Lecture Notes in Computer Science   **10249**

Commenced Publication in 1973
Founding and Former Series Editors:
Gerhard Goos, Juris Hartmanis, and Jan van Leeuwen

More information about this series at http://www.springer.com/series/7409

Eva Blomqvist · Diana Maynard
Aldo Gangemi · Rinke Hoekstra
Pascal Hitzler · Olaf Hartig (Eds.)

The Semantic Web

14th International Conference, ESWC 2017
Portorož, Slovenia, May 28 – June 1, 2017
Proceedings, Part I

 Springer

Editors

Eva Blomqvist
Linköping University
Linköping
Sweden

Diana Maynard
University of Sheffield
Sheffield
UK

Aldo Gangemi
Paris Nord University
Paris
France

Rinke Hoekstra
Vrije Universiteit Amsterdam
Amsterdam
The Netherlands

Pascal Hitzler
Wright State University
Dayton, OH
USA

Olaf Hartig
Linköping University
Linköping
Sweden

ISSN 0302-9743 ISSN 1611-3349 (electronic)
Lecture Notes in Computer Science
ISBN 978-3-319-58067-8 ISBN 978-3-319-58068-5 (eBook)
DOI 10.1007/978-3-319-58068-5

Library of Congress Control Number: 2017939119

LNCS Sublibrary: SL3 – Information Systems and Applications, incl. Internet/Web, and HCI

Preface

This volume contains the main proceedings of the ESWC 2017 conference. The ESWC conference is established as a yearly major venue for discussing the latest scientific results and technology innovations related to the Semantic Web and linked data. At ESWC, international scientists, industry specialists, and practitioners meet to discuss the future of applicable, scalable, user-friendly, as well as potentially game-changing solutions. This 14th edition took place from May 28 to June 1, 2017, in Portorož (Slovenia). Building on its past success, ESWC is also a venue for broadening the focus of the Semantic Web community to span other relevant research areas in which semantics and Web technology plays an important role. Thus, the chairs of ESWC 2017 organized two special tracks putting particular emphasis on usage areas where Semantic Web technologies are facilitating a leap of progress, namely: "Multilinguality" and "Semantic Web and Transparency."

Emerging from its roots in AI and Web technology, the Semantic Web today is mainly a Web of linked data, upon which a plethora of services and applications for all possible domains are being proposed. Some of the core challenges that the Semantic Web aims at addressing are the heterogeneity of content and its volatile and rapidly changing nature, its uncertainty, provenance, and varying quality. This in combination with more traditional disciplines — such as logical modelling and reasoning, natural language processing, databases and data storage and access, machine learning, distributed systems, information retrieval and data mining, social networks, Web science and Web engineering — shows the span of topics covered by this conference. The nine regular tracks, in combination with an in-use and applications track, and the two special tracks, constituted the main technical program of ESWC 2017.

The program also included three exciting invited keynotes. Lora Aroyo (Professor Human Computer Interaction Vrije Universiteit Amsterdam, The Netherlands, and Visiting Professor at Columbia University in the City of New York, USA) focused on the notion of ambiguity, discussing how ambiguity can be captured and even taken advantage of by capturing the diversity of interpretations. In particular, she discussed how to capture this diversity and how to allow machines to deal with it. Kevin Crosbie (Chief Product Officer, RavenPack) discussed the role of semantic intelligence in financial markets. In particular, he focused on the challenges of turning unstructured content into structured data, given that many new kinds of alternative data (social media, satellite imagery, etc.) are being used to complement traditional data for use in predictive modelling for financial trading algorithms. John Sheridan (Digital Director, The National Archives) discussed the use and benefits of Semantic Web technologies for digital archiving, in particular for managing heterogeneous metadata, dealing with uncertainty, and in areas such as provenance and trust.

The main scientific program of the conference comprised 51 papers: 40 research and 11 in-use and application papers, selected out of 183 reviewed submissions, which corresponds to an acceptance rate of 25% for the research papers submitted and 52%

for the in-use papers. A special thanks goes to our process improvement chair, Derek Doran, who helped us establish an improved quality assurance process during the paper selection, ensuring the originality and quality of the research papers that were accepted to the conference. This program was completed by a demonstration and poster session, in which researchers had the chance to present their latest results and advances in the form of live demos. In addition, the PhD Symposium program included ten contributions, selected out of 14 submissions.

This year's edition of ESWC's main scientific program presented a significant number of research papers with a focus on solving typical Semantic Web problems, such as entity linking, discoverability, etc., by using methods and techniques from areas such as machine learning and natural language processing, and reflecting in particular the current interest in deep learning. Work on both the fundamental development and use of Semantic Web technologies in relation to transparency is particularly interesting in view of current governmental and institutional open data initiatives.

The conference program also offered 12 workshops, six tutorials, and an EU Project Networking session. This year, an open call also allowed us to select and support five challenges. These associated events create an even more open, multidisciplinary, and cross-fertilizing environment at the conference, allowing for work-in-progress and practical results to be discussed. Workshops ranged from domain-focused topics, including the biomedical, scientific publishing, e-science, robotics, distributed ledgers, and scholarly fields, to more technology-focused topics ranging from RDF stream processing, query processing, data quality and data evolution, to sentiment analysis and semantic deep learning. Tutorial topics spanned NLP, ontology engineering, and linked data, including specific tutorials on knowledge graphs and rule-based processing of data. Proceedings from these satellite events are available in a separate volume.

The General and Program Committee chairs would like to thank the many people who were involved in making ESWC 2017 a success. First of all, our thanks go to the 24 track chairs and 360 reviewers, including 49 external reviewers, for ensuring a rigorous blind review process that led to an excellent scientific program and an average of four reviews per article. The scientific program was completed by an exciting selection of posters and demos chaired by Katja Hose and Heiko Paulheim.

Special thanks go to the PhD symposium chairs, Rinke Hoekstra and Pascal Hitzler, who managed one of the key events at ESWC, the PhD symposium. The brilliant PhD students will become the future leaders of our field, and deserve both encouragement and mentoring, which Rinke and Pascal made sure we could provide. We also had a great selection of workshops and tutorials, as mentioned earlier, thanks to the commitment of our workshop chairs, Agnieszka Ławrynowicz and Fabio Ciravegna, and tutorial chairs, Anna Lisa Gentile and Sebastian Rudolph.

Thanks to our EU Project Networking session chairs, Lyndon Nixon and Maria Maleshkova, we had the opportunity to facilitate meetings and exciting discussions between leading European research projects. Networking and sharing ideas between projects is a crucial success factor for such large research projects.

We are additionally grateful for the work and commitment of Monika Solanki, Mauro Dragoni, and all the individual challenges chairs, who successfully established a challenge track. The five challenges provided researchers and practitioners with the opportunity to compare their latest solutions in these challenge areas, ranging from topic-focused tasks such as question answering and semantic sentiment analysis, to practical tasks such as storage, semantic publishing, and open knowledge extraction.

We thank STI International for supporting the conference organization, and particularly Alexander Wahler as the conference treasurer. YouVivo GmbH deserve special thanks for the professional support of the conference organization, and for solving all practical matters. Our local chair Marko Grobelnik also deserves a special thanks for selecting the venue and for, together with his local organizers, Marija Kokelj, Monika Kropej, and Spela Sitar, arranging a great on-site experience for our conference attendees.

Further, we are very grateful to Ruben Verborgh, our publicity chair, who kept our community informed throughout the year, and Venislav Georgiev, who administered the website. Of course we also thank our sponsors, listed on the next pages, for their vital support of this edition of ESWC. We would like to stress the great work achieved by the Semantic Technologies coordinators Lionel Medini and Luigi Asprino, who maintained and updated our ESWC mobile app and published our conference dataset. A special thanks also to our proceedings chair, Olaf Hartig, who did an excellent job in preparing this volume with the kind support of Springer.

March 2017

Eva Blomqvist
Diana Maynard
Aldo Gangemi

Organization

Organizing Committee

General Chair

Eva Blomqvist — Linköping University, Sweden

Program Chairs

Diana Maynard — University of Sheffield, UK
Aldo Gangemi — Paris Nord University, France and ISTC-CNR, Italy

Workshops Chairs

Agnieszka Ławrynowicz — Poznan University of Technology, Poland
Fabio Ciravegna — University of Sheffield, UK

Poster and Demo Chairs

Katja Hose — Aalborg University, Denmark
Heiko Paulheim — University of Mannheim, Germany

Tutorials Chairs

Anna Lisa Gentile — University of Mannheim, Germany
Sebastian Rudolph — TU Dresden, Germany

PhD Symposium Chairs

Rinke Hoekstra — Vrije Universiteit Amsterdam, The Netherlands
Pascal Hitzler — Wright State University, USA

Challenge Chairs

Monika Solanki — University of Oxford, UK
Mauro Dragoni — Fondazione Bruno Kessler, Italy

Semantic Technologies Coordinators

Lionel Medini — University of Lyon, France
Luigi Asprino — University of Bologna, Italy

EU Project Networking Session Chairs

Lyndon Nixon — Modul Universität Vienna, Austria
Maria Maleshkova — Karlsruhe Institute of Technology (KIT), Germany

Process Improvement Chair

Derek Doran   Wright State University, USA

Publicity Chair

Ruben Verborgh   Ghent University, Belgium

Web Presence

Venislav Georgiev   STI International, Austria

Proceedings Chair

Olaf Hartig   Linköping University, Sweden

Treasurer

Alexander Wahler   STI, Austria

Local Organization Chair

Marko Grobelnik   Jožef Stefan Institute, Slovenia

Local Organization and Conference Administration

Katharina Vosberg   YouVivo GmbH, Germany
Marija Kokelj   PITEA, Slovenia
Monika Kropej   Jožef Stefan Institute, Slovenia
Spela Sitar   Jožef Stefan Institute, Slovenia

Program Committee

Program Chairs

Diana Maynard   University of Sheffield, UK
Aldo Gangemi   Paris Nord University, France and ISTC-CNR, Italy

Track Chairs

Vocabularies, Schemas, Ontologies

Helena Sofia Pinto   Universidade de Lisboa, Portugal
Silvio Peroni   University of Bologna, Italy

Reasoning

Uli Sattler   University of Manchester, UK
Umberto Straccia   ISTI-CNR, Italy

Linked Data

Jun Zhao   University of Oxford, UK
Axel Ngonga Ngomo   Universität Leipzig, Germany

Social Web and Web Science

Harith Alani   The Open University, UK
Wolfgang Nejdl   Leibniz Universität Hannover, Germany

Semantic Data Management, Big Data, Scalability

Maria Esther Vidal   University of Bonn, Germany and Universidad Simón
                              Bolívar, Venezuela
Jürgen Umbrich   Vienna University of Economics and Business, Austria

Natural Language Processing and Information Retrieval

Claire Gardent   CNRS, France
Udo Kruschwitz   University of Essex, UK

Machine Learning

Claudia d'Amato   University of Bari, Italy
Michael Cochez   Fraunhofer Institute for Applied Information
                              Technology FIT, Germany

Mobile Web, Sensors and Semantic Streams

Emanuele Della Valle   Politecnico di Milano, Italy
Manfred Hauswirth   TU Berlin, Germany

Services, APIs, Processes and Cloud Computing

Peter Haase   metaphacts GmbH, Germany
Barry Norton   Elsevier, UK

Multilinguality

Philipp Cimiano   Universität Bielefeld, Germany
Roberto Navigli   Sapienza University of Rome, Italy

Semantic Web and Transparency

Mathieu d'Aquin   The Open University, UK
Giorgia Lodi   CNR, Italy

In-Use and Industrial Track

Paul Groth   Elsevier Labs, The Netherlands
Paolo Bouquet   Trento University, Italy

Members (All Tracks)

Karl Aberer
Maribel Acosta
Alessandro Adamou
Nitish Aggarwal
Mehwish Alam
Harith Alani
Jose Julio Alferes
Muhammad Intizar Ali
Marjan Alirezaie
Carlo Allocca
Grigoris Antoniou
Alessandro Artale
Sören Auer
Nathalie Aussenac-Gilles
Franz Baader
Michele Barbera
Valerio Basile
Chris Biemann
Antonis Bikakis
Eva Blomqvist
Fernando Bobillo
Piero Bonatti
Kalina Bontcheva
Alex Borgida
Stefan Borgwardt
Paolo Bouquet
Charalampos Bratsas
John Breslin
Christopher Brewster
Carlos Buil Aranda
Paul Buitelaar
Davide Buscaldi
Joseph Busch
Raf Buyle
Elena Cabrio
Jean-Paul Calbimonte
Diego Calvanese
Andrea Calì
Matteo Cannaviccio
Amparo E. Cano
Pompeu Casanovas
Giovanni Casini
Michele Catasta

Irene Celino
Ismail Ilkan Ceylan
Pierre-Antoine Champin
Thierry Charnois
Gong Cheng
Christian Chiarcos
Key-Sun Choi
Philip Cimiano
Michael Cochez
Sergio Consoli
Olivier Corby
Oscar Corcho
Gianluca Correndo
Fabio Cozman
Christophe Cruz
Claudia d'Amato
Enrico Daga
Laura M. Daniele
Jérôme David
Brian Davis
Victor de Boer
Giuseppe De Giacomo
Gerard de Melo
Stefan Decker
Thierry Declerck
Makx Dekkers
Emanuele Della Valle
Tommaso Di Noia
Stefan Dietze
John Domingue
Derek Doran
Mauro Dragoni
Alistair Duke
Michel Dumontier
Mathieu d'Aquin
Maud Ehrmann
Henrik Eriksson
Vadim Ermolayev
Jérôme Euzenat
Nicola Fanizzi
Catherine Faron Zucker
Bettina Fazzinga
Miriam Fernandez

Javier D. Fernández
Sebastien Ferre
Besnik Fetahu
Giorgos Flouris
Antske Fokkens
Achille Fokoue
Muriel Foulonneau
Enrico Francesconi
Irini Fundulaki
Fabien Gandon
Aldo Gangemi
Roberto Garcia
Raúl García-Castro
Claire Gardent
Daniel Garijo
Anna Lisa Gentile
Chiara Ghidini
Alain Giboin
Claudio Giuliano
Birte Glimm
Jose Manuel Gomez-Perez
Julio Gonzalo
Rafael S. Gonçalves
Jorge Gracia
Alasdair Gray
Paul Groth
Tudor Groza
Francesco Guerra
Alessio Gugliotta
Giancarlo Guizzardi
Víctor Gutiérrez Basulto
Christophe Guéret
Asunción Gómez-Pérez
Peter Haase
Siegfried Handschuh
Andreas Harth
Olaf Hartig
Matthias Hartung
Oktie Hassanzadeh
Manfred Hauswirth
Conor Hayes
Johannes Heinecke
Benjamin Heitmann
Sebastian Hellmann
Pascal Hitzler
Rinke Hoekstra

Aidan Hogan
Laura Hollink
Matthew Horridge
Katja Hose
Geert-Jan Houben
Eero Hyvönen
Yazmin Angelica Ibanez-Garcia
Valentina Janev
Krzysztof Janowicz
Mustafa Jarrar
Anja Jentzsch
Clement Jonquet
Md. Rezaul Karim
Tomi Kauppinen
Takahiro Kawamura
Carsten Keßler
Ali Khalili
Sabrina Kirrane
Matthias Klusch
Craig Knoblock
Magnus Knuth
Boris Konev
Jacek Kopecky
Manolis Koubarakis
Adila A. Krisnadhi
Anastasia Krithara
Udo Kruschwitz
Patrick Lambrix
Christoph Lange
Michael Lauruhn
Alberto Lavelli
Agnieszka Lawrynowicz
Danh Le Phuoc
Domenico Lembo
David Lewis
Giorgia Lodi
Steffen Lohmann
Nuno Lopes
Vanessa Lopez
Nikolaos Loutas
Chun Lu
Markus Luczak-Roesch
Ioanna Lytra
Bernardo Magnini
Maria Maleshkova
Pierre Maret

Nicolas Matentzoglu
Andrea Maurino
Diana Maynard
Suvodeep Mazumdar
John P. Mccrae
Fiona McNeill
Nandana Mihindukulasooriya
Daniel Miranker
Riichiro Mizoguchi
Dunja Mladenic
Marie-Francine Moens
Pascal Molli
Elena Montiel-Ponsoda
Gabriela Montoya
Federico Morando
Yassine Mrabet
Claudia Müller-Birn
Hubert Naacke
Ndapandula Nakashole
Amedeo Napoli
Roberto Navigli
Wolfgang Nejdl
Axel Ngonga
Matthias Nickles
Andriy Nikolov
Olaf Noppens
Barry Norton
Andrea Giovanni Nuzzolese
Leo Obrst
Alessandro Oltramari
Magdalena Ortiz
Francesco Osborne
Matteo Palmonari
Jeff Z. Pan
Heiko Paulheim
Terry Payne
Laura Perez-Beltrachini
Silvio Peroni
Rafael Peñaloza
H. Sofia Pinto
Vassilis Plachouras
Axel Polleres
Simone Paolo Ponzetto
María Poveda-Villalón
Valentina Presutti
Laurette Pretorius

Cédric Pruski
Yuzhong Qu
Filip Radulovic
Diego Reforgiato Recupero
Georg Rehm
Achim Rettinger
Chantal Reynaud
Mikko Rinne
Petar Ristoski
Carlos R. Rivero
Giuseppe Rizzo
Riccardo Rosati
Marco Rospocher
Camille Roth
Marie-Christine Rousset
Ana Roxin
Sebastian Rudolph
Harald Sack
Hassan Saif
Sherif Sakr
Muhammad Saleem
Felix Sasaki
Ulrike Sattler
Luigi Sauro
Vadim Savenkov
Monica Scannapieco
Francois Scharffe
Ansgar Scherp
Stefan Schlobach
Jodi Schneider
Pavel Shvaiko
Gerardo Simari
Elena Simperl
Hala Skaf-Molli
Charese Smiley
Monika Solanki
Steffen Staab
Giorgos Stamou
Yannis Stavrakas
Luc Steels
Giorgos Stoilos
Audun Stolpe
Umberto Straccia
York Sure-Vetter
Mari Carmen Suárez-Figueroa
Vojtěch Svátek

Pedro Szekely
Anders Søgaard
Dhavalkumar Thakker
Martin Theobald
Allan Third
Keerthi Thomas
Ilaria Tiddi
Thanassis Tiropanis
Ioan Toma
David Toman
Alessandra Toninelli
Anna Tordai
Yannick Toussaint
Cassia Trojahn
Dmitry Tsarkov
Giovanni Tummarello
Juergen Umbrich
Jacopo Urbani
Ricardo Usbeck
Marieke Van Erp
Frank Van Harmelen

Jacco van Ossenbruggen
Paola Velardi
Chiara Veninata
Ruben Verborgh
Guido Vetere
Maria Esther Vidal
Carlos Viegas Damásio
Serena Villata
Holger Wache
Simon Walk
Shenghui Wang
Nick Webb
Chris Welty
Erik Wilde
Frank Wolter
Feiyu Xu
Peter Yeh
Amrapali Zaveri
Jun Zhao
Antoine Zimmermann

Additional Reviewers

Andrejs Abele
Manuel Atencia
Sotiris Batsakis
Julia Bosque-Gil
Elena Botoeva
Markus Brenner
Benjamin Cogrel
Zlatan Dragisic
Lukas Eberhard
Amosse Edouard
Mezghani Emna
Ronald Ferguson
Johannes Frey
Jinlong Guo
Amit Gupta
Amelie Gyrard
Lavdim Halilaj
Michael Hoffmann
Aidan Hogan
Filip Ilievski
Prateek Jain

Daniel Janke
Amit Joshi
Unmesh Joshi
Naouel Karam
Nazifa Karima
Amit Kirschenbaum
Ruediger Klein
Sarah Kohail
Philipp Koncar
Albert Meroño-Peñuela
Diego Moussallem
Frank Nack
Yaroslav Nechaev
Chifumi Nishioka
Inna Novalija
Jonas Oppenländer
Enrico Palumbo
Alexander Panchenko
George Papadakis
Peter Patel-Schneider
Viviana Patti

Minh Tran Pham
Guangyuan Piao
Andreas Pieris
Danae Pla Karidi
Gustavo Publio
José Luis Redondo-García
Sebastian Ruder
Anisa Rula
Hassan Saif
Ihab Salawdeh
Valerio Santarelli
Marvin Schiller
Lukas Schmelzeisen

Panayiotis Smeros
Tommaso Soru
Ilias Tachmazidis
Steffen Thoma
Trung-Kien Tran
Despoina Trivela
Sahar Vahdati
Massimo Vitiello
Binh Vu
Guohui Xiao
Benjamin Zarrieß
Lu Zhou

PhD Symposium Program Committee

Chairs

Rinke Hoekstra   Vrije Universiteit Amsterdam, The Netherlands
Pascal Hitzler   Wright State University, USA

Members

Chris Biemann   TU Darmstadt, Germany
Michelle Cheatham   Wright State University, USA
Philipp Cimiano   Bielefeld University, Germany
Claudia D'Amato   University of Bari, Italy
Chiara Di   Fondazione Bruno Kessler-IRST, Italy
  Francescomarino
Anna Lisa Gentile   IBM Research Almaden, USA
Chiara Ghidini   Fondazione Bruno Kessler-IRST, Italy
Pascal Hitzler   Wright State University, USA
Rinke Hoekstra   University of Amsterdam/VU University Amsterdam,
                        The Netherlands
Aidan Hogan   DCC, Universidad de Chile, Chile
Krzysztof Janowicz   University of California, Santa Barbara, USA
Agnieszka Lawrynowicz Poznan University of Technology, Poland
Matteo Palmonari   University of Milano-Bicocca, Italy
Axel Polleres   Vienna University of Economics and Business, WU Wien,
                        Austria
Stefan Schlobach   Vrije Universiteit Amsterdam, The Netherlands

Jodi Schneider	University of Illinois Urbana Champaign, USA
Juan Sequeda	Capsenta Labs, USA
Monika Solanki	University of Oxford, UK
Vojtěch Svátek	University of Economics, Prague, Czech Republic
Danai Symeonidou	INRA, France
Serena Villata	CNRS, Laboratoire d'Informatique, Signaux et Systèmes de Sophia-Antipolis, France

Steering Committee

Chair

| John Domingue | The Open University, UK and STI International, Austria |

Members

Claudia d'Amato	Universià degli Studi di Bari, Italy
Mathieu d'Aquin	Knowledge Media Institute KMI, UK
Philipp Cimiano	Bielefeld University, Germany
Oscar Corcho	Universidad Politécnica de Madrid, Spain
Fabien Gandon	Inria, W3C, Ecole Polytechnique de l'Université de Nice Sophia Antipolis, France
Valentina Presutti	CNR, Italy
Marta Sabou	Vienna University of Technology, Austria
Harald Sack	Hasso Plattner Institute (HPI), Germany

Sponsoring Institutions

Gold Sponsors

http://www.iospress.nl/   http://www.sti2.org/

Silver Sponsors

https://www.elsevier.com/

Abstract of Keynotes

Bringing Semantic Intelligence to Financial Markets

Kevin Crosbie

RavenPack, New York, US

Abstract. The most successful hedge-funds in today's financial markets are consuming large amounts of alternative data, including satellite imagery, point-of-sale data, news, social media and publications from the web. This new trend is driven by the fact that traditional factors have become less predictive in recent years, requiring sophisticated investors to explore new data sources. The majority of this new alternative content is unstructured and hence must first be converted into structured analytics data in order to be used systematically. Instead of building such capabilities themselves, financial firms are turning towards companies that specialize in this field. In this talk, Kevin will discuss some of the practical challenges of giving structure to unstructured content, how entities and ontologies may be used to link data and the ways in which semantic intelligence can be derived for use in financial trading algorithms.

Disrupting the Semantic Comfort Zone

Lora Aroyo[1,2]

[1] Vrije Universiteit Amsterdam, Amsterdam, The Netherlands
[2] Columbia University, New York, USA
lora.aroyo@vu.nl

Abstract. Ambiguity in interpreting signs is not a new idea, yet the vast majority of research in machine interpretation of signals such as speech, language, images, video, audio, etc., tend to ignore ambiguity. This is evidenced by the fact that metrics for quality of machine understanding rely on a ground truth, in which each instance (a sentence, a photo, a sound clip, etc) is assigned a discrete label, or set of labels, and the machine's prediction for that instance is compared to the label to determine if it is correct. This determination yields the familiar precision, recall, accuracy, and f-measure metrics, but clearly presupposes that this determination can be made. CrowdTruth is a form of collective intelligence based on a vector representation that accommodates diverse interpretation perspectives and encourages human annotators to disagree with each other, in order to expose latent elements such as ambiguity and worker quality. In other words, CrowdTruth assumes that when annotators disagree on how to label an example, it is because the example is ambiguous, the worker isn't doing the right thing, or the task itself is not clear. In previous work on CrowdTruth, the focus was on how the disagreement signals from low quality workers and from unclear tasks can be isolated. Recently, we observed that disagreement can also signal ambiguity. The basic hypothesis is that, if workers disagree on the correct label for an example, then it will be more difficult for a machine to classify that example. The elaborate data analysis to determine if the source of the disagreement is ambiguity supports our intuition that low clarity signals ambiguity, while high clarity sentences quite obviously express one or more of the target relations. In this talk I will share the experiences and lessons learned on the path to understanding diversity in human interpretation and the ways to capture it as ground truth to enable machines to deal with such diversity.

Keywords: Ambiguity · Crowdsourcing · Disagreement · Diversity · Perspectives · Opinions · Machine-crowd computation · Crowdsourcing ground truth

Semantic Web Technologies for Digital Archives

John Sheridan

The National Archives, Kew, UK

Abstract. What will people in the future know of today? As the homes for our collective memory archives have a special role to play. Semantic Web technologies address some important needs for digital archives and are being ever more embraced by the archival community.

Archives face a big challenge. The use of digital technologies has profoundly shaped what types of record are created, captured, shared and made available. Digital records are not just documents or email but all sorts of content such as websites, threaded discussions, video, websites, structured datasets and even computer code. Yet, in the digital era, when so much is encoded as 0s and 1s there is no long term solution to the challenge of preservation. All archives can do is make the institutional commitment to continue to invest, through generations of technological change, in the engineering effort required for records to continue to be available.

The National Archives is one of the world's leading digital archives. Our Digital Records Infrastructure, which makes extensive use of RDF and SPARQL, is capable of safely, securely and actively preserving large quantities of data. Our Web Archive provides a comprehensive record of government on the web. We also lead the maintenance of a register of file format signatures that is used relied on by archives and other memory institutions around the world.

As a digital archive we provide value by preserving digital records, keeping them safe for the future. We maintain the context for the records so their evidential value can be understood in the context of their creation and continuing use. We produce records so that they are available for others to access, and we also enable use.

Semantic Web technologies play a key role in each of these areas and are integral to our approach for preserving, contextualising, presenting and enable use of digital records. This presentation will explain why and how we have used semantic web technologies for digital archiving and the benefits we have seen, for managing heterogeneous metadata and also in areas such a provenance and trust. It will explore new opportunities for archives from using Semantic Web technologies in particular around contextual description, with digital records increasingly contextualising each other. This is part of a shift to a more fluid approach where context grows with an archives collection and in relation to other collections. Finally it will also look at the challenges for archives with using Semantic Web technologies in particular around how best to manage uncertainty in our data as we increasingly use probabilistic approaches.

Contents – Part I

Mobile Web, Sensors, and Semantic Streams Track

Natural Language Processing and Information Retrieval Track

Vocabularies, Schemas, and Ontologies Track

Reasoning Track

Social Web and Web Science Track

Semantic Web and Transparency Track

Contents – Part II

PhD Symposium

Semantic Data Management, Big Data, and Scalability Track

Traffic Analytics for Linked Data Publishers

Luca Costabello[1]([⊠]), Pierre-Yves Vandenbussche[1], Gofran Shukair[1],
Corine Deliot[2], and Neil Wilson[2]

[1] Fujitsu Ireland Ltd., Galway, Ireland
{luca.costabello,pierre-yves.vandenbussche,
gofran.shukair}@ie.fujitsu.com
[2] British Library, London, UK
{corine.deliot,neil.wilson}@bl.uk

Abstract. We present a traffic analytics platform for servers that publish Linked Data. To the best of our knowledge, this is the first system that mines access logs of registered Linked Data servers to extract traffic insights on daily basis and without human intervention. The framework extracts Linked Data-specific traffic metrics from log records of HTTP lookups and SPARQL queries, and provides insights not available in traditional web analytics tools. Among all, we detect visitor sessions with a variant of hierarchical agglomerative clustering. We also identify workload peaks of SPARQL endpoints by detecting heavy and light SPARQL queries with supervised learning. The platform has been tested on 13 months of access logs of the British National Bibliography RDF dataset.

Keywords: Linked data · Traffic analytics · Data publication · SPARQL

1 Introduction

Data providers have so far published hundreds of linked datasets, thus contributing to the birth of the Web of Data. Although technical and conceptual challenges of Linked Data publication have been largely discussed in literature [2,10], we believe that data publishers have still limited awareness of how datasets are accessed by visitors. In other words, data providers cannot easily find valuable insights into how triples are consumed and by whom. This has two consequences: first, publishers struggle to justify Linked Data investment with management. Second, they miss out technical benefits: poor access traffic knowledge hampers negotiating an affordable service level agreement with hosting suppliers; limited awareness of traffic spikes prevents predicting peaks during real-world events. While some works describe specific access metrics for linked datasets [13], no comprehensive analytics tool for Linked Data publishers has ever been proposed, and in most cases publishers have no choice but to manually browse through records stored in server access logs. Applications

© Springer International Publishing AG 2017
E. Blomqvist et al. (Eds.): ESWC 2017, Part I, LNCS 10249, pp. 3–18, 2017.
DOI: 10.1007/978-3-319-58068-5_1

for analysing traditional websites traffic exist, but none considers the specificities of Linked Data: none extracts insights by parsing SPARQL in query strings (e.g. how many resources occur in SPARQL queries, how many SPARQL endpoints workload peaks, the ratio of HTTP and SPARQL traffic, etc.). Besides, content negotiation with 303 URIs[1] is not interpreted as an atomic operation, thus overestimating the actual number of requests. No tools detect Linked Data visitors sessions, neither.

Bringing traffic analytics to Linked Data publishers requires solving three major challenges: first, we must choose meaningful metrics: are traditional web traffic analytics metrics sufficient? How should we tailor such metrics to Linked Data? Second, we must extract such metrics. This leads to additional subproblems: which data sources shall we mine, and how do we filter noise (e.g. search engine crawlers, robots)? Some metrics require knowledge of visitor *sessions*: how to detect such sessions in a Linked Data scenario? Being aware of workload peaks of SPARQL endpoints is important: how do we detect *heavy* SPARQL queries repeatedly issued in a short time? Last but not least, how do we deliver insights to end users without human intervention?

Research Contribution. Our contribution is twofold: first, we present a traffic analytics platform for Linked Data servers: we add novel Linked Data-specific metrics - to break down traffic by RDF content, capture SPARQL insights, and properly interpret 303 patterns. Second, we describe two mining tasks that the system adopts to extract some of the aforementioned metrics: (i) the reconstruction of Linked Data visitors sessions with time-based hierarchical agglomerative clustering, and (ii) the detection of workload peaks of SPARQL endpoints by predicting heavy and light SPARQL queries with supervised learning and SPARQL syntactic features. We evaluate our mining tasks on DBpedia 3.9 and British National Bibliography logs, and we publish the datasets to reproduce our results. We describe the insights from 13 months of access logs of the British National Bibliography dataset[2], and we show that our system reveals patterns that would otherwise require heavy manual intervention by dataset publishers.

The paper is organised as follows: Sect. 2 lists related works. Section 3 introduces the system architecture. Section 4 discusses the traffic metrics and their extraction. This section also describes how we identify visitor sessions, and how we detect heavy and light SPARQL queries. Results and reproducibility are described in Sect. 5. Section 6 proposes future extensions.

2 Related Work

Google Analytics[3] and other popular web analytics platforms[4] (e.g. Open Web Analytics, PIWIK[5]) are not designed for linked datasets (e.g. they do not provide

[1] https://www.w3.org/TR/cooluris.

[2] http://bnb.data.bl.uk.

[3] http://analytics.google.com.

[4] https://en.wikipedia.org/wiki/List_of_web_analytics_software.

[5] http://piwik.org, http://www.openwebanalytics.com.

access metrics for SPARQL). Many works over the last decade discuss access metrics for the traditional web [6]. Nevertheless, only a handful propose Linked Data-specific traffic metrics: Möller et al. [13] propose a list of Linked Data-specific metrics that cover HTTP and SPARQL access to RDF (e.g. ratio between 303 and 200 HTTP requests, number of *RDF-aware* agents, SPARQL query features, machine/vs human classification based on user-agent strings). The well-established USEWOD workshop series[6] is the reference for Linked Data usage mining (the workshop authors also publish a dataset of anonymised linked datasets access logs [11]). We reused and extended metrics defined in [6, 13].

Labeling SPARQL queries as light or heavy has been inspired by [5, 7, 9, 18]. Although formal analysis of the SPARQL language complexity has been extensively studied [16, 19], empirical works often adopt an informal notion of query complexity: for instance, Möller et al. [13] consider complexity as the ratio between the triple pattern types and the count of SPARQL queries. Others [4] refer to the count of classes and properties of a query. Other empirical studies propose syntactic features that distinguish low from high complexity queries, and confirm that non-conjunctive queries lead to higher execution times [5, 7, 18]. We trained our light/heavy SPARQL classifier with syntactic features proposed in these papers. Other works use supervised learning on SPARQL query logs: Hasan [9] goes as far as predicting SPARQL query execution time with k-nearest neighbours regression and support vector machines.

To the best of our knowledge, no existing work detects Linked Data client sessions. Traditional web sessions have been naively determined by identifying fixed-length inactivity gaps, but picking optimal values is awkward and error-prone [1]. Murray et al. present a variant of hierarchical agglomerative clustering (HAC) - later refined by [12] - that finds visitor-specific thresholds [14]. We adopt HAC in our Linked Data scenario. Other clustering approaches rely on model-based cluster analysis [15] and time-aware clustering [8, 17].

3   System Overview

The traffic analytics platform includes the following main components (Fig. 1):

Extract-Transform-Load (ETL) Unit. On a daily basis, for registered publishers, the Log Ingestion sub-component fetches and parses access logs from one or more linked dataset servers (see Fig. 2 for an example). Since data publishers adopt different access records formats, the platform relies on a flexible and customisable log parser. Records are filtered from search engines crawlers noise.

Metrics Extraction Unit. Extracts traffic metrics. It includes visitor session detection, and light/heavy SPARQL query classification (Sect. 4). Note that we do not support structured content embedded in HTML pages (e.g. microdata, RDFa). The system does not provide statistics on downloads of datasets dumps.

---

[6] http://usewod.org/.

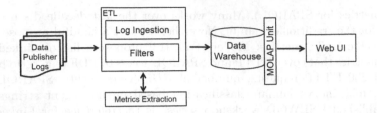

Fig. 1. Architecture of the traffic analytics platform for linked data publishers

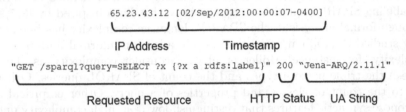

Fig. 2. A record of a linked data server access log (Apache commons logfile (https://httpd.apache.org/docs/trunk/logs.html#common)).

Data Warehouse and MOLAP Unit. Traffic metrics are stored in a data warehouse equipped with an SQL-compliant MOLAP[7] unit that answers queries with sub-second latency.

Web User Interface. The front end queries the RESTful APIs exposed by the MOLAP Unit, and generates a web UI that shows the metrics in Table 1 filtered by date, user agent type, and access protocol (see screenshots in Fig. 3).

4   Traffic Metrics

The traffic analytics platform features three categories of traffic metrics:

Content Metrics. How many times RDF resources have been accessed. We interpret content negotiation with 303 URIs as an atomic operation. We also count how many times resource URIs appear in SPARQL queries[8], and provide aggregates by *family* of RDF resource (i.e. instances, classes, properties, graphs).

Protocol Metrics. Information about the data access protocols used by visitors. Includes SPARQL-specific metrics such as the count of malformed queries and the detection of light and heavy SPARQL queries with supervised learning.

---

[7] Multidimensional Online Analytical Processing.
[8] This is a lower bound estimate. Access logs do not contain SPARQL result sets.

Table 1. Supported metrics.

Content Metrics	
Instances[a]	How many times RDF instances have been requested (HTTP requests and URIs in SPARQL queries)
Classes[a]	The count of URIs used as RDFS/OWL classes in SPARQL queries. URIs must be objects of rdf:type.
Properties[a]	The count of URIs used as predicates in SPARQL queries.
Graphs[a]	The count of URIs used as graphs in SPARQL queries (FROM/FROM NAMED, USING/USING NAMED, GRAPH).
Protocol Metrics	
Data access protocol[a]	The count of HTTP lookups and SPARQL queries, over a specific time frame.
SPARQL query type[a]	The count of SPARQL verbs over a specific time frame (e.g. how many SELECT, ASK, DESCRIBE, and CONSTRUCT).
Light/Heavy SPARQL[a]	Indicates the number of *light* and *heavy* SPARQL queries (see Sect. 4.2 for details).
HTTP methods	Indicates the count of HTTP verbs (e.g. GET, POST, HEAD).
303 Patterns[a]	The count of HTTP 303 patterns found in logs (303 URIs).
Misses	Keeping track of HTTP 404s is useful to understand whether visitors are looking for resources which are not currently included in the dataset.
Malformed queries[a]	The count of HTTP 400s shows how many malformed HTTP requests have been issued. We also show how many SPARQL queries contain syntax errors.
Other client-side errors	Useful, for example, to detect whether visitors attempt to access forbidden RDF resources (HTTP 403).
Server-side errors	The count of HTTP 5xxs is important to estimate whether error-triggering SPARQL queries have been repeatedly issued to a triplestore.
Audience Metrics	
Location	Continent, country, subdivision, and city of origin of a visitor.
Network provider	The name of the company or institute associated to a visitor's IP address, as returned by WHOIS lookups. If WHOIS data is not available, we provide the visitor's host network.
Language[a]	The preferred language requested by a visitor. Such information is extracted from the Accept-Language HTTP header (for HTTP lookups) and retrieved from SPARQL xsd language-tagged string literals (e.g. @en) or FILTER lang()s (e.g. lang(?l) = "en").
User agent	The user agent string provided by a visitor.
User agent type	To provide a clearer estimate of which clients are used to access a dataset, we group user agent strings into Software Libraries (e.g. Jena, Python SPARQL-client, etc.), Desktop Browsers, Mobile Browsers, and Others.
New vs Returning	New visitors versus visitors that have performed at least one visit before.
External referrer	HTTP Referer: headers. When dereferencing an RDF resource, the HTTP request might contain this optional header field that specifies the URI from which the request has been issued.
Sessions count	The count of all visitors sessions.
Session size	The number of requests sent by a visitor during a session (requests might be a mix of HTTP lookups and SPARQL queries).
Session depth[a]	The number of distinct RDF resources (graphs, classes, properties, instances) requested by a visitor during a session.
Session duration	The duration of a session.
Bounce rate	Indicates the percentage of sessions that contain only one resource request (whether this is an HTTP lookup or a SPARQL query).

[a] indicates linked data-specific entries

Audience Metrics. Besides traditional information about visitors (e.g. location, network provider), these measures include details of visitor sessions.

Table 1 describes the three categories, and highlights Linked Data-specific metrics (e.g. *light/heavy* SPARQL queries). In the following sections we describe how metric extraction is carried out by the system.

4.1 Content Metrics Extraction

Our traffic analytics platform supports Linked Data dual access protocol: it counts how many times an RDF resource is dereferenced with HTTP operations, but also how many times its URI is included in SPARQL queries. Unlike existing tools, we interpret content negotiation with 303 URIs as an atomic operation, thus counting each HTTP 303 pattern as a single request. Table 1 shows the list of aggregates by *family* (instances, classes, properties, graphs). While extraction of resources and their family is straightforward for HTTP operations (target URIs belong by default to the *instance* family), for SPARQL it requires parsing each query to determine the family of a resource (see Table 1 for parsing details).

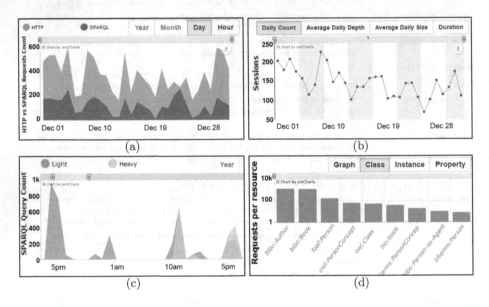

Fig. 3. Screenshots of the web UI: (a) HTTP vs SPARQL; (b) sessions insights; (c) light and heavy SPARQL queries spikes; (d) most popular RDF classes.

4.2 Protocol Metrics Extraction and SPARQL Queries Weight

Besides the count of HTTP lookups, 303 patterns, malformed queries, etc., this group includes the count of *light* and *heavy* SPARQL queries. We now describe the problem determined by this task, followed by our contribution.

```
                                 SELECT ?u ?c ?l
                                 WHERE {{
                                   ?u rdfs:comment ?c .
                                   FILTER(regex(str(?c), '(^|\\\\W)fr', 'i'))}
SELECT ?x                        UNION {
WHERE                              ?u rdfs:label ?l .
{?x a foaf:Person.}               OPTIONAL {?u skos:prefLabel ?l.}}}
```

(a)   (b)

Fig. 4. Sample SPARQL queries. *light* (a), and *heavy* (b). Note the presence of UNION, OPTIONAL, and FILTER regex in query (b).

Detecting Heavy and Light SPARQL Queries. We argue that identifying workload peaks of a SPARQL endpoint requires a rough estimate of how many *heavy* SPARQL queries have been sent to a SPARQL endpoint. A number of challenges arise: first, an (informal) definition of *weight* of a SPARQL query is required (i.e. when is a query *heavy* or *light?*). Second, a method to detect heavy and light queries is needed. Identifying *heavy* and *light* queries can be cast as a supervised binary classification problem; however, supervised learning requires a training and test dataset, and access logs collections usually do not include any information on execution time, least of all a manual annotation of *light/heavy* queries. Third, a comprehensive, established, and unified list of discriminative features for this specific classification task has not been defined yet, although features have been proposed for similar tasks (see related work in Sect. 2).

We use SPARQL syntactic features to train off-the-shelf supervised binary classifiers that label SPARQL queries in two categories: *light* and *heavy* (see Fig. 4 for an example). Although such categories can be more thoroughly explained by means of SPARQL complexity analysis, for the scope of this work we rely on the following qualitative definition:

Definition 1 (Heavy (light) SPARQL query). *Given a set Q of SPARQL queries, a query q ∈ Q is defined as heavy (light) if it requires considerable (little) computational and memory resources.*

We extract SPARQL 1.1 features from logs with Apache Jena 3.1.1: we combine what is proposed in recent literature [5,7,18] and experiment with five configurations (Table 2). Besides disjunctive statements [7], *Group 1* includes the number of joins, of FILTERs, and of regular expressions used in filters. *Group 2* adds the count of triple patterns, while *Group 3* also includes the count of * operators in SELECT statements (more than one might be present in case of nested queries). *Group 4* adds a number of other features, including FILTER-specific operators, projections, grouping operations, pagination, and the count of basic graph patterns. Finally, the complete set of features (*All*) adds the *type* of joins included in a query. Besides the six join configurations described in [7], we also keep track of the count of completely unbound and bound statements. In this paper we do not consider property paths, or negations (FILTER NOT EXISTS). We choose two binary classifier models, support vector machine classification

Table 2. Feature vectors configurations in our experiments (b=bound, u=unbound. Examples:`?s foaf:name ?n → ubu, ?s ?p?o → uuu`)

Group 1	• UNIONs count • OPTIONALs count • Joins count	• FILTERs count • FILTERs regex count	
Group 2	• Triple patterns count		∪ Group 1
Group 3	• SELECT * count		∪ Group 2
Group 4	• FILTERs expressions count • FILTERs strstarts count • FILTERs contains count • FILTERs in count • FILTERs or count • FILTERs lang count • Presence of ORDER BY • Projections count • Projection variables count	• REDUCEDs count • BINDs count • Basic graph patterns count • GRAPHs count • MINUS count • Presence of OFFSET/LIMIT • GROUP BY variables count • VALUES count	} ∪ Group 3
All	• Join types count (bub,bbb,uuu,ubu,ubb,bbu,uub,buu)		∪ Group 4

(SVC) and multinomial Naïve Bayes (NB), and we train such models on the training set described above (results presented in Sect. 5).

4.3   Audience Metrics Extraction and Visitor Session Identification

Most traffic analytics metrics described in this group (Table 1) rely on the notion of visitor *session*, and require query session analysis, i.e. the analysis of groups of queries. It is therefore crucial to (i) provide a Linked Data-compliant definition of session, and (ii) detect such sessions with acceptable precision and recall.

There is no general consensus on the definition of visitor session on the web [14,20]. Linked Data is no exception, as there seems to be no work in the Linked Data research community that tackles session detection identification. We adapt the definitions presented in [14] to Linked Data:

Definition 2 (Session). *A Session is defined as a sequence of requests issued with no significant interruptions by a uniquely identified visitor to a Linked Dataset. A session expires after a period of inactivity. Requests can either be HTTP lookups or SPARQL queries.*

Definition 3 (Visitor). *A visitor is a client uniquely identified by an IP address and a user agent string. Visitors can issue either HTTP operation or SPARQL queries.*

Besides problems shared with the traditional web (e.g. how to define a *period of inactivity* between sessions?), Linked Data session patterns might differ from traditional web sessions: Linked Data is meant for human consumption, *and* software agents: does this influence the shape of sessions? Does Linked Data dual protocol (HTTP lookups and SPARQL) lead to different patterns?

We adopted the variant of hierarchical agglomerative clustering (HAC) presented by Murray et al. in [14]. Although originally conceived for classic web logs analysis, we show that HAC has good performance also in a Linked Data scenario. HAC adopts a purely time-based gap detection, which is performed with hierarchical agglomerative cluster analysis. HAC is visitor-centered, and identifies sessions from the burstiness of a visitor's activity. HAC finds visitor-specific cut-off thresholds, and it does not need a global fixed cut-off threshold. For each visitor, HAC carries out two steps (Fig. 5a). First, it scans the gaps between HTTP requests to find a time interval that significantly increases the variance (Fig. 5a, line 3): such interval becomes the visitor-specific session cut-off. Then, it groups HTTP requests into sessions according to the cut-off (Fig. 5a, line 4). Figure 5b shows the cut-off threshold choice in detail: the algorithm loops through inter-session gaps in ascending order (lines 4–12); at each iteration it computes the time difference δ_g between the current gap g and the mean μ_I of the set of processed gaps I (line 6). If the ratio of δ_g and the standard deviation σ_I is greater than r_{max} (line 7), then r_{max} is replaced by such ratio and the cut-off c_v is set to g (lines 8,9). Figure 5c shows how sessions are built: if the gap between a request and the previous one is shorter than the cut-off c_v (line 4), the request is added to the current session (line 5). Otherwise, a new session is created (line 9).

Algorithm: HAC

Input   : The set $R = \{R_1, \ldots, R_n\}$ of requests for
           n visitors

Output : The sessions S of all n visitors

```
1 foreach visitor v do
2     Gᵥ ← get inter-requests time gaps from Rᵥ
3     cᵥ ← SetCutOff (Gᵥ)
4     Sᵥ ← BuildSessions (Rᵥ, cᵥ)
5     add Sᵥ to S
6 end
```

(a)

Function: SetCutOff

Input   : The gap intervals G_v for
        visitor v

Output : The visitor-specific cut-off
         threshold c_v

```
 1 I ← ∅
 2 rmax ← 0
 3 sort Gᵥ in ascending order
 4 foreach gap g ∈ Gᵥ do
 5     if I ≠ ∅ then
 6         δg ← (g − μI)
 7         if δg/σI > rmax then
 8             rmax ← δg/σI
 9             cᵥ ← g
10         end
11     end
12     I ← I ∪ {g}
13 end
```

(b)

Function: BuildSessions

Input   : The requests R_v for visitor v,
        the visitor-based threshold c_v

Output : The collection of sessions S_v
         for visitor v

```
 1 s ← new session
 2 foreach request rᵢ ∈ Rᵥ do
 3     gap g ← (t_{rᵢ} − t_{rᵢ₋₁})
 4     if g < cᵥ then
 5         add rᵢ to s
 6     end
 7     else
 8         Sᵥ ← Sᵥ ∪ {s}
 9         s ← new session
10     end
11 end
```

(c)

Fig. 5. HAC: (a) main function, (b) how per-visitor thresholds are set, (c) how HTTP/SPARQL requests are grouped in sessions.

5   Results

Heavy and Light SPARQL Queries. The light/heavy classifier must undergo ad-hoc training with dataset-specific access logs. To reproduce results, we describe how the classifier can be trained on public available USEWOD 2015 DBpedia 3.9 access logs [11] (we are not allowed to disclose the British National Bibliography access logs). The dataset used in this experiment is created as follows: we select random distinct queries from DBpedia 3.9 logs included in the USEWOD 2015 dataset. We execute such SPARQL queries on a local clone of DBpedia 3.9, and measure the response time of each query. Queries are issued from localhost. We run queries multiple times, to compute mean execution time μ_t and standard deviation ρ_t. We discard queries that led to HTTP and SPARQL errors, and queries with relative standard deviation $\rho_t/\mu_t > 0.8$. The resulting dataset contains 3,752 records (3,682 SELECT, 39 CONSTRUCT, and 31 ASK). We exclude DESCRIBE queries. We then choose a cut-off threshold $t_{cut-off}$=100ms, to separate *light* from *heavy* queries. Such value is chosen arbitrarily after manual inspection of the dataset: Fig. 6a shows the distribution of SPARQL against their execution time: the majority of queries falls within the first buckets. Note the extremely slow queries around 100 s. The chosen $t_{cut-off}$ leads to 3,192 *light* and 560 *heavy* queries (hence, the dataset is skewed towards *light* queries). Note that since we rely on a qualitative and scenario-dependent definition of *heavy* SPARQL query, the value of $t_{cut-off}$ can be replaced with another user-defined value, perhaps more meaningful under different conditions (e.g. different triplestore, server architecture, dataset). This paper only evaluates the quality of predictions with $t_{cut-off}$=100ms. Because supervised learning performance is influenced by $t_{cut-off}$, model re-training is required if such parameter is modified. We split the dataset in a training test and a test set. The two splits are generated with random stratified sampling, and account for 80% and 20% of records respectively. Training includes 2,553 *light* and 448 *heavy* queries; test includes 639 *light* and 112 *heavy* queries.

We run 10-fold cross-validated model selection with grid search over both support vector classifier and multinomial Naïve Bayes, using F1 score as scoring function: for SVC we iterate over $C = [0.1, 1, 10, 100]$ and linear versus radial basis function (RBF) kernels. For RBF kernels we try $\gamma = [0.1, 0.2, 0.5]$. Multinomial Naïve Bayes (NB) is trained with $\alpha = [0.5, 0.1, 0.2, 0.3, 0.4, 0.5, 1, 10]$, and learned class prior probabilities versus uniform prior. We fine-tune SVC model selection with randomized search over the ranges mentioned above, and find the best SVC estimator to have $C = 12.96$ and RBF kernel with $\gamma = 0.10$. The best NB classifier has $\alpha = 12.7$ and uniform prior. For SVC, we also train our models with each of the five groups of features described in Table 2. Figure 6b shows the quality of predictions by class when all features are used: although Naïve Bayes does not lead to reliable predictions of *heavy* queries (F1= 0.56), SVC reaches a F1 score of 0.66 ($P = 0.87$, but recall lags at $R = 0.54$, leaving space for future improvements). *Light* queries are labeled with F1= 0.95 by SVC and F1= 0.91 by Naïve Bayes. The average F1 score of SVC is 0.91, while NB yields to 0.86. Figures 6c and d show that AUC-ROC for SVC and Naïve Bayes are close

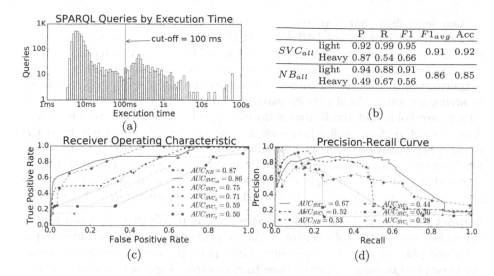

Fig. 6. *Heavy/light* SPARQL classification on USEWOD 2015: (a) queries distribution by execution time; (b) breakdown of performance (all features); (c) ROC and (d) PR curve for each feature configuration in Table 2.

(0.87 and 0.86 respectively). Nevertheless, due to class imbalance, we consider more reliable the area of the precision-recall curve (AUC-PR), where SVC shows the best result (AUC-PR= 0.67). Figures 6c and d also show that the entire set of features `All` is more effective than `Groups 1-4` (Table 2).

Visitor Session Identification. We assess the quality of the results of the session detection algorithm. We measure how well the adopted algorithm detects the beginning of a session. We compare precision, recall, and F1 score of HAC with sessions with fixed cut-offs of 15, 30, and 60 min respectively (these are common fixed-length session durations in literature [1, 14]).

Our evaluation dataset spans three consecutive random days of access records to the British National Bibliography Linked Dataset. The dataset includes 137 anonymised distinct visitors with at least 5 visits. Known search engine crawlers have been filtered out. Overall, the dataset includes 16,426 HTTP requests, 67.8% (11,128) of which have been issued by 10 distinct visitors that used software libraries, 31.6% (5,206) by 115 distinct visitors with desktop browsers, and 0.6% by other types of user agents (UA) - e.g. mobile browsers. Records include a timestamp, an HTTP (or SPARQL) request, and the user agent of the visitor. Records have been manually annotated by a domain expert as *session start* or *internal*: a total of 576 human-annotated session starts have been found, leading to 439 human-validated session gaps (the total count of session starts minus the 137 initial sessions for each visitor). As shown by Fig. 7a, session cut-offs are considerably spread out from their means μ. Data suggests that outliers greatly influence the means: software libraries are particularly affected with a standard deviation σ three times bigger than the mean. In this case, 50% of cut-offs are shorter than 29 m, but outliers bring the mean beyond 1 h.

Our tests show that HAC outperforms fixed-length cut-offs (Fig. 7b). When considering all user agents (*All UAs*), HAC's precision is always higher than the three fixed-length cut-offs (the 15 min cut-off leads to better recall, but precision is not satisfactory). Family-specific session detection shows that HAC is outperformed in the case of browsers (this presumably corresponds to human browsing sessions). Fixed cut-offs attempts show that precision increases with larger thresholds, but recall always decreases: this is because fixed-length sessions do not perform well when session gaps are highly spread out from the mean. Such behaviour is more prominent for software libraries than for desktop browsers, since session gaps of software libraries are more scattered: Fig. 7a shows that they have the highest coefficient of variation among the three categories ($c_v = \sigma/\mu = 2.57$), i.e. they have the highest relative standard deviation.

Impact of Mining Tasks and Lessons Learned. We show that our binary supervised classifier based solely on SPARQL syntactic features helps identifying workload peaks of SPARQL endpoints. This approach based on statistical learning overcomes missing query response time in access logs, as we cannot assume that logs include such information. Results show that SVC leads to encouraging results, with satisfactory detection of light queries. Heavy queries prediction comes with acceptable precision, although recall shows margin for improvement, namely through enhancements in feature extraction (e.g. adding property paths support) and training on larger datasets. The main disadvantage of our approach is the need for a adequately large and diverse training set. Class imbalance is also a limitation: *heavy* queries are rare - among more than 60k distinct queries from USEWOD logs which return 200 OK, only 560 led to mean execution times above the chosen cut-off with satisfactory relative standard deviation. Another limitation of our approach is the need to re-train the model if $t_{cut-off}$ is changed.

Session detection proves vital to deliver a more accurate estimate of the traffic on a Linked Data server: sessions enable understanding the duration of each visit, how often visitors come back, and how many distinct resources are requested during such time frame. They also serve as a course-grained alternative to the request count, thus helping gauging visitor engagement. The main benefit of HAC is its performance with intertwined human-machine traffic - a distinctive feature of Linked Data: we show that HAC outperforms fixed cut-offs on session gaps with high relative standard deviation. This is important for session analysis in Linked Data because, unlike traditional web, we must deal at the same time with *both* visits from desktop browsers (presumably by humans, hence with more condensed session gaps - see Fig. 7a), and from software libraries (that generate more scattered session durations). HAC also supports the dual access-protocol nature of Linked Data, since the time-based version of HAC presented in this paper is access protocol-independent. The heuristic can therefore be adopted as baseline, thus laying the foundations for content-based session detection heuristics that take into account SPARQL, HTTP, or mixed sessions. Note that, since we do not inject tracking code to identify single visitors, session detection with HAC should be considered a lower-bound estimate: we cannot circumvent intermediate components between visitors and datasets - e.g. caches (ISP, browsers,

	μ	σ	c_v	Median	95th perc	Support
All UAs	4h16m	6h16m	1.47	1h35m	18h20m	439
Desktop Browsers	5h43m	6h52m	1.20	2h57m	19h46m	259
Software Libraries	1h14m	3h10m	2.57	29m	3h18m	158

(a)

		P	R	F1
All UAs	HAC	0.91	0.72	0.80
	15 m	0.58	0.99	0.73
	30 m	0.72	0.75	0.73
	60 m	0.87	0.67	0.75
Desktop Browsers	HAC	0.88	0.63	0.73
	15 m	0.60	0.99	0.75
	30 m	0.77	0.96	0.86
	60 m	0.91	0.87	0.89
Software Libraries	HAC	0.96	0.93	0.95
	15 m	0.53	1.00	0.69
	30 m	0.43	0.23	0.30
	60 m	0.76	0.15	0.26

(b)

Fig. 7. (a) Gaps between manual annotated sessions in the evaluation dataset, and (b) session detection evaluation results: HAC algorithm vs fixed cut-offs (*All UAs* also includes 22 cut-offs from UA types not included in the table).

or others), proxy servers, NAT. Besides, visitors might fake user agent strings, thus leading to imprecise visitor identification.

Insights on the British National Bibliography Traffic Logs. We tested the the system on 13 months of logs of the British National Bibliography dataset (March 2014–April 2015). The dataset contains almost 100 million triples about books and serials, and is accessible with HTTP and SPARQL. The platform automatically filtered out 99.4% of records, as such requests were generated by search engines and malicious crawlers. Only 0.6% are genuine calls to the triplestore. The platform shows that the combined traffic of HTTP operations and SPARQL queries over the observed period increased by 30%. Interestingly, we identified a 95-fold increase in requests from software libraries, i.e. clients that interact with the triplestore by means of SPARQL queries. We identified 49 unique applications of this kind. Desktop browsers generated 62% of requests (either HTTP operations or SPARQL queries). SPARQL access accounts for 29% of total requests. 6% of queries have been classified as *heavy*. We identified 37 days that show unusual traffic spikes. One of these findings consists in a 1-h spike of 10 thousands *light* SELECT SPARQL queries. The traffic analytics platform also indicates the town and host network of origin, and shows that such queries come from a Java-based application. With this in mind, dataset administrators quickly found in logs that such spike consists in thousands of identical queries - probably originated from a bug in the client application. The platform shows that 33% of requests come from the United States, with the United Kingdom totalling 22%. Nevertheless, the most common city of origin of visitors is Frankfurt, in Germany. WHOIS lookups shows that the dataset has been accessed by more than 250 universities, at least 100 government-related host networks worldwide, and at least 20 other libraries. Our platform also measures user retention: over the analysed 13 months, bounce rate is 48%, meaning that almost half of visitors never came back after the first visit. The average monthly percentage of new visitors is 74%. Visitor session detection shows that visits

Instance	Freq (%)
resource/009910399	2.4
resource/007073756	1.9
person/LewisCS%28CliveStaples%291898-1963	0.6
person/TolkienJRR%28JohnRonaldReuel%291892-1973	0.6
resource/007073756/publicationevent/LondonHarperCollins1993c1978	0.6

(a)

Class	Freq (%)	Property	Freq (%)
dcterms:BibliographicResource	0.7	bibo:isbn10	8.7
bibo:Author	0.5	dcterms:title	4.9
bibo:Book	0.4	rdfs:label	3.2
resource/Author	0.2	dcterms:creator	2.4
foaf:Person	0.1	foaf:name	2.3

(b)   (c)

Fig. 8. Top-5 RDF instances, classes, and properties of the BNB dataset (HTTP operations *and* SPARQL queries). (a) instances, (b) classes, and (c) properties.

generated by software libraries have bigger, deeper, and longer sessions: the average daily size of software libraries sessions is 24 resources (against 2 resources for sessions originated from desktop browsers). The average daily depth of software libraries sessions is 11 unique resources (desktop browsers show on average 2 distinct resources per session). Software library sessions last in average 1 h and 3 min, while desktop browser sessions only 27 min. Figure 8 shows the most popular RDF instances, classes, and properties of the dataset.

Reproducibility and Public Demo. A public demo of the system is available at http://bit.ly/ld-traffic [3]. Datasets used for experiments with the SPARQL classifier and the session detector are available at http://bit.ly/traffic-ESWC2017. The heavy/light SPARQL classifier is implemented with `scikit-learn` 0.18.1 `SVC` and `MultinomialNaiveBayes`. Hyperparameters not listed in Sect. 4 are set to scikit-learn 0.18.1 defaults. Pseudo-code of the session detection algorithm is presented in Fig. 5. All experiments have been executed on an Intel Xeon E5-2420 v2 @ 2.20 GHz (1 socket, 6 cores, 12 threads), 128 GB RAM, Ubuntu 14.04 LTS, Virtuoso version 7.10.3211 (Virtuoso memory usage settings: `NumberOfBuffers=5450000`, `MaxDirtyBuffers=4000000`).

6 Conclusions and Future Work

We present a novel traffic analytics platform that relieves publishers from manual and time-consuming access log mining. We add novel Linked Data-specific metrics - to break down traffic by RDF content, capture SPARQL insights, and properly interpret 303 patterns. Platform aside, we also propose two mining tasks adopted by the system: the reconstruction of Linked Data visitors sessions - which we are the first to achieve with time-based hierarchical agglomerative clustering (establishing a baseline for future Linked Data-optimized heuristics), and the detection of workload peaks of SPARQL endpoints, achieved by predicting heavy and light SPARQL queries with a novel approach based on supervised learning and SPARQL syntactic features. The analysis of 13 months of access logs of the British National Bibliography dataset shows that our system effectively reveals visitors insights otherwise hidden to dataset publishers. These findings are useful, among all, to gauge SPARQL traffic spikes, and monitor trends (e.g. HTTP vs SPARQL traffic over time). They also help justifying investment in Linked Data, and enhancing the popularity of a dataset: for example, the awareness of decreasing user retention might prompt for better promotion (e.g. hackatons, spreading the word on community mailing lists, etc.). Also, if portions of a dataset are never accessed, perhaps better data documentation is required.

Future extensions will include additional metrics, such as statistics about noisy traffic that is now simply discarded by the platform (i.e. web crawlers). The heavy/light classifier feature set will be refined, and we will investigate whether this approach generalizes to additional linked datasets. We will enhance time-based session detection with content-based heuristics, such as relatedness of subsequent SPARQL queries, and structure and type of requested RDF entities.

Acknowledgements. This work has been supported by the TOMOE project funded by Fujitsu Laboratories Limited in collaboration with Insight Centre at NUI Galway.

References

1. Arlitt, M.: Characterizing web user sessions. ACM SIGMETRICS Perform. Eval. Rev. **28**(2), 50–63 (2000)
2. Buil-Aranda, C., Hogan, A., Umbrich, J., Vandenbussche, P.-Y.: SPARQL web-querying infrastructure: Ready for action? In: Alani, H., Kagal, L., Fokoue, A., Groth, P., Biemann, C., Parreira, J.X., Aroyo, L., Noy, N., Welty, C., Janowicz, K. (eds.) ISWC 2013. LNCS, vol. 8219, pp. 277–293. Springer, Heidelberg (2013). doi:10.1007/978-3-642-41338-4_18
3. Costabello, L., Vandenbussche, P., Shukair, G., Deliot, C., Wilson, N.: Access logs don't lie: Towards traffic analytics for linked data publishers. In: Proceedings of ISWC Posters & Demos Track (2016)
4. Demartini, G., Enchev, I., Wylot, M., Gapany, J., Cudré-Mauroux, P.: BowlognaBench—Benchmarking RDF analytics. In: Aberer, K., Damiani, E., Dillon, T. (eds.) SIMPDA 2011. LNBIP, vol. 116, pp. 82–102. Springer, Heidelberg (2012). doi:10.1007/978-3-642-34044-4_5
5. Dividino, R., Gröner, G.: Which of the following SPARQL queries are similar? why? In: Proceedings of LD4IE Workshop (2013)
6. Fasel, D., Zumstein, D.: A fuzzy data warehouse approach for web analytics. In: Lytras, M.D., Damiani, E., Carroll, J.M., Tennyson, R.D., Avison, D., Naeve, A., Dale, A., Lefrere, P., Tan, F., Sipior, J., Vossen, G. (eds.) WSKS 2009. LNCS (LNAI), vol. 5736, pp. 276–285. Springer, Heidelberg (2009). doi:10.1007/978-3-642-04754-1_29
7. Gallego, M.A., Fernández, J.D., Martínez-Prieto, M.A., de la Fuente, P.: An empirical study of real-world SPARQL queries. In: Proceedings of USEWOD (2011)
8. Halfaker, A., Keyes, O., Kluver, D., Thebault-Spieker, J., Nguyen, T., Shores, K., Uduwage, A., Warncke-Wang, M.: User session identification based on strong regularities in inter-activity time. In: Proceedings of WWW, pp. 410–418 (2015)

9. Hasan, R., Gandon, F.: A machine learning approach to SPARQL query performance prediction. In: Proceedings of WI, vol. 1, pp. 266–273. IEEE (2014)
10. Heath, T., Bizer, C.: Linked Data: Evolving the Web into a Global Data Space. Synthesis Lectures on the Semantic Web. Morgan & Claypool, Palo Alto (2011)
11. Luczak-Roesch, M., Berendt, B., Hollink, L.: USEWOD 2015 Research Dataset (2015). http://dx.doi.org/10.5258/SOTON/379407
12. Mehrzadi, D., Feitelson, D.G.: On extracting session data from activity logs. In: Proceedings of ISS, p. 3. ACM (2012)
13. Möller, K., Hausenblas, K., Cyganiak, R., Handschuh, S.: Learning from linked open data usage: Patterns & metrics. In: Proceedings of Web Science (2010)
14. Murray, G.C., Lin, J., Chowdhury, A.: Identification of user sessions with hierarchical agglomerative clustering. In: ASIS&T, vol. 43(1), 1–9 (2006)
15. Pallis, G., Angelis, L., Vakali, A.: Model-based cluster analysis for web users sessions. In: Hacid, M.-S., Murray, N.V., Raś, Z.W., Tsumoto, S. (eds.) ISMIS 2005. LNCS (LNAI), vol. 3488, pp. 219–227. Springer, Heidelberg (2005). doi:10.1007/11425274_23
16. Pérez, J., Arenas, M., Gutierrez, C.: Semantics and complexity of SPARQL. In: Cruz, I., Decker, S., Allemang, D., Preist, C., Schwabe, D., Mika, P., Uschold, M., Aroyo, L.M. (eds.) ISWC 2006. LNCS, vol. 4273, pp. 30–43. Springer, Heidelberg (2006). doi:10.1007/11926078_3
17. Petridou, S.G., Koutsonikola, V.A., Vakali, A.I., Papadimitriou, G.I.: Time-aware web users' clustering. IEEE Trans. Knowl. Data Eng. 20(5), 653–667 (2008)
18. Picalausa, F., Vansummeren, S.: What are real SPARQL queries like? In: Proceedings of SWIM, p. 7. ACM (2011)
19. Schmidt, M., Meier, M., Lausen, G.: Foundations of SPARQL query optimization. In: ICDT, pp. 4–33. ACM (2010)
20. Ye, C., Wilson, M.L., Rodden, T.: Develop, implement, and improve a web session detection model. In: Proceedings of IIiX, pp. 336–338. ACM (2014)

Explaining Graph Navigational Queries

Valeria Fionda[1] and Giuseppe Pirrò[2]([⊠])

[1] DeMaCS, University of Calabria, Rende, Italy
fionda@mat.unical.it
[2] Institute for High Performance Computing and Networking,
ICAR-CNR, Rende, Italy
pirro@icar.cnr.it

Abstract. Graph navigational languages allow to specify pairs of nodes in a graph subject to the existence of paths satisfying a certain regular expression. Under this evaluation semantics, connectivity information in terms of intermediate nodes/edges that contributed to the answer is lost. The goal of this paper is to introduce the GeL language, which provides query evaluation semantics able to also capture connectivity information and output graphs. We show how this is useful to produce query explanations. We present efficient algorithms to produce explanations and discuss their complexity. GeL machineries are made available into existing SPARQL processors thanks to a translation from GeL queries into CONSTRUCT SPARQL queries. We outline examples of explanations obtained with a tool implementing our framework and report on an experimental evaluation that investigates the overhead of producing explanations.

1 Introduction

Graph data pervade everyday's life; social networks, biological networks, and Linked Open Data are just a few examples of its spread and flexibility. The limited support that relational query languages offer in terms of recursion stimulated the design of query languages where *navigation* is a first-class citizen. Regular Path Queries (RPQs) [6], and Nested Regular Expressions (NREs) [16] are some examples. Also SPARQL has been extended with a (limited) navigational core called property paths (PPs). As for query evaluation, the only tractable languages are 2RPQs and NREs; adding conjunction (C2RPQs) makes the problem intractable (NP-complete) while evaluation of PPs still glitches (mixing set and bag semantics). Usually, queries in all these languages ask for *pairs of nodes* connected by paths conforming to a regular language over binary relations.

We research the problem of enhancing navigational languages with explanation functionalities and introduce the Graph Explanation Language (GeL in short). In particular, our goal is to *define formal semantics and efficient evaluation algorithms for navigational queries that return graphs useful to explain the results of a query.*

Part of this work was done while G. Pirrò was working at the WeST institute, University of Koblenz-Landau supported by the FP7 SENSE4US project.

E. Blomqvist et al. (Eds.): ESWC 2017, Part I, LNCS 10249, pp. 19–34, 2017.
DOI: 10.1007/978-3-319-58068-5_2

GeL is useful in many contexts where one needs to connect the dots [11]; from bibliographic networks to query debugging [8]. The practical motivation emerged from the SENSE4US FP7 project[1] aiming at creating a toolkit to support information gathering, analysis and policy modeling. Here, explanations are useful to enable users to find out previously unknown information that is of relevance for a query, understand how it is of relevance, and navigate it. For instance, a GeL query on DBpedia and OpenEI using concepts like Country and Vehicle (extracted from a policy document) allows to retrieve, for instance, the pair (Germany, ElectricCar) and its explanation, which includes the company ThyssenKrupp (intermediate node). This allows to deduce that ThyssenKrupp is potentially affected by policies about electric cars.

GeL by Example. We now give an example of what GeL can express (the syntax and semantics are introduced in Sect. 3).

Example 1 (Co-authors). *ISWC co-authors between 2002 and 2015.*

```
?x foaf:maker([swrc:series{=" swrc:semweb"}]&&[dc:issued({>2002}&&{<2015}]))/^foaf:maker ?y
```

The query uses path concatenation (/) nesting ([]), boolean combinations (&&) of (node) tests { } and backward navigation (^). The GeL syntax is purposely similar to previous navigational languages (e.g., NREs [16]). What makes a difference is the query evaluation semantics. Under the semantics of previous navigational languages, the evaluation would only look for pairs of co-authors (bindings of the variables ?x and ?y) connected by paths (in the graph) that satisfy the query. Under the GeL semantics, one can obtain both *pairs* of co-authors and a graph that gives an account of *why* each pair is an answer. Figure 1 shows the GUI of our explanation tool when evaluating the query on RDF data from DBLP. The tool allows to detail the explanation for each node in the answer. We only report explanations for ?x → S. Staab in Fig. 1(a) and ?x → C. utierrez in Fig. 1(b). One can see *why* S. Staab is linked with his co-authors; he had a paper with P. Mika, and R. Siebes and J. Brokestra are also authors of this paper. As for C. Gutierrez, we see that he had 8 ISWC papers, two of which with the same co-authors (i.e., M. Arenas and J. Pérez).   ◀

Contributions and Outline. We contribute: *(i)* GeL, which to the best of our knowledge is the first graph navigational language able to produce (visual) query explanations; *(ii)* formal semantics; *(iii)* efficient algorithms; *(iv)* a GeL2CONSTRUCT translation, which makes our framework readily usable on existing SPARQL processors; *(v)* an evaluation that investigates the overhead of the new explanation-based semantics.

The remainder of the paper is organized as follows. Section 3 provides some background, presents the GeL language, formalizes the notion of graph query explanation and introduces the formal semantics of the language. Section 4 presents the evaluation algorithms, a study of their complexity and outlines the GeL2CONSTRUCT translation. We discuss an experimental evaluation in Sect. 5, sketch future work and conclude in Sect. 6.

---

[1] http://www.sense4us.eu.

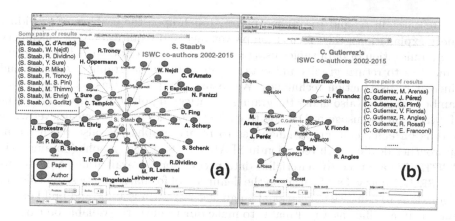

Fig. 1. ISWC co-authorship for S. Staab and C. Gutierrez.

2 Related Work

The core of graph query languages are Regular Path Queries (RPQs) that have been extended with other features, among which, conjunction (CRPQs) [6], inverse (C2RPQs) and the possibility to return and compare paths (EXPQs) [5]. Languages such as Nested Regular Expressions (NREs) [16] allow existential tests in the form of nesting, in a similar spirit to XPath. Finally, some languages have been proposed for querying RDF or Linked Data on the Web (e.g., [1,2,12,13]). There are some drawbacks that hinder the usage of these languages for our goal. The evaluation of queries in these languages (apart from ERPQs) returns set of pairs of nodes or set of (solution) mappings and no connectivity information is kept. Query evaluation in most of these languages (including ERPQs that return graphs) is not tractable (combined complexity); those languages that are tractable (e.g., NREs, RPQs) do not output graphs. We design efficient algorithms to reconstruct parts of the graph traversed to build to the answer.

There are approaches to retrieve subgraphs, querying for semantic associations and/or providing relatedness explanations. As for the first strand, we mention ρ-Queries [4] and SPARQ2L [3]; here the idea is to enhance RDF query languages to deal with semantic associations or path variables and constraints. Our work is different since we focus on navigational queries and our algorithms for query evaluation under the graph semantics are polynomial. As for the second strand, we mention RECAP [17], RelFinder [14], Explass [7] that generate relatedness explanation when giving as input two entities and a maximum distance k. The idea is to generate SPARQL queries (typically 2^k SPARQL queries) to retrieve paths of length k connecting the input pair; then, show paths after performing some filtering (e.g., only considering a subset of paths). The input of these approaches is a set of entities while in our case is a declarative navigational query; moreover, these approaches consider paths of a fixed length k (given as input) and require (SPARQL) queries to find these paths. Our work is also related to: *(i)* provenance (e.g., [9]); *(ii)* annotations (e.g., [22]) and *(iii)* module extraction (e.g., [19]).

Research in *(i)* and *(ii)* do not directly touch upon the problem of providing query explanations; their focus is on provenance and require complex machinery (e.g., annotation of data tuples using semirings). Our work defines formal query semantics to return graphs and obtain explanations via graph navigation and efficient reconstruction techniques. We focus on a precise class of queries that can be evaluated (also under the graph semantics) in polynomial time. The focus of *(iii)* is on the usage of Datalog to extract modules at ontological level while ours is on enhancing graph languages to return graphs. We also recall recent approaches dealing with recursion in SPARQL [18], where graphs (obtained via CONSTRUCT) are used to materialize data needed for the evaluation of the recursive SELECT query. Our focus is on the definition of formal semantics and efficient algorithms to enhance navigational languages to return graphs, and build explanations in an efficient way. Finally, to make our framework available on existing SPARQL processors we have devised a GEL2CONSTRUCT translation.

3 Building Query Explanations with GeL

We now provide some background information and then present the GeL language. We focus our attention on the Resource Description Framework (RDF). An RDF triple is a tuple of the form $\langle s, p, o \rangle \in \mathbf{I} \times \mathbf{I} \times \mathbf{I} \cup \mathbf{L}$, where $\mathbf{I}$ (IRIs) and $\mathbf{L}$ (literals) are countably infinite sets. Since we are interested in producing query explanations we do not consider bnodes. An RDF graph G is a set of triples. The set of terms of a graph will be $terms(G) \subseteq \mathbf{I} \cup \mathbf{L}$; $nodes(G)$ will be the set of terms used as a subject or object of a triple while $triples(G)$ is the set of triples in G. Since SPARQL property paths offer very limited expressive power, we will consider the well-known Nested Regular Expressions (NREs) as reference language. NREs [16] allow to express existential tests along the nodes in a path via nesting (in the same spirit of XPath) while keeping the (combined) complexity of query evaluation tractable. Each NRE $nexp$ over an alphabet of symbols Σ defines a *binary relation* $[\![nexp]\!]^G$ when evaluated over a graph G. The result of the evaluation of an NRE *is a set of pairs of nodes*. Other extensions (e.g., EPPs [10]) although adding expressive power to NREs (e.g., EPPs add path conjunction and path difference), all return pairs of nodes. This motivates the introduction of GeL, which tackles the problem of returning graphs from the evaluation of navigational queries that also help to explain query results.

3.1 Syntax of GeL

The syntax of GeL is defined by the following grammar:

$gexp := \tau \# exp \quad (\tau \in \{\text{FULL, FILTERED}, set\})$

$exp := a \; gtest(a \in \Sigma) \mid \hat{\ } a \; gtest(a \in \Sigma) \mid exp \, / \, exp \mid exp \mid exp \mid exp^* \mid exp\{l, h\}$

$gtest := gtest \; \&\& \; gtest \mid gtest \mid\mid gtest \mid (gtest) \mid [exp] \mid \{op \; val\}$

$op := > \mid < \mid = \mid \neq$

In the syntax, ˆ denotes backward navigation, / path concatenation, | path union, $\{l,h\}$ denotes repetition of an *exp* between l and h times; && and || conjunction and disjunction of *gtest*, respectively. Moreover, when the *gtest* is missing after a predicate, it is assumed to be the constant **true**. We kept the syntax of the language similar to that of NREs and other languages. We define novel query semantics and evaluation algorithms capable of: (i) returning graphs; (ii) keeping query evaluation tractable; (iii) building query explanations. The syntactic construct τ allows to output the answer either in the form of pairs of nodes (i.e., *set*) as usually done by previous navigational languages or in the form *of an explanation*. GeL can produce two types of explanations: one keeping the whole portion of the graph "touched" during the evaluation (FULL) and the other keeping only paths leading to results (FILTERED).

3.2 Semantics of GeL

Tiddi et al. [21] define explanations as *generalizations of some starting knowledge mapped to another knowledge under constraints of certain criteria*. We use GeL queries to define starting knowledge and Explanation Graphs (EGs) to formally capture the criteria that knowledge included into an explanation has to satisfy.

Definition 2 (Explanation Graph). *Given a graph G, a GeL expression e and a set of starting nodes $S \subseteq nodes(G)$, an EG is a quadruple $\Gamma{=}(V,E,S,T)$ where $V \subseteq nodes(G)$, $E \subseteq triples(G)$ and $T \subseteq V$ is a set of ending nodes, that is, nodes reachable from nodes in S via paths satisfying e.*

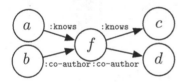

Fig. 2. An example graph.

Consider the graph G in Fig. 2 and the expression $e{=}(\texttt{:knows/:knows})|(\texttt{:co-author/:co-author})$. The answer with the semantics based on pairs of nodes (i.e., NREs) is the set of pairs of nodes: (a,c), (b,d). Under the GeL semantics, since there are two starting nodes a and b from which the evaluation produces results, one possibility would be to consider the EG capturing all results, that is, $\Gamma{=}(nodes(G), triple(G), \{a,c\}, \{b,d\})$. However, one may note that there exists a path (via the node f) from a to d in the EG even if the pair (a,d) does not belong to the answer. This could lead to misinterpretation of the query results and their explanation. To avoid these situations, we define G-soundness and G-completeness for EGs.

Definition 3 (G-Soundness). *Given a graph G and a GeL expression e, an EG is G-sound iff each ending node is reachable in EG from each starting node via a path satisfying e.*

Definition 4 (G-Completeness). *Given a graph G and a GeL expression e, an EG is G-complete iff all nodes reachable from some starting nodes, via a path satisfying e, are in the ending nodes.*

The EG Γ in the above example violates G-soundness because there exists only one path (via the node f) from a to d in Γ and such path does not satisfy the expression e. The following lemma guarantees G-soundness and G-completeness.

Lemma 1 (G-Sound and G-Complete EGs). *Explanation Graphs having a single starting node $v \in nodes(G)$ are G-sound and G-complete.*

Definition 5 (Query Explanation). *Given a GeL expression e and a graph G, a query explanation $\mathcal{E}^Q$ is a set of G-sound and G-complete EGs Γ_v.*

Returning to our example, the query explanation is the set $\{\Gamma_a, \Gamma_b\}$ s.t.:

1. $\Gamma_a = (\{a,f,c\}, \{\langle a, :\mathtt{knows}, f\rangle, \langle f, :\mathtt{knows}, c\rangle, a, \{a,c\})$
2. $\Gamma_b = (\{b,f,d\}, \{\langle b, :\mathtt{co\text{-}author}, f\rangle, \langle f, :\mathtt{co\text{-}author}, d\rangle\}, b, \{b,d\})$.

Note that query answering under the semantics returning pairs of nodes can be represented via the query explanation composed of the set of EGs: $\{\Gamma_v = (\emptyset, \emptyset, v, T) \mid v \in nodes(G)\}$. Since the formal semantics of GeL manipulates EGs, we now define: the counterpart for EGs of the composition ($\circ$) and union ($\cup$) operators used for binary relations, and operators to work with sets of EGs.

Definition 6 (EGs operators). *Let $\Gamma_i = (V_i, E_i, s_i, T_i)$, $i = 1, 2$ be EGs and $\Gamma_\perp = (\emptyset, \emptyset, \perp, \emptyset)$ denote the empty EG, where $\perp$ is a symbol not in the universe of nodes.*

Composition ($\circ$) and union ($\cup$) of EGs:

$$\Gamma_1 \circ \Gamma_2 = \begin{cases} \Gamma_\perp & \text{if } s_2 \notin T_1, \\ (V_1 \cup V_2, E_1 \cup E_2, s_1, T_2) & \text{if } s_2 \in T_1. \end{cases}$$

$$\Gamma_1 \cup \Gamma_2 = \begin{cases} (V_1 \cup V_2, E_1 \cup E_2, s_1, T_1 \cup T_2) & \text{if } s_1 = s_2, \\ \Gamma_1 & \text{if } s_1 \neq s_2 \wedge \Gamma_2 = \Gamma_\perp, \\ \Gamma_2 & \text{if } s_1 \neq s_2 \wedge \Gamma_1 = \Gamma_\perp, \\ \text{not defined} & \text{if } s_1 \neq s_2 \wedge \Gamma_1, \Gamma_2 \neq \Gamma_\perp. \end{cases}$$

The following definition formalizes extensions of the above operators over sets of EGs; here, the binary operator op $\in \{\circ, \cup\}$ is applied to all pairs Γ_1, Γ_2 such that Γ_1 belongs to the first set and Γ_2 to the second one.

Definition 7 (Operations over sets of EGs). *Let S_1 and S_2 be two sets of EGs.*

1. *For each op $\in \{\circ, \cup\}$ we define S_1 op $S_2 = \{\Gamma_1 \text{op} \Gamma_2 \mid \Gamma_1 \in S_1, \Gamma_2 \in S_2\}$.*
2. *(Disjoint union, direct sum): $S_1 \oplus S_2 = \{\Gamma \mid \Gamma \in S_1 \vee \Gamma \in S_2\}$.*

We now introduce the semantics of GeL in two variants: FULL (Λ) returning the portion of the graph *visited* during the evaluation and FILTERED (Φ), which only considers *successful* paths. Let G be a graph and e a GeL expression. Under the Λ semantics an explanation is the set of EGs where each EG Γ_v includes the nodes and edges of G *traversed* during the evaluation of e from $v \in nodes(G)$. For this semantics we introduce, in Table 1, the evaluation function $E_\Lambda[\![exp]\!]^G$.

Table 1. The FULL(Λ) and FILTERED(Φ) semantics of GeL EGs. (*): rule valid for both. Repetitions of GeL expressions are translated into unions of concatenations. In lines 5, 6, 10 and 11 if $gtest$ is not present then $\Lambda[\![gtest]\!]_{v}^{G} = \Phi[\![gtest]\!]_{v}^{G} = \texttt{true}$.

Construct	Semantics
$E[\![full\#exp]\!]^{G}$ $E[\![filt\#exp]\!]^{G}$	$E_{\Lambda}[\![exp]\!]^{G}$ $E_{\Phi}[\![exp]\!]^{G}$
$E_{\Lambda}[\![exp]\!]^{G}$	$\left\{ \bigoplus\limits_{v \in nodes(G)} \bigcup\limits_{\Gamma \in \Lambda[\![exp]\!]_{v}^{G}} \Gamma \right\}$
$E_{\Phi}[\![exp]\!]^{G}$	$\left\{ \bigoplus\limits_{v \in nodes(G)} \bigcup\limits_{\Gamma \in \Phi[\![exp]\!]_{v}^{G} \mid \Gamma.T \neq \emptyset} \Gamma \right\}$
$\Lambda[\![a\ gtest]\!]_{v}^{G}$	$\bigcup\limits_{(v,a,v') \in G \mid (\{v'\}, \emptyset, v', \{v'\}) \in \Lambda[\![gtest]\!]_{v'}^{G}} (\{v, v'\}, \{(v, a, v')\}, v, \{v'\})$
$\Lambda[\![\hat{}\ a\ gtest]\!]_{v}^{G}$	$\bigcup\limits_{(v',a,v) \in G \mid (\{v'\}, \emptyset, v', \{v'\}) \in \Lambda[\![gtest]\!]_{v'}^{G}} (\{v, v'\}, \{(v', a, v)\}, v, \{v'\})$
$\Lambda[\![exp_1 / exp_2]\!]_{v}^{G}$	$\Lambda[\![exp_1]\!]_{v}^{G} \circ \left(\bigoplus\limits_{v' \in \Gamma.T \mid \Gamma \in \Lambda[\![exp_1]\!]_{v}^{G}} (\Lambda[\![exp_2]\!]_{v'}^{G} \oplus (\{v'\}, \emptyset, v', \emptyset)) \right)$
$\Lambda[\![exp^{*}]\!]_{v}^{G}$	$(\{v\}, \emptyset, v, \{v\}) \cup \left(\bigcup\limits_{i=1}^{\infty} \Lambda[\![(exp_i)]\!]_{v}^{G} \mid exp_1 = exp \wedge exp_i = exp_{i-1}/exp \right)$
$\Lambda[\![exp_1 \mid exp_2]\!]_{v}^{G}$	$\Lambda[\![exp_1]\!]_{v}^{G} \cup \Lambda[\![exp_2]\!]_{v}^{G}$
$\Phi[\![a\ gtest]\!]_{v}^{G}$	$\bigoplus\limits_{(v,a,v') \in G \mid (\{v'\}, \emptyset, v', \{v'\}) \in \Phi[\![gtest]\!]_{v'}^{G}} (\{v, v'\}, \{(v, a, v')\}, v, \{v'\})$
$\Phi[\![\hat{}\ a\ gtest]\!]_{v}^{G}$	$\bigoplus\limits_{(v',a,v) \in G \mid (\{v'\}, \emptyset, v', \{v'\}) \in \Phi[\![gtest]\!]_{v'}^{G}} (\{v, v'\}, \{(v', a, v)\}, v, \{v'\})$
$\Phi[\![exp_1 / exp_2]\!]_{v}^{G}$	$\Phi[\![exp_1]\!]_{v}^{G} \circ \left(\bigoplus\limits_{v' \in \Gamma.T \mid \Gamma \in \Phi[\![exp_1]\!]_{v}(v)} \Phi[\![exp_2]\!]_{v'}^{G} \right)$
$\Phi[\![exp^{*}]\!]_{v}^{G}$	$(\{v\}, \emptyset, v, \{v\}) \oplus \left(\bigoplus\limits_{i=1}^{\infty} \Phi[\![(exp_i)]\!]_{v}^{G} \mid exp_1 = exp \wedge exp_i = exp_{i-1}/exp \right)$
$\Phi[\![exp_1 \mid exp_2]\!]_{v}^{G}$	$\Phi[\![exp_1]\!]_{v}^{G} \oplus \Phi[\![exp_2]\!]_{v}^{G}$
$\Phi[\![[exp]]\!]_{v}^{G}$	$\begin{cases} (\{v\}, \emptyset, v, \{v\}) \bigcup \Gamma & \text{if } \exists\ \Gamma \in \Phi[\![exp]\!]_{v}^{G} \wedge \Gamma.T \neq \emptyset \\ (\bot, \emptyset, \bot, \emptyset) & \text{otherwise} \end{cases}$
$(*)[\![[exp]]\!]_{v}^{G}$	$\begin{cases} (\{v\}, \emptyset, v, \{v\}) \bigcup \Gamma & \text{if } \exists\ \Gamma \in (\Lambda \mid \Phi)[\![exp]\!]_{v}^{G} \wedge \Gamma.T \neq \emptyset \\ \Gamma_{\bot} & \text{otherwise} \end{cases}$
$(*)[\![\{op\ val\}]\!]_{v}^{G}$	$\begin{cases} (\{v\}, \emptyset, v, \{v\}) & \text{if } Evaluate(v, op, val) = \texttt{true} \\ \Gamma_{\bot} & \text{otherwise} \end{cases}$
$(*)[\![gtest_1\ \&\&\ gtest_2]\!]_{v}^{G}$	$\begin{cases} (\{v\}, \emptyset, v, \{v\}) & \text{if } (\Lambda \mid \Phi)[\![gtest_1]\!]_{v}^{G} \neq \Gamma_{\bot} \wedge (\Lambda \mid \Phi)[\![gtest_2]\!]_{v}^{G} \neq \Gamma_{\bot} \\ \Gamma_{\bot} & \text{otherwise} \end{cases}$
$(*)[\![gtest_1 \mid\mid gtest_2]\!]_{v}^{G}$	$\begin{cases} (\{v\}, \emptyset, v, \{v\}) & \text{if } (\Lambda \mid \Phi)[\![gtest_1]\!]_{v}^{G} \neq \Gamma_{\bot} \vee (\Lambda \mid \Phi)[\![gtest_2]\!]_{v}^{G} \neq \Gamma_{\bot} \\ \Gamma_{\bot} & \text{otherwise} \end{cases}$

One may be only interested in the portion of G that actually contributed to build the answer; this gives the second semantics, where the query explanation is defined as the set of EGs such that each Γ_v only considers paths that start from $v \in nodes(G)$ and satisfy the expression (i.e., the successful paths). We introduce the evaluation function $E_{\Phi}[\![exp]\!]^{G}$ in Table 1. The difference between the semantics lays in the sets of nodes (V) and edges (E) included in the explanations graphs that form a query explanation. An expression is evaluated either via the rule at line 1 or 2, depending on the type of semantics (explanation) wanted.

<u>GeL expressiveness.</u> We chose NREs as reference language and added the possibility to test for node values reached when evaluating a nested expression and boolean combinations of tests. We added this type of tests since they allow to express queries like those in Example 1. Nevertheless, the focus of this paper is

on defining semantics and evaluation algorithms for navigational languages to output graphs besides pairs of nodes. This feature is not available in any existing navigational language (e.g., NREs [16], EPPs [10], SPARQL property paths).

4   Algorithms and Complexity

This section presents algorithms for the evaluation of GeL expressions under the novel semantics that also generate query explanations. The interesting result is that the evaluation of a GeL expression e in this new setting can be done efficiently. Let e be a GeL expression and G a graph. Let $|e|$ be the size of e, Σ_e the set of edge labels appearing in it, and $|G|=|nodes(G)| + |triples(G)|$ be the size of G. Algorithms that build explanations according to the FULL or FILTERED semantics are automata-based and work in two steps. The first step is shared and leverages *product automata*; the second step requires a *marking phase* only for the FILTERED semantics and is needed to include nodes and edges in the EGs that are relevant for the answer.

Building Product Automata. The idea is to associate to e (and to each *gtest* on the form of $[exp]$) a non deterministic finite state automaton with ϵ transitions $\mathcal{A}_e$ ($\mathcal{A}^{exp}$, resp.). Such automata can be built according to the standard Thomson construction rules over the alphabet $Voc(e)=\Sigma_e \cup \bigcup_{gtest \in e} gtest$, that is, by considering also *gtest* in e as basic symbols. The product automaton is a tuple $G \times \mathcal{A}_e = \langle Q^e, Voc(e), \delta^e, Q_0^e, F^e \rangle$ where Q^e is a set of states, $\delta^e : Q^e \times (Voc(e) \cup \epsilon) \to 2^{Q^e}$ is the transition function, $Q_0^e \subseteq Q^e$ is the set of initial states, and $F^e \subseteq Q^e$ is the set of final states. The building of the product automaton $G \times \mathcal{A}_e$ is based on an extension of the algorithm used by [16] based on the labeling of the nodes of G. In this phase, G is labeled wrt nested subexpressions in e, that is, for each node $n \in nodes(G)$ and nested subexpression exp in e, $exp \in label(n)$ if and only if there exists a node n' such that there is a path from n to n' in G satisfying exp. This allows to recursively label the graph G for each $[exp]$; hence, when the labeling wrt exp has to be computed, G has already been labeled wrt all the nested subexpressions $[exp']$ in exp.

Theorem 8 ([16]). *The product $G \times \mathcal{A}_e$ can be built in time $O(|G| \times |e|)$.*

Building Explanations. We now discuss algorithms that leverage product automata (of the GeL expression e and all nested subexpressions) to produce graph query explanations according to the FULL and FILTERED semantics. To access the elements of an explanation graph Γ (see Definition 2) we use the notation $\Gamma.x$, with $x \in \{V, E, S, T\}$. The main algorithm is Algorithm 1, which receives the GeL expression and the type of explanation to be built. In case of the FILTERED semantics the data structure **reached**, which maintains a set of states (n_i, q_j), is initialized via the procedure **mark** (line 3) reported in Algorithm 2; otherwise, it is initialized as the union of: *(i)* all the states of the product automaton $G \times \mathcal{A}_e$; *(ii)* all the states of the product automata $(G \times \mathcal{A}_{exp})$ of all the nested expressions in e (line 5). The procedure **mark** fills the set **reached** with all the states in all the product automata that contribute to obtain an answer; these are the states in a path from an initial state to a final state in the product automata. As shown in Algorithm 2, reached is populated by navigating the product automata backward from the final states to the initial ones. Then, the set of EGs composing a query explanation $\mathcal{E}^Q$ are initialized (lines 6–7; 9) by adding to $\mathcal{E}^Q$ an EG Γ_s for each initial state (s, q_0) of $G \times \mathcal{A}_e$. Moreover, the data structure seen is also initialized (line 8) by associating to each state (s, q_0) the node s (associated to the initial state (s, q_0)) from which it has been visited.

Input : GeL expression e, graph G, τ (*full* or *filt*)
Output: $\mathcal{E}^Q$: a query explanation as set of EG Γ_s
1. build the product automaton $G \times \mathcal{A}_e$
2. **if** *filt* /* filtered semantics */ **then**
3.   reached = $\underline{\text{mark}}(G \times \mathcal{A}_e, \emptyset)$
      /* reached keeps nodes in $G \times \mathcal{A}_e$ that are in a path to a final state */
4. **else**
5.   reached = $Q^e \cup \bigcup_{[exp] \text{ in } e} Q^{exp}$
6. **for all** $(s, q_o) \in Q_0^e$ **do**
7. $\Gamma_s = \langle \{s\}, \emptyset, s, \emptyset \rangle$
8.   seen$_{(s, q_o)} = \{s\}$
      /*seen for each state s_j keeps nodes in $G \times \mathcal{A}_e$ from which it has been reached*/
9. $\mathcal{E}^Q = \bigcup_{(s, q_o) \in Q_0^e} \{ \Gamma_s \}$
10. visit = $\bigcup_{(s, q_o) \in Q_0^e} \{ ((s, q_0), \{s\}) \}$
      /* visit keeps nodes to be visited */
11. $\underline{\text{buildE}}(G \times \mathcal{A}_e, \text{reached}, \mathcal{E}^Q, \text{visit})$
Algorithm 1. BuildExpl (e, G, τ)

Input: product automaton $G \times \mathcal{A}$, set of states reached
Build: set of states reached

1 reached = reached $\cup \bigcup_{(n, q_f) \in F^e} \{ (n, q_f) \}$
2 visit = $\bigcup_{(n, q_f) \in F^e} \{ (n, q_f) \}$ s.t. $q_f \in F$
3 visitN = $\emptyset$
4 **while** visit $\neq \emptyset$ **do**
5   **for all** (n, q) in visit **do**
6   **for all** transition $\delta((n', q'), x) \in G \times \mathcal{A}^e$ s.t. $(n, q) \in \delta((n', q'), x)$ **do**
7   **if** $(n', q') \notin$ reached **then**
8   visitN = visitN $\cup \{ (n', q') \}$
9   reached = reached $\cup \{ (n', q') \}$
10   **if** x is a *gtest* **then**
11   **for all** [exp] in x **do**
12                    $\underline{\text{mark}}(G \times \mathcal{A}^{exp}, \text{reached})$
13   visit = visitN
14   visitN = $\emptyset$
15 **return** reached
Algorithm 2. mark($G \times \mathcal{A}^e$, reached)

The data structure seen maintains for each state, reached while visiting the product automata, the starting nodes from which this state has already been visited. The usage of seen avoids to visit the same state more than once for each starting node. Finally, the data structure `visit` is also initialized with the initial states of $G \times \mathcal{A}_e$ (line 10); it contains all the states to be visited in the subsequent step plus the set of starting nodes for which these states have to be visited. Then, the EGs are built via **buildE** (Algorithm 3); all the states in `visit` are considered (line 2) only once for the entire set $B_{n,q}$, which keeps starting nodes for which states in `visit` have to be processed (line 3). Then, for each state $(n,q) \in$ `visit` all its transitions are considered (line 7); for each state $(n',q') \in$ `reached`, reachable from some $(n,q) \in visit$ via some transitions (line 8), the set of "new" starting nodes (D) for which (n',q') has to be visited in the subsequent step is computed with a possible update of the sets `visit` and `seen` (lines 9–12). If the transition is labeled with a predicate symbol in G (line 13), the EGs corresponding to nodes $s \in B_{n,q}$ are constructed by adding the corresponding nodes and edges (lines 14–16). If the transition is a *gtest* the building of the query explanation $\mathcal{E}^Q$ proceeds recursively by visiting the product automata associated to all nested (sub)expressions for *gtest* (lines 17–21).

Input: product $G \times \mathcal{A}^{\bar{e}}$, set of states *reached*, Explanation $\mathcal{E}^Q$, states to *visit*

```
1  visitN = ∅
2  for all (n, q) in visit do
3      B_{n,q} = ∪_{((n,q),S)∈visit} S
4      for all s ∈ B_{n,q} do
5          if q ∈ F^e then
6              add n to Γ_s.T
7      for all  transition δ^ē((n, q), x) do
8          for all  (n', q') ∈ δ^ē((n, q), x) s.t (n', q') ∈ reached do
9              D = B_{n,q} \ seen_{(n,q)}
10             if D ≠ ∅ then
11                 visitN = visitN ∪ {((n', q'), D)}
12                 seen_{(n',q')} = seen_{(n',q')} ∪ D
13             if x ∈ Σ_e then
14                 for all s ∈ B_{n,q} do
15                     add n' to Γ_s.V
16                     add (n, x, n') to Γ_s
17             else if x is a gtest then
18                 for all [exp] ∈ gtest do
19                     let (n, q_0) ∈ Q_0^{exp}
20                     seen_{(n,q_0)} = seen_{(n,q_0)} ∪ B_{n,q}
21                     buildE(G × A^{exp}, reached, E^Q, {(n, q), B_{n,q}})
22  buildE(A_e × G, reached, E^Q, visitN)
```
Algorithm 3. buildE($G \times \mathcal{A}$, reached, $\mathcal{E}^Q$, visit)

Theorem 9. *Given a graph G and a GeL expression e, the query explanation $\mathcal{E}^Q$ (according to both semantics) can be computed in time $\mathcal{O}(|nodes(G)| \times |G| \times |e|)$.*

Proof. The explanation $\mathcal{E}^Q$ built according to the FULL semantics can be constructed by visiting $G \times \mathcal{A}_e$ (Algorithm 3). In particular, for each starting state (n,q), the states and transitions of $G \times \mathcal{A}_e$ are all visited at most once (and the same also holds for the automata corresponding to the nested expressions of e). The starting and ending nodes of each EG are set during the visit of the product automaton. For each node s corresponding to a starting state $(s, q_o) \in Q_0^e$ an explanation graph Γ_s is created (Algorithm 1, lines 6–7); the set of nodes reachable from s is set to be $\Gamma_s.T = \{n \mid (n, q) \in F^e$ and (n, q) is reachable from $(s, q_o)\}$ (Algorithm 3 lines 5–6). Thus, each EG can be computed by visiting each transition and each node exactly once with a cost $O(|Q^e| + \sum_{[exp] \in e} |Q^{exp}| + |\delta^e| + \sum_{[exp] \in e} |\delta^{exp}|) = O(|G| \times |e|)$. Since the number of EGs to be constructed is bound by $|nodes(G)|$, the total cost of building the query explanation $\mathcal{E}^Q$, is $\mathcal{O}(|nodes(G)| \times |G| \times |e|)$. This bounds also take into account the cost of building product automata as per Theorem 8.

In the case of the FILTERED semantics, the marking phase does not increase the complexity bound; this is because the set `reachable`, which keeps reachable states, is built by visiting at most once all nodes and transitions in all the product automata, with a cost $O(|G| \times |e|)$.   □

Note that in Algorithm 3, the amortized processing time per node is lower than $|G| \times |e|$ when visiting the product automaton since the Breadth First Search(es) from each starting state are concurrently run according to the algorithm in [20]. Finally, the EGs in the $\mathcal{E}^Q$ built via Algorithm 1 are both G-sound and G-complete. It is easy to see by the definition of the product automaton, that there exists a starting state (n, q_0) that is connected to a final state (n', q_f) in $G \times \mathcal{A}_e$ and, thus, a path from n to n' in Γ_n if, and only if, there exists a path connecting n to n' in G satisfying e.

4.1 Translating GeL into SPARQL

The algorithms discussed in Sect. 4 are suitable for the implementation of GeL on a custom query processor. This has the advantage to guarantee a low complexity of query evaluation as we have formally proved. On the other hand, there is SPARQL, which is the standard for querying RDF data although offering limited navigational capabilities (via property paths). We wondered how the machineries developed for GeL *could be made available on existing SPARQL processors*. This will have the advantage of making GeL readily available for usage on the tremendous amount of RDF data accessible through SPARQL endpoints. We have devised a formal translation (GEL2CONSTRUCT) from GeL queries into CONSTRUCT SPARQL queries that produce RDF graphs as results of a SPARQL query. In particular, since current SPARQL processors can handle limited forms of recursive queries (as studied by Fionda et al. [10]) only a subset of GeL queries can actually be turned into CONSTRUCT queries. Such queries do not include closure operators (i.e., *). We have included in GeL path repetitions, that is, the possibility to express in a succinct way the union of concatenations of a GeL expression between l and h times. When translating GeL into CONSTRUCT queries one has to give up two main things. First, the complexity of query evaluation increases even if one can now rely on efficient and mature SPARQL query processors. Second, it is possible to only produce explanations

Table 2. Translating GeL into SPARQL (CONSTRUCT).

Construct	Translation
Θ^e(root)	'CONSTRUCT{'Θ^c(root.child(1))'} WHERE {'Θ^w(root.child(1))'}'
$\Gamma(n)$	n.s n.p n.o'.'
$\Theta^c(n^{/\ |\ |})$	Θ^c(n.child(1)) Θ^c(n.child(2))
$\Theta^c(n^{(u|^\wedge u)gtest})$	$\Gamma(n)$
$\Theta^w(n^{/})$	Θ^w(n.child(1)) Θ^w(n.child(2))
$\Theta^w(n^{|})$	'{'Θ^w(n.child(1))'} UNION{'Θ^w(n.child(2))'}'
$\Theta^w(n^{(u|^\wedge u)gtest})$	$\Gamma(n)$ 'FILTER('n.p'='u').' Θ^t(n.child(1))
$\Theta^t(n^{\&\&})$	Θ^t(n.child(1)) Θ^t(n.child(2))
$\Theta^t(n^{||})$	'{'Θ^t(n.child(1))'} UNION{'Θ^t(n.child(2))'}'
$\Theta^t(n^{[nexp]})$	'FILTER EXISTS'{'Θ^w(n.child(1))'}'
$\Theta^t(n^{\{op\ val\}})$	'FILTER('n.o op val')'

under the FILTERED semantics as SPARQL processors only provide parts of the graph that contribute to the answer while GeL relies on automata-based algorithms to also keep parts touched that do not contribute to the answer. Table 2 gives an overview of the translation. The translation algorithm, starts from the root of the parse tree of a GeL expression and applies translation rules recursively. Each GeL syntactic construct has associated a chuck of SPARQL code.

Theorem 10. *For every (non-recursive) GeL query* $\mathcal{P}=(\alpha, gel, \beta)$, $\alpha, \beta \in \mathcal{V} \cup \mathcal{I}$, *there exists a* CONSTRUCT *query* $Q_e = \mathcal{A}^t(\mathcal{P})$ *such that for every RDF graph G it holds that* $[\![\mathcal{P}]\!]_G = [\![Q_e]\!]_G$. *The* GEL2CONSTRUCT *algorithm* $\mathcal{A}^t$ *runs in time* $O(|\mathcal{P}|)$.

Proof (Sketch). The proof works by checking that the propagation of variable names (artificially generate) and terms along the parse tree is correct (see e.g., [10]).   □

5   Implementation and Evaluation

We implemented GeL and the explanation framework in Java. Beside our custom evaluator based on the algorithms discussed in Sect. 4, we have also implemented the GEL2CONSTRUCT translation to make available GeL's capabilities into existing SPARQL engines in an elegant and non-intrusive way.

Experimental Setting. We tested our approach using different datasets. The first is a subset of the FOAF network (~4M triples) obtained by crawling from 10 different seeds foaf:knows predicates up to distance 6 and then merging the graphs. The second one, is the Linked Movie Database (LMDB)[2], an RDF dataset containing information about movies and actors (~6M triples). We also considered data from YAGO (via the LOD cache[3]) (~22B triples) and DBpedia[4]

---

[2] http://linkedmdb.org.

[3] http://lod.openlinksw.com/sparql.

[4] http://dbpedia.org/sparql.

(~412M triples). The goal of the evaluation *is to measure the overhead of out-putting graphs as a result of navigational queries and build query explanations.* Because of the novelty of our approach it was not possible to compare it against other implementations, or run standard benchmarks to test the overhead of out-putting graphs instead of pairs of nodes. We tested the overhead of producing explanations both when using our custom processors and on SPARQL endpoints and also measured the size of the output returned. We used 6 queries per dataset for a total of 24 queries plus their SPARQL translation. Experiments have been run on a PC i5 CPU 2.6 GHz and 8 GB RAM; results are the average of 5 runs.

Overhead using the custom processor. We considered 6 queries (on FOAF data) including concatenations and *gtest* that ask for (pairs of) friends at increasing distance (from 1 to 6) with the condition that each friend (in the path) must have a link to his/her home page. For sake of space we report *the overhead* of generating explanations about Tim Berners-Lee (TBL) along with the size of the explanation (#nodes,#edges) generated under the FILTERED and FULL *semantics* (Tables 3 and 4). We observed a similar behavior when considering explanations related to other people in the FOAF network (e.g., A. Polleres, N. Lopes).

<table>
<tr><td colspan="3">Table 3. Overhead (secs).</td></tr>
<tr><th>FOAF</th><th>Filtered</th><th>Full</th></tr>
<tr><td>Q1</td><td>0.434</td><td>0.278</td></tr>
<tr><td>Q2</td><td>0.738</td><td>0.234</td></tr>
<tr><td>Q3</td><td>0.985</td><td>0.534</td></tr>
<tr><td>Q4</td><td>1.155</td><td>0.849</td></tr>
<tr><td>Q5</td><td>1.665</td><td>1.145</td></tr>
<tr><td>Q6</td><td>1.785</td><td>1.257</td></tr>
</table>

Table 3. Overhead (secs).

FOAF	Filtered	Full
Q1	0.434	0.278
Q2	0.738	0.234
Q3	0.985	0.534
Q4	1.155	0.849
Q5	1.665	1.145
Q6	1.785	1.257

Table 4. Size of the explanation.

FOAF	Filtered	Full
Q1	(6, 5)	(17, 17)
Q2	(18, 37)	(20, 45)
Q3	(18, 44)	(25, 53)
Q4	(23, 51)	(55, 90)
Q5	(36, 64)	(149, 236)
Q6	(177, 111)	(190, 139)

The evaluation of GeL queries under the explanation semantics does not have a significant impact on query processing time (the overhead is max. ~2 s) for friends at distance 6. This is not surprising as it confirms the complexity analysis discussed in Sect. 4 where we showed that our explanations algorithms run in polynomial time. The output of a GeL query clearly requires more space as it is a (explanation) graph. As one may expect, the FULL semantics produces larger graphs than the FILTERED semantics as it reports all parts of the graph touched (i.e., even paths that did not lead to any result). We can observe that for TBL, at distance 6 the explanation contains 190 nodes and 139 edges (resp., 177 and 111) under the FULL semantics (resp., FILTERED). The visual interface of the tool implementing GeL (see Fig. 1) allows to picking one node in the output and generate the corresponding explanation graph, zoom the graph, change the size of nodes/edges and perform free text search for nodes/edges. Running time for all queries were in the order of 6 seconds. Note that our algorithms work with the graphs loaded into main memory. In the next experiment we measure the overhead of generating explanations on large set of triples.

Overhead on SPARQL endpoints. Since we made available GeL's machinery also via CONSTRUCT queries, we tested the overhead of generating explanation (graphs) also on different datasets and SPARQL endpoints both local and remote. We set up a local BlazeGraph[5] instance where we loaded LMDB and accessed the other datasets via their endpoints. For each dataset we created 6 GeL queries and translated them into: (i) SELECT queries to mimic the semantics returning pairs of nodes and (ii) CONSTRUCT queries to mimic the explanation semantics. At this point, we need to make two important observations about generating explanations via translation into SPARQL. First, it is only possible to consider the FILTERED semantics as SPARQL engines do not keep track of the portions of the graph visited that did not contribute to the answer necessary for the FULL semantics. Second, explanations are only G-complete (see Definition 5) as it is not possible to keep separate the explanation for each node in the result of a CONSTRUCT while it can be done in GeL by using Explanation Graphs (see Definition 2). The overhead and size of results for DBpedia and YAGO are reported in the following figures.

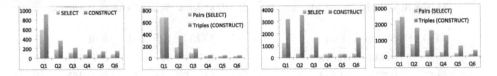

Fig. 3. DBpedia Time. **Fig. 4.** DBpedia Size. **Fig. 5.** YAGO Time. **Fig. 6.** YAGO size.

As it can be observed, running time for the CONSTRUCT (explanation) queries are always higher in DBpedia (Fig. 3) but always ∼1 s. The size of results (# triples) (Fig. 4) reaches 800 for Q1, which asks for (all pairs) fo people that have influenced each other (no filters). From Q2–Q6 each person in an influence path must be a scientist; this filter decreases at each step the size of the answer (∼100 for Q6). For YAGO (Fig. 5), accessed via LOD cache, we also observe that CONSTRUCT queries (asking for influences in YAGO among female people) require more time (<3 s) than SELECT queries, with an overhead of ∼2 s. Even in this case the overhead of generating explanations (considering the larger number of results generated) is bearable (Fig. 6). On LDMB (results not reported for sake of space) the overhead was of ∼1.5 s with average size of the explanation ∼700 triples. The GEL2CONSTRUCT translation (integrated in our tool) allows to obtain explanations from a variety of SPARQL endpoints online.

---

[5] https://www.blazegraph.com.

6   Concluding Remarks and Future Work

We have shown how current navigational languages (e.g., NREs) can be enhanced to return graphs besides pairs of nodes. Such kind of information is useful whenever one needs to connect the dots (e.g., bibliographic networks, exploratory search). We have described a language, formalized two semantics, and provided algorithms that use connectivity information to produce different types of query explanations. The interesting aspect is that query answering under the new explanation semantics is still tractable. We gave some examples of (visual) explanations generated with a tool implementing our framework and using real world data. There are several avenues for future research, among which: (i) studying explanations with negative information (e.g., which parts of a query failed); (ii) studying the expressiveness of GeL; (iii) assisting the user in writing queries [15]; (iv) including RDFS inferences.

References

1. Acosta, M., Vidal, M.-E.: Networks of linked data eddies: an adaptive web query processing engine for RDF data. In: Arenas, M., Corcho, O., Simperl, E., Strohmaier, M., d'Aquin, M., Srinivas, K., Groth, P., Dumontier, M., Heflin, J., Thirunarayan, K., Staab, S. (eds.) ISWC 2015. LNCS, vol. 9366, pp. 111–127. Springer, Cham (2015). doi:10.1007/978-3-319-25007-6_7
2. Alkhateeb, F., Baget, J.-F., Euzenat, J.: Extending SPARQL with regular expression patterns (for querying RDF). J. Web Sem. 7(2), 57–73 (2009)
3. Anyanwu, K., Maduko, A., Sheth, A.: SPARQ2L: towards support for subgraph extraction queries in RDF databases. In: WWW, pp. 797–806. ACM (2007)
4. Anyanwu, K., Sheth, A.: p-Queries: enabling querying for semantic associations on the semantic web. In: WWW, pp. 690–699. ACM (2003)
5. Barceló, P., Libkin, L., Lin, A.W., Wood, P.T.: Expressive languages for path queries over graph-structured data. ACM TODS 37(4), 31 (2012)
6. Calvanese, D., De Giacomo, G., Lenzerini, M., Vardi, M.Y.: Containment of conjunctive regular path queries with inverse. In: KR, pp. 176–185 (2000)
7. Cheng, G., Zhang, Y., Explass, Y.: Exploring associations between entities via top-k ontological patterns and facets. In: Proceedings of ISWC, pp. 422–437 (2014)
8. Consens, M.P., Liu, J.W.S., Rizzolo, F.: Xplainer: visual explanations of XPath queries. In: ICDE, pp. 636–645. IEEE (2007)
9. Dividino, R., Sizov, S., Staab, S., Schueler, B.: Querying for provenance, trust, uncertainty and other meta knowledge in RDF. J. Web Semant. 7(3), 204–219 (2009)
10. Fionda, V., Pirrò, G., Consens, M.P., Paths, E.P.: Writing more SPARQL queries in a succinct way. In: AAAI (2015)
11. Fionda, V., Gutierrez, C., Pirrò, G.: Building knowledge maps of web graphs. Artif. Intell. 239, 143–167 (2016)
12. Fionda, V., Pirrò, G., Gutierrez, C.: NautiLOD: a formal language for the web of data graph. ACM Trans. Web 9(1), 5:1–5:43 (2015)
13. Hartig, O., Pérez, J.: LDQL: a query language for the web of linked data. In: Arenas, M., Corcho, O., Simperl, E., Strohmaier, M., d'Aquin, M., Srinivas, K., Groth, P., Dumontier, M., Heflin, J., Thirunarayan, K., Staab, S. (eds.) ISWC 2015. LNCS, vol. 9366, pp. 73–91. Springer, Cham (2015). doi:10.1007/978-3-319-25007-6_5

14. Heim, P., Hellmann, S., Lehmann, J., Lohmann, S., Stegemann, T.: RelFinder: revealing relationships in RDF knowledge bases. In: Semantic Multimedia, pp. 182–187 (2009)
15. Lehmann, J., Bühmann, L.: AutoSPARQL: let users query your knowledge base. In: Antoniou, G., Grobelnik, M., Simperl, E., Parsia, B., Plexousakis, D., Leenheer, P., Pan, J. (eds.) ESWC 2011. LNCS, vol. 6643, pp. 63–79. Springer, Heidelberg (2011). doi:10.1007/978-3-642-21034-1_5
16. Pérez, J., Arenas, M., Gutierrez, C.: nSPARQL: a navigational language for RDF. J. Web Semant. 8(4), 255–270 (2010)
17. Pirrò, G.: Explaining and suggesting relatedness in knowledge graphs. In: Arenas, M., Corcho, O., Simperl, E., Strohmaier, M., d'Aquin, M., Srinivas, K., Groth, P., Dumontier, M., Heflin, J., Thirunarayan, K., Staab, S. (eds.) ISWC 2015. LNCS, vol. 9366, pp. 622–639. Springer, Cham (2015). doi:10.1007/978-3-319-25007-6_36
18. Reutter, J.L., Soto, A., Vrgoč, D.: Recursion in SPARQL. In: Arenas, M., Corcho, O., Simperl, E., Strohmaier, M., d'Aquin, M., Srinivas, K., Groth, P., Dumontier, M., Heflin, J., Thirunarayan, K., Staab, S. (eds.) ISWC 2015. LNCS, vol. 9366, pp. 19–35. Springer, Cham (2015). doi:10.1007/978-3-319-25007-6_2
19. Rousset, M.-C., Ulliana, F.: Extracting bounded-level modules from deductive RDF triplestores. In: Proceedings of the AAAI (2015)
20. Then, M., Kaufmann, M., Chirigati, F., Hoang-Vu, T., Pham, K., Kemper, A., Neumann, T., Vo, H.T.: The more the merrier: efficient multi-source graph traversal. VLDB Endowment 8(4), 449–460 (2014)
21. Tiddi, I., d'Aquin, M., Motta, E.: An ontology design pattern to define explanations. In: K-CAP, p. 3 (2015)
22. Zimmermann, A., Lopes, N., Polleres, A., Straccia, U.: A feneral framework for representing, reasoning and querying with annotated semantic web data. J. Web Semant. 11, 72–95 (2012)

A SPARQL Extension for Generating RDF from Heterogeneous Formats

Maxime Lefrançois$^{(\boxtimes)}$, Antoine Zimmermann, and Noorani Bakerally

Univ Lyon, MINES Saint-Étienne, CNRS, Laboratoire Hubert Curien UMR 5516,
42023 Saint-Étienne, France
{maxime.lefrancois,antoine.zimmermann,noorani.bakerally}@emse.fr

Abstract. RDF aims at being the universal abstract data model for structured data on the Web. While there is effort to convert data in RDF, the vast majority of data available on the Web does not conform to RDF. Indeed, exposing data in RDF, either natively or through wrappers, can be very costly. Furthermore, in the emerging Web of Things, resource constraints of devices prevent from processing RDF graphs. Hence one cannot expect that all the data on the Web be available as RDF anytime soon. Several tools can generate RDF from non-RDF data, and transformation or mapping languages have been designed to offer more flexible solutions (GRDDL, XSPARQL, R2RML, RML, CSVW, etc.). In this paper, we introduce a new language, SPARQL-Generate, that generates RDF from: (i) a RDF Dataset, and (ii) a set of documents in arbitrary formats. As SPARQL-Generate is designed as an extension of SPARQL 1.1, it can provably: (i) be implemented on top on any existing SPARQL engine, and (ii) leverage the SPARQL extension mechanism to deal with an open set of formats. Furthermore, we show evidence that (iii) it can be easily learned by knowledge engineers that know SPARQL 1.1, and (iv) our first naive open source implementation performs better than the reference implementation of RML for big transformations.

1 Introduction

We aim at lowering the overhead for web services and constrained things to embrace the Semantic Web formalisms and tool. A usual key step is to generate RDF from documents having various formats (or *triplify*). Indeed, companies and web services store and exchange documents in a multitude of data models and formats: the relational data model and XML (not RDF/XML) are still very present, data portals heavily rely on CSV, and web APIs on JSON. Furthermore, constrained things on the Web of things may be only able to support binary formats such as EXI or CBOR. Although effort has been made to define RDF data formats that are also compatible with the formats in use (e.g., RDF/XML is compatible with XML, JSON-LD is compatible with JSON, any EXI version

---

This paper has been partly financed by the ITEA2 12004 SEAS (Smart Energy Aware Systems) project, the ANR 14-CE24-0029 OpenSensingCity project, and a bilateral research convention with ENGIE R&D.

E. Blomqvist et al. (Eds.): ESWC 2017, Part I, LNCS 10249, pp. 35–50, 2017.
DOI: 10.1007/978-3-319-58068-5_3

of RDF/XML is compatible with EXI, etc.), it is unlikely that these formats will completely replace existing data formats one day. However, the RDF data *model* may still be used as a *lingua franca* to reach semantic interoperability and integration and querying of data having heterogeneous formats.

Several pieces of research and development focused on generating RDF from other models and formats, and sometimes led to the definition of standards. However, in the context of projects we participate in, we identified use cases and requirements that existing approaches satisfy only partially. These are reported in Sect. 2 and include:

- the solution must be expressive, flexible, and extensible to new data formats;
- the solution must generate RDF from several data sources with heterogeneous formats, potentially in combination with a RDF dataset;
- the solution should be easy to learn and to integrate in a typical semantic web engineering workflow, so that knowledge engineers can learn it easily to start prototyping triplifications.

Section 3 describes existing solutions and identify their limitations. In order to satisfy these requirements, we introduce SPARQL-Generate, an extension of SPARQL 1.1 that answers the aforementioned requirements and combines the following advantages: (1) it leverages SPARQL's expressivity and flexibility, including the standard extension mechanism for binding functions; (2) it may be implemented on top of any existing SPARQL engine.

The rest of this paper is organized as follows. Section 4 formally specifies the abstract syntax and semantics of the SPARQL-Generate language. These definitions enable to prove in Sect. 5.1 that it can be implemented on top of any existing SPARQL 1.1 engine, and propose a naive algorithm for this. Section 5.2 briefly describes a first open-source implementation on top of Apache ARQ, which has been tested on use cases from the related work and more. Finally, Sect. 5.3 proposes a comparative evaluation between SPARQL-Generate and RML on two aspects: performance of the reference implementations, and cognitive complexity of the query/mapping.

2   Use-Cases and Requirements

We identified two important use cases for generating RDF from heterogeneous data formats. They are originating from projects in which the stakeholders require strong interoperability in consuming and exchanging data, although data providers cannot afford the cost to move towards semantic data models.

Open Data. In the context of open data, organizations can rarely afford the cost of cleaning and reengineering their datasets towards more interoperable linked data. They sometimes also lack the expertise to do so. Therefore, data is published on a best effort basis in the formats that require least labour and resources. Yet, data consumers expect more uniform, self describing data sets that can be easily cross-related. In the case when a knowledge model has been agreed upon, it is important for the users to be able to prototype transformations to RDF from one or more of these data sources, potentially in different formats. In addition, the solution should be flexible enough to allow for fine-grained control on the generated RDF and the links between data sets, and should be able to involve contextual RDF data. The list of formats from which RDF may be generated must be easily extended. Finally, the solution must be easily used by knowledge engineers that know RDF and SPARQL.

Web of Things. In the emerging Web of Things, constrained devices must exchange lightweight messages due to their inherent bandwidth, memory, storage, and/or battery constraints. Yet, RDF formats have to encode a lot of textual information such as IRIs and literals with datatype IRIs. Although some research is led to design lightweight formats for RDF (such as a CBOR version of JSON-LD), it is likely that companies and device vendors will continue to use and introduce new binary formats that are optimized for their usage.

From these use cases, we identify the following requirements:

R1: transform several sources having heterogeneous formats;
R2: contextualize the transformation with an RDF Dataset;
R3: be extensible to new data formats;
R4: be easy to use by Semantic Web experts;
R5: integrate in a typical semantic web engineering workflow;
R6: be flexible and easily maintainable;
R7: transform binary formats as well as textual formats.

With these requirements in mind, the next section overviews existing solutions.

3 Related Work

Data publisher and consumer can implement *ad-hoc* transformation mechanisms to generate RDF from data with heterogeneous models and formats. Although this approach certainly leads to the most efficient solutions, it is also costly to develop and maintain, and inflexible. Several pieces of work aimed at simplifying this task.

Many *converters to RDF* have been listed by the W3C Semantic Web Education and Outreach interest group (SWEO): https://www.w3.org/wiki/ ConverterToRdf. Most of them target a specific format or specific metadata, such as ID3tag, BibTeX, EXIT, etc. Some like Apache Any23, datalift, or Virtuoso Sponger are designed to convert multiple data formats to RDF. Direct Mapping [1] describes a default transformation for relational data. These solutions are very ad hoc, implementation specific and barely allow the control of how RDF is generated. They do not provide a formal language that would allow to explicit and customize the conversion to RDF. As a result, the output RDF is often more geared towards describing the structure of the data rather than the data itself. It is still possible to compose these solutions with SPARQL construct rules that transform the generated RDF to the required RDF, but this requires to get familiar with the vocabulary used in the output of each of these tools. They hence do not satisfy most of the requirements listed in Sect. 2.

Other approaches propose to use a transformation or mapping language to tune the RDF generation. However, most of these solutions target only one or a few specific data models (e.g., the relational model) or formats (e.g., JSON). For instance GRDDL encourages the use of XSLT and targets XML inputs [2]. XSPARQL is based on XQuery and originally targeted XML [11], as well as the inverse transformation from RDF to XML, before being extended to the relational data model [10], then to JSON [4]. GRDDL and XSPARQL rely respectively on XSLT and XQuery, that have been proven to be Turing-complete. These languages are hence full-fledged procedural programming language with explicit algorithmic constructs to produce RDF.

Other formalisms have been designed to generate RDF from the relational data [7]. From these pieces of work originated R2RML [3], which proposes a RDF vocabulary to describe mappings to RDF. Finally, CSVW [12] also adopts this approach but targets the CSV data format.

One approach that stands out is RML [5], that extends the R2RML vocabulary to describe logical sources which are different from relational database tables. It generates RDF from JSON (exploiting JSONPath), XML (exploiting XPath), CSV[1], TSV, or HTML (exploiting CSS3 selectors). The approach is implemented on top of Sesame[2]. RML satisfies at least requirements R1, R3, R5. It would be possible to implement the support of binary data formats (R7), and ongoing research are led to integrate RDF sources on the Web of Linked Data (R2). Only RML and XSPARQL are specifically dedicated to the flexible generation of RDF from various formats.

In what follows, we propose an alternative to these approaches that is based on an extension of SPARQL 1.1, named SPARQL-Generate, that leverages its expressiveness and extensibility, and can be implemented on top of its engines.

4   SPARQL-Generate Specification

SPARQL-Generate is based on a query language that queries the combination of an RDF dataset and what we call a *documentset*, where each document is named and typed by an IRI. For illustration purposes, Fig. 1 is an example of a SPARQL-Generate query and the result of its execution on a RDF dataset that contains a default graph, and on a documentset that contains two documents identified by `<position.txt>` and `<measures.json>`. This query answers the question: *"What sensors are nearby, and what do they display?"*.[3] The concrete SPARQL-Generate syntax extends that of SPARQL 1.1 with three new clauses:

---

[1] RML is an implementation of the CSV on the Web standard [12].

[2] http://rdf4j.org/.

[3] Prefixes correspond to those registered at http://prefix.cc/ and are omitted to save space.

- The source clause is used to bind a variable to a document (here, ?pos and ?measures to the documents identified by <position.txt> and <measures.json>, respectively).
- The iterator clause allows to extract bits of documents using so-called *iterator functions*, duplicate a binding, and make a variable be successively bound to these extracted bits of documents (here, function sgiter:JSONListKeys is used to extract the set of keys of the JSON object that is bound to ?measures, and successively bind ?sensorId to these keys).
- Finally, the generate clause replaces and extends the construct clause with embedded SPARQL-Generate queries. This enables the modularization of queries and the factorization of the RDF generation.

Various data formats can be supported thanks to the extensible set of SPARQL 1.1 *binding* functions and SPARQL-Generate *iterator* functions.

Default graph (Turtle)

```
<s25> a :TempSensor ;
      geo:lat 38.677220 ;
      geo:long -27.212627 .
<s26> a :TempSensor ;
      geo:lat 37.790498 ;
      geo:long -25.501970 .
<s27> a :TempSensor ;
      geo:lat 37.780768;
      geo:long -25.496294 .
```

Document position.txt

```
37.780496,-25.495157
```

Document measures.json

```
{ "s25": 14.24,
  "s26": 18.18 }
```

Output (Turtle)

```
<s26> a :NearbySensor ;
      :temp 18.18 .
<s27> a :NearbySensor .
```

SPARQL-Generate query

```
GENERATE {
    ?sensor a :NearbySensor .

GENERATE {
    ?sensorIRI :temp ?temp .
}
ITERATOR sgiter:JSONListKeys(?measures) AS ?sensorId
WHERE {
    BIND( IRI( ?sensorId ) AS ?sensorIRI )
    FILTER( ?sensor = ?sensorIRI )
    BIND( CONCAT( "$." , ?sensorId ) AS ?jsonPath )
    BIND( sgfn:JSONPath( ?measures , ?jsonPath ) AS ?temp )
} .
}
SOURCE <position.txt> AS ?pos
SOURCE <measures.json> AS ?measures
WHERE {
    BIND( sgfn:SplitAtPosition(?pos,"(.*),(.*)",1) AS ?long )
    BIND( sgfn:SplitAtPosition(?pos,"(.*),(.*)",2) AS ?lat )
    ?sensor a :TempSensor .
    ?sensor geo:lat ?slat .
    ?sensor geo:long ?slong .
    FILTER( ex:distance(?lat, ?long, ?slat, ?slong) < 10 )
}
```

Fig. 1. Example of a SPARQL-Generate query execution on a default graph and two documents. This running example illustrates requirements **R1** and **R2**

4.1   SPARQL-Generate Concrete Syntax

The SPARQL-generate syntax is very close to the standard SPARQL 1.1 syntax with only slight additions to the EBNF [6, Sect. 19.8]:

```
[174] GenerateUnit ::= Generate
[175] Generate ::= Prologue GenerateQuery
[176] GenerateQuery ::= 'GENERATE' GenerateTemplate DatasetClause* IteratorOrSourceClause*
      WhereClause? SolutionModifier
[177] GenerateTemplate ::= '{' GenerateTemplateSub '}'
[178] GenerateTemplateSub ::= ConstructTriples? ( SubGenerateQuery ConstructTriples? )*
[179] IteratorOrSourceClause ::= IteratorClause | SourceClause
[180] IteratorClause ::= 'ITERATOR' FunctionCall 'AS' Var
```

```
[181] SourceClause ::= 'SOURCE' FunctionCall ('ACCEPT' VarOrIri )?'AS' Var
[182] SubGenerateQuery ::= 'GENERATE' ( SourceSelector | GenerateTemplate ) (
    IteratorOrSourceClause* WhereClause? SolutionModifier'.' )?
```

While the production of SPARQL Queries and SPARQL Updates respectively start at QueryUnit and UpdateUnit, the production of a SPARQL-Generate query starts at rule GenerateUnit. We wanted to not rewrite any of the SPARQL 1.1 production rules, this is why we do not use construct and introduce generate instead. This concrete syntax has two notable features.

Negotiating the Document Type. The first notable feature is in production rule [181]. The optional part ('ACCEPT' VarOrIri) allows to specify a type IRI for the document to bind in the source clause. If a SPARQL-Generate implementation chooses to look up the IRI of a document on the Web, they may retrieve different actual documents corresponding to different representations of the same resource. The optional accept component in the source clause is thought of as a hint for the implementation to choose how to negotiate the content of that resource. We chose to represent it as a IRI that identifies a document type, because the concept of content negotiation here goes beyond the usual HTTP Accept request header. It may also encompass other HTTP Accept-* parameters, and it may also describe other preferences to look up IRIs not related to the HTTP protocol. After negotiation with the server, the retrieved document type may be different from the requested document type.

Modularization and Reuse of Queries. The second feature is in production rule [182], and enables to modularize queries. A SPARQL-Generate sub-query (i.e., a query in the generate part of a parent query) may contain a generate template, including graph patterns and potentially other sub-queries. It can also refer to a IRI. As for the documentset, implementations are free to choose how this IRI must be looked up to retrieve the identified SPARQL-Generate query. This feature does not need to be described in the abstract syntax, but allows in practice (i) to publish queries on the Web and make them callable by other, and (ii) to modularize large queries and make them more readable. Of course, implementations need to take care about loops in query calls.

For now, SPARQL-Generate implementations are free to choose whether and how they use these informations. Section 5.2 describes the choices we made for our own implementation on top of Apache Jena. Let us now introduce the abstraction of the SPARQL-Generate syntax.

4.2 Abstract Syntax

We note $\mathbf{I}$, $\mathbf{B}$, $\mathbf{L}$, and $\mathbf{V}$ the pairwise disjoint sets of *IRIs*, *blank nodes*, *literals*, and *variables*. The set of *RDF terms* is $\mathbf{T} = \mathbf{I} \cup \mathbf{B} \cup \mathbf{L}$. The set of *triple patterns* is defined as $\mathbf{T} \cup \mathbf{V} \times \mathbf{I} \cup \mathbf{V} \times \mathbf{T} \cup \mathbf{V}$, and a *graph pattern* is a finite set of triple patterns. The set of all graph patterns is denoted $\mathcal{P}$. We denote $\mathbf{F_0}$ the set of SPARQL 1.1 function names,[4] which is disjoint from $\mathbf{T}$. We write $\mathcal{Q}$ the set of

---

[4] SPARQL 1.1 defines built-in functions with names IF, IRI, CONCAT, and so on.

SPARQL 1.1 query patterns. Finally, for any set X, we note $X^* = \bigcup_{n \geqslant 0} X^n$ the set of lists of X.

The set of *function expressions* is noted $\mathcal{E}$ and is the smallest set such that:

$$\mathbf{T} \cup \mathbf{V} \subseteq \mathcal{E} \qquad\qquad\qquad\qquad \text{(e.g., <position.txt>)} \quad (1)$$

$$(\mathbf{F_0} \cup \mathbf{I}) \times (\mathbf{T} \cup \mathbf{V})^* \subseteq \mathcal{E} \qquad \text{(e.g., CONCAT(\"\$.\",?id), sgiter:JSONListKeys(?m))} \quad (2)$$

$$\forall E \subseteq \mathcal{E}, (\mathbf{F_0} \cup \mathbf{I}) \times E^* \subseteq \mathcal{E} \quad \text{(i.e., the set of nested function expressions)} \quad (3)$$

The abstraction of production rule [181] is the set of *source clauses*, and enable to select a document in the documentset and bind it to a variable. For instance in the query above, variable ?pos is bound to the document identified by <position.txt>. Let us introduce a special element $\omega \notin \mathbf{T} \cup \mathbf{V}$, that represents *null*, and let us note $\hat{X} = X \cup \{\omega\}$ the *generalized set of X*.

Definition 1 (source clauses). *The set $\mathcal{S}$ of source clauses is defined by equation $\mathcal{S} = \mathcal{E} \times (\hat{\mathbf{I}} \cup \mathbf{V}) \times \mathbf{V}$. We use notation $v \xleftarrow{source} \langle e, a \rangle \in \mathcal{S}$ for a specific source clause, where $v \in \mathbf{V}$, $e \in \mathcal{E}$, and $a \in \hat{\mathbf{I}} \cup \mathbf{V}$.*

In most use cases, at some point one needs a given variable to iterate over several parts of the same document. For instance in the illustrating request, variable ?sensorId is successively bound to the keys of the JSON object bound to ?measures: "s25" and "s26". Other examples include the results of a XPath query evaluation over a XML document,[5] or the matches of a regular expression over a string.[6] In SPARQL, binding clauses involving binding functions are the only way through which one could extract a term from a literal. Yet, these functions output at most one RDF term. So they cannot be used to generate more solution bindings. Consequently, we introduce a second extension, the set of *iterator* clauses, which output a *set* of terms, and replace the current solution binding with as many solution bindings as there are elements in that set.

Definition 2 (iterator clauses). *The set of iterator clauses is defined as $\mathcal{I} = \mathbf{I} \times \mathcal{E}^* \times \mathbf{V}$. We use notation $v \xleftarrow{iterator} (u, e_0, \ldots, e_k) \in \mathcal{I}$ for a specific iterator clause, where $v \in \mathbf{V}$, $u \in \mathbf{I}$, $e_0, \ldots, e_k \in \mathcal{E}$, and $k \in \mathbb{N}$.*

We then extend the query pattern of SPARQL 1.1 queries $\mathcal{Q}$ with a list of source and iterator clauses, in any number and any order. We purposely do not change the definition of $\mathcal{Q}$ in order to facilitate the reuse of existing SPARQL implementations.

Definition 3 (SPARQL-Generate query patterns). *The set of SPARQL-Generate query patterns is defined as a sequence of source or iterator clauses followed by a query pattern: $\mathcal{Q}^+ = (\mathcal{S} \cup \mathcal{I})^* \times \mathcal{Q}$.*

---

[5] See test case *rmlproeg1* - http://w3id.org/sparql-generate/tests-reports.html.
[6] See test case *regexeg1* - http://w3id.org/sparql-generate/tests-reports.html.

Finally, the set of SPARQL-Generate queries augments Q^+ with a basic graph pattern, and potentially other SPARQL-Generate sub-queries.

Definition 4 (SPARQL-Generate queries). *The set of SPARQL-Generate queries is noted $\mathcal{G}$, and defined as the least set such that:*

$$\mathcal{P} \times Q^+ \subseteq \mathcal{G} \qquad \text{(simple SPARQL-Generate queries)} \qquad (4)$$

$$\forall G \subseteq \mathcal{G}, \mathcal{P} \times G^* \times Q^+ \subseteq \mathcal{G} \qquad \text{(nested SPARQL-Generate queries)} \qquad (5)$$

SPARQL-Generate queries defined by Eq. 4 are comparable to SPARQL CONSTRUCT queries, where a basic graph pattern will be instantiated with respect to a set of solution bindings. Those defined by Eq. 5 contain nested SPARQL-Generate queries, which are used to factorize the generation of RDF. For example, this enables to first generate RDF from the name of all the JSON object keys, and then iterate over the values for these keys, which may be arrays.

4.3 SPARQL-Generate Semantics

This section reuses some concepts from the SPARQL 1.1 semantics, that we redefine in an uncommon, yet equivalent, way for convenience in notations and definitions.

Definition of the SPARQL-Generate Data Model. A SPARQL-Generate query is issued against a data model that extends the one of SPARQL, namely RDF dataset. An *RDF dataset* is a pair $\langle D, N \rangle$ such that D is an RDF graph, called the *default graph*, and N is a finite set of pairs $\langle u, G \rangle$ where u is an IRI and G is an RDF graph, such that no two pairs contain the same IRI. In order to allow the querying of arbitrary data formats, we introduce the notion of a *documentset*, analogous to RDF datasets.

Definition 5 (Documentset). *A documentset is a finite set of triples $\Delta \subseteq \mathbf{I} \times \hat{\mathbf{I}} \times \mathbf{L}$. An element of Δ is a triple $\langle u, a, \langle d, t \rangle \rangle$ where: u is the name of the document; a is the requested type for the document; literal $\langle d, t \rangle$ models the document; and the literal datatype IRI t is the document type. Δ must be such that no pair of distinct triples share the same two first elements.*

In order to lighten formulas, we also note $\Delta : \hat{\mathbf{T}} \times \hat{\mathbf{T}} \to \hat{\mathbf{L}}$ the mapping that associates a pair $\langle u, a \rangle$ to a literal l if and only if $\langle u, a, l \rangle \in \Delta$, and to ω otherwise. A set of documents can hence be stored internally, or represent the Web: u represents where a look up (e.g., a series of HTTP GET following redirections) must be issued, a describes how the content must be negotiated, d is the content of the successfully obtained representation, and t describes the representation type (its media type, language, encoding, etc.).

Mappings. The set of *mappings* is noted $\mathcal{M}$, and is defined by Eq. (6) as a function from $\mathbf{T} \cup \mathbf{V}$ to the generalized set of terms. As opposed to standard SPARQL 1.1, we use a total function defined on the full set of terms and variables, and rely on the element ω to represent the image of unbound variables. As in SPARQL, The *domain* of a mapping is the set of variables that are bound to a term (see Eq. (7)).

$$\mu : \mathbf{T} \cup \mathbf{V} \to \hat{\mathbf{T}} \text{ s.t., } \forall t \in \mathbf{T}, \mu(t) = t \qquad (6)$$

$$\forall \mu \in \mathcal{M}, \mathsf{dom}(\mu) = \{v \in \mathbf{V} | \mu(v) \in \mathbf{T}\} \qquad (7)$$

We introduce a distinguished set of mappings called *substitution mappings*, whose domain is a singleton. i.e., $\forall v \in \mathbf{V}$ and $t \in \hat{\mathbf{T}}$, $[v/t]$ is a substitution mapping with:

$$\forall t' \in \mathbf{T}, [v/t](t') = t', \quad [v/t](v) = t, \quad \text{and } \forall x \in \mathbf{V}, x \neq v, [v/t](x) = \omega \qquad (8)$$

Then, the *left composition* operator $\mathring{,}$ is defined such that in $\mu_1 \mathring{,} \mu_2$, any variable that is commonly bound by μ_1 and μ_2 is finally bound to its value in mapping μ_1. In practice, this may be used to override bindings for variables in source or iterator clauses.

$$\mu_1 \mathring{,} \mu_2 : \begin{cases} x \mapsto \mu_1(x) & \text{if } x \in \mathsf{dom}(\mu_1) \\ x \mapsto \mu_2(x) & \text{if } x \in \mathsf{dom}(\mu_2) \setminus \mathsf{dom}(\mu_1) \\ x \mapsto \omega & \text{otherwise} \end{cases} \qquad (9)$$

Binding and Iterator Function Map. Each SPARQL engine recognizes a set of binding function IRIs F_b (e.g. here, at least `sgfn:JSONPath`, `sgfn:SplitAtPosition`, and `ex:distance`). A binding function maps function expressions used in binding clauses to their evaluation, i.e., a RDF term. Formally, for a given SPARQL engine, Eq. (10) defines a *binding functions map* f_b, that associates to any *recognized binding functions* its SPARQL binding function. The *SPARQL-Generate iterator functions map* is defined analogously for a SPARQL-Generate engine (e.g. here, it recognizes at least `sgiter:JSONListKeys`), except the evaluation of a function expression is a *set* of RDF terms. Given a set F_i of *recognized iterator functions*, Eq. (11) defines the *iterator functions map* f_i:

$$f_b : F_b \to \left(\hat{\mathbf{T}}^* \to \hat{\mathbf{T}}\right) \qquad (10)$$

$$f_i : F_i \to \left(\hat{\mathbf{T}}^* \to 2^{\hat{\mathbf{T}}}\right) \qquad (11)$$

Generalized Mappings. We generalize the definition of mappings so that their domains include the set of function expression. The set of *generalized mappings* is noted $\bar{\mathcal{M}}$. It contains the *generalization* $\bar{\mu}$ of every mapping $\mu \in \mathcal{M}$, where $\bar{\mu} : \mathbf{T} \cup \mathbf{V} \cup \mathcal{E} \to \hat{\mathbf{T}}$ is defined recursively as follows:

$$\forall t \in \mathbf{T} \cup \mathbf{V}, \bar{\mu}(t) = \mu(t) \qquad (12)$$

$$\forall \langle u, e_1, \ldots, e_n \rangle \in \mathcal{E} \text{ s.t. } u \in F_b, \ \bar{\mu}(\langle u, e_1, \ldots, e_n \rangle) = f_b(u)(\bar{\mu}(e_1), \ldots, \bar{\mu}(e_n)) \qquad (13)$$

Evaluation of source *and* iterator *Clauses.* A source clause $v \overset{\text{source}}{\longleftarrow} \langle e, a \rangle \in S$ is used to modify the binding μ so that variable v is bound to a document in Δ (e.g., ?pos is bound to "37.780496,-25.495157"). An iterator clause $v \overset{\text{iterator}}{\longleftarrow} \langle t, e_0, \ldots, e_k \rangle \in I$ is typically used to extract important parts of a document: from a binding μ, it enables, to generate several other bindings where variable v is bound to elements of the evaluation of $f_i(t)$ over $e_0, \ldots, e_k$ (e.g. here, ?sensorId will be successively bound to "s25" then to "s26"). Any number of source or iterator clauses can be combined in a list. Let $\Sigma \in (S \cup I)^n$, and $n \geqslant 1$. The set of solution mappings (i.e., the evaluation) for any list of source and iterator clauses $\llbracket \Sigma \rrbracket_\Delta^\mu$ can be defined by induction as follows:

$$\llbracket v \overset{\text{source}}{\longleftarrow} \langle e, a \rangle \rrbracket_\Delta^\mu = [v / \Delta(\bar{\mu}(e), a)] \,\natural\, \mu \tag{14}$$

$$\llbracket v \overset{\text{iterator}}{\longleftarrow} \langle t, e_0, \ldots, e_k \rangle \rrbracket_\Delta^\mu = \{ [v / t'] \,\natural\, \mu | t' \in f_i(t)(\bar{\mu}(e_0), \ldots, \bar{\mu}(e_k)) \} \tag{15}$$

$$\llbracket \langle \Sigma, v \overset{\text{source}}{\longleftarrow} e \rangle \rrbracket_\Delta^\mu = \{ \llbracket v \overset{\text{source}}{\longleftarrow} e \rrbracket_\Delta^{\mu'} | \mu' \in \llbracket \Sigma \rrbracket_\Delta^\mu \} \tag{16}$$

$$\llbracket \langle \Sigma, v \overset{\text{iterator}}{\longleftarrow} e \rangle \rrbracket_\Delta^\mu = \bigcup_{\mu' \in \llbracket \Sigma \rrbracket_\Delta^\mu} \llbracket v \overset{\text{iterator}}{\longleftarrow} e \rrbracket_\Delta^{\mu'} \tag{17}$$

Evaluation of SPARQL-Generate Query Patterns. Let $Q \in \boldsymbol{Q}$ be a SPARQL 1.1 query pattern, D be an RDF dataset, and $\llbracket Q \rrbracket_D^\mu$ be the set of solution mappings for Q that are compatible with a mapping μ, as defined by the SPARQL 1.1 semantics. Let also Σ be a list of source and iterator clauses. Then the evaluation of the SPARQL-Generate query pattern $Q^+ = \langle \Sigma, Q \rangle \in (S \cup I)^* \times \boldsymbol{Q}$ over D and a documentset Δ is defined by Eq. (18). We introduce a special *initial* mapping, $\mu_0 : v \mapsto \omega, \forall v \in \mathbf{V}$. Then, the set of solution mappings of any SPARQL Generate query Q^+ over Δ and D is defined by Eq. (19).

$$\llbracket Q^+ \rrbracket_{\Delta, D}^\mu = \bigcup_{\mu' \in \llbracket \Sigma \rrbracket_\Delta^\mu} \llbracket Q \rrbracket_D^{\mu'} \tag{18}$$

$$\llbracket Q^+ \rrbracket_{\Delta, D} = \llbracket Q^+ \rrbracket_{\Delta, D}^{\mu_0} \tag{19}$$

Generate Part of the SPARQL Generate Query. For any graph pattern $P \in \mathcal{P}$ and any mapping $\mu \in \mathcal{M}$, we note $\Bbbk^\mu(P)$ the RDF Graph generated by instantiating the graph pattern with respect to a mapping μ, following [6, Sect. 16.2.1]. We then define the evaluation of SPARQL-Generate queries recursively. Let be a simple SPARQL-Generate query $\langle P, Q \rangle \in \mathcal{P} \times \boldsymbol{Q}^+$, another query $G = \langle P, G_0, \ldots, G_j, Q \rangle \in \mathcal{P} \times \mathcal{G}^* \times \boldsymbol{Q}^+$, and a mapping μ. The following three equations define the RDF graph generated by G,

$$\Bbbk_{\Delta, D}^\mu(\langle P, Q \rangle) = \bigcup_{\mu' \in \llbracket Q \rrbracket_{\Delta, D}^\mu} \Bbbk^{\mu'}(P) \tag{20}$$

$$\Bbbk_{\Delta, D}^\mu(\langle P, G_0, \ldots, G_j, Q \rangle) = \bigcup_{\mu' \in \llbracket Q \rrbracket_{\Delta, D}^\mu} \left(\Bbbk^{\mu'}(P) \cup \bigcup_{0 \leq i \leq j} \Bbbk_{\Delta, D}^{\mu'}(G_i) \right) \tag{21}$$

$$\Bbbk_{\Delta, D}(G) = \Bbbk_{\Delta, D}^{\mu_0}(G) \tag{22}$$

5 Implementation and Evaluation

5.1 Generic Approach

It is advantageous to be able to implement SPARQL-Generate on top of any existing SPARQL 1.1 engine. In fact, such an engine already provides us with: (i) the binding functions map f_b (thus one can know for any mapping $\mu \in \mathcal{M}$ its generalization $\bar{\mu}$ to any binding function expression); (ii) a function SELECT that takes a SPARQL 1.1 query pattern as input, and returns a set of solution mappings; (iii) a function INSTANTIATE that takes a graph pattern $P \in \mathcal{P}$ and a mapping $\mu \in \mathcal{M}$ as input, and returns the RDF Graph corresponding to the instantiation of P with respect to μ; (iv) the management of RDF datasets D. Then an implementation of SPARQL-Generate would just need to provide: (1) the management of a documentset Δ, and (2) the iterator functions map f_i.

Let $\mathcal{V} = 2^{\mathcal{M}}$ be the set of *inline data blocks*. Then we note $\langle V, Q \rangle \in \mathcal{Q}$ the result of prefixing some SPARQL query $Q \in \mathcal{Q}$ by some inline data block $V \in \mathcal{V}$. Theorem 1 below allows us to design a naive algorithm[7] (Algorithm 1) that can be used to implement SPARQL-Generate on top of a SPARQL 1.1 engine.

Theorem 1. *Let be a SPARQL 1.1 query $Q \in \mathcal{Q}$, and a list of source and iterator clauses $\Sigma \in (\mathcal{S} \cup \mathcal{I})^*$. The evaluation of the SPARQL-Generate query pattern $\langle \Sigma, Q \rangle \in \mathcal{Q}^+$ is equal to the evaluation of $\langle [\![\Sigma]\!]_\Delta, Q \rangle$, where $[\![\Sigma]\!]_\Delta$ is the evaluation of Σ.*

Proof. First note that in the SPARQL 1.1 semantics, the evaluation of a SPARQL 1.1 query pattern Q prefixed by an inline data block V is a join between the evaluation of V (i.e., $[\![V]\!]_D = V$), and the evaluation of Q (i.e., $[\![Q]\!]_D$). With our notations, this translates to: $[\![\langle V, Q \rangle]\!] = \bigcup_{\mu \in V} [\![Q]\!]^\mu$. Substituting V by $[\![\Sigma]\!]_\Delta = [\![\Sigma]\!]_\Delta^{\mu_0}$ and combining with Eqs. 18 and 19 leads to the proof:

$$[\![\langle [\![\Sigma]\!]_\Delta^{\mu_0}, Q \rangle]\!]_{\Delta,D} = \bigcup_{\mu' \in [\![\Sigma]\!]_\Delta^{\mu_0}} [\![Q]\!]_{\Delta,D}^{\mu'} = [\![\langle \Sigma, Q \rangle]\!]_{\Delta,D}^{\mu_0} = [\![\langle \Sigma, Q \rangle]\!]_{\Delta,D} \qquad (23)$$

5.2 Implementation on Top of Apache Jena

This section overviews a first implementation of SPARQL-Generate with Algorithm 1 over the Jena ARQ SPARQL 1.1 engine.

---

[7] This algorithm is simplified and does not show subtleties in the management of blank nodes, which will be the focus of a future paper. On the other hand, the implementation already addresses this, see unit tests **bnode1** and **bnode2** at http://w3id.org/sparql-generate/tests-reports.html.

Algorithm 1. Naive implementation of SPARQL-Generate on top of any SPARQL 1.1 engine.

1: **procedure** GENERATE($\langle P, G_0, \ldots, G_j, \langle E_0, \ldots, E_n \rangle, Q \rangle, \mu$) ▷ See also Def. 4
2:     $M \leftarrow \{\mu\}$ ▷ M is a singleton containing one mapping
3:     **for** $0 \leq i \leq n$ **do**
4:         **if** $E_i = v \overset{\text{source}}{\longleftarrow} e$ **then** ▷ See also Def. 1
5:             **for all** $\mu \in M$ **do**
6:                 $\mu(v) \leftarrow \Delta(\bar{\mu}(e))$ ▷ See also Def. 5 and Eq. 12
7:             **end for**
8:         **else if** $E_i = v \overset{\text{iterator}}{\longleftarrow} \langle t, e_0, \ldots, e_k \rangle$ **then** ▷ See also Def. 2
9:             $M' \leftarrow \varnothing$
10:             **for all** $\mu \in M$ **do**
11:                 **for all** $t' \in f_i(t)(\bar{\mu}(e_0), \ldots, \bar{\mu}(e_k))$ **do** ▷ See also Eq. 11
12:                     $\mu' \leftarrow \mu$; $\mu'(v) \leftarrow t'$; and $M' \leftarrow M' \cup \{\mu'\}$
13:                 **end for**
14:             **end for**
15:             $M \leftarrow M'$ ▷ replace M by M'
16:         **end if**
17:     **end for**
18:     $M \leftarrow$ SELECT($\langle M, Q \rangle$) ▷ evaluate the query pattern prefixed by the computed inline data block
19:     $G \leftarrow \varnothing$ ▷ the empty RDF graph
20:     **for** $\mu \in M$ **do**
21:         $G \leftarrow G \cup$ INSTANTIATE(P, μ) ▷ operate a RDF graph union (not merge), i.e., do not merge blank nodes even if they share the same name
22:         **for** $0 \leq i \leq j$ **do**
23:             $G \leftarrow G \cup$ GENERATE(G_i, μ)
24:         **end for**
25:     **end for**
26:     **return** G
27: **end procedure**

Open-Source Code and Online Testbed. This implementation is open-source and available on GitHub,[8],[9] released as a Maven dependency,[10] can also be used as an executable jar, or as a Web API. SPARQL-Generate can also be tested online using a web form that calls the Web API.[11] The SPARQL-Generate editor uses the YASGUI library,[12] which has been modified to support the SPARQL-Generate syntax. Finally, one can load any of the library unit tests in this web form. These unit tests cover use cases from related work and more.[13]

---

[8] http://w3id.org/sparql-generate/get-started.html.
[9] https://github.com/thesmartenergy/sparql-generate.
[10] http://search.maven.org/#search|ga|1|sparql-generate.
[11] http://w3id.org/sparql-generate/language-form.html.
[12] http://yasqe.yasgui.org/.
[13] http://w3id.org/sparql-generate/tests-reports.html.

Supported Data Formats, and Extensibility. Binding and iterator functions are available for the following formats: JSON and CBOR (exploiting JSONPath, thus satisfying requirement **R7**), CSV and TSV (conforming to the RFC 4180, or custom), XML (exploiting XPath), HTML (exploiting CSS3 selectors), and plain text (exploiting regular expressions). A complete documentation of the available binding and iterator functions is available along with the documentation of the API.[14] The implementation relies on Jena's SPARQL binding function extension mechanism, and copies it for iterator functions. Therefore, covering a new data format in this implementation merely consists in implementing new binding and iterator functions like in Jena. This satisfies requirement **R3**. Even what is not covered by existing query languages can be implemented as an iterator function. For example, iterator function iter:JSONListKeys iterates on key names of a JSON object, which is not feasible using JSONPath. As another example, polymorphic binding function fn:CustomCSV enables to parse a CSV document with or without a header. Parameters guide the parsing and data extraction from CSV documents with sparse structures, but the function itself checks for the existence of a header. If present, it treats the parameter column as a string to refer to a column. Else, it treats is as the column index. This function hence covers the *Dialect Description* of CSVW.

Specific Implementation Choices (see Sect. 4.1*).* For the documentset Δ, this implementation uses the FileLocator Jena utility. It hence looks up a IRI depending on its scheme, except if a configuration file explicitly specifies a mapping to a local file. For now, the FileLocator does not look up for IRIs with schemes other than http and file. The implementation still covers these cases in two ways: (a) they may be explicitly mapped to local files, or (b) they may be provided to the engine through some initial binding. For instance, test case named *cbor-venueeg1*, featuring CBOR, uses option (a).

If the source clause accept option is set to some IANA media-type URI of the form http://www.iana.org/assignments/media-types/text/csv, then the library negotiates the specified media type with the server.[15] In any other case, the datatype of retrieved documents defaults to xsd:string.

Similarly, when a query calls another query with its IRI, the implementation uses the FileLocator Jena utility. If not explicitly mapped to a local file, then the implementation uses the SPARQL-Generate registered media type application/vnd.sparql-generate (file extension .rqg) as the Accepted media type to fetch it on the Web.[16]

5.3   Evaluation

As RML is the language that most closely satisfies the identified needs, we conducted a comparative evaluation of it and SPARQL-Generate. This evaluation

---

[14] http://w3id.org/sparql-generate/functions.html.

[15] There is no consensus on the mapping between URIs and Internet Media Types, althought this is the object of a W3C TAG finding [13].

[16] https://w3id.org/sparql-generate/language.html#IANA_considerations..

focuses on two aspects: performances of the reference implementations, and cognitive complexity of the query/mapping. For this purpose, we chose to focus on a very simple transformation from CSV documents generated by GenerateData.com to RDF. For every line, a few triples with the same subject, fixed predicates, and objects computed from one column, are generated. The report and the instructions to reproduce this experiment are available online.[17]

Performance of the Reference Implementations. Figure 2 shows that for this simple transformation, the execution time with sparql-generate-jena becomes faster than RML-Processor above ~1,500 rows, and linear. It is slightly above 3 min for 20,000 rows for sparql-generate-jena, when RML-Processor takes more than 6 min for 5,000 rows. Granted, comparing implementations does not necessarily highlight the true qualities of the approaches since optimizations, better choices of software libraries, and so on, could dramatically impact the results. Yet, with these experiments, we show that a straightforward and relatively naive implementation on top of Jena ARQ we achieve competitive performances. We argue that ease of implementation and use is the key benefit of our approach.

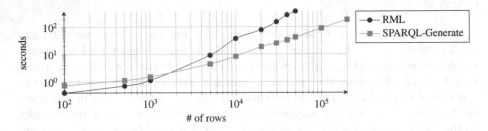

Fig. 2. Execution time for a simple transformation from CSV documents to RDF. Comparison between the current RML-Processor and sparql-generate-jena implementations.

Cognitive Complexity of the Query/Mapping. We conducted a limited study of the cognitive complexities of the languages we are comparing. On the experiment transformations, there are 12 terms from the R2RML and RML vocabularies, while SPARQL-Generate adds only 4 tokens to SPARQL 1.1 (source, iterator, sgiter:CSV and sgfn:CSV). Moreover, we realized that semantic web experts that have to carry on a triplification task usually observe the input data to identify the parts that have to be selected and formalize it with a selection pattern, such as a XPath or JSONPath query; then they draw an RDF graph or a graph pattern where they place the selected data from the input. This closely matches the structure of a SPARQL-generate query. The where clause contains the bindings that correspond to the select parts of the input documents; the generate clause contains the output graph patterns that reuse the extracted data. We also noticed

---

[17] https://w3id.org/sparql-generate/evaluation.html.

that when RML mappings get complex, they tend to grow to larger files than the equivalent SPARQL generate query, as can be witnessed by comparing our equivalent test cases.[18] These limitations in RML may be explained by the fact it extends R2RML whose triple maps are subject-centric. If one requires several triples to share the same object, then one must write several triple maps, that would have the same object map. This limitation impacts the cognitive complexity of the language. On the other hand, as the SPARQL-Generate concrete syntax is very close to that of SPARQL 1.1, we claim it makes it easy to learn and use by people that are familiar with the Semantic Web formalisms, satisfying requirement **R4** and **R5**. Nevertheless, from our own experience writing SPARQL-Generate queries, we identified some syntactic sugars that could strongly improve readability and conciseness of the queries. For instance one could use binding functions directly in the generate pattern, or use curly-bracket expressions instead of concatenating literals. Using such techniques, the running example query could be simplified as follows:

```
GENERATE {
  <http://example.com/person/{sgfn:CSV(?person,"PersonId")}> a foaf:Person ;
    foaf:name sgfn:CSV(?person, "Name" ) ;
    foaf:mbox <mailto:{sgfn:CSV(?person,"Email")}> ;
    foaf:phone <tel:{sgfn:CSV(?person,"Phone")}> ;
    schema:birthDate "{sgfn:CSV(?person,"Birthdate")}"^^xsd:dateTime ;
    schema:height "{sgfn:CSV(?person,"Height")}"^^xsd:decimal ;
    schema:weight "{sgfn:CSV(?person,"Weight")}"^^xsd:decimal .
} SOURCE <http://example.org/persons.csv> AS ?persons
  ITERATOR sgiter:CSV(?persons) AS ?person
```

Flexibility and Extensibility of the Languages. Work has been led to make RML be able to call external functions [8]. This is not necessary for SPARQL-Generate, and we believe that knowledge engineers are already familiar with SPARQL 1.1 functions, filtering capabilities, and solution sequence modifiers. This satisfies requirement **R6**.

6   Conclusion and Future Work

The problem of exploiting data from heterogeneous sources and formats is common on the Web, and Semantic Web technologies can help in this regard. However, adopting Semantic Web technologies does not automatically clear up those strong integration issues. Different solutions have been proposed to generate RDF from documents in heterogeneous formats. In this paper, we introduced a lightweight extension of SPARQL 1.1 called SPARQL-Generate, and compared it with the related work. We formally defined SPARQL-Generate and proved that it is (i) easily implementable on top of existing SPARQL engines; (ii) modular since extensions to new formats do not require a redefinition of the language (thanks to the use of SPARQL custom functions); (iii) easy to use by knowledge engineers because of its resemblance to normal SPARQL; and (iv) powerful

---

[18] See unit tests starting with RML⋆ at http://w3id.org/sparql-generate/tests-reports. html.

and flexible thanks to the custom function mechanism, the filtering capabilities, and the solution sequence modifiers of SPARQL 1.1. Our open-source implementation on top of Apache Jena covers many use cases, an is proven to be more efficient than the reference implementation of RML on a simple use case. Future plans consist of implementing more functions for more data formats, and extending the implementation to enable on the fly function integration (with an approach similar to [9]).

References

1. Arenas, M., Bertails, A., Prud'hommeaux, E., Sequeda, J.: A direct mapping of relational data to RDF. W3C Recommendation, W3C, 27 September 2012
2. Connolly, D.: Gleaning resource descriptions from dialects of languages (GRDDL). W3C Recommendation, W3C, 11 September 2007
3. Das, S., Sundara, S., Cyganiak, R.: R2RML: RDB to RDF mapping language. W3C Recommendation, W3C, 27 September 2012
4. Dell'Aglio, D., Polleres, A., Lopes, N., Bischof, S.: Querying the web of data with XSPARQL 1.1. In: Proceedings of the ISWC Developers Workshop 2014, Co-located with the 13th International Semantic Web Conference (ISWC 2014), Riva del Garda, Italy (2014)
5. Dimou, A., Sande, M.V., Colpaert, P., Verborgh, R., Mannens, E., Van de Walle, R.: RML: a generic language for integrated RDF mappings of heterogeneous data. In: Proceedings of the Workshop on Linked Data on the Web, Co-located with the 23rd International World Wide Web Conference (WWW 2014), Seoul, Korea (2014)
6. Harris, S., Seaborne, A.: SPARQL 1.1 query language. W3C Recommendation, W3C, 21 March 2013
7. Hert, M., Reif, G., Gall, H.C.: A comparison of RDB-to-RDF mapping languages. In: Proceedings the 7th International Conference on Semantic Systems, I-SEMANTICS 2011, Graz, Austria, pp. 25–32 (2011)
8. Junior, A.C., Debruyne, C., O'Sullivan, D.: Incorporating functions in mappings to facilitate the uplift of CSV files into RDF. In: Sack, H., Rizzo, G., Steinmetz, N., Mladenić, D., Auer, S., Lange, C. (eds.) ESWC 2016. LNCS, vol. 9989, pp. 55–59. Springer, Cham (2016). doi:10.1007/978-3-319-47602-5_12
9. Lefrançois, M., Zimmermann, A.: Supporting arbitrary custom datatypes in RDF and SPARQL. In: Sack, H., Blomqvist, E., d'Aquin, M., Ghidini, C., Ponzetto, S.P., Lange, C. (eds.) ESWC 2016. LNCS, vol. 9678, pp. 371–386. Springer, Cham (2016). doi:10.1007/978-3-319-34129-3_23
10. Lopes, N., Bischof, S., Polleres, A.: On the semantics of heterogeneous querying of relational, XML, and RDF data with XSPARQL. In: Proceedings of the 15th Portuguese Conference on Artificial Intelligence - Computational Logic with Applications Track (2011)
11. Polleres, A., Krennwallner, T., Lopes, N., Kopecký, J., Decker, S.: XSPARQL language specification. W3C Member Submission, W3C, 20 January 2009
12. Tandy, J., Herman, I., Kellogg, G.: Generating RDF from tabular data on the web. W3C Recommendation, W3C, 17 December 2015
13. Williams, S.: Mapping between URIs and internet media types. TAG Finding, W3C, 27 May 2002

Linked Data Track

Exploiting Source-Object Networks to Resolve Object Conflicts in Linked Data

Wenqiang Liu[1(✉)], Jun Liu[1], Haimeng Duan[1], Wei Hu[2], and Bifan Wei[1]

[1] MOEKLINNS Lab, Xi'an Jiaotong University, Xi'an, China
liuwenqiangcs@gmail.com, {liukeen,weibifan}@mail.xjtu.edu.cn,
duanhaimeng@gmail.com
[2] State Key Laboratory for Novel Software Technology,
Nanjing University, Nanjing, China
whu@nju.edu.cn

Abstract. Considerable effort has been exerted to increase the scale of Linked Data. However, an inevitable problem arises when dealing with data integration from multiple sources. Various sources often provide conflicting objects for a certain predicate of the same real-world entity, thereby causing the so-called *object conflict* problem. At present, object conflict problem has not received sufficient attention in the Linked Data community. Thus, in this paper, we firstly formalize the object conflict resolution as computing the joint distribution of variables on a heterogeneous information network called the *Source-Object Network*, which successfully captures three correlations from objects and Linked Data sources. Then, we introduce a novel approach based on network effects called *ObResolution* (object resolution), to identify a true object from multiple conflicting objects. ObResolution adopts a pairwise Markov Random Field (pMRF) to model all evidence under a unified framework. Extensive experimental results on six real-world datasets show that our method achieves higher accuracy than existing approaches and it is robust and consistent in various domains.

Keywords: Linked Data quality · Object conflicts · Truth discovery

1   Introduction

Considerable effort has been made to increase the scale of Linked Data. Especially, the number of available Linked Data sources in the Linking Open Data (LOD) project increases from 12 in 2007 to 1,146 in 2017.[1] In this paper, a Linked Data source refers to a dataset that has been published to the LOD project by individuals or organizations, such as YAGO. Linked Data resources are encoded in the form of ⟨*Subject, Predicate, Object*⟩ triples through the Resource Description Framework (RDF) format. The subject denotes the resource, and predicate is used to express a relationship between subject and object. Inevitably, errors

---

[1] http://lod-cloud.net/.

© Springer International Publishing AG 2017
E. Blomqvist et al. (Eds.): ESWC 2017, Part I, LNCS 10249, pp. 53–67, 2017.
DOI: 10.1007/978-3-319-58068-5_4

occur during such creation process given that many Linked Data sources on the web have been created from semi-structured datasets (e.g., Wikipedia) and unstructured ones through automatic or semi-automatic algorithms [4]. As a result, a predicate for the same real-world entity can have multiple inconsistent objects when dealing with data integration from multiple sources. For example, the objects of the *dbp:populationTotal* for *Beijing* in Freebase[2] and DBpedia[3] are "20,180,000" and "21,516,000" respectively. In this paper, this problem is called the *object conflict* problem. The concept of object conflicts can be defined as two objects are being in conflict only when their similarity is less than the defined threshold. According to this definition, it is also likely to regard two objects expressed in terms of different measure units as conflicts. But, the purpose of our study is to rank the trust values of all objects and provide the most common ones for users, rather than remove some objects directly. Therefore, people who use our methods can still see all objects.

A straightforward method to resolve object conflicts is to conduct the majority voting, which regards the object with the highest number of occurrences as the correct object. The drawback of this method is that it assumes that all Linked Data sources are equally reliable [12]. In reality, some Linked Data sources are more reliable than others and thus may produce inaccurate results in scenarios when there are some Linked Data sources provide untrustworthy objects. Many truth discovery methods have been proposed to estimate source reliability [11,12,20,23] in recent years to overcome the limitation of majority voting. The basic principle of these methods is that a source which provides trustworthy objects more often is more reliable, and an object from a reliable source is more trustworthy. Therefore, the truth discovery problem in these methods is formulated as an iterative procedure, which starts by assigning the same trustworthiness to all Linked Data sources, and iterates by computing the trust value of each object and propagating back to the Linked Data sources.

However, a major problem occurs in the aforementioned approaches. The iterative procedure in these methods is performed by simple weighted voting, which can result in that the rich getting richer over iterations [21]. Especially in Linked Data, data sharing between different Linked Data sources is common in practice. Therefore, errors can easily propagate and lead to wrong objects often appearing in many sources. As a result, methods based on an iterative procedure may derive a wrong conclusion. The situation is even worse for many predicates that are time sensitive, which the corresponding object tends to change over time (e.g., *dbo:populationTotal*), because many out-of-date objects often exist in more Linked Data sources than those up-to-date objects. The experimental results of [12] based on an iterative procedure also show the same conclusion, which obtained the lowest accuracy for the time-sensitive predicate.

To address this problem, we propose a new method, called *ObResolution* (Object Resolution), which utilizes the Source-Object network to infer the true object. This network successfully captures three correlations from objects and

---

[2] https://www.freebase.com/m/072p8.
[3] http://dbpedia.org/resource/Beijing.

Linked Data sources. For example, an object from a reliable source is more trustworthy and a source that provides trustworthy objects more often is more reliable. Thus, we build a message propagation-based method that exploits the network structure to infer the trust values of all objects and then the object with the maximum trust score is regarded as the true object. According to our evaluation, our method outperforms several existing truth discovery methods because these methods either model all clues by the iterative procedure, or do not take the sharing between Linked Data sources into consideration. We summarize the main contributions of our work as follows.

- We formalize the object conflict resolution problem as computing the joint distribution of variables in a heterogeneous information network called the *Source-Object Network*, which successfully captures three correlations from objects and Linked Data sources.
- We propose a novel truth discovery approach, ObResolution, to identify the truth in Linked Data. This approach leverages pairwise Markov Random Field (pMRF) to model the interdependencies from objects and sources, and a message propagation-based method is utilized that exploits the Source-Object Network structure to infer the trust values of all objects.
- We conducted extensive experiments on six real-world Linked Data datasets to validate the effectiveness of our approach. Our experimental results showed that our method achieves higher accuracy than several comparable methods.

The remainder of this paper is organized as follows. Section 2 formalizes our problem and the details of our method are discussed in Sect. 3. The evaluation of our method is reported in Sect. 4. Related work is discussed in Sect. 5. Section 6 presents the conclusion and future work.

2 Preliminaries

2.1 Basic Definitions

Definition 1 (RDF Triple) [13]. We let I denote the set of IRIs (Internationalized Resource Identifiers), B denote the set of blank nodes, and L denote the set of literals (denoted by quoted strings, e.g., *"Beijing City"*). An RDF triple can be represented by $\langle s, p, o \rangle \in (I \cup B) \times I \times (I \cup B \cup L)$, where s is called *subject*, p is *predicate*, and o is *object*.

Definition 2 (Trustworthiness of Sources). The trustworthiness $t(\omega_j)$ of a source ω_j is the average probability of the object provided by ω_j being true as defined as follows:

$$t(\omega_j) = \sum_{o_i \in F(\omega_j)} \tau(o_i)/|F(\omega_j)|, \tag{1}$$

where $F(\omega_j)$ is the set of objects provided by source ω_j and $\tau(o_i)$ denotes the trust value of an object o_i.

Definition 3 (Trust Values of Objects) [20]. The trust value $\tau(o_i)$ of an object o_i is the probability of being correct, which can be computed as

$$\tau(o_i) = \sum_{\omega_j \in \Omega(o_i)} t(\omega_j)/|\Omega(o_i)|, \tag{2}$$

where $\Omega(o_i)$ represents the set of sources that provide object o_i.

We let $O = \{o_i\}_m$ denote a set of conflicting objects for a certain predicate of a real-world entity. The process of object conflict resolution in Linked Data is formally defined as follows. Given a set of conflicting objects O, *ObResolution* assigns a trust score that lies in between 0 and 1 to each object. A score of object close to 1 indicates that we are very confident that this object is true. Therefore, the truth can be represented by $o^* = \arg\max_{o_i \in O} \tau(o_i)$.

2.2 Problem Analysis

Through the observation and analysis of the object conflicts in our sample Linked Data, we found three helpful correlations from Linked Data sources and objects to effectively distinguish between true and false objects.

- **Correlations among Linked Data Sources and Objects.** If an object comes from a reliable source, it will be assigned a high trust value. Thus a source that provides trustworthy objects often has big chance to be selected as a reliable source. For example, the object provided by DBpedia is more reliable than objects supported by many small sources because DBpedia is created from Wikipedia. This condition also serves as a basic principle for many truth discovery methods [8–11,18,23].
- **Correlations among Objects.** If two objects are similar, they should have similar trust values, which indicates that similar objects appear to have mutually support. For example, we assume that one source claims that the *dbp:height* of *Statue of Liberty* is "46.0248" and another says that it is "47". If one of these sources has a high trust value, the other should have a high trust value as well. Meanwhile, if two objects are mutually excluded, they cannot be both true. If one of them has a high trust value, the other should have a low trust value. For instance, if two different sources claim that the *dbp:height* of *Statue of Liberty* are "93" and "46.0248" respectively. If the true object is "46.0248", then "93" should be a wrong object.
- **Correlations among Linked Data Sources.** In many truth discovery methods, the trustworthiness of a source is formulated as the probability of the objects provided by this source being the truth. Therefore, the more same objects two different sources provide, the more similar is the trustworthiness of the two sources. Consider an extreme case when two sources provide the same objects for each predicate, and the trustworthiness of these two sources is the same.

As discussed, these three principles can be used to infer the trust values of objects. A key problem for object conflicts resolution is how to model these principles under a unified framework.

3   ObResolution Method

In this section, we formally introduce our proposed method called *ObResolution*, for discovering the most reliable objects from the set of conflicting objects. We first formulate the object conflict resolution problem as the Source-Object network analysis problem, which successfully captures all the correlations from objects and Linked Data sources. Subsequently, a message propagation-based method that exploits the Source-Object network structure is introduced to solve this problem. Finally several important issues that make this method practical are discussed.

3.1   Model Details

In general, the input to our problem includes three parts: (i) objects, which are the values of a certain predicate for the same real-world entity; (ii) Linked Data sources, which provide these objects, e.g., Freebase; and (iii) mappings between objects and Linked Data sources, e.g., which Linked Data sources provide which objects for the certain predicate of the same real-world entity. Thus, a set of objects and sources can be structured into a bipartite network. In this bipartite network, source nodes are connected to the object nodes, in which links represent the "provider" relationships. For ease of illustration, we present example network of six sources and four conflicting objects as shown in Fig. 1(a). According to the first principle, an object from a reliable source is more trustworthy, and thus a source that provides trustworthy objects than other sources. The "provided" relationship between a source and an object also indicates the interdependent relationship between the trust value of the object and the trustworthiness of the source. Besides the "provider" relationship between the source and object, among objects and among Linked Data sources also have correlations. For instance, because sources $\omega_1, \omega_3, \omega_5$ provide the same object o_1 in Fig. 1(a), they have a correlation for any two of these three sources. Therefore, the bipartite network in Fig. 1(a) can be converted to a heterogeneous information network called the Source-Object Network as shown in Fig. 1(b).

The Source-Object Network $G = (V, E)$ contains n Linked Data source nodes $\Omega = \{\omega_1, ..., \omega_n\}$ and m conflicting object nodes $O = \{o_1, ..., o_m\}$, $V = \Omega \cup O$, connected with edge set E. Owing to three types of correlations of objects and Linked Data sources, the Source-Object Network G has three types of edges $E = E_\Omega \cup E_O \cup E_{\Omega \rightarrow O}$, where $E_\Omega \subseteq \Omega \times \Omega$ represents the correlations between sources, $E_O \subseteq O \times O$ indicates the correlations among objects and $E_{\Omega \rightarrow O}$ represents the "provided" relationships between sources and objects.

Given a Source-Object Network, which successfully captures three correlations from objects and Linked Data sources, the task is to estimate the reliability of sources and the trust values of all conflicting objects. Each node in G is a random variable that can represent the trust values of objects and trustworthiness of sources. However, we find that the trust values of objects and trustworthiness of sources are assumed to be dependent on their neighbors and independent of all the other nodes in this network. This condition motivates us to select a method

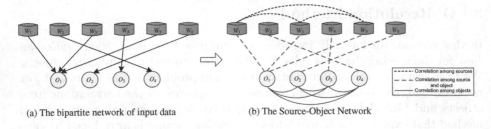

(a) The bipartite network of input data   (b) The Source-Object Network

Fig. 1. Illustration of an example source-Object Network

based on pMRF, which is a powerful formalism used to model real-world events based on the Markov chain and knowledge of soft constraints. Therefore, the Source-Object Network is represented by pMRF in this study. In fact, pMRF is mainly composed of three components: an unobserved field of random variables, an observable set of random variables, and the neighborhoods between each pair of variables. We let all the nodes $V = \Omega \cup O$ in G be observation variables. Thus, the unobserved variables $Y = Y_\Omega \cup Y_O$ have two types of labels.

(1) The unobserved variable y_i is the label of an object node. It indicates whether the corresponding object is the truth, which follows the Bernoulli distribution defined as follows.

$$P(y_i) = \begin{cases} \tau(o_i) & if\ o_i\ is\ true, i.e.,\ y_i = 1, \\ 1 - \tau(o_i) & if\ o_i\ is\ false, i.e.,\ y_i = 0. \end{cases} \quad (3)$$

(2) The unobserved variable y_i is the label of Linked Data source node which represents whether the corresponding source is a reliable source and also follows the Bernoulli distribution.

$$P(y_j) = \begin{cases} t(\omega_j) & if\ \omega_j\ is\ a\ reliable\ source, i.e.,\ y_j = 1, \\ 1 - t(\omega_j) & if\ \omega_j\ is\ a\ unreliable\ source, i.e.,\ y_j = 0. \end{cases} \quad (4)$$

The problem of inferring the trust values of conflicting objects and trustworthiness of sources can be converted to compute the joint distribution of variables in pMRF, which is factorized as follows:

$$P(y_1, ..., y_m, ..., y_{m+n}) = \frac{\prod_{c \in C} \psi_c(x_c)}{\sum_{x_c \in X} \prod_{c \in C} \psi_c(x_c)}, \quad (5)$$

where C denotes the set of all maximal cliques, the set of variables of a maximal clique is represented by x_c ($c \in C$), and $\psi_c(x_c)$ is a potential function in pMRF.

3.2   Inference Algorithms

In general, exactly inferencing the joint distribution of variables in pMRF is known to be a non-deterministic polynomial-time hard problem [17]. Loopy Belief Propagation (LBP) is an approximate inference algorithm that has been shown to perform extremely well for various of applications in the real word. In belief propagation, estimating the joint distribution of variables is a process of minimizing the graph energy. The key steps in the propagation process can be concisely expressed below.

– **Spreading the Belief Message.** The message from variable y_i to y_j is represented by $m_{i \rightarrow j}(y_j)$, $y_j \in \{0, 1\}$, which is defined as follows:

$$m_{i \rightarrow j}(y_j) = \sum_{y_i \in \{0,1\}} U(y_i, y_j) \psi_i(y_i) \prod_{y_k \in N(y_i) \cap Y \setminus \{y_j\}} m_{k \rightarrow i}(y_i), \tag{6}$$

where $N(v_i)$ indicates the set of neighbors of node y_i; $\psi_i(y_i)$ denotes the prior belief of $P(y_i)$, and $U(y_i, y_j)$ is a unary energy function.

– **Belief Assignment.** The marginal probability $P(y_i)$ of unobserved variable y_i is updated according to its neighbors, and is defined as follows:

$$P(y_i) = \psi_i(y_i) \prod_{y_j \in N(y_i) \cap Y} m_{j \rightarrow i}(y_i). \tag{7}$$

The algorithm updates all messages in parallel and assigns the label until the messages stabilizes, i.e. achieve convergence. Although convergence is not theoretically guaranteed, the LBP has been shown to converge to beliefs within a small threshold fairly quickly with accurate results [17]. After they stabilize, we compute the marginal probability $P(y_i)$. Thus, we can obtain the trust values of object and the trustworthiness of source. Given only one truth for a certain predicate of a real-world entity, the true object is o_i when $\tau(o_i)$ is the maximum. To date, we have described the main steps of LBP, but two problems occur in the algorithm, energy function and prior belief. These problems are discussed as follows.

Energy Function. The energy function $U(y_i, y_j)$ denotes the likelihood of a node with label y_i to be connected to a node with label y_j through an edge. The following three types of energy functions exist depending on the types of edges:

– The energy function between sources and objects. A basic principle between sources and objects is that the reliable source tends to provide true objects and unreliable sources to false objects. However, a reliable sources may also provide false objects as unreliable sources to true objects. In this study, we let β denote the likelihood between reliable sources and true objects, whereas δ denotes the likelihood between unreliable sources and false objects. Therefore, the energy function between sources and objects is shown in the first three columns of Table 1.
– The energy function among objects. The more similar the two objects are, the greater is the probability of them having the same trust values. Therefore, a positive correlation exists between the energy function and the similarity $S(o_i, o_j)$ between object o_i and o_j, as shown in Table 2.

Table 1. Energy function from sources and objects

Source	Object		Source	
	True	False	Reliable	Unreliable
Reliable	β	$1 - \beta$	ε	$1 - \varepsilon$
Unreliable	$1 - \delta$	δ	$1 - \varepsilon$	ε

Table 2. Energy function between objects

Object	Object	
	True	False
True	$S(o_i, o_j)$	$1 - S(o_i, o_j)$
False	$1 - S(o_i, o_j)$	$S(o_i, o_j)$

– The energy function among sources. We assume that the more same objects two different sources provide, the more similar the trustworthiness of the two sources are. The coefficient $\varepsilon = |F(\omega_i) \cap F(\omega_j)|/max(|F(\omega_i)|, |F(\omega_j)|)$ is used to denote the likelihood between sources ω_i and ω_j, where $F(\omega_i)$ is the set of objects provided by source ω_i as shown in the last two columns of Table 1.

The pseudo code of this method is shown in Algorithm 1.

Algorithm 1. ObResolution

Input : a set of conflicting objects $O = \{o_1, ..., o_m\}$, a set of Linked Data
sources $\Omega = \{\omega_1, ..., \omega_n\}$ and the mapping relations between O and Ω
Output: trust value $\tau(o_i)$, $o_i \in O$; trustworthiness of source $t(\omega_j)$, $\omega_j \in \Omega$
1 Initialize the prior belief of all nodes $\psi_i(y_i)$;
2 $\forall o_i, o_j \in O$: Calculating their similarity $S(o_i, o_j)$;
3 $\forall y_i, y_j \in Y$: $m_{i \rightarrow j}(y_i) = 1$; //Initialize the message
4 **repeat**
5   $\quad$ //Message propagation
6   $\quad$ **for** $j \leftarrow 1$ to $m + n$ **do**
7   $\quad\quad$ **for** $i \leftarrow 1$ to $m + n$ **do**
8   $\quad\quad\quad$ $m_{i \rightarrow j}(y_j) = \sum_{y_i \in \{0,1\}} U(y_i, y_j) \psi_i(y_i) \prod_{y_k \in N(y_i) \cap Y \setminus \{y_j\}} m_{k \rightarrow i}(y_i)$.
9   $\quad\quad$ **end**
10 $\quad$ **end**
11 **until** the convergence criterion is satisfied;
12 **for** $i \leftarrow 1$ to $m + n$ **do**
13 $\quad$ //Belief assignment
14 $\quad$ $P(y_i) = \psi_i(y_i) \prod_{y_j \in N(y_i) \cap Y} m_{j \rightarrow i}(y_i)$.
15 **end**
16 **return** $\tau(o_i), \forall o_i \in O$; $t(\omega_j), \omega_j \in \Omega$

3.3 Practical Issues

In this section, we discuss several important issues, including similarity functions and missing values, to ensure the practicality of our method.

Similarity functions. The energy function between objects depends on the similarity function. We respect the characteristic of each data type and adopt different similarity functions to describe the similarity degrees. We discuss two similarity functions for numerical and categorical data, which are the two most common data types.

For numerical data, the most commonly used similarity function is defined as:

$$S(o_i, o_k) = 1/1 + d(o_i, o_k), \tag{8}$$

$$d(o_i, o_k) = \begin{cases} 1 & if \ o_i = o_k = 0, \\ |o_i - o_k|/max(|o_i|, |o_k|) & others. \end{cases} \tag{9}$$

For string data, the Levenshtein distance [16] is adopted to describe the similarities of objects. The similarity function is defined as follows:

$$S(o_i, o_k) = 1 - ld(o_i, o_k)/max(len(o_i), len(o_k)), \tag{10}$$

where $ld(o_i, o_k)$ denotes the Levenshtein distance between objects o_i and o_k; $len(o_i)$ and $len(o_k)$ are the length of o_i and o_k respectively.

Apart from these functions, different similarity functions can be easily incorporated into our method to recognize the characteristics of various data types. We have obtained a few similarity functions in this study, which are selected based on data types. One of them is the Jaro-Winkler string similarity functions for names of people and strings that involve abbreviations [5].

Missing Values. Linked Data are built on the Open World Assumption, which states that what is not known to be true is simply unknown. Therefore, for the sake of simplicity in this study, we assume that all missing values are not known to be true.

4   Evaluation

4.1   The Datasets

Six Linked Data datasets were used in our experiments[4]. The first three datasets *persons, locations, organizations* are constructed based on the OAEI2011 New York Times dataset[5], which is a well-known and carefully created dataset of Linked Data. In order to draw more robust conclusions, three other domains, including *films, books* and *songs* are constructed through SPARQL queries over DBpedia. The construction process of datasets mainly involves the following two necessary steps.

1. **Identity Subjects**: We adopted a well-known tool, sameas.org[6], to identify subjects for the same real-world entities in the six dataset. Then, we crawled the data of every subject from BTC2014 [6], which is a comparatively complete LOD cloud tnat consists of 4 billion triples.

---

[4] For researchers who are interested in our method, six datasets and codes are available at https://github.com/xiaoqiangcs/ObResolution.
[5] http://data.nytimes.com/.
[6] http://sameas.org/.

2. **Schema Mapping**: We adopted a method combining automatic matching and manual annotation to produce more accurate schema mapping results. First, features (*Property Similarity*, *Value Overlap Ratio*, *Value Match Ratio* and *Value Similarity Variance*) are selected based on the description provided in [19]. These selected features can achieve good performance in Linked Data. Subsequently, we chose the Random Forest and Support Vector Machine model, which achieved the best F1-Measure in [19] as classifiers for schema matching. Manual annotation is used to break the tie when an agreement was unreachable on a predicate between the two classifiers.

The statistics of the six datasets are shown in Table 3. In this paper, an entity refers to something that has a real existence, such as a location. A subject is a URIs that identifies the entity and different sources adopt various subjects to denote the same real-world entity (e.g., *dbpedia:Statue of Liberty* and *freebase:m.072p8*). The "#ConflictingPredicates" refers to the number of subject/predicate pairs for which conflicting objects exist.

Table 3. Statistics of the six datasets

Datasets	# Entities	#Subjects	#ConflictingPredicates	#Triples
Persons	4,978	21,340	69,706	141,937
Locations	1,910	21,324	38,200	558,773
Organizations	2,000	4,529	14,000	15,928
Films	2,000	4,935	8,000	20,692
Books	9,081	15,644	45,405	71,532
Songs	2,000	2,872	10,000	7,170

One truth was selected from multiple conflicting objects for experimental verification. A strict process was established to ensure the quality of the annotation. This process mainly involved the following steps:

1. The annotators were provided annotated examples and annotation guidelines.
2. Every two annotators were asked to label the same predicate on the same entity independently.
3. The annotation results from two annotators were measured by using Cohen's kappa coefficient [1]. The agreement coefficient of the six datasets was set to be at least 0.75. When an agreement could not be reached, a third annotator was asked to break the tie.

The manually labeled results were regarded as the ground truth used in the evaluation.

4.2   Comparative Methods and Metrics

We compared our method with five well-known state-of-the-art truth discovery methods as competitors, which were modified, if necessary.

- *Majority Voting.* This method regards the object with the maximum number of occurrences as truth. Moreover, voting is a straightforward method.
- *Sums (Hubs and Authorities)* [7]. This method regards the object supported by the maximum number of reliable sources as true. In this study, a source is recognized as a reliable source if its trustworthiness score exceeds 0.5.
- *TruthFinder* [20]. This is a seminal work that is used to resolve conflicts based on source reliability estimation. It adopts Bayesian analysis to infer the trustworthiness of sources and the probabilities of a value being true.
- *ACCUCOPY* [2]. This method is a popular truth discovery algorithm that obtains the highest precision among all methods in [8]. ACCUCOPY considers the copying relationships between the sources, the accuracy of data sources, and the similarity between values.
- *F-Quality Assessment* [15]. This method is a popular algorithm used to resolve conflicts in Linked Data. Three factors, namely the quality of the source, data conflicts, and confirmation of values from multiple sources, are leveraged for deciding which value should be the true value.

In the experiments, accuracy as a unified measure is adopted and can be measured by computing the percentage of matched values between the output of each method and ground truths. The parameters of the baseline methods were set according to the authors' suggestions. We implemented all algorithms using Eclipse (Java) platform[7] by a single thread and conducted experiments on a windows sever computer with Intel Core E7-4820 CPU 2 GHz with 32 GB main memory, and Microsoft Windows 7 professional operating system.

4.3   Results

Accuracy Comparison. Figure 2 shows the performance of different methods on the six datasets in terms of accuracy. As shown in this figure, our method consistently achieves the best accuracy among all the methods. Majority voting achieves the lowest accuracy (ranging from 0.3 to 0.45) on the six datasets among all the methods. Majority voting performs poorly in Linked Data for two reasons. First, approximately 50% of predicates have no dominant object [12]. In this case, majority voting can only randomly select one object in order to break the tie. Second, majority voting assumes that all sources are equally reliable and does not distinguish them, which is not applicable to Linked Data as discussed in Introduction. Although source reliable estimation was taken into consideration in Sums, this method still achieves relatively low accuracy in all datasets because it only considers the correlation between sources and objects, but ignores the correlation between objects and the correlation between sources. TruthFinder, F-Quality Assessment and ACCUCOPY model all the clues by the iterative procedure, which easily leads to the problem of the rich getting richer over iterations. In this study, our proposed method utilizes the Source-Object network and successfully captures all the correlations from objects and Linked Data sources in a

---

[7] https://www.eclipse.org/.

unified framework to infer the true objects, which explains why our method consistently achieves the best accuracy among all the comparative methods.

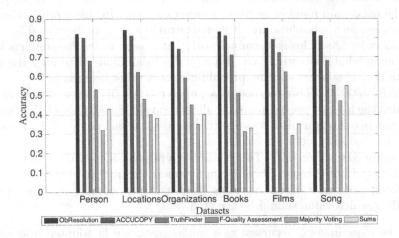

Fig. 2. Accuracy comparison in six datasets

Sensitive Analysis. We also studied the effect of the parameter β, δ on our methods. As discussed in Sect. 3, β indicates the likelihood between reliable source and true object, whereas δ denotes the likelihood between unreliable source and false object. Figure 3 shows that the accuracy of ObResolution varies in different values of β, δ in the same dataset, and ObResolution achieves best accuracy on six datasets with different values of β, δ ($\beta = 0.9, \delta = 0.7$ for *Persons*, for *Books* $\beta = 0.7, \delta = 0.9$). Therefore, parameters β, δ are sensitive to different datasets because different Linked Data datasets have different quality [22]. ObResolution uses different β, δ for different datasets to optimize the performance of our method.

5 Related Work

Resolving object conflicts is a key step for Linked Data integration and consumption. However, to the best of our knowledge, research on resolving object conflicts has not elicited enough attention in the Linked Data community. According to our survey, existing methods to resolve object conflicts in Linked Data can be grouped into three major categories of conflict handling strategies: conflict ignoring, conflict avoidance and conflict resolution.

- The conflict-ignoring strategy ignores the object conflicts and defers conflict resolution to users. For instance, Wang et al. [19] presented an effective framework to fuse knowledge cards from various search engines. In this framework, the fusion task involves card disambiguation and property alignment. For the value conflicts, this framework only adopted deduplication of the values and grouped these values into clusters.

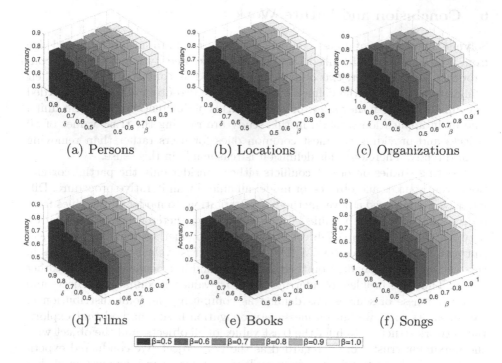

(a) Persons   (b) Locations   (c) Organizations

(d) Films   (e) Books   (f) Songs

■ β=0.5 ■ β=0.6 ■ β=0.7 ■ β=0.8 ■ β=0.9 ☐ β=1.0

Fig. 3. Sensitive analysis in six Linked Data datasets

- The conflict-avoidance strategy acknowledges the existence of object conflicts, but does not resolve these conflicts. Alternatively, they apply a unique decision to all data, such as manual rules. For instance, Mendes et al. [14] presented a Linked Data quality assessment framework called Sieve. In this framework, the strategy "Trust Your Friends," which prefers the data from specific sources, was adopted to avoid conflicts.
- The conflict-resolution strategy focuses on how to solve a conflict regarding the characteristics of all data and metadata. For example, Michelfeit et al. [15] presented an assessment model that leverages the quality of the source, data conflicts, and confirmation of values for determining which value should be the true value.

Previous work enlightens us on resolved object conflicts. In this paper, we propose a novel method that exploits the heterogeneous information network effect among sources and objects. Our approach is different in two aspects. First, we formalize the object conflict resolution problem through a heterogeneous information network, which successfully captures all the correlations from objects and Linked Data sources. Second, we adapt a message propagation-based method that exploits the network structure to infer the trust values of all objects. Our method has the following advantages: (1) it avoids the problem of the rich getting richer over iterations, and (2) it works in an unsupervised situation.

6 Conclusion and Future Work

Solving the problem of object conflicts is crucial to obtain insightful knowledge from a large number of Linked Data sources generated by numerous data contributors. In this paper, two objects are regarded as conflicts only when their similarity is less than the defined threshold. The two objects are still regarded as conflicts although they are expressed in terms of different measurement units. The main application scenario in this study are ranking the trust values of all objects and providing the most common ones for users rather than removing them directly. Therefore, this definition is reasonable in this sense.

Existing studies on object conflicts either consider only the partial correlations from sources and objects, or model all clues by an iterative procedure. Differently, we proposed a novel method, ObResolution, to model all the clues from sources and objects in a unified framework using a heterogeneous information network called the Source-Object Network. In our method, the Source-Object Network was represented by pMRF because the trust values of all the nodes in this network are dependent on their neighbors and independent of all the other nodes. Thus, the problem of inferring the trust values of conflicting objects and trustworthiness of sources was defined as computing the joint distribution of variables. As such, we built a message propagation-based method that exploits the network structure to infer the trust values of all objects, and the object with the maximum trust score is regarded as the true object. We conducted experiments on six datasets collected from multiple platforms and demonstrated that ObResolution exhibits higher accuracy than several comparative methods.

A potential direction for future research is to address more complicated conflict resolution scenarios, such as the situation involving copying relations of different sources. Another future research direction is to investigate the case where a certain predicate of a real-world entity has several true values.

Acknowledgments. This work is funded by the National Key Research and Development Program of China (Grant No. 2016YFB1000903), the MOE Research Program for Online Education (Grant No. 2016YB166) and the National Science Foundation of China (Grant Nos. 61370019, 61672419, 61672418, 61532004, 61532015).

References

1. Carletta, J.: Assessing agreement on classification tasks: the kappa statistic. Comput. Linguist. **22**(2), 249–254 (1996)
2. Dong, X.L., Berti-Equille, L., Srivastava, D.: Integrating conflicting data: the role of source dependence. PVLDB **2**(1), 550–561 (2009). Lyon, France
3. Dong, X.L., Gabrilovich, E., Murphy, K., Dang, V., Horn, W., Lugaresi, C., Sun, S., Zhang, W.: Knowledge-based trust: Estimating the trustworthiness of web sources. PVLDB **8**(9), 938–949 (2015). Hawaii, USA
4. Dutta, A., Meilicke, C., Ponzetto, S.P.: A probabilistic approach for integrating heterogeneous knowledge sources. In: Presutti, V., d'Amato, C., Gandon, F., d'Aquin, M., Staab, S., Tordai, A. (eds.) ESWC 2014. LNCS, vol. 8465, pp. 286–301. Springer, Cham (2014). doi:10.1007/978-3-319-07443-6_20

5. Jaro, M.A.: Advances in record-linkage methodology as applied to matching the 1985 census of tampa, florida. J. Am. Stat. Assoc. **84**(406), 414–420 (1989)
6. Käfer, T., Harth, A.: Billion Triples Challenge data set (2014). http://km.aifb.kit.edu/projects/btc-2014/
7. Kleinberg, J.M.: Authoritative sources in a hyperlinked environment. J. ACM **46**(5), 604–632 (1999)
8. Li, X., Dong, X.L., Lyons, K., Meng, W., Srivastava, D.: Truth finding on the deep web: is the problem solved? PVLDB **6**(2), 97–108 (2012). Istanbul, Turkey
9. Li, Q., Li, Y., Gao, J., Su, L., Zhao, B., Demirbas, M., Fan, W., Han, J.: A confidence-aware approach for truth discovery on long-tail data. PVLDB **8**(4), 425–436 (2014). Hangzhou, China
10. Li, Y., Li, Q., Gao, J., Su, L., Zhao, B., Fan, W., Han, J.: On the discovery of evolving truth. In: KDD, Sydney, Australia, pp. 675–684 (2015)
11. Li, Q., Li, Y., Gao, J., Zhao, B., Fan, W., Han, J.: Resolving conflicts in heterogeneous data by truth discovery and source reliability estimation. In: SIGMOD, Utah, USA, pp. 1187–1198 (2014)
12. Liu, W., Liu, J., Zhang, J., Duan, H., Wei, B.: Truthdiscover: a demonstration of resolving object conflicts on massive linked data. In: WWW, Perth, Australia (2017)
13. Manola, F., Miller, E., McBride, B.: RDF 1.1 primer. http://www.w3.org/TR/2014/NOTE-rdf11-primer-20140624/
14. Mendes, P.N., Mühleisen, H., Bizer, C.: Sieve: linked data quality assessment and fusion. In: EDBT/ICDT Berlin, Germany, pp. 116–123 (2012)
15. Michelfeit, J., Knap, T., Nečaský, M.: Linked data integration with conflicts. arXiv preprint arXiv:1410.7990 (2014)
16. Navarro, G.: A guided tour to approximate string matching. ACM Comput. Surv. **33**(1), 31–88 (2001)
17. Rayana, S., Akoglu, L.: Collective opinion spam detection: bridging review networks and metadata. In: KDD, Melbourne, Australia, pp. 985–994 (2015)
18. Vydiswaran, V., Zhai, C., Roth, D.: Content-driven trust propagation framework. In: KDD, San Diego, USA, pp. 974–982 (2011)
19. Wang, H., Fang, Z., Zhang, L., Pan, J.Z., Ruan, T.: Effective online knowledge graph fusion. In: Arenas, M., et al. (eds.) ISWC 2015. LNCS, vol. 9366, pp. 286–302. Springer, Cham (2015). doi:10.1007/978-3-319-25007-6_17
20. Yin, X., Han, J., Yu, P.S.: Truth discovery with multiple conflicting information providers on the web. IEEE Trans. Knowl. Data Eng. **20**(6), 796–808 (2008)
21. Yin, X., Tan, W.: Semi-supervised truth discovery. In: WWW, Lyon, France, pp. 217–226 (2011)
22. Zaveri, A., Rula, A., Maurino, A., Pietrobon, R., Lehmann, J., Auer, S., Hitzler, P.: Quality assessment methodologies for linked open data. Semant. Web J. **7**, 63–93 (2013)
23. Zhao, B., Rubinstein, B.I., Gemmell, J., Han, J.: A bayesian approach to discovering truth from conflicting sources for data integration. PVLDB **5**(6), 550–561 (2012). Istanbul, Turkey

Methods for Intrinsic Evaluation
of Links in the Web of Data

Cristina Sarasua[1(✉)], Steffen Staab[1,2], and Matthias Thimm[1]

[1] Institute for Web Science and Technologies,
University of Koblenz-Landau, Koblenz, Germany
{csarasua,staab,thimm}@uni-koblenz.de
[2] WAIS Research Group, University of Southampton, Southampton, UK

Abstract. The current Web of Data contains a large amount of inter-linked data. However, there is still a limited understanding about the quality of the links connecting entities of different and distributed data sets. Our goal is to provide a collection of indicators that help assess existing interlinking. In this paper, we present a framework for the intrinsic evaluation of RDF links, based on core principles of Web data integration and foundations of Information Retrieval. We measure the extent to which links facilitate the discovery of an extended description of entities, and the discovery of other entities in other data sets. We also measure the use of different vocabularies. We analysed links extracted from a set of data sets from the Linked Data Crawl 2014 using these measures.

Keywords: Data integration · Links · Quality · Monitoring · Semantic web

1 Introduction

Linked Data principles encourage data publishers to connect the resources in their data sets to other resources "so that more things can be discovered"[1]. With the increasing number of available data sets and links between them [8, 12], it becomes highly important to observe the extent to which existing links have desirable properties, as we need to ensure high quality to encourage the usage of Linked Data. Links should (i) follow the recommendations that apply to high quality data [14] (i.e. links should be accessible, syntactically valid, and semantically accurate), and (ii) links should enable the discovery of "more things", facilitating new insights from the data. Established data-driven quality assurance methodologies [10,11,14] suggest that the key steps for improving the status quo are: the definition of measures, the analysis of measurements and the subsequent monitoring of updates. So, to be able to analyse the quality of links, we need measures that help us assess all relevant quality aspects, including (i) and (ii).

---

[1] Berners-Lee, T. Linked Data Principles http://www.w3.org/DesignIssues/LinkedData.html.

© Springer International Publishing AG 2017
E. Blomqvist et al. (Eds.): ESWC 2017, Part I, LNCS 10249, pp. 68–84, 2017.
DOI: 10.1007/978-3-319-58068-5_5

Previous empirical studies on the adoption of Linked Data principles [6,12] report on the number of outgoing and incoming links of data sets, and the most frequently used predicates in RDF links. Recently, Hu et al. [7] studied degree distributions, as well as missing links in Bio2RDF based on symmetry and transitivity. Neto et al. [9] focused on the analysis of dead links in schema and entity link triples published in the Web of Data. While these studies, together with the findings provided by smaller evaluations of other link assessment methods focusing on (i) (e. g. Guéret et al. [4] and other quality dimensions like completeness [2] provide a characterization of existing links), they do not allow for assessing how many new things might be made discoverable thanks to the links (ii).

In this paper, we provide a framework for link analysis that takes into account principles of data integration in the Web of Data, addressing (ii). We suggest measures that focus on data quality dimensions inherent in the data, while extrinsic assessment would take into account the needs a user has in his specific context (cf. [14]). More specifically, our measures examine the effect that links have on entities (and consequently on data sets). We measure the extent to which links facilitate the discovery of an extended description of entities, and the discovery of other entities in other data sets. We also measure if they add different vocabularies (cf. Sect. 4.2) to the description of entities. Our measures are grounded on foundations of the field of Information Retrieval, as we acknowledge redundancy when we measure the gain in description, connectivity and number of used vocabularies. More precisely, the contributions of this paper are:

1. We identify a set of principles for data interlinking in the Web of Data (Sect. 3).
2. We define a set of measures to analyse available links in terms of these principles (Sect. 4).
3. We demonstrate the feasibility of the proposed framework with the implementation of the measures and carry out an empirical analysis of links extracted from the Linked Open Data Crawl [12] (Sect. 5).

2 Preliminaries

We introduce in this section the terminology and notation.

Definition 1. RDF Quadruple: *Given $\mathcal{U}$, a finite set of HTTP URIs, representing resources, $\mathcal{L}$ a finite set of literal values, and a finite set of blank nodes $\mathcal{B}$ where $U \cap \mathcal{L} = \mathcal{U} \cap \mathcal{B} = \mathcal{L} \cap \mathcal{B} = \emptyset$, a quadruple (s, p, o, c) is any element of the data space $Q = (\mathcal{U} \cup \mathcal{B}) \times \mathcal{U} \times (\mathcal{U} \cup \mathcal{L} \cup \mathcal{B}) \times \mathcal{U}$. s, p, o is a triple statement describing s, while c is the context (denoted by a URI) in which the triple is defined.*

Definition 2. RDF Data set: *An RDF data set D_c is a set of quadruples grouped by some context c $D_c \subseteq \{(s, p, o, c) \in Q\}$, where Q is the set of all quadruples.*

Definition 3. Home: *Given C the set of all contexts, and an entity (either a blank node or URI), home : $\mathcal{B} \cup \mathcal{U} \mapsto C$ is the function that maps the entity to the context c where the entity is defined. Note that when x is a vocabulary term (e. g. a class or a property), the c returned by home(x) is the identifier of the vocabulary where the term x was defined.*

The *home* function is customisable. For example, it can be defined to match the notion of data sets in the Linked Open Data literature [12], or it can be defined to match the graphs in data sets—the graphs in the SPARQL and N-Quads specifications. In this paper, we stick to the LOD cloud diagram[2] and analyse links on a data set basis.

For representing the relation between entities of different data sets, we define:

Definition 4. Link: *A link of D_c is a quadruple $(s,p,o,c) \in D_c$ such that $s \in \mathcal{U}, o \in \mathcal{U}, home(s) = c, home(o) \neq c$*

Definition 5. Interlinking: *The interlinking I_c of a data set D_c is the set of all links going out from D_c to any other data set: $I_c = \{(s,p,o,c) \in D_c|\ home(s) = c, home(o) \neq c\}$.*

To formally define our measures, we use a relational algebra-like notation. For this purpose we define selection σ, projection π and join $\bowtie$ as follows:

Definition 6. Selection: *Given $X \subseteq D_c$, a selection $\sigma_h(X)$ is the quadruples from X that satisfy a selection predicate h: $\sigma_h(X) = \{(s,p,o,c)|(s,p,o,c) \in X \wedge h(s,p,o,c) = true\}$*

Example 1. We can select the quadruples of the data set D_c that are `owl:sameAs` links by $\sigma_{p=owl:sameAs}(D_c) = \{(s,p,o,c)|(s,p,o,c) \in D_c, p = owl : sameAs\}$

Definition 7. Projection: *Given $X \subseteq D_c$, and Y a subset of the elements in the quadruples in X, a projection $\pi_Y(X)$ on attributes Y is the subset of X including the elements Y : $\pi_Y(X) = \{(s,p,o,c)[Y]|(s,p,o,c) \in X\}$*

Example 2. We can obtain the projection of all the entities appearing in the predicate and object positions of the quadruples of the data set D_c by $\pi_{p,o}(D_c) = \{(p,o)|(s,p,o,c) \in D_c\}$

Definition 8. EquiJoin: *Given $X_1 \subseteq D_1$ and $X_2 \subseteq D_2$, the Equi join of the two sets is the set of elements such that: $X_1 \bowtie_{X_1.o \theta X_2.s} X_2 = \{(X_1.s, X_1.p, X_1.o, X_2.s, X_2.p, X_2.o, c) \mid X_1.o = X2.s\}$*

Example 3. In Table 1 case (I), the equijoin of the two quadruples on the name and the link is the 7-tuple "d1:nn owl:sameAs d2:nn d2:nn rdfs:label "Natasha" d1.".

Now, we may re-state **our task** at hand as follows: Given a data set D_c containing the interlinking I_c, our task is to compare D_c and $D_c \backslash I_c$ and analyse the value that I_c gives to the data in terms of the principles for data interlinking in the Web of Data described in the following section.

---

[2] http://lod-cloud.net.

3   Principles for Data Interlinking in the Web of Data

The main reason to connect data sets is to enable their joint search, browsing or querying. As in any information system, when a user queries Linked Data it is important that she: (**n1**) finds all entities she is interested in (recall); (**n2**) finds only entities she is interested in (precision); (**n3**) is able to understand the relationship between entities in the Web; (**n4**) finds answers to all her questions no matter how heterogeneous in syntax, structure and semantics the questions are.

The existence of high quality links between entities can contribute to a better fulfilment of the aforementioned needs (n1–n4). In order to understand the way links can help, let us consider various interlinking examples (from (I) to (VII)) shown in Table 1. We analyse each of the examples, and derive from them desired properties for links (i. e. principles P1–P3).

Table 1. Examples of different interlinking cases.

Source data set	Target data set(s)
Entity Description	
(I) d1:nn foaf:name "Natasha Noy"	d2:nn foaf:name "Natalya F. Noy"
d1:nn dbo:affiliation d1:stanford	d2:nn dbo:affiliation d1:googleinc
d1:nn swrc:publication d1:p2012-1	d2:nn swrc:publication d2:p2015-1
d1:nn owl:sameAs d2:nn	
(II) d1:nn foaf:name "Natasha Noy"	d2:nn foaf:name "Natalya F. Noy"
d1:nn owl:sameAs d2:nn	d2:nn cito:likes d2:sfo
d2:ms foaf:name "Mark Smith"	d2:nn swc:holdsRole swc:Chair
(III) d1:nn foaf:name "Natasha Noy"	d2:nn foaf:name "Natasha Noy"
d1:nn owl:sameAs d2:nn	
Entity Connectivity	
(IV) d1:nn foaf:name "Natasha Noy"	d2:p1 foaf:name "Natasha Noy"
d1:nn dbo:affiliation	d2:p1 dbo:affiliation
dbr:Stanford_University	dbr:Stanford_University
d1:nn owl:sameAs d2:p1	d3:p5 foaf:name "Natasha Noy"
d1:nn owl:sameAs d3:p5	d3:p5 dbo:affiliation
	dbr:Stanford_University
d1:nn owl:sameAs d4:p1	d4:p1 dbo:affiliation
	dbr:Stanford_University
(V) d1:nn foaf:name "Natasha Noy"	d2:p1 foaf:name "Natasha Noy"
d1:nn dbo:affiliation	d2:p1 dbo:affiliation
dbr:Stanford_University	dbr:Stanford_University
d1:nn owl:sameAs d2:p1	d3:p5 foaf:name "Natasha Noy"
	d3:p5 dbo:affiliation
	dbr:Stanford_University
Vocabularies Involved in the Description	
(VI) d1:nn foaf:name "Natasha Noy"	d2:nn sioc:creator_of d2:post2
d2:nn rdf:type foaf:Person	d1:nn rdf:type proton:Human
d1:nn owl:sameAs d2:nn	d2:nn vivo:teachingOverview
	"Natasha Noy was a tutor in the
	SSSW08 summer school"
(VII) d1:nn foaf:name "Natasha Noy"	d2:nn foaf:name "Natasha Noy"
d1:nn owl:sameAs d2:nn	d2:nn foaf:currentProject d2:bioportal
	d2:nn foaf:pastProject d2:protege

Entity Description. In case (I) we see two entities linked via an `owl:sameAs` link. The two connected entities have different names, but represent the same person (Natalya F. Noy, also known as Natasha Noy informally). The source data set contains the publications that Natasha wrote when she worked at Stanford, and the target data set contains publications she has written while working at Google Inc. If we search for the publications written by Natasha and only consider the source data set, we exclusively see her Stanford publications. If we consider the link connecting the two entities referring to Natasha, we are able to also find her Google publications, giving us higher recall (**n1**).

In case (II) the two entities are also connected via an `owl:sameAs`. The target data set contains data about conferences and program committees, while the source data set does not contain this kind of data. If we look for persons who have been chairs of scientific events, and we only take into account the source data set, we are not able to find any person because we lack the information about the chairs of the events. In an Information Retrieval scenario, we would use query relaxation techniques, and the search query would be reformulated as a search for persons. The result would include the entities for Natasha Noy and Mark Smith (who is a student assistant and was never a chair). Conversely, if we consider the link, we have relevant information for the query and only Natasha is retrieved in the results. Therefore, in this case the link enables us to have higher precision (**n2**).

Observation: These two cases, have something in common: the links (s, p, o, c) extend the description of entities s. The description of an entity is the set of quadruples with s as subject, and literals, URIs and blank nodes as objects (cf. Sect. 4.2). When the linked data sets provide redundant information, links do not help in recall, nor in precision. Example (III) is a clear example of a scenario where we have redundant information and the description is not extended. Therefore, we formulate the first principle as:

> **P1: Try to create an interlinking that extends the description of entities of the source data set.**

Entity Connectivity. Case (IV) connects the entity referring to Natasha in d1 to the corresponding entities representing Natasha in data sets d2, d3 and d4. While these links do not extend description of the entity in d1 (i.e. they do not follow the Principle P1), they help in understanding the relationship between the entities in the Web of Data (**n3**). This understanding is necessary when for example, a change in the affiliation of Natasha is materialised in d1 to update her affiliation. The descriptions in d2, d3 and d4 could be subsequently changed, in order to keep the data up-to-date.

Observation: In (IV), we can see the importance of creating multiple links from the same entity to different external entities and data sets, increasing its connectivity (cf. Sect. 4.2). In Case (V), which is similar to case (IV) but without the links to d3 and d4, we see that if the links from d1:nn to the entities in d3 and d4 do not exist (as in case (V)), it is harder to reach the entities in other data sets that would need to be updated. This is similar in cases where the links are created to group entities, or to enable the browsing of different types of entities. We formulate the second principle as:

P2: Try to create an interlinking that increases the number of entities and data sets that source entities are connected to.

Heterogeneity of Descriptions. Case (VI) shows an example where the entity representing Natasha is connected via an `owl:sameAs` link to its corresponding entity in d2. The entity in d2 is described with vocabularies that are different from d1's vocabularies. In contrast, in case (VII) the entity in d2 contains a description that adds new information to the description of d1 (satisfies P1) but uses the same vocabulary as in d1 (i. e. FOAF).

Observation: in (VI), links help in answering a wider range of queries that might be formulated in different application contexts (**n4**). Using different vocabularies we are able to use and analyse entities from multiple perspectives. Hence, the third principle is:

P3: Try to create an interlinking that makes the source entities have a description with a higher number of vocabularies in their description.

These principles are not independent from each other. Principles P2 (entity connectivity) and P3 (vocabularies) are specializations of P1 (entity description). For some types of links (non-identity links), creating links to new entities in new data sets (P2), means that the description of the source entity is extended (P1). However, that does not necessarily happen the other way round. Analogously, if one uses further vocabularies in the links between entities (P3), the description of the source entity will be extended (P1). Therefore, when we analyse data in terms of these principles, we consider them as a three level test, in which having passed P1 is positive, but having passed P2 and P3, too, is even more positive. We do not claim that these principles are complete, and they may be extended.

4   Intrinsic Measures for Assessing the Quality of Links

The measures that we define do not provide an absolute assessment of the quality of links. That is, a particular measurement is not good or bad. Instead, we provide measures for a comparative assessment: we acknowledge that one interlinking is better than another in some dimension that we observe with regard to the principles in the previous section. It is up to the person or application inspecting the measurements to interpret its meaning, and make a decision based on it (e. g. a data publisher willing to improve her interlinking and using our measurements as a guide to decide where to start from).

We distinguish between descriptive statistics that give an overview of the size and the elements in I_c (see Sect. 4.1), and measures that assess the way the links in the interlinking I_c of the data set D_c follow the aforementioned principles (see Sect. 4.2).

4.1 Basic Descriptive Statistics

In order to describe basic properties of the interlinking of a data set, we use basic statistics proposed by related work (e. g. Void Vocabulary[3] and LOD Stats[4]), to compute the volume of the interlinking ($|I_c|$), and the distribution of linksets ($\{(x, |\sigma_{p=x}(I_c)|)\}$).

4.2 Principles-Based Measures

Since we would like to study the effect that links have on the entities of the source data set, our measures analyse links grouped by source entities. Note that in our analysis we focus on entities $e \in D_c$ such that $\nexists\,(e, rdf : type, rdfs : Class) \in D_c$. So, we look at the interlinking of individuals and not at vocabulary terms.

4.2.1 Two Views of the Quadruples About Entities

For each entity e, we distinguish two views of the set of quadruples that state something about e: the description view and the connectivity view of an entity.

Description View. This view focuses on all the quadruples in X describing the entity e.

We define the description of an entity e in $X \subseteq D_c$ as the projection that selects the predicates and objects from the set of quadruples of X about e, and entities defined to be identical to e (usually defined via the predicates `owl:sameAs` or `skos:exactMatch`).

$$desc(e, X) = \pi_{(p,o)}(\sigma_{s=e}(X)) \cup \pi_{(Q.p,Q.o)}(\sigma_{X.p=identity}((X \bowtie_{X.o=Q.s} Q))) \quad (1)$$

In order to have a more detailed view of the description, we differentiate between the entity's classification (i. e. the quadruples referring to the `rdf:type` of the entity):

$$classif(e, X) = \sigma_{p=\text{``}rdf:type\text{''}}(desc(e, X)) \quad (2)$$

and the rest of the description:

$$descm(e, X) = desc(e, X) \backslash classif(e, X) \quad (3)$$

---

[3] https://www.w3.org/TR/void/.
[4] http://stats.lod2.eu/links.

Example 4. In Table 1(VI), classif(d1:nn,D_1'= { (rdf:type, foaf:Person), (rdf:type, proton:Human)} and descm(d1:nn,D_1') = {(foaf:name, "Natasha Noy"), (owl:same AS, d2:nn), (foaf:name, "Natasha Noy"),(sioc:creator_of, d2:post2),(vivo:teaching Overview,"...")}

Additionally, we make a specification of $descm(e, X)$ and define $descmp$ to project only the predicates (instead of the predicates and values as in $descm(e, X)$).

$$descmp(e, X) = \pi_{(p)}(descm(e, X)) \tag{4}$$

To identify the vocabularies used in the description of an entity we define:

$$vocabd(e, X) = \{home(p)|(p, o) \in desc(e, X)\} \tag{5}$$

Connectivity View

This view focuses on the quadruples that state the connections between the entity e and other entities. Note that this view is a subview of the description view. Here, we ignore the quadruples about e, with literal values and quadruples describing identical entities to e.

We define the entity connectivity of an entity e in $X \subseteq D_c$ as the set containing the entities targeted from e:

$$econn(e, X) = \pi_o(\sigma_{s=e}(X)) \tag{6}$$

Analogously, we define the data set connectivity of an entity e in $X \subseteq D_c$ as the set containing the data sets targeted from e:

$$dconn(e, X) = \{home(o)| o \in econn(e, X)\} \tag{7}$$

Example 5. In Table 1(V), econn(d1:nn,D_1)={dbr:Google,d2:p1,d3:p5,d4:p1}, whereas dconn(d1:nn,D_1)={dbr,d2,d3,d4}

4.2.2   Measuring the Principles at an Entity and Data Set Level

Now that we have defined the sets for the description and the connectivity views (Sect. 4.2.1, let us look at the measures that are interesting to be applied on these sets, in order to state the extent to which the links going out of entity e follow principles P1, P2 and P3. We use the notation S to refer to any of the sets above.

Measure size. Measuring the size of data is a standard way of characterizing data. We measure the size of each of the sets above by calculating the cardinality of the corresponding set (i. e. $|S|$).

Measure diversity. When we observe if entities get their description (i. e. $classif(e, X)$ and $descm(e, X)$) extended when considering the links, we aim to identify redundancy. Furthermore, when we analyse the targeted entities and data sets, as well as the vocabularies used in the description and the links, we want to measure diversity both without and with links. In these two situations, we may encounter repetitions in the classification, the description,

Table 2. List of measures to analyse the fulfilment of data interlinking principles. Columns show: the name of the measure, the principle the measure belongs to, the random variables defined for the measure, and the formal definition of the measure.

ID	Principle/Description	Vars.	Definition
m11a	P1 #classes	-	$|classif(e, D_c^{\text{internal}})|, |classif(e, D_c)|$
m11c	P1 Classification Extension (entropy)	CS, CS'	$H(CS') - H(CS)$
m12a	P1 #predicate-objects	-	$|descm(e, D_c^{\text{internal}})|, |descm(e, D_c)|$
m12c	P1 Description Extension	DE, DE'	$H(DE') - H(DE)$
m13a	P1 #predicates	-	$|descmp(e, D_c^{\text{internal}})|, |descmp(e, D_c)|$
m13c	P1 Predicate Description Extension	DEP, DEP'	$H(DEP') - H(DEP)$
m21a	P2 #targeted entities	-	$|econn(e, D_c^{\text{internal}})|, |econn(e, D_c)|$
m21c	P2 Entity connectivity Extension	EC, EC'	$H(EC') - H(EC)$
m22a	P2 #targeted data sets	-	$|dconn(e, D_c^{\text{internal}})|, |dconn(e, D_c)|$
m22c	P2 Data set connectivity Extension	DC, DC'	$H(DC') - H(DC)$
m31a	P3 #Vocabularies in desc	-	$|vocabd(e, D_c^{\text{internal}})|, |vocabd(e, D_c)|$
m31c	P3 Increase #Vocabularies Used (entropy)	VD, VD'	$H(VD') - H(VD)$

the entity connectivity, the data set connectivity, and the vocabularies used in the description. Therefore, we extend the notion of our sets and model multisets (allowing repeated elements), counting the number of times each element appears in the multiset: (S, n) where n is $n : S \mapsto \mathbb{N}_{\geq 1}$, a function that given an $s \in S$ tells the number of times that s appears in S.

Diversity is a measure that takes into account the number of different (and non redundant) types of elements in a set, and at the same time takes into account how equally distributed the elements of each type are present in the set. For these two purposes, we use the Shannon Entropy [13], a standard measure used in Information Theory to measure diversity.

$$H(ELS) = -\sum_{s \in S} prob(ELS = s) \times \log prob(ELS = s) \qquad (8)$$

A low entropy value means that there is little diversity in the data. Note that $H(x) \geq 0$. In $classif(e, X)$, and $descm(e, X)$ repeated statements appear only when we consider the quadruples of the target data sets, because in one data set quadruples are supposed to be unique. Still, we calculate entropy to be able to signal redundancy when we compare the description with and without links.

Compare measurements. In order to accomplish our task of comparing measurements considering the links vs. not considering the links, we differentiate between the total set of quadruples in D_c, and the set of internal quadruples defined as:

$$D_c^{\text{internal}} = D_c \backslash I_c$$

We compare a measurement on D_c vs. the measurement on D_c^{internal} by subtracting the latter to the former.

Based on these three rationales, we define the following list of measures (cf. Table 2) to analyse the way links follow the principles. To measure the extension in classification, description, entity connectivity, data set connectivity and the increase in the number of vocabularies employed, we use the difference in entropy. For example, to check if the classification is extended, we define two random variables CS (in D_c^{internal}) and CS' (in D_c) and calculate $H(CS') - H(CS)$. The difference is zero when there is no information gain, negative when redundant information is gained, and positive otherwise.

5 Empirical Analysis

To demonstrate the feasibility of our approach for profiling the quality of links in the Linked Open Data cloud, we have implemented the measures in the SeaStar framework, which uses Java, the NxParser to parse N-Quads, and Jena for handling RDF data[5].

5.1 Data

We use data from the Linked Open Data Crawl[6], as it has been recognised as a sound snapshot of the LOD cloud in 2014 [12]. First we extracted the links from the crawled data, by parsing the dump line by line, and identifying each quadruple containing a subject and an object with different graph provenance, and therefore a different $home(x)$. While parsing the dump file, we excluded all syntactically invalid quadruples to work with clean data. Second, in order to analyse the links on a data set basis, we split the data crawl into individual data sets, taking as contexts the data set identifiers provided by Schmachtenberg et al.[7]. We selected a set of 35 data sets from the LOD2014 crawl (from different domains and containing several types of links), analysing a total of 1+ million links.

5.2 Methodology

We computed each of the measures listed in Table 2 for each of the linked entities in the data sets, for all types of links in the 35 data sets. Once we had all

---

[5] Source code: https://github.com/criscod/SeaStar.
[6] Linked Data Crawl http://goo.gl/lqxdgo.
[7] List of Data sets http://data.dws.informatik.uni-mannheim.de/lodcloud/2014/ISWC-RDB/tables/datasetsAndCategories.tsv.

the results, we first empirically validated the measures (Sect. 5.3). After that, we analysed the results on a data set basis (Sect. 5.4). We have published our experimental data and sources[8].

5.3   Measure Validation

Following standard practices in the literature of quality measures [3], we validate our measures by (i) checking that they do not provide the same measurement for all data sets D_i; and (ii) verifying that our measures are not all correlated with each other – otherwise having multiple measures would be of limited utility.

Discriminative Measures. We computed for each data set standard summary statistics such as the mean, standard deviation and quartiles considering all types of links simultaneously. As we see in the data files, the values for the measures vary across data sets, except for the classification extension (m11c) – where all data sets show a mean, standard deviation and quartiles of 0.0 for the difference in entropy. The other measures are discriminative.

Independent Measures. We computed the Spearman correlation of all the measurements within each data set, putting all types of links together. Table 3 shows the correlation values. The first row contains NaN values because the standard deviation(s) are equal to zero. Measures m21 and m22 are highly correlated (0.96), which makes sense, since m21 looks at the number of target entities and m22 at the number of target data sets. In theory, one may link to many target entities within a few data sets and viceversa; but the empirical analysis suggests that having both might not particularly interesting. Having only m21 seems to be sufficient.

Table 3. Correlation between measures, for all data sets and all types of links.

Measures	m11	m12	m13	m21	m22	m31
m11	NaN	NaN	NaN	NaN	NaN	NaN
m12		1.00	0.29	0.58	0.55	0.55
m13			1.00	−0.23	−0.22	0.76
m21				1.0	0.96	0.04
m22					1.0	0.02
m31						1.0

---

[8] Experimental data: extracted links http://141.26.208.201/links/ Measurements http://141.26.208.201/datameasures/ Python code and others https://github.com/criscod/SeaStar/tree/master/data.

5.4   Results

Let us first look at the types of links that exist in the data sets and second, at the adoption of the 3 core principles. We focus on identity links (e. g. `owl:sameAs`), relationship links (e. g. `wgs84:location`), classification links (e. g. `rdf:type`), similarity links (e. g. `skos:closeMatch`), and other more general links (e. g. `rdfs:seeAlso`).

Basic Descriptive Statistics. When we look at the type of links that is used the most in each of the data sets, in 17/35 data sets the type used at most is classification links (c), in 12/35 data sets it is relationship links (r), in 3/35 it is identity links (i) and in 3/35 it is other links (o). None of the data sets has similarity links (s). Table 4 shows the number of each type of link for each data set.

Table 4. Different types of links in the 35 data sets that we analysed.

Typelink	I	S	R	O	C	All
AEMET	0	0	96	0	57	153
BFS	1063	0	0	0	2862	3925
Bibbase	0	0	456	1401	0	1857
Bibsonomy	35646	0	2180	0	123080	160906
BNE	58	0	0	0	221	279
DNB	3577	0	8711	2278	55	14621
DWS Mannheim	71	0	296	39	926	1332
Eurostat	1182	0	2	0	1012	2196
Eye48	1	0	244	0	490	735
Fao	0	0	6	0	23	29
FigTrees	2	0	22	2	59	85
GeoVocab	11455	0	1759	113	7565	20892
GovWild	0	0	1998	0	0	1998
Harth	76	0	344	456	30	906
Icane	20	0	25	30	19	94
IMF	243	0	3	0	377	623
Korrekt	0	0	1174	0	7959	9133
L3S	1059	0	2478	1028	1089	5654

Typelink	I	S	R	O	C	All
LinkedGeoData	634	0	12	0	254	900
LOD2	26	0	282	50	180	538
NDLJP	1	0	178	60	267	506
Ontologi	0	0	5686	0	736	6422
Openei	6	0	323	0	203	532
Reegle	327	0	432	0	135	894
Revyu	1402	0	2145	1806	39772	45125
RodEionet	9	0	981	0	0	990
SemanticWeb	161	0	783	0	576295	577239
Sheffield	121	0	2189	1	27064	29375
Simia	6691	0	25113	0	38069	69873
Soton	50	0	352	0	160	562
SWCompany	2023	0	13473	421	43136	59053
TomHeath	7	0	34	4	6	51
Torrez	0	0	266	0	493	759
TWRPI	2	0	12	0	65	79
UKPostCodes	1	0	7	0	1	9

Principle-based Measurements. Since our user is a data publisher willing to improve the interlinking, for each measure we analyse the inequalities among entities of the same data set. For that, we generate multiple box plots (one per entropy-based measure and type of link)[9]. If a box plot suggests that there are entities that get their description less extended than other entities in the data set, the data publisher could think of generating further links from those entities to new target data sets. The important features of these plots are the medians (in red), the range and interquartile range—which can show big differences among the measurements of different entities when they are big—and the outliers, which in our case are relevant as they can be one of the weak spots to be improved.

Classification: for all data sets and all types of links, the difference in cardinality (m11a) and entropy (m11c) has a median of 0.0 and the range of boxes is [0.0,0.0]. That means that there are no cases in the data where entities have

---

[9] https://github.com/criscod/SeaStar/tree/master/data/plots.

been classified with classes defined in the source data set and the classification is inherited via identity links. However, given the number of links of type c, we see that data publishers do classify their entities with external classes.

Description: according to the m12a measurements, in all but two data sets the median of (p, o)-s gained is equal or below 2; the remaining two data sets show a median of 4 and 20. The median of new o-s gained instead (m13a) is 1 for 32 of the data sets (the other three have a median of 0). Observing the m12c measurements in the first row of box plots (Fig. 2), we notice that in links of type c the medians of the difference in entropy stay between 0 and 1, while in links of type i the medians vary among data sets and go up to 8. Also, in identity links there are way more outliers than in classification links (see the case of Bibsonomy). It makes sense that entities are not described homogeneously, and often publishers do not have the resources to review each generated identity link. Both things motivate that SeaStar shows the user source entities and other data sets as more positive references. In the case of m13c measurements, and for all types of links, we find data sets that have negative values for the difference in entropy. That means that the links add some redundancy by adding statements with predicates that were already in the source entity. However, the positive thing is that only a few data sets have the box in the negative area, and that happens for links of type relationship (r) and others (o). For example, that occurs when the data publisher adds multiple `rdfs:seeAlso` internal and external links. The medians are between −0.4 and 0.7. Comparing the box plots for identity links (type i) of the m12 and m13 measurements, we notice that in the former the range of the boxes is larger than the boxes in m13 measurements; in m12 the distance between the min and max is around 4 where as in m13 is around 0.2.

Connectivity: the medians for the number of new entities targeted (m21a) for three data sets are 3,4, and 11, and for the rest these are all equal or below 2 new entities targeted. In the difference of entropy (m21c), the box plots do not show redundancy, which would only be possible if we compared D_c with a basis of previously generated links and new links were added over the same target entity. This would be a positive thing, if those links managed to extend the description (P1). M21 measurements show medians between 0 and 8 as for links of type i, between 0.0 and 2.0 for links of type r and o, and between 0.0 and 1.0 for links of type c. The box plot with links of type i, shows a more skewed box (either to the left or to the right) than m12 measurements of the same type of links.

Heterogeneity: measurements m31a show that 27 data sets gain 1 vocabulary in their description, while the rest do not gain any new. The difference in entropy (m31c) is in several data sets negative (in outliers and in the interquartile range). For links of type c the medians in measurements m31c are between 0.0 and 1.0, while for links of type r medians are between −0.1 and 1.0.

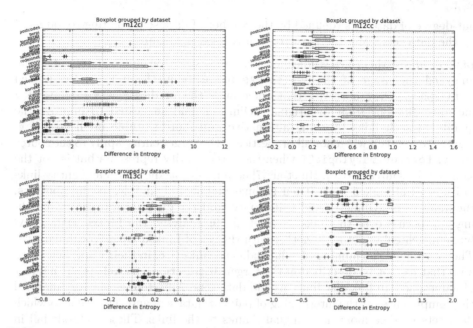

Fig. 1. Box plots showing m12c and m13c measurements for all data sets (m12c type c, m12c type, m13c type i, m13c type r).

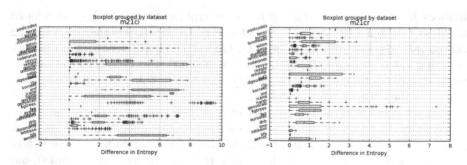

Fig. 2. Boxplots showing m21c measurements (links type i and type c) and box plots showing m31c measurements (links type r and type c).

6 Related Work

With the growth of Linked Data, there has been an increasing interest in assessing and monitoring the quality of available data [14].

Status of the Linked Data Web: while there were previous studies about the conformance of the Linked Data principles [6], the work by Schmachtenberg et al. [12] is the most recent study on the current adoption of Linked Data best practices. With regard to the linking principle, their analysis on data crawled from 1041 distinct data sets) showed descriptive statistics about the in- and out-degree of data sets (defined by the number of data sets pointing to/targeted by the data sets), and the most frequently used predicates.

Link Analysis: there are studies focusing exclusively on links. Halpin et al. [5] analyzed the usage of the `owl:sameAs` predicate in the links of the Linked Data space. They observed that sometimes the predicate was used with a meaning different from its original definition, and suggested to improve the quality of such links by using alternative and more suitable predicates (e.g. `skos:closeMatch` when not all properties of the target entity apply to the source entity; `foaf:primaryTopicOf` when the target entity represents but is not the same as the source entity). Hu et al. [7] empirically studied term and entity links in Biomedical Linked Data. Their findings include link and degree distributions, the analysis of symmetry and transitivity, and the evaluation of entity matching approaches over the links. Neto et al. [9] analysed the Linked Data crawl by Schmachtenberg et al., together with the set of Linked Open Vocabularies[10]. They examined the number of valid and dead links (i.e. in their work, links with an *o* that cannot be described in the target distribution), as well as the number of namespaces in link distributions and data sets. Albertoni et al. [1,2] analysed the completeness of the interlinking of pairs of data sets and the extent to which data sets become more multilingual thanks to the links. These methods fail in stating the extent to which links add value to the source data set in terms of the principles that we mention in this paper.

Methods for Assessing Accuracy of Links: several methods have been developed to assess the semantic accuracy of links (e.g. to decide whether `ch:koblenz owl:sameAs de:koblenz` holds or not). Guéret et al. [4] defined a framework including three measures from the area of network theory: degree, clustering coefficient and betweeness centrality of the entities in links; as well as two measures that the authors define: number of unclosed same as chains and description enrichment defined as the raw number of new statements gained by the source entity. While Guéret's et al. notion of description enrichment is related to ours, the main differences are that we are able to observe further dimensions (e.g. how the classification of entities and the connectivity is extended by the links), our approach is not only restricted to `owl:sameAs` links (as it applies to any link) and we are able to signal redundancy.

7   Conclusions and Future Work

We have presented a collection of measures whose goal is to help in gaining insights into the quality of existing links, and understanding the effect that links produce in the source data set. After analysing 35 data sets of the LOD cloud with these measures our findings show that source entities are not classified with internal classes, but with external classes via links, and identity links do not contribute to inheriting new classes. We also observed that there is certain redundancy in the properties and vocabularies used as for extending the description.

---

[10] LOV http://lov.okfn.org/dataset/lov/.

The differences between entities and data sets shown in the boxplots justify the need for our framework, which is able to pinpoint reference interlinked entities and data sets to data publishers.

As future work, we plan to extend our approach including mappings between classes and properties. We expect this add-on to help in identifying redundancy more precisely. Furthermore, we consider evaluating the usefulness of the measures with domain experts and observing the actions they take in data sets in response to the measurements.

Acknowledgements. The research leading to these results has received funding from the European Union's FP7 under grant agreement no. 611242—Sense4Us project. We also thank Thomas Gottron for our discussions at an initial phase of this research, and Leon Kastler for his feedback.

References

1. Albertoni, R., De Martino, M., Podestà, P.: A linkset quality metric measuring multilingual gain in skos thesauri. In: Linked Data Quality Co-located with ESWC 2015 (2015)
2. Albertoni, R., Pérez, A.G.: Assessing linkset quality for complementing third-party datasets. In: Proceedings of the Joint EDBT/ICDT 2013 Workshops, EDBT 2013 (2013)
3. Behkamal, B., Kahani, M., Bagheri, E., Jeremic, Z.: A metrics-driven approach for quality assessment of linked open data. J. Theor. Appl. Electron. Commer. Res. **9**(2), 64–79 (2014)
4. Guéret, C., Groth, P., Stadler, C., Lehmann, J.: Assessing linked data mappings using network measures. In: Simperl, E., Cimiano, P., Polleres, A., Corcho, O., Presutti, V. (eds.) ESWC 2012. LNCS, vol. 7295, pp. 87–102. Springer, Heidelberg (2012). doi:10.1007/978-3-642-30284-8_13
5. Halpin, H., Hayes, P.J., McCusker, J.P., McGuinness, D.L., Thompson, H.S.: When owl:sameAs isn't the same: an analysis of identity in linked data. In: Patel-Schneider, P.F., Pan, Y., Hitzler, P., Mika, P., Zhang, L., Pan, J.Z., Horrocks, I., Glimm, B. (eds.) ISWC 2010. LNCS, vol. 6496, pp. 305–320. Springer, Heidelberg (2010). doi:10.1007/978-3-642-17746-0_20
6. Hogan, A., Umbrich, J., Harth, A., Cyganiak, R., Polleres, A., Decker, S.: An empirical survey of linked data conformance. Web Semant. Sci. Serv. Agents World Wide Web **14**, 14–44 (2012)
7. Hu, W., Qiu, H., Dumontier, M.: Link analysis of life science linked data. In: Arenas, M., et al. (eds.) ISWC 2015. LNCS, vol. 9367, pp. 446–462. Springer, Cham (2015). doi:10.1007/978-3-319-25010-6_29
8. Schmachtenberg, M., Christian Bizer, A.J., Cyganiak, R.: Linking open data cloud diagram (2014). http://lod-cloud.net/
9. Neto, C.B., Kontokostas, D., Hellmann, S., Müller, K., Brümmer, M.: Assessing quantity and quality of links between linked data datasets (2016)
10. Pandian, C.R.: Software Metrics: A Guide to Planning, Analysis, and Application. CRC Press, Boca Raton (2003)
11. Rula, A., Zaveri, A.: Methodology for assessment of linked data quality. In: LDQ@ SEMANTICS (2014)

12. Schmachtenberg, M., Bizer, C., Paulheim, H.: Adoption of the linked data best practices in different topical domains. In: Mika, P., et al. (eds.) ISWC 2014. LNCS, vol. 8796, pp. 245–260. Springer, Cham (2014). doi:10.1007/978-3-319-11964-9_16
13. Shannon, C.E.: A mathematical theory of communication. ACM SIGMOBILE Mob. Comput. Commun. Rev. 5(1), 3–55 (2001)
14. Zaveri, A., Rula, A., Maurino, A., Pietrobon, R., Lehmann, J., Auer, S.: Quality assessment for linked open data: a survey. Semanti. Web J. (2015)

Entity Deduplication on ScholarlyData

Ziqi Zhang[1], Andrea Giovanni Nuzzolese[2], and Anna Lisa Gentile[3(✉)]

[1] Nottingham Trent University, Nottingham, UK
ziqi.zhang@ntu.ac.uk
[2] Semantic Technology Lab, ISTC-CNR, Rome, Italy
andrea.nuzzolese@istc.cnr.it
[3] IBM Research Almaden, San Jose, CA, USA
annalisa.gentile@ibm.com

Abstract. ScholarlyData is the new and currently the largest reference linked dataset of the Semantic Web community about papers, people, organisations, and events related to its academic conferences. Originally started from the Semantic Web Dog Food (SWDF), it addressed multiple issues on data representation and maintenance by (i) adopting a novel data model and (ii) establishing an open source workflow to support the addition of new data from the community. Nevertheless, the major issue with the current dataset is the presence of multiple URIs for the same entities, typically in persons and organisations. In this work we: (i) perform entity deduplication on the whole dataset, using supervised classification methods; (ii) devise a protocol to choose the most representative URI for an entity and deprecate duplicated ones, while ensuring backward compatibilities for them; (iii) incorporate the automatic deduplication step in the general workflow to reduce the creation of duplicate URIs when adding new data. Our early experiment focused on the person and organisation URIs and results show significant improvement over state-of-the-art solutions. We managed to consolidate, on the entire dataset, over 100 and 800 pairs of duplicate person and organisation URIs and their associated triples (over 1,800 and 5,000) respectively, hence significantly improving the overall quality and connectivity of the data graph. Integrated into the ScholarlyData data publishing workflow, we believe that this serves a major step towards the creation of clean, high-quality scholarly linked data on the Semantic Web.

1 Introduction

ScholarlyData [16] is the evolution of the Semantic Web Dog Food (SWDF) dataset[1]. So far it has taken care of refactoring data from the SWDF dataset at schema level, migrating data representation from the Semantic Web Conference (SWC) Ontology[2] to the new *conference-ontology*[3]. Moreover a workflow is in

---

[1] http://data.semanticweb.org.
[2] http://data.semanticweb.org/ns/swc/swc_2009-05-09.html.
[3] http://w3id.org/scholarlydata/ontology/.

© Springer International Publishing AG 2017
E. Blomqvist et al. (Eds.): ESWC 2017, Part I, LNCS 10249, pp. 85–100, 2017.
DOI: 10.1007/978-3-319-58068-5_6

place - cLODg[4] (conference Linked Open Data generator) [4] - to ease the tasks of data acquisition, conversion, integration, augmentation, which has already been used to add novel data to the portal[5].

Nevertheless, the problem of verifying data at instance level has not been tackled so far. Despite existing guidelines for populating the SWDF dataset in order to maximise the reuse of existing URIs and minimise the introduction of redundant URIs, these are, in practice, not carefully followed, resulting in a number of duplicate URIs in the dataset. This is especially common for people with multiple names and surnames, which sometime are reported inconsistently in different papers and in general for organisations, for which different variations of the name are often reported. In the initial refactoring from SWDF to ScholarlyData, while we tackled data model issues, we ignored the problems at instance level and we simply kept URIs as is.

In this work we address these issues and extend the cLODg workflow with data integration and verification steps. The aim is to have a procedure in place that when a new data graph is to be added to ScholarlyData, determines for each new URI in the new data graph, whether it refers to an existing entity in ScholarlyData with a different URI, or a truly unknown and new entity. We propose a three steps approach that for each input URI from the new data graph (i) creates a candidate pool of URIs that could potentially represent the same entity in the ScholarlyData dataset; (ii) discovers truly duplicate URIs (if any) for the input URI using supervised classification methods and (iii) finally resolves the duplicates and merges the new data graph into ScholarlyData.

Ultimately, we aim to ensure that each entity in ScholarlyData is represented by one unique URI, which is currently not the case. Therefore in this work, we also apply the proposed method to the existing ScholarlyData dataset in a one-off cleaning process to resolve and deprecate duplicate URIs. The contributions of this work are threefold. First, we expand the data publication workflow for ScholarlyData, adding automatic data integration capabilities. Second, we propose an efficient deduplication method for the scholar domain and show that it outperforms several state-of-the-art models significantly. Finally, we add a maintenance step in the general publication workflow that empirically determines how to resolve duplicate URIs, integrate information, while ensuring backward compatibility.

The paper is structured as follows. Section 2 examines related work; Sect. 3 describes the proposed method; Sect. 4 evaluates our method on a manually annotated dataset and discusses the deduplication results on the ScholarlyData dataset; and finally Sect. 5 concludes the paper and identifies future work.

2   Related Work

Scholarly data. The first considerable effort to offer comprehensive semantic descriptions of conference events is represented by the metadata projects at

---

[4] cLODG is an Open Source tool that provides a formalised process for the conference metadata publication workflow https://github.com/anuzzolese/cLODg2.

[5] The tool as been used for generating data for ISWC2016 and EKAW2016.

ESWC 2006 and ISWC 2006 conferences [13], with the Semantic Web Conference Ontology being the vocabulary of choice to represent such data. Increasing number of initiatives are pursuing the publication of data about conferences as Linked Data, mainly promoted by publishers such as Springer[6] or Elsevier[7] amongst many others. For example, the knowledge management of scholarly products is an emerging research area in the Semantic Web field known as Semantic Publishing [19]. Numerous initiatives are aimed at offering advanced exploration services over scholarly data, such as Rexplore [17], Open Citation[8], Google scholar[9] and DBLP[10].

Despite these continuous efforts, it has been argued that lots of information about academic conferences is still missing or spread across several sources in a largely chaotic and non-structured way [1]. Improving data quality remains a major challenge with scholarly data. This can include, amongst others, tasks of dealing with data-entry errors, disparate citation formats, enforcement of standards, ambiguous author names and abbreviations of publication venue titles [10]. The ScholarlyData dataset as well currently suffers from data quality issues, mainly due to duplicates, inconsistencies, misspelling and name variations. This work aims at filling this gap, by proposing a strategy to detect co-referent entities and resolve the duplicate URIs in scholarly datasets, specifically on ScholarlyData.

Entity deduplication. Entity deduplication looks for (nearly) duplicate records in relational data, usually amongst data points that belong to the same entity type. This is related to a wide range of literature, such as named entity coreference resolution in Natural Language Processing [22], and Link Discovery on the Semantic Web [14]. The typical approach depends on measuring a degree of 'similarity' between pairs of objects using certain metrics, then making a decision about whether a pair or groups of pairs are mutually duplicate. This is often done using strategies based on similarity threshold, classification (e.g., [2]) or clustering (e.g., [12]) methods. In [14], 10 most recent state-of-the-art systems are evaluated. Our work is most relevant to SILK [8] and LIMES [15], which include supervised models to overcome the arbitrary decision making of link-matching thresholds. Compared to SILK and LIMES, our method differs in terms of the learning algorithms and similarity metrics. In particular, our method deals with the situation where one URI is used much more frequently than its duplicates, which is found to be common in linked datasets [21] and making conventional set-overlap based measures ineffective. Also, the machine learning algorithms in LIMES (e.g., WOMBAT) only use positive examples for training while both SILK (i.e., the genetic algorithm in [8]) and our method can benefit from negative training examples. The 2013 KDD CUP [20] proposed an academic paper author linking task that drew 8 participating systems.

---

[6] http://lod.springer.com/wiki/bin/view/Linked+Open+Data/About.
[7] http://data.elsevier.com/documentation/index.html.
[8] http://opencitations.net/.
[9] https://scholar.google.com.
[10] http://dblp.uni-trier.de/.

However, the dataset used are very different and hence does not represent the problems in ScholarlyData. Duplicates are found very frequent, often as a result of parsing errors when importing data from different sources, and the winning system heavily relied on ad hoc pre-processing of author names. As we shall show later, duplicates in ScholarlyData are much less frequent and thus challenging to detect.

If addressed in a brute force fashion, the matching task has quadratic complexity as it requires pairwise comparisons of all the records. A common way to reduce complexity is the use of so-called *blocking strategies* (grouping records in blocks before comparing them) to reduce the search space. An approach is to exploit specific criteria in the data schema (the values of particular attributes) to split the data [7,18], as in the Sorted Neighborhood Method (SNM) [7] which performs sorting of the records according to a specifically chosen blocking key. Content based blocking strategies usually look for common tokens between two entities [18] and group entities that share some token(s). Finally, another solution to speed up the comparison step is to map the original entity representations in a lower dimensional space. The most prominent example is *Locality-Sensitive Hashing (LSH)*, which produces effective entity signatures [3]. In this work, we experiment with SNM [7] and a content based approach. We show that a simple content based technique [18] is as effective for this scenario, when searching for common tokens in name-like properties[11] of named entities.

Handling duplicate URIs. Once the sets of duplicate or co-referent URIs are detected, the next step is to determine how to resolve the co-reference and integrate information. A typical solution adopted in Linked Data is to link duplicates with `owl:sameAs` axioms. Glaser et al. [5] argue that `owl:sameAs` axioms are not suitable for co-referent URIs as they become indistinguishable, even though they may refer to different entities according to the context in which they are used. The Identity of Resources on the Web (IRW) ontology [6] formalises the distinction between information resources, such as webpages and multimedia files, and other kinds of Semantic Web non-information resources used in Linked Data and proposes a solution to link them. The solution adopted by DBpedia [11] for dealing with co-referent URIs follows the intuition of [5,6]: mirroring the structure of Wikipedia, DBpedia stores co-referent URIs in a separate graph called `redirects` and uses the HTTP response code `303 See Other` in order to (i) deal with the distinction between information and non-information resources and (ii) to implement the HTTP dereferencing.

Our solution encompasses those in [5,6,11] to allow to: (i) harmonise co-referent URIs by identifying information and non-information resources, and (ii) rely on HTTP redirect for dereferencing. We also extend it by enabling redirect in SPARQL queries by means of query rewriting. This enables backward compatibility to external clients that might refer to a certain resource by using any co-referent URIs.

---

[11] http://xmlns.com/foaf/0.1/name.

3   Deduplication

Given a set of URIs $E = \{e_1, e_2, ...e_n\}$ representing entities of the same type, the goal of deduplication is to: (i) identify sets of duplicate URIs that refer to the same real-world entity. We will call such URIs in each set *'co-referent'* to each other; (ii) determine in each subset one URI to keep (to be called the *'target URI'*), while deprecating the others (to be called the *'duplicates'*) and consolidating RDF triples from the duplicates into the target URI. In this work, we focus on two entity types that are found to be the dominant source of duplicates: **PERSON (PER)**, and **ORGANISATION (ORG)**.

For (i), we develop a process that identifies potential co-referent URI pairs $\langle e_i, e_j \rangle$ from E and submits each pair to a binary classifier to predict if they are truly co-referent. For (ii), we develop heuristics to deprecate duplicate URIs from co-referent pairs. Arguably, the first objective can also be achieved through the use of heuristic thresholds or clustering techniques. However, thresholds are often data-dependent [21], while our motivation for a classification-based approach is two-fold. First, as discussed before, our practical scenarios typically contain an existing, de-duplicated reference dataset D (i.e., ScholarlyData), and another data graph D' is generated from time to time and is to be merged into D (in our problem formalisation, both D and D' will be part of E). Hence the question to be asked is often classification-based: given a URI from D', determine if there exists one URI from D that refers to the same entity. Second, due to the nature of datasets, we expect clusters with 2 or more elements to be rather infrequent.

Beginning with an input set E, we first apply **blocking strategies** to identify pairs of URIs that are potentially co-referent. This should reduce the number of pairs to a number m such that $m \ll \binom{n}{2}$, where n is the number of URIs in E, $\binom{n}{2}$ is the number of *all* possible un-ordered pairs from E and will contain overwhelmingly negative elements as we expect true co-referent pairs to be rare. Next, each pair is passed to a **classification** process that predicts if the potential co-referent pair is positive. Finally, the positive co-referent pairs are submitted to the **URI harmonisation** process that identifies the target URI, removes the duplicates and merges RDF triples.

3.1   Blocking Strategies

Given the set of URIs E, the set P of all possible pair comparisons to perform is quadratic to the size of E. The blocking strategies aim at identifying a subset of comparisons P' such that $|P'| \ll |P|$, which achieves a good tradeoff between Reduction Ratio (RR) - the percentage of discarded comparisons from P - and Pair Completeness (PC) - the percentage of true positive (given a gold standard) that are covered by P'.

We experiment with two different solutions for this problem. First we use SNM [7]. We produce the list of all URIs in E with lexicographic ordering and we produce all combination for e_i with all e_j in a context window size of n (sliding window). We experiment with two kinds of ordering: one on the URIs

Table 1. Features for representing an organisation. $^{-1}$ indicates the inverse of a predicate.

Feature	Path	Target(s)	Example
Name	⟨conf:name⟩	Literals	'STLab-CNR', 'ISTC-CNR'
Members' names	⟨conf:withOrganisation^{-1}, conf:isAffiliationOf, conf:name⟩	Literals	'L. Page', 'S. Brin'
Members' URIs	⟨conf:withOrganisation^{-1}, conf:isAffiliationOf⟩	URIs	sdp:aldo-gangemi
Participated event URIs	⟨conf:withOrganisation^{-1}, conf:during⟩	URIs	sde:ESWC2009/ eswc/2009

themselves, one based on the literal value given the URI's predicate conf[12]:name for each e_i.

The second is a content based technique [18]. For each $e_i \in E$ we produce a candidate pair for comparisons for all e_j ($e_j \neq e_i$) that share at least one common token in their property values. Following [18], we choose name-like properties of entities (both persons and organisations). Details about experiments on the two blocking strategies are presented in Sect. 4.2.

3.2 Classification

Given a pair of URIs ⟨e_i, e_j⟩, we firstly build feature representations of e_i and e_j by traversing corresponding linked data graphs to gather their properties. Next, we derive a feature vector representation for the pair, which is then to be classified as either co-referent or not.

Features of URIs. These are generated by traversing paths on the linked data graph O, starting from e and following a series of predicates ⟨$p_1, p_2, ...p_m$⟩ that represent a particular semantic relation r, to reach another set of nodes, which can be either URIs or data literals. Depending on the semantic type of e, we define different semantic relations and paths. We use the ScholarlyData SPARQL endpoint[13] as the single point of access to the underlying linked data graph.

---

[12] The prefixes used in this paper are:

- conf: https://w3id.org/scholarlydata/ontology/conference-ontology.owl
- sdp: https://w3id.org/scholarlydata/person/
- sde: https://w3id.org/scholarlydata/event/
- sdo: https://w3id.org/scholarlydata/organisation
- sdi: http://www.scholarlydata.org/inproceedings/.

[13] https://w3id.org/scholarlydata/sparql/.

Table 2. Features for representing a person. The superscript $^{-1}$ indicates the inverse of a predicate.

Feature	Path	Target(s)	Example
Name	⟨conf:name⟩	Literals	'Tom Mitchell', 'T. Mitchell'
Affiliation names	⟨conf:hasAffiliation, conf:withOrganisation, conf:name⟩	Literals	'STLab-CNR', 'ISTC-CNR'
Affiliation URIs	⟨conf:hasAffiliation, conf:withOrganisation⟩	URIs	sdo:cnr-istc-italy
Participated event URIs	⟨conf:hasAffiliation, conf:during⟩	URIs	sde:ESWC2009/ eswc/2009
Published work URIs	⟨conf:hasContent^{-1}, conf:hasItem^{-1}, conf:hasAuthorList^{-1}⟩	URIs	sdi:ekaw2012/ paper/demos/109
Co-author URIs	⟨conf:hasContent^{-1}, conf:hasItem^{-1}, conf:hasItem, conf:hasContent⟩	URIs	sdp:aldo-gangemi, sdp:valentina-presutti
Title + abstract + keywords	starting from each each 'published work URI': ⟨conf:title⟩ ∨ ⟨conf:abstract⟩ ∨ ⟨conf:keyword⟩	Literals	'This paper describes a ...'

Table 3. Functions to measure similarity between two bag of features.

$$dice(e_i, e_j, t) = \frac{|f(e_i,t) \cap f(e_j,t)|}{|f(e_i,t) \cup f(e_j,t)|} \tag{1}$$

$$dice^{sqrt}(e_i, e_j, t) = \sqrt{sim_{dice}(f(e_i,t), f(e_j,t))} \tag{2}$$

$$cov(e_i, e_j, t) = \max\left\{ \frac{|f(e_i,t) \cap f(e_j,t)|}{|f(e_i,t)|}, \frac{|f(e_i,t) \cap f(e_j,t)|}{|f(e_j,t)|} \right\} \tag{3}$$

$$cov^{sqrt}(e_i, e_j, t) = \overline{cov(e_i, e_j, t)} \tag{4}$$

For an ORG, we gather features following the paths shown in Table 1. We then normalise URI values to lowercase, and normalise literal values by ASCII folding and replacing any consecutive non-alphanumeric characters with a single white space. For a PER, we gather features following the paths shown in Table 2. We apply the same normalisation to URI and literal values as that for ORG. Furthermore, for the feature 'title+abstract+keyword', we also tokenise the text and remove stopwords.

Features of a Pair of URIs. Next, given a pair of URIs, we create a vector representation of the pair based on the similarities between the features of each URI. Let $f(e, t)$ return a bag of features for the feature type t of the URI e. Depending on t, this will be either a duplicate-removed set ($fs(e, t)$) or a multiset ($fm(e, t)$). We then measure the similarity of each feature type for a pair of URIs in the range of $[0,1]$, using the functions shown in Table 3. Functions based on the Dice co-efficient ($dice$, $dice^{sqrt}$) evaluate the extent to which the information that the two URIs have in common (the top term) along a particular dimension (feature type t), can describe both of them (the bottom term). Functions based on the Coverage (cov, cov^{sqrt}) evaluate the maximum degree to which the common part of the two can describe either of them. This is to cope with the situation where one URI is used much more frequently than the other, which is found to be common in linked datasets [21]. As a result, conventional set-overlap based measures (e.g., Dice) tend to produce low similarity scores for such pairs, due to the lack of features for the minor URI.

Specifically, for a pair of ORG URIs, we use $fs(e,t)$ as the bag of feature function, and apply similarity functions (1) to (4) to each of the four feature types in Table 1. This produces a feature vector with a dimension of 16. For a pair of PER URIs, we use all similarity functions with the $fs(e,t)$ bag of feature representation on all feature types (Table 2), except 'title+abstract+keyword', for which we use all similarities with the $fm(e,t)$ bag of feature representation). This produces a feature vector for the pair with a dimension of 28.

Classification Models. Given a labelled dataset of PER or ORG URI pairs, each represented as a feature vector described above, we train a binary classifier for predicting new, unseen PER or ORG pairs. A plethora of classification models can be used for this kind of tasks. In this work, we select five models to experiment with, including: a Stochastic Gradient Descent (SGD) classifier, a Logistic Regression (LR) model, a Random Forest (RF) decision tree model, a linear Support Vector Machine (SVM-l) and a nonlinear SVM using a Radial Basis Function kernel (SVM-rbf). We use the implementation from the scikit-learn[14] library for all models. Details of the datasets and the training process are discussed in Sect. 4.3.

3.3   URI Harmonisation

The previous steps allow to identify a set of pairs consisting of duplicate URIs in the ScholarlyData dataset. The harmonisation is the task of identifying which are the URIs to keep after the data cleansing process. This task is not trivial, as it requires multiple activities: (i) closure identification; (ii) candidate selection; (iii) knowledge inheriting; (iv) recording of the harmonisation.

Closure Identification. Here we traverse the transitive chains of duplicates in order to identify the closures of duplicate URIs from available pairs. Closure identification is mandatory as the classification process returns predictions on potential co-referent pairs. Hence, it is possible to have transitive chains of co-referent pairs resulting from the classification in case a certain person or organisation is represented by more than two URIs in the dataset. This scenario

---

[14] http://scikit-learn.org/.

occurs when the lexicalisations provided for the names of people and organisation vary from one conference to another. For example, the individual 'Andrea Giovanni Nuzzolese' is associated with the URIs `sdp:andrea-giovanni-nuzzolese`, `sdp:andrea-nuzzolese`, and `sdp:andrea-g-nuzzolese`, which in turn are provided by the classifier as a set of pairs as follows.

```
<sdp:andrea-nuzzolese,sdp:andrea-giovanni-nuzzolese>
<sdp:andrea-g-nuzzolese,sdp:andrea-nuzzolese>
<sdp:andrea-giovanni-nuzzolese,sdp:andrea-g-nuzzolese>
```

Closure identification then produces a single tuple as:

```
<sdp:andrea-nuzzolese,sdp:andrea-giovanni-nuzzolese,sdp:andrea-g-nuzzolese>
```

Candidate Selection. The result of the previous activity is a set of tuples. Next, candidate selection aims at identifying for each tuple, a single candidate to keep as reference URI. The selection is performed by computing and comparing the degree centrality of each entity associated with a URI part of a tuple. The degree centrality is computed as the sum of indegree and outdegree, which count the number of incoming and outcoming ties of the node respectively. Accordingly, the selected URI is the one recording the highest degree centrality within the context of the tuple. To compute outdegree centrality, we do not take into account of datatype properties, as in ScholarlyData the only literals used for person and organisation are names. Hence, we assume that (i) an entity with alternative names is not more relevant with respect to another with a single name; (ii) the centrality is captured by object properties only.

Dataset Update. Once a single URI for each tuple has been selected, we associate the triples available through the other URIs of the tuple with the single selected URI. This is performed by means of SPARQL UPDATE queries.

Recording of the Harmonisation. This is carried out in parallel with the previous activity and its result is an RDF graph whose aim is twofold: (i) providing a description about how a final URI for a set of duplicates is derived and (ii) serving as background knowledge for enabling backward compatibility. The backward compatibility is needed in order to guarantee transparent access to any client application that relies on ScholarlyData without any client-side change being required. The RDF graph is modelled by using PROV-O[15] and describes the harmonisation in terms of provenance. Following the previous example, we show the resulting RDF graph generated below.

```
sdp:andrea-giovanni-nuzzolese a prov:Entity, sdo:Person ;
  prov:wasDerivedFrom sdp:andrea-nuzzolese, sdp:andrea-g-nuzzolese .
sdp:andrea-nuzzolese a prov:Entity, sdo:Person .
sdp:andrea-g-nuzzolese a prov:Entity, sdo:Person .
```

---

[15] https://www.w3.org/TR/prov-o/.

In this example, the entity `sdp:andrea-giovanni-nuzzolese` is typed as `prov:Entity` and associated with the entities `sdp:andrea-nuzzolese` and `sdp:andrea-g-nuzzolese` by means of the property `prov:wasDerivedFrom` that according to PROV-O allows to model the transformation of an entity into another, an update of an entity resulting in a new one or the construction of a new entity based on a pre-existing entity [9]. The objects of `prov:wasDerivedFrom` properties are recorded in the RDF graph along with their original triples[16] coming from the dataset before the update. On top of the RDF graph resulting from this activity we designed and implemented a software module that enables HTTP redirect and SPARQL query expansion in case deleted URIs are requested. The HTTP redirect is implemented by querying the provenance graph when a resource is not available in the default graph of ScholarlyData. Hence, if such a resource is available in the provenance graph the software module return a 301 HTTP status code, meaning that the resource has been moved permanently. For example, of response a HTTP/1.1 301 Moved Permanently to `sdp:andrea-giovanni-nuzzolese` is returned when the resource `sdp:andrea-nuzzolese` is requested.

The SPARQL query expansion is implemented by substituting all the occurrences of deleted URIs that appear as URI constants in a query with the valid ones. Again this is enabled by the provenance graph. The software is available on the GIT repository of cLODg[17]. For example, in the following, the first SPARQL SELECT statement is automatically converted to the second and then executed assuming that `sdp:andrea-giovanni-nuzzolese` is the candidate URI selected to inherit the knowledge from its duplicates and `sdp:andrea-nuzzolese` is one of the removed duplicates.

```
SELECT DISTINCT ?pred ?obj WHERE {sdp:andrea-nuzzolese ?pred ?obj}
SELECT DISTINCT ?pred ?obj WHERE {sdp:andrea-giovanni-nuzzolese
?pred ?obj}
```

4   Experiments

4.1   The Train/test Dataset

We manually labelled a dataset of pairs of PER URIs (*perD*) and pairs of ORG URIs (*orgD*). To select the PER pairs to annotate we retrieve all pairs that share at least a common value in one of their properties (we restrict these to `conf:name`, `conf:familyName`, `rdfs:label`; note that these properties can return multiple values for each entity). Similarly we select the ORG pairs (where the properties are restricted to `conf:name`, `rdfs:label`). For ORG we generate additional pairs by retrieving all affiliations for each person and generating all pair combinations when multiple affiliations are found for each person. We then manually annotated roughly 20% of all the candidates as positive or negative, discarding all the others. This resulted in 698 (148 positive, 550 negative) pairs for *perD* and 424 (188 positive, 236 negative) pairs for *orgD*.

---

[16] Due to space limitation those triples have been omitted in the example.

[17] https://github.com/anuzzolese/cLODg2.

4.2 Blocking

This part of experiment is aimed at identifying the optimal strategy for reducing pair comparisons. We evaluate this independently from the classification step by considering the *pair completeness* (PC), i.e. the fraction of true positive entity pairs contained in the list of candidate pairs returned by blocking, and the *reduction ratio* (RR), i.e. the percentage of pairs discarded after blocking is applied from the total number of possible pairs without blocking. The *harmonic mean* (HM) measures the tradeoff between PC and RR. These metrics give us some indication of the upper bound of performance for the classifiers, i.e. the percentage of duplicates that can be potentially identified, assuming perfect prediction. Given all PER or ORG URIs from ScholarlyData, we first run different blocking strategies to create candidate URI pairs. Next, we use the true positive pairs from the training data as reference to calculate the three metrics. Note that it is expected that this reference set may not include all true positives in the entire ScholarlyData dataset.

Tables 4 and 5 show the results of the blocking strategies for PER and ORG respectively. For the content based method, we keep all URI pairs that share at least a common value in their name-like properties. For the SNM method we produce the list of all URIs with the two orderings described in Sect. 3.1 and for each URI we generate all combinations in a context window n, from 5 to 90.

Results show that the content based method produces the best results for both PER and ORG. Note that although in training data preparation, we also used name-like properties as a proxy to find candidate pairs to annotate, the content based blocking uses name-like properties in a more general way (i.e., we tokenise property values and look for common tokens rather than values) and in practice generates far more candidate pairs.

We conclude to use the content based blocking for both PER and ORG to create candidate URI pairs. Specifically, we use multiple features L+N+S for PER (ref. Table 4) and P+N+L for ORG (ref. Table 5). We choose multiple

Table 4. Blocking results for PER. Results for (i) the SNM method (ordered by URI/name), at the variation of the window size, and (ii) the content based method with different features: N stands for `conf:name`, S for `conf:familyName`, L for `rdfs:label`.

SNM name	RR	PC	HM	SNM URI	RR	PC	HM	Content based	RR	PC	HM
5	≈ 1	0.73	0.84	5	≈ 1	0.6	0.75	L	≈ 1	0.38	0.55
10	≈ 1	0.8	0.89	10	≈ 1	0.71	0.83	N	≈ 1	0.38	0.55
20	≈ 1	0.91	0.95	20	≈ 1	0.8	0.89	S	≈ 1	1	≈ 1
30	0.99	0.91	0.95	30	0.99	0.84	0.91	L+N+S	≈ 1	1	≈ 1
50	0.99	0.93	0.96	50	0.99	0.84	0.91				
70	0.99	0.93	0.96	70	0.99	0.85	0.91				
90	0.98	0.95	0.97	90	0.98	0.89	0.93				

Table 5. Blocking results for ORG. Including: (i) the SNM method (ordered by URI/name), at the variation of the window size, and (ii) the content based method with different features: N stands for `conf:name`, P for affiliated person, L for `rdfs:label`.

SNM name	RR	PC	HM	SNM URI	RR	PC	HM	Content based	RR	PC	HM
5	≈ 1	1	≈ 1	5	≈ 1	0.49	0.66	P	≈ 1	0.84	0.91
10	0.99	1	≈ 1	10	0.99	0.59	0.74	N	≈ 1	1	≈ 1
20	0.99	1	0.99	20	0.99	0.65	0.78	L	≈ 1	1	≈ 1
30	0.98	1	0.99	30	0.98	0.66	0.79	P+N+L	≈ 1	1	≈ 1
50	0.97	1	0.99	50	0.97	0.68	0.8				
70	0.96	1	0.98	70	0.96	0.68	0.8				
90	0.95	1	0.98	90	0.95	0.71	0.81				

features over single features (i.e., S for PER, N or L for ORG) because in practice, using multiple features for content based blocking may result in better PC. Although the effects of using multiple as opposed to single features are indifferent for the experiment results, this may be partially attributed to the potentially incomplete reference set of true positive pairs, as discussed above.

Applied to the entire sections of PER and ORG URIs from the ScholarlyData dataset, this generates 1,468 PER and 3,717 ORG URIs to be classified, which is around 1% of the total possible pair combinations, and we expect this reasonably leads to a significant reduction in running time.

4.3   Classification

This part of experiment evaluates the performance of the binary classification models, using the datasets created in Sect. 4.1. The task addresses a typical problem in link discovery, therefore we compare against two state-of-the-art systems to be described below.

Training and testing. We split each labelled dataset (*perD* and *orgD*) randomly into a training set containing 75% of the data, and a testing set containing 25% of data. For each of our classifiers, we tune their hyper parameters by performing grid search using the training set with 10-fold cross validation. Both the labelled datasets and the optimised classification models are available for download[18]. The optimised classifiers are then applied to the test set. Both PER and ORG experiments are carried out independently from each other.

State-of-the-art. We compare our models against LIMES[19] and SILK[20], both of which offer supervised learning methods that can benefit from training data.

---

[18] https://github.com/ziqizhang/scholarlydata/tree/master/data/public/.
[19] Ver 1.1.2, https://github.com/AKSW/LIMES-dev/releases.
[20] Ver 2.6, https://github.com/silk-framework/silk/releases.

Table 6. Classification results for ORG.

		SGD	LR	RF	SVM-l	SVM-rbf
Positive examples	P	0.85	0.84	0.86	0.83	0.83
	R	0.8	0.8	0.82	0.86	0.86
	F1	0.83	0.82	0.84	0.85	0.85
Negative examples	P	0.83	0.82	0.84	0.87	0.87
	R	0.87	0.85	0.87	0.84	0.84
	F1	0.85	0.84	0.86	0.85	0.85
Total	P	0.84	0.83	0.85	0.85	0.85
	R	0.84	0.83	0.85	0.85	0.85
	F1	0.84	0.83	0.85	0.85	0.85

Table 7. Classification results for PER.

		SGD	LR	RF	SVM-l	SVM-rbf
Positive examples	P	0.78	0.88	0.92	0.77	0.88
	R	0.51	0.63	0.66	0.49	0.63
	F1	0.62	0.73	0.77	0.6	0.73
Negative examples	P	0.89	0.91	0.92	0.88	0.91
	R	0.96	0.98	0.99	0.96	0.98
	F1	0.93	0.94	0.95	0.92	0.94
Total	P	0.87	0.91	0.92	0.88	0.91
	R	0.87	0.91	0.92	0.87	0.91
	F1	0.86	0.90	0.91	0.86	0.90

For LIMES, we test *Wombat Simple* (**L-ws**) and *Wombat Complete* (**L-wc**) *supervised batch* models that are available at the time of writing. For a consistent experimental environment, the same set of URI features (Sect. 3.2) are used for all models. Both SILK and LIMES models are trained and tested on the train-test splits (75%–25%) for ORG and PER respectively. Their default configurations are used.

Analysis. Tables 6 and 7 show results of our five classifiers. The results show that detecting and resolving duplicates in scholarly linked datasets is not an easy task. When only positive examples are considered - as in the realistic scenarios, the performance of the classifiers is particularly weak for PER in terms of recall. Manual inspection of the training datasets reveals that, *on the one hand*, features of some URIs are very sparse. For example, URI sdp:gregoris-antoniou has no published work, while sdp:ghislain-atemezing has no affiliations. Consequently, the similarity between this URI and its true positive co-referent URI will be 0 in terms of this feature. To rectify this, an ensemble of classifiers could be employed. Different classifiers can be trained on different sub-sets of feature types. Then during testing, the optimal model is chosen dynamically depending on the availability of feature types. *On the other hand*, two URIs in a pair sometimes use features that, although are disjoint, often share certain implicit connections. For example, conference participants may use affiliations of different granularity from time to time, e.g., the names 'The Open University' and 'KMI'. Being able to measure the implicit connectivity between the two values could improve the model's recall. *Furthermore*, to improve recall, generalisation over certain features could also be helpful. For example, we derive generic event names (e.g., ESWC) based on their URIs, currently representing event series (e.g., ESWC2009, ESWC2011). All of these will be explored in the future work.

Overall, the RF model offers the best trade-off for both PER and ORG, especially on positive examples. Therefore in Tables 8 and 9 we compare the

Table 8. SoA results for ORG.

Table 9. SoA results for PER.

		RF	L-ws	L-wc	SILK			RF	L-ws	L-wc	SILK
Positive examples	P	0.86	0.63	0.73	0.59	Positive examples	P	0.92	0.34	0.78	0.48
	R	0.82	0.64	0.49	1.0		R	0.66	0.63	0.17	0.80
	F1	0.84	0.64	0.59	0.74		F1	0.77	0.44	0.28	0.6
Negative examples	P	0.84	0.72	0.69	1.0	Negative examples	P	0.92	0.85	0.79	0.94
	R	0.87	0.71	0.86	0.47		R	0.99	0.61	0.98	0.71
	F1	0.86	0.72	0.77	0.64		F1	0.95	0.71	0.88	0.81
Total	P	0.85	0.68	0.70	0.70	Total	P	0.92	0.62	0.79	0.79
	R	0.85	0.68	0.70	0.70		R	0.92	0.62	0.79	0.73
	F1	0.85	0.68	0.70	0.70		F1	0.91	0.62	0.79	0.76

results of this model against the best results of SILK and LIMES models[21]. As the results show, our method makes significant improvement in both tasks on F1, and strikes much better balance between precision and recall. This could be partially attributed to the usage of the 'Coverage' functions that may cope with infrequently used URIs more effectively. However, readers should note that on the one hand, our method is specifically tailored to the problem while both LIMES and SILK are general purpose matching systems; on the other hand, we did not extensively test all configurations of the two systems but used their default settings.

Finally, we use the trained RF model to label the sets of PER and ORG URI pairs identified by content based blocking, as discussed before. The output is then passed to URI harmonisation.

4.4 URI Harmonisation

The aim of this experiment is twofold: (i) assessing the classification output, and (ii) updating the dataset. First we manually checked all pairs labelled as positive. For PER we recorded 101 correct resolutions over 118, i.e. 0.86 of precision. For ORG we recorded 884 correct resolutions out of 1,262 pairs, i.e. 0.7 of precision. We then used the resulting cleaned output to harmonise the URIs on the ScholarlyData dataset. Consequently, from 101 pairs of PER URIs we derived 94 unique individuals each of them harmonising on average 2.05 distinct individuals. The average size of the graph associated with each individual

---

[21] LIMES allows setting a threshold for predicted mappings. We tested different thresholds from 0.1 to 1.0 with increment of 0.1 and found that LIMES-wc is insensitive to the threshold while LIMES-ws is. For complete results and optimal thresholds see: https://github.com/ziqizhang/scholarlydata/tree/master/data/public/soa_results.

involved in the harmonisation for PER counted of 20.3 distinct RDF triples. Similarly, for ORG we kept 531 correct resolutions out of 884 pairs harmonising 2.67 URIs on average. The graph associated with each individual involved in the harmonisation for ORG counted of 10.2 distinct RDF triples on average.

5   Conclusions

In this work, we introduced an approach to address a key issue in the publication of scholarly linked data on the Semantic Web, i.e., the presence of duplicate URIs for the same entities. Using the ScholarlyData dataset as a reference, our approach uses *blocking techniques* to narrow down a list of candidate duplicate URI pairs, exploits *supervised classification methods* to label the true positives and then devises a protocol to choose the most representative URI for an entity to keep in ScholarlyData and to make sure that we preserve all facts from duplicated URIs. To our knowledge, this is by far the first attempt to solve such issues on the largest conference dataset in the Semantic Web community. Future work will be carried out in a number of directions. Firstly, we will look into the issue of other types of URIs, such as events. Next, in terms of the classification process, we will explore the possibilities of improvement discussed before. We will also develop methods that exploit the dependency between the different types of URIs, where the solution to one task can feed into that of another (e.g., the de-duplication of ORG URIs could potentially address the disjoining issue of 'affiliation URI' and 'affiliation names' features of PER). Finally, we will explore the inclusion of human in the loop, in an 'active-learning' fashion to both minimise the human effort on annotation and improve the accuracy of our method.

References

1. Bryl, V., Birukou, A., Eckert, K., Kessler, M.: What is in the proceedings? combining publishers and researchers perspectives. In: SePublica 2014 (2014)
2. Clark, K., Manning, C.: Entity-centric coreference resolution with model stacking. In: Association for Computational Linguistics (2015)
3. Duan, S., Fokoue, A., Hassanzadeh, O.: Instance-Based Matching of Large Ontologies Using Locality-Sensitive Hashing. pp. 49–64 (2012)
4. Gentile, A.L., Acosta, M., Costabello, L., Nuzzolese, A.G., Presutti, V., Reforgiato Recupero, D.: Conference live: accessible and sociable conference semantic data. In: Proceedings of WWW Companion, pp. 1007–1012 (2015)
5. Glaser, H., Jaffri, A., Millard, I.: Managing co-reference on the semantic web. In: Linked Data on the Web (LDOW 2009) (2009)
6. Halpin, H., Presutti, V.: The identity of resources on the web: an ontology for web architecture. Appl. Ontol. **6**(3), 263–293 (2011)
7. Hernandez, M.A., Stolfo, S.J.: The merge/purge problem for large databases. In: Proceedings of SIGMOD 1995. ACM (1995)
8. Isele, R., Bizer, C.: Learning expressive linkage rules using genetic programming. Proc. VLDB Endow. **5**(11), 1638–1649 (2012)
9. Lebo, T., Sahoo, S., McGuinness, D.: Prov-o: The prov ontology. W3C recommendation, W3C, April 2013. https://www.w3.org/TR/prov-o/

10. Lee, D., Kang, J., Mitra, P., Giles, C.L., On, B.-W.: Are your citations. Commun. ACM **50**(12), 33–38 (2007)
11. Lehmann, J., Isele, R., Jakob, M., Jentzsch, A., Kontokostas, D., Mendes, P.N., Hellmann, S., Morsey, M., van Kleef, P., Auer, S., Bizer, C.: DBpedia - a large-scale, multilingual knowledge base extracted from Wikipedia. Semant. Web J. **6**, 167–195 (2013)
12. Mamun, A.-A., Aseltine, R., Rajasekaran, S.: Efficient record linkage algorithms using complete linkage clustering. PLoS ONE **11**(4), e0154446 (2016)
13. Möller, K., Heath, T., Handschuh, S., Domingue, J.: Recipes for semantic web dog food — the ESWC and ISWC metadata projects. In: Aberer, K., Choi, K.-S., Noy, N., Allemang, D., Lee, K.-I., Nixon, L., Golbeck, J., Mika, P., Maynard, D., Mizoguchi, R., Schreiber, G., Cudré-Mauroux, P. (eds.) ASWC/ISWC - 2007. LNCS, vol. 4825, pp. 802–815. Springer, Heidelberg (2007). doi:10.1007/978-3-540-76298-0_58
14. Nentwig, M., Hartung, M., Ngomo, A.-C.N., Rahm, E.: A survey of current link discovery frameworks. Semant. Web (Preprint):1–18 (2015)
15. Ngomo, A.-C.N., Auer, S.: LIMES: a time-efficient approach for large-scale link discovery on the web of data. In: Proceedings of IJCAI 2011, pp. 2312–2317 (2011)
16. Nuzzolese, A.G., Gentile, A.L., Presutti, V., Gangemi, A.: Conference Linked data: the scholarlydata project. In: Groth, P., Simperl, E., Gray, A., Sabou, M., Krötzsch, M., Lecue, F., Flöck, F., Gil, Y. (eds.) ISWC 2016. LNCS, vol. 9982, pp. 150–158. Springer, Cham (2016). doi:10.1007/978-3-319-46547-0_16
17. Osborne, F., Motta, E., Mulholland, P.: Exploring scholarly data with rexplore. In: Alani, H., Kagal, L., Fokoue, A., Groth, P., Biemann, C., Parreira, J.X., Aroyo, L., Noy, N., Welty, C., Janowicz, K. (eds.) ISWC 2013. LNCS, vol. 8218, pp. 460–477. Springer, Heidelberg (2013). doi:10.1007/978-3-642-41335-3_29
18. Papadakis, G., Niederée, C., Fankhauser, P.: Efficient entity resolution for large heterogeneous information spaces. pp. 535–544 (2011)
19. Shotton, D.: Semantic publishing: the coming revolution in scientific journal publishing. Learned Publishing **22**(2), 85–94 (2009)
20. Solecki, B., Silva, L., Efimov, D.: KDD cup 2013: author disambiguation. In: Proceedings of the 2013 KDD Cup 2013 Workshop, KDD Cup 2013, pp. 9:1–9:3. ACM, New York (2013)
21. Zhang, Z., Gentile, A.L., Blomqvist, E., Augenstein, I., Ciravegna, F.: An unsupervised data-driven method to discover equivalent relations in large linked datasets. Semant. web **8**(2), 197–223 (2017)
22. Zheng, J., Chapman, W., Crowley, R., Savova, G.: Coreference resolution: a review of general methodologies and applications in the clinical domain. Biomed. Inform. **44**(6), 1113–1122 (2011)

Machine Learning Track

WOMBAT – A Generalization Approach for Automatic Link Discovery

Mohamed Ahmed Sherif[1]([✉]), Axel-Cyrille Ngonga Ngomo[1,2],
and Jens Lehmann[3,4]

[1] R&D Department II, Computing Center,
University of Leipzig, 04109 Leipzig, Germany
{sherif,ngonga}@informatik.uni-leipzig.de
[2] Data Science Group, University of Paderborn,
Pohlweg 51, 33098 Paderborn, Germany
ngonga@upb.de
[3] Computer Science Institute, University of Bonn, Römerstr. 164,
53117 Bonn, Germany
jens.lehmann@cs.uni-bonn.de
[4] Fraunhofer IAIS, Schloss Birlinghoven, 53757 Sankt Augustin, Germany
jens.lehmann@iais.fraunhofer.de

Abstract. A significant portion of the evolution of Linked Data datasets lies in updating the links to other datasets. An important challenge when aiming to update these links automatically under the open-world assumption is the fact that usually only positive examples for the links exist. We address this challenge by presenting and evaluating WOMBAT, a novel approach for the discovery of links between knowledge bases that relies exclusively on positive examples. WOMBAT is based on generalisation via an upward refinement operator to traverse the space of link specification. We study the theoretical characteristics of WOMBAT and evaluate it on 8 different benchmark datasets. Our evaluation suggests that WOMBAT outperforms state-of-the-art supervised approaches while relying on less information. Moreover, our evaluation suggests that WOMBAT's pruning algorithm allows it to scale well even on large datasets.

1 Introduction

The Linked Open Data Cloud has grown from a mere 12 datasets at its beginning to a compendium of more than 9,000 public RDF data sets.[1] In addition to the number of the datasets published growing steadily, we also witness the size of single datasets growing with each new edition. For example, *DBpedia* has grown from 103 million triples describing 1.95 million things (DBpedia 2.0) to 583 million triples describing 4.58 million things (DBpedia 2014) within 7 years. This growth engenders an increasing need for automatic support when maintaining evolving datasets.

---

[1] http://lodstats.aksw.org.

© Springer International Publishing AG 2017
E. Blomqvist et al. (Eds.): ESWC 2017, Part I, LNCS 10249, pp. 103–119, 2017.
DOI: 10.1007/978-3-319-58068-5_7

One of the most crucial tasks when dealing with evolving datasets lies in updating the links from these data sets to other data sets. While supervised approaches have been devised to achieve this goal, they assume the provision of both positive and negative examples for links [1]. However, the links available on the Data Web only provide positive examples for relations and no negative examples, as the open-world assumption underlying the Web of Data suggests that the non-existence of a link between two resources cannot be understood as stating these two resources are not related. Consequently, state-of-the-art supervised learning approaches for link discovery can only be employed if the end users are willing to provide the algorithms with information that is generally not available on the Linked Open Data Cloud, i.e., with negative examples.

We address this drawback by proposing the first approach for learning links based on positive examples only. Our approach, dubbed WOMBAT, is inspired by the concept of generalisation in quasi-ordered spaces. Given a set of positive examples we aim to find a classifier that covers a large number of positive examples (i.e., achieves a high recall on the positive examples) while still achieving a high precision. We use Link Specifications (LS, see Sect. 2) as classifiers [1,7,14]. We are thus faced with the challenge of using various similarity metrics, acceptance thresholds and nested logical combinations of those when learning. The contributions of this paper are: (1) We provide the first approach for learning LS that is able to learn links from positive examples only. (2) Our approach is based on an upward refinement operator for which we analyse its theoretical characteristics. (3) We use the characteristics of our operator to devise a pruning approach and improve the scalability of WOMBAT. (4) We evaluate WOMBAT on 8 benchmark datasets and show that in addition to needing less training data, it also outperforms the state of the art in most cases.

2   Preliminaries

The aim of link discovery (LD) is to discover the set $\{(s,t) \in S \times T : Rel(s,t)\}$ provided an input relation Rel and two sets S (source) and T (target) of RDF resources. To achieve this goal, declarative LD frameworks rely on LS, which describe the conditions under which $Rel(s,t)$ can be assumed to hold for a pair $(s,t) \in S \times T$. Several grammars have been used for describing LS in previous works [6,15,19]. In general, these grammars assume that LS consist of two types of atomic components: *similarity measures* m, which allow comparing property values of input resources and *operators* op, which can be used to combine these similarities to more complex LS. Without loss of generality, we define a similarity measure m as a function $m : S \times T \rightarrow [0,1]$. An example of a similarity measure is the edit similarity dubbed edit[2] which allows computing the similarity of a pair $(s,t) \in S \times T$ with respect to the properties p_s of s and p_t of t. We use *mappings* $M \subseteq S \times T$ to store the results of the application of a similarity function to $S \times T$ or subsets thereof. We denote the set of all mappings as $\mathcal{M}$ and the set

---

[2] We define the edit similarity of two strings s and t as $(1 + lev(s,t))^{-1}$, where lev stands for the Levenshtein distance.

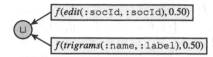

Fig. 1. Example of a complex LS. The filter nodes are rectangles while the operator nodes are circles. :socID stands for social security number.

Table 1. Link specification syntax and semantics

LS	$[[LS]]_M$
$f(m, \theta)$	$\{(s,t)\|(s,t) \in M \wedge m(s,t) \geq \theta\}$
$L_1 \sqcap L_2$	$\{(s,t)\|(s,t) \in [[L_1]]_M \wedge (s,t) \in [[L_2]]_M\}$
$L_1 \sqcup L_2$	$\{(s,t)\|(s,t) \in [[L_1]]_M \vee (s,t) \in [[L_2]]_M\}$
$L_1 \backslash L_2$	$\{(s,t)\|(s,t) \in [[L_1]]_M \wedge (s,t) \notin [[L_2]]_M\}$

of all LS as $\mathcal{L}$. We define a *filter* as a function $f(m, \theta)$. We call a specification *atomic* when it consists of exactly one filtering function. A complex specification can be obtained by combining two specifications L_1 and L_2 through an *operator* that allows merging the results of L_1 and L_2. Here, we use the operators $\sqcap$, $\sqcup$ and $\backslash$ as they are complete and frequently used to define LS. An example of a complex LS is given in Fig. 1.

We define the semantics $[[L]]_M$ of a LS L w.r.t. a mapping M as given in Table 1. Those semantics are similar to those used in languages like SPARQL, i.e., they are defined extensionally through the mappings they generate. The mapping $[[L]]$ of a LS L with respect to $S \times T$ contains the links that will be generated by L. A LS L is *subsumed* by L', denoted by $L \sqsubseteq L'$, if for all mappings M, we have $[[L]]_M \subseteq [[L']]_M$. Two LS are *equivalent*, denoted by $L \equiv L'$ iff $L \sqsubseteq L'$ and $L' \sqsubseteq L$. Subsumption ($\sqsubseteq$) is a partial order over $\mathcal{L}$.

3 Constructing and Traversing Link Specifications

The goal of our learning approach is to learn a specification L that generalizes a mapping $M \subseteq S \times T$ which contains a set of pairs (s,t) for which $Rel(s,t)$ holds. Our approach consists of two main steps. First, we aim to derive initial atomic specifications A_i that achieve the same goal. In a second step, we combine these atomic specifications to the target complex specification L by using the operators $\sqcap$, $\sqcup$ and $\backslash$. In the following, we detail how we carry out these two steps.

3.1 Learning Atomic Specifications

The goal here is to derive a set of initial atomic specifications $\{A_1, \ldots, A_n\}$ that achieves the highest possible F-measure given a mapping $M \subseteq S \times T$ which contains all known pairs (s,t) for which $Rel(s,t)$ holds. Given a set of similarity functions m_i, the set of properties P_s of S and the set of properties P_t of T, we begin by computing the subset of properties from S and T that achieve a coverage above a threshold $\tau \in [0,1]$, where the coverage of a property p for a knowledge base K is defined as

$$coverage(p) = \frac{|\{s : (s,p,o) \in K\}|}{|\{s : \exists q : (s,q,o) \in K\}|}. \tag{1}$$

Now for all property pairs $(p, q) \in P_s \times P_t$ with $coverage(p) \geq \tau$ and $coverage(q) \geq \tau$, we compute the mappings $M_{ij} = \{(s, t) \in S \times T : m_{ij}(s, t) \geq \theta_j\}$, where m_{ij} compares s and t w.r.t. p and q and M_{ij} is maximal w.r.t. the F-measure it achieves when compared to M. To this end, we apply an iterative search approach. Finally, we select M_{ij} as the atomic mapping for p and q. Thus, we return as many atomic mappings as property pairs with sufficient coverage. Note that this approach is not quintessential for WOMBAT and can thus be replaced with any approach of choice which returns a set of initial LS that is to be combined.

3.2   Combining Atomic Specifications

After deriving atomic LS as described above, WOMBAT computes complex specifications by using an approach based on generalisation operators. The basic idea behind these operators is to perform an iterative search through a solution space based on a score function. Formally, we rely on the following definitions:

Definition 1 ((Refinement) Operator). *In the quasi-ordered space* $(\mathcal{L}, \sqsubseteq)$, *we call a function from* $\mathcal{L}$ *to* $2^{\mathcal{L}}$ *an (LS) operator. A downward (upward) refinement operator* ρ *is an operator, such that for all* $L \in \mathcal{L}$ *we have that* $L' \in \rho(L)$ *implies* $L' \sqsubseteq L$ $(L \sqsubseteq L')$. L' *is called a* specialisation *(generalisation) of* L. $L' \in \rho(L)$ *is usually denoted as* $L \rightsquigarrow_\rho L'$.

Definition 2 (Refinement Chains). *A refinement chain of a refinement operator* ρ *of length* n *from* L *to* L' *is a finite sequence* $L_0, L_1, \ldots, L_n$ *of LS, such that* $L = L_0, L' = L_n$ *and* $\forall i \in \{1 \ldots n\}, L_i \in \rho(L_{i-1})$. *This refinement chain goes through* L'' *iff there is an* i $(1 \leq i \leq n)$ *such that* $L'' = L_i$. *We say that* L'' *can be reached from* L *by* ρ *if there exists a refinement chain from* L *to* L''. $\rho^*(L)$ *denotes the set of all LS which can be reached from* L *by* ρ. $\rho^m(L)$ *denotes the set of all LS which can be reached from* L *by a refinement chain of* ρ *of length* m.

Definition 3 (Properties of refinement operators). *An operator* ρ *is called (1) (locally) finite iff* $\rho(L)$ *is finite for all LS* $L \in \mathcal{L}$; *(2)* **redundant** *iff there exists a refinement chain from* $L \in \mathcal{L}$ *to* $L' \in \mathcal{L}$, *which does not go through (as defined above) some LS* $L'' \in \mathcal{L}$ *and a refinement chain from* L *to* L' *which does go through* L''; *(3)* **proper** *iff for all LS* $L \in \mathcal{L}$ *and* $L' \in \mathcal{L}$, $L' \in \rho(L)$ *implies* $L \not\equiv L'$. *An LS upward refinement operator* ρ *is called* **weakly complete** *iff for all LS* $\perp \sqsubseteq L$ *we can reach a LS* L' *with* $L' \equiv L$ *from* $\perp$ *(most specific LS) by* ρ.

We designed two different operators for combining atomic LS to complex specifications: The first operator takes an atomic LS and uses the three logical connectors to append further atomic LS. Assuming that $(A_1, \ldots, A_n)$ is the set of atomic LS found, φ can be defined as follows:

$$
\varphi(L) = \begin{cases} \bigcup_{i=1}^n A_i & \text{if } L = \perp \\ (\bigcup_{i=1}^n L \sqcup A_i) \cup (\bigcup_{i=1}^n L \sqcap A_i) \cup (\bigcup_{i=1}^n L \backslash A_i) & \text{otherwise} \end{cases}
$$

$$\psi(L) = \begin{cases} \{A_{i_1} \setminus A_{j_1} \sqcap \cdots \sqcap A_{i_m} \setminus A_{j_m} \mid A_{i_k}, A_{j_k} \in \mathbf{A} \\ \quad \text{for all } 1 \le k \le m\} & \text{if } L = \bot \\ \{L \sqcup A_i \setminus A_j \mid A_i \in \mathbf{A}, A_j \in \mathbf{A}\} & \text{if } L = A \ (\textit{atomic}) \\ \{L_1\} \cup \{L \sqcup A_i \setminus A_j \mid A_i \in \mathbf{A}, A_j \in \mathbf{A}\} & \text{if } L = L_1 \setminus L_2 \\ \{L_1 \sqcap \cdots \sqcap L_{i-1} \sqcap L' \sqcap L_{i+1} \sqcap \cdots \sqcap L_n \mid L' \in \psi(L_i)\} \\ \quad \cup \{L \sqcup A_i \setminus A_j \mid A_i \in \mathbf{A}, A_j \in \mathbf{A}\} & \text{if } L = L_1 \sqcap \cdots \sqcap L_n (n \ge 2) \\ \{L_1 \sqcup \cdots \sqcup L_{i-1} \sqcup L' \sqcup L_{i+1} \sqcup \cdots \sqcup L_n \mid L' \in \psi(L_i)\} \\ \quad \cup \{L \sqcup A_i \setminus A_j \mid A_i \in \mathbf{A}, A_j \in \mathbf{A}\} & \text{if } L = L_1 \sqcup \cdots \sqcup L_n (n \ge 2) \end{cases}$$

Fig. 2. Definition of the refinement operator ψ.

This naive operator is not a refinement operator (neither upward nor downward). Its main advantage lies in its simplicity allowing for a very efficient implementation. However, it cannot reach all specifications, e.g., a specification of the form $(A_1 \sqcup A_2) \sqcap (A_3 \sqcup A_4)$ cannot be reached. Examples of chains generated by φ are as follows:

1. $\bot \leadsto_\varphi A_1 \leadsto_\varphi A_1 \sqcup A_2 \leadsto_\varphi (A_1 \sqcup A_2) \setminus A_3$
2. $\bot \leadsto_\varphi A_2 \leadsto_\varphi A_2 \sqcap A_3 \leadsto_\varphi (A_2 \sqcap A_3) \setminus A_4$

The second operator, ψ, uses a more sophisticated expansion strategy in order to allow learning arbitrarily nested LS and is shown in Fig. 2. Less formally, the operator works as follows: It takes a LS as input and makes a case distinction on the type of LS. Depending on the type, it performs the following actions:

- The $\bot$ LS is refined to the set of all combinations of $\setminus$ operations. This set can be large and will only be built iteratively (as required by the algorithm) with at most approx. n^2 refinements per iteration (see the next section for details).
- In LS of the form $A_1 \setminus A_2$, ψ can drop the second part in order to generalise.
- If the LS is a conjunction or disjunction, the operator can perform a recursion on each element of the conjunction or disjunction.
- For LS of any type, a disjunction with an atomic LS can be added.

Below are two example refinement chains of ψ:

1. $\bot \leadsto_\psi A_1 \setminus A_2 \leadsto_\psi A_1 \leadsto_\psi A_1 \sqcup A_2 \setminus A_3$
2. $\bot \leadsto_\psi A_1 \setminus A_2 \sqcap A_3 \setminus A_4 \leadsto_\psi A_1 \sqcap A_3 \setminus A_4 \leadsto_\psi A_1 \sqcap A_3 \leadsto_\psi (A_1 \sqcap A_3) \sqcup (A_5 \setminus A_6)$

ψ is an upward refinement operator with the following properties.

Proposition 1. ψ *is an upward refinement operator.*

Proof. For an arbitrary LS L, we have to show for any element $L' \in \psi(L)$ that $L \sqsubseteq L'$ holds. The proof is straightforward by showing that L' cannot generate less links than L via case distinction and structural induction over LS:

- $L = \bot$: Trivial.
- L is atomic: Adding a disjunction cannot result in less links (this also holds for the cases below).
- L is of the form $L_1 \setminus L_2$: $L' = L_1$ cannot result in less links.
- L is a conjunction / disjunction: L' cannot result in less links by structural induction.   □

Proposition 2. *ψ is weakly complete.*

Proof. To show this, we have to show that an arbitrary LS L can be reached from the $\bot$ LS. First, we convert everything to negation normal form by pushing $\setminus$ inside, e.g. LS of the form $L_1 \setminus (L_2 \sqcap L_3)$ are rewritten to $(L_1 \setminus L_2) \sqcup (L_1 \setminus L_3)$ and LS of the form $L_1 \setminus (L_2 \sqcup L_3)$ are rewritten to $(L_1 \setminus L_2) \sqcap (L_1 \setminus L_3)$ exhaustively. We then further convert the LS to conjunction normal including an exhaustive application of the distribute law, i.e., conjunctions cannot be nested within disjunctions. The resulting LS is dubbed L' and equivalent to L. We show that L' can always be reached from $\bot$ via induction over its structure:

- $L' = \bot$: Trivial via the empty refinement chain.
- $L' = A$ (atomic): Reachable via $\bot \rightsquigarrow_\psi A \setminus A' \rightsquigarrow_\psi A$.
- $L' = A_1 \setminus A_2$ (atomic negation): Reachable directly via $\bot \rightsquigarrow_\psi A_1 \setminus A_2$.
- L' is a conjunction with m elements: $\bot \rightsquigarrow_\psi A_{i_1} \setminus A_{j_1} \sqcap \cdots \sqcap A_{i_m} \setminus A_{j_m}$ where an element $A_{i_k} \setminus A_{j_k}$ is chosen as follows: Let the k-th element of conjunction L' be L''.
  - If L'' is an atomic specification A, then $A_{i_k} = A$ (A_{j_k} can be arbitrarily).
  - If L'' is an atomic negation $A_1 \setminus A_2$, then $A_{i_k} = A$ and $A_{j_k} = A_2$.
  - If L'' is a disjunction, the first element of this disjunction falls into one the above two cases and A_{i_k} and A_{j_k} can be set as described there.
  Each element of L'' is then further refined to L' as follows:
  - If L'' is an atomic specification A: $A \setminus A_{j_k}$ is refined to A.
  - If L'' is an atomic negation $A_1 \setminus A_2$: No further refinements are necessary.
  - If L'' is a disjunction. The first element of the disjunction is first treated according to the two cases above. Subsequent elements of the disjunction are either atomic LS or atomic negation and can be added straightforwardly as the operator allows adding disjunctive elements to any non-$\bot$ LS.

Please note that the case distinction is exhaustive as we assume L' is in conjunctive negation normal form, i.e., there are no disjunctions on the outer level, negation is always atomic, conjunctions are not nested within other conjunction and elements of disjunctions within conjunctions cannot be conjunctions.   □

Proposition 3. *ψ is finite, not proper and redundant.*

Proof. Finiteness: There are only finitely many atomic LS. Hence, there are only finitely many atomic negations and, consequently, finitely many possible conjunctions of those. Consequently, $\psi(\bot)$ is finite. The finiteness of $\psi(L)$ with $L \neq \bot$ is straightforward.

Properness: The refinement chain $\bot \rightsquigarrow_\psi^* A_1 \sqcap A_2 \rightsquigarrow_\psi^* (A_1 \sqcup A_2) \sqcap A_2$ is a counterexample.

Redundancy: The two refinement chains $A_1 \sqcap A_3 \rightsquigarrow_\psi^* (A_1 \sqcup A_2) \sqcap A_3 \rightsquigarrow_\psi^* (A_1 \sqcup A_2) \sqcap (A_3 \sqcup A_4)$ and $A_1 \sqcap A_3 \rightsquigarrow_\psi^* A_1 \sqcap (A_3 \sqcap A_4) \rightsquigarrow_\psi^* (A_1 \sqcup A_2) \sqcap (A_3 \sqcup A_4)$ are a counterexample. □

Naturally, the restrictions of ψ (being redundant and not proper) raise the question whether there are LS refinement operators satisfying all theoretical properties:

Proposition 4. *There exists a weakly complete, finite, proper and non-redundant refinement operator in $\mathcal{L}$.*

Proof. Let C be the set of LS in $\mathcal{L}$ in conjunctive negation normal form without any LS equivalent to $\bot$. We define the operator α as $\alpha(\bot) = C$ and $\alpha(L) = \emptyset$ for all $L \neq \bot$. α is obviously complete as any LS has an equivalent in conjunctive negation normal form. It is finite as S can be shown to the finite with an extended version of the argument in the finiteness proof of ψ. α is trivially non-redundant and it is proper by definition. □

The existence of an operator which satisfies all considered theoretical criteria of a refinement operator is an artifact of only finitely many semantically inequivalent LS existing in $\mathcal{L}$. This set is however extremely large and not even small fractions of it can be evaluated in all but very simple cases. For example, the operator α as $\alpha(\bot) = C$ and $\alpha(L) = \emptyset$ for all $L \neq \bot$ is trivially non-redundant and it is proper by definition. Such an operator α is obviously not useful as it does not help *structuring the search space*. Providing a useful way to structure the search space is the main reason for refinement operators being successful for learning in other complex languages as it allows to gradually converge towards useful solutions while being able to prune other paths which cannot lead to promising solutions (explained in the next section). This is a reason why we sacrificed properness and redundancy for a better structure of the search space.

4 The WOMBAT Algorithm

We have now introduced all ingredients necessary for defining the WOMBAT algorithms. The first algorithm, which we refer to as *simple* version, uses the operator φ, whereas the second algorithm, which we refer to as *complete*, uses the refinement operator ψ. The complete algorithm has the following specific characteristics: First, while ψ is finite, it would generate a prohibitively large number of refinements when applied to the $\bot$ concept. For that reason, those refinements will be computed stepwise as we will illustrate below. Second, as ψ is an upward refinement operator it allows to prune parts of the search space, which we will also explain below. We only explain the implementation of the complex WOMBAT algorithm as the other is a simplification excluding those two characteristics.

Algorithm 1 shows the individual steps of WOMBAT complete. Our approach takes the source dataset S, the target dataset T, examples $E \subseteq S \times T$ as well as

---

Algorithm 1. WOMBAT Learning Algorithm

Input: Sets of resources S and T; examples $E \subseteq S \times T$; property coverage
threshold τ; set of similarity functions $\mathbf{F}$

1 $\mathbf{A} \longleftarrow null$ (the list of initial atomic metrics);
2 $i \longleftarrow 1$;
3 **foreach** *property $p_s \in S$* **do**
4 | **if** *coverage(p_s) $\geq \tau$* **then**
5 | | **foreach** *property $p_t \in T$* **do**
6 | | | **if** *coverage(p_t) $\geq \tau$* **then**
7 | | | | Find atomic metric $m(p_s, p_t)$ that leads to highest F-measure;
8 | | | | Optimize similarity threshold for $m(p_s, p_t)$ to find best mapping A_i;
9 | | | | Add A_i to $\mathbf{A}$;
10 | | | | $i \longleftarrow i + 1$;

11 $\Gamma \longleftarrow \bot$ (initiate search tree Γ to the root node $\bot$);
12 $F_{best} \longleftarrow 0$, $L_{best} \longleftarrow null$;
13 **while** *termination criterion not met* **do**
14 | Choose the node with highest scoring LS L in Γ;
15 | **if** $L == \bot$ **then**
16 | | **foreach** $A_i, A_j \in \mathbf{A}$*, where $i \neq j$* **do**
17 | | | Only add refinements of form $A_i \setminus A_j$;
18 | **else**
19 | | Apply operator to L;
20 | | **if** *L is a refinement of $\bot$* **then**
21 | | | **foreach** $A_i, A_j \in \mathbf{A}$*, where $i \neq j$* **do**
22 | | | | In addition to refinements, add conjunctions with specifications of the form $A_i \setminus A_j$ as siblings;
23 | **foreach** *refinement L'* **do**
24 | | **if** *L' is not already in the search tree Γ* **then**
25 | | | Add L' to Γ as children of the node containing L;
26 | Update F_{best} and L_{best};
27 | **if** *F_{best} has increased* **then**
28 | | **foreach** *subtree $t \in \Gamma$* **do**
29 | | | **if** $F_{best} > F_{max}(t)$ **then**
30 | | | | Delete t;

31 Return L_{best};

---

the property coverage threshold and the set of considered similarity functions as input. In Line 3, the property matches are computed by optimizing the threshold for properties that have the minimum coverage (Line 7) as described in Sect. 3.1. The main loop starts in Line 13 and runs until a termination criterion is satisfied, e.g. (1) a fixed number of LS has been evaluated, (2) a certain time has elapsed,

(3) the best F-score has not changed for a certain time or (4) a perfect solution has been found. Line 14 states that a heuristic-based search strategy is employed. By default, we employ the F-score directly. More complex heuristics introducing a bias towards specific types of LS could be encoded here. In Line 15, we make a case distinction: Since the number of refinements of $\bot$ is extremely high and not feasible to compute in most cases, we perform a stepwise approach: In the first step, we only add simple LS of the form $A_i \setminus A_j$ as refinements (Line 17). In Line 22, we add more complex conjunctions if the simpler forms are promising. Apart from this special case, we apply the operator directly. Line 24 updates the search tree by adding the nodes obtained via refinement. For redundancy elimination, we only add those nodes to the search tree which are not already contained in it.

The subsequent part starting from Line 26 defines our *pruning procedure*: Since ψ is an upward refinement operator, we know that the set of links generated by a child node is a superset of or equal to the set of links generated by its parent. Hence, while both precision and recall can improve in subsequent refinements, they cannot rise arbitrarily. Precision is bound as false positives cannot disappear during generalisation. Furthermore, the achievable recall r_{max} is that of the most general constructable LS, i.e., $\mathcal{A} = \bigcup A_i$ This allows to compute an upper bound on the achievable F-score. In order to do so, we first build a set S' with those resources in S occurring in the input examples E as well as a set T' with those resources in T occurring in E. The purpose of those is to restrict the computation of F-score to the fragment $S' \times T' \subseteq S \times T$ relevant for example set E. We can then compute an upper bound of precision of a LS L as follows:

$$p_{max}(L) = \frac{|E|}{|E| + |\{(s,t) \mid (s,t) \in [[L]], s \in S' \text{ or } t \in T'\} \setminus E|}$$

F_{max} is then computed as the F-measure obtained with recall r_{max} and precision p_{max}, i.e., $F_{max} = \frac{2 p_{max} r_{max}}{p_{max} + r_{max}}$. It is an upper bound for the maximum achievable F-measure of any node reachable via refinements. We can disregard all nodes in the search tree which have a maximum achievable F-score that is lower than the best F-score already found. This is implemented in Line 28. The pruning is conservative in the sense that no solutions are lost. In the evaluation, we give statistics on the effect of pruning. WOMBAT ends by returning L_{best} as the best LS found, which is the specification with the highest F-score. In case of ties, we prefer shorter specifications over long ones. Should the tie persist, then we prefer specifications that were found early.

Proposition 5. WOMBAT *is complete, i.e., it will eventually find the LS with the highest F-measure within* $\mathcal{L}$.

Proof. This is a consequence of the weak completeness of ψ and the fact that the algorithm will eventually generate all refinements of ψ. For the latter, we have to look at the refinement of $\bot$ as a special case since otherwise a straightforward application of ψ is used. For the refinements of $\bot$ it is easy to show via induction over the number of conjunctions in refinements that any element in $\psi(\bot)$ can be reached via the algorithm. (The pruning is conservative and only prunes nodes never leading to better solutions.)   □

5   Evaluation

We evaluated our approach using 8 benchmark datasets. Five of these benchmarks were real-world datasets while three were synthetic. The real-world interlinking tasks used were those in [9]. The synthetic datasets were from the OAEI 2010 benchmark[3]. All experiments were carried out on a 64-core 2.3 GHz PC running *OpenJDK* 64-Bit Server 1.7.0_75 on *Ubuntu* 14.04.2 LTS. Each experiment was assigned 20 GB RAM.

For testing WOMBAT against the benchmark datasets in both its simple and complete version, we used the jaccard, trigrams, cosine and qgrams similarity measures. We used two termination criteria: Either a LS with F-measure of 1 was found or a maximal depth of refinement (10 resp. 3 for the simple resp. complete version) was reached. This variation of the maximum refinement trees sizes between the simple and complete version was because WOMBAT complete adds a larger number of nodes to its refinement tree in each level. The coverage threshold τ was set to 0.6. A more complete list of evaluation results are available at the project web site.[4] Altogether, we carried out 6 sets of experiments to evaluate WOMBAT.

In *the first set of experiments*, we compared the average F-Measure achieved by the simple and complete versions of WOMBAT to that of four other state-of-the-art LS learning algorithms within a 10-fold cross validation setting. The other four LS learning algorithms were EAGLE [15] as well as the *linear, conjunctive* and *disjunctive* versions of EUCLID [16]. EAGLE was configured to run 100 generations. The mutation and crossover rates were set to 0.6 as in [15]. To address the non-deterministic nature of EAGLE, we repeated the whole process of 10-fold cross validation 5 time and present the average results. EUCLID's grid size was set to 5 and 100 iterations were carried out as in [16]. The results of the evaluation are presented in Table 2. The simple version of WOMBAT was able to outperform the state-of-the-art approaches in 4 out of the 8 data sets and came in the second position in 2 datasets. WOMBAT complete was able to achieve the best F-score in 4 data sets and achieve the second best F-measure in 3 datasets. On average, both versions of WOMBAT were able to achieve an F-measure of 0.9, by which WOMBAT outperforms the three version of EUCLID by an average of 11%. While WOMBAT was able to achieve the same performance of EAGLE in average, WOMBAT is still to be preferred as (1) WOMBAT only requires positive examples and (2) EAGLE is indeterministic by nature.

---

[3] http://oaei.ontologymatching.org/2010/.
[4] https://github.com/AKSW/LIMES/tree/master/evaluationsResults/wombat.

Table 2. 10-fold cross validation F-measure results.

Dataset	WOMBAT Simple	WOMBAT Complete	EUCLID Linear	EUCLID Conjunction	EUCLID Disjunction	EAGLE
Person 1	**1.00**	**1.00**	0.64	0.97	**1.00**	0.99
Person 2	**1.00**	0.99	0.22	0.78	0.96	0.94
Restaurants	**0.98**	0.97	0.97	0.97	0.97	0.97
DBLP-ACM	0.97	**0.98**	**0.98**	**0.98**	**0.98**	**0.98**
Abt-Buy	0.60	0.61	0.06	0.06	0.52	**0.65**
Amazon-GP	0.70	0.67	0.59	0.71	**0.73**	0.71
DBP-LMDB	0.99	**1.00**	0.99	0.99	0.99	0.99
DBLP-GS	**0.94**	**0.94**	0.90	0.91	0.91	0.93
Average	**0.90**	**0.90**	0.67	0.80	0.88	**0.90**

For *the second set of experiments*, we implemented an evaluation protocol based on the assumptions made at the beginning of this paper. Each input dataset was split into 10 parts of the same size. Consequently, we used 3 parts (30%) of the data as training data and the rest 7 parts (70%) for testing. This was to implement the idea of the dataset growing and the specification (and therewith the links) for the new version of the dataset having to be derived by learning from the old dataset. During the learning process, the score function was the F-measure achieved by each refinement of the portion of the training data related to $S \times T$ selected for training (dubbed $S' \times T'$ previously). The F-measures reported are those achieved by LS on the test dataset. We used the same settings for EAGLE and EUCLID as in the experiments before. The results (see Table 3) show clearly that our simple operator outperforms all other approaches in this setting. Moreover, the complete version of WOMBAT reaches the best F-measure on 2 datasets and the second-best F-measure on 3 datasets. This result of central importance as it shows that WOMBAT is well suited for the task for which it was designed. Interestingly, our approach also outperforms the approaches that rely on negative examples (i.e. EUCLID and EAGLE). The complete version of WOMBAT seems to perform worse than the simple version because it can only explore a tree of depth 3. However, this limitation was necessary to test both implementations using the same hardware.

In the *third set of experiments*, we measured the effect of increasing the amount of training data on the precision, recall and F-score achieved by both simple and complete versions of WOMBAT. The results are presented in Fig. 3. Our results suggest that the complete version of WOMBAT is partly more stable in its results (see ABT-Buy and DBLP-Google Scholar) and converges faster towards the best solution that it can find. This suggests that once trained on a dataset, our approach can be used on subsequent versions of real datasets, where a small number of novel resources is added in each new version, which

Table 3. A comparison of WOMBAT F-measure against 4 state-of-the-art approaches on 8 different benchmark datasets using 30% of the original data as training data.

Dataset	WOMBAT Simple	WOMBAT Complete	EUCLID Linear	EUCLID Conjunction	EUCLID Disjunction	EAGLE
Person 1	**1.00**	**1.00**	0.95	0.96	0.99	0.92
Person 2	**0.99**	0.79	0.80	0.82	0.88	0.69
Restaurants	**0.97**	0.88	0.87	0.84	0.89	0.88
DBLP-ACM	**0.95**	0.91	0.88	0.89	0.91	0.85
Abt-Buy	**0.44**	0.40	0.29	0.29	0.29	0.27
Amazon-GP	**0.54**	0.41	0.31	0.30	0.32	0.32
DBP-LMDB	**0.98**	**0.98**	0.97	0.96	0.97	0.89
DBLP-GS	**0.91**	0.74	0.83	0.76	0.74	0.69
Average	**0.85**	0.76	0.74	0.73	0.75	0.69

is the problem setup considered in this paper. On the other hand, the simple version is able to find better LS as it can explore longer sequences of mappings.

In the *fourth set experiments*, we measured the learning time for each of the benchmark datasets. The results are also presented in Fig. 3. As expected, the simple approach is time-efficient to run even without any optimization. While the complete version of WOMBAT without pruning is significantly slower (up to 1 order of magnitude), the effect of pruning can be clearly seen as it reduces the runtime of the algorithm while also improving the total space that the complete version of WOMBAT can explore. These results are corroborated by our *fifth set of experiments*, in which we evaluated the pruning technique of the complete version of WOMBAT. In those experiments, for each of aforementioned benchmark datasets we computed what we dubbed as *pruning factor*. The pruning factor is the number of searched nodes (search tree size plus pruned nodes) divided by the maximum size of the search tree (which we set to 2000 nodes in this set of experiments). The results are presented in Table 5. Our average *pruning factor* of 2.55 shows that we can discard more than 3000 nodes while learning specifications.

In *a final set of experiments*, we compared the two versions of WOMBAT against the 2 systems proposed in [8]. To be comparable, we used the same evaluation protocol in [8], where 2% of the gold standard was used as training data and the remaining 98% of the gold standard as test data. The results (presented in Table 4) suggests that WOMBAT is capable of achieving better or equal performance in 4 out of the 6 evaluation data sets. While WOMBAT achieved inferior F-measures for the other 2 data sets, it should be noted that the competing

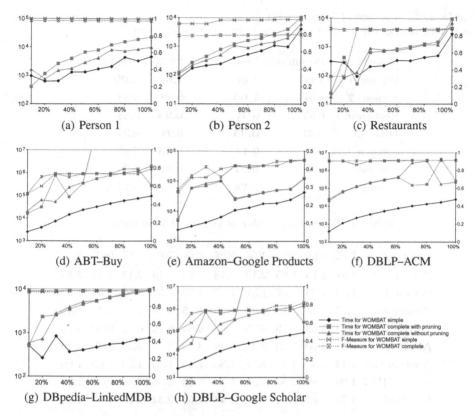

(a) Person 1 (b) Person 2 (c) Restaurants

(d) ABT–Buy (e) Amazon–Google Products (f) DBLP–ACM

(g) DBpedia–LinkedMDB (h) DBLP–Google Scholar

Fig. 3. Runtime and F-measure results of WOMBAT. The x-axis represents the fraction of positive examples from the gold standard used for training. The left y-axis represents the learning time in milliseconds with time out limit of 10^7 ms, processes running above this upper limit were terminated, all time plots are in log scale. The right y-axis represents the F-measure values.

systems are optimised for a low number of examples and they also get negative examples as input. Overall, these results suggest that our approach can generalise a small number of examples to a sensible LS.

Overall, our results show that ψ and φ are able to learn high-quality LS using only positive examples. When combined with our pruning algorithm, the complete version of ψ achieves runtimes that are comparable to those of φ. Given its completeness, ψ can reach specifications that simply cannot be learned by φ (see Fig. 4 for an example of such a LS). However, for practical applications, φ seems to be a good choice.

Table 4. Comparison of WOMBAT F-measure against the approaches proposed in [8] on 6 benchmarks using 2% of the original data as training data.

Dataset	Pessimistic	Re-weighted	Simple	Complete
Persons 1	**1.00**	**1.00**	**1.00**	**1.00**
Persons 2	0.97	**1.00**	0.80	0.84
Restaurants	0.95	0.94	**0.98**	0.88
DBLP-ACM	0.93	**0.95**	0.94	0.94
Amazon-GP	0.39	0.43	**0.53**	0.45
Abt-Buy	0.36	**0.37**	**0.37**	0.36
Average	0.77	**0.78**	0.77	0.74

Table 5. The *pruning factor* of the benchmark datasets.

Dataset	10%	20%	30%	40%	50%	60%	70%	80%	90%	100%
Person 1	1.57	2.13	1.85	2.13	2.13	2.13	2.13	2.13	2.13	2.13
Person 2	1.29	1.29	1.57	1.57	1.57	1.57	1.57	1.57	1.57	1.57
Restaurant	1.17	1.45	1.17	1.45	1.45	1.45	1.45	1.45	1.45	1.45
DBLP-ACM	6.23	5.58	6.79	6.85	6.85	6.85	6.79	6.79	6.93	6.79
Abt-Buy	3.38	3.00	3.00	3.39	3.39	3.39	1.79	3.39	3.39	3.39
Amazon-GP	1.14	1.38	1.33	1.37	1.38	1.45	1.54	1.59	1.60	1.60
DBP-LMDB	1.00	1.86	2.86	1.86	1.86	2.33	2.36	2.36	2.36	2.36
DBLP-GS	1.79	1.93	2.01	2.36	2.45	1.66	2.44	2.26	1.97	2.05

6   Related Work

There is a significant body of related work on *positive only learning*, which we can only briefly cover here. The work presented by [13] showed that logic programs are *learnable* with arbitrarily low expected error from positive examples only. [18] introduced an algorithm for learning from labeled and unlabeled documents based on the combination of Expectation Maximization (EM) and a naive Bayes classifier. [2] provides an algorithm for learning from positive and unlabeled examples for statistical queries. The pLSA algorithm [24] extends the original unsupervised probabilistic latent semantic analysis, by injecting a small amount of supervision information from the user.

For learning with *refinement operators*, significant previous work exists in the area of Inductive Logic Programming and more generally concept learning which we only briefly sketch here. A milestone was the Model Inference System in [20]. Shapiro describes how refinement operators can be used to adapt a hypothesis to a sequence of examples. Afterwards, refinement operators became widely used as a learning method. In [23] some general results regarding refinement operators in quasi-ordered spaces were published. In [3] and later [4], algorithms for learning in description logics (in particular for the language $\mathcal{ALC}$) were created

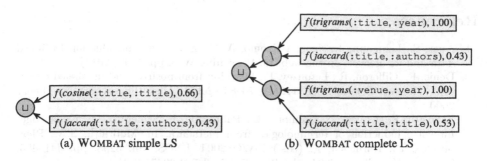

(a) WOMBAT simple LS   (b) WOMBAT complete LS

Fig. 4. Best LS learned by WOMBAT for the *DBLP-GoogleScholar* data set.

which also make use of refinement operators. Recent studies of refinement operators include [11,12] which analysed properties of $\mathcal{ALC}$ and more expressive description logics. A constructive existence proof for ideal operators in the lightweight $\mathcal{EL}$ description logics has been shown in [10]. DEER [21] uses refinement operators for automatic datasets enrichment.

Most LD approaches for *learning LS* developed are supervised. One of the first approaches to target this goal was presented in [5]. While this approach achieves high F-measures, it also requires large amounts of training data. Hence, methods based on active learning have also been developed (see, e.g., [7,17]). In general, these approaches assume some knowledge about the type of links that are to be discovered. For example, unsupervised approaches such as PARIS [22] aim to discover exclusively owl:sameAs links. Newer unsupervised techniques for learning LS include approaches based on probabilistic models [22] and genetic programming [16,19], which all assume that a 1-to-1 mapping is to be discovered. To the best of out knowledge, this paper presents the first LD approach designed to learn from positive examples only.

7   Conclusions and Future Work

We presented the (to the best of our knowledge) first approach to learn LS from positive examples via generalisation over the space of LS. We presented a simple operator φ that aims to achieve this goal as well as the complete operator ψ. We evaluated φ and ψ against state-of-the-art link discovery approaches and showed that we outperform them on benchmark datasets. We also considered scalability and showed that ψ can be brought to scale similarly to φ when combined with the pruning approach we developed. In future work, we aim to parallelize our approach as well as extend it by trying more aggressive pruning techniques for better scalability.

Acknowledgments. This work has been supported by H2020 projects SLIPO (GA no. 731581) and HOBBIT (GA no. 688227) as well as the DFG project LinkingLOD (project no. NG 105/3-2) and the BMWI Project GEISER (project no. 01MD16014).

References

1. Auer, S., Lehmann, J., Ngonga Ngomo, A.-C., Zaveri, A.: Introduction to linked data and its lifecycle on the web. In: Reasoning Web, pp. 1–90 (2013)
2. Denis, F., Gilleron, R., Letouzey, F.: Learning from positive and unlabeled examples. Theoret. Comput. Sci. **348**(1), 70–83 (2005). Algorithmic Learning Theory 2000
3. Esposito, F., Fanizzi, N., Iannone, L., Palmisano, I., Semeraro, G.: Knowledge-intensive induction of terminologies from metadata. In: McIlraith, S.A., Plexousakis, D., Harmelen, F. (eds.) ISWC 2004. LNCS, vol. 3298, pp. 441–455. Springer, Heidelberg (2004). doi:10.1007/978-3-540-30475-3_31
4. Iannone, L., Palmisano, I., Fanizzi, N.: An algorithm based on counterfactuals for concept learning in the semantic web. Appl. Intell. **26**(2), 139–159 (2007)
5. Isele, R., Bizer, C.: Learning linkage rules using genetic programming. In: Sixth International Ontology Matching Workshop (2011)
6. Isele, R., Jentzsch, A., Bizer, C.: Efficient multidimensional blocking for link discovery without losing recall. In: WebDB (2011)
7. Isele, R., Jentzsch, A., Bizer, C.: Active learning of expressive linkage rules for the web of data. In: Brambilla, M., Tokuda, T., Tolksdorf, R. (eds.) ICWE 2012. LNCS, vol. 7387, pp. 411–418. Springer, Heidelberg (2012). doi:10.1007/978-3-642-31753-8_34
8. Kejriwal, M., Miranker, D.P.: Semi-supervised instance matching using boosted classifiers. In: Gandon, F., Sabou, M., Sack, H., d'Amato, C., Cudré-Mauroux, P., Zimmermann, A. (eds.) ESWC 2015. LNCS, vol. 9088, pp. 388–402. Springer, Cham (2015). doi:10.1007/978-3-319-18818-8_24
9. Köpcke, H., Thor, A., Rahm, E.: Evaluation of entity resolution approaches on real-world match problems. Proc. VLDB Endow. **3**(1–2), 484–493 (2010)
10. Lehmann, J., Haase, C.: Ideal downward refinement in the EL description logic. In: 19th International Conference on Inductive Logic Programming, Leuven, Belgium (2009)
11. Lehmann, J., Hitzler, P.: Foundations of refinement operators for description logics. In: Blockeel, H., Ramon, J., Shavlik, J., Tadepalli, P. (eds.) ILP 2007. LNCS (LNAI), vol. 4894, pp. 161–174. Springer, Heidelberg (2008). doi:10.1007/978-3-540-78469-2_18
12. Lehmann, J., Hitzler, P.: Concept learning in description logics using refinement operators. Mach. Learn. J. **78**(1–2), 203–250 (2010)
13. Muggleton, S.: Learning from positive data. In: Muggleton, S. (ed.) ILP 1996. LNCS, vol. 1314, pp. 358–376. Springer, Heidelberg (1997). doi:10.1007/3-540-63494-0_65
14. Ngonga Ngomo, A.-C.: Link discovery with guaranteed reduction ratio in affine spaces with minkowski measures. In: Cudré-Mauroux, P., Heflin, J., Sirin, E., Tudorache, T., Euzenat, J., Hauswirth, M., Parreira, J.X., Hendler, J., Schreiber, G., Bernstein, A., Blomqvist, E. (eds.) ISWC 2012. LNCS, vol. 7649, pp. 378–393. Springer, Heidelberg (2012). doi:10.1007/978-3-642-35176-1_24
15. Ngonga Ngomo, A.-C., Lyko, K.: EAGLE: efficient active learning of link specifications using genetic programming. In: Simperl, E., Cimiano, P., Polleres, A., Corcho, O., Presutti, V. (eds.) ESWC 2012. LNCS, vol. 7295, pp. 149–163. Springer, Heidelberg (2012). doi:10.1007/978-3-642-30284-8_17
16. Ngonga Ngomo, A.-C., Lyko, K.: Unsupervised learning of link specifications: deterministic vs. non-deterministic. In: Proceedings of the Ontology Matching Workshop (2013)

17. Ngomo, A.-C.N., Lyko, K., Christen, V.: COALA – correlation-aware active learning of link specifications. In: Cimiano, P., Corcho, O., Presutti, V., Hollink, L., Rudolph, S. (eds.) ESWC 2013. LNCS, vol. 7882, pp. 442–456. Springer, Heidelberg (2013). doi:10.1007/978-3-642-38288-8_30
18. Nigam, K., McCallum, A.K., Thrun, S., Mitchell, T.: Text classification from labeled and unlabeled documents using EM. Mach. Learn. **39**(2–3), 103–134 (2000)
19. Nikolov, A., dAquin, M., Motta, E.: Unsupervised learning of link discovery configuration. In: Simperl, E., Cimiano, P., Polleres, A., Corcho, O., Presutti, V. (eds.) ESWC 2012. LNCS, vol. 7295, pp. 119–133. Springer, Heidelberg (2012). doi:10.1007/978-3-642-30284-8_15
20. Shapiro, E.Y.: Inductive inference of theories from facts. In: Lassez, J.L., Plotkin, G.D. (eds.) Computational Logic: Essays in Honor of Alan Robinson. The MIT Press (1991)
21. Sherif, M.A., Ngomo, A.-C.N., Lehmann, J.: Automating RDF dataset transformation and enrichment. In: Gandon, F., Sabou, M., Sack, H., dAmato, C., Cudré-Mauroux, P., Zimmermann, A. (eds.) ESWC 2015. LNCS, vol. 9088, pp. 371–387. Springer, Cham (2015). doi:10.1007/978-3-319-18818-8_23
22. Suchanek, F.M., Abiteboul, S., Senellart, P.: PARIS: probabilistic alignment of relations, instances, and schema. PVLDB **5**(3), 157–168 (2011)
23. Laag, P.R.J., Nienhuys-Cheng, S.-H.: Existence and nonexistence of complete refinement operators. In: Bergadano, F., Raedt, L. (eds.) ECML 1994. LNCS, vol. 784, pp. 307–322. Springer, Heidelberg (1994). doi:10.1007/3-540-57868-4_66
24. Zhou, K., Gui-Rong, X., Yang, Q., Yu, Y.: Learning with positive and unlabeled examples using topic-sensitive PLSA. IEEE Trans. Knowl. Data Eng. **22**(1), 46–58 (2010)

Actively Learning to Rank Semantic Associations for Personalized Contextual Exploration of Knowledge Graphs

Federico Bianchi[✉], Matteo Palmonari, Marco Cremaschi,
and Elisabetta Fersini

University of Milan - Bicocca, Viale Sarca 336, Milan, Italy
{federico.bianchi,palmonari,cremaschi,fersiniel}@disco.unimib.it

Abstract. Knowledge Graphs (KG) represent a large amount of Semantic Associations (SAs), i.e., chains of relations that may reveal interesting and unknown connections between different types of entities. Applications for the contextual exploration of KGs help users explore information extracted from a KG, including SAs, while they are reading an input text. Because of the large number of SAs that can be extracted from a text, a first challenge in these applications is to effectively determine which SAs are most interesting to the users, defining a suitable ranking function over SAs. However, since different users may have different interests, an additional challenge is to personalize this ranking function to match individual users' preferences. In this paper we introduce a novel active learning to rank model to let a user rate small samples of SAs, which are used to iteratively learn a personalized ranking function. Experiments conducted with two data sets show that the approach is able to improve the quality of the ranking function with a limited number of user interactions.

1 Introduction

Knowledge Graphs (KG) represent entities of different types, their properties and binary relations that interconnect these entities. KGs are today frequently used to support interoperability among applications also in the industry, while languages like RDF support the publication of KGs as open linked data. A problem that has recently gained attention is how to exploit the vast amount of knowledge available in proprietary or open KGs to deliver useful information to the users. While query answering is aimed at satisfying specific information needs, knowledge exploration provides mechanisms to deliver information that is estimated to be interesting for the users in a proactive fashion [1].

One approach to support knowledge exploration is to push content from KGs while users are carrying out familiar tasks, such as querying a search engine, watching media content [2], or reading a text of interest [3,4]. We refer to the latter approaches as *contextual KG exploration*, where an input text (possibly a description of media content) is used as a entry point to let users explore information extracted from the KG. By using well-known entity linking techniques [3],

E. Blomqvist et al. (Eds.): ESWC 2017, Part I, LNCS 10249, pp. 120–135, 2017.
DOI: 10.1007/978-3-319-58068-5_8

entities mentioned in the text are linked to a KG. If more than one entity is found, semi-walks in the KG that connect the two entities, i.e., **Semantic Associations** (SAs) of finite length between the two entities [4,5] reveal connections between entities, which may provide new and interesting insights into the topic of the input text. For example, a SA found in DBpedia revealing that Clinton and Trump have been both members of the Democratic Party has been found interesting by many Italian students who read about US Election 2016.

The main problem arising in contextual KG exploration is that a very large amount of SAs can be found between a set of entities extracted from even relatively short texts. For example, from an input article[1], as many as 40.107 SAs are found in DBpedia with DaCENA[2], a data journalism prototype for contextual KG exploration [4]. The crucial research challenge to exploit such a large amount of SAs represented in KGs is to provide effective methods to identify those few SAs that are more interesting for the users. Several approaches have been proposed that use measures based on graph analytics to rank and filter SAs [5,6]. However, *different users may be interested in different kinds of SAs*, which suggests that the ranking function should adapt to the preferences of individual users. One approach proposes to personalize a ranking function by learning from explicit user preferences [7], but does not address the problem of minimizing the labels collected from the users, which is crucial when exploring very large sets of SAs. Starting from these observations in this paper we address the following research questions: (RQ1) Can we learn to rank SAs by iteratively collecting a small number of labels from a user, so that we can personalize the content delivered to her based on her preferences? (RQ2) Do we need personalization in contextual exploration of KGs with SAs, or can we assume that different users are interested in the same content?

To answer to Q1, we propose an active learning to rank model to reduce the number of SAs that need to be labeled by the user. The model comprises: (1) a workflow to iteratively collect labels from a user and learn to rank the SAs based on her preferences using the RankSVM algorithm; (2) algorithms to actively select the SAs that the user has to label; (3) different approaches to select the first set of SAs that the user has to label, thus solving a cold start problem affecting the above mentioned active sampling algorithms, (4) a set of features based on KG analytics to represent SAs and support the model. To evaluate the effectiveness of the proposed model under different configurations and against different baselines, we have built two data sets consisting of ratings given by different users on a complete set of SAs extracted for different pieces of news articles. Results show that the proposed approach is feasible and provides a consistent improvement of the ranking quality with a limited number of interactions. To answer to Q2, we measure the agreement among ratings given by different users to SAs found for the same articles. Results clearly show that different users are interested in different content, thus confirming the need for personalization methods in contextual KG exploration.

---

[1] https://goo.gl/RFvqZh.

[2] http://www.dacena.org/article/84.

To the best of our knowledge this is the first attempt to use active sampling to learn to rank SAs, thus improving on state of the art approaches that require a large number of labels to learn a ranking function over SAs. The paper is organized as follows: in Sect. 2, we further motivate the proposed approach by discussing contextual KG exploration, with an example of application; in Sect. 3, we explain our active learning to rank model; in Sect. 4 we describe the experiments conducted to evaluate our model; in Sect. 5 we discuss related work, while in Sect. 6 we draw some conclusions and discuss future work.

2   Contextual KG Exploration

Applications for the contextual exploration of KGs enrich the experience of a user who is accessing textual or multimedia content by delivering information extracted from a KG [2–4]. The input content tell us something about current interests of the user, thus providing a starting point to select pieces of valuable information, because helpful to expand her knowledge or to better understand the content itself. Named Entity Recognition and Linking (NEEL) techniques [3] can be applied to an input text to extract a first a set of entities from the KG, which can be subsequently used to retrieve additional information, e.g., SAs. In the following, we refer to SAs as *semi-walks in a KG* [5], *which do not contain loops and whose nodes are entities.* We discuss how SAs can be used as source of additional information in these application by referring to DaCENA [4] (Data Context for News Articles), an application that supports exploration of KGs to address one of the missions of data journalism, i.e., *"[...] to provide context, clarity and, [...] find truth in the expanding amount of digital content"* [8]

DaCENA presents a set of SAs extracted from a KGs as additional information (a data context) to a user who is reading a news article. End users can read the article and explore the extracted SAs from an interactive interface. Figure 1 shows a screenshot of the interface, and readers can play with the same example online[3]. The graph shows the k-most interesting SAs (40.107 for this article), where k can be set by the user. When the user clicks on an entity node, e.g., *Separatism*, SAs from/to such node are shown in the lower panel and ordered by estimated interest. SAs are extracted with a process controlled through a back-end. In DaCENA we are currently using TextRazor[4] as NEEL tool and DBpedia[5] as reference KG. Once entities are extracted we make several queries to the DBpedia SPARQL endpoint to extract the SAs that connect all these entities (we use a SPARQL endpoint to ensure that we retrieve up-to-date information). While we started with extracting SAs of maximum length equal to three [4], now we consider only SAs of maximum length equal to two, because we found - with preliminary user studies - that SAs of length greater than two are seldom considered interesting by the users. Otherwise, while we previously found every SA from a principal entity to every other entity, we now consider

---

[3] http://www.dacena.org/article/84.
[4] https://www.textrazor.com/.
[5] https://dbpedia.org/.

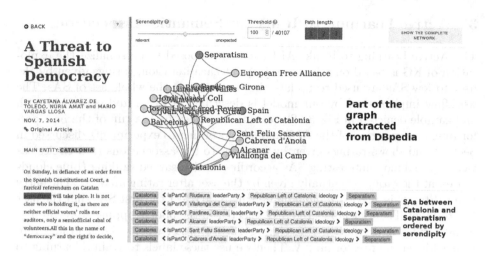

Fig. 1. DaCENA interface

shorter SAs between every entity extracted from the text. Processing an article may require significant amount of time (up to thirty minutes) if semantic data are fetched by querying a SPARQL endpoint as we currently do. Therefore, texts and data are processed off-line so as to make the interactive visualization features as much fluid as possible. DaCENA currently uses a measure to evaluate *interestingness* of SAs named **Serendipity** [4]. Serendipity is defined as a parametric linear combination of **Relevance**, a measure that evaluates the relevance of a SA with respect to a text, and **Rarity**, a measure that evaluates how much a SA may be unexpected for the users. A SA is relevant if the virtual document built by concatenating the abstracts of each entity occurring in the SA is similar to the given text (we compute the cosine similarity between word vectors weighted using TF-IDF). Instead, a SA is unexpected, or rare, when it is composed by properties that are not frequently used in the KG (see [4] for the formula). Let α be a parameter used for balancing the weight of each measure, and *text* be the input text; the serendipity $S(\pi)$ of an SA π, is computed by $S(\pi, text) = \alpha relevance(\pi, text) + (1 - \alpha) \, rarity(\pi)$. In the interface shown in Fig. 1 the user can adjust the serendipity parameter to favor relatedness or unexpectedness.

By analyzing several articles with DaCENA, we could observe that thousands or even dozens of thousands of SAs can be extracted from an article (see, e.g., the example shown in Fig. 1), while preliminary user studies suggest that users do not want to look at more than 100 SAs. Thus, the ranking function used to push the most interesting SAs upfront and filter out other SAs is crucial to help users effectively explore the KG content. Moreover, while some user may be interested in finding out information about small municipalities associated with separatism, other users may be more interested in information about more important cities. If different users have different interests (an hypothesis validated in our experiments), mechanisms to personalize KG exploration are needed.

3　Active Learning to Rank for Semantic Associations

The Active Learning to Rank (ALR) model proposed to personalize the exploration of KG is based on a learning loop: at each iteration, ratings given by the user to few SAs are used to update the ranking over the whole set of SAs. The workflow implemented by our model is described in Fig. 2 and explained with an example depicted in Fig. 3. The example is taken from a run of the best performing configuration of the ALR model (according to experiments discussed in Sect. 4) and shows ratings given by one user to few associations (red circles) as well as the 3 most interesting SAs according to the learned ranking (blue clouds represent the ratings eventually given by the user after rating all SAs). The entry point (Step 1) is a *bootstrapping* phase where we select the first SAs that the user has to label. The user labels the SAs selected in the bootstrapping step (Step 2) using ratings in a graded scale, e.g., $<1, 2, 3, 4, 5, 6>$, where higher grades represent higher interest for an SA. Then we use these labels to train a learning to rank algorithm (step 3), which ranks all the SAs by assigning them a score. If the user decides that she is satisfied with the ranking obtained so far, the loop stops. Else, we proceed to further improve the ranking by collecting more labels using active sampling (Step 5). In active sampling, observations are selected with the aim of optimizing the ranking function with as few labels as possible. To find the observations for which labels are estimated to be more informative, active sampling algorithms use the scores determined by the learned ranking function. This prerequisite, motivates the need for introducing a bootstrapping step (Step 1) where labels are not selected using active sampling. After Step 5, we close the loop by repeating Step 2. Observe that after the first iteration, the user always labels SAs selected with active sampling. In Fig. 3, it can be noticed that the ranking improves after the second iteration (Step 3). The main steps of the loop, i.e., steps 1, 3 and 5, are explained in details here below.

Step 1: Bootstrap Active Learning. Two approaches are proposed: the first one (alternative 1a) is based on clustering algorithms while the second one uses an heuristic ranking function (alternative 1b). The latter has the advantage that

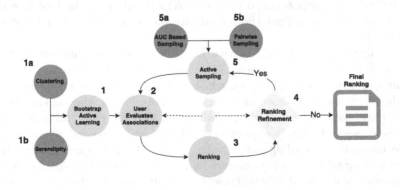

Fig. 2. Workflow of the active learning to rank model

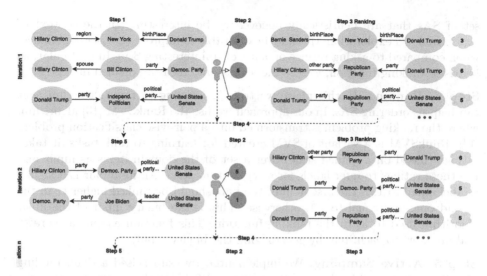

Fig. 3. Example of iterative ranking refinement with the ALR model (Color figure online)

we have an ordered set of SAs (hence, a small set of interesting SAs) to present to the user even before she provides any label.

Alternative 1a. The assumption at the basis of this approach is the following: observations that would be rated in a similar way by a user are spatially near in the feature space used to represent the SAs, while observations rated in different ways should be distant in this space. Based on this assumption, the best way to quickly collect the training data is to cluster the data set and take the most representative observation for each cluster (the observation nearest to the mean of the cluster). Approaches similar to this have been already considered in active learning settings [9,10], in which clustering is used to find the first observations for machine learning models. We test two clustering algorithms: the first one is the Dirichlet Process Gaussian Mixture Model [11] (Dirichlet) that has been chosen for its ability to automatically find the best number of cluster inside the data set; this is useful because we cannot know a-priori which could be the correct number of cluster for a given set of SAs. We also evaluate a second clustering algorithm, the Gaussian Mixture Model (Gaussian) [12]. It is important to notice that these algorithms select SAs that are representative of a data set, but that does not mean that these SAs are also meaningful or interesting for a user.

Alternative 1b. Another approach that we propose to use for bootstrapping the model (also in consideration of the latter remark) is to use an heuristic ranking measure. In this way, not only we can show to a user a set of SAs even before she provides any rating, but we can also ask their ratings on a set of SAs that are heuristically believed to be interesting. In the context of contextual KG exploration, this may be desirable to improve the user experience, when compared to asking ratings on a set of uninteresting SAs. In particular, as heuristic ranking function, we use the *serendipity* measure defined in Sect. 2.

Step 3: Ranking. In this phase we train a learning to rank algorithm that can help us ordering SAs. In our approach we use the RankSVM [13] algorithm, where the ranking problem is transformed into a pair-wise classification problem [14]. RankSVM is a variant of SVM created for learning to rank tasks, it takes ratings (based on a graded scale) over a set of the domain items as input and use these ratings to infer labels for a set of item pairs. An item pair is assigned a label equal to 1, if the first item of the pair should be ranked higher than the second one, and equal to −1, otherwise. This is the binary input of the inner algorithm used to learn the ranking function. This function assigns a score (a real number) to each item by generalizing the binary input.

Step 5: Active Sampling. We implemented two supervised active sampling algorithms proposed in the document retrieval field. Both the algorithms use a pairwise approach, meaning that they can be directly used on pairwise learning to rank algorithms, like RankSVM.

Alternative 5a: the first algorithm [15] (denominated *AUC-Based Sampling*, or, shortly, *AS*, in the next sections) tries to optimize the Area Under the Curve (AUC), by selecting individual observations for labeling, without explicitly comparing every pair of domain items. The algorithm is thus known to be sub-optimal, runs efficiently. It is essentially based on the computation of the estimated probability of a binary class for an observation (thus, it was used in a binary setting). The algorithm uses a parameter λ to calibrate the weight of two different probability estimations.

Alternative 5b: the second algorithm [16] (denominated *Pairwise Sampling*, or, shortly, *PS*, in the next sections) explicitly compares pairs of SAs to select the most informative pairs. The most informative pairs are the ones that maximize two measures: Local Uncertainty (LU), which estimates the uncertainty of the relative order within the pair, and Global Uncertainty (GU), which estimates the uncertainty of the position of each element of the pair within the global ranking. A parameter p is used to tune the weight assigned to the LU measure. In this case, users are then asked to rate each of the most informative pairs of SAs. With this approach, we can evaluate if the uncertainty score used to select the pair is incoherent with user ratings, thus providing more informative labels to RankSVM. The explicit generation of the observation-pairs makes this algorithms less efficient than AUC-Based Sampling, which may prevent its application to the exploration of a large number of SAs.

3.1 Features

To represent the SAs inside our platform we used different measures. In this way, we define feature vectors for the active learning to rank algorithms. We normalize data extracted with these measures using a standard normalization techniques by removing the mean and scaling to unit variance.

Global PageRank. We use the data in [17] to collect a global score of the PageRank inside DBpedia. In this, way we are able to get an overall value of the importance of an entity inside the KG. The Global Pagerank of a SA is computed as the average global pagerank of every entity occurring in that SA.

Local PageRank. We compute PageRank on the sub-graph, defined by the SAs extracted from an input text, to measure the *centrality importance* of each entity.

Local HITS. We ran the HITS (Hyperlink-Induced Topic Search) [18] algorithm to compute two scores for each node of the local graph. Authority score, indicates how much a node is *important*, while hub score indicates nodes that point to nodes with a high authority score. The algorithm gives two scores for each SA: one for the average of the authority values and one for the average of the hub values.

Temporal Relevance. Using the Wikimedia API we extract the number of times a Wikipedia entity (page) as been accessed in a specific date (date of the publication of a given text, for example). In this way we are able to measure the value of importance related to timing. For example, if we consider Wikipedia access[6] on the page Paris, we see that the entity has been accessed 8.331 times on 12-11-2015 and 171.988 times on 14-11-2015, when on 13-11-2015 there have been terrorist attacks in Paris. The temporal relevance for a SA is given by the average temporal relevance of all the entities in the SA.

Path Informativeness. We use a measure defined in [5], which is based on the concept of Predicate Frequency Inverse Triple Frequency (PF-ITF).

Path Pattern Informativeness We use a measure on path patterns, defined in [5], to get the informativeness of patterns extracted from paths.

Relevance and Rarity. We use the two measures explained in Sect. 2 as components of our Serendipity measure [4].

4   Experiments

The purpose of the experimental evaluation is to validate the hypothesis that personalization is important in KG exploration, to evaluate the performance of the proposed model, and to compare alternative approaches proposed for different steps of the model. The targets of the application are fairly educated users familiar with IT technologies. So far, we used in our experiments master students from Computer Science, Mechanical Engineering and Communication Sciences with good English reading skills. All the data sets used in our expriments are available online[7]. The experiments were run on a machine with a Intel Core i5 (4th Gen, 1.6 Ghz).

---

[6] http://tools.wmflabs.org/pageviews/.
[7] https://github.com/vinid/semantic-associations-survey.

4.1 Experimental Settings

To test our model we built two different datasets, each one consisting of triples $<text_i, A_i, ratings_{u,i}>$, where $text_i$ is a text extracted from an article retrieved from online news platforms like NYT and The Guardian, A_i is the set of all SAs extracted from $text_i$ with our tool DaCENA, and $ratings_{u,i}$ contains the labels assigned by a user u to every SA in A_i. From each triple in a dataset, we can derive a complete ranking of the retrieved SAs for one user, i.e., a personal ideal ranking. We describe the creation of each dataset here below.

Short Articles Many Users (SAMU). We collected user ratings for this data set using an online form. For ratings we choose a graded scale from 1 to 6, following guidelines suggested in a recent study [19]. Differently from a five-valued ordinal scale, this scale provides a symmetric range that clusters scores in two sets: scores with a negative tendency (1, 2 and 3) and scores with a positive tendency (4, 5 and 6). Each user had to evaluate the complete set of associations extracted from one article, thus we had to choose articles small enough to let users perform their task without being subject to fatigue bias [19]. We thus selected the first self-contained paragraphs of articles from NYT and Guardian with the following features: articles topic concern politics, is reasonably well-known and engaging for foreign (Italian) educated users; the number of associations extracted by DaCENA is comprised between 50 and 100 SAs. The average task completion time resulted in 12 min - little below the fatigue bias threshold mentioned in [19]. We also wanted to have preferences of different users on a same article to measure inter-user agreement and validate the "personalization hypothesis": we needed a number of articles small enough to collect at least 3 evaluations from different users. Articles were assigned randomly to each user to avoid any bias. After evaluating the first article, a user could stop or evaluate more articles. We stopped searching users for the evaluation when we collected evaluations by at least 3 users on each article, which resulted in a total of 14 different users, and 25 gold standards (personal rankings).

Long Articles Few Users (LAFU). We wanted also to evaluate if results obtained over small SA sets are comparable with results obtained with (and thus generalizable to) large SA sets. To this end, two users were asked to rate thousands of SAs extracted for two full-length articles, with the goal of evaluating heuristic functions used in an early version of DaCENA. In this case, we used a three-valued scale for ratings, from 1 to 3. The two users involved in the evaluation of the longer articles were Communication Sciences students with no background in Computer Science. They were granted several days for completing the task, and asked to complement their task with a qualitative analysis.

Using the ideal rankings in the two gold standards, we measure the quality of the rankings returned by our model at different iterations using Normalized Discounted Cumulative Gain (nDCG) computed over the top-10 ranked SAs, denoted by nDCG@10. In addition, we compute the Area Under the nDCG@10 Curve (AUNC) as an aggregate performance measure, the curve is based on the nDCG@10 values at each iteration. We carry out experiments in two different settings. In *Contextual Exploration Settings*, we consider the workflow as implemented in a system that supports contextual exploration: the set from which we select the observations to label is the same set used to evaluate the performance of the model. In these settings, we make sure that observations labeled during previous iterations are not labeled a second time by the user. In *Cross Validation Settings*, which was used also in previous work [15], active sampling always picks SAs from the training set. Although not amenable in contextual exploration, this approach is helpful to evaluate the robustness of the model. In fact, we can use 2Fold-Stratified Cross Validation to make sure that results can be reasonably generalized and do not depend on specific data.

4.2 Configurations and Baselines

We evaluate different configurations of the model, based on the alternative algorithms proposed in the two steps of the loop. Direct comparison with other state-of-the-art approaches is difficult because we could not find an active learning to rank approach for SAs. For Bootstrap Active Learning, we consider three approaches: two clustering algorithms (Gaussian vs. Dirichlet), and the Serendipity heuristic function, for which we set $\alpha = 0, 5$. For Active Sampling, we consider two algorithms: AUC Based Sampling [15] (AS) and Pairwise Sampling [16] (PS). Parameters of these two algorithms have been determined experimentally, and set to $\lambda = 0.8$ and $p = 1$. The six configurations of the active learning to rank workflow described above are compared also against three different baselines:

- *Random + Random*: RankSVM is still used to learn a ranking function, but is trained using ratings assigned to SAs that are randomly selected, both in the bootstrap and active sampling steps.
- *Serendipity No-AL*: we consider the ranking determined with Serendipity, which is not based on active learning and does not change across iterations.
- *Random No-AL*: we consider random rankings of SAs, which are not based on active learning and do change across iterations.

Random algorithm are run multiple times to stabilize values (100 hundred thousands of them).

Configuration Details. In the SAMU data sets Dirichlet and Gaussian Clustering, in the first iterations, selected an average number of clusters equal to 3 (and thus, an average number of 3 SAs are selected from this two methods in the first iteration); for this reason, to feed the model with a balanced number of observations, on the average, for both Serendipity and Random Active Learning we choose to select 3 SAs when using Serendipity and Random in the bootstrapping step. Finally for this data set we select 2 SAs to be evaluated at each active sampling step. The number of observations collected at each iteration was increased in the LAFU data set. The clustering algorithms in this data set selected an average number of cluster equal to 5, leading to 5 SAs to be labeled when using Serendipity and Random for bootstrapping. In the active sampling phase, for the LAFU data set, we selected 6 observation to be labeled (since we have more data) at each iteration. In this data set we could not use the *Temporal Relevance* as a feature because articles in this data set are not recent. We used a RankSVM with polynomial kernel on the LAFU data set that was able to output the results of a single iteration in what we considered interactive time (less then 2 s).

4.3   Results and Discussion

User interests and personalization. We have measured Inter-Rater Reliability [20] (IRR) to assess the usefulness of personalization within this context. Our idea is based on the assumption that different users are interested in different things. IRR was computed on the SAMU, which had the same SAs rated by different users. We used two measures: Krippendorff's alpha, which gave in output a value of 0.06154, weighted using an ordinal matrix, and Kendall's W, which gave a score of 0.2608. We can see that for all the five texts used in this experiment, IRR is low and distant from 1, the value that usually represents unanimity between the raters. We also show the distribution of the ordinal ratings for the data sets in Table 1.

Table 1. Rating distribution in the two data sets

Rating	1	2	3	4	5	6
SAMU	23.7%	14.5%	22.3%	20.9%	10.1%	5.5%
LAFU	67.4%	30.1%	2.5%	NaN	NaN	NaN

Contextual Exploration Settings. In this setting the active learning algorithms Fig. 4 were able to perform better than the baseline considered (we can notice that active learning approaches completely outperform the non active learning ones). AUNC values can be seen in Table 3, the algorithm that performs better is the one that uses serendipity for the bootstrap step and AS for the active sampling step (Fig. 2).

Cross Validation Settings. The results we obtained in the cross validation settings provide further evidence that the best result is obtained by the Serendipity heuristic, combined with the use of AS, However, in this setting we achieve a worse performance, due to the fact that the active learning algorithms are not able to access to the test data. The plots can be found in Fig. 4 while the computed areas are in Table 3.

Bootstrap Active Learning Analysis. To construct a ranking model we need at least two examples with different labels, is not always possible in the first iteration (the user, in that step, can assign to each observation the same degree of interest). We evaluated the *time for first iteration* value, that corresponds to the average number of iterations needed for each method to have a training useful

Table 2. AUNC in cross validation

Configurations	SAMU	LAFU
Gaussian AS	3.0168	2.6455
Dirichlet AS	3.0011	2.6872
Gaussian PS	2.9975	NaN
Dirichlet PS	3.0009	NaN
Serendipity AS	**3.0742**	**2.711**
Serendipity PS	3.0302	NaN
Random Random	2.976	2.6013

Table 3. AUNC in contextual exploration

Configurations	SAMU	LAFU
Gaussian AS	3.0242	2.747
Dirichlet AS	3.0711	2.7174
Gaussian PS	2.9629	NaN
Dirichlet PS	3.019	NaN
Serendipity AS	**3.2018**	**3.0817**
Serendipity PS	3.1399	NaN
Random Random	2.9359	2.673
Serendipity No-AL	2.7199	2.734
Random No-AL	2.3199	1.7971

for training the learning to rank algorithm. The worst case of these algorithms is represented by a user who gives the same score to every observation. The results are visible in Table 4. We show data for both Cross Validation (CV) and Contextual Exploration Setting (CE). We can see that with the LAFU data set finding the first observations needed to train the model becomes more difficult for methods with the except of Dirichlet that can probably adapt itself to the dataset in an easier way.

Table 4. #Iterations to find the first training set for the ranking model

Algorithm	#Iter. SAMU CV	#Iter. LAFU CV	#Iter. SAMU CE	#Iter. LAFU CE
Dirichlet	1.16	1.11	1.010	1.063
Gaussian	1.08	1.16	1.066	1.381
Serendipity	1.08	1.5	1.346	1.1363
Random	1.25	1.5	1.1866	1.229

Discussion. Based on the evidence collected through the above experiments, we can provide answers to research questions introduced in Sect. 1. In relation to RQ2, we observed that different users have different interests, which motivates the need for personalization in KG exploration approaches. In relation to RQ1, the ALR model introduced in this paper supports shows remarkable improvement over ranking methods that do not use active sampling. In this context and for the selected set of features, we found that AUC-based sampling performs better than pairwise sampling, both in terms of effectiveness and efficiency, and that the combination of Serendipity (for bootstrapping) and AUC-based active sampling outperforms every alternative configurations on both data sets.

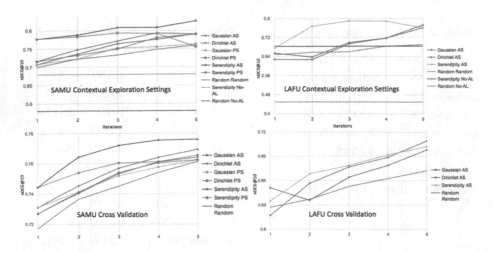

Fig. 4. nDCG@10 in contextual exploration and cross validation settings

5 Related Work

We compare our work to previous work in the field of interactive KG exploration and of learning to rank approaches for KG exploration.

Interactive Knowledge Graph Exploration. Several methods, described and compared in a recent survey [1], combine navigation, filtering, sampling and visualization to let users explore large data sets. One approach to entity expansion provides an example of contextual KG exploration, but does not focus on the retrieval of SAs like our approach [2]. RelFinder is a web application that finds SAs between two entities selected by a user [21]. Other applications similar to RelFinder also incorporate measures to evaluate and explain SAs [5,6,22]. Refer [3] is a Wordpress Plugin that help a user enrich an article with additional information extracted from KBs like Wikipedia. The plugin finds entities in the article and recommends SAs that are estimated to by unknown to the user. Refer is an example of contextual exploration of KG; the main difference between their approach and our approach is that we introduce a model to order all SAs, introducing a machine learning model to personalize the exploration. None of the approaches mentioned above or surveyed in [1] introduces methods to learn information to show to the users based on their explicit feedback. An interesting approach seen in the literature [23] uses genetic programming to find strong relationships in linked data; in their experiments, eight judges were asked to evaluate the relationships, but relationships with low inter-user agreement were not considered positive examples for training because not interesting for all users. Since different users have different interests, we train our model based on the preferences of individual users using ALR.

Learning to Rank and Active Learning for KG Exploration. Learning to rank has been extensively applied in document retrieval [14] but only in one approach to KG exploration [7]. This approach use a variant of SVM to rank SAs extracted from Freebase, but does not try to minimize the inputs needed to learn the ranking function through active learning. In addition, some of their features are specifically tailored on the Freebase structure while our features can be easily applied to any KG (with the exception of Temporal Relevance, which requires bridges from the KG to Wikipedia). Active learning to rank introduces techniques to select the most informative observations to train the model. In our approach, we have implemented and tested two different techniques proposed for document retrieval. A first approach collects labels over individual observations (SAs in our case) and solves the cold-start problem by randomly selecting positive and negative instances from a subset of the data reserved for training [15]. In our interactive approach we pick the SAs that are labeled by the user form the same set that has to be ranked, which is coherent with contextual KG exploration scenarios. However, we have also conducted tests with data split in a training and a test set to show the robustness of the model. In addition, we provided a principled approach to solve the cold-start problem in our domain. The second approach, which collects labels over pairs of observations [16], seems to be not only less efficient, but also less effective for ranking SAs. To the best of our knowledge, ours is the first attempt to apply active learning to rank to the problem of exploring SAs.

6   Conclusion

Experimental results presented in this paper suggest that active learning approaches can be effectively used to optimize the ranking of SAs extracted from KGs, thus supporting personalized exploration of complex relational knowledge made available in these graphs. Such personalization mechanisms have also shown to be important for knowledge exploration, since different users are interested in different content. We have also found that, an approach that combines our Serendipity measure [4] and AUC-based active sampling outperforms different alternative configurations. In future work, we plan to analyze the impact of individual features on the performance of an active learning to rank model for SAs, and evaluate the use of additional measures. In addition, we want to incorporate our active learning to rank model into the DaCENA application, by tackling the challenge of designing human-data interaction patterns that can engage the users. We would also like, in the future, to improve the performance of our application. So far, we preferred to have fresher information via a SPARQL endpoint despite the longer processing time, because processing is performed offline. In journalism, freshness of information is relevant and we plan to further investigate methods to refresh/update SAs after processing in the future.

Acknowledgement. We thank our colleague Federico Cabitza for his knowledgeable advises about the creation of the SAMU data set.

References

1. Bikakis, N., Sellis, T.: Exploration and visualization in the web of big linked data: a survey of the state of the art. preprint arXiv:1601.08059 (2016)
2. Redondo-García, J.L., Hildebrand, M., Romero, L.P., Troncy, R.: Augmenting TV newscasts via entity expansion. In: Presutti, V., Blomqvist, E., Troncy, R., Sack, H., Papadakis, I., Tordai, A. (eds.) ESWC 2014. LNCS, vol. 8798, pp. 472–476. Springer, Cham (2014). doi:10.1007/978-3-319-11955-7_69
3. Tietz, T., Jäger, J., Waitelonis, J., Sack, H.: Semantic annotation and information visualization for blogposts with Refer. In: VOILA 2016, vol. 1704, pp. 28–40 (2016)
4. Palmonari, M., Uboldi, G., Cremaschi, M., Ciminieri, D., Bianchi, F.: DaCENA: serendipitous news reading with data contexts. In: Gandon, F., Guéret, C., Villata, S., Breslin, J., Faron-Zucker, C., Zimmermann, A. (eds.) ESWC 2015. LNCS, vol. 9341, pp. 133–137. Springer, Cham (2015). doi:10.1007/978-3-319-25639-9_26
5. Pirrò, G.: Explaining and suggesting relatedness in knowledge graphs. In: Arenas, M., et al. (eds.) ISWC 2015. LNCS, vol. 9366, pp. 622–639. Springer, Cham (2015). doi:10.1007/978-3-319-25007-6_36
6. Cheng, G., Zhang, Y., Qu, Y.: Explass: exploring associations between entities via Top-K ontological patterns and facets. In: Mika, P., et al. (eds.) ISWC 2014. LNCS, vol. 8797, pp. 422–437. Springer, Cham (2014). doi:10.1007/978-3-319-11915-1_27
7. Chen, N., Prasanna, V.K.: Learning to rank complex semantic relationships. IJSWIS 8(4), 1–19 (2012)
8. Gray, J., Chambers, L., Bounegru, L.: The Data Journalism Handbook. O'Reilly Media Inc., Sebastopol (2012)
9. Kang, J., Ryu, K.R., Kwon, H.-C.: Using cluster-based sampling to select initial training set for active learning in text classification. In: Dai, H., Srikant, R., Zhang, C. (eds.) PAKDD 2004. LNCS (LNAI), vol. 3056, pp. 384–388. Springer, Heidelberg (2004). doi:10.1007/978-3-540-24775-3_46
10. Yuan, W., Han, Y., Guan, D., Lee, S., Lee, Y.K.: Initial training data selection for active learning. In: ICUIMC, p. 5. ACM (2011)
11. Gershman, S.J., Blei, D.M.: A tutorial on bayesian nonparametric models. J. Math. Psychol. 56(1), 1–12 (2012)
12. Tan, P.N., et al.: Introduction to Data Mining. Pearson Education India, Upper Saddle River (2006)
13. Lee, C.-P., Lin, C.-J.: Large-scale linear ranksvm. Neural Comput. 26(4), 781–817 (2014)
14. Liu, T.-Y.: Learning to rank for information retrieval. Found. Trends Inf. Retr. 3(3), 225–331 (2009)
15. Donmez, P., Carbonell, J.G.: Active sampling for rank learning via optimizing the area under the ROC curve. In: Boughanem, M., Berrut, C., Mothe, J., Soule-Dupuy, C. (eds.) ECIR 2009. LNCS, vol. 5478, pp. 78–89. Springer, Heidelberg (2009). doi:10.1007/978-3-642-00958-7_10
16. Qian, B., Li, H., Wang, J., Wang, X., Davidson, I.: Active learning to rank using pairwise supervision. In: SIAM International Conference Data Mining, pp. 297–305. SIAM (2013)
17. Thalhammer, A., Rettinger, A.: PageRank on Wikipedia: towards general importance scores for entities. In: Sack, H., Rizzo, G., Steinmetz, N., Mladenić, D., Auer, S., Lange, C. (eds.) ESWC 2016. LNCS, vol. 9989, pp. 227–240. Springer, Cham (2016). doi:10.1007/978-3-319-47602-5_41

18. Kleinberg, J.M.: Authoritative sources in a hyperlinked environment. JACM **46**(5), 604–632 (1999)
19. Cabitza, F., Locoro, A.: Questionnaires in the design and evaluation of community-oriented technologies. Int. J. Web-Based Commun. **13**(1), 4–35 (2017)
20. Gwet, K.L.: Handbook of inter-rater reliability: the definitive guide to measuring the extent of agreement among raters. Advanced Analytics, LLC (2014)
21. Heim, P., Hellmann, S., Lehmann, J., Lohmann, S., Stegemann, T.: RelFinder: revealing relationships in RDF knowledge bases. In: Chua, T.-S., Kompatsiaris, Y., Mérialdo, B., Haas, W., Thallinger, G., Bailer, W. (eds.) SAMT 2009. LNCS, vol. 5887, pp. 182–187. Springer, Heidelberg (2009). doi:10.1007/978-3-642-10543-2_21
22. Fang, L., Sarma, A.D., Yu, C., Bohannon, P.: Rex: explaining relationships between entity pairs. Proc. VLDB **5**(3), 241–252 (2011)
23. Tiddi, I., d'Aquin, M., Motta, E.: Learning to assess linked data relationships using genetic programming. In: Groth, P., Simperl, E., Gray, A., Sabou, M., Krötzsch, M., Lecue, F., Flöck, F., Gil, Y. (eds.) ISWC 2016. LNCS, vol. 9981, pp. 581–597. Springer, Cham (2016). doi:10.1007/978-3-319-46523-4_35

Synthesizing Knowledge Graphs for Link and Type Prediction Benchmarking

André Melo[(✉)] and Heiko Paulheim

University of Mannheim, B6 26, 68159 Mannheim, Germany
{andre,heiko}@informatik.uni-mannheim.de
http://dws.informatik.uni-mannheim.de

Abstract. Despite the growing amount of research in link and type prediction in knowledge graphs, systematic benchmark datasets are still scarce. In this paper, we propose a synthesis model for the generation of benchmark datasets for those tasks. Synthesizing data is a way of having control over important characteristics of the data, and allows the study of the impact of such characteristics on the performance of different methods. The proposed model uses existing knowledge graphs to create synthetic graphs with similar characteristics, such as distributions of classes, relations, and instances. As a first step, we replicate already existing knowledge graphs in order to validate the synthesis model. To do so, we perform extensive experiments with different link and type prediction methods. We show that we can systematically create knowledge graph benchmarks which allow for quantitative measurements of the result quality and scalability of link and type prediction methods.

Keywords: Knowledge graphs · Link prediction · Type prediction · Benchmarking

1 Introduction

Benchmarking is an important way of evaluating and comparing different methods for a given task. Having datasets with various characteristics is a crucial part of designing good benchmarking tests, allowing to thoroughly analyze the performance of a method under various conditions.

With the growing adoption and usage of Web-scale knowledge graphs, the data quality of those graphs has drawn some attention, and methods for improving the data quality, e.g., by predicting missing types and links, have been proposed. While there are a few benchmarking datasets for other tasks in the Semantic Web community, like SPARQL query performance [18,30], ontology matching [6], entity linking [8], machine learning [28], and question answering [15], benchmarks for the task of type and link prediction are still missing. In contrast, the majority of approaches is only tested on one or few datasets, most prominently different versions of DBpedia, which makes it difficult to compare the approaches [24]. Thus, it would be desirable to have benchmarking datasets

© Springer International Publishing AG 2017
E. Blomqvist et al. (Eds.): ESWC 2017, Part I, LNCS 10249, pp. 136–151, 2017.
DOI: 10.1007/978-3-319-58068-5_9

with different characteristics, such as the number of entities, relation assertions, number of types, the taxonomy of types, the density of the knowledge graph, etc. Furthermore, it would be interesting to be able to have some control over these characteristics, vary them if necessary, and generate a knowledge graph following defined settings.

Generating data artificially for evaluation purposes is not something new. Data synthesizers have been used in various research areas. IBM Quest Synthetics Data Generator[1] is probably the most famous of them. It generates transaction tables for frequent pattern mining. There are also generators, e.g., for spatial-temporal data [31], clustering and outlier detection [5], data for information discovery and analysis systems [29], and high-dimensional datasets [1].

The overall goal is to synthesize a multitude of knowledge graphs to design benchmarkings for the tasks of link and type prediction. A first step to achieve this goal is to be able to replicate already existing datasets. In this paper, we propose knowledge graph models, and a synthesis process that is able to generate data based on the models. To show the validity of the synthesis approach, our main goal is to replicate the performance measures obtained for evaluation measures when performing link and type prediction with various state-of-the-art methods. We want to minimize the distance between the original dataset and the synthesized replicas for these measures, and also preserve method rankings. In our case, we select five methods for each task.

In order to be able to run systematic scalability tests with different approaches, we also explore the possibility to generate replicas of different sizes (number of entities and facts). The results should be preserved when varying the size of the synthesized data.

The rest of this paper is structured as follows. Section 2 discusses related work. We introduce our model for knowledge graphs in Sect. 3, and discuss the synthesis approach in Sect. 4. In a set of experiments, we discuss the validity of our approach in Sect. 5, and conclude with a summary and an outlook on future work.

2   Related Work

There have been works which address the synthesis of knowledge graphs for benchmarking purposes. However, most efforts were focused on synthesizing A-box assertions for a specific T-box. Moreover, these works generate benchmarking datasets for different tasks in the Semantic Web, but none of them focus on link and type prediction.

Guo et al. [12] propose a method for benchmarking Semantic Web knowledge base systems on large OWL applications. They present the Lehigh University Benchmark (LUBM), which has an ontology for the university domain and includes the Univ-Bench artificial data generator (UBA), as well as a set of queries and performance measures for evaluation. The data generator synthesizes A-boxes of arbitrary size to evaluate scalability. The data contains information

---

[1] http://www.philippe-fournier-viger.com/spmf/datasets/IBM_Quest_data_generator.zip.

about universities, which are artificially created based on some predefined restrictions, e.g. minimum and maximum number of departments, student/faculty ratio, which are based on arbitrary defined ranges.

SP2Bench [30] is a SPARQL performance benchmark based on DBLP data. It features a data generator, which can create arbitrarily large datasets. Similarly to UBA, the authors synthesize the A-box based on an existing T-box, in this case the DBLP ontology, and a dataset specific model used to generate the synthetic data. The model uses logistic curves and simple intervals to describe characteristics of the DBLP data, such as the number and types of publications, distribution of citations, and level of incompleteness over years.

Morsey et al. [18] created a SPARQL query benchmark based on DBpedia to evaluate knowledge base storage systems. They gather a set of real world queries extracted with query log mining, and run them on datasets of different sizes generated from DBpedia. Their "data generation" process consists of sampling the original DBpedia dataset and changing the entities namespace. Two sampling methods are considered: *rand*, which basically randomly selects a fraction of the triples, and *seed*, which first sample a subset of the classes, then instances of these classes are also sampled and added to a queue. This process is iterated until the target dataset size is reached.

Linked Data Benchmark Council (LDBC) [2] developed the social network benchmark (SNB) and the semantic publishing benchmark (SPB). The SNB which includes a data generator that enables the creation of synthetic social network data representative of a real social network. The data generated includes properties occurring in real data, e.g. irregular structure, structure/value correlations and power-law distributions. The benchmark covers main aspects of social network data management, including interactive, business intelligence and graph analytics workload. The SPB is similar to the SNB, but it concerns the scenario of a media organization that maintains RDF descriptions of its catalogue of creative works.

3   Knowledge Graph Model

We define a knowledge graph $\mathcal{K} = (\mathcal{T}, \mathcal{A})$, where $\mathcal{T}$ is the T-box and $\mathcal{A}$ is the A-box containing relations assertions $\mathcal{A}_R$ and type assertions $\mathcal{A}_C$. We define N_C as the set of concepts (types), N_R as the set of roles (object properties) and N_I as the set of individuals (entities which occur as subject or object in relations). The set of relation assertions is defined as $\mathcal{A}_R = \{p(s,o)|p \in N_R \wedge s,o \in N_I\}$ and the set of type assertion as $\mathcal{A}_C = \{C(s)|C \in N_C \wedge s \in N_I\}$.

In our proposed model, we learn the joint distribution of types over instances. To that end, we compute $P(T)$, which is the probability of an individual having a set of types T. We define the set of types $\tau(s)$ of a given individual s as $\tau(s) = \{C|C(s) \in \mathcal{A}_C\}$ and the set of individuals of given set of types T as $E_T = \{s|\tau(s) = T\}$. This is important because most knowledge graphs allow instances to have multiple types, and by modeling the distribution of instances over sets of types we can capture the dependencies between types, which is relevant for the problem described in this paper. It is important to notice that, e.g., Arnold_Schwarzenegger with set of types $T = \{$Actor, Politician, BodyBuilder$\}$ is not considered to belong to $\{$Actor, Politician$\}$ when computing the distributions. With that, we make sure that $\sum_{T \in \mathcal{P}(N_C)} P(T) = 1$, where $\mathcal{P}(N_C)$ is the powerset of types containing all possible combinations of types

$$P(T) = \frac{|\{s | \tau(s) = T\}|}{|N_I|} \tag{1}$$

We also model the joint distribution of relations and the type set of their subject (T_s) and object (T_o), which we call $P(r, T_s, T_o)$. This distribution allows us to model how different types are related, and capture domain and range restrictions of relations in a fine grained way. For example, we can model not only that the relation playsFor has domain Athlete and range SportsTeam, but also how athletes are distributed over more specific types (e.g., FootballPlayer, BasketballPlayer, etc.) and how teams are distributed over subclasses of SportsTeam (e.g., FootballTeam, BasketballTeam, etc.), and most importantly, we can model that FootballPlayer playsFor FootballTeam and BasketballPlayer playsFor BasketballTeam.

We model the joint distribution $P(r, T_s, T_o)$ with the chain rule (3). We decompose it into the distribution of relations over facts $P(r)$, conditional distributions of subject type set given relation $P(T_s|r)$ and a conditional distributions of object type set given subject type set and relation $P(T_o|r, T_s)$.

$$P(r, T_s, T_o) = P(r)P(T_s|r)P(T_o|r, T_s) \tag{2}$$

$$P(r) = \frac{|\{p(s, o) \in \mathcal{A}_R | p = r\}|}{|\mathcal{A}_R|} \tag{3}$$

$$P(T_s|r) = \frac{|\{p(s, o) \in \mathcal{A}_R | p = r \wedge \tau(s) = T_s\}|}{|\{p(s, o) \in \mathcal{A}_R | p = r\}|} \tag{4}$$

$$P(T_o|r, T_s) = \frac{|\{p(s, o) \in \mathcal{A}_R | p = r \wedge \tau(s) = T_s \wedge \tau(o) = T_o\}|}{|\{p(s, o) \in \mathcal{A}_R | p = r \wedge \tau(s) = T_s\}|} \tag{5}$$

It is important to note that in case there are inconsistencies in the knowledge graph, such as domain/range violations or the assignment of inconsistent types, they are also captured in the distribution $P(r, T_s, T_o)$, and can be later replicated with their respective probabilities.

Besides the probability distributions of types and relations, individuals also follow a certain probability distribution, and not all relations have a uniform distribution w.r.t. their subjects and objects. In many cases, when selecting the individuals from E_{T_s} and E_{T_o}, there might be some bias which we should take into account. For instance, if we select $r =$ livesIn, $T_s = \{$Person$\}$ and $T_o = \{$Country$\}$, we should not select the individual for Country based on an uniform distribution. The distribution should be biased towards more populous countries, e.g., the probability of selecting China should be much higher that of Vatican. At the same time, for the $r =$ capitalOf with $T_o = \{$Country$\}$, the distribution of countries should be uniform since all the countries are equally likely to have a capital.

After selecting the relation r and type set of subject T_s and object T_o, we then need to select the subject and object individuals. Since in our synthesis process we first generate the individuals and their type assertions and then generate the relations assertions, there exist a limited number of individuals belonging to a given type set T which we define as $n_T = |E_T|$.

Following those considerations, we compute the conditional distributions of subject and object individuals given a relation and type set of subject and object, which we call $P(e|r, T_s)$ and $P(e|r, T_o)$, respectively. To that end, we count the occurrences of subject individuals for all relations r and subject type set T_s, and occurrences of object individuals for all r and T_o. We then sort the individuals by frequency in descending order and fit a distribution model.

We need to select an instance from a finite set E_T, and we should be able to vary the size n_T in order to be able to scale the knowledge base up and down. Therefore we consider the use of uniform and exponential truncated distributions (c.f. Eqs. 6 and 7).

$$f(x,b) = \begin{cases} \frac{1}{b} & \text{, if } 0 \le x < b \\ 0 & \text{, otherwise} \end{cases} \quad (6) \qquad f(x,b) = \begin{cases} \frac{e^{-x}}{1-e^{-b}} & \text{, if } 0 \le x < b \\ 0 & \text{, otherwise} \end{cases} \quad (7)$$

In truncated distributions, occurrences are limited to values which lie inside a given range. In the case of Eqs. 6 and 7, that interval is $0 \le x < b$. It is important to use truncated functions, because when synthesizing relation assertions and selecting the individual for a given type, we can set $b = n_T$, and select an individual amongst the limited number of individuals that have the required type.

All distributions presented earlier in this section can effectively replicate some characteristics of a knowledge graph, such as in and out degree and density of the graph, however, they are not able to replicate more complex patterns involving paths in the graph. An example of such pattern in a knowledge graph containing data about families is that people who are married to the parent of a given child are also the parent of that child with some confidence. This pattern can be represented with the horn rule below.

$$\texttt{marriedTo(x,y)} \wedge \texttt{childOf(x,z)} \Rightarrow \texttt{childOf(y,z)} \quad [conf = 0.93]$$

Horn rules are basis of inductive logic programming (ILP) systems, such as ALEPH [19], WARMR [11], DL-Learner [14], and AMIE [9]. There are also ILP extensions with probabilistic methods [27] and that can efficiently handle numerical attributes [17]. We choose to use AMIE especially because of its better scalability in comparison to ALEPH and WARMR.

As most ILP systems, AMIE uses techniques to restrict the search space. AMIE mines only *closed* and *connected* rules. A rule is connected if all of its atoms are connected transitively to every other atom of the rule, and two atoms are connected if they share a variable or a constant. A rule is closed if every variable in the rule appears at least twice. Such rules do not predict merely the existence of a fact (e.g. diedIn(x,y) $\Rightarrow$ wasBornIn(x,z), which is connected rule, but not closed), but the concrete arguments for it (e.g. diedIn(x,y) $\Rightarrow$ wasBornIn(x,y)).

We use the horn rules learned by AMIE in our KB model in order to represent more complex patterns and use their associated PCA (partial close-world assumption) confidence value in the synthesis. In our model, we are able to ensure various relation characteristics. The RDF Schema domain and range restrictions can be ensured by the joint distribution $P(r, T_s, T_o)$. The horn rules can model symmetric, transitive, equivalent, and inverse properties.

To cover even more complex schemas, we additionally learn functionality, inverse functionality and non-reflexiveness from the data. All relations which do not have any same individual as both subject and object of a triple are considered non-reflexive, all relations with object cardinality of 1 are considered functional, and with subject cardinality of 1 are considered inverse functional. Learning these characteristics from data allows us to detect relations which might not have been conceived as, or not defined as such in the schema, but which in the available data present the characteristics. For instance, a dataset with the childOf relation, which is not functional, might contain data about people which have exclusively one child, and with our approach we ensure this characteristic is replicated.

4   Synthesis Process

Algorithm 1 summarizes the process of synthesizing a knowledge graph. As input, it uses the probability distributions $P(T)$, $P(r, T_s, T_o)$, $P(e|r, T_s)$, and $P(e|r, T_o)$, a set of horn rules $\mathcal{H}$, as well as the desired number of individuals n_e and relation assertions n_f to be synthesized.

The function VERIFY_TRIPLE first verifies if the exact same triple is already present in the synthesized KG. Then it checks whether functionality, inverse functionality, and non-reflexiveness are satisfied. That is, it verifies if there is no assertion with the given subject already present in the KG for functional relations, no assertion with the given object for inverse functional relations, and the given subject and object are different individuals for non-reflexive relations.

The function CHECK_HORN_RULES ensures that the patterns learned with the horn rules are replicated in the synthesized data. It checks if a newly synthesized fact triggers any of the learned horn rules. If a rule is triggered, the rule will produce a new fact with a probability equal to that of its confidence. The new facts produced by rules also need to be checked against the horn rules again, which means that the CHECK_HORN_RULES function is called recursively until it does not produce any new facts.

The function UPDATE_DISTRIBUTION makes sure that the original distribution $P(r, T_s, T_o)$ is not distorted by the production of new facts from horn rules, which may not follow $P(r, T_s, T_o)$. Therefore, it is necessary to adjust the joint distribution in order to compensate this effects. We do that by simply keeping counts for the relations, subject and object type sets, and based on the number of facts to be synthesized and the distribution of already synthesized facts we can adjust $P(r, T_s, T_o)$.

Algorithm 1. Knowledge base synthesis process

```
 1: function GEN_KB(n_e, n_f, P(T), P(r), P(r, T_s, T_o), H)
 2:     A ← ∅                                              ▷ Create empty A-Box
 3:     E ← {}                                    ▷ Map of type sets and their entities
 4:     for i ← 1 to n_e do                                ▷ synthesize entities
 5:         T_i ← randomly choose from P(T)
 6:         E[T_i] ← E[T_i] ∪ {e_i}
 7:         for C ∈ T_i do
 8:             A ← A ∪ {C(e_i)}
 9:         end for
10:     end for
11:     i ← 0
12:     while i < n_f do                          ▷ synthesize relation assertions
13:         r_i, T_{s_i}, T_{o_i} ← randomly choose from P(r, T_s, T_o)    ▷ use chain rule
14:         s_i ← SELECT_ENTITY(E[T_{s_i}], P(e|r_i, T_{s_i}))
15:         o_i ← SELECT_ENTITY(E[T_{o_i}], P(e|r_i, T_{o_i}))
16:         if VERIFY_TRIPLE(s_i, r_i, o_i) then
17:             A ← A ∪ {r_i(s_i, o_i)}
18:             CHECK_HORN_RULES(A, (s_i, r_i, o_i), H)
19:             UPDATE_DISTRIBUTION(P(r, T_s, T_o))
20:             i ← i + 1
21:         end if
22:     end while
23:     return A
24: end function
```

Another detail not shown in Algorithm 1 is the use of a pool of subjects for functional and pool of objects for inverse functional relations. We do that in order to avoid generating facts which violate the functionality and inverse functionality restrictions. If no pools are considered, the probability of generating violating facts for a given relation increases linearly with the number of already existent facts. With the pools, all individuals of a given type are initially in the pool, and whenever an individual is picked to generate a new fact, this individual is removed from the pool and cannot be picked again, therefore preventing the violations.

In the synthesis process some characteristics can be easily changed. Noise can be introduced by smoothing the distribution $P(r, T_s, T_o)$, making the probability for invalid combinations of relations, subject and object types non-zero. The density of the knowledge graph can be altered by modifying the ratio n_f/n_e between number of facts and individuals. It is possible to change the scale of the synthetic knowledge graphs by simply multiplying the original number of individuals n_e and facts n_f by a constant. That is, assuming that the number of relations in the knowledge graph are linear, i.e., the number of relation assertions grows linearly with the number of individuals.

However, some knowledge graphs might have relations which are quadratic, e.g. owl:differentFrom that indicates individuals that are not the same. Therefore, for the quadratic relations of a knowledge base, we need to scale the number of relation assertions quadratically with the number of individuals. This kind of relations are rather rare, and they can be difficult to automatically detect. We use a simple heuristic based on thresholds for the average number of different objects per subject and different subjects per object. If both thresholds are reached, we assume the relation to be quadratic.

One important characteristic is that the synthesis process is based on pseudo random number generators (PRNG), therefore, the process is deterministic and identical datasets can be generated if the same seed is used. By using different seeds, it is also possible to generate different datasets from the same model and with similar characteristics, allowing us to test the stability of methods.

5   Experiments

The link prediction task consists of predicting the existence (or probability of correctness) of edges in the graph (i.e., triples). This is important since existing knowledge graphs are often missing many facts, and some of the edges they contain are incorrect. Nickel et al. [20] present a review of multirelational models, many of which have been used for the link prediction task. In this paper, we select five popular methods to be used in our experiments: Path Ranking algorithm [13], SDValidate [26], Holographic embeddings (HolE) [21], Translation embeddings (TransE) [4] and RESCAL [22]. In our experiments, we evaluate the prediction of relation assertions only. All the measurements reported were obtained using 5-fold cross-validation. The test set consists of the 20% of positive positive triples selected in the cross-validation, plus negative examples. There are the same number of positive and negative examples in the test set, and the negative examples are generated by corrupting each of the positive triples following the method described by Bordes et al. [4].

Type prediction can be considered a subtask of link prediction where we are interested on prediction links for the relation rdf:type. There are several type prediction approaches which rely on external features [3,10,23,32], however, in this paper, we concentrate on methods which rely on features extracted from the knowledge graph. The methods used in the experiments are SDType [25] and SLCN [16], as well as the state-of-art multilabel classifiers MLC4.5 [7], MLP [33] and MLkNN [34] – multilabel versions of decision tree, multilayer perceptron and k-nearest neighbors – with ingoing and outgoing links used as features as described in [16].

As input knowledge graphs, we use Wikidata, DBpedia (2015-10), and NELL. We use the following smaller domain specific datasets: Thesoz[2], Semantic Bible[3] AIFB portal[4], Nobel Prize[5] and Mutagenesis. We also select four of the largest

---

[2] http://www.gesis.org/fileadmin/upload/dienstleistung/tools_standards/thesoz_skos_turtle.zip.

[3] http://www.semanticbible.com/.

[4] http://www.aifb.kit.edu/web/Web_Science_und_Wissensmanagement/Portal.

[5] http://www.nobelprize.org/nobel_organizations/nobelmedia/nobelprize_org/developer/manual-linkeddata/terms.html.

Table 1. Statistics about the datasets used in the experiments

Dataset	Entities	Types	Rels	Type ass.	Relation ass.	Density
Wikidata	19060716	474	482	40198183	18955236	$1.082 \cdot 10^{-10}$
DBpedia	4940352	1027	646	31521734	14747048	$9.353 \cdot 10^{-10}$
NELL	1475674	276	248	5565472	174621	$3.233 \cdot 10^{-10}$
AIFB	27100	63	82	59613	59349	$9.855 \cdot 10^{-7}$
Mutagenesis	14157	91	4	48111	26533	$3.310 \cdot 10^{-5}$
SemanticBible	789	71	31	2563	2482	$1.286 \cdot 10^{-4}$
Thesoz	48540	10	16	109960	275430	$7.306 \cdot 10^{-6}$
NobelPrize	10013	23	18	19506	30148	$1.671 \cdot 10^{-5}$
ESWC2015	1285	16	25	1285	4062	$9.840 \cdot 10^{-5}$
ISWC2013	2548	20	39	2545	9992	$3.946 \cdot 10^{-5}$
WWW2012	3836	22	43	3907	15406	$2.435 \cdot 10^{-5}$
LREC2008	3502	7	24	3502	16514	$5.611 \cdot 10^{-5}$

conference datasets from the Semantic Web dog food corpus[6], i.e., LREC2008, WWW2012, ISWC2013, and ESWC2015. Some relevant statistics about the datasets used in the experiments are shown in Table 1.

For every input KG, we synthesize replicas of three different sizes increased by factors of 10. For smaller datasets we also scale the replicas up. On the Semantic Web dog food datasets we synthesize replicas of sizes 10%, 100% and 1000%. For large datasets we scale the replicas down (DBpedia and Wikidata replicas are of sizes 0.01%, 0.1% and 1%, and the remaining datasets 1%, 10% and 100%).

We use the scikit-kge[7] implementation of HolE, TransE and RESCAL, and the scikit-learn implementation of MLkNN, MLC4.5 and MLP. We implemented the remaining methods ourselves. The proposed synthesis process code is available to download.[8]

The evaluation measures used in the link experiments are the area under the precision-recall curve (PR AUC) and area under the ROC curve (ROC AUC). For the type prediction experiments we use micro-averaged F_1-score and accuracy. We compute the distance of these evaluation measures between the results on the original datasets, and their synthetic replicas. In order to compare the ranking of methods, we use the Spearman-ρ rank correlation coefficient. All the results reported in this paper were obtained with 5-fold cross-validation.

In order to evaluate how the different parts of the proposed knowledge base model affect the results on link and type prediction tasks, we use 6 different models in our evaluation: M1, M2, M3, e(M1), e(M2) and e(M3):

---

[6] http://data.semanticweb.org/dumps/conferences/.
[7] https://github.com/mnick/scikit-kge.
[8] https://github.com/aolimelo/kbgen.

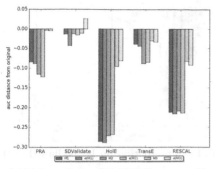

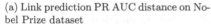

(a) Link prediction PR AUC distance on Nobel Prize dataset

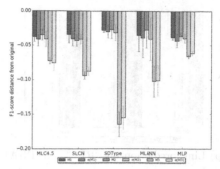

(b) Type prediction F_1-score distance on
Wikidata

Fig. 1. Distances of performance measures to original datasets

- M1 is the simplest version, which considers only the distributions $P(T)$ and $P(r, T_s, T_o)$. The bias to selection of individuals is not considered, and individuals are always selected from an uniform distribution. No relation characteristics (apart from domain and range restrictions covered by $P(r, T_s, T_o)$) are considered.
- M2 is M1 plus functionality, inverse functionality and non-reflexiveness of relations.
- M3 is M2 plus the horn rules learned with AMIE.
- The models e(Mi) are the model Mi plus the biases to selection of individuals $P(e|r, T_s)$ and $P(e|r, T_o)$.

We use AMIE with its default parameter settings (i.e., no rules with constants, maximum rule length = 3, confidence computed with PCA, minimum support = 100 examples, minimum head coverage = 0.01).

We use PRA with maximum path length of 3 for all datasets. For HolE, TransE and RESCAL we learn embeddings with 20 dimensions and maximum of 100 epochs. While this may not be the optimal settings for most datasets, we consistently use the same settings throughout all of our experiments, since our aim is not to achieve optimal results, but to show that the benchmark synthesis works as desired.

Figure 1a shows an example of PR AUC distance on link prediction from for the Nobel Prize datasets between original and replica (100% size) with the 5 selected methods. It is clear that the use of horn rules significantly improves the results, as M3 and e(M3) performs better than the other methods, except from SDValidate, which relies exclusively on distributions of relations and object types and does not exploit more complex path patterns.

Figure 1b shows an example of F_1-score distance on type prediction for Wikidata between original and replica (0.1% size). It is noticcable that horn rules do not improve the results, as M1, M2, e(M1) and e(M2) perform better than M3 and e(M3). This is explained by the fact that most of the evaluated type prediction methods rely solely on ingoing and outgoing links of entities. Moreover, as

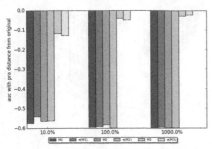   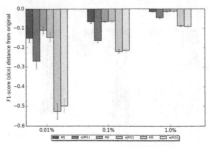

(a) Link prediction on LREC2008 (PRA)   (b) Type prediction on DBpedia (SLCN)

Fig. 2. Effect of scaling the replica sizes up and down

Table 2. Summary of the link prediction results

	PR AUC						ROC AUC					
	M1	e(M1)	M2	e(M2)	M3	e(M3)	M1	e(M1)	M2	e(M2)	M3	e(M3)
ρ_{all}	0.527	0.643	0.567	0.607	0.643	**0.653**	0.647	0.613	**0.657**	0.613	0.610	0.577
ρ_{large}	0.640	0.740	0.590	0.580	0.730	**0.800**	0.650	0.610	0.640	0.630	**0.670**	0.620
d_{all}	0.243	0.247	0.247	0.245	**0.112**	0.115	0.231	0.230	0.231	0.231	**0.109**	0.111
d_{large}	0.215	0.228	0.216	0.219	**0.082**	0.089	0.211	0.215	0.208	0.211	**0.087**	0.095

explained in Sect. 3, horn rules can disturb the original distribution $P(r, T_s, T_o)$, which is crucial for the replication of ingoing and outgoing links.

Tables 2 and 3 show a summary of the results obtained over all datasets for type prediction and link prediction, respectively. The values with subscript *all* report the average of the results over all different sizes of replicas, while those with subscript *large* report the averages over the largest size of replicas only. We do that because different models, especially M3 and e(M3), perform relatively worse than others for smaller replica sizes (c.f. Fig. 2), and we also want to know how the models perform when ruling out this effect.

The results of Table 2 indicate that in terms of distance, M3 is the best method overall, however, when it comes to preserving the rankings, the results become more mixed. It is clear that introducing the horn rules does have a positive effect on the model, especially for the distances which are reduced to less than half of that of other models. In Table 3 we can see that M2 is the best overall in terms of distance for both PR and ROC AUC, while for the rankings, the use of horn rules again have a positive impact with M3 being the best method overall. The link prediction results were reported for all datasets apart from DBpedia and Wikidata. Because of the large size of these two datasets and the complexity of the approaches, the experiments did not finish in less than a week.

We also perform the Nemenyi test in order to find how significant the differences of the evaluated models is, both in terms of distance and ranking. Figure 3 shows the critical distance diagrams. For the distances d the models on the left side are the best performers, since lower distances are desired, while for Spearman's rank correlations ρ the models on the right side are the best performers, since higher correlations are desired.

Table 3. Summary of the type prediction results

	F_1-score						Accuracy					
	M1	e(M1)	M2	e(M2)	M3	e(M3)	M1	e(M1)	M2	e(M2)	M3	e(M3)
ρ_{all}	0.208	0.221	0.195	0.259	**0.362**	0.265	0.290	0.334	0.315	0.343	**0.406**	0.307
ρ_{large}	0.343	0.273	0.251	0.400	**0.456**	0.410	0.470	0.420	0.357	0.498	**0.502**	0.433
d_{all}	0.086	0.098	**0.082**	0.083	0.131	0.130	0.061	0.064	**0.057**	0.061	0.065	0.066
d_{large}	0.059	0.065	**0.055**	0.057	0.083	0.084	0.056	0.060	**0.054**	0.057	0.061	0.062

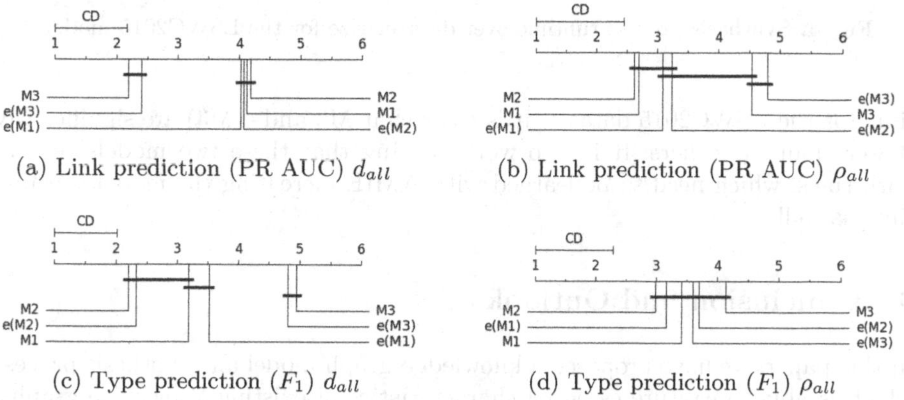

(a) Link prediction (PR AUC) d_{all}

(b) Link prediction (PR AUC) ρ_{all}

(c) Type prediction (F_1) d_{all}

(d) Type prediction (F_1) ρ_{all}

Fig. 3. Nemenyi critical distance diagrams for link and type prediction

In Fig. 3a we can see that PR AUC distances on link prediction between the models with horn rules (M3 and e(M3)) and the others is very significant, while the differences in terms of Spearman-ρ from Fig. 3b are closer to the critical distance (CD). We can also observe that the difference between M3 and e(M3) is not significant, indicating that the use of bias to selection of instances does not have a great impact. One possible explanation for that is the fact that, in order to simplify our model and abstract from specific instances, we assume that, for a given type set, the most frequent instances are always the same. That is, if we consider the type set {Country} as object of livesIn and beatifiedPlace, we assume that the most frequent country in both cases is the same individual, while in reality the most frequent country for livesIn would be China and for beatifiedPlace Italy. Since the computation of the bias can be very expensive, especially for larger datasets with high number of types and individuals, M3 would be a more reasonable choice than e(M3).

When analyzing Fig. 3c, we notice that, in terms of F_1-score distance, the M2, e(M2) and M1 are not significantly different from each other, and the use of horn rules has a significant negative effect. The Spearman-ρ from Fig. 3d values are very close to each other, without any significant difference between the evaluated models.

We illustrate the difference in runtime for the synthesis processes with different methods with Fig. 4. The plot shows the number of facts generated over

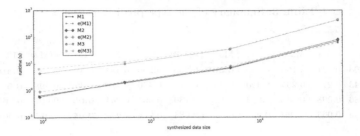

Fig. 4. Synthesis process runtime over dataset size for the ESWC2015 model

time for the ESWC2015 dataset. It is clear that M3 and e(M3) are significantly slower than the others. It is also worth noting that these two models require horn rules, which need to be learned with AMIE increasing the model learning time as well.

6   Conclusion and Outlook

In this paper, we have proposed a knowledge graph model and synthesis process which is able to capture essential characteristics of existing knowledge graphs, which allows us to create replicas of those graphs at different scales.

Extensive experiments comparing the replicas and original datasets in the link and type prediction tasks were conducted. We have performed evaluations with five different methods for each tasks and comparisons of distances and methods rankings between replicas and original datasets. Overall, the model M3 was the best performer, and the use of horn rules significantly improved the results. The use of a bias to selection of subject and object individuals did not show any significant improvement. In general, we recommend the use of M3, unless the objective is to replicate the results of type prediction on a single methods, without performing any comparisons. In that case M2, which does not include horn rules, would be the best option.

In the future, we intend to start synthesizing knowledge graphs from scratch, which would involve the synthesis of whole schemas. We plan to create a system which enables users to synthesize data based on a set of parameters that gives control on important characteristics of a knowledge base, such as number of entities, types, relations, assertions of types and relations, density, connectivity. Finally, we want to synthesize a set of knowledge bases of different characteristics to create a larger collection of benchmarks for link prediction and type prediction.

Acknowledgements. The work presented in this paper has been partly supported by the Ministry of Science, Research and the Arts Baden-Württemberg in the project SyKo^{2}W^2 (Synthesis of Completion and Correction of Knowledge Graphs on the Web).

References

1. Albuquerque, G., Löwe, T., Magnor, M.: Synthetic generation of high-dimensional datasets. IEEE Trans. Vis. Comput. Graph. **17**(12), 2317–2324 (2011). (TVCG, Proc. Visualization / InfoVis)
2. Angles, R., Boncz, P., Larriba-Pey, J., Fundulaki, I., Neumann, T., Erling, O., Neubauer, P., Martinez-Bazan, N., Kotsev, V., Toma, I.: The linked data benchmark council: a graph and rdf industry benchmarking effort. SIGMOD Rec. **43**(1), 27–31 (2014)
3. Palmero Aprosio, A., Giuliano, C., Lavelli, A.: Automatic expansion of DBpedia exploiting wikipedia cross-language information. In: Cimiano, P., Corcho, O., Presutti, V., Hollink, L., Rudolph, S. (eds.) ESWC 2013. LNCS, vol. 7882, pp. 397–411. Springer, Heidelberg (2013). doi:10.1007/978-3-642-38288-8_27
4. Bordes, A., Usunier, N., Garcia-Duran, A., Weston, J., Yakhnenko, O.: Translating embeddings for modeling multi-relational data. In: Burges, C.J.C., Bottou, L., Welling, M., Ghahramani, Z., Weinberger, K.Q. (eds.) Advances in Neural Information Processing Systems, vol. 26, pp. 2787–2795. Curran Associates, Inc. (2013)
5. Chawla, S., Gionis, A.: k-means-: a unified approach to clustering and outlier detection. In: Proceedings of the 13th SIAM International Conference on Data Mining, Austin, Texas, USA, pp. 189–197. SIAM (2013)
6. Cheatham, M., Dragisic, Z., Euzenat, J., Faria, D., Ferrara, A., Flouris, G., Fundulaki, I., Granada, R., Ivanova, V., Jiménez-Ruiz, E., et al.: Results of the ontology alignment evaluation initiative 2015. In: 10th ISWC Workshop on Ontology Matching (OM), pp. 60–115 (2015)
7. Clare, A., King, R.D.: Knowledge discovery in multi-label phenotype data. In: Raedt, L., Siebes, A. (eds.) PKDD 2001. LNCS (LNAI), vol. 2168, pp. 42–53. Springer, Heidelberg (2001). doi:10.1007/3-540-44794-6_4
8. van Erp, M., Mendes, P., Paulheim, H., Ilievski, F., Plu, J., Rizzo, G., Waitelonis, J.: Evaluating entity linking: an analysis of current benchmark datasets and a roadmap for doing a better job. In: Proceedings of the Language Resources and Evaluation Conference, ELRA (2016)
9. Galárraga, L.A., Teflioudi, C., Hose, K., Suchanek, F.M.: AMIE: association rule mining under incomplete evidence in ontological knowledge bases. In: WWW 2013, Rio de Janeiro, Brazil, pp. 413–422. ACM (2013)
10. Gangemi, A., Nuzzolese, A.G., Presutti, V., Draicchio, F., Musetti, A., Ciancarini, P.: Automatic typing of DBpedia entities. In: Cudré-Mauroux, P., et al. (eds.) ISWC 2012. LNCS, vol. 7649, pp. 65–81. Springer, Heidelberg (2012). doi:10.1007/978-3-642-35176-1_5
11. Goethals, B., Bussche, J.: Relational association rules: getting WARMER. In: Hand, D.J., Adams, N.M., Bolton, R.J. (eds.) Pattern Detection and Discovery. LNCS (LNAI), vol. 2447, pp. 125–139. Springer, Heidelberg (2002). doi:10.1007/3-540-45728-3_10
12. Guo, Y., Pan, Z., Heflin, J.: LUBM: a benchmark for owl knowledge base systems. Web Semant. **3**(2–3), 158–182 (2005)
13. Lao, N., Cohen, W.W.: Relational retrieval using a combination of path-constrained random walks. Mach. Learn. **81**(1), 53–67 (2010)
14. Lehmann, J.: Dl-learner: learning concepts in description logics. J. Mach. Learn. Res. **10**, 2639–2642 (2009)

15. Lopez, V., Unger, C., Cimiano, P., Motta, E.: Evaluating question answering over linked data. Web Semant. Sci. Serv. Agents World Wide Web **21**, 3–13 (2013)
16. Melo, A., Paulheim, H., Völker, J.: Type prediction in RDF knowledge bases using hierarchical multilabel classification. In: Proceedings of the International Conference on Web Intelligence, Mining and Semantics, WIMS 2016, Nîmes, France, pp. 14:1–14:10 (2016)
17. Melo, A., Theobald, M., Völker, J.: Correlation-based refinement of rules with numerical attributes. In: Proceedings of the International Florida Artificial Intelligence Research Society Conference, FLAIRS, Pensacola, Florida (2014). http://www.aaai.org/ocs/index.php/FLAIRS/FLAIRS14/paper/view/7819
18. Morsey, M., Lehmann, J., Auer, S., Ngonga Ngomo, A.-C.: DBpedia SPARQL benchmark – performance assessment with real queries on real data. In: Aroyo, L., Welty, C., Alani, H., Taylor, J., Bernstein, A., Kagal, L., Noy, N., Blomqvist, E. (eds.) ISWC 2011. LNCS, vol. 7031, pp. 454–469. Springer, Heidelberg (2011). doi:10.1007/978-3-642-25073-6_29
19. Muggleton, S.: Learning from positive data. In: Muggleton, S. (ed.) ILP 1996. LNCS, vol. 1314, pp. 358–376. Springer, Heidelberg (1997). doi:10.1007/3-540-63494-0_65
20. Nickel, M., Murphy, K., Tresp, V., Gabrilovich, E.: A review of relational machine learning for knowledge graphs. Proc. IEEE **104**(1), 11–33 (2016)
21. Nickel, M., Rosasco, L., Poggio, T.A.: Holographic embeddings of knowledge graphs. CoRR abs/1510.04935 (2015)
22. Nickel, M., Tresp, V., Kriegel, H.P.: A three-way model for collective learning on multi-relational data. In: Proceedings of the 28th International Conference on Machine Learning (ICML 2011), pp. 809–816. ACM (2011)
23. Nuzzolese, A.G., Gangemi, A., Presutti, V., Ciancarini, P.: Type inference through the analysis of Wikipedia links. In: WWW 2012 Workshop on Linked Data on the Web, Lyon, France. CEUR Workshop Proceedings, vol. 937 (2012)
24. Paulheim, H.: Knowledge graph refinement: a survey of approaches and evaluation methods. Semant. Web **8**(3), 489–508 (2017)
25. Paulheim, H., Bizer, C.: Type inference on noisy RDF data. In: Alani, H., et al. (eds.) ISWC 2013. LNCS, vol. 8218, pp. 510–525. Springer, Heidelberg (2013). doi:10.1007/978-3-642-41335-3_32
26. Paulheim, H., Bizer, C.: Improving the quality of linked data using statistical distributions. Int. J. Semant. Web Inf. Syst. **10**(2), 63–86 (2014)
27. Raedt, L., Frasconi, P., Kersting, K., Muggleton, S. (eds.): Probabilistic Inductive Logic Programming. LNCS (LNAI), vol. 4911. Springer, Heidelberg (2008)
28. Ristoski, P., Vries, G.K.D., Paulheim, H.: A collection of benchmark datasets for systematic evaluations of machine learning on the semantic web. In: Groth, P., Simperl, E., Gray, A., Sabou, M., Krötzsch, M., Lecue, F., Flöck, F., Gil, Y. (eds.) ISWC 2016. LNCS, vol. 9982, pp. 186–194. Springer, Cham (2016). doi:10.1007/978-3-319-46547-0_20
29. Samadi, B., Cipolone, A., Lin, P.J., Xiao, R., Jeske, D.R., Holt, D., Rend, C., Cox, S.: Development of a synthetic data set generator for building and testing information discovery systems. In: Third International Conference on Information Technology, pp. 707–712 (2006)
30. Schmidt, M., Hornung, T., Lausen, G., Pinkel, C.: Sp2bench: a SPARQL performance benchmark. CoRR abs/0806.4627 (2008)
31. Theodoridis, Y., Nascimento, M.A.: Generating spatiotemporal datasets on the WWW. SIGMOD Rec. **29**(3), 39–43 (2000)

32. Yosef, M.A., Bauer, S., Hoffart, J., Spaniol, M., Weikum, G.: HYENA: hierarchical type classification for entity names. In: COLING 2012, 24th International Conference on Computational Linguistics, Proceedings of the Conference: Posters, Mumbai, India, pp. 1361–1370 (2012)
33. Zhang, M.L., Zhou, Z.H.: Multilabel neural networks with applications to functional genomics and text categorization. IEEE Trans. Knowl. Data Eng. **18**(10), 1338–1351 (2006)
34. Zhang, M.L., Zhou, Z.H.: ML-KNN: a lazy learning approach to multi-label learning. Pattern Recogn. **40**(7), 2038–2048 (2007)

Online Relation Alignment for Linked Datasets

Maria Koutraki[1,2]([✉]), Nicoleta Preda[3], and Dan Vodislav[4]

[1] FIZ Karlsruhe – Leibniz Institute for Information Infrastructure,
Karlsruhe, Germany
[2] Institute AIFB, Karlsruhe Institute of Technology (KIT), Karlsruhe, Germany
maria.koutraki@kit.edu
[3] Universtiy of Paris-Saclay, Versailles, France
nicoleta.preda@uvsq.fr
[4] ETIS CNRS, University of Cergy-Pontoise, Cergy-Pontoise, France
dan.vodislav@u-cergy.fr

Abstract. The large number of linked datasets in the Web, and their diversity in terms of schema representation has led to a fragmented dataset landscape. Querying and addressing information needs that span across disparate datasets requires the alignment of such schemas. Majority of *schema* and *ontology alignment* approaches focus exclusively on class alignment. Yet, relation alignment has not been fully addressed, and existing approaches fall short on addressing the dynamics of datasets and their size.

In this work, we address the problem of *relation alignment* across disparate linked datasets. Our approach focuses on two main aspects. First, *online relation alignment*, where we do not require full access, and sample instead for a minimal subset of the data. Thus, we address the main limitation of existing work on dealing with the large scale of linked datasets, and in cases where the datasets provide only *query access*. Second, we learn supervised machine learning models for which we employ various features or *matchers* that account for the diversity of linked datasets at the instance level. We perform an experimental evaluation on real-world linked datasets, DBpedia, YAGO, and Freebase. The results show superior performance against state-of-the-art approaches in schema matching, with an average relation alignment accuracy of 84%. In addition, we show that relation alignment can be performed efficiently at scale.

1 Introduction

In the recent years, the number of datasets exposed as linked data has grown continuously. Estimates show that there are more than 1000 datasets, with roughly 30 billion facts in the form of triples [1]. The decentralized nature of these datasets and furthermore, the lack of mechanisms and proper documentation for reusing existing schemas has led to a large number of schemas, 650 in LOD [25]. In many cases classes and relations across such schemas are redundant and not aligned (with only 2% of schemas aligned [25]). As a consequence this leads to a disintegrated dataset landscape [9].

© Springer International Publishing AG 2017
E. Blomqvist et al. (Eds.): ESWC 2017, Part I, LNCS 10249, pp. 152–168, 2017.
DOI: 10.1007/978-3-319-58068-5_10

This particular problem has been partly addressed by *ontology alignment* approaches, which almost exclusively have focused on class alignment across disparate ontologies [27]. In contrast to the class alignment, *relation alignment* has not seen such progress. Yet an equally important task, helping address information needs that span across datasets. Through both class and relation alignment we can rewrite and perform federated queries across disparate datasets.

The few existing relation alignment approaches, like the state of the art [28] fail to incorporate two main aspects of linked datasets. First, [28] is a blocking solution that requires several iterations over full datasets and which takes several hours to complete. As shown in [25] full dataset access is not always possible. Many datasets limit to only *query access* and to a few thousands of triples per query. Secondly, it does not take into account the *evolving nature* of datasets. Upon updates, only some of the alignments may require to be modified. Yet, the existing solutions need to be run from scratch on the full datasets, thus making such approaches inefficient.

We propose *SORAL – Supervised Ontology Relation ALignment*, a supervised machine learning approach for *relation alignment* across disparate schemas and datasets. Similar to [17], we work under the assumption that `owl:sameAs` statements exist between entities across datasets [25]. Additionally we operate in an online setting requiring only SPARQL query access to datasets, thus allowing for large scale alignment, independent of dataset size.

Figure 1 shows *SORAL*, consisting of two main steps. First, for a *source relation* and source dataset, we generate *candidate relations* for alignment to a *target dataset*. As candidates we consider relations that have entity instances in common (based on `owl:sameAs` statements). In this step we employ *sampling strategies* to cope with the scale of datasets and ensure the efficiency of *SORAL*. Second, for a relation pair, we compute features from the common entity instances, and information we extract from the relations, e.g. *domain/range* of relations etc. *SORAL* optimizes for *efficiency* by performing the alignment in an online setting, and *accuracy* by providing qualitative relation alignments. To this end, we make the following contributions:

– We propose an approach for finding relation subsumptions and equivalences;
– We perform the relation alignment in an *online* setting that ensures *efficiency*;
– We conducted an extended experimental evaluation using real-world knowledge bases (KBs).

2 Related Work

Scope. Our goal is to discover *subsumption* and *equivalence* relationships between the relations of two KBs exported using SPARQL endpoints. This problem is also known as *ontology alignment*. Ontology alignment includes: *instance alignment*, *class alignment*, and *relation alignment*. To align relations, we rely on existing work on aligning instances [5,6,20]. Hence, we see this effort as complementary to ours.

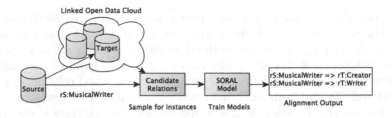

Fig. 1. SORAL approach overview.

Related work in this field is usually categorized in *schema-based* and *instance-based* approaches depending on the data that is used to produce the alignments. The majority of the work focuses on *class alignment* and much less on *relation alignment*, which is the objective of this work.

Schema-Based. Schema matching systems like in [3,8,22,26] align classes by relying on schema constrains, which are often unavailable in LOD. COMA++ [3] expects as input OWL descriptions and ignores data instances. While in [8] the approach relies on user assistance for the alignment process. Other systems like in [26] serve as proxy over existing schema matching tools, thereby providing means on combining the different functionalities of the individual platforms. BLOOMS [15], aligns classes in different KBs where the schema definitions are available. In our vision, the alignment of KBs should happen even in cases where explicit schema information or other database constraints are not available. For that reason we choose an instance-based approach.

Instance-Based: Class Alignment. Many instance-based approaches are proposed for class alignment in ontologies [12,16,31], contrary to our work, that focuses on relation alignment. Movshovitz-Attias et al. [23] infer *subconcept* and *superconcept* hierarchies from a KB. [24] produces equivalence and subsumption relationships between classes. Contrary to the class alignment work, the difference in the case of relations lies on two main aspects. First, classes and relations represent two different constructs in schemas. For classes it is sufficient to simply measure set overlap in terms of entities and other schematic information. In the case of relations we have the domain and range which should map between two relations, and even in cases they match, two distinct relations might have different semantics (e.g. `bornIn` and `presidentOf`).

Instance-Based: Relation Alignment. A large corpus of works focus on the discovery of *equivalence* or *similarity* relationships between relations, with early examples in relational databases [11,21], and Web services [18,19]. In the case of KBs, the schemas are in large numbers and vary heavily. For example, YAGO and DBpedia vary greatly in their representation, where YAGO has 37 object relations, while DBpedia has 688, whereas in terms of entity instances they have a high overlap. Thus, by considering only *equivalence* alignments we cannot map relations that have similar semantics or subsume another. This is due to the fact that many relations map to more than one relation in a target KB (e.g. YAGO to DBpedia). In itself finding *subsumption* relationships is more challenging, yet, it reflects the true nature of this problem. In the following we describe only the approaches that are capable of handling subsumption relationships [28,30,32].

ILIADS [30] and PARIS [28] serve for the dual purpose of aligning instances and schema elements between two KBs. In ILIADS as in [7], the authors acknowledge the difficulty of the relation alignment problem, where systems designed for class and instance alignment perform poorly. In the case of PARIS, it is acknowledged as a state of the art approach by related work [32], thus, we compare against it, and show the advantages of *SORAL* achieving significant improvement over PARIS. Galárraga et al. [13] propose *ROSA*, a relation alignment approach. *ROSA* operates on a similar setting as ours. However, as we show in our experimental evaluation in Sect. 7, relying solely on *association rule mining* measure PCA [14] is insufficient. Finally, the aforementioned works [13,28,30,32] operate on KB snapshots, contrary to our approach, which operates on small data samples that are queried from the respective KBs, thus, tackling one of the major problems of *efficiency* of existing works.

3 Preliminaries

Knowledge Bases. We assume that knowledge bases (KB) are represented in RDFS[1]. A KB consists of a set of triples $\mathcal{K} \subseteq \mathcal{E} \times \mathcal{R} \times (\mathcal{E} \cup \mathcal{L})$, where $\mathcal{E}$ is a set of resources that we refer as entities, $\mathcal{L}$ a set of literals, and $\mathcal{R}$ a set of relations. An *entity* is identified by a URI and represents a real-world object or an abstract concept. A *relation* (or property) is a binary predicate. A *literal* is a string, date or number.

Given a triple (x, r, y) (aka a statement), x is known as subject, r as relation, and y as object. Based on the nature of the objects, we classify relations in two categories: (i) *entity-entity* relations with x and y being entities, and (ii) *entity-literal* relations where y is a literal.

We use $r(x, y)$ to refer to the triple (x, r, y), and the tuple (x, y) as an *instantiation* of r. Without loss of generality, for an entity-entity relation $r \in \mathcal{R}$ we denote its inverse as $r^- \in \mathcal{R}$, and that the triples of r^- are also contained by $\mathcal{K}$ $(\forall r(x, y) \in \mathcal{K} \Rightarrow r^-(y, x) \in \mathcal{K})$.

Classes and Instances. A *class* is an entity that represents a set of objects, e.g., the class of all politicians. An entity that is a member of a class is called an *instance* of that class. The rdf:type relation connects an instance to a class. For example, *Barack_Obama* is a member of the class of politicians: rdf:type*(Barack_Obama, Politician)*.

Equivalence of Instances. Equivalence between two entities is expressed through owl:sameAs statements. For example, *yago:US* and *dbpedia:USA* (both referring to the same country), the equivalence of the two instances is expressed as owl:sameAs*(yago:US, dbpedia:USA)*.

---

[1] https://www.w3.org/RDFS/.

4 Problem Statement

We consider a KB pair (a source $\mathcal{K}_S$ and a target $\mathcal{K}_T$) that can be queried through SPARQL endpoints. Next, we assume that owl:sameAs statements exist between entity instances of $\mathcal{K}_S$ and $\mathcal{K}_T$. For two equivalent instances $e_s \equiv e_t$, from $\mathcal{K}_S$ and $\mathcal{K}_T$, respectively, we assume that, $\mathcal{K}_S$ stores owl:sameAs(e_s, e_t) and $\mathcal{K}_T$ stores owl:sameAs(e_t, e_s).

Definition 1 (Relation Subsumption). *For two relations r_S and r_T, we say that r_S is* subsumed *by r_T, or that r_T subsumes r_S and write $r_S \subseteq r_T$ or $r_S \Rightarrow r_T$ iff*

$$\forall x, y, r_S(x, y) \text{ is true} \rightarrow r_T(x, y) \text{ is also true.}$$

The notion of subsumption is independent of the relation extensions/name in the two KBs. We can only rely on the facts stored in the KBs to learn the subsumption relationships.

Goal. *For two knowledge bases, a source $\mathcal{K}_S(\mathcal{E}_S, \mathcal{R}_S, \mathcal{E}_S \cup \mathcal{L}_S)$ and a target $\mathcal{K}_T(\mathcal{E}_T, \mathcal{R}_T, \mathcal{E}_T \cup \mathcal{L}_T)$, and a relation $r_S \in \mathcal{R}_S$, find the relations $r_T \in \mathcal{R}_T$ s.t. $r_S \subseteq r_T$.*

The relationship of *equivalence* between two relations is expressed as two way subsumption relationship: $r_S \equiv r_T$ iff $r_S \subseteq r_T$ and $r_T \subseteq r_S$. In this way, we support the computation of both *subsumption* and *equivalence* relationships.

In this work, we consider only entity-entity relations. We do not consider entity-literal relations, because the equivalences between literals are managed differently and are not materialized in KBs. However, once such equivalences are established, one can use our approach to align entity-literal relations.

5 SORAL: Relation Alignment

We propose *SORAL*, an online relation alignment approach, which for a pair of knowledge bases, a source $\mathcal{K}_S$ and a target $\mathcal{K}_T$, works under two assumptions: (i) each KB is accessible through a SPARQL endpoint, (ii) for a KB pair we assume that their entity instances are partially aligned.

The process of discovering *subsumption* and *equivalence* relation alignments in *SORAL* is outlined in two main steps:

(1) **Candidates Generation.** For a relation $r_S \in \mathcal{R}_S$, find all overlapping relations $r_T \in \mathcal{R}_T$ i.e. $\exists x, y : r_S(x, y) \in \mathcal{K}_S \wedge r_T(x, y) \in \mathcal{K}_T$.
(2) **Supervised Model.** For every candidate pair $\langle r_S, r_T \rangle$, we classify it as correct or incorrect depending if the subsumption relationship holds or not. We propose a set of features (Sect. 5.3) which we use to learn a supervised machine learning model.

5.1   Candidate Generation

We observe that for a relation r_S its super-relations r_T must have at least one instantiation in common. Thus, they must satisfy the following constraint:

$$\exists x_S, y_S, x_T, y_T : r_S(x_S, y_S) \wedge x_T \equiv x_S \wedge y_T \equiv y_S \wedge r_T(x_T, y_T)$$

Since, the relations reside at different endpoints, for the candidate generation we perform a *federated* query. The conjunction of the first three terms of the expression can be evaluated at $\mathcal{K}_S$. However, for the last term the query is evaluated at $\mathcal{K}_T$. More precisely, we first evaluate the following query at $\mathcal{K}_S$:

> Q_1: SELECT DISTINCT ?x_T ?y_T FROM <K_S> WHERE
> { ?x_S r_S ?y_S. ?x_S owl:sameAs ?x_T. ?y_S owl:sameAs ?y_T. }

From the query result-set $S_{r_S} = \{(x_1, y_1) \ldots\}$ of Q_1 we discover the *candidate relations* from $\mathcal{K}_T$ for which these pairs are *instantiations*. Note that a tuple (x_T, y_T) can be the instantiation of the inverse of a relation at $\mathcal{K}_T$. Hence, we discover both *direct* and *inverse* relations. For this purpose, we use the *VALUES* feature from SPARQL 1.1 and compute using only one query both inverse and direct relations.

> Q_2: SELECT DISTINCT ?r_T ?d FROM <K_T> WHERE {
> VALUES (?x ?y) {(x₁ y₁) ...}
> { SELECT ?x ?r_T ?y ?d WHERE {?x ?r_T ?y. VALUES ?d {"d"}}}
> UNION
> { SELECT ?x ?r_T ?y ?d WHERE {?y ?r_T ?x. VALUES ?d {"i"}}} }

The result-set of query $Q2$, $C(r_S)$, denotes the set of candidate relations r_T. For simplicity, we assume that all relations are direct. This is because we have assumed (Sect. 3) that each KB defines for each relation, its inverse.

We avoid the transfer of the entire instantiations of a relation (as it introduces a bottleneck), and instead sample for a minimal set of tuples (x_T, y_T) that are transferred from the source to the target. We discuss the sampling strategies in the following

5.2   Sampling Strategies

In many cases SPARQL endpoints place limitations on the number of triples one can transfer. Even when full access is provided, transferring the full data is expensive (both in terms of time and network bandwidth consumption).

To overcome these issues we suggest three *sampling* strategies for entity instance selection, which we use to generate *relation candidates*. For r_S, the objective is to sample for a minimal set of *representative* instantiations S_{r_S}, such that they provide optimal coverage on discovering super-relations r_T.

First–N Sampling. It is an efficient way to query for the first N returned entity samples (x_T, y_T). The drawback is that it does not provide *representative samples*, due to the fact that it is subject to how the data is added into the

KBs. This represents a baseline to show the impact of sampling strategies on generating accurate relation alignments.

Random Sampling. In this case we use the RAND() feature of SPARQL 1.1 to query for representative samples for S_{r_S}. Contrary to first-N, ordering the matching triples through the RAND() function is expensive and introduces a significantly higher overhead in the query-execution when performing the relation alignment. Moreover, dependent on the sample size, the samples might be biased towards the classes with more instances, and as such it has an impact in the candidate relations we generate from $\mathcal{K}_T$.

Stratified Sampling. Here, we account for the possibly missing candidate relations from $\mathcal{K}_T$ due to the sample size, for relations whose domain/range belongs to fine grained entity types. Through stratified sampling we can achieve a better coverage in terms of uncovered relation alignments for r_S.

We group entities into homogeneous groups/strata (a group is represented by an entity type) before sampling. Additionally, the groups are further defined at various depths at the type taxonomy levels in order to find out the optimal groups which yield the highest coverage in terms of candidate relations. To ensure the *disjointness* (a prerequisite for stratified sampling), from the *transitive type closure* for an entity, we associate it to its most specialized type (dependent at what level in the type taxonomy we are interested to group entities), thus, ensuring disjointness of groups.

At high levels in the taxonomy, the sampling is similar to random sampling, whereas for deeper levels, we have more groups, and as a consequence more representative samples, leading to an increased coverage of candidate relations from $\mathcal{K}_T$. The query to retrieve the samples is shown below, where N is the sample size for each strata, proportional to the number of entities in it and the given total sample size.

```
Q₃: SELECT DISTINCT ?xT ?yT WHERE {
      ?xS rs ?yS. ?xS owl:sameAs ?xT.
      ?yS owl:sameAs ?yT. ?xS rdf:type ?type.
   } ORDER BY RAND() LIMIT N
```

5.3 Features

In this section, we describe the features we use to learn the supervised model in *SORAL* to predict whether the subsumption relationship for a relation pair holds. We consider features which we categorize into two main categories. Firstly, we consider inductive logic programming (ILP) approaches that work under the *open* and *closed* world assumptions w.r.t the instantiations of relations r_S and r_T. In the second group (GRS), we look into general statistics extracted from relations, specifically we assess the domain/range class distribution of the relations r_S and r_T. It is important to note that the features are computed over the sampled entity instances S_{r_S} and for the candidate relation in $C(r_S)$.

ILP – Features

Closed World Assumption – CWA. Proposed in [10], cwa_{conf} is used to mine association rules and was first used for relation alignment between two KBs in state of the art [28]. It works under the *closed world assumption*, where the KBs are assumed to be complete, hence there are no missing statements. The cwa_{conf} score is computed as in Eq. 1. It measures the overlap in terms of the number of instantiations in common between r_S and r_T, normalised by the number of instantiations of r_S.

$$cwa_{conf}(r_S \Rightarrow r_T) := \frac{overlap(r_S \wedge r_T)}{r_S} \tag{1}$$

We gather such counts by querying the respective SPARQL endpoints of $\mathcal{K}_S$ and $\mathcal{K}_T$.

The cwa_{conf} measure provides strong signal for the alignment of relations in the case when the relations are *complete* (i.e., KBs have all statements for a given relation). However, due to the fact that KBs are constructed from different sources, complementary statements for an entity in different KBs are considered as counter-examples.

Partial Completeness Assumption – PCA. The second feature, pca_{conf}, is an ILP measure proposed in [14] with the purpose of mining *Horn rules* for an input KB. It is able to handle incompleteness in KBs, and works under the *partial completeness assumption*. For a relation r_S and instance x, it is assumed that the KB contains either *all* or *none* of the r_S-triples with x as a *subject*. As counter-examples for the alignment $r_S \Rightarrow r_T$, is considered any pair (x, y) that is an instantiation of r_S but not of r_T. More formally, $counter(r_S \Rightarrow r_T)$ $:= \#(x, y) : r_S(x, y) \wedge \exists y_2, y_2 \neq y : r_T(x, y_2) \wedge \neg r_T(x, y)$.

$$pca_{conf}(r_S \Rightarrow r_T) := \frac{overlap(r_S \wedge r_T)}{overlap(r_S \wedge r_T) + counter(r_S \Rightarrow r_T)} \tag{2}$$

Similar to cwa_{conf}, to compute pca_{conf} we extract the counts for r_S and r_T through SPARQL queries on $\mathcal{K}_S$ and $\mathcal{K}_T$. Note that, pca_{conf} can have maximal score even when the overlap between relation instantiations consists of few tuples with no counter-examples. This is helpful in detecting alignments with a small overlap in the two KBs. On the other hand, it increases the likelihood of getting false positives. These are caused by erroneous facts and incomplete data.

However, considering jointly cwa_{conf} and pca_{conf} features has the effect of *regularizing* each other; for high pca_{conf} scores but low cwa_{conf} the chance of having a correct alignment is low, and vice-versa. We make use of this assumption and compute a *joint probability* score for a relation pair being correct by considering the pca_{conf} and cwa_{conf} scores jointly.

Partial Incompleteness Assumption – PIA. The pca_{conf} may hold for *functional relations* (see Eq. 4), but is less probable for one-to-many relations. The intuition is that if the average number of r_S-triples per subject is high, then it is more likely that not all r_S-triples of some subject x have been extracted.

The same observation holds also for the triples of r_T. Note, that we have yet to prove that r_S is subsumed in r_T. Hence, the counter-examples of less functional relations should be weighted less than the counter-examples of more functional relations. We measure the pia_{conf} score as:

$$pia_{conf}(r_S \Rightarrow r_T) := \frac{overlap(r_S \wedge r_T)}{overlap(r_S \wedge r_T) + (counter(r_S \Rightarrow r_T) * func(r_S))} \tag{3}$$

where, $func(r_S)$ measures the probability that a counter example may not exist.

General Relation Statistics (GRS) – Features. In the GRS feature group, we propose features that take into account the cardinality of the two relations, and the types extracted from the entity instances in the *domains* and *ranges* of a relation.

Functionality of a Relation. A relation is called *functional*, if there are no two distinct facts that share the relation and the first argument. Real life KBs may be noisy, and contain distinct facts for relations that should be functional (such as *bornIn*). Therefore, we make use of the functionality [28]:

$$func(r) := \frac{\#x : \exists y : r(x,y)}{\#(x,y) : r(x,y)} \tag{4}$$

Perfect functional relations have a functionality of 1.

From this intuition we assume that if r_S is subsumed by r_T, then the number of facts per subject in r_S should be lower than in r_T. In other words, the functionality of r_S should be grater or equal to the functionality of r_T, that is if $r_S \Rightarrow r_T \rightarrow func(r_S) \geqslant func(r_T)$. Finally, the functionality feature we consider here is the functionality difference between r_S and r_T, $func_{diff} := func(r_S) - func(r_T) \in (-1,1)$.

Weighted Jaccard Similarity. We measure the similarity of the type distributions $\mathcal{D}(\cdot)$ between the relations r_S and r_T. We denote with $\mathcal{D}^s(r_S)$ and $\mathcal{D}^s(r_T)$ the type distribution for *subject* entities for r_S and r_T, respectively; each type is represented by the proportion of entities (out of the total entities for a relation).

Since we are interested to find the subsumption $r_S \Rightarrow r_T$, the intuition behind this feature is that the respective type distributions should be similar, or $\mathcal{D}(r_S)$ should be entailed in $\mathcal{D}(r_T)$. In other words, the target relation r_T, should be able to represent entity instances from r_S, namely through its *domain* entity type definition[2].

To compute $\mathcal{D}(\cdot)$, first we unify the type representation of entities in the two KBs by picking one of the taxonomies in either $\mathcal{K}_S$ or $\mathcal{K}_T$. This is necessary, due to the fact that two KBs use different schemas. We are able to do this due to the restriction that we consider only entities that are linked across KBs through `owl:sameAs` statements. The similarity is computed as in Eq. 5.

$$WJ(\mathcal{D}(r_S), \mathcal{D}(r_T)) := \frac{\sum_{t \in \mathcal{D}(r_S) \wedge \mathcal{D}(r_T)} min(\sigma(t^S), \sigma(t^T))}{\sum_{t \in \mathcal{D}(r_S) \wedge \mathcal{D}(r_T)} max(\sigma(t^S), \sigma(t^T))} \tag{5}$$

---

[2] Each relation in a RDFS schema has two properties denoting the *domain* and *range*.

where, $\mathcal{D}(\cdot)$ represents either the type distribution for the subject or object values based on that if the relation is a direct or inverse relation. $\sigma(t)$ represents the score of a type in the distribution for a given relation.

Weighted Jaccard Dissimilarity. Contrary to the previous feature which is computed on overlapping types, in this case, we compute the *weighted dissimilarity score* (WDS) between r_S and r_T. Specifically, if a specific entity type from r_S does not exist in r_T, this accounts for a dissimilarity between the two relations, and lowers the likelihood of r_T subsuming r_S.

$$WDS(\mathcal{D}(r_S), \mathcal{D}(r_T)) := \frac{\sum_{t \in \mathcal{D}(r_S) \wedge t \notin \mathcal{D}(r_T)} \sigma(t^S)}{|\mathcal{D}(r_S) \setminus \mathcal{D}(r_T)|} \tag{6}$$

The intuition is that if r_T cannot represent types from r_S which account to a large proportion of entities, then it is unlikely for $r_S \Rightarrow r_T$ to hold.

ILP Score Relevance Probability. In the case of ILP features, the scores are subject to KB pairs. For specific relations we can achieve nearly perfect pca_{conf}score with low overlap in terms of entity instantiations between relations (e.g. single entity x), yet, with the subsumption relationship unlikely to hold. On the other hand, low pca_{conf}scores may result even if the overlap of instantiations is high in absolute numbers, but due to the complementary nature of datasets might yield to many counter examples.

We circumvent such shortcomings of the ILP features, by assessing the likelihood of the $pca_{conf}$and cwa_{conf}scores for a subsumption relationship to hold between a relation pair. We measure the *prior* probabilities (in our training data) and use these probabilities as features in our learning algorithm. To do so, we first discretize the $pca_{conf}$and cwa_{conf}scores into the ranges with a cut-off point of 0.1, $\{0, 0.1, ..., 1.0\}$. For a given pca_{conf}or cwa_{conf}score, we denote with $\langle r_S, r_T \rangle^c$ the set of relation alignment candidates that are correct, and with $\langle r_S, r_T \rangle$ the complete set of relation alignment candidates.

ILP Prior. We measure the *prior probabilities*, $p(correct|pca_{conf})$ or $p(correct|cwa_{conf})$, where the subsumption relationship holds for a pca_{conf}or cwa_{conf}score as the ratio of the cases where the relation holds, divided by the total relation pairs having that discretized score.

Joint PCA and CWA Probability. Here we address the shortcomings of $cwa_{conf}$and pca_{conf}score, by learning a *joint probability* score in which the subsumption relationship holds. We compute the joint relevance probability as following.

$p(correct|pca_{conf}, cwa_{conf}) :=$

$$\frac{\#\langle r_S, r_T \rangle^c : pca_{conf}(r_S \Rightarrow r_T) = pca_{conf} \wedge cwa_{conf}(r_S \Rightarrow r_T) = cwa_{conf}}{\#\langle r_S, r_T \rangle : pca_{conf}(r_S \Rightarrow r_T) = pca_{conf} \wedge cwa_{conf}(r_S \Rightarrow r_T) = cwa_{conf}} \tag{7}$$

where for a given discretized $pca_{conf}$and cwa_{conf}score we simply count the number of relation pairs whose scores match and the corresponding alignment is relevant, over the total number of relation pairs with the respective pca_{conf} and cwa_{conf} scores. We expect this feature to be sparse, hence, we use the simple priors as fall-back features.

6 Experimental Setup

Here we present the evaluation setup of *SORAL*. In our setup we host the evaluation datasets in the Virtuoso Universal Server[3] with SPARQL 1.1 support.

Datasets. We evaluate *SORAL* on the following real-world knowledge bases which are the most commonly used datasets from LOD and serve as general purpose datasets.

YAGO (Y). From the YAGO2 [29] dataset we use the *core* and *transitive type* facts, excluding the entity labels given that we consider only entity–entity relations. The size of YAGO is approximately 900 MB.

DBpedia (D). For DBpedia [2] we consider the *entity types* and the *mapping-based properties*, with a size of 5.5 GB.

Freebase (F). For Freebase[4] we take a subset of entities that have owl:sameAs links to DBpedia, with a total of 30 GB of data.

Entity-Entity Relations. From the aforementioned KBs, we extract all possible entity-entity relations, and filter out those with less than 50 triples. After filtering we are left with 36 relations from YAGO, 563 from DBpedia, and 1666 from Freebase.

Entity Links owl:sameAs. We are in hold of owl:sameAs links between the pairs DBpedia – YAGO (2.8 MM links), and DBpedia – Freebase (3.8 MM links), in the case of YAGO and Freebase we infer the owl:sameAs links (2.7 MM links) through the DBpedia links.

Sampling Strategies Setup. In order to measure the right samples size to have an optimal coverage and maintain the efficiency of our approach, we vary the sample sizes between {100, 500, 1000}. We evaluate all three sampling strategies, and in the case of stratified sampling we construct the strata based on the DBpedia type taxonomy[5]. We take advantage of the fact that all sampled entities from the various KBs have equivalent entities in DBpedia. We opt for the DBpedia taxonomy due to the fact that the types form a hierarchy.

Ground-Truth Construction. Due to the limited resources, the authors of the paper have served as expert annotators and we manually construct the ground-truth for the relation alignment process. The ground-truth is constructed for each KB pair. We guide the process of ground-truth creation by displaying the individual pairs $\langle r_S, r_T \rangle$. For each pair, apart form the labels of the relations,

---

[3] http://virtuoso.openlinksw.com.

[4] http://www.freebase.com.

[5] http://mappings.dbpedia.org/server/ontology/classes/.

the expert annotators need to assess a sample of instantiations for each relation in order to assess correctly whether a relation subsumes another.

Learning Framework. In $SORAL$ we feed in all the computed features into a supervised model. In our case we use a *logistic regression* (LR) model [4] (any other model can be used), and learn a binary classifier. For each relation pair $\langle r_S, r_T \rangle$ it outputs either *'correct'* or *'incorrect'* depending whether the subsumption relationship holds in our ground-truth dataset. Finally, we evaluate the performance of our model by considering a 5-fold cross-validation strategy.

Evaluation Metrics. We evaluate $SORAL$ on two main aspects: (i) accuracy and (ii) efficiency. In terms of accuracy, we compute standard evaluation metrics, like precision (**P**), recall (**R**) and F-measure (**F1**). For *efficiency* we measure the overhead in the query-execution process in terms of: (i) time (**t**), as the amount of time taken to sample entities for the set S_{r_S}, and (ii) network bandwidth usage (**b**) as the amount of bytes we transfer over the network for the different sample sizes in S_{r_S}. Note here, that a traditional schema-matching approach operating on a dataset snapshot, and as such they introduce an incomparably higher overhead in terms of bytes and time.

Competitors. We compare $SORAL$ against two existing approaches. First is a *state of the art* approach in relation alignment, a system called *PARIS* [28], which implements the cwa_{conf} measure. The second competitor, is called *ROSA* [13] and makes use of the pca_{conf} measure. It must be noted that the competitors are unsupervised approaches, and as such do not need training data. However, as it will be shown in our evaluation applying such measures in an unsupervised manner results in poorer performance.

To ensure fairness in our comparison, since the two baselines use specific thresholds for the alignment process, we select their best threshold parameter such that it maximizes the F1 score. The best configuration for *PARIS* is a threshold of $cwa_{conf} = 0.1$, whereas for *ROSA* a threshold of $pca_{conf} = 0.3$.

Table 1. The performance of $SORAL$ for the different sampling strategies and sizes.

Sampling	100			500			1000		
	P	R	F1	P	R	F1	P	R	F1
firstN	.80	.45	.56	.83	.48	.60	.82	.48	.59
random	.80	.45	.56	.81	.44	.56	.82	.44	.56
str.lvl–2	.80	.40	.51	.81	.44	.57	.77	.40	.49
str.lvl–3	.80	.42	.54	**.84**	**.52**	**.64**	.82	.50	.61
str.lvl–4	.79	.44	.55	.82	.49	.61	.80	.49	.59
str.lvl–5	.77	.44	.55	.82	.47	.59	.80	.48	.60

7   Results and Discussion

In this section we present the evaluation results of *SORAL* by assessing two main aspects: (i) the performance for the relation alignment task and its comparison against the competitors, (ii) the *efficiency* by measuring the introduced overhead in terms of time and network bandwidth overload through the sampling strategies.

7.1   Relation Alignment Accuracy

Sampling Configurations. Here, we show the impact of the sampling strategies on the candidate generation process. Through sampling we can *efficiently* perform the alignment process in an online setting, and at the same time provide accurate results.

Table 1 shows the sampling strategies presented in Sect. 5.2. In the case of the *stratified sampling* we construct the strata based on the DBpedia type taxonomy with a depth level from 2 up to 5 (DBpedia has a maximum depth level of 7). The results are averaged across the different KB pairs. Here we aim at finding the best parameter configuration for *SORAL*, specifically for: (i) sampling strategy, and (ii) sample size.

From Table 1 we see that the best performing strategy is based on the *stratified sampling*, with strata constructed at the depth level of three, and with a sample size of 500. With these two parameters, *SORAL* yields an F1 score of 0.64. We note a small fluctuation in terms of precision for the sample sizes of 500 and 1000. We check for statistical significance on the resulting F1 scores, however, the difference proved to be insignificant with a $p - value = .17$. On the other hand, we find statistical significance between the F1 scores for 100 and 500 entity samples ($p - value = .01$).

Table 2. The performance of *SORAL* in comparison to the competing baselines.

	Sampled data									Full data								
	SORAL			ROSA			PARIS			SORAL			ROSA			PARIS		
				$pca_{th} = 0.3$			$cwa_{th} = 0.1$						$pca_{th} = 0.3$			$cwa_{th} = 0.1$		
$\mathcal{K}_S \to \mathcal{K}_T$	P	R	F1	P	R	F1	P	R	F1	P	R	F1	P	R	F1	P	R	F1
YAGO → DBpedia	1.0	.68	**.81**	.17	.75	.28	.71	.66	.68	.92	.73	**.81**	.06	.68	.11	.42	.54	.48
YAGO → Freebase	.87	.67	**.76**	.11	.78	.20	.55	.59	.57	.82	.82	**.82**	.03	1.0	.05	.40	1.0	.57
DBpedia → Freebase	.72	.36	**.48**	.10	.67	.18	.31	.50	.40	.69	.38	**.49**	.05	.85	.09	.31	.65	.42
DBpedia → YAGO	.86	.60	**.71**	.30	.72	.43	.70	.66	.68	.57	.49	**.53**	.18	.55	.27	.40	.45	.43
Freebase → DBpedia	.88	.51	.64	.27	.79	.41	.65	.65	.65	.87	.66	**.75**	.34	.93	.50	.72	.57	.64
Freebase → YAGO	.72	.34	**.46**	.22	.39	.28	.42	.37	.39	.69	.74	**.71**	.61	.86	.71	.73	.60	.66
Average	.84	.52	**.64**	.19	.68	.29	.55	.57	.56	.76	.64	**.69**	.21	.81	.29	.49	.63	.53

SORAL Effectiveness. Table 2 shows the results for the best configuration of *SORAL* (with stratified sampling at depth level 3) for the individual KB pairs. In the "Sampled Data" results group we compare *SORAL* against the competitors *ROSA* and *PARIS*. The results are highly significant when comparing *SORAL* against the competitors for the F1 score. In terms of precision, we have a relative improvement of 336% when compared to *ROSA*, whereas in the case of *PARIS* we have a 51% improvement. In terms of recall, we perform worse, however, the low pca_{conf} and cwa_{conf} thresholds for our baselines ($pca_{th} = 0.3$ and $cwa_{th} = 0.1$), lead to a low precision which makes such results hardly usable. Increasing the thresholds results an increase of precision, however, strongly penalizing the recall of the competitors.

In the group of results in column "Full Data", we show the results of *SORAL* and the competitors when computed without sampling for the relation candidate generation. Here too we outperform the competitors to a large extent, with an average difference of 16% in terms of F1 for *PARIS* and even higher difference of 40% to *ROSA*.

It is interesting to note that for *SORAL* on average we miss 13% of aligned relations from the real coverage performance of *SORAL* when comparing the results with sampling and without. In terms of F1 measure we have only a small difference of 6%, which presents an optimal results when comparing the advantages we get in terms of *efficiency* through sampling.

Finally, we note varying performance of *SORAL* (with and without sampling) across the different KB pairs. For instance, in the case of YAGO → DBpedia we have perfect precision and the best F1 score. It is evident that DBpedia and YAGO represent high quality KBs, whereas Freebase is more noisy, thus, leading to a poorer performance. The main reason for such noise is that relations in Freebase are created by its users, and as such are not unique, in the sense that there are multiple relations representing similar concepts in Freebase.

From Table 2 we further notice that in the case of DBpedia to YAGO we have worse results when comparing the full against results obtained through sampling. This may be caused due to the difference in the number of statements per relation in the two KBs. DBpedia has a larger set of entity instances and statements per relation, and as such the ILP and GRS features are penalized more heavily in the full dataset. We believe that in this case the sampling plays a normalizing factor for such differences in terms of statements and the impact on the feature computation.

Feature Ablation. In the feature ablation we show as to what is attributed the performance of *SORAL*. Figure 2 shows the feature ablation results for the different feature groups in Sect. 5.3. The results are averaged across the different KB pairs and are computed based on models trained on the full data, such that we can assess the true power of the different groups.

The highest impact is attributed to the *GRS* feature group. This follows our intuition where we hypothesized that for a relation to be subsumed in a target relation, one of the important factors is the ability of r_T to represent the entity

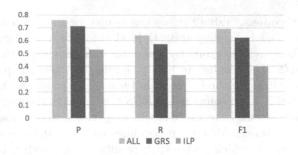

Fig. 2. Feature ablation for the different feature groups.

instances. In other words, the entity types for two relations should be similar. The next assumption we made was that *ILP* feature scores, depending on the KB pairs, sometimes are insufficient, hence, the corresponding likelihood scores of the ILP measures can provide additional information on predicting correctly the label of a relation pair.

7.2   Query-Execution Overhead

Here we show the overhead in terms of *time* and *network bandwidth* usage for the *sampling* process. These represent important aspects considering that our relation alignment is setup in an online setting.

Time overhead. With respect to the *time* efficiency factor, we measured the time it takes to perform the different sampling strategies. The most efficient sampling strategy is *first–N* taking the least amount of time. The *stratified* sampling strategy requires the most amount of time, with a maximum of 3 *s* (for 1000 instances); a significantly higher execution time compared to other strategies. However, such an overhead for the task at hand is arguably not high.

Network bandwidth overhead. Understandably the amount of overhead in terms of bandwidth is uniformly distributed across the different sampling strategies. That is, considering that we sample for the same amount of entity instances. The highest amount of bandwidth overhead is introduced when we sample for 1000 entity instances, with a maximum 140 *kb*. In the performance evaluation of *SORAL* in Table 1, we found the optimal results to be with 500 entity sample instances, resulting in bandwidth overhead of 60 *kb*. This contrary to existing approaches that require the full content of a KB, represents a highly efficient way to compute such alignments.

8   Conclusion

We proposed *SORAL*, a supervised machine learning approach for relation alignment. We perform the specific task where for a given source relation r_S we find relation subsumption in a target KB. Furthermore, we employ *SORAL* in an online setting, where we ensure the *efficiency* of the proposed approach by extracting minimal samples of entity instances for a relation from the respective SPARQL endpoints. To ensure the accuracy of *SORAL* we computed features that range from *association rule mining* features, to *general statistics* from relations.

References

1. Datasets by topical domain. http://linkeddatacatalog.dws.informatik.uni-mannheim.de/state/
2. Auer, S., Bizer, C., Kobilarov, G., Lehmann, J., Cyganiak, R., Ives, Z.: DBpedia: a nucleus for a web of open data. In: Aberer, K., et al. (eds.) ASWC/ISWC - 2007. LNCS, vol. 4825, pp. 722–735. Springer, Heidelberg (2007). doi:10.1007/978-3-540-76298-0_52
3. Aumueller, D., Do, H.-H., Massmann, S., Rahm, E.: Schema and ontology matching with coma++. In: SIGMOD (2005)
4. Bishop, C.M.: Pattern Recognition and Machine Learning, vol. 1. Springer, Heidelberg (2006)
5. Bizer, C., Heath, T., Idehen, K., Berners-Lee, T.: Linked data on the Web. In: WWW (2008)
6. Böhm, C., de Melo, G., Naumann, F., Weikum, G.: Linda: distributed web-of-data-scale entity matching. In: CIKM (2012)
7. Cheatham, M., Hitzler, P.: String similarity metrics for ontology alignment. In: Alani, H., et al. (eds.) ISWC 2013. LNCS, vol. 8219, pp. 294–309. Springer, Heidelberg (2013). doi:10.1007/978-3-642-41338-4_19
8. Cruz, I.F., Antonelli, F.P., Stroe, C.: Agreementmaker: efficient matching for large real-world schemas and ontologies. PVLDB 2, 1586–1589 (2009)
9. d'Aquin, M., Adamou, A., Dietze, S.: Assessing the educational linked data landscape. In: WebSci (2013)
10. Dehaspe, L., Toivonen, H.: Discovery of frequent datalog patterns. Data Min. Knowl. Discov. 3, 7–36 (1999)
11. Dhamankar, R., Lee, Y., Doan, A., Halevy, A.Y., Domingos, P.: imap: discovering complex mappings between database schemas. In: SIGMOD (2004)
12. Doan, A.-H., Madhavan, J., Domingos, P., Halevy, A.: Ontology matching: a machine learning approach. In: Staab, S., Studer, R. (eds.) Handbook of ontologies. International Handbooks on Information Systems, pp. 385–403. Springer, Heidelberg (2004). doi:10.1007/978-3-540-24750-0_19
13. Galárraga, L., Preda, N., Suchanek, F.M.: Mining rules to align knowledge bases. In: AKBC (2013)
14. Galárraga, L., Teflioudi, C., Hose, K., Suchanek, F.M.: Amie: association rule mining under incomplete evidence in ontological knowledge bases. In: WWW (2013)
15. Jain, P., Hitzler, P., Sheth, A.P., Verma, K., Yeh, P.Z.: Ontology alignment for linked open data. In: Patel-Schneider, P.F., Pan, Y., Hitzler, P., Mika, P., Zhang, L., Pan, J.Z., Horrocks, I., Glimm, B. (eds.) ISWC 2010. LNCS, vol. 6496, pp. 402–417. Springer, Heidelberg (2010). doi:10.1007/978-3-642-17746-0_26
16. Kirsten, T., Thor, A., Rahm, E.: Instance-based matching of large life science ontologies. In: Cohen-Boulakia, S., Tannen, V. (eds.) DILS 2007. LNCS, vol. 4544, pp. 172–187. Springer, Heidelberg (2007). doi:10.1007/978-3-540-73255-6_15

17. Koutraki, M., Preda, N., Vodislav, D.: Sofya: Semantic on-the-fly relation alignment. In: EDBT (2016)
18. Koutraki, M., Vodislav, D., Preda, N.: Deriving intensional descriptions for web services. In: CIKM (2015)
19. Koutraki, M., Vodislav, D., Preda, N.: Doris: discovering ontological relations in services. In: ISWC (2015)
20. Lacoste-Julien, S., Palla, K., Davies, A., Kasneci, G., Graepel, T., Ghahramani, Z.: Sigma: simple greedy matching for aligning large knowledge bases. In: KDD (2013)
21. Madhavan, J., Bernstein, P.A., Doan, A., Halevy, A.: Corpus-based schema matching. In: ICDE (2005)
22. Miller, R.J., Haas, L.M., Hernández, M.A.: Schema mapping as query discovery. In: VLDB (2000)
23. Movshovitz-Attias, D., Whang, S.E., Noy, N., Halevy, A.: Discovering subsumption relationships for web-based ontologies. In: Proceedings of the 18th International Workshop on Web and Databases (2015)
24. Parundekar, R., Knoblock, C.A., Ambite, J.L.: Linking and building ontologies of linked data. In: Patel-Schneider, P.F., Pan, Y., Hitzler, P., Mika, P., Zhang, L., Pan, J.Z., Horrocks, I., Glimm, B. (eds.) ISWC 2010. LNCS, vol. 6496, pp. 598–614. Springer, Heidelberg (2010). doi:10.1007/978-3-642-17746-0_38
25. Schmachtenberg, M., Bizer, C., Paulheim, H.: Adoption of the linked data best practices in different topical domains. In: Mika, P., et al. (eds.) ISWC 2014. LNCS, vol. 8796, pp. 245–260. Springer, Cham (2014). doi:10.1007/978-3-319-11964-9_16
26. Seligman, L., Mork, P., Halevy, A.Y., Smith, K.P., Carey, M.J., Chen, K., Wolf, C., Madhavan, J., Kannan, A., Burdick, D.: Openii: an open source information integration toolkit. In: SIGMOD (2010)
27. Shvaiko, P., Euzenat, J.: Ontology matching: state of the art and future challenges. IEEE Trans. Knowl. Data Eng. 25, 158–176 (2013)
28. Suchanek, F.M., Abiteboul, S., Senellart, P.: Paris: probabilistic alignment of relations, instances, and schema. PVLDB 5(3), 157–168 (2011)
29. Suchanek, F.M., Kasneci, G., Weikum, G.: YAGO: a core of semantic knowledge - unifying WordNet and Wikipedia. In: WWW (2007)
30. Udrea, O., Getoor, L., Miller, R.J.: Leveraging data and structure in ontology integration. In: SIGMOD (2007)
31. Wang, S., Englebienne, G., Schlobach, S.: Learning concept mappings from instance similarity. In: Sheth, A., Staab, S., Dean, M., Paolucci, M., Maynard, D., Finin, T., Thirunarayan, K. (eds.) ISWC 2008. LNCS, vol. 5318, pp. 339–355. Springer, Heidelberg (2008). doi:10.1007/978-3-540-88564-1_22
32. Wijaya, D.T., Talukdar, P.P., Mitchell, T.M.: Pidgin: ontology alignment using web text as interlingua. In: CIKM (2013)

Tuning Personalized PageRank
for Semantics-Aware Recommendations
Based on Linked Open Data

Cataldo Musto(✉), Giovanni Semeraro, Marco de Gemmis, and Pasquale Lops

Department of Computer Science, University of Bari Aldo Moro, Bari, Italy
{cataldo.musto,giovanni.semeraro,marco.degemmis,pasquale.lops}@uniba.it

Abstract. In this article we investigate how the knowledge available in the Linked Open Data cloud (LOD) can be exploited to improve the effectiveness of a semantics-aware graph-based recommendation framework based on Personalized PageRank (PPR).

In our approach we extended the classic *bipartite* data model, in which only user-item connections are modeled, by injecting the *exogenous* knowledge about the items which is available in the LOD cloud. Our approach works in two steps: first, all the available items are automatically mapped to a DBpedia node; next, the resources gathered from DBpedia that describe the item are connected to the item nodes, thus enriching the original representation and giving rise to a *tripartite* data model. Such a data model can be exploited to provide users with recommendations by running PPR against the resulting representation and by suggesting the items with the highest PageRank score.

In the experimental evaluation we showed that our semantics-aware recommendation framework exploiting DBpedia and PPR can overcome the performance of several state-of-the-art approaches. Moreover, a proper tuning of PPR parameters, obtained by better distributing the weights among the nodes modeled in the graph, further improved the overall accuracy of the framework and confirmed the effectiveness of our strategy.

Keywords: Graphs · Recommender systems · Linked open data · PageRank

1 Introduction

According to recent statistics[1], 150 billions of RDF[2] triples and almost 10.000 linked datasets are now available in the so-called Linked Open Data (LOD) cloud [6]. The *nucleus* of this set of interconnected semantic datasets is represented by DBpedia [1], the RDF mapping of Wikipedia that acts as a huge *hub* for most of the RDF triples made available in the LOD cloud.

---

[1] http://stats.lod2.eu/.

[2] http://www.w3.org/TR/rdf-concepts/.

© Springer International Publishing AG 2017
E. Blomqvist et al. (Eds.): ESWC 2017, Part I, LNCS 10249, pp. 169–183, 2017.
DOI: 10.1007/978-3-319-58068-5_11

The collaborative effort behind the LOD initiative is bringing several benefits: first, the rise of these *data silos* is contributing to push forward the vision of the *Web of Data*; second, the tremendous growth of semantically-annotated data is leading practitioners and researchers to investigate *whether* and *how* data gathered from the LOD cloud can be exploited to develop new services or to improve the performance of intelligent and knowledge-intensive applications.

As an example, graph-based recommender systems [13] can benefit of the information available in the LOD cloud. In this case the classic bipartite user-item representation can be easily extended by injecting in the graph the resources available in the LOD cloud which are connected to the properties describing the item. This can help to discover surprising connections between the movies: as an example, by mining the information available in the LOD cloud it emerges that both *The Matrix* and *The Lost World: Jurassic Park* movies have been shot in Australia, and this can in turn help to generate better (and unexpected) recommendations by exploiting such new information (Fig. 1).

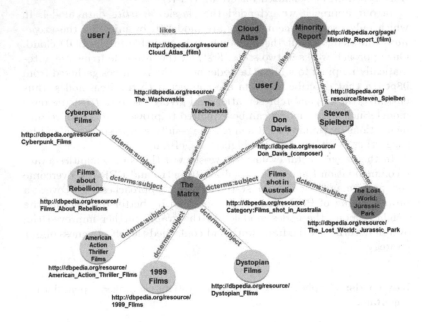

Fig. 1. A (tiny) portion of the connections between users, items and entities encoded in Linked Open Data cloud. Purple nodes represent users, blue nodes represent items, yellow nodes represent entities. (Color figure online)

According to these insights, it immediately emerges that recommender systems (RSs) may tremendously benefit of the data points available in the LOD cloud. To this end, in this article we investigate the impact of such *exogenous knowledge* on the performance of a graph-based recommendation framework. We focused our attention on graph-based approaches since they use a uniform formalism to represent both collaborative and LOD-based features. Indeed, in the first case users and items are represented as *nodes* and preferences are represented as *edges*. Similarly, entities from the DBpedia are represented as *nodes* while the connections between them (expressed through RDF properties) are represented as *edges*. It is very straightforward that through a simple mapping of the items to be recommended with the URIs available in the LOD cloud, both representations can be connected and merged in a unique and powerful formalism. Given such a representation, we adopted Personalized Page Rank (PPR) [11] as recommendation algorithm and we suggested the items with the highest PageRank score. Specifically, we biased PageRank by modifying the personalization vector p (as shown in [11]) through the adoption of different combinations to distribute the weights.

In the experimental evaluation we showed that our approach is able to overcome the predictive accuracy obtained by several state of the art baselines. Moreover, in this work we made one step further with respect to our previous research [14], and we investigated how the distribution of the weights in PPR algorithm can influence the overall performance of the system. The results emerged from the experiments showed that a proper tuning of the weights, which also considers the nodes coming from the LOD cloud, may further improve the accuracy of the framework. To sum up, with this article we provide the following contributions:

1. We propose a methodology to feed a graph-based recommendation algorithm with features gathered from DBpedia;
2. We investigate whether different weights of the features gathered from DBpedia influence the recommendation accuracy;
3. We validated our methodology through several experiments, and we showed that our approach obtains higher results with respect to all the baselines we took into account.

The rest of the paper is organized as follows: Sect. 2 analyzes the related literature. The description of the graph-based recommender system and the overview the different methodologies for distributing weights in PPR are provided in Sect. 3. The details of the experimental evaluation on two state of the art datasets are described in Sect. 4, while conclusions and future work are drawn in Sect. 5.

2   Related Work

This work investigates two different research lines: graph-based recommender systems and LOD-based recommender systems. In the following we present the current literature in both areas.

Graph-based Recommender Systems. Most of the literature in the area of graph-based RS is inspired by PageRank [18] and random walk. As an example, FolkRank [12] is an adaptation of PageRank used for tag recommendation, that relies on a representation in which resources are modeled along with the tags the community used to annotate them. Next [7], Bogers proposed ContextWalk, a movie recommender system relying on PageRank, modeling in a graph tags, genres and actors. Similarly, Baluja et al. [2] present a recommender system for YouTube videos based on random walks on the bipartite user-video graph. Finally, de Gemmis et al. [10] recently evaluated the applicability of Random Walk with Restart to Linked Data and investigated to what extent such graph-based representation can lead to *serendipitous* recommendations. A distinguishing aspect of this article with respect to the current literature is that none of the previously mentioned approaches investigated the integration of LOD-based data points. A similar attempt has been presented in [17], in which the paths connecting users and items via LOD-based properties are extracted and are used to train a classifier in a pure machine learning-based approach. Differently from this work, we encoded LOD-based features along with collaborative ones in a hybrid graph-based representation and we exploited Personalized PageRank as recommendation technique. Moreover, as previously introduced, in this article we continued carrying on our previous research [14] by better investigating how the distribution of the weights in PPR algorithm can influence the overall performance of the system.

LOD-based Recommender Systems. An updated and detailed review of the literature on recommendation approaches leveraging Linked Open Data is presented in [9]. In that survey, those approaches are classified as *top-down semantic approaches*, i.e. relying on the integration of external knowledge sources. In most of the current literature, properties gathered from DBpedia are exploited to define new similarity measures, as in [19]. The use of DBpedia for similarity calculation is also the core of the work presented by Musto et al. [15]: in that paper music preferences are extracted from Facebook and similarity measures are exploited to build a personalized music playlist. Moreover, a relevant research line investigated Linked Open Data to generate new descriptive features for the items. This is done in [8], where the authors exploit DBpedia to gather one or more labels describing the genre played by each artist the user liked, and by Baumann et al. [5], who extract features from Freebase[3] to describe artists. The use of Freebase as knowledge base to feed recommendation algorithms is also investigated by Nguyen et al. [16], who assessed the effectiveness of such a source for music recommendations. Another interesting attempt is due to Basile et al. [3,4] who obtained the best results in the ESWC 2014 Recommender Systems Challenge[4] by proposing an ensemble of several widespread algorithms running on diverse sets of features gathered from the LOD cloud. Differently from the current literature, our work aim to exploit features gathered from the Linked Open Data to build a *hybrid model of user preferences* that merges both collaborative and LOD-based data points in a unique graph-based representation which exploits PPR as recommendation algorithm.

---

[3] https://www.freebase.com/.
[4] http://challenges.2014.eswc-conferences.org/index.php/RecSys.

3   Methodology

In this section we describe our graph-based recommendation methodology. First, we show how we extended the original *bipartite representation* based on user-item connections by introducing features gathered from DBpedia, next we provide some basics of PPR algorithm and we introduce the different approaches for distributing the weight in the tripartite LOD-enriched representation.

3.1   Graph-Based Representations

The main idea behind our graph-based model is to represent *users* and *items* as *nodes* in a graph. Formally, given a set of users $U = \{u_1, u_2, \ldots u_n\}$ and a set of items $I = \{i_1, i_2, \ldots i_m\}$, a graph $G = \langle V, E \rangle$ is instantiated. It is worth to note that G is a bipartite graph, since it models two different kind of entities (that is to say, *users* and *items*). Next, an edge connecting a user u_i to an item i_j is created for each positive feedback expressed by that user ($likes(u_i, i_j)$), thus $E = \{(u_i, i_j)|likes(u_i, i_j) = true\}$. Clearly, if each user and each item have at least a *positive* rating, then $|V| = |U| + |I|$. This is a very basic formulation, built on the ground of simple *collaborative* data points, since we just modeled user-item couples, as in collaborative filtering algorithms.

As previously explained, in this work we enriched this basic graph by introducing some *extra* nodes and *extra* edges, by exploiting the data points available in the LOD cloud. However, before performing this *enriching* process, it is mandatory to carry out a *mapping* step. The goal of the mapping procedure is to identify, for each item in the dataset, the corresponding DBpedia node the item refers to. As an example, we associate the book *The Shining* with its corresponding URI in the LOD cloud[5]. It is worth to emphasize that the mapping is a necessary and mandatory step to get an *entry point* to the LOD cloud. Once the mapping is completed, it is possible to gather all the extra features describing our items and to model our *tripartite graph* accordingly. Further details about the mapping procedure will be provided in Sect. 4.

Formally, after the mapping process we define an extended graph $G_{LOD} = \langle V_{ALL}, E_{ALL} \rangle$, where $V_{ALL} = V \cup V_{LOD}$ and $E_{ALL} = E \cup E_{LOD}$. In this case, E_{LOD} is the set of the new connections resulting from the properties encoded in the LOD cloud (e.g. *writer*, *subject*, *genre*, etc.), while V_{LOD} is the new set of nodes representing the resources gathered from the LOD cloud (e.g. *Stephen_King* or *Gothic_Novels*) that are connected to the items $i_1 \ldots i_m \in I$.

As previously stated, G_{LOD} is a tripartite graph, since beyond users and items also the resources gathered from the LOD cloud describing the items are now modeled. However, one could argue that some of the properties used to gather new resources from the LOD cloud may not be relevant for the recommendation task, and should be filtered out from the representation. This issue can be solved by applying *features selection* (FS) techniques on the complete *tripartite* representation and to maintain in the graph only those labeled as

---

[5] http://dbpedia.org/resource/The_Shining_(film).

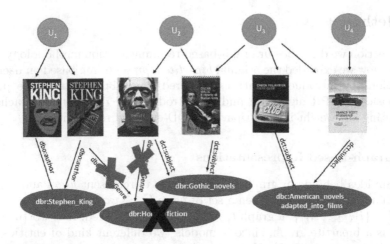

Fig. 2. The extended *tripartite* graph-based data model, including also (some of) the resources coming from DBpedia.

relevant by the algorithm. Due to space reasons, the discussion regarding the application of FS techniques on such a representation is out of the scope of the paper. However, in our previous research [14] we showed that the application of FS techniques can significantly improve the accuracy of the recommendation algorithm.

A partial example of such a representation is shown in Fig. 2. In this case, the resources connected to the properties available that are not deemed as relevant by a generic FS techniques (in this case, dbr:Horror_fiction) are filtered out.

3.2   Running Personalized PageRank

Given such *bipartite* and *tripartite* representations, we need an algorithm to provide each node $i \in I$ with a relevance score, in order to rank the available items and provide users with recommendations. To calculate the relevance of each item, we used a well-known variant of the PageRank called *Personalized PageRank (PPR)* [11]. In the original formulation of the PageRank an evenly distributed prior probability is assigned to each node ($\frac{1}{N}$, where N is the number of nodes). Differently from PageRank, PPR adopts a non-uniform personalization vector assigning different weights to different nodes to get a bias towards some nodes (specifically, the preferences of a specific user). As an example, Fig. 3 shows a configuration of PPR where 80% of the total weight is evenly distributed among items liked by user U_2 and 20% is evenly distributed among the remaining nodes. In this case, these values are set through a simple heuristic.

The main issue of the above described configuration is that the distribution of the weights does not consider the importance of the information coming from the LOD cloud. Indeed, all the nodes gathered from DBpedia (*red* nodes) are assigned with a very low probability, thus they poorly influence the random walks in the

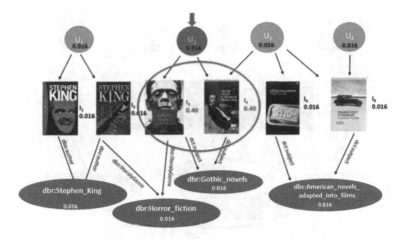

Fig. 3. Personalized PageRank distributing 80% of the weight to the items liked by User U_2 and 20% to the remaining nodes. (Color figure online)

graph and, in turn, the recommendations generated by the algorithm. This is a very important issue, since it is straightforward that a preference for a certain item also reflects some kind of preference for the *properties* of that item, as well (*e.g.*, the director, the genre of a movie, etc.), and the basic formulation of PPR ignores this aspect.

Accordingly, in this work we investigated whether different distribution of the weights in the *extended graph* may lead to an improvement of the accuracy of our recommendation strategy. As an example, in Fig. 4 we distributed half of the weight assigned to the items the user liked to the resources coming from DBpedia. In this way, it is more likely that the recommendations generated by the algorithms are also influenced by the preference the user expressed towards specific characteristics (resources) of the items she liked. In the experimental evaluation we tried to empirically validate this hypothesis by defining *different schemas* for distributing the weights among the available nodes and by analyzing their effectiveness in *a top-N recommendation task*.

4   Experimental Evaluation

In order to validate our approach, we carried out three experiments. First, we investigated whether graph-based recommender systems benefit of the introduction of LOD-based features *(Experiment 1)*. Next, we evaluated whether a different distribution of the weights in Personalized PageRank may lead to better recommendations as well *(Experiment 2)*. Finally, we perform a comparison with the performance of several state of the art techniques *(Experiment 3)*.

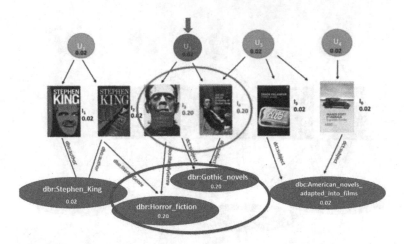

Fig. 4. Personalized PageRank distributing 40% of the weight to the items liked by User U_2 and 40% to the resources directly connected to the items. Finally, 20% is assigned to the remaining nodes.

4.1 Experimental Protocol

The evaluation was performed on two state of the art datasets, as MovieLens (ML1M)[6], a dataset for movie recommendation, and DBbook[7], a dataset for book recommendations which comes from the previously mentioned ESWC 2014 challenge. Some statistics about the datasets are provided in Table 1.

Experiments were performed by adopting a 5-fold cross validation as regards MovieLens, while a single training/test split for DBbook[8]. In the first case, we

Table 1. Description of the datasets.

	MovieLens	DBbook
Users	6040	6181
Items	3883	6733
Ratings	1,000,209	72372
Sparsity	96.42%	99.83%
Positive ratings	57.51%	45.85%
Avg. ratings/user $\pm\ \sigma$	165.59 $\pm$ 192.74	11.70 $\pm$ 5.85
Median/mode per user	96/21	11/5
Avg. ratings/item $\pm\ \sigma$	269.88 $\pm$ 384.04	10.74 $\pm$ 27.14
Median/mode per item	124/1	4/1

---

[6] http://grouplens.org/datasets/movielens/.
[7] http://www.di.uniba.it/%7Eswap/datasets/dbbooks_data.zip.
[8] http://challenges.2014.eswc-conferences.org/index.php/RecSys.

built the splits on our own, while for DBBOOK we used the split used during the challenge. We face a Top-N recommendation task: given that MovieLens preferences are expressed on a 5-point Likert scale, we deem as *positive* ratings only those equal to 4 and 5. On the other side, DBbook is already available as *binarized*, thus no further processing was needed.

In order to enrich the graph G with LOD-based features, each item in the dataset was mapped to a DBpedia entry. As regards MovieLens, 3,300 movies were successfully mapped (85% of the items) while 6,600 items (98.02%) from DBbook dataset were associated to a DBpedia node. The mapping was performed by querying a DBpedia SQL Endpoint by using the *title* of the movie or the *name* of the book. Non-exact mappings were managed by exploiting a similarity measure based on Levenshtein distance[9]. We made our mappings available online[10]. The items for which a DBpedia entry was not found were only represented by using *collaborative* data points. MovieLens entries were described through 60 different DBpedia properties, while DBbook ones using 70 properties.

As recommendation algorithm we used the Personalized PageRank (damping factor equal to 0.85, as in [18]) and we compared the effectiveness of the graph topologies we previously introduced:

- G, which models the basics *collaborative* information about user ratings;
- G_{LOD}, which enrichs G by introducing all the features gathered from DBpedia;
- G_{LOD-FS}, which filters the enriched graph by only considering a subset of the resources connected to the properties selected by a feature selection technique FS.

As regards G_{LOD-FS} configuration, we used Principal Component Analysis (PCA) and Information Gain (IG) as feature selection techniques, since in our previous research [14] they emerged as the best-performing ones for the datasets we took into account. We made available the list of the properties chosen by each algorithm[11]. It is worth to note that also some *datatype* property (as movie length) is included in the list. For each of the above mentioned tripartite representations we evaluated *four different strategies* to distribute the weights in Personalized PageRank algorithm. The three values in the configurations describe the weight assigned to the *items* the user liked, to the *nodes* gathered from the LOD and to all the other nodes in the graph, respectively.

- 80/0/20, the original formulation which does not consider the resources gathered from DBpedia, thus only distributing 80% of the weights among the items the user liked, as in Fig. 3;
- 60/20/20, which distributes 60% of the weight among the items the user liked and gives the remaining 20% to the resources directly connected to the items;
- 40/40/20, which equally distributes the weights among the items and the resources describing the items, as in Fig. 4;

---

[9] https://en.wikipedia.org/wiki/Levenshtein_distance.
[10] https://tinyurl.com/hj4q5go.
[11] http://tinyurl.com/zmfulol.

– 20/60/20, which emphasizes the importance of the resources gathered from the LOD, giving them 60% of the weight.

To sum up, we evaluated different configurations with a fixed increase (20%) of the weight assigned to the nodes gathered from DBpedia. For each tripartite representation G_{LOD} and G_{LOD-FS} four different run of the experiments were carried out. Given that we evaluated two different FS techniques, twelve different configurations were compared in the experiments. Moreover, it is also worth to emphasize that Personalized PageRank has to be executed for each user in the dataset, since the distribution of the weights change as the preferences of the user change as well.

For each experimental session, the performance of each graph topology was evaluated by exploiting state-of-the-art ranking metrics as *F1-Measure, Mean Average Precision (MAP) and Normalized Discounted Cumulative Gain (NDCG)*. In order to ensure the reproducibility of the results, metrics were calculated by exploiting the Rival evaluation framework[12]. Statistical significance was assessed by exploiting Wilcoxon and Friedman tests, chosen after running the Shapiro-Wilk test[13], which revealed the non-normal distribution of the data. Finally, the source code of our graph-based recommendation framework has been published on GitHub[14].

4.2 Discussion of the Results

Experiment 1. In the first Experiment we evaluated the effectiveness of our *tripartite* graph-based representation with respect to the original *bipartite* representation that did not include any feature gathered from the LOD cloud. Results are presented in Table 2.

As emerged from the results, the injection of the knowledge extracted from DBpedia significantly improves the accuracy of the basic bipartite representation (reported as G) for all the metrics we took into account. This validates our conjecture that the introduction of LOD-based data points positively affects the overall effectiveness of our recommendation algorithm. As expected, our approach also benefits of the adoption of feature selection techniques (with the exception of F1@5 on MovieLens data), since configurations based on PCA and IG obtained the best results on MovieLens and DBbook, respectively. The improvement is statistically significant ($p < 0.05$) only for G_LOD and $G\_LOD - IG$ while it is not for $G\_LOD - PCA$. Even if the increase is typically tiny, the adoption of a nonparametric test as Mann-Whitney made the difference resulting as statistically significant.

It is worth to note that even without feature selection our enriched representation tend to overcome the baseline (with the exception of F1@5 on DBbook data). In this first experiment we did not modify the distribution of the weights

---

[12] http://rival.recommenders.net/.

[13] http://en.wikipedia.org/wiki/Shapiro-Wilk_test.

[14] https://github.com/cataldomusto/lod-recsys/.

Table 2. Results of Experiment 1. Configurations overcoming the baseline are reported in bold.

MovieLens	Weight	F1@5	MAP	NDCG
G		0.5396	0.7842	0.9003
G_LOD	80/0/20	**0.5406**	**0.7845**	**0.9010**
G_LOD-PCA	80/0/20	**0.5398**	**0.7853**	**0.9014**
G_LOD-IG	80/0/20	**0.5404**	**0.7852**	**0.9013**

DBbook	Weight	F1@5	MAP	NDCG
G		0.5507	0.7018	0.7861
G_LOD	80/0/20	0.5494	**0.7040**	**0.7886**
G_LOD-PCA	80/0/20	**0.5508**	**0.7056**	**0.7897**
G_LOD-IG	80/0/20	**0.5528**	**0.7062**	**0.7898**

in PPR algorithm, which is run by using the classic proportion (80% of the weights to the items the user liked, 20% to all the other nodes in the graph).

To sum up, we can conclude this first experiment by confirming our hypothesis that the exogenous knowledge encoded in DBpedia contributes to improving the accuracy of our graph-based recommender system.

Experiment 2. In the *second experiment* we validated our insight that a different weights distribution in PPR, which emphasizes the importance of nodes gathered from DBpedia, may lead to better recommendations. Our results, presented in Tables 3 and 4 for MovieLens and DBbook, respectively, gave interesting insights.

The first outcome of the experiment is the connection between the sparsity of the dataset and the optimal distribution of the weights. Indeed, the configuration that distributes most of the weight to the resources directly connected to the items the user liked (reported as 20/60/20) never overcame the baseline for all the graph topologies. This means that the when the datasets is not sparse (as for MovieLens) it is not necessary to distribute weights to the nodes gathered from the LOD, since random walks have to be mainly driven by *collaborative data points*. Moreover, a different distribution of the weights neither improve our metrics on G-LOD configurations. This is probably due to the fact that some of the movie-related resources injected in the graph are noisy, thus by distributing some of the weight to that resources it is likely that poorly relevant movies may be recommended.

The only configurations that benefit of a different weight distribution are those that use feature selection to filter out the resources connected to non-relevant properties. Indeed, all the metrics we reported got an statistically significant improvement ($p < 0.05$) on both G-LOD-PCA and G-LOD-IG graphs. Overall, the best-performing configuration is that based on IG which equally distributes the weight among item nodes and resource nodes (40/40/20). In this case we obtained a significant improvement over the baseline we took into account, thus confirming our experimental hypothesis.

Table 3. Results of Experiment 2 on `MovieLens` data. The baselines running PPR with the original weight distribution are highlighted with a grey background. The configurations overcoming the baselines are reported in bold, the overall best-performing configuration is highlighted with a dark grey background.

MovieLens	Weight	F1@5	MAP	NDCG
G-LOD	80/0/20	*0.5406*	*0.7845*	*0.9010*
G-LOD	60/20/20	0.5399	0.7833	0.9003
G-LOD	40/40/20	0.5389	0.7825	0.8999
G-LOD	20/60/20	0.5380	0.7813	0.8994
G-LOD-PCA	80/0/20	*0.5398*	*0.7853*	*0.9014*
G-LOD-PCA	60/20	**0.5403**	**0.7855**	**0.9015**
G-LOD-PCA	40/40/20	**0.5401**	**0.7856**	**0.9016**
G-LOD-PCA	20/60/20	0.5393	0.7847	0.9011
G-LOD-IG	80/0/20	*0.5404*	*0.7852*	*0.9013*
G-LOD-IG	60/20/20	**0.5407**	0.7852	0.9013
G-LOD-IG	40/40/20	**0.5409**	**0.7857**	**0.9018**
G-LOD-IG	20/60/20	0.5400	0.7851	0.9012

Our conjecture about the connection between the sparsity of the dataset and the distribution of the weights is also confirmed on `DBbook`. As shown in Table 4, a different distribution of the weights improve our metrics also on the complete *tripartite* graph `G-LOD`. This behavior is due to the high sparsity of the dataset (see Table 1), which makes collaborative data points coming from user-item connections poorly significant to properly model user preferences, thus also item-properties connections gathered from `DBpedia` have to be modeled as well.

In this case the best-performing configuration is again that based on Information Gain as feature selection technique and a distribution of the weights which assigns a small part of the weight (60/20/20) to the resources connected to the nodes extracted from the LOD cloud. However, also results emerging from `DBbook` confirmed our hypothesis since a different distribution of the weights improved the accuracy of our recommendation strategy on these data as well. All the improvements are statistically significant for $p < 0.05$.

Experiment 3. Finally, in the third experiment we compared our best-performing configuration to the results obtained by several baselines. To calculate the results, we exploited MyMediaLite Recommender System library[15]. As baselines, User-to-User Collaborative Filtering (U2U-KNN), Item-to-Item Collaborative Filtering (I2I-KNN), and a simple popularity-based approach suggesting the items the users (positively) voted the most were used. Moreover, we also compared our approach to Bayesian Personalized Ranking Matrix Factorization (BPRMF) presented in [20], and its extended version which also models side information, since they both emerged as the best-performing baselines in related

---

[15] http://www.mymedialite.net/.

literature [17]. Due to space reasons, we only report the best performing configuration of each baseline: as regards U2U-KNN and I2I-KNN, neighborhood size was set to 80., while BPRMF was run by setting the number of latent factors equal to 100. Results are reported in Table 5. Unfortunately, it was not possible to compare our approach to other methodologies based on Linked Open Data (as [17] or [19]) since most of these approaches did not use publicly available datasets or were based on user studies.

Table 4. Results of Experiment 2 on DBbook data. The baselines running PPR with the original weight distribution are highlighted with a grey background. The configurations overcoming the baselines are reported in bold, the overall best-performing configuration is highlighted with a dark grey background.

DBbook	Weight	F1@5	MAP	NDCG
G-LOD	80/0/20	*0.5494*	*0.7040*	*0.7886*
G-LOD	60/20/20	**0.5498**	**0.7041**	**0.7887**
G-LOD	40/40/20	**0.5504**	**0.7042**	**0.7888**
G-LOD	20/60/20	**0.5496**	**0.7042**	**0.7889**
G-LOD-PCA	80/0/20	0.5507	0.7056	0.7897
G-LOD-PCA	60/20/20	0.5503	0.7054	0.7896
G-LOD-PCA	40/40/20	0.5499	0.7055	**0.7899**
G-LOD-PCA	20/60/20	0.5504	0.7054	0.7897
G-LOD-IG	80/0/20	0.5528	0.7062	0.7898
G-LOD-IG	60/20/20	**0.5533**	**0.7065**	**0.7905**
G-LOD-IG	40/40/20	0.5526	**0.7064**	**0.7903**
G-LOD-IG	20/60/20	0.5524	0.7060	**0.7901**

Table 5. Experiment 3. Comparison to baselines. Best-performing configurations are highlighted in grey.

MovieLens	F1@5	MAP	NDCG
G-LOD-IG (40/40)	0.5409	0.7857	0.9018
U2U	0.5220	0.7512	0.8721
I2I	0.5043	0.7230	0.8445
Popularity	0.514	0.7271	0.8602
BPRMF	0.5179	0.7338	0.8631
BPRMF+Side	0.5218	0.7498	0.8704

DBbook	F1@5	MAP	NDCG
G-LOD-IG (60/20)	0.5533	0.7065	0.7905
U2U	0.5193	0.6588	0.7417
I2I	0.5111	0.6448	0.7328
Popularity	0.5296	0.6549	0.7562
BPRMF	0.5290	0.6655	0.7589
BPRMF+Side	0.5304	0.6701	0.7631

As shown in the table, our approach based on PPR and LOD significantly outperforms all the baselines took into account. Results are particularly interesting since our methodology is able to overcome also the accuracy of widespread and well-performing baselines based on matrix factorization as BPRMF, which are commonly considered as *state-of-the-art techniques* in recommender systems community. These results definitely confirmed the soundness of our methodology as well as the insight of better distributing weights in PPR to emphasize the importance of the resources gathered from DBpedia.

5   Conclusions and Future Work

In this work we proposed a *semantics-aware recommendation methodology* based on Personalized PageRank algorithm, and we evaluated different techniques to distribute the weights in our graph-based representation. Experimental results showed that our graph-based recommender can benefit of the infusion of novel knowledge coming from the LOD cloud. Interestingly, the impact of LOD-based features is particularly positive when data are more sparse, making our framework very suitable for also for challenging recommendation settings. Finally, our methodology was able to overcome several baselines on two state of the art datasets. A publicly available implementation of the framework as well as of the splits used for the evaluation guarantee the reproducibility of the experimental results.

As future work we will investigate the impact of resources not directly connected with the items the user liked, in order to assess about the usefulness of injecting more knowledge in our representation, and we will also evaluate how a different distribution of the weights impacts on different recommender systems metrics as the *diversity* or the *serendipity* of the recommendations.

References

1. Auer, S., Bizer, C., Kobilarov, G., Lehmann, J., Cyganiak, R., Ives, Z.: DBpedia: a nucleus for a web of open data. In: Aberer, K., Choi, K.-S., Noy, N., Allemang, D., Lee, K.-I., Nixon, L., Golbeck, J., Mika, P., Maynard, D., Mizoguchi, R., Schreiber, G., Cudré-Mauroux, P. (eds.) ASWC/ISWC -2007. LNCS, vol. 4825, pp. 722–735. Springer, Heidelberg (2007). doi:10.1007/978-3-540-76298-0_52
2. Baluja, S., Seth, R., Sivakumar, D., Jing, Y., Yagnik, J., Kumar, S., Ravichandran, D., Aly, M.: Video suggestion and discovery for YouTube: taking Random Walks through the view graph. In: Proceedings of the 17th International Conference on World Wide Web, pp. 895–904. ACM (2008)
3. Basile, P., Musto, C., de Gemmis, M., Lops, P., Narducci, F., Semeraro, G.: Aggregation strategies for linked open data-enabled recommender systems. In: European Semantic Web Conference (ESWC) (2014)
4. Basile, P., Musto, C., Gemmis, M., Lops, P., Narducci, F., Semeraro, G.: Content-based recommender systems + DBpedia knowledge = semantics-aware recommender systems. In: Presutti, V., Stankovic, M., Cambria, E., Cantador, I., Iorio, A., Noia, T., Lange, C., Reforgiato Recupero, D., Tordai, A. (eds.) SemWebEval 2014. CCIS, vol. 475, pp. 163–169. Springer, Cham (2014). doi:10.1007/978-3-319-12024-9_21

5. Baumann, S., Schirru, R.: Using linked open data for novel artist recommendations. In: Proceedings of the 13th International Society for Music Information Retrieval Conference, ISMIR (2012)
6. Bizer, C.: The emerging web of linked Data. IEEE Intell. Syst. **24**(5), 87–92 (2009)
7. Bogers, T.: Movie recommendation using Random walks over the contextual graph. In: Proceedings of the 2nd International Workshop on Context-Aware Recommender Systems (2010)
8. Bostandjiev, S., O'Donovan, J., Höllerer, T.: TasteWeights: a visual interactive hybrid recommender system. In: Proceedings of the Sixth ACM Conference on Recommender Systems, pp. 35–42. ACM (2012)
9. de Gemmis, M., Lops, P., Musto, C., Narducci, F., Semeraro, G.: Semantics-aware content-based recommender systems. In: Ricci, F., Rokach, L., Shapira, B. (eds.), Recommender Systems Handbook, pp. 119–159. Springer (2015)
10. de Gemmis, M., Lops, P., Semeraro, G., Musto, C.: An investigation on the serendipity problem in recommender systems. Inf. Process. Manage. **51**(5), 695–717 (2015)
11. Haveliwala, T.H.: Topic-sensitive PageRank: a context-sensitive ranking algorithm for web search. IEEE Trans. Knowl. Data Eng. **15**(4), 784–796 (2003)
12. Hotho, A., Jäschke, R., Schmitz, C., Stumme, G., Althoff, K.-D.: FolkRank: A ranking algorithm for folksonomies. In: LWA, vol. 1, pp. 111–114 (2006)
13. Kantor, P.B., Rokach, L., Ricci, F., Shapira, B.: Recommender Systems Handbook. Springer, New York (2011)
14. Musto, C., Lops, P., Basile, P., de Gemmis, M., Semeraro, G.: Semantics-aware graph-based recommender systems exploiting linked open data. In: Proceedings of the 2016 Conference on User Modeling Adaptation and Personalization, UMAP 2016, pp. 229–237. ACM, New York (2016)
15. Musto, C., Semeraro, G., Lops, P., Gemmis, M., Narducci, F.: Leveraging social media sources to generate personalized music playlists. In: Huemer, C., Lops, P. (eds.) EC-Web 2012. LNBIP, vol. 123, pp. 112–123. Springer, Heidelberg (2012). doi:10.1007/978-3-642-32273-0_10
16. Nguyen, P.T., Tomeo, P., Noia, T., Sciascio, E.: Content-based recommendations via DBpedia and freebase: a case study in the music domain. In: Arenas, M., Corcho, O., Simperl, E., Strohmaier, M., d'Aquin, M., Srinivas, K., Groth, P., Dumontier, M., Heflin, J., Thirunarayan, K., Staab, S. (eds.) ISWC 2015. LNCS, vol. 9366, pp. 605–621. Springer, Cham (2015). doi:10.1007/978-3-319-25007-6_35
17. Ostuni, V.C., Di Noia, T., Di Sciascio, E., Mirizzi, R.: Top-N recommendations from implicit feedback leveraging linked open data. In: Proceedings of the ACM Conference on Recommender Systems, pp. 85–92. ACM (2013)
18. Page, L., Brin, S., Motwani, R., Winograd, T.: The PageRank Citation Ranking: Bringing Order to the Web (1999)
19. Passant, A.: dbrec — Music recommendations using DBpedia. In: Patel-Schneider, P.F., Pan, Y., Hitzler, P., Mika, P., Zhang, L., Pan, J.Z., Horrocks, I., Glimm, B. (eds.) ISWC 2010. LNCS, vol. 6497, pp. 209–224. Springer, Heidelberg (2010). doi:10.1007/978-3-642-17749-1_14
20. Rendle, S., Freudenthaler, C., Gantner, Z., Schmidt-Thieme, L.: BPR: Bayesian personalized ranking from implicit feedback. In: Proceedings of the Twenty-Fifth Conference on Uncertainty in Artificial Intelligence, pp. 452–461. AUAI Press (2009)

Terminological Cluster Trees for Disjointness Axiom Discovery

Giuseppe Rizzo[✉], Claudia d'Amato, Nicola Fanizzi, and Floriana Esposito

LACAM – Dipartimento di Informatica,
Università degli Studi di Bari "Aldo Moro", Bari, Italy
{giuseppe.rizzo1,
claudia.damato,nicola.fanizzi,floriana.esposito}@uniba.it
http://lacam.di.uniba.it

Abstract. Despite the benefits deriving from explicitly modeling concept disjointness to increase the quality of the ontologies, the number of disjointness axioms in vocabularies for the *Web of Data* is still limited, thus risking to leave important constraints underspecified. Automated methods for discovering these axioms may represent a powerful modeling tool for knowledge engineers. For the purpose, we propose a machine learning solution that combines (unsupervised) distance-based clustering and the divide-and-conquer strategy. The resulting *terminological cluster trees* can be used to detect candidate disjointness axioms from emerging concept descriptions. A comparative empirical evaluation on different types of ontologies shows the feasibility and the effectiveness of the proposed solution that may be regarded as complementary to the current methods which require supervision or consider atomic concepts only.

1 Introduction

With the growth of the Web of Data, along with the *Linked Data* initiative [13], a large number of datasets are being published using a standard data model that connects lots of knowledge bases within a uniform semantic space. In this context, Web ontologies are used as formal vocabularies that support many important services based on *automated reasoning*, such as *classification*, *query answering*, *population* and *enrichment*, or *reconciliation* (*instance matching*). In this perspective, ontologies represent a means to ensure the quality of data.

Debugging strategies may be employed to prevent the introduction of conflicting assertions that hinder the employment of reasoning services. However many ontologies still represent simplified data models for the targeted domain failing to capture some underlying intended constraints [21]. A common problem is the lack of an explicit representation of negative knowledge, usually expressed in the form of *disjointness axioms*. Conversely, the acquisition of such axioms may enhance the mentioned rich services.

In the literature, sundry approaches for discovering disjointness axioms have been proposed. Recent methods apply association rule mining [18,19]. However,

© Springer International Publishing AG 2017
E. Blomqvist et al. (Eds.): ESWC 2017, Part I, LNCS 10249, pp. 184–201, 2017.
DOI: 10.1007/978-3-319-58068-5_12

they can capitalize on the available of intensional knowledge only to some marginal extent. In these works, heterogeneous sources are exploited, with most of the features being lexical, also based on external corpora. Additionally, most of the approaches move from the assumption that disjointness may hold among concepts when the sets of their instances, that can be thought of as empirical approximations of their extensions, do not tend to *overlap* [11,20]. Hence a more data-driven approach could be exploited: it may derive from finding partitions of similar individuals occurring in the knowledge base according to some criterion of choice, by maximizing the separation (i.e. minimizing the overlap) among different partitions. This objective boils down to a *clustering problem* [1] which is a classic topic in *machine learning*. In the context of the *Semantic Web* (SW), clustering methods for individuals described by ontologies have been proposed, extending classic algorithms such as K-MEANS or K-MEDOIDS [9] with the ability of taking into account intensional knowledge. They have been mainly employed for concept learning or for the automated detection of *concept drift* or *novelty* [10].

In the line of the unsupervised statistical approaches, ours relies on methods that produce hierarchical clustering structures, while taking into account the intensional knowledge provided by the ontology. The goal is to derive potential disjointness axioms, by exploiting the background knowledge on the schema, similarly to *relational learning* frameworks [15]. Indeed, we adopt an approach based on *conceptual clustering*, that aims at learning intensional descriptions of emerging clusters of individuals that may involve even complex concept descriptions, differently from other unsupervised approaches. Our solution is based on a novel form of logical tree model [7], dubbed *terminological cluster tree* [16], that can be regarded as an extension of the terminological decision tree [8]. Both are produced through *divide-and-conquer* algorithms, but while the latter are essentially classifiers that solve supervised concept learning problems exploiting information-based heuristics (*information gain* or other *purity* measures), the former rely on specific metrics and on a notion of *cluster prototype* [9,10].

Unlike other (unsupervised) clustering models, the proposed solution also aims at intensional definitions of the clusters, i.e. concept descriptions that describe their individuals. Another advantage is that the number of clusters – which has a strong impact on the quality of the clustering structure – is not a required parameter; conversely, descends from the number of dense data regions found in the given instance space. Indeed, in the induction of (terminological) cluster trees this number depends on a notion of *purity*, that determines the stop condition for further branchings to prevent compromising the clusters *separation*. Once the tree is grown, groups of (disjoint) clusters located at sibling nodes identify concepts involved in candidate disjointness axioms to be derived. The discovered axioms can be validated by a domain expert/ontology engineer and/or may even be automatically involved in a debugging process for eliciting cases of inconsistency which cannot arise when disjointness axioms are lacking.

Summary. In the next section, related works are briefly surveyed. In Sect. 3, disjointness axiom discovery problem is formalized as a clustering problem for individuals in a knowledge base. Section 4 illustrates the approach to the induction of terminological cluster trees and their application to the targeted problem. Section 5 presents a comparative experimental evaluation of the proposed solution on common ontologies. Finally, Sect. 6 concludes this work delineating future research directions.

2   Related Work

The problem of discovering the disjointness axioms to enrich and improve the quality of ontological knowledge bases has been receiving a growing attention. In early works, the mentioned *strong disjointness assumption* [5] (SDA), which states that the children of a common parent in the subsumption hierarchy should be disjoint, has been exploited in a pinpointing algorithm for semantic *clarification* (i.e. the process of automatically enriching ontologies by appropriate disjointness statements [17]. Focusing first on text and successively on RDF datasets, unsupervised methods for mining axioms, including disjointness axioms, have been proposed [12,20]. Their main limitation is the inability to exploit background knowledge, which on the contrary may help in increasing the number of axioms discovered while filtering out unnecessary or wrong axioms. The main limitations of supervised methods is the necessity of axioms for training that may demands costly work by domain experts.

Besides, methods based on *relational learning* [15] and *formal concept analysis* [3] have been proposed, but none specifically aimed at assessing the quality of the induced axioms. This is pointed out also in [19] and additional approaches [11,18] based on *association rule mining* have been introduced to better address this limitation. The goal was studying the correlation between classes. Specifically, *(negative) association rules* and the use of a *correlation coefficient* have been considered. Also in these cases, background knowledge is not explicitly exploited. In [15], a tool for *repairing* various types of ontology modeling errors is described; it uses methods from the DL-LEARNER framework [14] to enrich ontologies with axioms induced from existing instances.

Our solution is based on an unsupervised approach, deriving from previous works on concept learning and inductive classification [6]. Specifically, we propose a hierarchical conceptual clustering method that is able to provide intensional cluster descriptions. It exploits a novel form of a family of semidistances for the individuals in knowledge bases [6] which involves reasoning on the knowledge base. The method is grounded on the notion of medoid as cluster prototype to give a topological structure to the representation of the instance space [1]. Related, but partitive clustering approaches [1], such as the BISECTING K-MEDOIDS [9] or the *partition around medoids*, combined with evolutionary programming [10] have been proposed. They cluster individuals in Web ontologies by exploiting metrics that are related to those adopted in this work. They can be easily extended for producing hierarchical structures of clusters. However, the derivation of related concepts as intensional definitions for the clusters requires the adoption of additional and suitable concept learning algorithms.

Specifically, the method proposed in this paper relies on logic tree models [4] which essentially adopt a divide-and-conquer strategy to derive a hierarchical structure. The learning method can work both in supervised and unsupervised mode, depending on the availability of information about the instance classification to be exploited for separating sub-groups of instances. Terminological decision trees were derived [8] in the former case, while for the latter case, first-order logic clustering trees [7] were proposed to induce concepts expressed in *clausal logics* for the clusters. The C0.5 system, which is integrated in the TILDE framework [4], is able to induce concepts as conjunctions of literals (clause bodies) installed at inner nodes. Almost all these exiting methods are grounded on the exploitation of an heuristic based on the *information gain*, employed in the supervised case. Differently, our approach tends to maximize the separation between cluster medoids according to a semi-distance measure that can be also computed before the learning phase [6] making a more efficient use of the computationally expensive reasoning services.

3 Disjointness Discovery as a Conceptual Clustering Problem

In this section, we formalize the problem of *discovering concept disjointness axioms* from an ontological knowledge base in terms of a *clustering* task. We will borrow notation and terminology from *Description Logics* (DLs) [2], being the theoretical foundation of the standard representation languages for the SW. Hence, we will use the terms *concept (description)* and *role* as synonyms of *class* and *property*, respectively and we will denote a *knowledge base* (KB) with $\mathcal{K} = \langle \mathcal{T}, \mathcal{A} \rangle$, where $\mathcal{T}$ is the *TBox* (containing terminological axioms regarding concepts and roles) and $\mathcal{A}$ is the *ABox* (containing concept/role assertions regarding individuals). $\mathsf{Ind}(\mathcal{A})$ will denote the set of individuals (resource names) occurring in $\mathcal{A}$. *Subsumption, equivalence* and logic *entailment* will be denoted with the usual symbols.

Before formalizing the problem of discovering concept disjointness axioms, for the sake of clarity and completeness, we recall some basic *clustering* methods. Clustering is an unsupervised learning task aiming at grouping a collection of objects into subsets or *clusters*, such that those within each cluster are more closely related/similar to one another, than objects assigned to different clusters [1]. In the general setting, an object is usually described in terms of features from a selected set $\mathcal{F}$; a measure of similarity between objects is expressed in terms of a distance function, e.g. in the case of attribute-value datasets of objects that are often described by tuples of numeric features, the *Euclidean* distance (or its extensions) is typically adopted. A more complex goal is to move from flat to natural *hierarchical clustering* structures. Another difference among the various clustering models is related to the form of membership of the objects with respect to the clusters. In the simplest (*crisp*) case, e.g. K-MEANS, cluster membership *can be exclusive*: each object belongs to one cluster. Extensions, such as FUZZY C-MEANS or EM [1], admit overlapping clusters as the objects exhibit a graded membership (*responsibility*) w.r.t. the clusters. An interesting class of methods is represented by the *conceptual clustering* approaches. In the resulting clustering structures, the objects are arranged into clusters that are intentionally, rather than extensionally, defined. Differently from other methods, conceptual clustering algorithms may exploit available background knowledge for building descriptions for each cluster. Besides the propositional data representations (or the equivalent vector spaces mentioned above) more expressive data representation, e.g. through richer logic languages may be necessary. When such expressive representations are considered, suitable (dis)similarity measures have to be adopted.

Moving from the observation that a disjointness axiom, involving two or more concepts, may hold if their extensions do not overlap (as introduced in Sect. 1), the task of discovering disjointness axioms may be regarded as an unsupervised conceptual clustering problem aiming at finding separate partitions of individuals of the KB (such that each subset consists of similar individuals, according to a given similarity criterion) and producing intensional descriptions for them. The problem is defined as follows:

Definition 3.1 (Disjointness axiom discovery as a conceptual clustering problem)

Given
- *a knowledge base* $\mathcal{K} = \langle \mathcal{T}, \mathcal{A} \rangle$
- *a set of training individuals* $\mathbf{I} \subseteq \mathsf{Ind}(\mathcal{A})$

Find
- *a partition* Π *of* $\mathbf{I}$ *in a set of pairwise disjoint clusters* $\Pi = \{\mathbf{C}_1, \ldots, \mathbf{C}_{|\Pi|}\}$
- *for each* $i = 1, \ldots, |\Pi|$, *a concept description* D_i *that describes* $\mathbf{C}_i$, *so that:*
    $\forall a \in \mathbf{C}_i : \ \mathcal{K} \models D_i(a)$ *and*
    $\forall b \in \mathbf{C}_j, j \neq i : \ \mathcal{K} \models \neg D_i(b)$.
    Hence $\forall D_i, D_j, i \neq j : \ \mathcal{K} \models D_j \sqsubseteq \neg D_i$.

Note that the number of clusters (say $K = |\Pi|$) is not a required parameter to be provided tentatively. Also note that, the problem of discovering disjointness axioms resorting to machine learning methods can be formalized in different ways, depending on the type of approach (supervised or unsupervised) to be employed. In the next section, a solution to the formalized problem is presented.

4 Terminological Cluster Trees for Disjointness Learning

The proposed approach is grounded on a two-steps process. In the first step, given a knowledge base, clusters and the related concepts that describe them are discovered and organized in a tree structure. In the second step, the induced structure is exploited for learning a set of disjointness axioms. The model can be formally defined as follows:

Definition 4.1 (Terminological cluster tree). *Given a knowledge base* $\mathcal{K}$, *a terminological cluster tree (TCT) is a binary logical tree where each node stands for a cluster of individuals* $\mathbf{C}$ *and such that:*

Algorithm 1. Routines for inducing a TCT

```
 1 const ν, δ: thresholds                          19 function SPLIT(I, E): pair of sets of individuals
 2                                                  20 input I: set of individuals
 3 function INDUCETCT(I, C): TCT                     21        E: concept description
 4 input I: set of individuals                      22 begin
 5        C: concept description                     23    ⟨P, N⟩ ← RETRIEVEPOSNEG(I, E, δ)
 6 begin                                             24    b ← p(P) { prototype }
 7    T ← new TCT                                    25    c ← p(N) { prototype }
 8    if STOPCONDITION(I, ν) then                    26    I_left ← ∅
 9       T ← ⟨null, I, null, null⟩                   27    I_right ← ∅
10    else                                           28    for each a ∈ I
11       S ← ρ(C) { set of candidate specializations } 29       if d(a, b) ≤ d(a, c) then
12       E* ← SELECTBESTCONCEPT(S, I)                30          I_left ← I_left ∪ {a}
13       ⟨I_left, I_right⟩ ← SPLIT(I, E*)            31       else
14       T_left ← INDUCETCT(I_left, E*)              32          I_right ← I_right ∪ {a}
15       T_right ← INDUCETCT(I_right, ¬E*)           33
16       T ← ⟨E*, I, T_left, T_right⟩                34    return ⟨I_left, I_right⟩
17    return T                                       35 end
18 end
```

- *each node contains a concept D (defined over the signature of $\mathcal{K}$) describing $\mathbf{C}$;*
- *each edge departing from an internal node corresponds to the outcome of the membership test of individuals with respect to D.*

A tree-node is represented by a quadruple $\langle D, \mathbf{C}, T_{left}, T_{right} \rangle$ with the left and right subtrees connected by either departing edge.

The construction of the model combines elements of logical decision trees induction (recursive partitioning and refinement operators for specializing concept descriptions) with elements of *instance-based learning* (a distance measure over the instance space). The details of the algorithms for growing a TCT (*step 1.*) and deriving intensional definitions of disjoint concepts (*step 2.*) are reported in the sequel.

4.1 Growing Terminological Cluster Trees

A TCT T is induced by means of a recursive strategy (see Algorithm 1), which follows the schema proposed for terminological decision trees (TDTs) [8]. The ultimate goal is to find a partition of *pure* clusters.

The main routine INDUCETCT is to be invoked passing $\mathbf{I}$ and $\top$ as parameters. In this recursive function, the base case tests the STOPCONDITION predicate, i.e. whether the measure of cohesion of the cluster $\mathbf{I}$ exceeds a given threshold ν. Further details about the heuristics and the stop condition will be reported later on.

In the inductive step, which occurs when the stop condition does not hold, the current (parent) concept description C has to be specialized using a refinement operator (ρ) that spans over a search space of concepts subsumed by C. A set of candidate specializations $\mathbf{S} = \rho(C)$ is obtained. For each $E \in \mathbf{S}$, the sets of positive and negative individuals, i.e. the instances of E and of $\neg E$, respectively denoted by $\mathbf{P}$ and $\mathbf{N}$, are retrieved by RETRIEVEPOSNEG. A tricky situation may occur when either **N** or **P** is empty for a given E (e.g. in the absence of disjointness axioms). In such a case, RETRIEVEPOSNEG assigns individuals in **I\P** to **N** (resp. in **I\N** to **P**) when the distance between them and the prototype of **P** (resp. **N**) exceeds the threshold δ. A representative element for **P** and **N** is determined as a prototype, i.e. their *medoid*, a central element with the minimal average distance w.r.t. the other elements in the cluster. Then, function SELECTBESTCONCEPT evaluates the candidate specializations in terms of the *cluster separation* computed through a heuristic (see Eq. 1) and returns the best concept $E^* \in \mathbf{S}$, that is the one for which the distance between the medoids of the related positive and negative sets is maximized. Then E^* is installed in the current node. Hence, the individuals are partitioned by SPLIT to be routed along the left or right branch of E^*. Differently from TDTs, the routine does not decide the branch where the individuals will be sorted according to a concept membership test (instance check): it splits individuals according to the distance w.r.t. the tow prototypes, i.e. the medoids of **P** and **N**.

This divide-and-conquer algorithm is applied recursively until the instances routed to a node satisfy the stop condition. Note that, the number of the clusters is not required as an input but it depends on the number of branches grown: it is naturally determined by the algorithm according to the data distribution in the regions of the instance space.

The proposed approach relies on a downward refinement operator that can generate the concepts to be installed in child-nodes performing a specialization process on the concept, say C, installed in a parent-node or its complement:

ρ_1 by adding a concept atom (or its complement) as a conjunct: $C' = C \sqcap (\neg)A$;

ρ_2 by adding a general existential restriction (or its complement) as a conjunct: $C' = C \sqcap (\neg)(\exists)R.\top$;

ρ_3 by adding a general universal restriction (or its complement) as a conjunct: $C' = C \sqcap (\neg)(\forall)R.\top$;

ρ_4 by replacing a sub-description C_i in the scope of an existential restriction in C with one of its refinements: $\exists R.C'_i \in \rho(\exists R.C_i) \wedge C'_i \in \rho(C_i)$;

ρ_5 by replacing a sub-description C_i in the scope of a universal restriction with one of its refinements: $\forall R.C'_i \in \rho(\forall R.C_i) \wedge C'_i \in \rho(C_i)$.

Note that the cases of ρ_4 and ρ_5 are recursive. Please, also note that the refinement operator take the KB (and particularly the TBox) strictly into account. Indeed refinements that are consistent with the KB are always returned.

The algorithms for growing TCTs and TDTs share a common structure but differ on the criterion for selecting the test concepts installed in the nodes: while *information gain* is adopted by the latter, the procedure for TCTs resorts to a measure of distance defined over the individuals occurring in the knowledge base. Specifically, the heuristic for selecting the best refinement of the parent concept is defined as follows:

$$E^* = \operatorname*{argmax}_{D \in \rho(C)} d\left(p(\mathbf{P}), p(\mathbf{N})\right) \tag{1}$$

where $\mathbf{P}$ and $\mathbf{N}$ are sub-clusters output by RETRIEVEPOSNEG$(\mathbf{I}, D, \delta)$, $d(\cdot, \cdot)$ is a distance measure between individuals and $p(\cdot)$ is a function that maps a cluster to its *prototype*. As previously mentioned, the adopted $p(\cdot)$ computes the *medoid* of the cluster.

The required measure for individuals should capture aspects of their semantics in the context of the KB. We resort to variation of a language-independent dissimilarity measure proposed in previous works [6,9,10]. Given the knowledge base $\mathcal{K}$, the idea is to compare the behavior of the individuals w.r.t. a set of concepts $\mathcal{C} = \{C_1, C_2, \ldots, C_m\}$ that is dubbed *context* or *committee* of features. For each $C_i \in \mathcal{C}$, a *projection function* $\pi_i : \mathsf{Ind}(\mathcal{A}) \to [0, 1]$ is defined as a simple mapping:

$$\forall\, a \in \mathsf{Ind}(\mathcal{A}) \qquad \pi_i(a) = \begin{cases} 1 & \text{if } \mathcal{K} \models C_i(a) \\ 0 & \text{if } \mathcal{K} \models \neg C_i(a) \\ 0.5 & \text{otherwise} \end{cases} \tag{2}$$

where the third value (0.5) represents a case of maximal uncertainty on the membership. As an alternative, the estimate of the likelihood for a generic individual a of being an instance of C_i could be considered. Especially with densely populated ontologies (as those forming the Web of Data) the probability $\Pr[\mathcal{K} \models C_i(a)]$ may be estimated by $|r_{\mathcal{K}}(C_i)|/|\mathsf{Ind}(\mathcal{A})|$, where $r_{\mathcal{K}}()$ denotes the *retrieval* of a concept w.r.t. $\mathcal{K}$, i.e. the set of individuals of $\mathsf{Ind}(\mathcal{A})$ that (can be proven to) belong to C_i [2].

Hence, a family of distance measures $\{d_n^{\mathcal{C}}\}_{n \in \mathbb{N}}$ can be defined as follows: $d_n^{\mathcal{C}} : \mathsf{Ind}(\mathcal{A}) \times \mathsf{Ind}(\mathcal{A}) \to [0, 1]$ with

$$d_n^{\mathcal{C}}(a, b) = \left[\sum_{i=1}^{m} w_i \left[1 - \pi_i(a)\pi_i(b)\right]^n \right]^{1/n} \tag{3}$$

Non uniform values for $\boldsymbol{w}$ can be considered to reflect the specific importance of each feature. For example it may be set according to an entropic measure [6,10] based on the average information brought by each concept:

$$\forall i \in \{1, \ldots, m\} \qquad w_i = - \sum_{k \in \{-1,0,+1\}} \mu_i(k) \log \mu_i(k) \tag{4}$$

where, given a generic $a \in \mathsf{Ind}(\mathcal{A})$, the following estimates can be used: $\mu_i(+1) \approx \Pr[\mathcal{K} \models C_i(a)]$, $\mu_i(-1) \approx \Pr[\mathcal{K} \models \neg C_i(a)]$ and $\mu_i(0) = 1 - \mu_i(+1) - \mu_i(-1)$.

The growth of a TCT can be stopped by resorting to a heuristic that is similar to the one employed for selecting the best concept description. This requires the employment of a threshold $\nu \in [0, 1]$ for the value of $d(\cdot, \cdot)$. If the value is lower than the threshold, the branch growth is stopped.

4.2   Extracting Candidate Disjointness Axioms from TCTs

The procedure for discovering/extracting disjointness axioms requires a TCT as input. It is reported in Algorithm 2.

Algorithm 2. Routines for deriving disjointness axioms from a TCT

```
1 function DERIVECANDIDATEAXIOMS(T): set of        12 function COLLECT(C, T): set of concepts
      axioms                                        13 input C: concept description
2 input T: TCT                                      14        T: TCT
3 begin                                             15 begin
4    A ← ∅                                          16    let T = ⟨D, I, T_left, T_right⟩
5    CS ← COLLECT(T, T)                             17    if T_left = T_right = null then    { leaf }
6    for each C ∈ CS do                             18       return {C}
7       for each D ∈ CS do                          19    else
8          if (D ⊑ ¬C) ∉ A and C ≢ D then           20       CS_left ← COLLECT(C ⊓ D, T_left)
9             A ← A ∪ {D ⊑ ¬C}                       21       CS_right ← COLLECT(¬C ⊓ ¬D, T_right)
10   return A                                       22       return (CS_left ∪ CS_right)
11 end                                              23 end
```

Given a TCT T, function DERIVECANDIDATEAXIOMS traverses the tree to collect the concept descriptions that are used as parents of the leaf-nodes. In this phase, it generates a set of concept descriptions **CS**, collecting the concepts installed in leaf-nodes (see COLLECT). Then, it considers all pairs of elements in **CS** that are not equivalent and adds a disjointness axiom $D ⊑ ¬C$, if it does not already occur in it.

The set of concept descriptions **CS** is obtained traversing the tree. The procedure COLLECT tries to find the concept descriptions for which disjointness axioms may hold by exploring the paths from the root the leaves by collecting the concepts description installed in the internal nodes.

Note that the hierarchical nature of the approach allows one to generalize this function, controlling the maximum depth of the tree traversal with a parameter. This would produce fewer and more general axioms than the case reported above.

5   Experiments

Two experimental evaluation sessions have been performed to assess the feasibility of discovering disjointness axioms through our approach based on TCTs[1]. We also compared our method with two related statistical methods that have been recently proposed (see Sect. 2).

In the first session of experiments, we considered Web ontologies, freely available, containing disjointness axioms and describing various domains, namely: BIOPAX, NEW TESTAMENT NAMES (NTN), FINANCIAL, GEOSKILLS, MONETARY, and DBPEDIA. In the second experimental session we also considered MUTAGENESIS and VICODI (that originally lack of disjointness axioms). Their principal characteristics are summarized in Table 1. The distance measure d_2^C was employed by our method, with a context of features C made up of the atomic concepts in each KB.

The method has been tested on the problem of (re)discovering disjointness axioms previously removed from the KB: (a) in the first session single axioms are targeted; (b) in the second session comparative experiments on versions of the ontologies enriched with further disjointness axioms (by virtue of the SDA) have been performed.

---

[1] Code and test ontologies are available at: http://github.com/Giuseppe-Rizzo/TCT.

Table 1. Ontologies employed in the experimental sessions

Ontology	DL language	#Concepts	#Roles	#Individuals	#Disj. Ax.'s
BioPax	$\mathcal{ALCIF}(D)$	74	70	323	85
NTN	$\mathcal{SHIF}(D)$	47	27	676	40
Financial	$\mathcal{ALCIF}(D)$	60	16	1000	113
GeoSkills	$\mathcal{ALCHOIN}(D)$	596	23	2567	378
Monetary	$\mathcal{ALCHIF}(D)$	323	247	2466	236
DBPedia3.9	$\mathcal{ALCHI}(D)$	251	132	16606	11
Mutagenesis	$\mathcal{AL}(D)$	86	5	14145	0
Vicodi	$\mathcal{ALHI}(D)$	196	10	16942	0

For each experimental session, the targeted problem and the parameter setup are described, then the outcomes are discussed.

5.1 Re-discovery of a Target Disjoint Axiom

Settings. In this session, a copy of each ontology was created removing a *target disjointness axiom*. Each copy was employed to extract a training set: given the target axiom, say $C \sqsubseteq \neg D$, we considered only individuals that belong to C and D to induce the TCTs. Table 2 lists C and D of the removed axioms for each ontology.

The experiment was repeated picking various values for the threshold ν controlling the tree growth in Algorithm 1; we report the results for $\nu = 0.9, 0.8, 0.7$. The value for δ set to 0.6. The effectiveness of the method was evaluated in terms of the *number of cases of inconsistency that were due to the addition of discovered axioms*.

Table 2. Summary of the axioms ($C \sqsubseteq \neg D$) removed from each ontology in the first experiment

Ontology	C	D
BioPax	bioSource	xref
NTN	Man	Populace ⊔ Woman ⊔ Supernatural Being
Financial	PermanentOrder	Account ⊔ Region
GeoSkills	Educational Level	Educational Pathway
Monetary	ISO3166-Country Code	ISO31813-Market Identifier Code ⊔ ISO4217-Currency Code
DBpedia3.9	Activity	Person

Outcomes. Table 3 illustrates the results of this session. Preliminarily it is worthwhile to note that the new method was able to rediscover the target disjointness axioms for all ontologies but DBPEDIA3.9, as discussed in the following, and it could also determine new axioms regarding concept descriptions that were equivalent to those considered in the target axioms. This case depended on the definition of ρ and on the presence of equivalence axioms in the knowledge bases: ρ assumes that all the concept names are distinct regardless of the existence of equivalence and subsumption axioms. As a result, the operator could produce potentially redundant intensional definitions. For example, in the case of BIOPAX bioSource in the target axiom is alternatively described by ExternalReferenceUtilityClass ⊓ ∃TAXONREF.⊤ (with ExternalReferenceUtilityClass ⊒ bioSource and being bioSource the domain of the role TAXONREF) and xref that is equivalent to ¬ExternalUtilityClass ⊓ PublicationXRef ⊓ ¬dataSource. Also, in the experiments with NTN, the proposed method suggested the disjointness between ¬SupernaturalBeing ⊓ Person ⊓ hasSex.Male (≡ Man) and SupernaturalBeing ⊓ God (≡ God, since God ⊑ SupernaturalBeing).

Table 3. Number of inconsistencies (#inc.) and total number of discovered axioms (#ax's) in the first experimental session (with varying values of ν)

Ontology	TCT 0.9		TCT 0.8		TCT 0.7	
	#inc	#ax's	#inc	#ax's	#inc	#ax's
BIOPAX	2	53	2	53	3	52
NTN	10	70	9	73	10	75
FINANCIAL	0	125	0	126	0	127
GEOSKILLS	2	345	1	347	4	347
MONETARY	0	432	0	432	0	433
DBPEDIA3.9	45	45	44	44	43	43

Moreover, the number of inconsistencies caused by the addition of axioms derived from the TCTs was quite small, especially when compared to the number of axioms predicted (see Table 3). Noticeably, in the cases of BIOPAX and NTN most of the instances were routed to two leaves, while the others were empty, yielding a large number of further new axioms. In such cases, the disjointness axioms involving concept descriptions that correspond to the empty clusters can be added to the knowledge base with no risk of making it inconsistent. In the experiments with larger ontologies, e.g. FINANCIAL, MONETARY and GEOSKILLS, very few empty clusters were observed because of the larger training sets, increasing the quality of clustering and of the derived axioms.

By way of ρ the algorithm can exploit all the concepts (and roles) defined in the signature of $\mathcal{K}$, but it generally considers only a subset of individuals that occur in the ABox, $\mathbf{I} \subseteq \mathsf{Ind}(\mathcal{A})$. Therefore the heuristic wrongly favored refinements that tended to generate poor splits, sorting most of the individuals along one branch. This represents a sort of *small disjunct problem* that typically affects concept learning (with TDTs). However, we noted that this issue seldom occurred with larger training sets. Decreasing the threshold ν (that controls the depth of the branches), no significant difference was observed, because of the homogeneity of the individuals routed to the nodes.

The case of DBPEDIA3.9 deserves a deeper analysis. We observed that instances of a concept like Person hardly discernible in terms of the distance measure (given the choice of $\mathcal{C}$). However, as an inner concept Person was further refined into a number of sibling specializations leading to disjointness axioms involving more specific concepts. Thus, the method ended up finding disjointness axioms between pairs of more specific concepts like Activity and Person⊓∃nationality.United_states owing to the presence in the training set of various individuals that describe American citizens. Also, the new method allowed to elicit a disjointness axiom between Activity and a concept describing non-American artists. This is a potential drawback of data-driven methods that consider general axioms including complex concepts instead of mere concept names. In our case this issue can be avoided with a more careful tuning of the parameters (thresholds) that control the growth of the tree, to prevent the involvement of overly specific concepts.

The time required for learning TCTs was quite limited spanning from few seconds (on BIOPAX) up to one hour (on MONETARY), especially depending on the number of concepts and axioms in the ontologies. Further factors that affected the efficiency of the proposed approach were the inference services required by ρ (e.g. checking the satisfiability of computed specializations) and the computation of the medoids.

5.2 Comparison to Other Approaches Under SDA

Settings. In the second experiment, we considered two further knowledge bases (excluded from the previous experiments due to the mentioned lack of disjointness axioms): MUTAGENESIS and VICODI. We considered extended versions of the ontologies reported in Table 1. Specifically, in order to test our method in comparison with two described in Sect. 2, in a scenario where the ontologies feature non trivial numbers of disjointness axioms, new versions were produced by adding disjointness axioms that involve sibling concepts in the hierarchy, according to the SDA, provided that they would keep the ontology consistent. Then, for each ontology, a fraction f of disjointness axioms was randomly removed. To determine unbiased estimates of the performance indices (i.e. independent of the specific selection of axioms removed), the empirical evaluation procedure was repeated 10 times per ontology also varying f: 20%, 50%, 70%.

Adopting the same parameter setup of the previous session, our method was compared against two related methods (see Sect. 2), respectively, one based on *Pearson's correlation coefficient* (PCC) and another exploiting *negative association rules* (NAR). As for the latter, rules were mined using APRIORI; the required parameters, *minimum support rate, minimum confidence rate* and *maximum rule length* were set, respectively, to 10%, 50% and 3 (also in consideration of the sparseness of the instance distributions w.r.t. the concepts in the considered ontologies). The effectiveness of the methods was evaluated in terms of *the average number of inconsistencies caused by the addition of discovered axioms* (the less the better) and of *the average number of axioms that were discovered* and *rate of removed axioms re-discovered* (the larger the better).

Outcomes. In general, the method based on TCTs produced good clusterings of the training sets: the clusters were well-described by the concept descriptions in the TCTs.

As expected, the number of discovered axioms generally decreased with larger fractions of axioms removed since the resulting trees showed generally a less complex structure. Moreover the experiments showed that a nonnegligible impact on the effectiveness can come from properly tuning the threshold ν. Also, a sort of *horizon effect* was observed: the heuristic based on distance measure acted as a sort of *pre-pruning criterion* that stopped the growth of the tree too prematurely. In addition, in some case, TCTs with (nearly) empty clusters were produced. This was due to the mentioned cases of imbalanced instance distributions w.r.t. the various concepts: for example concepts with few instances are frequent in FINANCIAL, but such cases occur also in the other ontologies. However, this phenomenon was mitigated by the presence of an overall larger number of individuals to be clustered w.r.t. the previous experiment.

The outcomes reported in Table 4 show that, in absolute terms, more axioms were generally discovered by our method (considering all three choices for ν) compared with the two other methods. Moreover, the number of inconsistencies introduced (in case of direct addition to the knowledge bases) was quite limited in proportion to the number of axioms produced: for example, with MONETARY and VICODI less than the 3.5% in almost 20000 discovered axioms. This is interesting in the perspective of an integration in an ontology enrichment process: a larger variety of possibly redundant axioms may be proposed for validation with a very limited chance of introducing errors. On the other hand, the table shows that the compared approaches exhibited a more stable behavior with respect to the fractions of removed axioms f because they could discover axioms involving exclusively named concepts of the knowledge base signature whose instances are more likely to be available. Moreover, a weak correlation between two concepts is unlikely to depend on the presence of a disjointness axiom involving them. This led them also not to introduce further inconsistencies in the experiments. Inspecting sampled TCTs to gain a deeper insight into the outcomes, we could note that, for ontologies with a smaller number of concepts, such as BioPAX and NTN, the refinement operator tended to introduce the same concept in more branches. As a consequence, a large number of axioms were discovered due to the replication of some sub-trees.

We also noted that TCTs for ontologies like DBPEDIA, VICODI and MONETARY presented concept descriptions installed at inner nodes that could be generally considered as disjoint but for few cases represented by specific

Table 4. Experimental comparison of the various approaches: average numbers of cases of inconsistency (#inc.) and total numbers of discovered axioms (#ax's)

Ontology	f	TCT 0.9		TCT 0.8		TCT 0.7		PCC		NAR	
		#inc	#ax's	#inc	#ax's	#inc	#ax's	#inc	#ax's	#inc	#ax's
BioPax	20%	235	3859	357	4235	365	4256	257	280	352	2990
	50%	125	3576	357	4176	432	4115				
	70%	125	3432	235	3875	417	4154				
NTN	20%	312	3128	343	3126	354	3124	32	957	376	3766
	50%	234	3023	234	3034	235	3034				
	70%	156	2987	176	2679	123	2675				
Financial	20%	76	165	87	325	96	276	124	1112	542	5366
	50%	37	143	56	307	53	259				
	70%	33	143	43	276	40	221				
GeoSkills	20%	234	14289	357	14297	432	14345	456	13384	456	13299
	50%	231	14123	356	14154	417	14256				
	70%	234	14122	358	14154	377	14187				
Monetary	20%	535	13456	573	13453	623	13460	543	13384	423	13456
	50%	315	13236	432	13236	532	13236				
	70%	247	13127	231	13127	312	13127				
Mutagenesis	20%	34	14753	43	14847	43	14978	20	2264	45	14832
	50%	23	14753	31	14753	32	14978				
	70%	23	14753	32	14753	32	14978				
Vicodi	20%	431	18231	485	18432	502	18432	475	15518	472	18721
	50%	142	18231	345	18432	467	18431				
	70%	141	18231	345	18432	312	18432				
DBPedia3.9	20%	1345	29730	1432	30143	1432	30567	1243	30470	1243	30365
	50%	1346	29730	1431	30143	1433	30567				
	70%	1343	19730	1432	30143	1432	30567				

individuals. For example, this situation applied to the concepts Actor and President in DBPEDIA, sharing the instance RONALD_REAGAN. The resulting axioms cannot be considered as wrong; they are intended to be submitted to the validation of a domain expert.

Considering the performance in terms of rate of rediscovered axioms (a sort of *recall* index), Table 5 shows very high rates of recovery using the TCT-based method. Even more so, it allows to express more general disjointness axioms than those obtained through the other algorithms: these tackle only the disjointness between concept names whereas from the TCTs axioms involving arbitrarily complex concept descriptions can be derived as a product of the refinement operator adopted in learning procedure. This explains why, in most cases, the number of discovered axioms, but also the number of inconsistencies, was larger with respect to those observed with the compared methods. As reported in Table 5, on average an amount of axioms could not be rediscovered by the

Table 5. Average rates (and standard deviations) of removed axioms re-discovered using the various approaches (with standard and pre-pruned TCTs)

Ontology	f	TCT – *standard mode*			TCT – *early stopping*		
		TCT 0.9	TCT 0.8	TCT 0.7	TCT 0.9	TCT 0.8	TCT 0.7
BioPax	20%	0.90 ± 0.12	0.76 ± 0.13	0.74 ± 0.13	0.80 ± 0.23	0.65 ± 0.23	0.70 ± 0.13
	50%	0.85 ± 0.13	0.74 ± 0.13	0.74 ± 0.13	0.63 ± 0.23	0.63 ± 0.23	0.63 ± 0.23
	70%	0.85 ± 0.13	0.74 ± 0.12	0.74 ± 0.14	0.69 ± 0.13	0.67 ± 0.13	0.66 ± 0.14
NTN	20%	0.99 ± 0.08	0.95 ± 0.06	0.95 ± 0.08	0.70 ± 0.15	0.67 ± 0.15	0.67 ± 0.14
	50%	0.97 ± 0.03	0.93 ± 0.10	0.93 ± 0.01	0.55 ± 0.13	0.54 ± 0.13	0.54 ± 0.15
	70%	0.90 ± 0.10	0.89 ± 0.11	0.89 ± 0.10	0.55 ± 0.13	0.55 ± 0.13	0.55 ± 0.13
Financial	20%	0.99 ± 0.08	0.99 ± 0.08	0.99 ± 0.08	0.60 ± 0.10	0.59 ± 0.11	0.59 ± 0.11
	50%	0.97 ± 0.03	0.97 ± 0.03	0.97 ± 0.03	0.56 ± 0.10	0.56 ± 0.10	0.56 ± 0.10
	70%	0.95 ± 0.05	0.95 ± 0.05	0.95 ± 0.05	0.56 ± 0.10	0.56 ± 0.10	0.56 ± 0.10
GeoSkills	20%	0.99 ± 0.08	0.99 ± 0.08	0.99 ± 0.08	0.70 ± 0.15	0.69 ± 0.11	0.69 ± 0.11
	50%	0.92 ± 0.10	1.00 ± 0.00	1.00 ± 0.00	0.65 ± 0.23	0.65 ± 0.23	0.65 ± 0.23
	70%	0.92 ± 0.10	0.92 ± 0.10	0.92 ± 0.10	0.65 ± 0.23	0.63 ± 0.22	0.62 ± 0.23
Monetary	20%	0.99 ± 0.08	1.00 ± 0.00	1.00 ± 0.00	0.65 ± 0.23	0.63 ± 0.20	0.62 ± 0.23
	50%	0.94 ± 0.13	1.00 ± 0.00	1.00 ± 0.00	0.63 ± 0.12	0.66 ± 0.15	0.65 ± 0.11
	70%	0.94 ± 0.13	0.91 ± 0.14	0.91 ± 0.13	0.62 ± 0.12	0.60 ± 0.13	0.60 ± 0.12
Mutagenesis	20%	1.00 ± 0.00	1.00 ± 0.00	1.00 ± 0.00	0.56 ± 0.13	0.53 ± 0.14	0.51 ± 0.12
	50%	1.00 ± 0.00	1.00 ± 0.00	1.00 ± 0.00	0.52 ± 0.14	0.51 ± 0.15	0.50 ± 0.11
	70%	1.00 ± 0.00	1.00 ± 0.00	1.00 ± 0.00	0.50 ± 0.15	0.50 ± 0.16	0.50 ± 0.15
Vicodi	20%	0.95 ± 0.02	0.90 ± 0.08	0.90 ± 0.08	0.65 ± 0.03	0.63 ± 0.01	0.62 ± 0.01
	50%	0.95 ± 0.02	0.90 ± 0.08	0.90 ± 0.08	0.62 ± 0.03	0.62 ± 0.04	0.62 ± 0.03
	70%	0.95 ± 0.02	0.90 ± 0.08	0.90 ± 0.08	0.58 ± 0.04	0.57 ± 0.05	0.58 ± 0.04
DBPedia3.9	20%	1.00 ± 0.00	1.00 ± 0.00	1.00 ± 0.00	0.70 ± 0.12	0.68 ± 0.13	0.67 ± 0.12
	50%	1.00 ± 0.00	1.00 ± 0.00	1.00 ± 0.00	0.65 ± 0.23	0.68 ± 0.13	0.64 ± 0.12
	70%	0.96 ± 0.08	0.90 ± 0.08	0.90 ± 0.08	0.65 ± 0.22	0.68 ± 0.13	0.64 ± 0.12

new method. The TCT-based approach assumes the availability of the instances belonging the concepts involved in a target axiom. But in the evaluation, especially with Financial and Vicodi, some concepts were endowed with a very small number of instances (less than 10). As a result, the proposed approach could not detect cases of disjointness due to the lack of good cluster medoids. For example, in the experiments with Financial, our method was unable to discover the disjointness of the concepts Loan and Sex. For this ontology, Sex was used to model the customer's gender but no specific assertion was available: the gender was instead designed as a subconcept (Male/Female). A similar case was observed also in the experiments on Vicodi: the disjointness of Actor and Artefact could not be discovered.

Regarding the ability of discovering axioms also preventing the cases of inconsistency due to the process, one may argue that growing taller trees, thus involving very specific concept descriptions like those produced in these experiments, may turn out to be time-consuming and error-prone. To test this aspect, we checked the quality of the axioms obtained by prematurely stopping the growth of TCTs at a certain level instead of reaching the leaves namely, it was stopped if the induced axioms would introduce a case of inconsistency. Hence, Table 5 reports also the results of the experiments run with this policy implemented in the method (*early stopping* columns). As expected, the proportions of detected axioms were lower than those obtained using the standard method: no more than 80% of the removed axioms were rediscovered. This means that, although the progressive specialization of the concepts included in the disjointness axioms may lead to cases of inconsistency, working in *standard mode* the method can help elicit relevant correct axioms (inconsistencies may be avoided applying a *post-pruning* strategy aimed at preventing the production of defective axioms). Besides, we noted that the aforementioned exploration procedure was often stopped too early: in most cases, the consistency of the knowledge base was preserved by adding axioms that can be found within the 10th and the 15th level in the TCTs. One of the benefits deriving from the proposed approach concerns the ability to overcome one of the downsides of association rules: they cannot be considered as logical rules, rather they merely denote statistical correlations between two or more features (that hold with a degree of uncertainty).

Finally, our new method showed it could be more efficient than the one based on APRIORI, that was especially slow in the step of generating the frequent patterns. The running-time for inducing TCTs spanned from less than a minute to hours for the various ontologies, while the time required by the association rule mining was larger in most cases. Note that this also depends on the maximum length of the rules: mining longer rules (to discover more axioms) would make the method infeasible.

6   Conclusions and Outlook

In this work, we have illustrated the terminological cluster trees, an extension of terminological decision trees [8] that, unlike these supervised classification models, aims at solving an unsupervised problem: clustering individual resources occurring in Web ontologies. As an application, we have cast the task of discovering disjointness axioms as the mentioned clustering problem and have proposed a solution that exploits the new models. In the presented empirical evaluation the effectiveness of the proposed approach was tested. Compared to related unsupervised approaches, the new method proved to be able to discover disjointness axioms involving complex concept descriptions exploiting the ontology as a source of background knowledge, unlike the other methods based on the statistical correlation between instances.

This work can be extended along various directions. One regards the investigation of other distances measures for individuals and different notions of separation between clusters. In addition, this approach can be integrated with other machine learning-based frameworks for ontology engineering, such as DL-LEARNER [14], as a service for enriching the terminology of lightweight ontologies.

The method can be further improved by introducing a post-pruning step for better tackling the problem of empty-clusters. Finally, the empirical evaluation may be extended considering further methods and well-modeled ontologies endowed with disjointness axioms which seem hard to find.

References

1. Aggarwal, C.C., Reddy, C.K.: Data Clustering: Algorithms and Applications, 1st edn. Chapman & Hall/CRC, Boca Raton (2013)
2. Baader, F., Calvanese, D., McGuinness, D., Nardi, D., Patel-Schneider, P. (eds.): The Description Logic Handbook, 2nd edn. Cambridge University Press, Cambridge (2007)
3. Baader, F., Ganter, B., Sertkaya, B., Sattler, U.: Completing description logic knowledge bases using formal concept analysis. In: Veloso, M. (ed.) Proceedings of IJCAI 2007, pp. 230–235. AAAI Press, Menlo Park (2007)
4. Blockeel, H., De Raedt, L.: Top-down induction of first-order logical decision trees. Artif. Intell. **101**(1–2), 285–297 (1998)
5. Cornet, R., Abu-Hanna, A.: Usability of expressive description logics - a case study in UMLS. In: Kohane, I. (ed.) Proceedings of AMIA 2002, pp. 180–184. AMIA (2002)
6. d'Amato, C., Fanizzi, N., Esposito, F.: Query answering and ontology population: an inductive approach. In: Bechhofer, S., Hauswirth, M., Hoffmann, J., Koubarakis, M. (eds.) ESWC 2008. LNCS, vol. 5021, pp. 288–302. Springer, Heidelberg (2008). doi:10.1007/978-3-540-68234-9_23
7. De Raedt, L., Blockeel, H.: Using logical decision trees for clustering. In: Lavrač, N., Džeroski, S. (eds.) ILP 1997. LNCS, vol. 1297, pp. 133–140. Springer, Heidelberg (1997). doi:10.1007/3540635149_41
8. Fanizzi, N., d'Amato, C., Esposito, F.: Induction of concepts in web ontologies through terminological decision trees. In: Balcázar, J.L., Bonchi, F., Gionis, A., Sebag, M. (eds.) ECML PKDD 2010. LNCS (LNAI), vol. 6321, pp. 442–457. Springer, Heidelberg (2010). doi:10.1007/978-3-642-15880-3_34
9. Fanizzi, N., d'Amato, C.: A hierarchical clustering method for semantic knowledge bases. In: Apolloni, B., Howlett, R.J., Jain, L. (eds.) KES 2007. LNCS (LNAI), vol. 4694, pp. 653–660. Springer, Heidelberg (2007). doi:10.1007/978-3-540-74829-8_80
10. Fanizzi, N., d'Amato, C., Esposito, F.: Evolutionary conceptual clustering based on induced pseudo-metrics. Int. J. Semant. Web Inf. Syst. **4**(3), 44–67 (2008)
11. Fleischhacker, D., Völker, J.: Inductive learning of disjointness axioms. In: Meersman, R., et al. (eds.) OTM 2011. LNCS, vol. 7045, pp. 680–697. Springer, Heidelberg (2011). doi:10.1007/978-3-642-25106-1_20
12. Haase, P., Völker, J.: Ontology learning and reasoning: dealing with uncertainty and inconsistency. In: da Costa, P., et al. (eds.) Uncertainty Reasoning for the Semantic Web I. LNCS, vol. 5327, pp. 366–384. Springer, Heidelberg (2008)
13. Heath, T., Bizer, C.: Linked Data: Evolving the Web into a Global Data Space. Synthesis Lectures on the Semantic Web. Morgan & Claypool Publishers, San Rafael (2011)
14. Hellmann, S., Lehmann, J., Auer, S.: Learning of OWL class descriptions on very large knowledge bases. Int. J. Semant. Web Inf. **5**(2), 25–48 (2009)
15. Lehmann, J., Bühmann, L.: ORE - a tool for repairing and enriching knowledge bases. In: Patel-Schneider, P.F., et al. (eds.) ISWC 2010. LNCS, vol. 6497, pp. 177–193. Springer, Heidelberg (2010). doi:10.1007/978-3-642-17749-1_12

16. Rizzo, G., d'Amato, C., Fanizzi, N., Esposito, F.: Induction of terminological cluster trees. In: Bobillo, F., et al. (ed.) Proceedings of URSW 2016. CEUR Workshop Proceedings, vol. 1665, pp. 49–60. CEUR-WS.org (2016)
17. Schlobach, S.: Debugging and semantic clarification by pinpointing. In: Gómez-Pérez, A., Euzenat, J. (eds.) ESWC 2005. LNCS, vol. 3532, pp. 226–240. Springer, Heidelberg (2005). doi:10.1007/11431053_16
18. Völker, J., Fleischhacker, D., Stuckenschmidt, H.: Automatic acquisition of class disjointness. J. Web Semant. **35**(P2), 124–139 (2015)
19. Völker, J., Niepert, M.: Statistical schema induction. In: Antoniou, G., Grobelnik, M., Simperl, E., Parsia, B., Plexousakis, D., Leenheer, P., Pan, J. (eds.) ESWC 2011. LNCS, vol. 6643, pp. 124–138. Springer, Heidelberg (2011). doi:10.1007/978-3-642-21034-1_9
20. Völker, J., Vrandečić, D., Sure, Y., Hotho, A.: Learning disjointness. In: Franconi, E., Kifer, M., May, W. (eds.) ESWC 2007. LNCS, vol. 4519, pp. 175–189. Springer, Heidelberg (2007). doi:10.1007/978-3-540-72667-8_14
21. Wang, T.D., Parsia, B., Hendler, J.: A survey of the web ontology landscape. In: Cruz, I., et al. (eds.) ISWC 2006. LNCS, vol. 4273, pp. 682–694. Springer, Heidelberg (2006). doi:10.1007/11926078_49

Embedding Learning for Declarative Memories

Volker Tresp[1,2(✉)], Yunpu Ma[1,2], Stephan Baier[2], and Yinchong Yang[1,2]

[1] Siemens AG, Corporate Technology, Munich, Germany
Volker.Tresp@Siemens.com
[2] Ludwig-Maximilians-Universität München, Munich, Germany

Abstract. The major components of the brain's declarative or explicit memory are *semantic memory* and *episodic memory*. Whereas semantic memory stores general factual knowledge, episodic memory stores events together with their temporal and spatial contexts. We present mathematical models for declarative memories where we consider semantic memory to be represented by triples and episodes to be represented as quadruples i.e., triples in time. E.g., *(Jack, receivedDiagnosis, Diabetes, Jan1)* states that Jack was diagnosed with diabetes on January 1. Both from a cognitive and a technical perspective, an interesting research question is how declarative data can efficiently be stored and semantically be decoded. We propose that a suitable data representation for episodic event data is a 4-way tensor with dimensions subject, predicate, object, and time. We demonstrate that the 4-way tensor can be decomposed, e.g., using a 4-way Tucker model, which permits semantic decoding of an event, as well as efficient storage. We also propose that semantic memory can be derived from the episodic model by a marginalization of the time dimension, which can be performed efficiently. We argue that the storage of episodic memory typically requires models with a high rank, whereas semantic memory can be modelled with a comparably lower rank. We analyse experimentally the relationship between episodic and semantic memory models and discuss potential relationships to the corresponding brain's cognitive memories.

1 Introduction

The main components of the brain's declarative or explicit memory are *semantic memory* and *episodic memory*. Both are considered long-term memories and store information potentially over the life-time of an individual [1,4,9,34]. The semantic memory stores general factual knowledge, i.e., information we *know*, independent of the context where this knowledge was acquired. Episodic memory concerns information we remember and includes the spatiotemporal context of events [38]. There is evidence that these main cognitive categories are partially dissociated from one another in the brain, as expressed in their differential sensitivity to brain damage [10]. However, there is also evidence indicating that the different memory functions are not mutually independent and support one another [13].

© Springer International Publishing AG 2017
E. Blomqvist et al. (Eds.): ESWC 2017, Part I, LNCS 10249, pp. 202–216, 2017.
DOI: 10.1007/978-3-319-58068-5_13

In this paper we discuss technical models for semantic and episodic memories and compare them with their biological counterparts. In particular, we consider that a technical realization of a semantic memory is a knowledge graph (KG) which is a triple-oriented knowledge representation. Popular technical large-scale KGs are DBpedia [2], YAGO [35], Freebase [5], NELL [7], and the Google KG [32]. There exist reliable KGs with more than a hundred billion triples that support search, text understanding and question answering [32].

In our approach we model episodes as events in time that can be represented by quadruples. Thus whereas the triple *(Jack, hasDiagnosis, Diabetes)* might reflect that Jack has diabetes, the quadruple *(Jack, receivedDiagnosis, Diabetes, Jan1)* would represent the diagnostic event.

We propose that biologically plausible representations of both semantic and episodic memories can be achieved by a decomposition of the adjacency tensors describing the memories. The decomposition leads to a highly compressed form of the memories and exhibits a form of memory generalization or inductive learning, in form of a generalization to new triples and quadruples [26]. If each entity and each predicate has a unique latent representation, information is shared across all memory functions.

We propose that semantic memory is a long-term storage for episodic memory where the exact timing information is lost, and that both memories rely on the same latent representations. In particular we propose that semantic memory can be derived from episodic memory by marginalizing the time dimension, an operation which can be performed elegantly when nonnegative tensor decompositions are used as memory models. Whereas the storage of episodic memory typically requires decomposition models with a high rank, semantic memory can be stored with a comparably lower rank.

The paper is organized as follows. In the next section we introduce the unique representation hypothesis which postulates latent representations of generalized entities that are shared between memory functions. Section 3 covers latent representation models for semantic and episodic memory and Sect. 4 describes the tensor models. In Sect. 5, we discuss relationships between episodic and semantic memories and discuss memory querying. Section 6 discusses the biological relevance of the proposed model and Sect. 7 presents experimental results. Section 8 contains our conclusions.

2   Unique-Representation Hypothesis

A technical realization of a semantic memory is a knowledge graph (KG) which is a triple-oriented knowledge representation. Here we consider a slight extension to the subject-predicate-object triple form by adding the value in the form $(e_s, e_p, e_o; \ Value)$ where $Value$ is a function of e_s, e_p, e_o and, e.g., can be a Boolean variable (*True* for *1*, *False* for *0*) or a real number. Thus *(Jack, likes, Mary; True)* states that Jack (the subject or head entity) likes Mary (the object or tail entity). Note that e_s and e_o represent the entities for subject index s and object index o. To simplify notation we also consider e_p to be a generalized

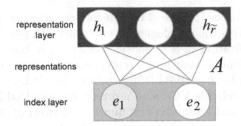

Fig. 1. A graphical view of the unique-representation hypothesis. The model can operate bottom-up and top-down. In the first case, index node e_i activates the representation layer via its latent representation, implemented as weight vector. In the figure, e_1 is activated, all other index nodes are inactive and the representation layer is activated with the pattern $\mathbf{h} = \mathbf{a}_{e_1}$. In top-down operation, a representation layer can also activate index nodes. The activation of node e_i is then equal to the inner product $\mathbf{a}_{e_i}^\top \mathbf{h}$. We consider here formalized nodes which might actually be implemented as ensembles of distributed neurons or as stable activation patterns of distributed neurons. Here and in the following we assume that the matrix A stores the latent representations of all generalized entities. The context makes it clear if we refer to the latent representations of entities, predicates, or time indices.

entity associated with predicate type with index p. For the episodic memory we introduce e_t, which is a generalized entity for time t.

The *unique-representation hypothesis* assumed in this paper is that each entity or concept e_i, each predicate e_p and each time step e_t has a unique latent representation —$\mathbf{a}_i$, $\mathbf{a}_p$, or $\mathbf{a}_t$, respectively— in form of a set of real numbers, represented as a vector or a matrix. The assumption is that the representations are shared among all memory functions, which permits information exchange and inference between the different memories. For simplicity of discussion, we assume that the latent representations form vectors and that the dimensionalities of these latent representations for entities and predicates are $\tilde{r}$ such that $\mathbf{a}_i \in \mathbb{R}^{\tilde{r}}$, $\mathbf{a}_p \in \mathbb{R}^{\tilde{r}}$, and for time is $\tilde{r}_T$ such that $\mathbf{a}_t \in \mathbb{R}^{\tilde{r}_T}$. Figure 1 shows a simple network realization.

3   Semantic and Episodic Knowledge Graph Models

3.1   Semantic Knowledge Graph

We now consider an efficient representation of a KG. First, we introduce the three-way semantic adjacency tensor $\mathcal{X}$ where the tensor element $x_{s,p,o}$ is the associated *Value* of the triple (e_s, e_p, e_o). Here $s = 1, \ldots, S$, $p = 1, \ldots, P$, and $o = 1, \ldots, O$. One can also define a companion tensor Θ with the same dimensionality as $\mathcal{X}$ and with entries $\theta_{s,p,o}$. It contains the natural parameters of the model and the connection to $\mathcal{X}$ for Boolean variables is

$$P(x_{s,p,o}|\theta_{s,p,o}) = \text{sig}(\theta_{s,p,o}) \tag{1}$$

where $\text{sig}(arg) = 1/(1+\exp(-arg))$ is the logistic function (Bernoulli likelihood), which we use in this paper. If $x_{s,p,o}$ is a real number then we can use a Gaussian distribution with $P(x_{s,p,o}|\theta_{s,p,o}) \sim \mathcal{N}(\theta_{s,p,o}, \sigma^2)$.

As mentioned, the key concept in embedding learning is that each entity and predicate e has an $\tilde{r}$-dimensional latent vector representation $\mathbf{a} \in \mathbb{R}^{\tilde{r}}$. In particular, the embedding approaches used for modeling a semantic KGs assume that

$$\theta_{s,p,o}^{sem} = f^{sem}(\mathbf{a}_{e_s}, \mathbf{a}_{e_p}, \mathbf{a}_{e_o}). \tag{2}$$

Here, the function $f^{sem}(\cdot)$ predicts the value of the natural parameter $\theta_{s,p,o}^{sem}$. In the case of a KG with a Bernoulli likelihood, $\text{sig}(\theta_{s,p,o}^{sem})$ represents the confidence that the triple (e_s, e_p, e_o) is true and we call the function an *indicator mapping function*. We discuss examples in the next section.

Latent representation approaches have been used very successfully to model large KGs. It has been shown experimentally that models using latent factors perform well in these high-dimensional and highly sparse domains. Since an entity has a unique representation, independent of its role as a subject or an object, the model permits the propagation of information across the KG. For example if a writer was born in Munich, the model can infer that the writer is also born in Germany and probably writes in the German language [24, 25]. For a recent review, please consult [26].

Due to the approximation, $\text{sig}(\theta_{Jack,marriedTo,e}^{sem})$ might be smaller than one for the true spouse. The approximation also permits inductive inference: We might get a large $\text{sig}(\theta_{Jack,marriedTo,e}^{sem})$ also for persons e that are *likely* to be married to *Jack* and $\text{sig}(\theta_{s,p,o}^{sem})$ can, in general, be interpreted as a confidence value for the triple (e_s, e_p, e_o). More complex queries on semantic models involving existential quantifiers are discussed in [19].

3.2 An Event Model for Episodic Memory

Whereas a semantic KG model reflects the state of the world, e.g., of a clinic and its patients, observations and actions describe discrete events, which, in our approach, are represented by an episodic event tensor. In a clinical setting, events might be a prescription of a medication to lower the cholesterol level, the decision to measure the cholesterol level and the measurement result of the cholesterol level; thus events can be, e.g., actions, decisions and measurements.

The episodic event tensor is a four-way tensor $\mathcal{Z}$ where the tensor element $z_{s,p,o,t}$ is the associated *Value* of the quadruple (e_s, e_p, e_o, e_t). The indicator mapping function then is

$$\theta_{s,p,o,t}^{epi} = f^{epi}(\mathbf{a}_{e_s}, \mathbf{a}_{e_p}, \mathbf{a}_{e_o}, \mathbf{a}_{e_t}),$$

where we have added a representation for the time of an event by introducing the generalized entity e_t with latent representation $\mathbf{a}_{e_t} \in \mathbb{R}^{\tilde{r}_t}$. This latent representation compresses all events that happen at time t. As discussed, an example from a clinical setting could be *(Jack, receivedDiagnosis, Diabetes, Jan1)* which states that Jack was diagnosed with diabetes on January 1.

4 Tensor Decompositions

4.1 Tensor Decompositions

There are different options for modelling the indicator mapping functions $f^{epi}(\cdot)$ and $f^{sem}(\cdot)$. In this paper we will only consider multilinear models derived from tensor decompositions. Tensor decompositions have shown excellent performance in modelling KGs [26].

Specifically, we consider a 4-way Tucker model for episodic memory in the form

$$f^{epi}(\mathbf{a}_{e_s}, \mathbf{a}_{e_p}, \mathbf{a}_{e_o}, \mathbf{a}_{e_t}) = \sum_{r_1=1}^{\tilde{r}} \sum_{r_2=1}^{\tilde{r}} \sum_{r_3=1}^{\tilde{r}} \sum_{r_4=1}^{\tilde{r}_T} a_{e_s,r_1} \, a_{e_p,r_2} \, a_{e_o,r_3} \, a_{e_t,r_4} \, g^{epi}(r_1, r_2, r_3, r_4). \quad (3)$$

Here, $g^{epi}(r_1, r_2, r_3, r_4) \in \mathbb{R}$ are elements of the core tensor $\mathcal{G}^{epi} \in R^{\tilde{r} \times \tilde{r} \times \tilde{r} \times \tilde{r}_T}$.

4.2 Inner Product Formulation of Tensor Decompositions

Note that we can rewrite Eq. 3 as

$$f^{epi}(\mathbf{a}_{e_s}, \mathbf{a}_{e_p}, \mathbf{a}_{e_o}, \mathbf{a}_{e_t}) = \mathbf{a}_{e_o}^{\top} \mathbf{h}^{object}$$

where

$$\mathbf{h}^{object} = \sum_{r_1=1}^{\tilde{r}} \sum_{r_2=1}^{\tilde{r}} \sum_{r_4=1}^{\tilde{r}_T} a_{e_s,r_1} \, a_{e_p,r_2} \, a_{e_t,r_4} \, g(r_1, r_2, :, r_4).$$

Thus if we consider subject, predicate, and time as inputs, we can evaluate the likelihood for different objects by an inner product between the latent representation of the object $\mathbf{a}_{e_o}$ with a vector $\mathbf{h}^{object}$ derived from the latent representations of the subject, the predicate, the time and the core tensor. Similarly we can calculate likely subjects, predicates, and time instances. We propose that this formulation is biologically more plausible, since inner products are operations that are easily performed by formalized neurons [31]. Also the representation is suitable for a sampling approach in querying (see the next section).

5 Querying Memories

5.1 Probabilistic Querying

In many applications one is interested in retrieving triples with a high likelihood, conditioned on some information, thus we are essentially faced with an optimization problem. To answer a query of the form *(Jack, receivedDiagnosis, ?, Jan1)* we need to solve

$$\arg\max_{e_o} f^{epi}(\mathbf{a}_{Jack}, \mathbf{a}_{receivedDiagnosis}, \mathbf{a}_{e_o}, \mathbf{a}_{Jan1}).$$

Of course one is often interested in a set of likely answers. In [37] it was shown that likely triples can be generated by defining a Boltzmann distribution derived from an energy function. By enforcing non-negativity of the factors and the core tensor entries, the energy function for a Tucker model becomes $\mathbb{E}(s, p, o, t) = -\log f^{epi}(\mathbf{a}_{e_s}, \mathbf{a}_{e_p}, \mathbf{a}_{e_o}, \mathbf{a}_{e_t})$ and the quadruple probability becomes

$$
P(s, p, o, t) \propto \left(\sum_{r_1=1}^{\tilde{r}} \sum_{r_2=1}^{\tilde{r}} \sum_{r_3=1}^{\tilde{r}} \sum_{r_4=1}^{\tilde{r}_T} a_{e_s, r_1} \, a_{e_p, r_2} \, a_{e_o, r_3} \, a_{e_t, r_4} \, g^{epi}(r_1, r_2, r_3, r_4), \right)^{\beta}
$$

(4)

where β plays the role of an inverse temperature: A large β would put all probability mass to triples with high functional values whereas a small β would assign probability mass also to triples with lower functional values. Note that for querying, we obtain a probability distribution over s, p, o, t and we can define marginal queries like $P(s, p, o)$ and conditional queries like $P(o|s, p, t)$. It is straightforward to generate likely samples from these distributions (see Fig. 2).

Since the Tucker decomposition is an instance of a sum-product network [27], conditionals and marginals can easily be computed: A conditioning means that the index nodes are simply clamped to their respective values and marginalization means that the index nodes of the marginalized variables are all active, indicated by a vector of ones. Figure 2 shows some examples.

5.2 Semantic Memory Derived from Episodic Memory

Note that we can derive semantic queries from the episodic memory. As an example, the probability for the statement *(Jack, receivedDiagnosis, Diabetes, Jan1)* can be queried from the episodic memory directly, the probability for the statement *(Jack, receivedDiagnosis, Diabetes)* can also be derived from the episodic memory if we assume that semantic memory simply aggregates episodic memory slices. In particular we get for a semantic memory (with $\beta = 1$),

$$
P(s, p, o) \propto \sum_{r_1=1}^{\tilde{r}} \sum_{r_2=1}^{\tilde{r}} \sum_{r_3=1}^{\tilde{r}} a_{e_s, r_1} \, a_{e_p, r_2} \, a_{e_o, r_3} \, g^{sem}(r_1, r_2, r_3),
$$

(5)

where

$$
g^{sem}(r_1, r_2, r_3) = \sum_{t} \sum_{r_4=1}^{\tilde{r}_T} a_{e_t, r_4} \, g^{epi}(r_1, r_2, r_3, r_4).
$$

(6)

Technically, semantic memory can be derived from episodic memory by setting all index nodes for time to active, as shown in Fig. 2D. Thus, if we accept that the semantic memory is a long-term storage for episodic memory, *we do not need to model semantic memory separately, since it can be derived from episodic memory!*

As part of a consolidation process we propose that $g^{sem}(r_1, r_2, r_3)$ is stored explicitly. The main reason is that $\tilde{r}_T$ is typically quite large (see discussion in Subsect. 6.2) and realizing the summation in Eq. 6 for each recall of semantic memory could be quite expensive. With $\tilde{r} \ll \tilde{r}_T$, the semantic memory has a small footprint and can be calculated efficiently. We also propose that the

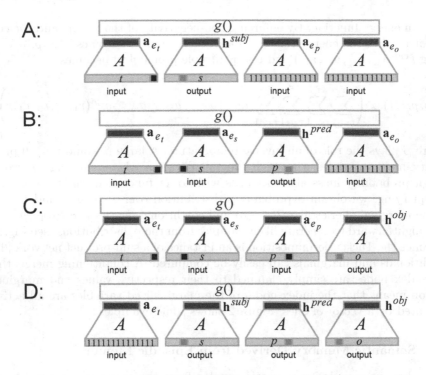

Fig. 2. The semantic decoding using a 4-dimensional Tucker tensor model for episodic memory. A: To sample a subject s given time t, we marginalize predicate p and object o. B: Here, o is marginalized and one samples a predicate p, given t, s. C: Sampling of an object o, given t, s, p. D: By marginalizing the time dimension, we obtain a semantic memory.

assumption that $\tilde{r} \ll \tilde{r}_T$ is quite plausible from a biological view point, as discussed in the following section.

Note that we implicitly assume that a fact that was encountered as an event is true forever. In the example above we would conclude that a diagnosed diabetes would be valid for lifetime. This would also agree with the weak expressiveness of standards like the Resource Description Framework (RDF) which do not model negations. Implementing negations and expressive constraints would require stronger ontologies. Temporal RDF graphs are discussed in [14,15]. Some diseases, such as infections, on the other hand, can be cured. There are a number of ways of how this can be handled, for example by considering the relationships between *hasDisease* and *wasCuredFromDisease*. The latter could be implemented as an event *hasDisease* but with $Value = -1$.

6   Relationships to Human Memories

This section speculates about the relevance of the presented models to human memory functions. In particular we present several concrete hypotheses.

6.1   Unique-Representation Hypothesis for Entities and Predicates

The *unique-representation hypothesis* states that each generalized entity e is represented by an index node and a unique (rather high-dimensional) latent representation $\mathbf{a}_e$ that is stored as weight patterns connecting the index nodes with nodes in the representation layer (see Fig. 1). Note that the weight vectors might be very sparse and in some models non-negative. The latent representations integrate all that is known about a generalized entity, they are the basis for episodic memory and semantic memory, and they can be instrumented for prediction and decision support in working memory. Among other advantages, a unique representation would explain why background information about an entity is seemingly effortlessly integrated into both sensor scene understanding and decision support, at least for entities familiar to the individual.

We proposed formalized nodes which might actually be implemented as ensembles of distributed neurons or as stable activation patterns of distributed neurons. Neurons which very selectively respond to specific entities and concepts have been found in the medial temporal lobe (MTL). In particular, researchers have reported on a remarkable subset of MTL neurons that are selectively activated by strikingly different pictures of given individuals, landmarks or objects and in some cases even by letter strings with their names [28]. For example, some neurons have been shown to selectively respond to prominent actors like "Jennifer Aniston" or "Halle Berry". These are called concept cells by the authors.

In the consolidation theory of human memory it is assumed that, after some period of time, semantic memory, and possibly also episodic memory, is consolidated in cerebral cortex. Often neurons with similar receptive fields are clustered in sensory cortices and form a topographic map [12]. Topological maps might also be the organizational form of neurons representing entities. Thus, entities with similar latent representations might be topographically close. A detailed atlas of semantic categories has been established in extensive fMRI studies showing topographically sorted local representations of semantic concepts [16].

6.2   Perception and Memory Formation

It is well established that new episodic memories are formed in the hippocampus, which is part of the MTL. We propose that the hippocampus is the region where index nodes for generalized entities are formed and that these index nodes establish a presence in the cortex during memory consolidation. The nodes in the representation layer might be in higher order sensory layers and in association cortex. The hippocampal memory index theory [36] agrees with this model and proposes that, in particular, time indices are established in the hippocampus. These are linked to the representations formed as responses to an episodic sensory input in the higher order sensory and association cortices. This model would also support the idea that $\tilde{r}_T$ must be large since $\mathbf{a}_{e_t}$ would need to represent all processed sensory information. The semantic decoding of $\mathbf{a}_{e_t}$ by the episodic memory then corresponds to the semantic understanding of a sensory input, i.e., would be the essence of perception. Note that a recall of a past episode would simply mean that the corresponding node e_t is activated which then activates $\mathbf{a}_{e_t}$ in the representation layer. $\mathbf{a}_{e_t}$ can be semantically decoded, enabling an individual to semantically describe the past episode, and could activate the corresponding past sensory impressions, providing an individual with a sensory impression of the past episode.

6.3   Tensor Memory Hypothesis

The hypothesis states that semantic memory and episodic memory are implemented as functions applied to the latent representations involved in the generalized entities which include entities, predicates, and time. Thus neither the knowledge graphs nor the tensors ever need to be stored explicitly!

6.4   Semantic Memory and Episodic Memory

In our interpretation, semantic memory is a long-term storage for episodic memory. This is biologically attractive since no involved transfer from episodic to semantic memory is required. We propose that this is supported by cognitive studies on brain memory functions: It has been argued that semantic memory is information we have encountered repeatedly, so often that the actual learning episodes are blurred [8,12]. Similarly, it has been speculated that a gradual transition from episodic to semantic memory can take place, in which episodic memory reduces its sensitivity and association to particular events, so that the information can be generalized as semantic memory. Thus some theories speculate that episodic memory may be the "gateway" to semantic memory [3,20,22,33,34,39]. [23] is a recent overview on the topic.

Our model supports inductive inference in form of a probabilistic materialization. As an example, consider that we know that Max lives in Munich. The probabilistic materialization that happens in the factorization should already predict that Max also lives in Bavaria and in Germany. Thus both facts and inductively inferred facts about an entity are represented in its local environment. There is a certain danger in probabilistic materialization, since it might lead to overgeneralizations, reaching from national prejudice to false memories. In fact in many studies it has been shown that individuals produce false memories but are personally absolutely convinced of their truthfulness [21,29].

7   Experiments

The goal of our experiments was to investigate the quality of the tensor decompositions for the semantic and episodic memory. The 3-way and 4-way tensors were factorized using a Tucker decomposition with unique latent representations for entities (as subjects and objects), predicates and time. We considered three model settings. The first setting was unconstraint using a binary cross-entropy (Bernoulli cost function) with additional $l2$ norm penalty on all parameters. In the second setting we constrained all parameters to be non-negative and used a mean squared error cost function and in the third setting we enforced non-negativity as well and used an $l1$ norm penalty on all parameters to encourage sparse solutions.

7.1   Data Set

Our experiments are based on the open Freebase KG, since it contains relatively many predicate types.[1] Triples in the Freebase KG have been extracted from Wikipedia, WordNet, and many other web resources. In our experiments we extracted a subset which includes $10k$ entities, 285 relation types, and in total $141k$ positive triples. Most other triples were treated as unknown and only a small number of triples, generated by a corruption of observed positive triples, is treated as negative. The protocol for sampling negative triples follows Bordes *et al.* [6]: For each true triple (s, p, o) in the data set, we generated 5 negative triples by replacing the object o with corrupted entities o' drawn from the set of objects.

To generate a data set for episodic memory, we assigned a time index to each triple, in a way that all triples with the same subject obtained the same time index. Overall we used 40 different time indices.[2] Similar to the corruption of semantic triples, the negative samples of episodic quadruples (s, p, o, t) are drawn by corrupting the objects o to o' or the temporal index t to t', meaning that (s, p, o', t) services as a negative evidence of the episodic memory at instance t, and (s, p, o, t') is a true fact which cannot be correctly recalled at instance t'. The cost function is composed of cross-entropy and additional $L2$ or $L1$ norm and can be written for episodic quadruples as

$$\mathcal{L} = - \sum_{(s,p,o,t) \in \mathcal{T}} \log \theta^{epi}_{s,p,o,t} - \sum_{(s',p',o',t') \in \mathcal{C}} \log(1 - \theta^{epi}_{s',p',o',t'}) + \lambda ||A||_{1or2},$$

where (s, p, o, t) are true quadruples in the data set, and (s', p', o', t') are the corresponding corrupted quadruples.

7.2   Evaluation and Implementation

All the latent models were implemented using the open source libraries TensorFlow and Keras.[3] The latent representations of all entities, predicates, time

---

[1] This is crucial for investigating human's semantic memory since the system of memory includes "words and verbal symbols, their meanings and the relations between them" according to Tulving [38]. Large number of relation types contained in the knowledge graph enrich the hidden structure and enhance the complexity of the knowledge graph, and this will give a more realistic and accurate simulation of human's semantic memory.

[2] This assignment resembles the active learning process of humans since, during the early stage of learning, we observe the true effects related to a certain subject during a fixed period of time, and store these facts in the form of episodic memory. In our experiment, we highly abstract this learning procedure.

[3] https://en.wikipedia.org/wiki/Comparison_of_deep_learning_software. The code for the experiments is given in https://github.com/Yunpu/Episodic-Memory.

indices and the core tensor from the Tucker tensor decomposition are initialized with Glorot uniform initialization [11]. All the models are optimized using mini-batch adaptive gradient descent using the Adam update rule [18].

We split the data set, both semantic and episodic, into three subsets, where 70% were treated as the training set, 20% as the testing set, and the remaining 10% as the cross-validation set.

7.3   Experimental Results

Figure 3 shows area under precision recall (AUPRC) scores for the training and the test sets for the three settings as a function of the rank of the model. We report results for a semantic memory ("semantic"), for the episodic memory ("episodic") and for the semantic memory derived form the episodic memory by marginalization or projection ("projection"). Note that episodic is evaluated on the episodic data ("remember") whereas semantic and projection is evaluated on the semantic data ("know"). We see that the episodic experiment typically requires a higher rank to obtain good performance. The reason is that the episodic tensor is even sparser than the semantic tensor and contains fewer clear global patterns. The figures also show that we can obtain a semantic memory by projecting the episodic memory, confirming that episodic memory is a "gateway" to semantic memory. In general, the unconstrained model gives better scores. But note that for the non-negative models we performed the projection as discussed in Fig. 2 by entering a vector of ones for the time indices, whereas for the unconstrained setting, we first fully reconstructed the tensor entries and then did the summation on the reconstructed entries over time dimension. The latter procedure becomes infeasible for large episodic KGs.

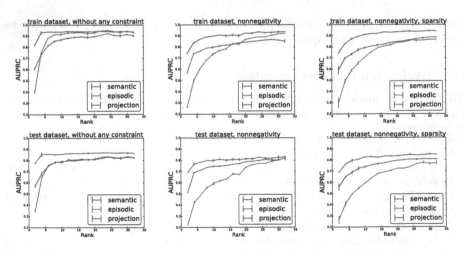

Fig. 3. AUPRC scores of the training and testing data sets for different model settings as a function of the rank.

In Fig. 4 we plot recall score vs rank. In setting 3 with both non-negativity and sparsity constraints (see the third panel of Fig. 4), the projection almost overlaps with the curve of semantic memory for the train and test data set. This observation explicitly indicates that the projected episodic memory function possesses the same memory capacity and quality as the semantic memory function, and this is the central result of our experiment.

Sparsity is an important feature of biological brain functions. Biological experiments indicate that the dentate gyrus (DG) and the CA3 subregion of the hippocampus sustain active neurons, which are connected by sparse projections from DG to CA3 through mossy fibers [17]. CA3 is considered to be crucial for establishing a memory trace during memory consolidation [30]. In the first unconstrained setting we obtained a sparsity of 3%, in the second setting with non-negativity constraints we obtained a sparsity of 30%, and in the third setting with non-negativity constraints and $l1$ norm to encourage sparsity, we obtained 58% sparsity.

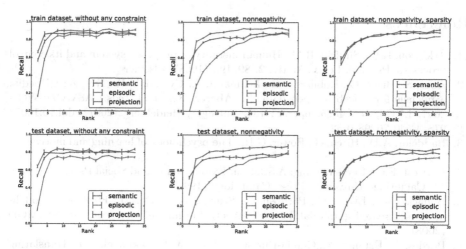

Fig. 4. Recall scores of the training and testing data set for different model settings as a function of the rank.

8   Conclusions

We have derived technical models for episodic memory and semantic memory based on a decomposition of the corresponding adjacency tensors. Whereas semantic memory only depends on the latent representations of subject, predicate and object, episodic memory also depends on the latent representation of the time of an event. We also proposed that semantic memory can be directly derived from episodic memory by marginalizing the time dimension.

As has been shown by several studies, the test set performances of tensor decomposition approaches are state-of-the-art [26]. If we want to use the models for memory recall, we are, however, mostly interested in reproducing stored memories accurately. Currently our models do not perform sufficiently well here (the training scores are around 0.9 instead of 1.0 on the training set). We attribute this to the limited ranks $\tilde{r}$ and $\tilde{r}_T$ of the models. As discussed, $\mathbf{a}_{e_t}$ must encode all processed sensory information from various modalities and must be extremely high-dimensional to be able to do so, thus a large $\tilde{r}_T$ is necessary. On the other hand, the rank for the latent representations for entities and predicates can be somewhat smaller, since the semantic memory, being formed by an integration process over the episodic memory model, requires a smaller rank for a good approximation, as confirmed in the experiment. Another issue is that the training data also contains triples and quadruples that do not follow regular patterns. It is correct to smooth over these triples and quadruples, since one cannot generalize from those, but one would want a truthful memory system to be able to recall also events that do not follow any regular patterns. Finding suitable solutions here is part of future work.

References

1. Atkinson, R.C., Shiffrin, R.M.: Human memory: a proposed system and its control processes. Psychol. Learn. Motiv. **2**, 89–195 (1968). Elsevier
2. Auer, S., Bizer, C., Kobilarov, G., Lehmann, J., Cyganiak, R., Ives, Z.: DBpedia: a nucleus for a web of open data. In: Aberer, K., et al. (eds.) ASWC/ISWC - 2007. LNCS, vol. 4825, pp. 722–735. Springer, Heidelberg (2007). doi:10.1007/978-3-540-76298-0_52
3. Baddeley, A.D., Hitch, G., Bower, G.H.: The psychology of learning and motivation (1974)
4. Bartlett, F.C.: Remembering: A Study in Experimental and Social Psychology, vol. 14. Cambridge University Press, Cambridge (1995)
5. Bollacker, K., Evans, C., Paritosh, P., Sturge, T., Taylor, J.: Freebase: a collaboratively created graph database for structuring human knowledge. In: SIGMOD (2008)
6. Bordes, A., Usunier, N., Garcia-Duran, A., Weston, J., Yakhnenko, O.: Translating embeddings for modeling multi-relational data. In: Advances in Neural Information Processing Systems 26 (2013)
7. Carlson, A., Betteridge, J., Kisiel, B., Settles, B., Hruschka Jr., E.R., Mitchell, T.M.: Toward an architecture for never-ending language learning. In: AAAI, vol. 5, p. 3 (2010)
8. Conway, M.A.: Episodic memories. Neuropsychologia **47**(11), 2305–2313 (2009). Elsevier
9. Ebbinghaus, H.: Über das Gedächtnis: Untersuchungen zur experimentellen Psychologie. Duncker & Humblot (1885)
10. Gazzaniga, M.S., Ivry, R.B., Mangun, G.R.: Cognitive Neuroscience: The Biology of the Mind. Norton, New York (2013)
11. Glorot, X., Bengio, Y.: Understanding the difficulty of training deep feedforward neural networks. In: AISTATS, vol. 9, pp. 249–256 (2010)
12. Gluck, M.A., Mercado, E., Myers, C.E.: Learning and Memory: From Brain to Behavior. Palgrave Macmillan, New York (2013)

13. Greenberg, D.L., Verfaellie, M.: Interdependence of episodic and semantic memory: evidence from neuropsychology. J. Int. Neuropsychological Soc. **16**(05), 748–753 (2010). Cambridge Univ Press
14. Gutierrez, C., Hurtado, C.A., Vaisman, A.: Introducing time into RDF. IEEE Trans. Knowl. Data Eng. **19**(2) (2007). IEEE
15. Hoffart, J., Suchanek, F.M., Berberich, K., Lewis-Kelham, E., De Melo, G., Weikum, G.: Yago2: exploring and querying world knowledge in time, space, context, and many languages. In: WWW. ACM (2011)
16. Huth, A.G., de Heer, W.A., Griffiths, T.L., Theunissen, F.E., Gallant, J.L.: Natural speech reveals the semantic maps that tile human cerebral cortex. Nature **532**(7600), 453–458 (2016). Nature Publishing Group
17. Kesner, R.P., Rolls, E.T.: A computational theory of hippocampal function, and tests of the theory: new developments. Prog. Neurobiol. **79**(1), 1–48 (2006). Elsevier
18. Kingma, D., Ba, J.: Adam: a method for stochastic optimization. arXiv:1412.6980 (2014)
19. Krompaß, D., Jiang, X., Nickel, M., Tresp, V.: Probabilistic latent-factor database models. In: ECML PKDD (2014)
20. Kumar, A., Irsoy, O., Su, J., Bradbury, J., English, R., Pierce, B., Ondruska, P., Gulrajani, I., Socher, R.: Ask me anything: dynamic memory networks for natural language processing. arXiv:1506.07285 (2015)
21. Loftus, E., Ketcham, K.: The Myth of Repressed Memory: False Memories and Allegations of Sexual Abuse. Macmillan, New York (1996)
22. McClelland, J.L., McNaughton, B.L., O'Reilly, R.C.: Why there are complementary learning systems in the hippocampus and neocortex. Psychol. Rev. **102**(3), 419 (1995). American Psychological Association
23. Morton, N.W.: Interactions between episodic and semantic memory. Technical report, Vanderbilt Computational Memory Lab (2013)
24. Nickel, M., Tresp, V., Kriegel, H.-P.: A three-way model for collective learning on multi-relational data. In: ICML (2011)
25. Nickel, M., Tresp, V., Kriegel, H.-P.: Factorizing YAGO: scalable machine learning for linked data. In: WWW (2012)
26. Nickel, M., Murphy, K., Tresp, V., Gabrilovich, E.: A review of relational machine learning for knowledge graphs: from multi-relational link prediction to automated knowledge graph construction. Proc. IEEE **104**(1), 11–33 (2016). IEEE
27. Poon, H., Domingos, P.: Sum-product networks: a new deep architecture. In: ICCV (2011)
28. Quiroga, R.Q.: Concept cells: the building blocks of declarative memory functions. Nat. Rev. Neurosci. **13**(8), 587–597 (2012). Nature Publishing Group
29. Roediger, H.L., McDermott, K.B.: Creating false memories: remembering words not presented in lists. J. Exp. Psychol. Learn. Mem. Cogn. **21**(4), 803 (1995). American Psychological Association
30. Rolls, E.T.: A computational theory of episodic memory formation in the hippocampus. Behav. Brain Res. **215**(2), 180–196 (2010). Elsevier
31. Rosenblatt, F.: The perceptron: a probabilistic model for information storage and organization in the brain. Psychol. Rev. **65**(6), 386 (1958). American Psychological Association
32. Singhal, A.: Introducing the knowledge graph: things, not strings, May 2012. http://googleblog.blogspot.com/2012/05/introducing-knowledge-graph-things-not.html

33. Socher, R., Gershman, S., Sederberg, P., Norman, K., Perotte, A.J., Blei, D.M.: A Bayesian analysis of dynamics in free recall. In: Advances in Neural Information Processing Systems (2009)
34. Squire, L.R.: Memory and Brain. Oxford University Press, New York (1987)
35. Suchanek, F.M., Kasneci, G., Weikum, G.: Yago: a core of semantic knowledge. In: WWW (2007)
36. Teyler, T.J., DiScenna, P.: The hippocampal memory indexing theory. Behav. Neurosci. **100**(2), 147 (1986). American Psychological Association
37. Tresp, V., Esteban, C., Yang, Y., Baier, S., Krompaß, D.: Learning with memory embeddings. arXiv:1511.07972 (2015)
38. Tulving, E.: Episodic and semantic memory 1. In: Organization of Memory. Academic, London (1972)
39. Yee, E., Chrysikou, E.G., Thompson-Schill, S.L.: The cognitive neuroscience of semantic memory. In: Oxford Handbook of Cognitive Neuroscience. Oxford University Press (2014)

Mobile Web, Sensors, and Semantic Streams Track

Spatial Ontology-Mediated Query Answering over Mobility Streams

Thomas Eiter[1], Josiane Xavier Parreira[2], and Patrik Schneider[1,2(✉)]

[1] Vienna University of Technology, Vienna, Austria
[2] Siemens AG Österreich, Vienna, Austria
patrik@kr.tuwien.ac.at

Abstract. The development of (semi)-autonomous vehicles and communication between vehicles and infrastructure (V2X) will aid to improve
road safety by identifying dangerous traffic scenes. A key to this is the
Local Dynamic Map (LDM), which acts as an integration platform for
static, semi-static, and dynamic information about traffic in a geographical context. At present, the LDM approach is purely database-oriented
with simple query capabilities, while an elaborate domain model as captured by an ontology and queries over data streams that allow for semantic concepts and spatial relationships are still missing. To fill this gap, we
present an approach in the context of ontology-mediated query answering that features conjunctive queries over DL-Lite$_A$ ontologies allowing
spatial relations and window operators over streams having a pulse. For
query evaluation, we present a rewriting approach to ordinary DL-Lite$_A$
that transforms spatial relations involving epistemic aggregate queries
and uses a decomposition approach that generates a query execution
plan. Finally, we report on experiments with two scenarios and evaluate
our implementation based on the stream RDBMS PipelineDB.

1 Introduction

The development of (semi)-autonomous vehicles needs extensive communication
between vehicles and the infrastructure, which is covered by Cooperative Intelligent Transport Systems (C-ITS). These systems collect temporal data (e.g.,
traffic light signal phases) and geospatial data (e.g., GPS positions), which is
exchanged by vehicle-to-vehicle, vehicle-to-infrastructure, and combined communications (V2X). V2X aids to improve road safety by analyzing traffic scenes
that could lead to accidents (e.g. red light violation). A key technology for this
is the Local Dynamic Map (LDM) [1], which acts as an integration platform for
static, semi-static, and dynamic information in a geographical context.

Current approaches for an LDM, however, are purely database-oriented with
simple query capabilities. Our aim is to enable spatial-stream conjunctive queries
(CQs) over a semantically enriched LDM for safety applications, such as detection of red light violations on complex intersections managed by a roadside
C-ITS station. To realize spatial query answering (QA) over mobility streams,
spatial and streaming data must be lifted to the setting of ontology-mediated

© Springer International Publishing AG 2017
E. Blomqvist et al. (Eds.): ESWC 2017, Part I, LNCS 10249, pp. 219–237, 2017.
DOI: 10.1007/978-3-319-58068-5_14

QA with the frequently used ontology language DL-Lite$_A$. However, bridging the gap between stream processing and ontology-mediated QA is not straightforward, as the semantics of DL-Lite$_A$ must be extended with spatial relations and stream queries using window operators. For this, we build on the work on spatial QA in [12] and extend ontology-mediated QA with epistemic aggregate queries (EAQs) [10] to detemporalize the streams. The extension preserves first-order rewritability, which allows us to evaluate a CQ with spatial atoms over a stream RDBMS. Our contributions are briefly summarized as follows:

- we outline the field of V2X integration using LDMs in the mobility context (Sect. 2);
- we introduce a data model and query language suited for mobility streams (Sects. 3 and 4);
- we present a spatial-stream QA approach for DL-Lite$_A$ defining its semantics with the focus of preserving FO-rewritability. The QA approach is based on CQ over DL-Lite$_A$ ontologies, which combines window operators over streams having a pulse and spatial relations over spatial objects (Sects. 4 and 5);
- we provide a technique for query rewriting taking the above into account. For query evaluation, we extend and apply the known techniques of (a) epistemic aggregate queries, e.g., average, for a "detemporalization" of the streams; and (b) provide a technique for query decomposition using hypertrees (Sects. 5 and 6);
- we have implemented a prototype and performed experiments in two scenarios to evaluate its applicability (Sect. 7).

In the final Sect. 8, we discuss related work and conclude with ongoing and future work.

2   V2X Integration using a Local Dynamic Map

The base communication technologies (i.e., the IEEE 802.11p standard) allow wireless access in vehicular environments, which enables messaging between vehicles themselves and the infrastructure, called V2X communication. Traffic participants and roadside C-ITS stations broadcast every 100 ms messages for informing others about their current state such as position, speed, and traffic light signal phases [1]. The main types of

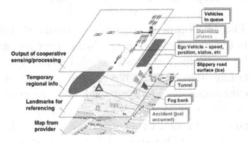

Fig. 1. The four layers of a LDM [1] (Color figure online)

V2X messages are *Cooperative Awareness Messages* (CAM) that provide high frequency status updates of a vehicle's position, speed, vehicle type, etc.; *Map Data Messages* (MAP) that describe the detailed topology of an intersection, including its lanes and their connections; *Signal Phase and Timing Messages*

(SPaT) that give the projected signal phases (e.g., green) for a lane; and *Decentralized Environmental Notification Messages* (DENM) that inform if specific events like road works occur in a designated area.

The *Local Dynamic Map* (LDM) is a comprehensive integration effort of V2X messages; the SAFESPOT project [1] introduced the concept of an LDM as an integration platform to combine static geographic information system (GIS) maps with data from dynamic environmental objects (e.g., vehicles, pedestrians). This was motivated by advanced safety applications (e.g. detect red light violation) that need an "overall" picture of the traffic environment. The LDM has the following four layers (see Fig. 1):

- *Permanent static*: the first layer contains static information obtained from GIS maps and includes roads, intersections, and points-of-interest;
- *Transient static*: the second layer extends the static map by detailed local traffic informations such as fixed ITS stations, landmarks, and intersection features like lanes;
- *Transient dynamic*: the third layer contains temporary regional information like weather, road or traffic conditions (e.g., traffic jams), and signal phases;
- *Highly dynamic*: the fourth layer contains dynamic information of road users detected by V2X messages, in-vehicle sensors like the GPS module.

Current research (e.g., [19]) on architectures of an LDM identified that it can be built on top of a spatial RDBMS enhanced with streaming capabilities. As recognized by [19], an LDM should be represented by a world model, world objects, and data sinks on the streamed input. However, an elaborate domain model, captured by an LDM ontology, and extended queries over data streams allowing spatial relations are still missing. The ontology represents an integration schema modeled in DL-Lite$_A$ and captures the layers of an LDM. Likewise, the LDM ontology must represent the content of the V2X messages and more general GIS objects (e.g., parking or petrol stations) (cf. [11]).

Safety applications on intersections. "Road intersection safety" is an important application for improving road safety [1]. Intersections are the most complex environments and need special attention, where hazardous situations like *obstructed view* or *red-light violation* might lead to accidents. We take the latter as a motivation and running example.

Example 1. The following query detects red-light violations on intersections by searching for vehicles y with speed above 30 km/h on lanes x whose signals will turn red in 4 s:

$$q_1(x,y) : LaneIn(x) \wedge hasLocation(x,u) \wedge intersects(u,v) \wedge pos_{(line,\,4s)}(y,v)$$
$$\wedge Vehicle(y) \wedge speed_{(avg,\,4s)}(y,r) \wedge (r > 30) \wedge isManaged(x,z)$$
$$\wedge SignalGroup(z) \wedge hasState_{(first,\,-4s)}(z, Stop)$$

Query q_1 exhibits the different dimensions which need to be combined: (a) $Vehicle(y)$ and $isManaged(x,z)$ are ontology atoms, which have to be unfolded in respect to the ITS domain modelled in the LDM ontology; (b) $intersects(u,v)$ and $hasLocation(x,u)$ are spatial atoms, where the first checks spatial intersection and the second the assignment of a geometry to an object; (c) $speed_{(Avg\ 2s)}(y, v)$ defines a window operator that aggregates the average speed of the vehicles over the stream and $hasState_{(first,\ -4s)}$ gives us the upcoming traffic light state.

3 Streams, Pulses, and Spatial Databases

We now introduce the data model and sources that are used in our spatial-stream QA.

Streams and pulses. Our data model is *point-based* (vs. *interval-based*) and captures the *valid time* (vs. *transaction time*) saying that some data item is valid at that time point. We extend this validity of time, and say that a data item is valid *from its time point until the next data item is added* to the stream. To capture streaming data, we introduce the *timeline* $\mathbb{T}$, which is a *closed* interval of $(\mathbb{N}, \leq)$. A (data) *stream* is a triple $F = (\mathbb{T}, v, P)$, where $\mathbb{T}$ is a timeline, $v : \mathbb{T} \to \langle \mathcal{F}, \mathcal{S}_\mathcal{F} \rangle$ is a function that assigns to each element of $\mathbb{T}$, called *timestamp* (or time point), data items (called *membership assertions*) of $\langle \mathcal{F}, \mathcal{S}_\mathcal{F} \rangle$, where $\mathcal{F}$ (resp. $\mathcal{S}_\mathcal{F}$) is a *stream (resp. spatial with streams) database*, and P is an integer called *pulse* defining the general interval of consecutive data items on the timeline (cf. [6,20]). A pulse generates a stream of data items with the frequency derived from the interval length. We always have a *main pulse* $P_\mathbb{T}$ with a fixed interval length (usually 1) that defines the lowest granularity of the validity of data items. The pulse also aligns the data items, which arrive asynchronously in the database (DB), to the timeline.

Extending [20], we allow additional *larger pulses* that generate streams with a lower frequency allowing larger intervals. Larger pulses also imply that their generated data items are valid longer than items from the main pulse, thus allowing us to resize the window size of a query and perform optimizations such as caching. Furthermore, pull-based queries are executed at any single time point i denoted as $\mathbb{T}_i$. Push-based queries are evaluated asynchronously where the lowest granularity is given by $P_\mathbb{T}$.

Example 2. For the timeline $\mathbb{T} = [0, 100]$, we have the stream $F_{CAM} = (\mathbb{T}, v, 1)$ of vehicle positions and speed at the assigned time points $v(0) = \{speed(c_1, 30),\ pos(c_1, (5, 5)),\ speed(b_1, 10),\ pos(b_1, (1, 1))\}$, $v(1) = \{speed(c_1, 29), pos(c_1, (6, 5))\ speed(b_1, 5),\ pos(b_1, (2, 1))\}$, and $v(2) = \{speed(c_1, 34),\ pos(c_1, (7, 5))\}$ for the individuals c_1 and b_1. A second "slower" stream $F_{SPaT} = (\mathbb{T}, v, 5)$ captures the next signal state of a traffic light: $v(0) = \{hasState(t_1, Red)\}$ and $v(5) = \{hasState(t_1, Green)\}$. As F_{SPaT} has a pulse of $p = 5$, we know $v(4) = \emptyset$ but under an alternative semantics with an *inertia assumption*, we could conclude $v'(4) = \{hasState(t_1, Red)\}$. Further, the static ABox contains the assertions $Car(c_1)$, $Bike(b_1)$, and $SignalGroup(t_1)$.

Spatial databases and topological relations. We recall the essential idea based on *Point-Set Topological Relations* (see [12]). Spatial relations are defined via pure set theoretic operations on a point set $P_E \subseteq \mathbb{R}^2$ in the plane. An *admissible geometry* $g(s)$ is a sequence $p = (p_1, \ldots, p_n)$ of points over P_F, where $P_F \subseteq P_E$ is the set of explicit points. We define a *spatial DB* over Γ_S as a pair $\mathcal{S} = (P_F, g)$ of a point set P_F and a mapping $g : \Gamma_S \rightarrow \bigcup_{i \geq 1} P_F{}^i$, where Γ_S is a set of spatial objects. The *extent* of a geometry p (full point set) is given by the function $points(p)$ as a (possibly infinite) subset of P_E. For a spatial object s, we let $g(s)$ be its geometry and let $points(s) := points(g(s))$. For our KB, we consider the following admissible geometries p over P_F, and let $P_E = \bigcup_{s \in \Gamma_S} points(s)$ (see [12] for further ones):

- *points* are the sequences $p = (p_1)$, where $points(p_1) = \{p_1\}$;
- *line segments* are sequences $p = (p_1, p_2)$, and $points(p) = \{\alpha p_1 + (1-\alpha)p_2 \mid \alpha \in \mathbb{R}, 0 \leq \alpha \leq 1\}$;

We use *points* to evaluate the spatial relations of two spatial objects via their respective geometries and define the relations in terms of *pure set operations* (see [12] for more):

- *Inside*(x, y): $points(x) \subseteq points(y)$ and *Outside*(x, y): $points(x) \cap points(y) = \emptyset$;
- *Contains*(x, y): $points(y) \subseteq points(x)$, *Intersect*$(x, y)$: $points(x) \cap points(y) \neq \emptyset$.

A spatial relation $S(s, s')$ with $s, s' \in \Gamma_S$ holds on a spatial DB $\mathcal{S}$, written $\mathcal{S} \models S(s, s')$, if $S(g(s), g(s'))$ is true. Relative to *points*, this is easily captured by a first-order (FO) formula over $(\mathbb{R}^2, \leq)$, and on geo-spatial RDBMS rewritable into FO queries.

Combining spatial and stream databases. Following an ontology-mediated QA approach, the LDM ontology is the global schema called the TBox $\mathcal{T}$, whereon we link normal, spatial, and stream DBs. We distinguish between a (standard) static ABox $\mathcal{A}$, a stream ABox $\mathcal{F}$, a static-spatial ABox $\mathcal{S}_{\mathcal{A}}$, and a spatial ABox with stream support $\mathcal{S}_{\mathcal{F}}$. These ABoxes can be stored in respective DBs, and combined in different ways. We focus on a stream DB with limited support for spatial data, which acts also as a storage for $\mathcal{S}_{\mathcal{A}}$.

4   Syntax, Semantics, and Query Language of DL-Lite$_A$ (S,F)

We start from previous work in [12], which introduced spatial CQ answering for DL-Lite$_A$, and lift the semantics from the spatial DL-Lite$_A$ KB to the spatial-stream KB.

Syntax and semantics of DL-Lite$_A$. We consider a vocabulary of individual names Γ_I, domain values Γ_V (e.g., $\mathbb{N}$), and spatial objects Γ_S. Given atomic concepts A, atomic roles P, and atomic attributes E, we define (a) *basic* concepts B, *basic roles* Q, and *basic value-domains* E (attribute ranges); (b) *complex* concepts C, *complex* role expressions R, and *complex attributes* V; and (c) value-domain expressions D:

$$
\begin{array}{ll}
B ::= A \mid \exists Q \mid \delta(U_C) & C ::= \top_C \mid B \mid \neg B \mid \exists Q.C' \\
E ::= \rho(U_C) & D ::= \top_D \mid D_1 \mid \ldots \mid D_n \\
Q ::= P \mid P^- & R ::= Q \mid \neg Q \qquad V ::= U \mid \neg U
\end{array}
$$

where P^- is the *inverse* of P, $\top_D$ is the *universal* value-domain and $\top_C$ is the *universal* concept; furthermore, U_C is a given attribute with domain $\delta(U_C)$ (resp. range $\rho(U_C)$). A DL-Lite$_A$ *knowledge base* (KB) is a pair $\mathcal{K} = (\mathcal{T}, \mathcal{A})$ where the TBox $\mathcal{T}$ and the ABox $\mathcal{A}$ consist of finite sets of axioms as follows:

- *inclusion assertions* of the form $B \sqsubseteq C$, $Q \sqsubseteq R$, $E \sqsubseteq D$, and $U \sqsubseteq V$; respectively
- *functionality assertions* of the form $funct\ Q$ and $funct\ U$;
- *membership assertions* of the form $A(a)$, $D(c)$, $P(a, b)$, and $U(a, c)$, where a, b are individual names in Γ_I and c is a value in Γ_V.

The semantics of DL-Lite$_A$ is in terms of FO interpretations $\mathcal{I} = (\Delta^{\mathcal{I}}, \cdot^{\mathcal{I}})$, where the domain $\Delta^{\mathcal{I}} \neq \emptyset$ is the disjoint union of $\Delta_I^{\mathcal{I}}$ of $\Delta_V^{\mathcal{I}}$ and $\cdot^{\mathcal{I}}$ is an *interpretation function* as usual (see [9]). Satisfaction of axioms and logical implication are denoted by $\models$. We assume the *unique name assumption* (UNA) for different individuals resp. domain values and adopt the *constant domain assumption*, saying that all models share the same domain.

Syntax DL-Lite$_A$ (S,F). Let $\mathbb{T}$ be a timeline and let Γ_S, Γ_I, and Γ_V be pairwise disjoint sets as above. A *spatial-stream knowledge base* is a tuple

$$\mathcal{K}_{S\mathcal{F}} = \langle \mathcal{T}, \mathcal{A}, \mathcal{S}_A, \langle \mathcal{F}, \mathcal{S}_{\mathcal{F}} \rangle, \mathcal{B} \rangle,$$

where $\mathcal{T}$ (resp. $\mathcal{A}$) is a DL-Lite$_A$ TBox (resp. ABox), $\mathcal{S}_A$ is a spatial DB, and $\langle \mathcal{F}, \mathcal{S}_{\mathcal{F}} \rangle$ is a stream DB with support for spatial data. Furthermore, $\mathcal{B} \subseteq \Gamma_I \times \Gamma_S$ is a partial function called the *spatial binding* from $\mathcal{A}$ to $\mathcal{S}_A$ and $\mathcal{F}$ to $\mathcal{S}_{\mathcal{F}}$. If we restrict to a spatial KB resp. stream KB, we drop $\mathcal{F}$ (resp., $\mathcal{S}$) and have

$$\mathcal{K}_S = \langle \mathcal{T}, \mathcal{A}, \mathcal{S}_A, \mathcal{B} \rangle \quad \text{resp.} \quad \mathcal{K}_{\mathcal{F}} = \langle \mathcal{T}, \mathcal{A}, \mathcal{F} \rangle.$$

We introduce for DL-Lite$_A$ the possibility to specify the *localization* of atomic concepts and roles. For this, we extend their syntax similar as in [12] as follows:

$$C ::= \top_C \mid B \mid \neg B \mid \exists Q.C' \mid (\textbf{\textit{loc }} A) \mid (\textbf{\textit{loc}}_s\ A)$$
$$R ::= Q \mid \neg Q \mid (\textbf{\textit{loc }} Q) \mid (\textbf{\textit{loc}}_s\ Q),$$

where $s \in \Gamma_S$ and the concept and roles are as before. Intuitively, $(loc\ A)$ is the set of individuals in A that can have a spatial extension (e.g., $(loc\ Parks)$), and $(loc_s\ A)$ is the subset where it is s (e.g., $(loc_{(48.20, 16.37)}\ Vienna)$).

The extension with streaming is captured by the following axiom schemes:

$$(\textbf{\textit{stream}}_F\ C) \quad \text{and} \quad (\textbf{\textit{stream}}_F\ R),$$

where F is a particular stream over either complex concepts C or roles R in $\langle \mathcal{F}, \mathcal{S}_{\mathcal{F}} \rangle$.

Example 3. For Example 2, a TBox may contain $(stream_{F_{CAM}}\ speed)$, $(stream_{F_{CAM}}\ (loc\ pos))$, $(stream_{F_{CAM}}\ Vehicle)$, and $(stream_{F_{SPaT}}\ hasState)$, and we have further axioms $Car \sqsubseteq Vehicle$, $Bike \sqsubseteq Vehicle$, and $Ambulance \sqsubseteq \exists hasRole.Emergency$.

Semantics DL-Lite$_A$ (S,F). We give a semantics to the localization $(loc\ Q)$ and $(loc_s\ Q)$ for individuals of Q with some spatial extension resp. located at s, such that a KB $\mathcal{K}_S = \langle \mathcal{T}, \mathcal{A}, \mathcal{S}, \mathcal{B} \rangle$ can be readily transformed into an ordinary DL-Lite$_A$ KB $\mathcal{K}_O = \langle \mathcal{T}', \mathcal{A}' \rangle$, using the fresh spatial top concept $C_{\mathcal{S}_T}$ and spatial concepts C_s. An *interpretation of* $\mathcal{K}_S$ is a structure $\mathcal{I}_S = \langle \Delta^{\mathcal{I}}, \cdot^{\mathcal{I}}, b^{\mathcal{I}} \rangle$, where $\langle \Delta^{\mathcal{I}}, \cdot^{\mathcal{I}} \rangle$ is an interpretation of $\langle \mathcal{T}, \mathcal{A} \rangle$ and $b^{\mathcal{I}} \subseteq \Delta^{\mathcal{I}} \times \Gamma_S$ is a partial function that assigns some individuals a location, such that for every $a \in \Gamma_I$, $(a, s) \in \mathcal{B}_A$ implies $b^{\mathcal{I}}(a^{\mathcal{I}}) = s$. We extend the semantics with $(loc\ Q)$ and $(loc_s\ Q)$, where Q is an atomic role in $\mathcal{T}$ by $((loc\ A)$ and $(loc_s\ A)$ are accordingly):

$$(loc\ Q)^{\mathcal{I}_S} \supseteq \{(a_1, a_2) \mid (a_1, a_2) \in Q^{\mathcal{I}} \wedge \exists s \in \Gamma_S : (a_2, s) \in b^{\mathcal{I}}\},$$
$$(loc_s\ Q)^{\mathcal{I}_S} = \{(a_1, a_2) \mid (a_1, a_2) \in Q^{\mathcal{I}} \wedge (a_2, s) \in b^{\mathcal{I}}\}.$$

The transformation of $\mathcal{K}_S$ to an ordinary DL-Lite$_A$ KB $\mathcal{K}_O$ is described in [12,13].

The idea of an initial streaming semantics is by interpreting the stream over the full timeline, which can be captured by a finite sequence $\mathcal{F}_A = (\mathcal{F}_i)_{\mathbb{T}_{min} \leq i \leq \mathbb{T}_{max}}$ of temporal ABoxes, which is obtained via the evaluation function v on $\mathcal{F}$ and $\mathbb{T}$ (cf. [7,15]). Hence, we define the interpretation of the point-based model over $\mathbb{T}$ as a sequence $\mathcal{I}_F = (\mathcal{I}_i)_{\mathbb{T}_{min} \leq i \leq \mathbb{T}_{max}}$ of interpretations $\mathcal{I}_i = \langle \Delta^{\mathcal{I}}, \cdot^{\mathcal{I}_i} \rangle$; $\mathcal{I}_F$ is a model of $\mathcal{K}_F$, denoted $\mathcal{I}_F \models \mathcal{K}_F$ iff $\mathcal{I}_i \models \mathcal{F}_i$ and $\mathcal{I}_i \models \mathcal{T}$, for all $i \in \mathbb{T}$.

The semantics of the $(stream_F\ C)$ and $(stream_F\ R)$ axioms is along the same line. A stream axiom is satisfied, if a complex concept C (resp. role R) holds over all the time points of stream $F = (\mathbb{T}, v, P)$; thus we restrict our models such that:

$$(stream_F\ C)^{\mathcal{I}} = \bigcap\nolimits_{i \in tp(\mathbb{T}, P)} C^{\mathcal{I}_i} \text{ and } (stream_F\ R)^{\mathcal{I}} = \bigcap\nolimits_{i \in tp(\mathbb{T}, P)} R^{\mathcal{I}_i},$$

where $tp(\mathbb{T}, P)$ is a set of time points determined by the segmentation of $\mathbb{T}$ by P. This allows us to check for the *satisfiability* of a KB and gives us a global consistency, which is of theoretical nature, since we would need to know the full timeline.

Spatial-stream query language over DL-Lite$_A$ (S,F). We next define spatial-stream conjunctive queries over $\mathcal{K}_{S\mathcal{F}}$. Such queries may contain ontology, spatial, and stream predicates. An spatial-stream CQ $q(\mathbf{x})$ is a formula:

$$\bigwedge\nolimits_{i=1}^{l} Q_{O_i}(\mathbf{x}, \mathbf{y}) \wedge \bigwedge\nolimits_{j=1}^{n} Q_{S_j}(\mathbf{x}, \mathbf{y}) \wedge \bigwedge\nolimits_{k=1}^{m} Q_{F_k}(\mathbf{x}, \mathbf{y}) \tag{1}$$

where $\mathbf{x}$ are the *distinguished* (*answer*) variables, $\mathbf{y}$ are either *non-distinguished* (*existentially quantified*) variables, individuals from Γ_I, or values from Γ_V and

- each $Q_{O_i}(\mathbf{x}, \mathbf{y})$ has the form $A(z)$ or $P(z, z')$, where A is a concept name, P is a role name and z, z' are from $\mathbf{x} \cup \mathbf{y}$;
- each atom $Q_{S_j}(\mathbf{x}, \mathbf{y})$ is from the vocabulary of spatial relations (see Sect. 3) and of the form $S(z, z')$, with z, z' from $\mathbf{x} \cup \mathbf{y}$;

- $Q_{F_j}(\mathbf{x}, \mathbf{y})$ is similar to $Q_{O_i}(\mathbf{x}, \mathbf{y})$ but adds the vocabulary for stream operators, which are taken from [6] and relate to CQL operators [3]. Moreover, we have a window $\boxplus$ over a stream F_j that is derived from L (in $\mathbb{Z}^+$ for past, or in $\mathbb{Z}^-$ for future) time units resp. $\mathbb{T}_i$, and an *aggregate function* $agr \in \{count, sum, first, \ldots\}$ (see Sect. 5 for details) that is applied to the data items in the window:[1]
  - $\boxplus_T^L agr$ represents the aggregate of last/next L time units of stream F_j;
  - $\boxplus_T$ represents the current tuples of F_j with $L = 0$;
  - $\boxplus_T^O agr$: represents the aggregate of all previous L time units of F_j;

Example 4. We modify $q_1(x, y)$ of Example 1 and use the stream operators instead:

$$q_1(x, y) : LaneIn(x) \wedge hasLocation(x, u) \wedge intersects(u, v) \wedge position_{\boxplus_T^4 line}(y, v)$$
$$\wedge\, Vehicle(y) \wedge speed_{\boxplus_T^4 avg}(y, r) \wedge (r > 30) \wedge isManaged(x, z)$$
$$\wedge\, SignalGroup(z) \wedge hasState_{\boxplus_T^{-4} first}(z, Stop)$$

Certain answer semantics with spatial atoms. In the streamless setting, due to the OWA, queries are evaluated over all (possibly infinitely many) models. *Certain answers* retain the tuples that are answers in all possible models. More formally, a *match* for $q(\mathbf{x})$ in an interpretation $\mathcal{I} = \langle \Delta^{\mathcal{I}}, \cdot^{\mathcal{I}} \rangle$ of $\mathcal{K}$ is a function $\pi : \mathbf{x} \cup \mathbf{y} \to \Delta^{\mathcal{I}}$ such that $\pi(c) = c^{\mathcal{I}}$, for each constant c in $\mathbf{x} \cup \mathbf{y}$, and for each $i = 1, \ldots n$ and $j = 1, \ldots, m$:

(i) $\pi(z) \in A^{\mathcal{I}}$, for $Q_{O_i}(\mathbf{x}, \mathbf{y}) = A(z)$ (concept atoms);
(ii) $(\pi(z), \pi(z')) \in P^{\mathcal{I}}$, for $Q_{O_i}(\mathbf{x}, \mathbf{y}) = P(z, z')$ (role atoms); and
(iii) $\exists s, s' \in \Gamma_S : (\pi(z), s) \in b^{\mathcal{I}} \wedge (\pi(z'), s') \in b^{\mathcal{I}} \wedge \mathcal{S} \models S(s, s')$,
   for $Q_{S_j}(\mathbf{x}, \mathbf{y}) := S(z, z')$ (spatial atoms).

A tuple $\mathbf{c} = c_1, \ldots, c_k$ over Γ_I is a *(certain) answer* for $q(\mathbf{x})$ in $\mathcal{I}$, $\mathbf{x} = x_1, \ldots, x_k$, if $q(\mathbf{x})$ has some match π in $\mathcal{I}$ where $\pi(x_i) = c_i$, $i = 1, \ldots, k$; and $\mathbf{c}$ is an answer for $q(\mathbf{x})$ over $\mathcal{K}$, if it is an answer in every model $\mathcal{I}$ of $\mathcal{K}$. The *result* $Cert(q(\mathbf{x}), \mathcal{K})$ of $q(\mathbf{x})$ over $\mathcal{K}$ is the set of all its answers. If we drop $\mathcal{T}$, we obtain a DB setting and let $Eval(q(\mathbf{x}), \mathcal{I})$ be the set of matches of $q(\mathbf{x})$ over the single model $\mathcal{I}$ of $\mathcal{A}$ under closed world assumption.

Regarding spatial atoms, as shown in [12,13] the semantic correspondence between $\mathcal{K}_{\mathcal{O}}$ and $\mathcal{K}_{\mathcal{S}}$ guarantees that we can rewrite $q(\mathbf{x})$ into an equivalent query $uq(\mathbf{x})$ over $\mathcal{K}_{\mathcal{S}}' = \langle \mathcal{T}', \mathcal{A}', \mathcal{S}_{\mathcal{A}} \rangle$. Using the rewriting and the semantic correspondence of $\mathcal{K}_{\mathcal{O}}$ and $\mathcal{K}_{\mathcal{S}}$, spatial atoms can be rewritten into a "standard" DL-Lite$_A$ UCQ, thus, answering spatial CQs is still FO-rewritable (details in [12,13]).

5   Query Rewriting by Stream Aggregation

We aim at answering queries at *a single* time point $\mathbb{T}_i$ with stream atoms that define *aggregate functions* on different windows sizes relative to $\mathbb{T}_i$. For this, we

---

[1] This would be represented in CQL as `R[Range L]`, `R[Now]`, `R[Range L Slide D]`, etc.

consider a semantics based on *epistemic aggregate queries* (EAQ) over ontologies [10] by dropping the order of time points for the membership assertions and handle the (streamed) assertions as *bags*, which is similar to "classic" stream processing approaches.

Epistemic aggregate queries. As described in [10], EAQ are defined over bags of numeric and symbolic values, called *groups* and denoted as $\{| \cdot |\}$. Aggregates cannot be directly transferred to DL-Lite, since with the certain answer semantics each model has different groups due to unknown individuals, which leads to empty answers. [10] extended database semantics for aggregates with an epistemic operator $\mathbf{K}$ and a two-layer evaluation using the completion w.r.t $\mathcal{T}$. The basic idea is to close the aggregate query, so only known individuals are grouped and aggregated. More formally, an EAQ is defined as[2]

$$q_a(\mathbf{x}, agr(y)) : \mathbf{K} \; \mathbf{x}, y, \mathbf{z}. \; \phi,$$

where $\mathbf{x}$ are the *grouping variables*, $agr(y)$ is the *aggregate function and variable*, and ϕ is a CQ called *main conditions*; $\mathbf{z}$ are the disjoint existential variables of ϕ. We call $\mathbf{w} := \mathbf{x} \cup y \cup \mathbf{z}$ the $\mathbf{K}$-variables of ϕ. The definition of a group was extended in [10] by a multiset $H_{\mathbf{d}}$ of groups $\mathbf{d}$, called $\mathbf{K}$-group, as:

$$H_{\mathbf{d}} := \{| \; \pi(y) \mid \pi \in KSat_{\mathcal{I},\mathcal{K}}(\mathbf{z}; \phi) \text{ and } \pi(\mathbf{x}) = \mathbf{d} \; |\},$$

where $KSat$ are the *satisfying* $\mathbf{K}$-matches of ϕ *for* the model $\mathcal{I}$ of $\mathcal{K}$ and given by:

$$KSat_{\mathcal{I},\mathcal{K}}(\mathbf{w}; \phi) := \{\pi \in Eval(\phi, \mathcal{I}) \mid \pi(\mathbf{w}) \in Cert(aux_{q_a}, \mathcal{K})\},$$

where $aux_{q_a}(\mathbf{w}) \leftarrow \phi$ is the auxiliary atom used to map $\mathbf{w}$ only to *known solutions*. The set of $\mathbf{K}$-answers for an EAQ query q over $\mathcal{I}$ and $\mathcal{K}$ can now be derived as:

$$q_a^{\mathcal{I}} := \{(\mathbf{d}, agr(H_{\mathbf{d}})) \mid \mathbf{d} = \pi(\mathbf{x}), \text{ for some } \pi \in KSat_{\mathcal{I},\mathcal{K}}(\mathbf{w}; \phi)\}.$$

The *epistemic certain answers* $ECert(q_a, \mathcal{K})$ for a query q_a over $\mathcal{K}$ is the set of $\mathbf{K}$-answers that are answers in every model $\mathcal{I}$ of $\mathcal{K}$. To compute $ECert(q_a, \mathcal{K})$, [10] gave a "general algorithm" GA that (1) computes the certain answers, (2) projects on the $\mathbf{K}$-variables, and (3) aggregates the resulting tuples. Importantly, evaluating EAQs reduces to standard CQ evaluation over DL-Lite$_A$ with LOGSPACE data complexity.

Filtered and merged temporal ABoxes. Our approach is to evaluate the EAQ over one or more *filtered and merged temporal* ABoxes. The filtering and merging, relative to the window size and $\mathbb{T}_i$, creates several *windowed* ABoxes $\mathcal{A}_{\boxplus_{\phi}}$, which are the union of the static ABox $\mathcal{A}$ and the filtered stream ABoxes from $\mathcal{F}$. The EAQ aggregates are applied on each *windowed* ABox $\mathcal{A}_{\boxplus_{\phi}}$ by aggregating normal objects, concrete values, and spatial objects. More formally, a stream atom $\phi \boxplus_T^L agr$ is evaluated as an EAQ over ontologies

$$q_{\phi}(\mathbf{x}, agr(y)) : \mathbf{K} \; \mathbf{x}, y, \mathbf{z}. \; \phi \boxplus_T^L,$$

---

[2] We simplified EAQs of [10] by omitting ψ and consider only aggregates with a single variable.

where $\mathbf{x}$ are the grouping variables and y is the aggregate variable, $\mathbf{z}$ are the disjoint existential variables, and ϕ is a subquery of q with atoms in the same *scope* of the window operator $\boxplus_T^L$ and aggregate functions agr.

Example 5. For query $q_1(x, y)$ of Example 4, we have three EAQs represented as:

$$q_{pos}(y, line(v)) : \mathbf{K}\, y, v.\, Vehicle(y) \wedge position(y, v);$$
$$q_{speed}(y, avg(r)) : \mathbf{K}\, y, r.\, Vehicle(y) \wedge speed(y, r);$$
$$q_{state}(z, first(m)) : \mathbf{K}\, z, m.\, hasState(z, m)$$

We extend the evaluation of EAQs for the stream setting, such that an EAQ is evaluated over the window relative to $\mathbb{T}_i$, the window operator $\boxplus_T^L$, and the pulse P. $KSat_{\mathcal{I}_\boxplus, \mathcal{K}_\boxplus}(\mathbf{w}; \phi)$ is now the set of $\mathbf{K}$-matches of ϕ for a model $\mathcal{I}_\boxplus$ of $\mathcal{K}_\boxplus$, where the *windowed* ABox $\mathcal{A}_\boxplus$ is defined as $\mathcal{A}_\boxplus = \mathcal{A} \cup \bigcup \{\mathcal{A}_i \mid w_s \leq i \leq w_e\}$. We have four cases for the window size L and a pulse P, where P enlarges L according to its interval length:

– a current window with $L = 0$, i.e. $w_s = w_e = \mathbb{T}_i$;
– a past window with $L > 0$ leading to $w_s = (\mathbb{T}_i - L)$ and $w_e = \mathbb{T}_i$;
– a future window with $L < 0$ that is $w_s = \mathbb{T}_i$ and $w_e = (\mathbb{T}_i + L)$; and
– the entire history with O resulting in $w_s = 0$ and $w_e = \mathbb{T}_i$.

We obtain KB $\mathcal{K}_\boxplus = \langle \mathcal{T}, \mathcal{A}_\boxplus \rangle$ as above; the *epistemic (certain) answers* for q_ϕ over $\mathcal{K}_\boxplus$ are naturally defined as $ECert_\boxplus(q_\phi, \mathcal{K}_\boxplus) = \bigcap_{\mathcal{I}_\boxplus \models \mathcal{K}_\boxplus} q_\phi^{\mathcal{I}_\boxplus}$, where

$$q_\phi^{\mathcal{I}_\boxplus} = \{(\mathbf{d}, agr(H_{\mathbf{d}})) \mid \mathbf{d} = \pi(\mathbf{x}), \text{ for some } \pi \in KSat_{\mathcal{I}_\boxplus, \mathcal{K}_\boxplus}(\mathbf{w}; \phi)\}$$

are the $\mathbf{K}$-matches that are answers in the model $\mathcal{I}_\boxplus$ of $\mathcal{K}_\boxplus$. In $ECert_\boxplus$, we did not yet address the *validity* of an assertion, say in $\mathcal{A}_{\boxplus_1}$, until the next assertion in $\mathcal{A}_{\boxplus_3}$. Two semantics are suggestive: the first ignores intermediate time points, and thus $\mathcal{A}_{\boxplus_2}$ will be unknown. The second fills the missing gaps with the previous assertion, i.e. copies it from $\mathcal{A}_{\boxplus_1}$ to $\mathcal{A}_{\boxplus_2}$. For specific aggregate functions, e.g., *max*, *min*, or *last*, the two semantics coincide, but for *sum*, *avg*, and *count*, they are different.

Example 6. We pose the query $q_1(x, y)$ at $\mathbb{T}_1$ and replace the stream atoms with auxiliary atoms related to the EAQ of Example 5:

$$q_1(x, y) : LaneIn(x) \wedge hasLocation(x, u) \wedge intersects(u, v) \wedge q_{pos}(y, v)$$
$$\wedge q_{speed}(y, r) \wedge (r > 30) \wedge isManaged(x, z) \wedge q_{state}(z, Stop)$$

The queries are computed using the ABoxes $\mathcal{A}_{\boxplus[0,1]} = \mathcal{A} \cup \mathcal{A}_0 \cup \mathcal{A}_1$ and $\mathcal{A}_{\boxplus[1,4]} = \mathcal{A} \bigcup_{1 \leq i \leq 4} \mathcal{A}_i$. This leads under $ECert_\boxplus$ for q_{speed} to the groups $G_{c_1} = \{|30, 29, 34|\}$ and $G_{b_1} = \{|10, 5|\}$, which results in $q_{speed} = \{(c_1, 31), (b_1, 7.5)\}$. The results for the other EAQ are $q_{state} = \{(t_1, Red)\}$ and $q_{pos} = \{(c_1, ((5, 5), (6, 5), (7, 5))), (b_1, ((1, 1), (2, 1)))\}$.

$ECert_\boxplus$ gives the certain answers for a single EAQ including the ontology atoms in the same scope as the stream atoms. Answering the full CQ q can be done by answering each EAQ q_{ϕ_k} separately and joining the answers, i.e.,

$$ECertAll(q, \mathcal{K}_\mathcal{F}, \mathbb{T}_i) = ECert_\boxplus(q_{\phi_1}, \mathcal{K}_{\boxplus_{w_1}}) \bowtie \cdots \bowtie ECert_\boxplus(q_{\phi_j}, \mathcal{K}_{\boxplus_{w_j}}),$$

Algorithm 1. NSQ - Answer Naive Stream Query

Input: A stream conjunctive query q, time point $\mathbb{T}_i$, and a KB $\mathcal{K}_\mathcal{F}$
Output: Set of tuples O
/* Step 1: Detemporalize   */
foreach Q_{F_i} *of* q **do**

$\quad$ $\mathcal{A}_{\boxplus_i} \leftarrow \mathcal{A} \bigcup_{w_s \leq j \leq w_e} \mathcal{A}_j$ according to $\boxplus_T^L$ and $\mathbb{T}_i$;

$\quad$ $\mathcal{K}_{\boxplus_i} \leftarrow \langle \mathcal{T}, \mathcal{A}_{\boxplus_i} \rangle$;

$\quad$ build $aux_i(\mathbf{x}, y, \mathbf{z})$ from $\phi \boxplus_T^L agr$ of Q_{F_i} ;

$\quad$ build $q_{i,1}(\mathbf{x}, y, \mathbf{z}^\phi)$ from PerfectRef$(aux_i, \mathcal{T})(\mathbf{x}, y, \mathbf{z})$;

$\quad$ build $q_{i,2}(x, agr(y))$ from $q_{i,1}(\mathbf{x}, y, \mathbf{z}^\phi)$ and $\phi \boxplus_T^L agr$;

$\quad$ $R_{i,1} :=$ evaluate Answer$(aux_i, \mathcal{K}_{\boxplus_i})$ /* certain answers */ ;

$\quad$ $R_{i,2} :=$ evaluate $q_{i,1}$ over $R_{i,1}$ /* $\mathbf{K}$ projection */ ;

$\quad$ $R_i \;\; :=$ evaluate $q_{i,2}$ over $R_{i,2}$ /* aggregation */ ;

$\quad$ Add R_i to $\mathcal{A}_{aux}$ and replace Q_{F_i} in q with $R_i(x, y)$;

/* Step 2: Standard evaluation   */
$O :=$ evaluate Answer$(q, \langle \mathcal{T}, \mathcal{A} \cup \mathcal{A}_{aux} \rangle)$;

---

where the $w_k = w(\phi_k, \mathbb{T}_i)$ are the computed window sizes and $A \bowtie B = \{t$ over $sig(A) \cup sig(B) \mid t[sig(A)] \in A, \; t[sig(B)] \in B\}$ is the join (cf. [18]) of sets A, B of $\mathbf{K}$-answers, where $sig()$ is the relational signature of a $\mathbf{K}$-answer set. The new $\mathbf{K}$-answers are also answers in every model $\mathcal{I}$ of $\mathcal{K}$. More details on deriving the q_{ϕ_j} are given in Sect. 6.

We now introduce the algorithm NSQ (see Algorithm 1), where $\mathbf{z}^\phi$ are the non-distinguished variables in ϕ and PerfectRef (resp. Answer) is the "standard" query rewriting (resp. evaluation) as in [9]. NSQ extends the GA of [10] to compute the answers for stream CQs as follows: (1) calculate the epistemic answer for each stream atom over the different windowed ABoxes and store the result in an *auxiliary ABox* using new atoms. Furthermore, replace each stream atom with a new auxiliary atom; (2) calculate the certain answers over $\mathcal{A}$ and the auxiliary ABox, using "standard" DL-Lite$_A$ query evaluation. A proof sketch for correctness of NSQ is given in [13], viz. that for every stream CQ q, KB $\mathcal{K}_\mathcal{F}$, and time point $\mathbb{T}_i$, we have NSQ$(q, \mathcal{K}_\mathcal{F}, \mathbb{T}_i) = ECertAll(q, \mathcal{K}_\mathcal{F}, \mathbb{T}_i)$. It considers that q must be constrained by $\mathcal{T}$ and that aggregate functions must obey conditions as in [10]; it exploits that answering each EAQ (Step 1) can be decoupled from answering the full CQ.

Standard aggregates. Different aggregate functions for use in $ECert(q, \mathcal{K})$ were already discussed in [10]. For *last* and *first*, we extend the definition of $H_\mathbf{d}$, as the sequence of time points is lost. By iteratively checking if we have a match in one of the ABoxes $\mathcal{A}_{\boxplus_i} w_s \leq i \leq w_e$, we can determine the first resp. last match. The extension of $H_\mathbf{d}$ for *first* and *last* is by checking each model for match (details in [13]). In an implementation, the first/ last match can be simply cached while processing the stream.

Spatial aggregates. For *spatial objects*, we define geometric aggregate functions on the multiset of $H_\mathbf{d}$. As the order of assertions (i.e., points) is lost, we need to rearrange them to create an admissible geometry $g(s)$ that is a sequence $p = (p_1, \ldots, p_n)$. We add new aggregates on $H_\mathbf{d}$ to create new admissible geometries $g(s_\mathbf{d})$:

- agr_{point}: we evaluate *last* to get the last available position p_1 and set $g(s_{\mathbf{d}}) := (p_1)$;
- agr_{line}: we create $p = (p_1, \ldots, p_n)$, where $p_1 \neq p_n$ and determine a total order on the bag of points in each **K**-group, such that we have a starting point using last and iterate backwards finding the next point;
- agr_{angle}: This aggregate function determines angles (in degrees) in a geometry by applying (1) agr_{line}, then (2) obtain a simplified geometry using smoothing, and (3) calculate the angles between the lines of the geometry.

Besides the above aggregate functions, more functions such as computing the convex hull or minimum spanning tree can be applied. In contrast to numerical aggregates, spatial aggregates introduce for each **K**-group $(\mathbf{d}, agr(H_{\mathbf{d}}))$ a new spatial object $s_{\mathbf{d}}$ of Γ_S and an admissible geometry $g(s_{\mathbf{d}})$ with $agr(H_{\mathbf{d}}) = (p_1, \ldots, p_n)$. This is achieved by (a) adding a binding $(\mathbf{d}, s_{\mathbf{d}})$ to $\mathcal{B}$ and (b) creating a new mapping $g : s_{\mathbf{d}} \rightarrow (p_1, \ldots, p_n)$ in $\mathcal{S}_{aux}$. For simplicity, we assume that Γ_S is static and contains (candidates for) $s_{\mathbf{d}}$ already.

Combining spatial and stream queries. We combine the spatial and temporal elements of a query q and KB $\mathcal{K}$ as follows: (1) detemporalize the stream atoms using EAQs; (2) transform q and $\mathcal{K}$ to an ordinary UCQ and KB as in Sect. 4, where in Step 2 of Algorithm 1 $Cert(q, \langle \mathcal{T}, \mathcal{A} \cup \mathcal{A}_{aux}, \mathcal{S}_A \cup \mathcal{S}_{aux}, \mathcal{B} \rangle)$ is changed to $Cert_S(uq, \langle \mathcal{T}', \mathcal{A}' \cup \mathcal{A}_{aux} \rangle)$. We still keep LOGSPACE data complexity, which follows from the data complexity of single EAQs and the fact that the number of aggregate atoms bounds the number of EAQs. As shown before, spatial binding and relations do not increase the data complexity.

6   Query Evaluation by Hypertree Decomposition

We focus on pull-based evaluation of spatial-stream CQs, which is already challenging, as we must deal with three different types of query atoms that need different evaluation techniques over possibly separate DBs. Ontology atoms are evaluated over the static ABox $\mathcal{A}$ using the "standard" DL-Lite$_A$ query rewriting, i.e., *PerfectRef* [9]. For spatial query atoms, we need to dereference the bindings by joining the binding $\mathcal{B}$ and the spatial ABox $\mathcal{S}_A$, where we evaluate the spatial relations (e.g., *Inside*) over the spatial objects of the join; Stream query atoms are computed as described in Algorithm 1 over the stream ABox $\mathcal{F}$ and the spatial ABox with stream support $\mathcal{S}_\mathcal{F}$.

Evaluation strategies. In [12], we introduced spatial CQ evaluation based on the assumptions that *no bounded variables* occur in spatial atoms and the CQ $q_S(\mathbf{x})$ has to be *acyclic*. This allows an evaluation in two stages:

(1) evaluate the ontology part of $q_S(\mathbf{x})$ by dropping all spatial atoms over $\mathcal{K}_S'$. For this, we can apply the standard query rewriting and evaluate the resulting UCQ over $\mathcal{A}$;
(2) filter the result of Step (1), by evaluating the spatial atoms on the matches π (for the distinguished variables $\mathbf{x}$) taking the bindings $\mathcal{B}$ to $\mathcal{S}_A$ into account.

As shown in [12], one evaluation strategy is based on the hypergraph of q_S and the derived join plan, while another is based on compiling $q_S(\mathbf{x})$ into a single, large UCQ with spatial joins. The hypergraph-based strategy is well suited for lifting to spatial-stream CQs as the partial EAQ results are already stored. Hence, we merge it with the two-stage evaluation of Sect. 5 (detemporalization). For this, we aim to find large subqueries of combined stream and ontology atoms, and an efficient evaluation order (the join plan), which allows the partial evaluation and merging of the intermediate results to obtain the final result. In our opinion, the hypergraph-based strategy has the advantage of allowing fine-grained caching, full control over the evaluation, and possibly different DBs.

Hypergraphs and join trees. Many works have been dedicated to connecting hypergraphs, (acyclic) DB schemes, and join trees (see [18] for an overview). For decomposing a query q, the query hypergraph $H(q) = (V, E)$ is popular, where the vertices V represent the variables in q and the hyperedges in E capture the atoms in q with shared variables. In case of an *acyclic conjunctive query* (ACQ), which is defined in terms acyclicity of $H(q)$, a *join tree* can be generated from $H(q)$ that yields a plan for computing the query q. We focus here on α-acyclicity, which can be efficiently tested by the *GYO-reduction* (cf. [18]). A specific join tree J_H can be found via the *maximum-weight spanning tree* T_S of the *intersection graph* I_H of H, where edge weights of T_S are edge counts of V in I_H.

Details on the query evaluation. The combined evaluation extends our spatial evaluation strategy with hypertree decomposition of a hypergraph, by keeping intermediate results of each step in memory. The main steps of our query evaluation algorithm are:

(1) construct the α-acyclic hypergraph H_q from q and label each hyperedge in H_q with l_O, l_S, and l_F if it represents an ontology, spatial, or stream atom, resp. the combination of them; l_F gets the window size assigned, e.g., $l_{F,2}$ for $speed_{\boxplus_T^2 avg}$.

(2) build the join tree J_q of H_q and extract the subtrees J_{ϕ_i} in H_q, such that each node is covered by the same label $l_{F,n}$. The intention is to extract subtree CQs that share the same window size L (where static queries have $L = 0$), so they can be evaluated together and cached for future query evaluations.

(3) apply detemporalization as in Algorithm 1, where for each subtree J_{ϕ_i} the stream CQ q_{ϕ_i} is extracted and computed. The results are stored in a (virtual) relation R_{ϕ_i}, and each J_{ϕ_i} is replaced with a query atom pointing to R_ϕ.

(4) traverse J_q bottom up, left-to-right, to evaluate the CQ q_{ϕ_i} for each subtree J_{ϕ_i} (now without stream atoms) and keep the results in memory for future steps. Ontology and spatial atoms are evaluated as described before.

Example 7. The subqueries and join order of query $q3(x, v)$ in Table 1 is as follows:

(1) $q_{3,F1}(x, y) : Vehicle(x) \wedge position_{\boxplus_T^{10} line}(x, y);$

(2) $q_{3,N1}(v,u) : LaneIn(v) \wedge hasLocation(v,u)$; and

(3) $q_3(x,v) : q_{3,F1}(x,y) \wedge intersects(y,u) \wedge q_{3,N1}(v,u)$.

Caching for future queries is achieved by storing the intermedia results in memory with an expiration time according to L and the pulse. Static results never expire.

7 Implementation and Experimental Evaluation

We have implemented a prototype for our spatial-stream QA approach in JAVA 1.8 using the open-source PIPELINEDB 9.6.1 (https://www.pipelinedb.com/) as the spatial-stream RDBMS. The hypertree decomposition for each query is computed once using the implementation at https://www.dbai.tuwien.ac.at/proj/hypertree/. Based on it, each subquery is evaluated separately and (spatial) joined in-memory. For the FO-rewriting of DL-Lite$_A$, we used the implementation of

Table 1. Benchmark queries (windows size in seconds)

$q_1(x,y,z) :$	$Car(x), speed(x,y)[avg, 10],$ $vehicleMaker(x,z), y > 30$	cars w/ brands, travelling above 30 km/h
$q_2(x,y) :$	$LaneIn(x), hasSignal(x,y),$ $SignalGroup(y),$ $signalState(y,z)[last, 15], z = \text{"R"}$	lanes and signal groups switched to red
$q_3(x,v) :$	$Vehicle(x), pos(x,y)[line, 10],$ $inside(y,u), hasGeo(v,u), LaneIn(v)$	vehicles on incoming lanes
$q_4(x,y) :$	$Vehicle(x), pos(x,w)[line, 30],$ $intersects(w,z), pos(y,z)[line, 30], Car(y)$	vehicles with crossed paths
$q_5(x,y) :$	$Vehicle(x), speed(x,z)[avg, 15],$ $pos(x,y)[line\_angle, 15], z > 30,$ $y > -10, y < 10$	vehicles above 30 km/h heading straight
$q_6(x,y) :$	Taken from Example 1	Detection of red-light violation
$q_7(x,z) :$	$LaneIn(x), isPartof(x,u), Intersection$ $(u), u = \text{"I1"}, hasSignal(x,y),$ $SignalGroup(y), signalState(y,r)[last,$ $15], r = \text{"R"}, connect(x,q), connect(q,v),$ $Lane(v), hasSignal(v,z), SignalGroup(z),$ $signalState(z,s)[last, 15], s = \text{"R"}$	synthetic, testing many ontology atoms
$q_8(x,y) :$	$Vehicle(x), pos(x,y)[line, 20], intersects$ $(y,u), LaneIn(r), hasGeo(r,u),$ $intersects(y,v), LaneIn(s), hasGeo(s,v),$ $intersects(y,w), LaneIn(t),$ $hasGeo(t,w), within(y,z), hasGeo(q,z),$ $Intersection(q)$	synthetic, testing many spatial atoms
$q_9(x,q,r,s,t,u) :$	$Vehicle(x), speed(x,q)[avg, 1], speed(x,r)$ $[avg, 5], speed(x,s)[avg, 10],$ $speed(x,t)[avg, 25], speed(x,u)[avg, 50]$	synthetic, testing many stream atoms

PerfectRef in OWLGRES 0.1 [24] for now; more recent (and more efficient) implementations for query rewriting (e.g., [23]) are available.

Our experiment is based on two scenarios of monitoring vehicles and traffic lights (a) on a single intersection and (b) on a network of locally connected intersections, both managed by a single roadside C-ITS station. The ontology, queries (see Table 1), the experimental setup with logs, and the implementation are available on http://www.kr.tuwien.ac.at/research/projects/loctrafflog/eswc2017. We use a custom DL-Lite$_A$ LDM ontology with 119 concepts (with 113 inclusion assertions); 34 roles and 28 data roles (with 31 inclusion assertions). The LDM ontology models the C-ITS domain in a layered approach, separating concepts like ITS features (e.g., intersection topology), geo-features (e.g., POIs), geometries (e.g., polygon), actors (e.g., vehicles), events (e.g., accidents); and roles like partonomies (e.g., isPartOf), spatial relations, and generic roles (e.g., speed).

For (a), we have a T-shaped intersection as shown in Fig. 2 that represents a real-world deployment of a C-ITS station in Vienna. It connects two roads with 13 lanes and 3 signal groups that are linked to the lanes. We developed a synthetic data generator that simulates the movement of 10, 100, 500, 1000, 2500, and 5000 vehicles on a single intersection updating the streams averagely 50 ms. This allows us to generate streams with up to 10000 data points per sec. and stream.[3] We chose random starting points and simulated linear movements on a constant pace, creating a stream of vehicle positions. We also simulated simple signal phases for each traffic light that toggle between red and green every 3 s. The aim of this scenario is to show for simple driving patterns the scalability of our approach in the number of vehicles. For (b), we use a realistic traffic simulation of 9 intersections in a grid, developed with the microscopic traffic simulation PTV VISSIM (http://vision-traffic.ptvgroup.com/en-us/products/ptv-vissim/), allowing us to simulate realistic driving behavior and signal phases. The intersection structure, driving patterns and signal phases are more complex, but the number of vehicles is lower (max. 300) than in (a), as we quickly have traffic jams. We developed an adapter to extract the actual state of each simulation step, allowing us to replay the simulation from the logs. To vary data throughput, we ran the replay with 0 ms, 100 ms (real-time), 250 ms and 500 ms delay.

Results. We conducted our experiments on a Mac OS X 10.6.8 system with an Intel Core i7 2.66 GHz, 8 GB of RAM, and a 500GB HDD. The average of 11 runs for query rewriting time and evaluation time was calculated, where the largest outlier was dropped. The results are shown in Table 2 presenting query type (O for ontology, F for stream, and S for spatial atoms), number of subqueries $\#Q$, size of rewritten atoms $\#A$, and t as the average evaluation time (AET) in seconds for n vehicles or the delay in ms.

The baseline spatial-stream query is q_3 for 500 vehicles, where we have a time-to-load (TOL) of 0.22 s, an evaluation time for the stream (resp. ontology) atom of 0.54 s (resp. 0.03 s), and a spatial join time of 0.05 s. Clearly, 50% of the

---

[3] The intervals vary due to the number of vehicles, so we scale the DB updates up to 12 generators.

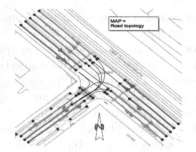
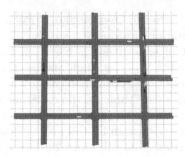

Fig. 2. Schematic representation of scenario (a) and (b) (Color figure online)

Table 2. Results (t in secs) for scenario (a) and (b), marked with * are signal streams

Type		#Q	#A	(a) t with #vehicles						(b) t with ms sim. delay			
				10	100	500	1000	2500	5000	0	100	250	500
q_1	O, F	1	1	0.85	0.82	0.91	1.05	1.22	1.58	0.78	0.74	0.73	0.71
q_2	O, F^*	1	6	0.83	0.83	0.83	0.83	0.83	0.83	0.77	0.77	0.72	0.71
q_3	O, S, F	3	23	0.89	0.87	1.00	1.25	1.39	1.74	0.83	0.81	0.77	0.75
q_4	O, S, F	3	22	1.10	1.09	1.24	1.53	1.81	2.32	1.02	1.00	0.95	0.93
q_5	O, S, F	3	42	1.11	1.10	1.26	1.39	1.90	1.92	1.05	1.00	0.98	0.96
q_6	O, S, F	7	52	1.39	1.39	1.49	1.69	2.36	2.28	1.40	1.28	1.26	1.25
q_7	O, F^*	6	69	1.16	1.16	1.16	1.16	1.16	1.16	1.15	1.12	1.11	1.09
q_8	O, S	9	73	0.92	0.94	1.30	1.43	1.72	2.19	0.99	0.98	0.92	0.91
q_9	O, F	9	105	1.67	1.73	1.99	2.06	2.49	2.97	1.71	1.68	1.66	1.63

AET is use for the stream atom (including rewriting steps). The TOL could be reduced by pre-compiling the program; this shortens evaluation by roughly 0.2 s. Initial evaluation of the queries q_4, q_5, q_6 and q_9 show that with each new stream subquery the number of results dropped down to zero, which seems an implementation issue of PIPELINEDB with *Continuous Views* on the same stream with different window sizes. We found a workaround by adding a delay of 0.2 s which again increases the number of results. This delay increases the AET, e.g. by 0.76 s in q_9, and might be ignored with future versions of PIPELINEDB and other stream RDBMS. The synthetic queries with mostly ontology (q_6), spatial (q_7), and stream atoms (q_9) clearly show that the challenging part of query evaluation are the stream aggregates. The good performance of PIPELINEDB allows us to work on condensed results (reducing the join sizes); however, stream aggregates could be further accelerated by calculated continuously inline aggregates on the DB, which are skimmed by our queries. Notably, PIPELINEDB keeps not always the order of inserted data points; this does not affect our bag semantics.

In general, our approach is designed to retain *complete* results; however, completeness might be lost as (1) the underlying spatial-stream RDBMS loses results as described above; (2) evaluation of a subquery is slow and subsequent queries start too late. One can solve (1) at the level of the spatial-stream RDBMS, and (2) can be overcome by continuous inline aggregates and query parallelization. In conclusion, the experiments show that the AET of our experimental prototype is for up to 500 vehicles below 1.5 s (except q_9). This suggests that with optimizations, e.g. quick detection of red-light violations on complex intersections is feasible.

8 Related Work and Conclusion

Data stream management systems (DSMSs) such as STREAM [3], were built supporting streaming applications by extending RDMBS [14]. More recently, RDF stream processing engines, such as C-SPARQL [5], SPARQLstream [8], and CQELS [17], were proposed for processing RDF streams integrated with other Linked Data sources. Besides C-SPARQL, most of them follow the DSMSs paradigm and do not support stream reasoning. EP-SPARQL [2] resp. LARS [6] proposes a language that extends SPARQL resp. CQ with stream reasoning, but translates KBs into expressive (less efficient) logic programs. Regarding spatio-temporal RDF stream processing, a few SPARQL extensions were proposed, such as SPARQL-ST [21] and st-SPARQL [16]. Closest to our work are (i) [22], which supports spatial operators as well as aggregate functions over temporal features (ii) [8], which allows evaluating OQA queries over stream RDBMS, and (iii) [20], which extends SPARQL with aggregate functions (using advanced statistics) evaluated over streamed and ordered ABoxes. This work differs regarding (a) the evaluation approach using EAQ with aggregates on the query and not ontology level, (b) hypergraph-based query decomposition, and (c) the main focus of querying streams of spatial data in an OQA setting.

Our approach is situated in-between "classical" stream processing approaches that handle the streaming data as bags in windows, and temporal QA over DL-Lite using temporal operators like LTL in [4], which are evaluated over a (two-sorted) model separating the object and temporal domain. We believe that detemporalization with its bag semantics suffices for the C-ITS case, since the order of V2X messages is not guaranteed, and for most of the normal as well as spatial aggregates it can be ignored (e.g., sum) or is implicit in the data (e.g., Euclidian distance of points). Besides [4], similar temporal QA is investigated in [7,15], which are all on the theoretical side and provide no implementation yet. Finally, we build on the results for EAQs in [10], but we introduce spatial streams and more complex queries.

This work is sparked by the LDM for V2X communications, which serves as an integration effort for streaming data (e.g., vehicle movements) in a spatial context (e.g., intersections) over a complex domain (e.g., a mobility ontology). We introduced a suitable approach using ontology-mediated QA for realizing the LDM. For spatial-streaming queries, bridging the gap between stream processing and ontology-mediated QA is not straightforward; we extended previous work in [12] and used epistemic aggregate queries to detemporalize the stream sources. The latter preserves FO-rewritability, which allows us to evaluate conjunctive queries with spatial atoms over existing stream RDBMSs. We also provided a technique to construct query execution plans using hypergraph decomposition, and we have implemented a proof-of-concept prototype to assess the feasibility of our approach on two experiments with mobility data. The results are encouraging, as the evaluation time appeared to be moderate already without optimization.

Future work. Our ongoing and future research is directed to advance the theoretical and practical aspects of our approach. On the theoretical side, a detailed correctness proof for the algorithm that accounts for all different aggregate functions is needed. So far, consistency for QA is neglected and could be enforced in different ways by repairs. The query language could be lifted to SPARQL, but epistemic aggregates, query decomposition, and spatial relations would need reevaluation. On the practical side, our implementation should be extended to pull-based QA with extensive caching and inline aggregates on the DB, along with other optimizations, such as using the pulse for pre-caching resp. window size optimizations, and different query rewriting techniques. Also more complex spatial aggregates, i.e., trajectories, should be considered. Furthermore, cyclic queries need to be handled. The implementation could be tested in more complex scenarios like event detection (e.g., bus delays) with public transport data.

Acknowledgements. Supported by the Austrian Research Promotion Agency project Industrienahe Dissertationen and the Austrian Science Fund projects P26471 and P27730.

References

1. Andreone, L., Brignolo, R., Damiani, S., Sommariva, F., Vivo, G., Marco, S.: Safespot final report. Technical report D8.1.1 (2010)
2. Anicic, D., Fodor, P., Rudolph, S., Stojanovic, N.: EP-SPARQL: a unified language for event processing and stream reasoning. In: Proceedings of WWW 2011, pp. 635–644 (2011)
3. Arasu, A., Babu, S., Widom, J.: The CQL continuous query language: semantic foundations and query execution. VLDB J. **15**(2), 121–142 (2006)
4. Artale, A., Kontchakov, R., Kovtunova, A., Ryzhikov, V., Wolter, F., Zakharyaschev, M.: First-order rewritability of temporal ontology-mediated queries. In: IJCAI 2015, pp. 2706–2712 (2015)
5. Barbieri, D.F., Braga, D., Ceri, S., Valle, E.D., Grossniklaus, M.: C-sparql: a continuous query language for rdf data streams. Int. J. Semant. Comput. **4**(1), 3–25 (2010)
6. Beck, H., Dao-Tran, M., Eiter, T., Fink, M.: LARS: A logic-based framework for analyzing reasoning over streams. In: Proceedings of AAAI 2015, pp. 1431–1438 (2015)
7. Borgwardt, S., Lippmann, M., Thost, V.: Temporalizing rewritable query languages over knowledge bases. J. Web Sem. **33**, 50–70 (2015)
8. Calbimonte, J.-P., Mora, J., Corcho, O.: Query rewriting in RDF stream processing. In: Sack, H., Blomqvist, E., d'Aquin, M., Ghidini, C., Ponzetto, S.P., Lange, C. (eds.) ESWC 2016. LNCS, vol. 9678, pp. 486–502. Springer, Cham (2016). doi:10.1007/978-3-319-34129-3_30

9. Calvanese, D., Giacomo, G.D., Lembo, D., Lenzerini, M., Rosati, R.: Tractable reasoning and efficient query answering in description logics: the dl-lite family. J. Autom. Reasoning **39**(3), 385–429 (2007)
10. Calvanese, D., Kharlamov, E., Nutt, W., Thorne, C.: Aggregate queries over ontologies. In: Proceedings of ONISW 2008, pp. 97–104 (2008)
11. Eiter, T., Füreder, H., Kasslatter, F., Parreira, J.X., Schneider, P.: Towards a semantically enriched local dynamic map. In: Proc. 23rd World Congress on Intelligent Transport Systems (ITSWC-2016), Melbourne, October 10–14, 2016 (2016)
12. Eiter, T., Krennwallner, T., Schneider, P.: Lightweight spatial conjunctive query answering using keywords. In: Cimiano, P., Corcho, O., Presutti, V., Hollink, L., Rudolph, S. (eds.) ESWC 2013. LNCS, vol. 7882, pp. 243–258. Springer, Heidelberg (2013). doi:10.1007/978-3-642-38288-8_17
13. Eiter, T., Parreira, J.X., Schneider, P.: Towards spatial ontology-mediated query answering over mobility streams. In: Proceedings of Stream Reasoning Workshop 2016, pp. 13–24 (2016)
14. Golab, L., Özsu, M.T.: Issues in data stream management. SIGMOD Rec. **32**(2), 5–14 (2003)
15. Klarman, S., Meyer, T.: Querying temporal databases via OWL 2 QL. In: Kontchakov, R., Mugnier, M.-L. (eds.) RR 2014. LNCS, vol. 8741, pp. 92–107. Springer, Cham (2014). doi:10.1007/978-3-319-11113-1_7
16. Koubarakis, M., Kyzirakos, K.: Modeling and querying metadata in the semantic sensor web: the model stRDF and the query language stSPARQL. In: Aroyo, L., Antoniou, G., Hyvönen, E., Teije, A., Stuckenschmidt, H., Cabral, L., Tudorache, T. (eds.) ESWC 2010. LNCS, vol. 6088, pp. 425–439. Springer, Heidelberg (2010). doi:10.1007/978-3-642-13486-9_29
17. Le-Phuoc, D., Dao-Tran, M., Xavier Parreira, J., Hauswirth, M.: A native and adaptive approach for unified processing of linked streams and linked data. In: Aroyo, L., Welty, C., Alani, H., Taylor, J., Bernstein, A., Kagal, L., Noy, N., Blomqvist, E. (eds.) ISWC 2011. LNCS, vol. 7031, pp. 370–388. Springer, Heidelberg (2011). doi:10.1007/978-3-642-25073-6_24
18. Maier, D.: The Theory of Relational Databases. Computer Science Press, Rockville (1983)
19. Netten, B., Kester, L., Wedemeijer, H., Passchier, I., Driessen, B.: Dynamap: A dynamic map for road side its stations. In: Proceedings of ITS World Congress 2013 (2013)
20. Özçep, Ö.L., Möller, R., Neuenstadt, C.: Stream-query compilation with ontologies. In: Pfahringer, B., Renz, J. (eds.) AI 2015. LNCS (LNAI), vol. 9457, pp. 457–463. Springer, Cham (2015). doi:10.1007/978-3-319-26350-2_40
21. Perry, M., Jain, P., Sheth, A.P.: SPARQL-ST: extending SPARQL to support spatiotemporal queries. Geospatial Semant. Semant. Web **12**, 61–86 (2011)
22. Quoc, H.N.M., Le Phuoc, D.: An elastic and scalable spatiotemporal query processing for linked sensor data. In: Proceedings of Semantics 2015, pp. 17–24. ACM (2015)
23. Rodríguez-Muro, M., Kontchakov, R., Zakharyaschev, M.: Ontology-based data access: *Ontop* of databases. In: Alani, H., et al. (eds.) ISWC 2013. LNCS, vol. 8218, pp. 558–573. Springer, Heidelberg (2013). doi:10.1007/978-3-642-41335-3_35
24. Stocker, M., Smith, M.: Owlgres: a scalable owl reasoner. In: Proceedings of OWLED 2008 (2008)

Optimizing the Performance of Concurrent RDF Stream Processing Queries

Chan Le Van$^{(\boxtimes)}$, Feng Gao, and Muhammad Intizar Ali

INSIGHT Centre for Data Analytics, National University of Ireland, Galway, Ireland
{chan.levan,feng.gao,ali.intizar}@insight-centre.org

Abstract. With the growing popularity of Internet of Things (IoT) and sensing technologies, a large number of data streams are being generated at a very rapid pace. To explore the potentials of the integration of IoT and semantic technologies, a few RDF Stream Processing (RSP) query engines are made available which are capable of processing, analyzing and reasoning over semantic data streams in real-time. This way, RSP mitigates data interoperability issues and promotes knowledge discovery and smart decision making for time-sensitive applications. However, a major hurdle in the wide adoption of RSP systems is their query performance. Particularly, the ability of RSP engines to handle a large number of concurrent queries is very limited which refrains large scale stream processing applications (e.g. smart city applications) to adopt RSP. In this paper, we propose a shared-join based approach to improve the performance of an RSP engine for concurrent queries. We also leverage query federation mechanisms to allow distributed query processing over multiple RSP engine instances in order to gain performance for concurrent and distributed queries. We apply load balancing strategies to distribute queries and further optimize the concurrent query performance. We provide a proof of concept implementation by extending CQELS RSP engine and evaluate our approach using existing benchmark datasets for RSP. We also compare the performance of our proposed approach with the state of the art implementation of CQELS RSP engine.

Keywords: Linked Data · RDF Stream Processing · Query optimization

1 Introduction

A merger of semantic technologies and stream processing has resulted into RDF Stream Processing (RSP). RSP engines refer to the stream processing engines which have the capability to process semantically annotated RDF data streams. Over the past few years, different RSP engines have been proposed, such as CQELS [16], C-SPARQL [4], and SPARQLStream [6] etc. RSP engines continuously consume streaming RDF data as an input and generate query results as an output. An RSP query is registered once and executed constantly. The continuous query execution is a resource intensive task and most of the existing RSP engines suffer from scalability and performance issues while processing

ⓒ Springer International Publishing AG 2017
E. Blomqvist et al. (Eds.): ESWC 2017, Part I, LNCS 10249, pp. 238–253, 2017.
DOI: 10.1007/978-3-319-58068-5_15

multiple concurrent RSP queries. The inability of the RSP engines to handle a large number of concurrent queries is a major performance bottleneck, hence, a main obstacle to the wider adoption of RSP engines. This RSP performance bottleneck has different causes. For example, in CQELS, each input query is separately compiled and parsed into an individual execution plan with a designated data buffer assigned to that particular query. Therefore, the performance of the engine would definitely degrade whenever multiple concurrent queries are registered to the engine. Many real-time applications consuming continuous data streams often share input data streams for a large number of concurrent queries. Therefore, we foresee a great opportunity to address the performance and scalability issues of RSP engines by proposing a resource sharing strategy where multiple concurrent queries could share the resources required to process common input data streams. These resource sharing strategies will certainly ease the burden over RSP engines and consequently improve the performance.

In this paper, we propose resource sharing and load balancing techniques to optimize the performance of concurrent RSP queries. We take the concept of the shared join operator (also known as multiple join operator) introduced in [16], to support sharing memory and computation resources. Taking the advantage of queries having the shared input streams, the multiple join operator uses a join component to produce shared results for all queries, and a routing component/router is used to route the output items to the corresponding query. We extend the approach of shared join operator and introduce the creation of join sequences using a *reutilization* metric. This way, a network of shared join operators are formed for multiple queries. The network contains different shared join operators consuming the same input streams, and in order to produce the query results a query evaluation process involves traversing a path in the network, we called this process a join sequence. In our approach, some intermediate results can also propagate to other queries, if they contain triple patterns referring to the same stream data buffers. *Reutilization* metric is used to choose optimal join sequences. We extended the existing implementation of CQELS engine, and named it as CQELS+. We evaluated CQELS+ in comparison to the existing state-of-the-art CQELS implementation, which showcases that CQELS+ is capable of handling a larger number of concurrent queries. We further evaluated CQELS+ performance for different load balancing techniques for multiple instances of RSP engines deployed in a distributed environment. We evaluated the performance of multiple instances of CQELS+ and compare it with the performance of a single engine instance. The evaluation results have revealed that we can have lower query latency by deploying parallel engine instances.

The remainder of the paper is organized as follows; Sect. 2 lays the theoretical foundations for our approach. Section 3 elaborates the details for the shared join operations and load balancing strategies for parallel RSP engines. Section 4 presents the experimental design to evaluate our approach. This section also provides experimental results analysis. Section 5 reviews the literature before we conclude and discuss future plans in Sect. 6.

2   Foundations

In this section, we introduce the basic concepts of RSP, including multi-way join, shared join operator and network of multiple join operators, which forms the basis of our proposed approach [16].

Following the principles of Linked Data [5], the concept of Linked Stream Data [25] was introduced in order to bridge the gap between data streams and Linked Data. Linked stream data helped to simplify the data integration process among heterogeneous streaming sources as well as between streaming and static sources. Below we present the basic concepts of RDF Stream Processing model, which are also followed in CQELS data model.

- **RDF stream** is a bag of elements $\langle (s, p, o) : [t] \rangle$, where (s, p, o) is an *RDF triple* [22] and t is a timestamp.
- **Window operators** are inspired by the time-based and triple-based window operators of CQL data stream management system [3].
- **Stream Graph Pattern** is supplemented into the *GraphPatternNotTriples* pattern[1] to represent window operators on RDF Stream.

2.1   Multi-way Join

The concept to multi-way join was introduced in [19], same concept was re-used for CQELS [16], where authors propose a single multi-way join that works over more than two stream data buffers (also known as window buffers). Window buffers generate and propagate results in a single step rather than creating the join results by passing through a multiple-stage binary joins in the query execution pipeline. Suppose we have n window data buffers $W_1, .., W_n$ involved in a join operation. A multi-way join operates as follows:

1. When a new stream data item (tuple) comes to any window buffers, it is used to recursively probe other window buffers to generate the join output, i.e., when a data item M arrives at W_1, the join processing starts to look for next join candidate in $W_2, W_3, ..., W_n$.
2. Let us assume W_2 is the next joining candidate, the algorithm will check data inside W_2 to see if there are any existing tuples which are compatible with M.
3. If there is any compatible data item available, the intermediate result sets are created and the algorithm recursively continues until all of the involved window buffers are visited or stops if no compatible data items found.

2.2   Shared Join Operator and Network of Shared Join Operators

Chapter 7 in [16] introduces the *shared join operator* to share the computation among multi-way join operations from queries with the same set of window buffers of RDF streams. As discussed in [16], let us assume we have k multi-way join operators, each involves m window buffers i.e., we have k multi-way

---

[1] SPARQL 1.1 Grammar: https://www.w3.org/TR/sparql11-query/#grammar.

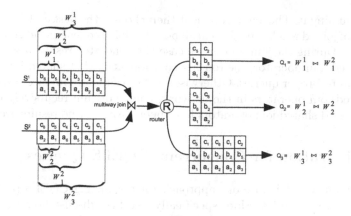

Fig. 1. Shared join operator example

join operators: $(W_1^1 \cdots \bowtie W_1^m)$, ..., $(W_k^1 \cdots \bowtie W_k^m)$. Additionally, let us assume with any $1 \leq i \leq m$, we always have W_1^i, W_2^i, ..., W_k^i referring to the same window buffer (regardless to the window alignment), then the equation below (containment property [12]) is always true:

$$W_j^1 \cdots \bowtie W_j^m \subseteq W_{max}^1 \cdots \bowtie W_{max}^m \tag{1}$$

In Eq. 1, W_{max}^i refers to the window buffer in the set $\{W_1^i, W_2^i, ..., W_k^i\}$) with the maximum size. We call W_{max}^i is a shared window buffer.

Intuitively, Eq. 1 tells us that the join results of any multi-way join operator for a fixed set of window buffers is included in the join results of the operator for the set of windows with the maximum length. For example, Fig. 1 shows a shared join operator for three queries Q_1, Q_2 and Q_3 containing 3 multi-way joins $W_1^1 \bowtie W_1^2$, $W_2^1 \bowtie W_2^2$ and $W_3^1 \bowtie W_3^2$, respectively. These queries receive input from two RDF streams: S^1 and S^2. As we can see in the picture, W_{max}^1 will be W_3^1 and W_{max}^2 will be W_3^2. By applying Eq. 1, we have:

$$\forall j \in \{1, 2, 3\}, W_j^1 \bowtie W_j^2 \subseteq W_3^1 \bowtie W_3^2 \tag{2}$$

Leveraging the containment property, the approach in [16] designed a shared join operator. This operator targets in queries sharing the same set of window buffers. It consists of two components: join component and routing component. The join component has the duty of generating the shared results that contain results of all involved queries. The routing component is responsible for filtering the shared results and routing the filtered results to proper queries.

It has been discussed in the literature that in practice the queries registered to the processing engine often share only subsets of input streams. In [16], authors introduced the concept of *network of shared join operators* (NSJO). NSJO can share the execution for sub-queries (part of the whole queries) from a group of queries. NSJO includes a set of shared join operators, where each shared join operator consumes the same window buffers. When a new tuple comes to a shared window buffer, it triggers the join components of related shared join operators in the

NSJO consuming it. The join component then chooses the best cost join sequence (the order of joined window buffers) in all possible join sequences and generates the join output. During the join process, in case the generated results are the output of any shared join operator, the routing component of that shared join operator routes them to proper queries. Otherwise, if the results are further consumed by other shared join operators in the network, the algorithm recursively continues until all related shared join operators have been visited or no results are created.

3   Optimization for Concurrent CQELS Queries

In this section, we elaborate our approach for optimizing CQELS performance under concurrent queries. More specifically, we first discuss how we facilitate NSJO in CQELS+ (the extended CQELS), including how to create the join sequences (the order of joined window buffers) in a join graph, how to calculate the reutilization metric, and provide an example of the join graph. Then, we elaborate how we deploy multiple CQELS+ engines in parallel and distribute the workload among different engine instances in order to obtain better performance.

Algorithm 1. *create_Join_Sequences(curr_Join_Vertex, queries)*

---

1 **while** *coverage < number_of_queries* **do**
2 | HAQ ← 0; considering_Patterns ← ∅ ;
3 | **for** $j \in [1..number\_of\_physical\_windows]$ **do**
4 | | cal ← cal_Reuse_Metric(cur_join_Vertex, physical_Windows[j]);
5 | | patterns = get_Considering_Patterns(cur_join_Vertex, physical_Windows[j]) ;
6 | | **if** *cal > HAQ* **then**
7 | | | HAQ = cal; next_Buffer = physical_Windows[j];
8 | | | considering_Patterns ← patterns;
9 | | **end**
10 | **end**
11 | **if** *HAQ > 0* **then**
12 | | next_Join_Vertex ← create_Join_Vertex(next_Buffer) ;
13 | | next_Join_Vertex.set_Considered_Patterns(considering_Patterns) ;
14 | | curr_Join_Vertex.add_Nexts(next_Join_Vertex) ;
15 | | prev_Join_Vertex ← curr_Join_Vertex ;
16 | | curr_Join_Vertex ← next_Join_Vertex ;
17 | | coverage += count_Covered_Queries(curr_Join_Vertex) ;
18 | | create_Join_Sequences(current_Join_Vertex, queries) ;
19 | | curr_Join_Vertex ← prev_Join_Vertex ;
20 | **end**
21 | **else**
22 | | return;
23 | **end**
24 **end**

3.1   CQELS+: Network of Shared Join Operators

According to [16], when a new tuple comes to the shared window buffer(a physical window in our approach) in the NSJO, join sequences in the join graph of this physical window are evaluated to generate the join results for involved queries. In our approach, this join graph is built at the query registration time based on a metric called *reutilization*. This metric is defined to create the optimal join sequences in the graph. More specifically, in each step of forming a join sequence, one physical window is chosen to be the next join candidate if the generated join results can be consumed by the most queries. The process of creating a join graph has two steps including *create_Join_Sequences* as described in Algorithm 1 and supplement join sequences for *special queries* containing multiple stream patterns referring to the same physical window.

Each Shared Window Buffer (SWB) from the involved queries triggers Algorithm 1 to create the join graph. The *curr_Join_Vertex* parameter for the first call is the vertex containing the SWB itself and SWB's graph patterns. The *coverage* variable in line 1, initialized by 0, holding the number of queries has been considered, is used to escape the algorithm. Line 2 initializes the *HAQ* and *considering_Patterns* variables. HAQ is intended to hold the *H*ighest *A*mount of *Q*ueries reusing the generated join results. The *considering_Patterns* keeps the considering stream patterns in the queries sharing join variables with the current join vertex. From line 3 to line 8, the algorithm iterates over all of the involved physical windows and calculates the *reutilization metric* detailed in the next section. The HAQ and *considering_Patterns* variables are updated when we found a higher calculated value. If the condition in line 6 evaluates to *true* (i.e. the next join candidate is found), we reset the pointers for the current vertex and previous vertexes based on the newly found candidate as from line 12 to 16. With the chosen join candidate, we count the number of covered queries and update the coverage variable (line 17). At line 18, we recursively call the algorithm with the input parameter as the found join candidate. Line 19 assigns the previous state of the considering join candidate to create other sequences (if any). Finally, the algorithm terminates in line 23 if we are not able to find any join candidate, or all the queries have been already considered, i.e. *coverage* equals or is greater than *number_of_queries*.

Calculating the Reutilization Metric: Given n queries $(Q_1,..,Q_n)$, m join sequences $(S_1, .., S_m)$ that can share the join results for a set of sub-queries in the current vertex C, t patterns $(P_1, .., P_t)$ referring to the physical window in C. We define $J_j^i(k,l,v)$ as a counter for calculating the reutilization for variable v, on sequence S_i and pattern P_j, where $v \in V = \{$set of variables in all queries$\}$, $1 \leq i \leq m, 1 \leq j \leq t$, $l =$ index of join variable in P_j, e.g., in the first triple pattern in PW1 (Fig. 2), the index of variable (?person_q1) and (?loc_q1) are 1 and 3, respectively[2]. $k =$ index of the joining variable in S_i, $J_j^i(k,l,v)$ is given in the equation below:

---

[2] The index of join variable in join sequence is similarly defined.

$$J_j^i(k,l,v) \begin{cases} = 1 \text{ , if } P_j \notin S_i \wedge S_i \in Q_y \wedge p_j \in Q_y \\ \qquad \wedge var(S_i, k) = v \wedge var(P_j, l) = v \\ = 0 \text{ , if otherwise.} \end{cases} \tag{3}$$

The functions $var(S_i, k)$ and $var(P_j, l)$ allow to retrieve the variable of a sequence S_i at index k or the variable of a pattern P_j at index l. We calculate the metric based on the Eq. 4:

$$R = Max(\sum_{v \in V} \sum_{k \in [1,|S_i|]} \sum_{l \in [1,|P_j|]} J_j^i(k,l,v)), \tag{4}$$

where $|S_i|$ and $|P_j|$ gives the size of S_i and P_j, respectively.

Supplement Join Sequences: Special queries mentioned in Sect. 2.2 requires more than one join sequences to generate the join results because any arrived tuple in the shared window buffer is matched with multiple stream patterns inside one query. As Algorithm 1 only considers one join sequence for each query, we need to supplement the rest of join sequences. To do this, we count the number of special queries and rerun Algorithm 1 with the condition that created join sequences are not taken into account.

Join Graph Example: Now we show an example for building the join graph for 4 queries: Q1, Q2, Q3 and Q4[3] with 5 data buffers shown in Fig. 2. In this example, we need to build the join graph for the physical window PW1 because this one stores the streaming data. Figure 3 demonstrates how the graph is constructed. Starting from physical window PW1, the algorithm calculates the highest reutilization metric and chooses the corresponding physical window. As a result, PW2 is chosen with a reutilization value of 3, the highest value among all candidates. There are three shared join variables in three join sub-sequences. In this case, the join results are reused by three queries, which are Q1, Q3 and Q4. Continuing on this path, PW3 is visited next because it has highest reutilization of 2, and the join results on this path are re-used by Q1 and Q3. At this point, all patterns in Q1 are visited and Q1 is covered. After this buffer, PW1, PW4 and PW4 (again) are respectively chosen and Q3 is also covered. Similar processes are repeated to cover all involved queries, then the algorithm stops. Q2, Q3 and Q4 are covered more than once as shown in the dashed branches in Fig. 3. These three special queries require multiple join sequences to generate enough join results. Therefore, we have to run the supplementation step to create join sequences starting from these patterns to make sure, that we do not miss the generated join results. In Fig. 3, the red color indicates the queries covered after running Algorithm 1, while the blue color demonstrates the queries covered after the *supplement join sequence* step.

---

[3] Q4 in this paper is actually Q5 in [20], we do not use Q4 in [20] because its join sequence is too long for the demonstration.

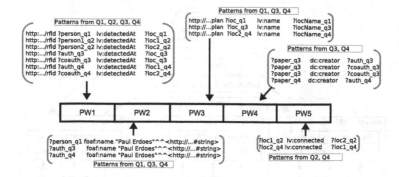

Fig. 2. Shared window buffers for patterns extracted from 4 queries

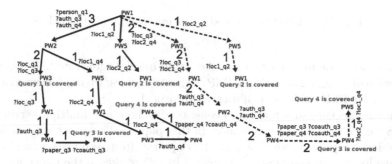

Fig. 3. Join graph example (Color figure online)

3.2   Load Balancing for Parallel CQELS+ Instances

In Sect. 2.2 we discussed the means for optimizing RSP performance leveraging shared join operations. However, the described approach is still a centralized one, i.e., the join graphs and all queries are managed by a single machine. While this is feasible for small-scale applications, it cannot satisfy the need for large-scale systems, where multiple servers are deployed, possibly at different geographical locations, to handle the excessively high concurrency. Evidence for the concurrency limitations for RSP can be found in CityBench [1]. Notice that in CityBench only duplicated queries are tested over a single engine instance. To cope with large-scale applications, a distributed way of processing RDF streams is necessary. In [9,11], a service-oriented approach was proposed to realize RSP federation via service composition. This way, multiple engine instances, even with different engine types, can be deployed in parallel to collaboratively answer an RSP query. In this paper, we follow this principle and develop means for an efficient distributed RSP.

In [10], the problem of multi-modal QoS optimization for the event service composition plan is studied, i.e., it provides means for creating optimal RSP query federation. However, it does not address the problem of the overall system performance. In particular, it does not provide heuristics on which specific engine

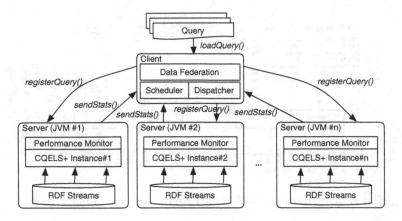

Fig. 4. Architecture for load balancing over parallel CQELS+

instance should be used when a new federated query is generated. To determine the workload assigned to multiple RSP engine instances, we implemented a query scheduler for RSP engines, as illustrated in Fig. 4.

The architecture shown in Fig. 4 consists of a client and multiple servers hosting RSP engines (CQELS+ instances in this paper). The client is a centralized controlling component, which accepts RSP queries from applications and utilizes the data federation component (described in [9]) to create the query federation plan. Then, the scheduler determines which engine instance should be picked for evaluating the plan (which is also a CQELS query) at runtime. The scheduler communicates with the performance monitors implemented on the servers and continuously receives information on the status of the servers. Currently, we implemented three different heuristics (evaluated in Sect. 4.2) for the load balancing strategy:

- **Rotation:** each instance takes its turn to deploy a new query in a rotation, i.e., we distribute queries evenly among the engine instances;
- **Minimal average latency:** the engine instance with the lowest average query latency is picked to deploy the new query. For the latency monitoring mechanism, we follow the approach provided by CityBench [1], and
- **Minimal data buffer size:** this strategy chooses the current instance which has the minimum total number of elements in data buffers, to register the next query, as the processing time of the join operator depends significantly on the join selectivity of the data buffers. Heuristically, smaller data buffers typically have smaller join selectivity.

4 Evaluation

In this section, we conduct three experiments. First we compare our shared join approach in the CQELS+ engine with the original CQELS. Then, we show the performance of the load balancing over CQELS+ engines. Finally, we evaluate the query registration time for multiple queries. All experiments are deployed on a machine running Ubuntu 12.04 with Intel Quad-Core i7-3520M CPU (2.90 GHz) and 16 GB RAM. The tests are compiled for 64-bit Java Virtual Machine (JVM build 1.7.0_80b15). All experimentation results are reproducible[4]. In the following, we present the detailed design of the experiments and analyze the results.

4.1   Evaluating Shared Joins in CQELS+

In this experiment, we reproduce the experiments in LS-Bench[5] and compare the performance between the publicly available version of CQELS[6] with our implementation of CQELS+. The performance is measured in throughput, i.e., the number of triples processed per unit time. We choose 4 queries: Q2, Q3, Q5 and Q6 from LS-Bench for this evaluation, since they involve the streaming and static data. For each query, we increase the number of duplicated queries registered to the engines.

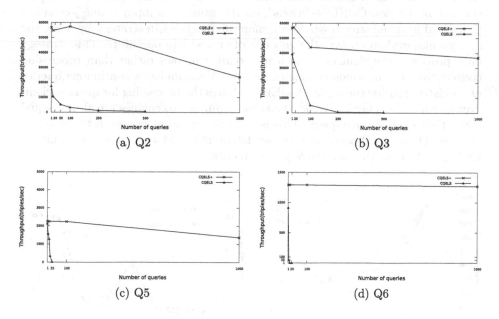

(a) Q2   (b) Q3

(c) Q5   (d) Q6

Fig. 5. Throughput comparison between CQELS+ and CQELS

Figure 5 shows the results of this experiment. CQELS+ outperforms CQELS in throughput as well as in handling concurrent queries. The results in Fig. 5a and b

---

[4] CQELS+: https://github.com/chanlevan/CqelsplusLoadBalancing.
[5] LS-Bench: https://code.google.com/archive/p/lsbench/.
[6] CQELS engine: https://code.google.com/archive/p/lsbench/wikis/howto_cqels.wiki.

point out that for Q2 and Q3, CQELS is not able to process more than 500 queries and the throughput is about 20000 and 40000 triples per second, respectively. As indicated in Fig. 5c and d, CQELS responses if the number of queries is not higher than 25 and the maximum throughput is about 2200 triples per second for Q5 and 800 for Q6. The results show that the throughput of CQELS decreases drastically when we increase the number of concurrent queries. Conversely, CQELS+ can handle one thousand queries with higher throughput, about 60000 triples per second for Q2 and Q3, nearly 2300 for Q5 and 1300 for Q5. The performance of CQELS+ reduces slowly when we increase the number of queries.

These results are because of two main factors: the join operator and the scheduling mechanism of the JVM. In the CQELS version we evaluated, the join operators are repeatedly triggered by the arrivals of data elements and the consecutive stream windows and data arriving at those windows are processed independently. Furthermore, if the input data coming to stream windows has the high join selectivity and a large number of involving stream windows, a lot of intermediate results are generated. The resources used by those intermediate results have to be released frequently, which causes considerable overhead for the JVM. All of these factors have limited the capability of CQELS for handling concurrent queries. CQELS+ instead, on the same data input (same join selectivity and involved stream windows), manipulates the join strategy by using the incremental evaluation over the network of shared join operators. This strategy only processes the changes of data in stream windows rather than processing consecutive stream windows independently. When multiple continuous queries are registered inside the engine, CQELS+ shares the processing for queries whenever possible. Notably in Fig. 5c, the throughput of two engines are almost equal when they evaluate one query. This is because the window size in the query is too long (1 day) and contains very big data buffers, which makes it difficult for CQELS+ to look for compatible join elements.

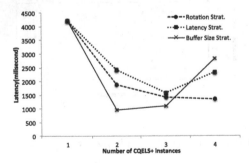

Fig. 6. Latency with increasing CQ-ELS+ instances

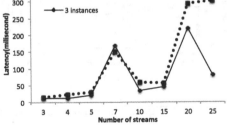

Fig. 7. Latency with increasing number of streams

4.2   Evaluating Load Balancing over CQELS+

In this experiment, we run two different tests. We show the results of the load balancing model presented in Sect. 3.2 in the first test. Then, we choose the best policy and test the scalability of the system with increasing number of streams. To demonstrate the both tests, we implement the client-server architecture as shown in Fig. 4. A new query is registered only after the previous query has been successfully registered to the engine. After all the queries are deployed, we keep the system running for 15 min and monitor the latency. For the input data, we use the CityBench [1] as they support end-to-end query latency monitoring. The latency is captured as follows: each query in this test has some joins over sensor data streams, and the queries are slightly adjusted (e.g., removing numerical filters) so that each sensor observation (in the form of a group of triples) can produce at least one result, then, the latency can be derived by comparing the timestamp of the first result produced by each sensor observation and the timestamp of observation entering the triple streams. Notice that all sensor streams produce observations simultaneously with a same frequency.

In the first test, we deploy different CQELS+ engines in different JVMs and apply the aforementioned load balancing strategies, i.e. rotation, minimum average latency and total buffers size. We choose 4 different queries: Q1, Q2, Q3 and Q4 (from CityBench). We register 400 queries (randomly picked from Q1 to Q4) to the system. Figure 6 illustrates the results. With a single instance, the latency for 400 queries is up to 4.5 s. The latency decreases about 4 times when we deploy 2 and 3 CQELS+ instances. With this configuration, the buffer size load-balancing strategy is the most efficient one, while the other two strategies also improve the latency for 2 and 3 instances. The reason behind this is perhaps that the number of data buffer size is more accurate in representing the workload, and the latency observed may fluctuate in time. When we use 4 instances, the latency tends to increase. This indicates that the overhead of multiple JVMs and CQELS+ instances starts to outweigh the benefit of load balancing.

The second test aims to check the latency of the multiple CQELS+ instances when we vary the number of streams. Previous experiments (including the LS-Bench results) tested only a limited number of streams. Now we increase the number of streams and monitor the latency over 2 and 3 CQELS+ instances with the data buffer size policy. These streams are picked in turns by 400 randomly generated different queries with Q1 as a template. Figure 7 shows the experiment results. Generally, the latency increases when more streams are used, and with more than 7 streams some unstable processing states can be observed (e.g., latency "spikes" for 7 and 20+ streams). We also observed that with more than 30 streams, the engine often stops responding. The latency increases when we scale the number of streams from 3 to 25 streams in the both configurations (2 and 3 instances). The more streams are involved, the more concurrent threads are created and invoked to stream the data into the system. This makes the system response slower when the number of streams increases. However, the abnormal increase in latency appears with the 7 streams and 25 streams in Fig. 7.

This abnormal behavior is perhaps caused by an incorrect engine implementation, which may also be the cause for the system failure when the number of streams is higher than 30.

4.3   Evaluating the Query Registration Time

In the first experiment, we showed that the CQELS+ outperforms CQELS at runtime. However, this is because we build the join probing graphs before query execution. This introduces an overhead for registering new queries: the graph must be updated constantly. Also, the more queries are deployed, the more time it takes to update the graph. Figure 8 shows the time taken for the query registration using 400 random queries from the template of Q1 in the second experiment, using different load balancing strategies. From the results, we can see that for the data buffer strategy, it takes more time than the other two, possibly because this strategy has a higher chance of resulting in the different number of queries deployed on the engine instances.

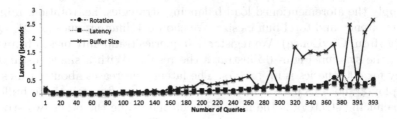

Fig. 8. Query registration time for 4 CQELS+ instances

5   Related Work

Sharing query results is not a novel for database community, e.g. optimization for multiple queries for static database systems by sharing the query operations have been discussed in [7,8,24]. However, for stream processing systems more complicated query semantics are required to be incorporated and additionally the dynamicity of data buffers and input streams have to be considered. Existing RSP engines like C-SPARQL [4] uses a black box approach for handling the semantic queries, thus not able to provide much optimization for multiple queries. Benchmark results like CityBench [1] also showed that with a native join operator implementation, CQELS is better than C-SPARQL at handling multiple queries. Although an adaptive join routing is discussed in [16], where an estimated cost for the join sequences is calculated by monitoring the index of data buffers at run-time. However, it is currently not implemented in CQELS, thus we are not able to compare the performance difference to our approach. Also, we argue that a static join graph has less overhead at run-time while we acknowledge that we are relocating the complexity to query registration time.

Distributed query processing has long been discussed in database community [15] and also in linked data community [2,23]. Regarding distributed RSP, In [13] a C-SPARQL version with parallel streaming is developed mainly to optimize RDFS reasoning by splitting and filtering sub-streams. In CQELS Cloud [17], where extensions for CQELS are made for utilizing the elasticity of cloud environments and allow processing nodes to join and leave the network on-demand. However, the load shedding in CQELS Cloud relies on existing DSMS systems (e.g., Storm[7]), whereas in our approach different strategies are designed and tested. Load balancing techniques have been studied extensively in the literature [14,18,21]. Various metrics, from basic execution latency and bandwidth usage [14] to sophisticated service correlations [21] and network path analysis [18] have been proposed to evaluate the load and determine the shedding strategy.

6 Conclusions and Future Work

In this paper, we realized shared join operations for CQELS in order to improve its performance when handling multiple concurrent queries. Particularly, we have discussed when and how stream and static inputs can share the data. We provided a solution to share join operators. Our approach pre-processes the queries and builds a join graph before constructing the network of shared join operators. The join graph is constructed based on the heuristics that each vertex should generate join results reusable by as many queries as possible. Our experiments showed that CQELS+ can handle more concurrent queries with higher throughputs, compared to the original CQELS. In order to further improve the performance for concurrent RSP queries, we followed the principle of RSP query federation and applied load balancing strategies over distributed RSP engines. Evaluations for the load balancing strategies showed that deploying multiple CQELS+ instances can lower the latency, but the memory overhead for too many parallel instances may outweigh its benefit. Also, we found that while the minimal data buffer size strategy performs best at reducing the query latency, it has longer query registration time.

In future work, we plan to investigate on implementing the probing graph based on both the statistics of data in window buffers and join variable. Real-time data from the stream sources come to the system unpredictably, which means the data inside the window buffers is changing and consequently changes the join selectivity. Therefore, the join probing graph built based merely on join variable in patterns is not able to guarantee the optimal join probing sequences. On the other hand, we must consider the overhead when monitoring the dynamics of the data buffers. We also plan to investigate more sophisticated load balancing strategies, e.g., defining a more precise cost model for the query processing pipeline and use it to estimate the total cost after a new query registration. This may involve checking the similarity between queries using metrics like the

---

[7] Twitter Storm: http://storm.apache.org/.

number of shareable data buffers, the graph edit distance between the join probing graphs before and after the query registration etc.

Acknowledgment. Authors are extremely thankful to John Breslin, Alessandra Mileo and Danh-Le Phouc for their valuable feedback and guidance. The work conducted during this study is supported by Science Foundation Ireland (SFI) under grant No. SFI/12/RC/2289.

References

1. Ali, M.I., Gao, F., Mileo, A.: CityBench: a configurable benchmark to evaluate RSP engines using smart city datasets. In: Arenas, M., et al. (eds.) ISWC 2015. LNCS, vol. 9367, pp. 374–389. Springer, Cham (2015). doi:10.1007/978-3-319-25010-6_25
2. Ali, M.I., Pichler, R., Truong, H.-L., Dustdar, S.: DeXIN: an extensible framework for distributed XQuery over heterogeneous data sources. In: Filipe, J., Cordeiro, J. (eds.) ICEIS 2009. LNBIP, vol. 24, pp. 172–183. Springer, Heidelberg (2009). doi:10.1007/978-3-642-01347-8_15
3. Arasu, A., Babu, S., Widom, J.: The CQL continuous query language: semantic foundations and query execution. VLDB J. **15**(2), 121–142 (2006)
4. Barbieri, D.F., Braga, D., Ceri, S., Della Valle, S.E., Grossniklaus, M.: C-SPARQL: SPARQL for continuous querying. In: Proceedings of WWW, pp. 1061–1062. ACM (2009)
5. Bizer, C., Heath, T., Berners-lee, T.: Linked data - the story so far. Int. J. Semant. Web Inf. Syst. **5**, 1–22 (2009)
6. Calbimonte, J.-P., Jeung, H., Corcho, O., Aberer, K.: Enabling query technologies for the semantic sensor web. Proc. IJSWIS **8**(1), 43–63 (2012)
7. Deen, S.M., Al-Qasem, M.: A query subsumption technique. In: Bench-Capon, T.J.M., Soda, G., Tjoa, A.M. (eds.) DEXA 1999. LNCS, vol. 1677, pp. 362–371. Springer, Heidelberg (1999). doi:10.1007/3-540-48309-8_34
8. Diao, Y., Franklin, M.J.: High-performance XML filtering: an overview of YFilter. IEEE Data Eng. Bull. **26**, 41–48 (2003)
9. Gao, F., Ali, M.I., Mileo, A.: Semantic discovery and integration of urban data streams. In: Proceedings of the 13th International Semantic Web Conference (ISWC 2014), Workshop on Semantics for Smarter Cities (2014)
10. Gao, F., Curry, E., Ali, M.I., Bhiri, S., Mileo, A.: QoS-aware complex event service composition and optimization using genetic algorithms. In: Franch, X., Ghose, A.K., Lewis, G.A., Bhiri, S. (eds.) ICSOC 2014. LNCS, vol. 8831, pp. 386–393. Springer, Heidelberg (2014). doi:10.1007/978-3-662-45391-9_28
11. Gao, F., Curry, E., Bhiri, S.: Complex event service provision and composition based on event pattern matchmaking. In: Proceedings of the 8th ACM International Conference on Distributed Event-Based Systems, Mumbai, India. ACM (2014)
12. Hammad, M.A., Franklin, M.J., Aref, W.G., Elmagarmid, A.K.: Scheduling for shared window joins over data streams. In: VLDB. VLDB Endowment (2003)
13. Hoeksema, J., Kotoulas, S.: High-performance distributed stream reasoning using S4. In: Ordering Workshop at ISWC (2011)
14. Koerner, M., Kao, O.: Multiple service load-balancing with openflow. In: 2012 IEEE 13th International Conference on High Performance Switching and Routing (HPSR), pp. 210–214. IEEE (2012)

15. Kossmann, D.: The state of the art in distributed query processing. ACM Comput. Surv. (CSUR) **32**(4), 422–469 (2000)
16. Le-Phuoc, D.: A native and adaptive approach for linked stream data processing. Ph.D. thesis, National University of Ireland Galway, IDA Business Park, Lower Dangan, Galway, Ireland (2012)
17. Le-Phuoc, D., Nguyen Mau Quoc, H., Le Van, C., Hauswirth, M.: Elastic and scalable processing of linked stream data in the cloud. In: Alani, H., et al. (eds.) ISWC 2013. LNCS, vol. 8218, pp. 280–297. Springer, Heidelberg (2013). doi:10.1007/978-3-642-41335-3_18
18. Matsuba, H., Joshi, K., Hiltunen, M., Schlichting, R.: Airfoil: a topology aware distributed load balancing service. In: 2015 IEEE 8th International Conference on Cloud Computing (CLOUD), pp. 325–332. IEEE (2015)
19. Naughton, V.J.F., Burger, J.: Maximizing the output rate of multi-way join queries over streaming information sources. In: VLDB. VLDB Endowment (2003)
20. Le-Phuoc, D., Xavier Parreira, J., Hauswirth, M.: Linked stream data processing. In: Eiter, T., Krennwallner, T. (eds.) Reasoning Web 2012. LNCS, vol. 7487, pp. 245–289. Springer, Heidelberg (2012). doi:10.1007/978-3-642-33158-9_7
21. Porter, G., Katz, R.H.: Effective web service load balancing through statistical monitoring. Commun. ACM **49**(3), 48–54 (2006)
22. Prud'hommeaux, E., Seaborne, A.: SPARQL query language for RDF. W3C Recommendation **4**, 1–106 (2008)
23. Schwarte, A., Haase, P., Hose, K., Schenkel, R., Schmidt, M.: FedX: optimization techniques for federated query processing on linked data. In: Aroyo, L., et al. (eds.) ISWC 2011. LNCS, vol. 7031, pp. 601–616. Springer, Heidelberg (2011). doi:10.1007/978-3-642-25073-6_38
24. Sellis, T.K.: Multiple-query optimization. ACM Trans. Database Syst. **13**(1), 23–52 (1988)
25. Sequeda, J.F., Corcho, O.: Linked stream data: a position paper. In: SSN (2009)

AGACY Monitoring: A Hybrid Model
for Activity Recognition
and Uncertainty Handling

Hela Sfar[(✉)], Amel Bouzeghoub, Nathan Ramoly, and Jérôme Boudy

CNRS Paris Saclay, Telecom SudParis, SAMOVAR, Évry, France
{hela.sfar,amel.bouzeghoub,
nathan.ramoly,jerome.boudy}@telecom-sudparis.eu

Abstract. Acquiring an ongoing human activity from raw sensor data is a challenging problem in pervasive systems. Earlier, research in this field has mainly adopted data-driven or knowledge based techniques for the activity recognition, however these techniques suffer from a number of drawbacks. Therefore, recent works have proposed a combination of these techniques. Nevertheless, they still do not handle sensor data uncertainty. In this paper, we propose a new hybrid model called AGACY Monitoring to cope with the uncertain nature of the sensor data. Moreover, we present a new algorithm to infer the activity instances by exploiting the obtained uncertainty values. The experimental evaluation of AGACY Monitoring with a large real-world dataset has proved the viability and efficiency of our solution.

Keywords: Smart home · Uncertainty · Ontology · Machine learning

1 Introduction

Nowadays, we face an emerging use of context-aware systems in various domains in order to ensure of end-user well being and quality of life. The smart homes are a trending context aware applications. In smart homes, context information about the user context is gathered and used to monitor and track their activities. Smart environments rely on sensors to monitor the interaction between the users, objects, and the environment. However, in real world applications, sensor data are not always well-aimed and precise [10] due to hardware failure, energy depletion, etc. Therefore, a context-aware system should be sensitive enough to the missing or imprecise sensor data in order to make the right decisions. To recognize activities and handle sensors' uncertainty, the main solutions can be generally classified as *data driven*, *knowledge based*, and *hybrid* approaches. Data driven methods apply different supervised machine learning techniques to classify sensor data into activities based on the given training data. For example, Hidden Markov Models (HMM) [12] and Support Vector Machines (SVM) [13] are two well-known classifiers. Although this group of methods is suited to handle uncertainty and to deal with a broad range of sensors, it needs a large amount

© Springer International Publishing AG 2017
E. Blomqvist et al. (Eds.): ESWC 2017, Part I, LNCS 10249, pp. 254–269, 2017.
DOI: 10.1007/978-3-319-58068-5_16

of training data to set up a model and estimate its parameters [10]. On the other hand, knowledge based methods use ontologies and reasoning engines to infer proper activities from current sensors input. Despite their powerful semantic representation of real world data and their reasoning capabilities, their use is restricted to a limited number of sensors: when big number of sensors are used then the manual creation of the ontology will be hard and painful. Therefore, more the number of sensor increased more the conception of the ontology and the reasoning part becomes difficult. The number of sensors in which the ontology is limited can be set in empirically. Given the limitations of both data driven and knowledge based approaches, combining them is a promising research direction as it was stated in the analysis done in [10]. Intuitively, a hybrid approach takes the "best of both worlds" by using a combination of methods. Such an approach is able to provide a formal, semantic and extensible model with the capability of dealing with uncertainty of sensor data and reasoning rules [10]. Therefore, proposing hybrid models has been the motivation of recent works including [3,11,14]. Nevertheless, the lack of sensor data uncertainty consideration is the main drawback of the aforementioned approaches. To overcome this limitation, this paper proposes a new hybrid model combining data driven and knowledge based methods for activity recognition and sensor data uncertainty handling. The main contributions of the paper are as follows:

1. Introduction of the AGACY Monitoring hybrid model, available online, that integrates knowledge based and data driven techniques for activity recognition. This novel approach handles uncertain sensor data and exploits these uncertainty values to compute the produced activities uncertainty values.
2. Improvement of an existing method for feature extraction [8] in order to deal with more time distant actions and their uncertainty values.
3. Invention of new algorithm for current activity instances inferring called AGACY.

The rest of this paper is organized as follows: Sect. 2 discusses related work about hybrid models for activity recognition, Sect. 3 presents the architecture of AGACY Monitoring model. Sections 4 and 5 provide more details about each layer in the architecture. Section 6 reports experimental results. Finally, Sect. 7 concludes the paper.

2 Related Work

The combination of data driven and knowledge based methods for activity recognition has been a recent topic of interest. Therefore, few hybrid activity recognition systems have been proposed in the literature.

COSAR [4,15] is a context-aware mobile application that combines machine learning techniques and an ontology. As a first step, the machine learning method is triggered in order to predict the most probable activities based on a provided training data. Then, an ontological reasoner is applied to refine the results by selecting the set of possible activities performed by a user based on his/her location acquired by a localization server. Despite the fact that the sensor data are supposed to be certain, COSAR deals with the uncertainty of the transformation of the localization from a physical format to a symbolic one. Another hybrid model that combines a machine learning technique, an ontology, and a log-linear system has been proposed in [14]. The aim of this approach is to recognize a multilevel activity structure that holds 4 levels: atomic gesture (Level 4), manipulative gesture (Level 3), simple activity (Level 2), and complex activity (Level 1). The atomic gestures are recognized through the application of a machine learning technique. Moreover, using a probabilistic ontology defined by the log-linear, and standard ontological reasoning tasks, the manipulative gestures, simple activities, and complex activities are inferred. Each level is deduced based on a time window that contains elements from the previous level. Even though the work in [14] is similar to the previous one regarding the absence of sensor data uncertainty's handling, the inference of the 4 levels activities is based on a probabilistic reasoning that represents a sort of uncertainty.

FallRisk [11] is another pervasive system that combines data driven and knowledge based methods. Its main objective is to detect a fall of an elderly person living independently in a smart home. FallRisk is a platform that integrates several systems that use machine learning methods for fall detection. It filters the results of these systems thanks to the use of an ontology that stores the contextual information about the elderly person. The main advantage of this system is that it is extensible to integrate several fall detection systems. Moreover, the contextual information of the elderly is taken into account. However, this work does not consider any kind of uncertainty.

FABER [16] is a pervasive system used to detect abnormal behavior of a patient. Firstly, it deduces events and actions from the acquired sensor data. This is done based on simple ontological inference methods. Then, these events and actions are sent to a Markov Logic Network (MLN) as a machine learning method to analyze the event logs and infer the start/end time of activities. The inferred activity boundaries are communicated together with actions and events to the knowledge based inference engine. This engine evaluates the rules modeling abnormal behaviors and detected abnormal behaviors are communicated to the hospital center for further analysis by the doctors. Nevertheless, similarly to previous works this system does not handle uncertainty of sensor data.

SmartFABER [3] system is an extension and an improvement of FABER [16]. These two frameworks share the same aims. Regarding SmartFABER [3], instead of communicating the inferred events and actions to MLN classifier, the system sends them to a module that is in charge of building vectors of features based on the received events. Then, these features are communicated to a machine learning module for the classification of activities. Next, a proposed algorithm called SmartAggregation is applied to infer current activity instances. For deducing an activity's instance from a sequence of events classified to an activity, the algorithm verifies whether each event satisfies a set of conditions. These conditions are defined by a human expert after a deep analysis of the semantic of activity. If all events satisfy all conditions, then an activity instance could be inferred. This work is proved to outperform FABER [16]. However, it suffers from two main drawbacks: (1) There is no uncertainty handling for sensor data; (2) The performance of the SmartAggregation algorithm depends heavily on the defined conditions. It can suffer from time consuming if there is a huge number of conditions that need semantic verification.

3 The AGACY Monitoring Architecture Overview

The overall architecture is composed of two layers: the Knowledge based layer and the Data driven layer, as depicted in Fig. 1. The multimodal fusion process aims to homogenize the sensor data and compute their uncertainty's value (e.g. FSCEP system [1]). This process is out of the scope of this paper. We assume that the homogeneous sensor data together with their uncertainty values are provided to the Knowledge based layer. The knowledge based layer then represents semantically the incoming sensor data together with their uncertainty values (Ontological modeling). Afterwards, it infers actions and events from the modeled sensor data (Semantic reasoning) and computes their uncertainty's values. The obtained actions and the computed uncertainty values are sent to the data driven layer. The layer is responsible for: (1) classification of the actions into features (Time & Uncertainty-based features extraction), (2) classification of the features and actions into activities (Dempster Shafer theory for activity classification), and finally (3) inferring of activities' instances (Activities instances inferring under uncertainty). In the following, further explanation of each layer is provided.

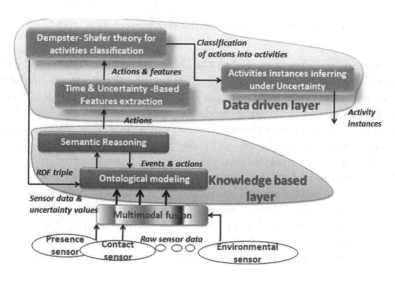

Fig. 1. The architecture of the AGACY Monitoring

4 Knowledge Based Layer

4.1 Ontological Modeling

The ontological modeling module allows sensor data to be formally conceptualized. This conceptualization, as the ontology is defined for, serves to provide a semantic model from the data. However, the native ontological representation is known to poorly handle uncertainty. Therefore, some attempts for uncertainty integration into ontological models have been realized [5,6].

In this work, we adopt the model proposed in [6] since it could be attached to any existing ontology without the need for redesigning it. However, the problem of this model is the lack of temporal representation. In order to distinguish two similar sensor data that have the same uncertainty value but come in two different timestamps, it is highly important to associate an uncertainty value (assigned to a sensor data) with the unique time. Hence, to overcome this problem, we extended the model in [6] by adding a temporal element (the class Time). Our proposed model is depicted in Fig. 2.

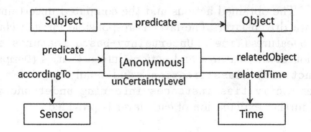

Fig. 2. The extended uncertainty representation model

Anonymous resource is related to the class: *Subject* through the ObjectProperty: *predicate* and to the *Object* through the ObjectProperty: *relatedObject*. It holds the uncertainty level of the triple < *Subject*; *predicate*; *Object* >, according to the model in Fig. 2, using its dataProperty: *unCertaintyLevel*. It is also linked to the source of uncertainty – namely a sensor or a set of sensors that derive uncertainty– with the objectProperty: *accordingTo*, and to class: *Time* via the objectProperty: *relatedTime*.

Through this model, the uncertainty can be easily integrated into any ontology. To do so, we have designed an ontology to represent sensors, sensor data, sensor data uncertainty values, actions, events, persons, time, and so on. Figure 3 shows an excerpt of this ontology, in which the model from Fig. 2 is used in order to represent the uncertainty. As Fig. 3 depicts, among the basic high level concepts in the ontology we find: *Sensor, Action, Event, Activity, Person, Object,* and *Time.*

Events and actions have also values of uncertainty that are represented as dataProperties. Next subsection will describe how the events and actions are inferred and how their uncertainty values are computed.

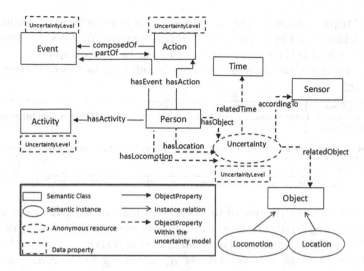

Fig. 3. An excerpt of the ontology with uncertainty integration

4.2 Semantic Reasoning

This module is mainly responsible for the inference of events and actions and their uncertainty values. Events are deduced from sensor data and actions are inferred from events. Formally we define an event as follows:

Definition 1. *Event Instance: Let E be the set of event labels; T a set of all possible time-stamps; and C a set of uncertainty values. An event instance is a triple ev(e_i, t_i, c_i) where $e_i \in E$, $t_i \in T$ and $c_i \in C$.*

The events are deduced by combining the *logic with possibilities* [7] with the *modified semantic logic rules* in [6]. On the one hand, in [6] the traditional semantic logic rules are modified in order to integrate the uncertainty in a clause's definition. This is done according to the ontological model in Fig. 2 (without the concept Time). Then, this modified rule is able to propagate the uncertainty value from the premise to the conclusion of the rule. However, the propagation is done from only one clause within the premise. Thus, the case when the rule contains more than one clause is omitted. On the other hand, the *logic with possibilities* has the advantage to propagate values of possibilities (in our case values of uncertainty (denoted Unc in Eq. 1) in a rule from a premise with multiple clauses to the conclusion. This is done through the following rules:

$$Unc(p \wedge q) = min(Unc(p), Unc(q)); \ Unc(p \vee q) = max(Unc(p), Unc(q))$$
$$Unc(\neg p) = 1 - Unc(p); \ Unc(p \longrightarrow q) = max(1 - Unc(p), Unc(q)) \tag{1}$$

Accordingly, in order to deduce the events, we combine the *logic with possibilities* with the *modified semantic logic rules* where the premise is a set of clauses linked with logic operators. A clause is a RDF triple translated in logic [6].

This RDF triple models the sensor data with their uncertainty according to the ontology depicted in Fig. 3. The conclusion is an event. To propagate the clauses' uncertainty towards the conclusion, the rules in Eq. 1 are applied.

In this example, we demonstrate how to infer an instance of event labeled *sitOnChair*:

$\forall$ se$_1$, se$_2$ $\in$ *{Sensors}*, t$_1$, t$_2$ $\in$ *{Time}*, and p $\in$ *{Person}*

(p hasLocomotion [**a Uncertainty; uncertaintyLevel** n$_1$; **relatedObject SitOn; relatedTime** t$_1$; **accordingTo se$_1$**]

$\wedge$

(p hasObject [**a Uncertainty; uncertaintyLevel** n$_2$; **relatedObject Chair; relatedTime** t$_2$; **accordingTo se$_2$**])

$\longrightarrow$ ev(SitOnChair, max(t$_1$,t$_2$), min(n$_1$, n$_2$))

Example of rule for inferring the event instance sitOnChair

As we can see, the premise of this rule contains 2 clauses. Each one is a RDF triple representing an uncertain sensor data according to the ontology depicted in Fig. 3. The first clause means *"the resident p is observed to have the locomotion SitOn at time t_1 with uncertaintyLevel n_1 according to the sensor se_1"* while the second clause translates the information *"the resident p is observed to get the object Chair at time t_2 with uncertaintyLevel n_2 according to the sensor se_2"*. The rule's output always creates a new event with an uncertainty value deduced according to rules described in Eq. 1. Then, a production of the event with label *sitOnChair* is deduced. Since the premise is a conjunction of clauses, its uncertainty value is the minimum between both values of clauses' uncertainty. Accordingly, this new event will be modeled in the ontology as an instance of the concept *Event* and the uncertainty value as a dataProperty of the instance.

After deducing events, the actions are inferred. An action is defined as follows:

Definition 2. *Action Instance: Let A be the set of actions labels, T the set of all possible timestamps, C the set of uncertainty values. An action instance is a triple $ac(a_i, t_i, c_i)$ where $a_i \in A$, $t_i \in T$ and $c_i \in C$.*

The actions and their uncertainty values are inferred by simply applying the logic with possibilities rules. The action is the conclusion of a rule, the clauses of its premise are the events or some defined restrictions.

This example is to illustrate how we can deduce the production of an action's instance with label *SitOnChairAtKitchenTable*:

$$ev(SitOnChair, t_1, n_1) \wedge ev(PresenceAtKitchenTable, t_2, n_2) \wedge (t_1 \geq t_2) \wedge ((t_1 - t_2) \leq 5s) \longrightarrow (SitOnChairAtKitchenTable, t_2, \min(n_1, n_2))$$

The previous rules are examples of rules that are managed by the system. However, a set of required rules are defined by experts according to the semantic of the activities to be monitored and the types of sensors in the smart environment. Other rules can be inferred through an ontological inference engine according to the axioms defined in the ontology.

The deduced actions are then communicated to the module **Time & Uncertainty-based features extraction** described in the following section.

5   Data Driven Layer

5.1   Time and Uncertainty-Based Features Extraction

For each received action $ac(a_i, t_i, c_i)$, this module is in charge of building a feature vector representing the sequence S of the recent actions in a time window with a size equal to n. $S = \langle ac(a_{i-n+1}, t_{i-n+1}, c_{i-n+1}), ..., ac(a_{i-1}, t_{i-1}, c_{i-1}), ac(a_i, t_i, c_i)\rangle$.

In this work, we improve the technique proposed in [8] for features extraction. We have chosen this method since it is proved to be effective in recognizing activities based on streams of sensor events or actions instead of streams of sensor data compared to traditional features extraction techniques. This technique builds a vector of features for each events sequence S. The produced vector of feature of a sequence of events S_i holds these information: (i) The label of the feature, K_i; (ii) The time t_0 of the first event in S_i; (iii) The time t_i of the last event in the sequence S_i; (iv) The list of events under S_i; (v) A weight value fine-tunes the contribution of each event in S_i, so that recent events participate more than the older ones. This weight value is computed as follows (Eq. 2):

$$F_{k_i}(S_i) = \sum_{ev(e_j, t_j) \in S_i} exp(-\chi(t_i - t_j)) \times f_{k_i}(ev(e_j, t_j)) \qquad (2)$$

where the factor χ determines the time-based decay rate of the events in S_i; t_i-t_j is expressed in seconds and $f_{k_i}(ev(e_j, t_j))$ is the time-independent participation of $ev(e_j, t_j)$ in the computation of the F_{k_i} value. The other way around if $ev(e_j, t_j)$ participates in the execution of k_i then $f_{k_i}(ev(e_j, t_j)) = 1$ else 0.

As we can see in Eq. 2, the greater difference is of the time distance between $ev(e_j, t_j)$ and $ev(e_i, t_i))$, the less $ev(e_j, t_j)$ participates in the computation of F_{k_i}.

In our understanding, this may be true when the approximate duration of the feature is short (e.g. in terms of seconds or some minutes). In contrast, when it is long (e.g. in terms of hours), this hypothesis is not always valid. Let us have the example of the feature "stove usage" in [3]. The duration of the execution of this feature, for a particular recipe, may be equal to a number of hours. We suppose that the vector of this feature contains the following two events *openStove* and *closeStove*. Since the duration of the feature "stove usage" is in terms of hours, the value of $(t_{closeStove} - t_{openStove})$ may be equal to 1 h or 2 h. Accordingly, by applying Eq. 2 the event *openStove* does not participate in the weight calculus of the feature "stove usage". Therefore, incorrect value of weight may be obtained. Accordingly, the execution of the event *openStove* must have a high impact in the execution of the feature. Intuitively, the stove can not be used if it is not opened. Moreover, this technique assumes that the only factor that may have impact on the computation of the feature's weight is the time distance between events. However, when information about actions and events uncertainty values is provided, this information should be taken into consideration in the computation of the feature's weight. Therefore, we made the following assumptions:

A1. The uncertainty values must have an impact on the calculus of the feature weight: the higher uncertainty value of event or action is, the more the weight of the feature increases.

A2. A product containing the uncertainty value of an event or action, must be higher or equal than this uncertainty value: The uncertainty value must not be decreased when it is multiplied.

To overcome the problem mentioned above, we propose a new version of Eq. 2 that meets the assumptions A1 and A2. We formally define the notion of Short Time Feature (STF) and Long Term Feature (LTF) as follows:

Definition 3. *Short Time Features (STF) is a set of features that holds only features having duration less than ten minutes. Let $Dur(f)$ be the duration of feature f: $f \in STF \Leftrightarrow Dur(f) <= 600\,s$.*

Definition 4. *Long Term Features (LTF) is a set of features that holds only features having duration more than ten minutes. Let $Dur(f)$ be the duration of feature f: $f \in LTF \Leftrightarrow Dur(f) > 600\,s$.*

Firstly, it is important to note that This value of ten minutes is chosen intuitively and it is variable according to the experiments. Then, to compute the weight of the feature, we propose the following Eq. 3.

$$F_{k_i}(S_i) = \sum_{ac(a_j,t_j,c_j) \in S_i} c_j \times Fact_{\chi,\delta t_{ij}} \times f_{k_i}(ac(a_j,t_j,c_j)) \tag{3}$$

$$Fact_{\chi,\delta t_{ij}} = \begin{cases} exp(-\chi * \delta^h t_{ij}) & If\ (k_i \in LTF) \\ \frac{1}{\chi * \delta^s t_{ij}} & If((\chi * \delta^s t_{ij} \succ 1)\ \&\ (k_i \in STF)) \\ 1 & Otherwise \end{cases}$$

It is worth mentioning that in our work the features vectors are built from actions instead of events, in contrast to [8] and the size n of the time window is chosen according to the nature of the feature (STF or LTF). We note that $c_j \times Fact_{\chi,\delta t_{ij}}$, in Eq. 3, is a sort of uncertainty where $c_j \times Fact_{\chi,\delta t_{ij}} \leq c_j$ (Assumption A2). $\delta t_{ij} = t_i - t_j$. $\delta^h t_{ij}$ means that δt_{ij} is expressed in hours and $\delta^s t_{ij}$ means that δt_{ij} is expressed in seconds. To fix the problem of the time delay in Eq. 2, we distinguish three cases: (1) The feature is a LTF ($k_i \in LTF$). Then, to compute $Fact_{\chi,\delta t_{ij}}$ the same function ($exp(-\chi * \delta t_{ij})$) in Eq. 2 is used. However, δt_{ij} is expressed in hours ($\delta^h t_{ij}$) instead of seconds. Accordingly, the actions that happened earlier could participate in the weight computation in contrast to Eq. 2; (2) The feature is a STF ($k_i \in STF$) and $\chi * \delta t_{ij} \succ 1$. Then, $Fact_{\chi,\delta t_{ij}} = \frac{1}{\chi * \delta^s t_{ij}}$. The quotient function is chosen to compute $Fact_{\chi,\delta t_{ij}}$ since it has a less decreasing shape than the exponentiation function. (3) If $(\chi * \delta^s t_{ij}) \prec 1$, then $\frac{1}{\chi * \delta^s t_{ij}} \succ 1$ and accordingly $c_j \times Fact_{\chi,\delta t_{ij}} \succ c_i$ that does not correspond to Assumption A2. Therefore, $Fact_{\chi,\delta^s t_{ij}} = 1$. This is due to the fact that since t_i and t_j are very close, $ac(a_j,t_j,c_j)$ must have the highest impact on the weight computation, i.e. the value 1 in [0..1] is chosen. As a result, computed weights of the feature are used as their uncertainty values. Afterwards, the actions and the features together with their uncertainty values are sent to `Dempster-Shafer theory for activity classification` module.

Algorithm 1. AGACY

Input: G, minUncert, tolUncert, β_{delay}, $maxDelay_A$, A
Output: Inst: a set of activity instances

```
 1:  ℵ ⟵ segmentation(G, maxDelay_A);
 2:  Inst ⟵ φ;
 3:  LowUncert ⟵ ℵ;
 4:  for each g in ℵ do
 5:      if (∀ac(a_i, t_i, c_i) ∈ g ⟶ c_i ≻ minUncer) then
 6:          inst ⟵ activity instance of A that is generated for
                   the observation g;
 7:          Inst ⟵ Inst ∪ inst;
 8:          LowUncert ⟵ lowUncert\g;
 9:          G ⟵ G\all actions in g;
10:      else
11:          missClass ⟵ φ;
12:          ratioUncert ⟵ ComputeRatioCert(g);
13:          if ratioUncert ≻ tolUncert then
14:              missClass ⟵ missClass ∪ GetLowUncert(g);
15:              K ⟵ 0;
16:              toBeReplaced ⟵ φ;
17:              while (k ≺ (card(G))&(card(toBeReplaced) ≺ card(missClass)) do
18:                  if (ac(a_k, t_k, ck) ∈ G)&(ac(a_k, t_k, ck) ∉ g)&(c_k ≻ minUncert)&(∀ac(a_j, t_j, c_j) ∈ g ⟶
                         |t_j − t_k| ≺ maxDelay_A + β_delay) then
19:                      toBeReplaced ⟵ toBeReplaced ∪ ac(a_k, t_k, c_k);
20:                  end if
21:                  K ⟵ K + 1;
22:              end while
23:          end if
24:      end if
25:      if card(toBeReplaced) = card(missClass) then
26:          g ⟵ g\missClass;
27:          g ⟵ g ∪ toBeReplaced;
28:          inst ⟵ activity instance that is generated
                   from the observation g;
29:          Inst ⟵ Inst ∪ inst;
30:          lowUncert ⟵ lowUncert\g;
31:          Partition(missClass, ℵ, maxDelay_A);
32:          G ⟵ G\ {all actions in g};
33:      end if
34:  end for
```

5.2 Dempster-Shafer Theory for Activity Classification

In order to classify activities, we use the well known Dempster Shafer theory (DS) [2]. It has proven to provide decent result in comparison to other techniques such as J48 Decision Tree [9]. The model is used through a directed acyclic graph, where actions in AGACY Monitoring are equivalent to the named evidences in DS, the called hypothesis in DS are features in AGACY Monitoring, and outputs are activities with their uncertainty values. We note here that activities are composed of set of features and as sated in the previous section features are composed of set of actions. The uncertainty values of activities are computed based on the uncertainty values of the features and the mass function value defined in [9].

5.3 Activities Instances Inferring Under Uncertainty

In this section, we present $AGACY$, our algorithm for activity instances inference. The algorithm aims to improve an existing one called $SmartAggregation$ [3].

We improve this method since it is accurate and it improves the well known algorithm Naive aggregation [3]. The basic idea of the *SmartAggregation* algorithm is the following: if two consecutive events that occurred respectively at t_i and t_{i+1} are classified with the same activity's class $A_i = A_{i+1}$ and verify the defined conditions of the activity A_i, $C^{(A)}$, they are considered as an observation[1] generated by the same activity instance of the class A_i. Otherwise, if an event does not satisfy all the conditions, the algorithm tries to integrate it into an observation of a previous inferred instance providing that it satisfies all the conditions defined by human experts. Hence, this algorithm, despite its accuracy, suffers from scalability and time consuming. Assuming a big number of events, the algorithm must check each event for satisfying all the conditions, which is demanding and time consuming. To tackle these problems, we refine our recognition method by the proposed Algorithm 1. It takes advantage of the obtained uncertainty values of the activities to infer the current instances of activity. The basic idea of Algorithm 1 is the following: a group of actions could be considered as an observation of an activity instance providing that (1) all the actions have uncertainty values beyond a defined minimum value, e.g. MINUNCERT, (Lines 2–9) and (2) the time distance between every pair of consecutive actions within the group must be lower than $maxDelay_A$. The second condition is always true since the first step done by the algorithm is a segmentation over the output of the DS (Line 1): All the actions associated with the same activity class A and temporally close (according to $maxDelay_A$) are grouped together to obtain a set of groups $\aleph$. G is a set of all actions without segmentation. In this case, there are no conditions to be verified for the actions compared to *SmartAggregation* algorithm. Indeed, the uncertainty value of each action is replaced, before the execution of the algorithm, with the uncertainty value of its classified activity class. Therefore, when the uncertainty value c of an action ac is high, it is almost sure that the action ac is really produced and belongs to the correct class of activity act, hence, must belong to an instance of act.

This process is less time consuming than *SmartAggregation* algorithm. On the other hand, when at least one action in the group has an uncertainty value lower than MINUNCERT, the algorithm accepts some time distance shift between actions regarding the uncertainty degree (Lines 10–32). To do so, it checks firstly if the ratio of the number of actions (RATIOUNCERT) in the group that have lower uncertainty value than MINUNCERT is higher than a defined value called TOLUNCERT (Line 13). If so, the algorithm attempts to replace the actions that have lower uncertainty values with other actions from other groups. A new action can be added if it has higher uncertainty value than MINUNCERT and the time distance between the action and each action in the group is lower than the $MaxDelay_A + \beta_{delay}$ (Lines 18 and 19). β_{delay} represents a sort of tolerance about the time distance between two actions. Thus, we assume that uncertainty value is more important than the time distance. Finally, when all uncertain actions are replaced with actions having high uncertainty values with some time distance tolerance, this new group can generate a new activity instance (Line 27). Afterwards, the removed actions having lower uncertainty values than

---

[1] An observation is a sequence of events that is generated by an activity instance.

MINUNCERT in the old group are distributed to the rest of the groups according to $maxDelay_A$ (Line 31).

6 Evaluation and Discussion

For this experiments, we evaluate the proposed AGACY Monitoring system. In this section, we first describe the dataset used for the experiments. Then, the experiments and the achieved results are presented.

6.1 DataSet

We used real-life data collected in highly rich smart environments. The dataset[2] [17] was obtained as a part of the Opportunity project[3]. The dataset focus on activities concerning breakfast that holds (from AGACY Monitoring perspective) homogeneous sensor data (level 4), events (level 3), actions (level 2), and activities (level 1) that have been done by three persons S10, S11, and S12 with three different routines each (ADL1-3). In order to test AGACY Monitoring system, this dataset does not contain information about sensor data uncertainty values (level 4). Therefore, we have randomly annotated the level 4 in the dataset with high uncertainty values in [0.8..1]. Moreover, we have injected a set erroneous sensor data in the dataset annotated with low uncertainty values in [0..0.4].

6.2 Implementation and Experimental Setup

We have implemented the AGACY Monitoring system using JAVA. Regarding data driven layer, for this dataset, we have considered the set of features depicted in Table 1 to be treated in the *Time & Uncertainty based features extraction* module.

Table 1. List of considered features. STF: Short Time Feature, LTF: Long Term Feature

Feature name	STF/LTF	Duration (s)
PrepareCoffee	STF	600
Drink	STF	120
GatherCutlery	STF	600
GatherFood	STF	600
Eat	STF	1200
PutAwayFood	STF	600
PutAwayCutl	STF	600
Dishwasher	STF	300
Resting	STF	600

For the *Activity classification* module, the experiments are not limited only to the use of DS – we also have the SVM for activity classification. The results obtained from this module are compared with those obtained by Rim et al. [14]. Finally, regarding the *Activities' instances inferring under uncertainty* the algorithm AGACY has 4 variables: MINCERT, TOLCERT, MAXDELAY$_A$, β_{DELAY}. Due to the lack of space, we only present the most adequate values regarding this experiments: MINCERT $= 0.5$; TOLCERT $= 1$; MAXDELAY$_A = 30$ s; $\beta_{\text{DELAY}} = 10$ s;

---

[2] The dataset: http://webmind.dico.unimi.it/care/annotations.zip.
[3] http://www.opportunity-project.eu.

6.3 Evaluation and Results

Activity Recognition Evaluation. As described in the previous sections, the *Activity classification* module outputs the predicted activity class. It is worth mentioning that the system has no False Negative result (FN = 0): it outputs always at least one activity. Figure 4 depicts the average precision measure over the three routines for all subjects by varying the value n of the time window (Tw) (see Subsect. 4.1) in [60 s..300 s] with one erroneous sensor data for five correct ones (e.g. 1/5 erroneous sensor data). The code source of AGACY Monitoring is available online[4]. As it is clearly shown in Fig. 4, the DS with n = 180 s reaches 91% of precision recognition rate. For time windows shorter or longer than 180, DS tends to become less efficient: DS is efficient where time window are properly proportional to the activities: when the time window is too big, there may be conflict between activities, while when it is too small, DS does not have enough data to be efficient. On the other hand, SVM gives better results for short time window (n <= 120 s), but with the increase of n value, the accuracy of classification gets worse. This maybe explained by the fact that SVM is not as dependent on features weights as DS. In general, DS provides better results than SVM. Figure 5 shows the average precision measure over the three routines for all subjects by varying the proportion of the introduced uncertain sensor data compared to the correct sensor data with n = 180 s. As it is clearly shown in the figure the DS is more efficient than SVM: The precision values of DS are in range [0,74..0,91], however that of SVM are in range [0,65..0,77]. Both methods have a decreasing precision values when the number of uncertain sensor data in the dataset increases. This is an expected result since the methods will have less certain data to make the right decision. However, despite the dataset half contains uncertain sensor data, the system is able to predict the activity with a good precision level (74%).

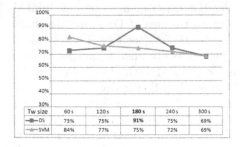

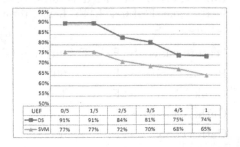

Tw size	60 s	120 s	180 s	240 s	300 s
DS	73%	75%	91%	75%	69%
SVM	84%	77%	75%	72%	69%

UEF	0/5	1/5	2/5	3/5	4/5	1
DS	91%	91%	84%	81%	75%	74%
SVM	77%	77%	72%	70%	68%	65%

Fig. 4. Average recognition precision for all subjects over the three routines with different values of the size n of the time window (Tw)

Fig. 5. Average recognition precision for all subjects over the three routines for varying frequency of uncertain event (UEF). The value 1/5 means there is one uncertain sensor data for five correct ones. 1 means there is one uncertain sensor data for one correct

---

[4] https://github.com/Nath-R/AGACY-monitoring.

Table 2. Average recognition results over three routines for the three subjects obtained by AGACY Monitoring (with DS) and the system proposed in [14] for n = 180 s.

All users	AGACY monitoring	[14]
Precision	0,91	0,91
recall	1	0,65

We have compared the obtained results with another system proposed in [14]. The system has been applied with the same dataset (without erroneous sensor data and uncertainty values).

As it is shown in Table 2 the two systems have the same average precision recognition. However, AGACY Monitoring outperforms the second system within recall value. This can be explained by the fact that the system in [14] returns null if it can not infer an activity. However, AGACY Monitoring has the advantage to always predict an activity. Moreover, the proposed AGACY Monitoring system is effective despite the introduced erroneous sensor data in the dataset. This confirms the ability of AGACY Monitoring to effectively handle uncertainty and predict activities.

Evaluation of AGACY Algorithm for Activities Instances Inference. In this paragraph, we show the evaluation of the AGACY algorithm for inferring the activities instances. The accuracy evaluation is based on the following metrics: precision, recall, and F1 score. Furthermore, we have evaluated the system for time consuming. We tested the AGACY algorithm, on the output of the activities classification module with DS (n = 180 s and the value 1/5 of erroneous sensor data in the dataset).

Moreover, we have implemented and tested the Smart Aggregation algorithm [3] with the same dataset without uncertainty annotations. The average time consuming results over the three routines of both AGACY and SmartAggregation are presented in Fig. 6. The results regarding average accuracy of activities instances detection are depicted in Table 3.

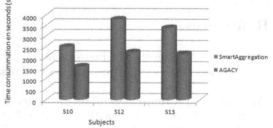

Fig. 6. Average time consuming of activity instances recognition for all subjects over the three routines

Table 3. Average recognition results for activities instances detection over three routines for subjects S10, S11, and S12.

All users	AGACY	SmartAgg.
Precision	0.916	0.854
Recall	0.830	0.872
F1	0.871	0.862

Using above-mentioned experimentation, we obtained two major results: firstly, as it is highlighted in Fig. 6, AGACY has significantly reduced the time execution for activities' instances recognition regarding *SmartAggregation*. Approximately, for the used dataset, AGACY reduces 40% of *SmartAggregation*'s time consummation.

Secondly, in terms of performance both algorithms are efficient with similar F1 score value. As a result, the experiments have proved that AGACY is fast and efficient.

7 Conclusion and Future Work

In this paper, we proposed the AGACY Monitoring hybrid model for activity recognition and sensor data uncertainty handling. The main novelty of AGACY Monitoring is that it combines knowledge based with data driven methods. Thus, several modules contribute to compute the uncertainty value of the expected output. Unlike the related work, AGACY Monitoring supports the inherent uncertain nature of sensor data and exploits it to compute the uncertainty value of each module's output. Besides, the system is able to integrate background knowledge with the data driven method for the recognition of current activity instances. Moreover, the experimental results confirm the viability and the performance of the proposed system and its high level precision even during the presence of uncertain sensor data.

Our future work involves the reuse of an existed upper ontology such DOLCE ontology.

Acknowledgements. This work has been partially supported by the project COCAPS (https://agora.bourges.univ-orleans.fr/COCAPS/) funded by Single Inter-ministrial Fund N20 (FUI N20).

References

1. Jarraya, A., Ramoly, N., Bouzeghoub, A., Arour, K., Borgi, A., Finance, B.: A new model for context perception in smart homes. In: OTM Conferences: COOPIS (2016)
2. Lotfi, A.Z.: A simple view of the Dempster-Shafer theory of evidence, its implication for the rule of combination. AI Mag. **7**, 85–90 (1986). ACM
3. Riboni, D., Bettini, C., Civitares, G., Janjua, Z.H.: SmartFABER: recognizing fine-grained abnormal behaviors for early detection of mild cognitive impairment. Artif. Intell. Med. **67**, 57–74 (2016). Elsevier
4. Riboni, D., Bettini, C.: COSAR: hybrid reasoning for context-aware activity recognition. Pers. Ubiquit. Comput. **3**, 271–289 (2011). Springer
5. Singh, A., Juneja, D., Sharma, A.: A fuzzy integrated ontology model to manage uncertainty in semantic web: the FIOM. Int. J. Comput. Sci. Eng. **3**, 1057–1062 (2011). Citeseer
6. Aloulou, H., Mokhtari, M., Tiberghien, T., Endelin, R., Biswas, J.: Uncertainty handling in semantic reasoning for accurate context understanding. Knowl.-Based Syst. **77**, 16–28 (2015). Elsevier
7. Dubois, D., Lang, J., Prade, H.: Automated reasoning using possibilistic logic: semantics, belief revision, and variable certainty weights. IEEE Trans. Knowl. Data Eng. **6**, 64–71 (1994). IEEE
8. Krishnan, N.C., Cook, D.J.: Activity recognition on streaming sensor data. Pervasive Mob. Comput. **10**, 138–154 (2014). Elsevier

9. Sebbak, D., Benhammadi, F., Chibani, A., Amirat, Y., Mokhtari, A.: Dempster-Shafer theory-based human activity recognition in smart home environments. Ann. Telecommun. **69**, 171–184 (2014). Springer
10. Ye, J., Dobson, S., McKeever, M.: Situation identification techniques in pervasive computing: a review. Pervasive Mob. Comput. **9**, 36–66 (2012). Elsevier
11. De Backere, F., Ongenae, F., Van den Abeele, F., Nelis, J., Bonte, P., Clement, E., Philpott, M., Hoebeke, J., Verstichel, S., Ackaert, A., De Turck, F.: Towards a social and context-aware multi-sensor fall detection and risk assessment platform. Comput. Biol. Med. **64**, 307–320 (2015). Elsevier
12. Patterson, D.J., Fox, D., Kautz, H., Philipose, M.: Fine-grained activity recognition by aggregating abstract object usage. In: ISWC (2005)
13. Hearst, M.A., Dumais, S.T., Osuna, E., Platt, J., Scholkopf, B.: Support vector machines. IEEE Intell. Syst. Appl. **13**, 18–28 (1998)
14. Helaoui, R., Riboni, D., Stuckenschmidt, H.: A probabilistic ontological framework for the recognition of multilevel human activities. In: UbiComp (2013)
15. Riboni, D., Bettini, C.: Context-aware activity recognition through a combination of ontological and statistical reasoning. In: Zhang, D., Portmann, M., Tan, A.-H., Indulska, J. (eds.) UIC 2009. LNCS, vol. 5585, pp. 39–53. Springer, Heidelberg (2009). doi:10.1007/978-3-642-02830-4_5
16. Riboni, D., Bettini, C., Civitarese, G., Janjua, Z.H., Helaoui, R.: Fine-grained recognition of abnormal behaviors for early detection of mild cognitive impairment. In: PerCom (2015)
17. Lukowicz, P., Pirkl, G., Bannach, D., Chavarriaga, R.: Recording a complex, multi modal activity data set for context recognition. In: ARCS, 1st Workshop on Context-Systems Design, Evaluation and Optimisation (2010)

Natural Language Processing and Information Retrieval Track

Mapping Natural Language to Description Logic

Bikash Gyawali, Anastasia Shimorina, Claire Gardent$^{(\boxtimes)}$, Samuel Cruz-Lara,
and Mariem Mahfoudh

CNRS/LORIA, Nancy, France
{bikash.gyawali,anastasia.shimorina,claire.gardent,
samuel.cruz-Lara,mariem.mahfoudh}@loria.fr

Abstract. While much work on automated ontology enrichment has focused on mining text for concepts and relations, little attention has been paid to the task of enriching ontologies with complex axioms. In this paper, we focus on a form of text that is frequent in industry, namely system installation design principle (SIDP) and we present a framework which can be used both to map SIDPs to OWL DL axioms and to assess the quality of these automatically derived axioms. We present experimental results on a set of 960 SIDPs provided by Airbus which demonstrate (i) that the approach is robust (97.50% of the SIDPs can be parsed) and (ii) that DL axioms assigned to full parses are very likely to be correct in 96% of the cases.

Keywords: Natural language processing · OWL · Quality checks

1 Introduction

While there has been much work on enriching ontologies from texts [14,23], most of this work focuses on concepts and relations. As noted in [21], "ontology learning from text is mostly restricted to inexpressive ontologies while the acquisition of complex axioms involving logical connectives, role restrictions and other expressive features of OWL remains largely unexplored."

There are several reasons why addressing this bottleneck is important.

First, manually creating ontologies is a difficult and time consuming task which requires a high level of domain knowledge and technical expertise. Being able to automate or semi-automate the enrichment of ontologies with complex axioms would help diminish the time and expertise required for building the knowledge bases required by semantic applications.

Second, being able to enrich an ontology using complex axioms derived from text would permit semantic reasoning on the content of that text. In particular, it would permit querying the content of that text in the context of the background knowledge encoded in the ontology (using e.g., conjunctive tree queries on the enriched ontology). It would also allow for consistency checking. Is the text consistent with the knowledge contained in the initial ontology? Is a set of text consistent both internally (are all texts consistent with each other?) and externally (is each text in this set consistent with the ontology?).

© Springer International Publishing AG 2017
E. Blomqvist et al. (Eds.): ESWC 2017, Part I, LNCS 10249, pp. 273–288, 2017.
DOI: 10.1007/978-3-319-58068-5_17

Third, it would help bridge the gap between a document-centric and a model-centric view of information. In industries and in governmental services and European organizations, a great number of technical documents (i.e. documents meant to pass information in a way as less ambiguous as possible) are manipulated. Alternatively to this text-based document approach, there is also a use of multiple kinds of "models" (Description Logic Knowledge Bases, UML diagrams, Enterprise Architecture diagrams, etc.), which have increasing popularity. Document-centric and model-centric approaches are, nowadays, largely disconnected however. Being able to automatically map a text to the corresponding model would facilitate the task of technical authors. This would mean that not only the human-friendly text is distributed, but also the associated computer-processable equivalent (model) of the text content.

In this paper, we consider the task of enriching an existing OWL DL (Description Logic) ontology with complex axioms derived from text. We focus on a form of text that is frequent in industry, namely system installation design principles (SIDP) such as (1a-b).

(1) a. Pipes shall be identified with labels.
    b. Spacer shall be used only with attachment device to increase distance with regards to structure, bundles or other systems.

System installation design principles are regulations and directives about how to install a system or a set of systems in a functional area (e.g., electrical and optical system or Water Waste System). For instance, at Airbus, for each aircraft project, a set of SIDPs is produced to ensure that planes comply with all system requirements and take into consideration applicable regulations and procedures. In what follows, we distinguish between *simple SIDPs* consisting of a single clause (e.g., 1a) and *complex SIDPs* which involve more than one clause (e.g., 1b).

Our contribution is threefold.

– **Semantic Parsing.** We propose a framework for mapping SIDPs to OWL DL axioms which combines an automatically derived lexicon, a small hand-written grammar, a parser and a surface realiser. This framework is modular, robust and reversible. It is modular in that, different lexicons or grammars may be plugged to meet the requirements of the semantic application being considered. For instance, the lexicon (which relates words and concepts) could be built using a concept extraction tool, i.e. a text mining tool that extracts concepts from text (e.g., [2]). And the grammar could be replaced by a grammar describing the syntax of other document styles such as cooking recipes. It is robust in that, in the presence of unknown words, the parser can skip words and deliver a connected (partial) parse. And, it is reversible in that the same grammar and lexicon can be used both for parsing and for generation.
– **Quality Assessment.** We provide a method for assessing the system output. A chief difficulty when mapping text to semantic representations is the lack of accepted criteria for assessing the correctness of the semantic representations derived by the system. We tackle this issue by exploiting techniques

from both natural language generation and automated reasoning. We evaluate our approach on a dataset of 960 System Installation Design Principles and report results on coverage (proportion of SIDP parsed) and quality assessment (BLEU score for re-generated sentences, statistics on syntactic well-formedness). In particular, we show that the system provides a DL formula for 97.50% of the input SIDPs and that the DL formulae assigned to full parses are very likely to be correct in 96% of the cases.

– **Ontology Enrichment.** We show that the output of our semantic parser can be used to enrich an existing ontology.

2 Related Work

[21] propose a method for converting natural language definitions to complex DL axioms by first, parsing definitions and second, applying ad hoc transformation rules to parse trees. One main difference with our approach is that we use a generic framework for semantic parsing instead. This allows for a more modular and principled approach (for instance, different semantics could be experimented with). This permits the exploitation of well-understood, highly optimised parsing algorithms (we use a standard CKY algorithms extended to increase robustness to unknown words). And this supports a mapping between natural language and logical formulae that is both direct and reversible – by contrast, [21]'s two step approach (dependency parsing followed by the application of transformation rules) makes it more difficult to identify sources of errors as re-generation cannot be used to "visualise" the natural language content of the derived semantic formulae.

More recently, [17] proposed a neural semantic parser which derives DL formulae from natural language definitions. One important difference with our approach is that the data they are training on has been synthesised. In particular, the input text has been authored using ACE (Attempto Controlled English [13]) and grammar based generation. In contrast, we work on human authored text.

Our work also differs from approaches such as [1,20] which support the authoring of complex DL axioms. In these approaches, DL axioms are authored in an interactive fashion using controlled languages. The main difference with our work is that our approach does not require the use of a control language. Instead, we provide a framework for automatically parsing uncontrolled natural language as can be found for instance, in industrial documentation.

Finally, there has been much work on extracting knowledge from dictionaries [3,18] but as mentioned in the introduction, in these approaches, the knowledge derived is restricted to the lower layer of the "ontology-learning layer cake" [3] and thus focuses mostly on extracting concepts and relations rather than complex axioms.

3 Approach Overview

A central bottleneck when developing semantic parsers and text generators is the lack of parallel corpus aligning text and semantic representations. Such corpora are not generally available and are costly and difficult to build (humans find it difficult to associate a text with a logical formula representing its meaning). Moreover, as noted in [22], even if such corpora can be built, their logical coverage is often restricted and so any semantic parser trained on them will fail to analyse logical structures missing from the training data.

We therefore explore an alternative approach in which we combine a symbolic grammar-based approach with an automatically acquired lexicon and a robust parsing algorithm which can skip unknown words to produce connected, possibly partial, parses. The approach also integrates a surface realiser which given a DL formula can produce a text expressing the meaning of that formula. Figure 1 outlines our approach showing the interaction of various components. The lexicon maps verbs and noun phrases to complex and simple concepts respectively. The grammar provides a declarative specification of how text relates to meaning (as represented by OWL DL formulae). The parser and the generator exploit the grammar and the lexicon to map natural language to OWL DL formulae (semantic parsing) and OWL DL formulae to natural language (generation and more specifically, surface realisation), respectively.

In the following sections, we describe each of these components in detail. Section 4 summarises the results obtained when processing 960 SIDPs from Airbus to enrich an existing ontology developed by Airbus engineers.

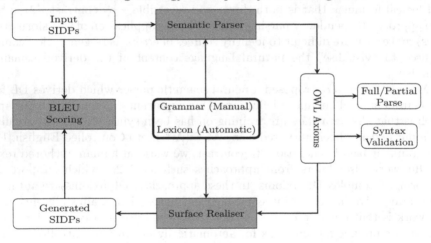

Fig. 1. Parsing and generation of airbus SIDPs.

3.1 Grammar

The grammar provides a declarative specification of the relation between natural language phrases and Description Logic formulae. We use a Feature-Based Lexicalised Tree Adjoining Grammar augmented with a unification-based flat semantics (FB-LTAG, [7]). We start by introducing FB-LTAG. We then define the semantic representation language it integrates and its mapping to OWL. Finally, we show how lexical and grammatical knowledge can be dissociated thereby allowing for increased genericity.

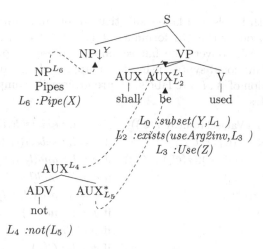

Fig. 2. Example FB-LTAG with unification-based semantics. The variables decorating the tree nodes (e.g., X) abbreviate feature structures of the form $[idx : X]$ where X is a unification variable.

FB-LTAG. Figure 2 shows an illustrating FB-LTAG. In essence[1], an FB-LTAG with unification semantics consists of a set of (trees, semantics) pairs where tree nodes may be labelled with non recursive feature structures (sets of feature-value pairs where values can be constants or unification variables) and semantics may contain unification variables shared with variables occurring in the corresponding tree. During parsing, trees are combined using the grammar operations (adjunction and substitution) and unification is applied to both the feature structures in the tree and the literals in the semantic representation. The semantics of a derived tree is the union of the semantics of the trees contributing to its derivation modulo unification. For instance, given the sentence *Pipes shall not be used*, the combination of the three trees shown in Fig. 2 will yield the derived tree and the semantics shown in Fig. 3. That is, the grammar assigns to sentence (2), a flat semantic formula which is equivalent to the DL formula (2a) whose interpretation (b)[2] can be glossed as (2c) or more simply, (2d).

(2) Pipes shall not be used
    a. $Pipe \sqsubseteq \neg \exists useArg2^-.(Use)$
    b. $\{x \mid x \in Pipe\} \subseteq D^I \setminus \{y \mid (x, y) \in useArg2 \wedge x \in Use\}$
    c. *Pipes are not in the set of things that are the arg2 participant of a Use event*
    d. *Pipes are not things that are used*

Semantic Representation Language. In the grammar, the semantic representation language used is a flattened version of description logic where subformulae

---

[1] See [7] for a more precise definition of the FB-LTAG framework.
[2] D^I is the domain of interpretation.

are associated with labels and labels substituted for subformulae. For instance, the flattened version of the DL formula $C1 \sqsubseteq \exists R.C2$ is $l_0:subset(l_1,l_2)$, $l_1:C1$, $l_2:exists(l_3)$, $l_3:C2$. We convert the flat semantic representations output by the parser to OWL functional syntax using the mapping shown below where $\tau(X)$ is the DL conversion of X, l_i are labels, C_i are arbitrarily complex DL concepts, and R are DL roles.

$$\tau(\phi) = \begin{cases} \text{ObjectSomeValuesFrom}(:R\ \tau(C)) & \text{if } \phi = l_i : exists(R,l_j)\ \ l_j : C \\ \text{SubClassOf}(\tau(C_1)\ \tau(C_2)) & \text{if } \phi = l_i : subset(l_j,l_k)\ \ l_j : C_1\ \ l_k : C2 \\ \text{ObjectIntersectionOf}(\tau(C_1)\ \tau(C_2)) & \text{if } \phi = l_i : and(l_j,l_k)\ \ l_j : C_1\ \ l_k : C_2 \\ (\tau(C1) \sqcap \tau(C2)) & \text{if } \phi = l_i : and(l_j,l_k)\ \ l_j : C1\ \ l_k : C2 \\ (\tau(C1) \sqcup \tau(C2)) & \text{if } \phi = l_i : or(l_j,l_k)\ \ l_j : C1\ \ l_k : C2 \\ not(\tau(C)) & \text{if } \phi = l_i : not(l_j)\ \ l_j : C \\ R^- & \text{if } \phi = Rinv \\ C & \text{if } \phi = l_i : C(x) \end{cases}$$

Further examples of the DL translations assigned by our grammar to natural language sentences are shown in Table 1. Semantically, the grammar makes use of the following DL constructs: $\top$ (the most general concept), disjunction, conjunction, negation, role inverse, universal and existential restrictions. Syntactically, the grammar covers simple and complex SIDPs (i.e., SIDP consisting of more than one clause). Note that temporal or spatial relations such as *after* and *near* are not given any special semantics. They are simply DL roles i.e., binary relations. Also numerical restrictions have not been modelled yet and should be added for a finer grained semantics of nominal phrases (E.g., *at least 3 cables*).

Dissociating Grammar and Lexicon. Figure 2 shows an FB-LTAG in which trees and semantic representations are *lexicalised* in that each tree is associated with

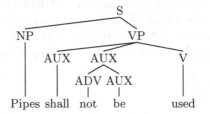

$L_6:Pipe(X)\ L_0 :subset(L_6,L_4)\ L_4:not(L_5)$
$L_5:exists(useArg2inv,L_3)\ L_3:Use(Z)$

$Pipe \sqsubseteq \neg\exists useArg2^-.(Use)$

Fig. 3. Derived tree. The flat semantics representation produced by the grammar is equivalent to the description logic formula shown.

lexical items and with a specific semantic predicate. In practice though, grammar and lexicons are kept separate and the grammar contains trees and semantic schemas which are instantiated during parsing or generation using a lexicon. Figure 4 shows an illustrating example with a lexical entry on the left and the corresponding grammar unit on the right. During generation/parsing, the semantic literals listed in the lexicon (here, *Use* and *useArg2inv*) are used to instantiate the variables (here, *A2* and *Rel*) in the semantic schema (here, L_0:*subset(X,L_1)* L_2:*exists(A2,L_3)* L_3:*Rel(Y)*). Similarly, the Anchor value (*used*) is used to label the terminal node marked with the anchor sign (◇) and each coanchor is used to label the terminal node with corresponding name. Thus, the strings *shall* and *be* will be used to label the terminal nodes $V1$ and $V2$ respectively.

Semantics:
Rel = Use
A2 = useArg2inv
Tree: nx0V
Anchor: *used*
Coanchor: V1 → *shall/V*
Coanchor: V2 → *be/V*

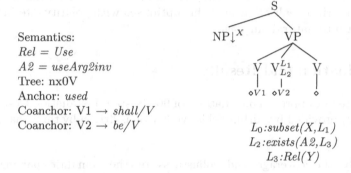

Fig. 4. Example lexical entry and grammar unit

Importantly, this separation between grammar and lexicon supports modularity in that e.g., different lexicons and/or grammars could be plugged into the system.

In essence, the lexicon provides a mapping between natural language phrases and (simple or complex) DL concepts. For the work presented here, we built the lexicon in a rather ad hoc fashion by applying regular expressions and a customised NP chunker[3] to extract verbal and nominal lexical entries from SIDPs. This lexicon could instead be built in a more principled way e.g., by using tools for automatically identifying and extracting concepts and relations (cf. e.g., [4–6,19]).

Similarly, to parse text whose syntax and semantics differs from those of SIDPs, another grammar could be used and mapping SIDPs to a semantic representation language other than OWL DL could be done by simply modifying the semantic component of the grammar.

3.2 Semantic Parser and Surface Realiser

Given a grammar G, a lexicon L and an input sentence S, the *semantic parser* derives from the input sentence S, the parse tree and the description logic formula

---
[3] We use the NLTK regular expression chunker.

associated by G and L to S. For a given sentence, the automatically extracted lexicon can produce a very large number of derivations. The parser uses a CKY algorithm [9] augmented with a simple heuristic to prune the initial search space (lexical entries whose co-anchors are not present in the input string are not selected) and a robustness mechanism for skipping unknown words, i.e. words present in the input that are absent from the lexicon.

Conversely, the *surface realiser* takes as input a grammar G, a lexicon L and a DL formula ϕ and outputs a sentence S associated by G and L to ϕ. The differences with the semantic parser are twofold. First, grammar trees are selected based on their associated semantics (rather than their associated lexical items for parsing). Second, tree combinations are not constrained by word order (during parsing, only trees whose yield matches the linear order of the input string are tried out for combination). We use the GenI surface realiser[4], a tabular bottom-up surface realisation algorithm optimised with polarity filtering[5] to map DL formulae to natural language.

4   Evaluation and Results

We evaluate our approach on a dataset of 960 System Installation Design Principles (SIDP) provided by Airbus. This evaluation is driven by three main research objectives:

- to study the coverage and robustness of the semantic parsing module (Sect. 4.1)
- to assess the correctness of the derived DL formulae (Sect. 4.2)
- to analyse the impact of our semantic parsing on the ontology learning task (Sect. 4.3).

In average, each SIDP sentence consists of 19.88 tokens (min: 5, max: 87). While their syntax is relatively simple (as illustrated in Example 1, an SIDP usually consists of a main clause which may be complemented with a subordinate clause expressing a condition), SIDPs have a complex compositional semantics resulting from the interaction of word order, logical operators, sentence structure and functor arity. Table 1 shows some of the semantic patterns that need to be derived. To capture this semantic variability, we manually specify a grammar consisting of 52 trees. As mentioned above, the lexicon (10 781 lexical entries) associating word and terms (sequences of tokens) to grammar units is constructed automatically by applying regular expressions and NP chunking to the input SIDPs.

---

[4] See [8] for a more detailed definition of the GenI Surface Realiser.

[5] Polarity filtering filters the initial search space by eliminating all combinations of grammar trees which cannot possibly lead to a successful derivation either because it can be calculated that a given tree will not be able to combine with the other trees (a resource will not be used) or, conversely, that some tree(s) are missing to yield a syntactically valid sentence (a resource will be missing).

Table 1. Text and meaning variations

Logical Operators	
Only S shall be used by O	$\neg S \sqsubseteq \neg \exists useA2^-.(Use \sqcap \exists by.O)$
S should be used by <u>all</u> O	$O \sqsubseteq \exists by^-.(Use \sqcap \exists useA2.S)$
S shall <u>not</u> be used by O	$S \sqsubseteq \neg \exists useA2^-.(Use \sqcap \exists by.O)$
Word Order	
S shall be used by O <u>only</u>	$S \sqsubseteq \neg \exists useA2^-.(Use \sqcap \exists by.\neg O)$
<u>All</u> S shall be used by O	$S \sqsubseteq \exists useA2^-.(Use \sqcap \exists by.O)$
Arity	
S shall be used	$S \sqsubseteq \exists useA2^-.(Use)$
S shall be used by O	$S \sqsubseteq \exists useA2^-.(Use \sqcap \exists by.O)$
S shall be used by O on PO	$S \sqsubseteq \exists useA2^-.(Use \sqcap \exists by.O \sqcap \exists on.PO)$
Sentence Structure	
S shall be used by O <u>before</u> entering connections	$(Use \sqcap \exists useA2.S \sqcap \exists by.O) \sqsubseteq$
	$\exists before.(Enter \sqcap \exists enterA2.Connections)$
Modifiers	
S shall be used <u>directly</u> by O	$S \sqsubseteq \exists useA2^-.(Use \sqcap \exists directly.(\exists by.O))$
S shall be used by O <u>between</u> C and D	$S \sqsubseteq \exists useA2^-.(Use \sqcap \exists useA3.(O \sqcap$
	$\exists betweenA1^-.(Between \sqcap$
	$\exists betweenA2.C \sqcap \exists betweenA3.D)))$

4.1 Mapping SIDPs to Complex Axioms

Using the grammar, the lexicon and the parser described in the preceding sections, we obtain the results shown in Table 2. Recall that simple SIDPs are rules consisting of a single clause (1a) while complex SIDPs are rules including a condition, usually expressed by a subordinate or an infinitival clause (1b). Table 2 shows that most (97.50%) SIDPs can be parsed. Manual examination of the results reveals that the few cases where parsing fails (2.50 %) are mainly due to missing lexical entries resulting in a syntactically incomplete parse tree (typically, a verbal argument is missing because the corresponding lexical entry is missing).

Table 2. Parsing results (Coverage)

	Full parse	Partial parse	Failure	Total & Ratio
Simple SIDP	155	290	11	456 (47.50 %)
Complex SIDP	48	443	13	504 (52.50 %)
Total & Ratio	203 (21.15 %)	733 (76.35 %)	24 (2.50 %)	960 (100 %)

Table 2 also shows that a high proportion (76.35%) of the parses are partial parses that is, parses where some of the input words are ignored. Partial parses may be more or less partial though. A partial parse may simply ignore (skip over) a single word or it may ignore a whole subordinate clause. It is thus important to have some criteria for evaluating the correctness of the semantic representations derived by the parser. We show how this issue can be addressed in the following section.

4.2 Assessing Correctness

One key difficulty when using semantic parsing for ontology enrichment is that there is no known metrics for automatically checking the correctness of the semantic representations derived by the parser.

Checking Syntactic Well-Formedness. It is possible however to check the *well-formedness* of the semantic representations. If the semantic representation derived by the parser fails to convert to a well-formed description logic formula, we know that either an incorrect semantics has been assigned to some lexical entry or there is an inconsistency in the output of the syntactic component. To impose this well-formedness check, we first convert the flat semantic representations output by the parser to OWL functional syntax as explained in Sect. 3.1. We then input the resulting OWL formulae to the OWL Functional Syntax parser provided by OWL API 4.1.3 [10] and check that they are subclass axioms (because SIDP are rules which, in our modelling, translate to DL axioms of the form $C_1 \sqsubseteq C_2$). The results are as follows (Table 3).

Table 3. Well-formedness results

Full parses		Partial parses		All parses	
well-formed	ill-formed	well-formed	ill-formed	well-formed	ill-formed
203	0	695	38	898	38
100%	0%	94.8%	5.2%	96%	4%

These results show that most parses produce a subclass axiom for both full and partial parses. While this does not guarantee that the derived semantics correctly captures the meaning of the input SIDP, the well-formedness check allows us to quickly identify parses which are definitely incorrect. These are few (4% of all parses) and closer examination of the data reveals that these are mostly due to syntactically complex SIDP such as (3) which are insufficiently covered by the grammar and/or the lexicon.

(3) Pipes shall be defined by considering red zones for repair in order to allow the removal of the channel without having to modify the rest of the setup or to use specific procedures and tools.[6]

In sum, the well-formedness check allows us to quickly identify cases where semantic parsing yields a definitely incorrect semantic representation.

Comparing Input and Re-Generated Text. We can further evaluate the correctness of full and partial parses by exploiting the fact that the grammar is declarative and can therefore be used both for parsing and for generation. Given an input SIDP S with derived semantics ϕ, we input ϕ to an existing FB-LTAG surface realiser (namely, GenI [8]) and we compare the sentence generated by this surface realiser to the initial input S. To measure the degree to which the re-generated sentence resembles the input sentence, we use BLEU [16], a precision metric standardly used in Natural Language Processing (in particular, in machine translation) for assessing the similarity between two sentences. By re-generating from partial parse semantics and comparing the resulting text with the input using a sentence similarity metrics, we can get a more precise assessment of the quality of the semantic parser output – a low *BLEU score* suggests that indeed the partial parse is very partial and that the derived semantics is likely incorrect while a high one will point to examples where e.g., a single word has been skipped. Table 4 shows the results.

Table 4. Measuring the similarity between input and re-generated sentences. Low: BLEU $\leq$ 32%, Medium: 33% $\geq$ BLEU $\leq$ 66%, High: BLEU $\geq$ 67%

		Low	Medium	High	Total (Ratio)
F-Parse	S-SIDP	0	0	155	155 (16.55%)
	C-SIDP	0	0	48	48 (5.12%)
P-Parse	S-SIDP	105	122	63	290 (30.98%)
	C-SIDP	296	102	45	443 (47.32%)
Total (Ratio)		401 (42.84%)	224 (23.93%)	311 (33.22%)	936 (100%)

For full parses, the BLEU scores are high thereby indicating that re-generated sentences are either identical or very similar to the input SIDP and suggesting that the derived DL formulae adequately capture the meaning of the input SIDP (since regenerating from it produces a sentence identical or highly similar to it). Note that, because the grammar captures some paraphrastic relations (e.g., X *will be used only with* Y / X *will be used with* Y *only*), a re-generated sentence may be different from the input SIDP even if the parse is complete.

---

[6] Because the SIDPs we are working on are confidential data, we modified the lexical items contained in that example. The syntactic structure was preserved however and illustrates the type of syntactic examples our grammar fails to cover.

For partial parses, the proportion of low BLEU scores is noticeably higher for complex SIDPs reflecting the fact that the syntax and semantics of conditions is only partially covered by the grammar. Low BLEU scores for simple SIDPs with partial parses are mainly due to missing or incorrect lexical entries. For instance, our lexicon does not include lexical entries for references to tables and figures. Hence sentence (4a) yields the partial parse covering the words in (4b).

(4) a. En6049 split conduit shall be attached to open backshell as defined in figure below.

b. En6049 split conduit shall be attached to open backshell.

4.3 Ontology Enrichment

Given a DL formula produced by the semantic parser, we go on to enrich an existing ontology from Airbus with that formula. The Airbus ontology describes plane components and encompasses about 650 classes, 1300 individuals, 200 object and data properties, and more than 7400 various logical axioms. To enrich the ontology, we consider the following subtasks:

1. Identifying classes and properties contained in the DL formula which already exist in the ontology;
2. Enriching the ontology with classes and properties contained in the derived DL formulae and which do not exist in the ontology;
3. Checking for consistency and for unsatisfiable classes when adding the full DL formulae derived through parsing.

Once the ontology enrichment module[7] receives a new formula along with the lists of classes and properties it contains, the following steps are performed (see Fig. 5).

Classes and Properties. First, we check if classes presented in the formula exist in the ontology. As classes do not have IRIs in formulae after the semantic parsing, an IRI must be prepended to it. Along with that, we normalise a string underlying the class name: it is set to the lower-case letters, and punctuation signs are removed. Then, by a complete enumeration of all the IRIs existing in the ontology, we try to establish a match between a new class and existing classes. If an exact match was found, we consider that the class used in the axiom is already present in the ontology and should therefore not be added to it.

If the class was not found in the ontology, we resort to lemmatisation. We use the Stanford CoreNLP Toolkit [15] to lemmatise and POS-tag each token making up the class names. We also remove determiners such as *the, a* which might be present in the class names produced by the parser. For the latter, we make use of the morphological tag DT with which tokens were labelled during the POS-tagging phase. For instance, *the green pipes* becomes a class with the name *Green_Pipe*. Once the class name derived by parsing has been lemmatised,

---

[7] For interaction with the ontology, we use OWL API 4.1.3 and HermiT 1.3.8.

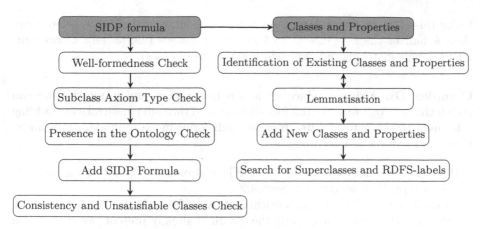

Fig. 5. Ontology enrichment component.

we proceed to compare it to the set of existing classes in the ontology. If a match is detected, we mark the class as an existing one. Otherwise, the class is listed as a new class, and it is added to the ontology using the base IRI, the class name being its lemmatised form.

Table 5. Ontology enrichment statistics

New classes	935
Existing classes	89
New object properties	84
Existing object properties	0
superclasses found	498
RDFS-label matches found	7
new added SIDP formulae	798 (85.3%)
rejected SIDP formulae due to syntax errors	38 (4.0%)
rejected SIDP formulae due to redundancy	91 (9.7%)
rejected SIDP formulae due to inconsistency	9 (1.0%)

We also apply the matching procedure used for class names to the list of properties contained in the DL formulae produced by parsing.

Statistics of the class and property identification are shown in the first section of Table 5. The number of existing classes (89 cases) we found is relatively small compared to the amount of new classes (935 cases). To better link those new classes with the existing concepts, we searched (i) for RDFS-labels which bear the same name as the derived class and (ii) for possible superclasses. We hypothesise that a class C is a superclass of another class C_i if C contains a substring of C_i.

Using this strategy with superclasses was successful: we managed to relate more than a half of the new classes to their super classes (498 cases). Conversely, RDFS-labels matching did not yield many links (7 cases).

Complex DL Axioms. Once we added new classes and properties, we can enrich the ontology with the full formulae derived through parsing. Before adding a formula to the ontology, we check for well-formedness, redundancy, inconsistency and un-satisfiability:

- If the formula does not follow the OWL syntax and it is not of the subclass axiom type, it is rejected (cf. Sect. 4.2).
- At each step of the ontology enrichment, we refresh the list of axioms which are currently in the ontology. If the axiom is already present, we do not add it.
- If the axiom was successfully added, we perform the ontology consistency check and verify that the ontology does not have unsatisfiable classes.

The results of the ontology enrichment procedure are presented in the third section of Table 5. In total, 14.7% of formulae were rejected due to various reasons. Some of the formulae (4.0%) were not well-formed (see Sect. 4.2), some of them were already present in the ontology (9.7%), others were rejected as they led to inconsistency or unsatisfiable classes (1.0%). Mainly the formula rejection occurred with partial parses. Commonly, only the main clause of an SIDP rule was parsed, leaving out the subordinate part. In such a way, different rules (S *shall be used when Y, S shall be used if Z*) were either reduced to the same partial parse (*S shall be used*), or, in case of a negative sentence, to two parses with contrasting meaning (*S shall be used, S shall not be used*). The former accounts for redundant formulae, the latter for inconsistent formulae.

Since we check the consistency and search for unsatisfiable classes after each addition of a new SIDP formula, it enables us to identify inconsistent formulae on the fly. Thus, a strong point of our approach is the immediate detection of incompatible SIDP rules without having to compute justifications, which is a challenging issue for real-world ontologies (see [11], for example).

5   Conclusion

In industries, governmental services and European organizations, system requirements are key information which need to be queried and checked for consistency. Mostly however, these are listed in documents and disconnected from formal models. Being able to translate text to model would permit e.g., to check consistency (analysis). Conversely, being able to translate models to text (generation) would allow technical authors to update the documentation to reflect changes in the model. In this paper, we showed how techniques from natural language processing could be used to provide a reversible framework for mapping natural language requirements to description logic and vice versa. We further argued that such a reversible framework helps address one key issue for semantic parsing namely, how to detect incorrect output.

In future work, we plan to explore how our reversible framework could be used to build a text-semantics corpus using data extension and recombination as suggested in [12]; and how to use this corpus to train a more robust, more generic semantic parser for system requirements.

Acknowledgments. This research has been partially supported by the ITEA 2 Eureka Cluster Programme under the ModelWriter project (Grant 13028).

References

1. Bernstein, A., Kaufmann, E., Göhring, A., Kiefer, C.: Querying ontologies: A controlled english interface for end-users. In: Gil, Y., Motta, E., Benjamins, V.R., Musen, M.A. (eds.) ISWC 2005. LNCS, vol. 3729, pp. 112–126. Springer, Heidelberg (2005). doi:10.1007/11574620_11
2. Bozsak, E., et al.: KAON — Towards a large scale semantic web. In: Bauknecht, K., Tjoa, A.M., Quirchmayr, G. (eds.) EC-Web 2002. LNCS, vol. 2455, pp. 304–313. Springer, Heidelberg (2002). doi:10.1007/3-540-45705-4_32
3. Buitelaar, P., Cimiano, P., Magnini, B.: Ontology Learning From Text: Methods, Evaluation and Applications. IOS press, Amsterdam (2005)
4. Charlet, J., Szulman, S., Pierra, G., Nadah, N., Téguiak, H.V., Aussenac-Gilles, N., Nazarenko, A.: Dafoe: A multimodel and multimethod platform for building domain ontologies. 2e Journées Francophones sur les Ontologies. ACM, Lyon, France (2008)
5. Cimiano, P., Völker, J.: Text2Onto: A framework for ontology learning and data-driven change discovery. In: Montoyo, A., Muñoz, R., Métais, E. (eds.) NLDB 2005. LNCS, vol. 3513, pp. 227–238. Springer, Heidelberg (2005). doi:10.1007/11428817_21
6. Frantzi, K.T., Ananiadou, S., Tsujii, J.: The *C-value/NC-value* method of automatic recognition for multi-word terms. In: Nikolaou, C., Stephanidis, C. (eds.) ECDL 1998. LNCS, vol. 1513, pp. 585–604. Springer, Heidelberg (1998). doi:10.1007/3-540-49653-X_35
7. Gardent, C., Kallmeyer, L.: Semantic construction in feature-based TAG. In: Proceedings of EACL. pp. 123–130. Association for Computational Linguistics (2003)
8. Gardent, C., Kow, E.: A symbolic approach to near-deterministic surface realisation using tree adjoining grammar. In: Proceedings of ACL, pp. 328–335. ACL (2007)
9. Harrison, M.A.: Introduction to Formal Language Theory. Addison-Wesley Longman Publishing Co., Inc., Boston (1978)
10. Horridge, M., Bechhofer, S.: The OWL API: A java API for OWL ontologies. Semant. Web **2**(1), 11–21 (2011)
11. Horridge, M., Parsia, B., Sattler, U.: Explaining inconsistencies in OWL ontologies. In: Godo, L., Pugliese, A. (eds.) SUM 2009. LNCS (LNAI), vol. 5785, pp. 124–137. Springer, Heidelberg (2009). doi:10.1007/978-3-642-04388-8_11
12. Jia, R., Liang, P.: Data recombination for neural semantic parsing (2016). arXiv preprint. arXiv:1606.03622
13. Kaljurand, K., Fuchs, N.E.: Verbalizing OWL in attempto controlled english. In: OWLED, vol. 258 (2007)

14. Maedche, A., Staab, S.: Semi-automatic engineering of ontologies from text. In: Proceedings of the 12th International Conference on Software Engineering and Knowledge Engineering, pp. 231–239. Citeseer (2000)

15. Manning, C.D., Surdeanu, M., Bauer, J., Finkel, J.R., Bethard, S., McClosky, D.: The stanford CoreNLP natural language processing toolkit. In: ACL (System Demonstrations), pp. 55–60 (2014)

16. Papineni, K., Roukos, S., Ward, T., Zhu, W.J.: BLEU: A method for automatic evaluation of machine translation. In: Proceedings of ACL, pp. 311–318. ACL (2002)

17. Petrucci, G., Ghidini, C., Rospocher, M.: Ontology learning in the deep. In: Blomqvist, E., Ciancarini, P., Poggi, F., Vitali, F. (eds.) EKAW 2016. LNCS (LNAI), vol. 10024, pp. 480–495. Springer, Cham (2016). doi:10.1007/978-3-319-49004-5_31

18. Ruiz-Casado, M., Alfonseca, E., Castells, P.: Automatic extraction of semantic relationships for wordnet by means of pattern learning from wikipedia. In: Montoyo, A., Muñoz, R., Métais, E. (eds.) NLDB 2005. LNCS, vol. 3513, pp. 67–79. Springer, Heidelberg (2005). doi:10.1007/11428817_7

19. Szulman, S., Aussenac-Gilles, N., Charlet, J., Nazarenko, A., Sardet, E., Teguiak, H.: DAFOE: A platform for building ontologies from texts. In: EKAW (2010)

20. Tablan, V., Polajnar, T., Cunningham, H., Bontcheva, K.: User-friendly ontology authoring using a controlled language. In: Proceedings of LREC (2006)

21. Völker, J., Hitzler, P., Cimiano, P.: Acquisition of OWL DL axioms from lexical resources. In: Franconi, E., Kifer, M., May, W. (eds.) ESWC 2007. LNCS, vol. 4519, pp. 670–685. Springer, Heidelberg (2007). doi:10.1007/978-3-540-72667-8_47

22. Wang, Y., Berant, J., Liang, P.: Building a semantic parser overnight. In: Proceedings of ACL 2015, pp. 1332–1342 (2015)

23. Zouaq, A., Nkambou, R.: Building domain ontologies from text for educational purposes. IEEE Trans. Learn. Technol. 1(1), 49–62 (2008)

Harnessing Diversity in Crowds and Machines for Better NER Performance

Oana Inel[(✉)] and Lora Aroyo

Vrije Universiteit Amsterdam, Amsterdam, The Netherlands
{oana.inel,lora.aroyo}@vu.nl

Abstract. Over the last years, information extraction tools have gained a great popularity and brought significant performance improvement in extracting meaning from structured or unstructured data. For example, named entity recognition (NER) tools identify types such as people, organizations or places in text. However, despite their high F1 performance, NER tools are still prone to brittleness due to their highly specialized and constrained input and training data. Thus, each tool is able to extract only a subset of the named entities (NE) mentioned in a given text. In order to improve *NE Coverage*, we propose a hybrid approach, where we first aggregate the output of various NER tools and then validate and extend it through crowdsourcing. The results from our experiments show that this approach performs significantly better than the individual state-of-the-art tools (including existing tools that integrate individual outputs already). Furthermore, we show that the crowd is quite effective in *(1)* identifying mistakes, inconsistencies and ambiguities in currently used ground truth, as well as in *(2)* a promising approach to gather ground truth annotations for NER that capture a multitude of opinions.

Keywords: Crowdsourcing · Disagreement · Diversity · Perspectives · Opinions · Named entity extraction · Named entity typing · Hybrid machine-crowd workflow · Crowdsourcing ground truth

1 Introduction

Named entity recognition (NER) is a powerful information extraction (IE) technique for identifying named entities (NEs) such as people, places, organizations, events and, to some extent, numerical values or time periods. Nowadays, there is an abundance of off-the-shelf NER tools [1]. When compared however, their output significantly varies in terms of: *(1)* the existence of an entity, *(2)* the entity surface form (*i.e.*, entity span) and the entity type, *(3)* the knowledge base used for disambiguation, or *(4)* the confidence scores given for an entity. This makes it difficult to choose the best NER tool as they all seem to have a partially good and partially not so good performance.

Even though some NER tools have reached human-like performance, they are still highly dependent on the input type and ground truth (gold standard) [2]. For example, a NER tool trained on particular input types or entity types performs

© Springer International Publishing AG 2017
E. Blomqvist et al. (Eds.): ESWC 2017, Part I, LNCS 10249, pp. 289–304, 2017.
DOI: 10.1007/978-3-319-58068-5_18

well only on similar data. Research [3] has shown that NER tools trained on English news articles achieve an accuracy of 85%–90% on this type of data, but perform very poor on short, ill-formed texts, such as microblogs. Similarly, the quality and the size of the ground truth could bias NER towards a particular annotation perspective. In [2], the authors show that many NER tools have very low performance when dealing with the diversity of miscellaneous entity types.

The mainstream approach of gathering ground truth for NER is still by means of experts, who typically follow over-specified annotation guidelines to increase the *inter-annotator agreement* between experts. Such guidelines are known to be prone to denying the intrinsic language ambiguity and its multitude of perspectives and interpretations [4]. Thus, ground truth datasets might not always be 'gold' or 'true' in terms of capturing the real text meaning and interpretation diversity. More recent work has been focusing on capturing the *inter-annotator disagreement* [5] to provide a new type of ground truth, where language ambiguity is considered. As crowdsourcing has proven to be a reliable method for IE in various domains, *e.g.*, news [6], tweets [7] and more specialized tasks such as entity typing [8], there is an increasing number of hybrid NER approaches that combine machine and crowd-based IE [9]. However, they all suffer from the same *'lack of understanding of ambiguity'* as the traditional NER tools.

This paper aims to answer the following research question: *can we leverage the machine and crowd diversity to improve NER performance?*. We propose a hybrid multi-machine-crowd approach where state-of-the-art NER tools are combined and their aggregated output is validated and improved through crowdsourcing. We perform the crowdsourcing experiments in the context of the CrowdTruth approach [10] and methodology [5] that aims at capturing the inherent language ambiguity by means of disagreement. Thus, we argue that:

- **H1:** Aggregating the output of NER tools by harnessing the inter-tool disagreement (Multi-NER) performs better than the individual NERs (Single-NER); we experiment with existing Wikipedia sentence-based ground truth datasets and show that disagreement among NER improves their performance; we also show that the crowd is effective in spotting NER mistakes;
- **H2:** NER performance is influenced by the rigidness of the ground truth;
- **H3:** Crowdsourced ground truth by harnessing inter-annotator disagreement produces diversity in annotations and thus, improves the aggregated output of NER tools; we show that the crowd can produce a better ground truth.

The main contributions of this paper, besides addressing the above mentioned results, are: *(1)* a hybrid workflow for NER that improves significantly current NER by means of disagreement-based aggregation and crowdsourcing; *(2)* a method for improving ground truth datasets through fostering disagreement among the machines and crowd; *(3)* a data and NER tool agnostic method to improve the NE coverage, *i.e.*, can be used with any type or any number of NER tools and can be applied on any number and type of entities; *(4)* a disagreement-aware approach that effectively mitigates the issues of NER tools.

The paper is structured as follows. Section 2 introduces the use case and the datasets, while Sect. 3 covers the state of the art. Section 4 contains the comparative analysis of multiple NER tools and their aggregated output. Section 5 outlines the crowdsourcing experimental setup. Further, Sect. 6 presents the crowdsourcing results, while Sect. 7 discusses the results. Finally, Sect. 8 concludes and introduces the future work.

2 Use Case and Datasets

We performed named entity extraction with five state-of-the-art NER tools: NERD-ML[1], TextRazor[2], THD[3], DBpediaSpotlight[4], and SemiTags[5]. NERD-ML [11] is an extension of NERD [12], a NER tools unifier, that uses machine learning for improved results. We performed a comparative analysis of *(1)* their performance (output) and *(2)* their combined performance (output), on two ground truth (GT) evaluation datasets used during Task 1 of the Open Knowledge Extraction (OKE) semantic challenge at ESWC in 2015[6] (*OKE*2015) and 2016[7] (*OKE*2016) respectively. Table 1 presents the summary of the datasets: in total, there are 156 Wikipedia sentences with 1007 annotated named entities of types *place*, *person*, *organization* and *role*.

Table 1. Datasets overview

OKE2015			*OKE2016*		
Sentences	Named entities		Sentences	Named entities	
101	Place	120	55	Place	44
	Person	304		Person	105
	Organization	139		Organization	105
	Role	103		Role	86
Total	101	664[a]		55	340

[a]The sum per type is not equal to 664 because 2 entities have 2 distinct types.

3 Related Work

3.1 Open Knowledge Extraction Systems

The systems proposed during the OKE challenges have been evaluated on datasets described in Sect. 2. The ADEL system [13] had the best performance

---

[1] http://nerd.eurecom.fr.

[2] https://www.textrazor.com.

[3] http://ner.vse.cz/thd/.

[4] http://dbpedia-spotlight.github.io/demo/.

[5] http://nlp.vse.cz/SemiTags/.

[6] https://github.com/anuzzolese/oke-challenge.

[7] https://github.com/anuzzolese/oke-challenge-2016.

in 2015, with an F1-score of 0.60, by implementing a hybrid 3-steps approach that combines an off-the-shelf NER model together with POS-tagging, a linking step through DBpedia and Wikipedia, and a pruning step for removing the entities that are out of scope. A second system, FRED [14], had a micro F1-score of 0.34 and a macro F1-score of 0.22. However, the lower performance is due to the fact that the system was used with its default settings, without being adapted for this challenge. Similarly, the third participating system, FOX [15], is an off-the-self system. The system is not able to recognize the type *role*, thus, the F1-score is around 0.49. The enhanced version of ADEL [16] combines different models to improve the entity recognition and entity linking. The system described in [17] applies filtering and merging heuristics on the combined output of NER tools and semantic annotators. It outperforms ADEL with an F1-score above 0.65.

3.2 Crowdsourcing Named Entities

Crowdsourcing proved to be effective in gathering data semantics for various tasks, such as medical relation extraction [18], temporal events ordering [19,20], entity salience [21]. State-of-the-art NER tools have good performance when tested on news articles, but perform very poor on microblogs [3]. Thus, crowdsourcing has been used as an alternative to identify named entities in tweets [7,22]. When dealing with crowdsourced data, the quality plays an important role. Typical solutions for assessing the quality of crowdsourced data are based on the hypothesis [23] that there is only one right answer. However, we operate under the assumption that the disagreement among workers is not noise, but a signal [5,24] of *(i)* input ambiguity, *(ii)* worker quality and *(iii)* task clarity. Therefore, we run our crowdsoucing experiments on the CrowdTruth [10] framework.

3.3 Multi-NER, Hybrid Named Entity Recognition

Harnessing the agreement among NER tools proved to be effective in [25], since entities missed by one NER can be extracted by another NER. Agreement among NER tools is well captured by majority vote systems [26]. However, this could cut off relevant information such as, information supported by only one extractor and cases with more than one solution. When dealing with data on heterogeneous topics and domains, the accuracy of extracting named entities has been shown to increase when NER tools are combined [25,27].

In [9] the need of designing hybrid approaches for NER pipelines is stressed, based on the reliable crowd performance when identifying named entities in tweets. Systems that integrate machines and crowd have been already developed [6,28]. On the one hand, in [6], the authors propose a probabilistic model to choose the most relevant data that needs to be annotated by the crowd, in a hybrid machine-crowd approach. On the other hand, the crowdsourcing component has been integrated as a plugin in the GATE framework [28], but they still assume there is only one correct answer. Hybrid expert-crowd approaches [29] have also been envisioned. The authors optimize in time and cost the process of gathering expert annotations by involving the crowd: the experts mark the named entities, while the crowd provides the type of the entities.

4 Single-NER vs. Multi-NER Comparison

In this section we introduce the **Multi-NER approach**, an approach that combines the output of five state-of-the-art NER tools. The NER tools whose output we combine are mentioned in Sect. 2. On the one hand, by performing a comparative analysis of the five individual NER tools and their combined output (Multi-NER), we aim to validate **H1**. On the other hand, by performing an empirical analysis of the cases where NER tools perform poorly, we aim to identify the factors that influence their performance (**H2**).

4.1 Single-NER vs. Multi-NER - Entity Surface

According to [2], the performance of each state-of-the-art NER differs on a dataset due to the fact that each NER tool uses different training data and different learning algorithms. However, evaluating the disagreement among them [25] proves to be effective in generating better outcomes. First, we compare the five Single-NERs and Multi-NER on the GT in Table 1, by looking at the entity surface. For this analysis we use all the NEs in the GT and all their alternatives, *i.e.*, all the surface forms for each entity in the GT, extracted by any NER tool. Considering this, we measure the following:

- *true positive (TP)*: the NE has the same surface form and the same offsets as the NE in the GT;
- *false positive (FP)*: the NE is only a partial overlap with the NE in the GT;
- *false negative (FN)*: the NEs in the GT that were not extracted by any NER, nor the Multi-NER.

The comparison is shown in Table 2. Overall, there are high differences in terms of the number of TP, FP and FN cases for each state-of-the-art NER, but their performance in F1-score is still very similar. Although it seems that

Table 2. NER evaluation at the level of entity surface

	$OKE2015$						$OKE2016$					
	TP	FP	FN	P	R	F1	TP	FP	FN	P	R	F1
NERD-ML	401	93	263	0.812	0.604	**0.693**	209	37	131	0.85	0.615	**0.713**
SemiTags	366	**37**	298	**0.908**	0.551	0.686	161	**14**	179	**0.92**	0.474	0.625
THD	199	114	465	0.636	0.3	0.407	122	73	218	0.626	0.359	0.456
DBpediaSpotlight	411	234	253	0.637	0.619	0.628	**228**	119	**112**	0.657	**0.671**	0.664
TextRazor	**431**	177	**232**	0.709	**0.65**	0.678	207	105	133	0.663	0.609	0.635
Multi-NER	**555**	403	109	0.579	**0.836**	0.684	299	218	41	0.578	0.879	0.698

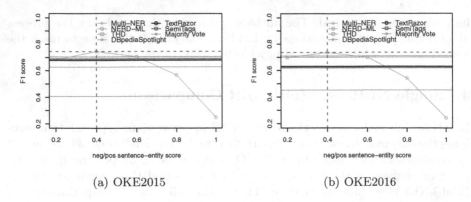

(a) OKE2015   (b) OKE2016

Fig. 1. Annotation quality F1 per negative/positive sentence-entity threshold

NERD-ML performs the best in F1-score across the two datasets, when looking at the exact numbers, we observe that the Multi-NER approach covers a significantly larger pool of entities, *i.e.*, has a significantly higher number of TPs and also a significantly lower number of FNs. However, the combined output of the NER tools also introduces a lot of FPs, but this only slightly decreases its performance.

The reason for the increased number of FPs is due to the high disagreement between the NER tools on the surface form of the entity (*i.e.*, the NER tools do not agree on the exact entity span). On the one side, Multi-NER has a higher recall (with about 30%) on both datasets compared to TextRazor, the tool with the highest recall. This proves that the Multi-NER approach retrieves a higher number of relevant entities. On the other side, the low precision indicates the fact that many entities retrieved are not correct. Thus, our focus should be on improving the precision of the Multi-NER approach, while keeping a high recall.

To show that combining NER output and harnessing the diversity among them is beneficial, we applied the CrowdTruth methodology [10]. First, we introduce a core metric, the *sentence-entity score* which shows the likelihood of an entity to be in the GT based on how many NER tools extracted it. The *sentence-entity score* is equal to the ratio of NER tools that extracted the entity. In Fig. 1a and b we plotted the F1-score values for each NER and the F1-score of the Multi-NER approach for each sentence-entity score threshold. We use the sentence-entity score as a threshold for differentiating between a positive and a negative named entity. At the threshold of 0.4, Multi-NER outperforms the rest of the tools. Using McNemar's test, the results show that the difference in performance between NERD-ML and Multi-NER at its best performing threshold is statistically significant (*OKE2015*: $p < 2.2e^{-16}$, *OKE2016*: $p < 3.247e^{-11}$).

We have also plotted the F1-score for the majority vote approach, a mainstream approach when combining multiple NER tools. In our case, the majority vote includes all the entities that were extracted by at least 3 NER (sentence-entity score >=0.6). The difference of performance is also statistically significant for majority vote vs. Multi-NER ($OKE2015$: $p < 2.2e^{-16}$, $OKE2016$: $p < 7.025e^{-12}$). Overall, the Multi-NER outperforms the state-of-the-art NER tools at a sentence-entity score $>=0.4$, which fosters the idea that disagreement is beneficial, and it also outperforms the majority vote approach.

4.2 Single-NER vs. Multi-NER - Entity Surface and Entity Type

To better understand the results of our Multi-NER approach, we focus on analyzing the cases where the NER tools underperform. Table 3a and b contain the combined NER evaluation at the entity surface based on the entity type. We show how the TP, FP and FN cases from Table 2 are distributed across the types of interest: *person*, *place*, *organization* and *role*. The remaining of the section focuses on analyzing the FN and FP cases.

Table 3. NER evaluation at the level of entity surface and entity type

(a) OKE2015

	TP					FP					FN				
	Place	People	Org	Role	Total	Place	People	Org	Role	Total	Place	People	Org	Role	Total
NERD-ML	90	142	106	65	403	22	21	42	17	102	30	162	33	38	263
SemiTags	100	168	100	0	368	16	2	19	2	39	20	136	39	103	298
THD	62	35	55	49	201	17	17	62	29	125	58	269	84	54	465
DBpedia-Spotlight	99	156	81	77	413	26	62	124	26	238	21	148	58	26	253
TextRazor	110	174	109	40	433	31	14	118	24	187	9	130	30	63	232
Multi-NER	116	219	130	92	558	**54**	**91**	**214**	**66**	425	**4**	**85**	**9**	**11**	**108**

(b) OKE2016

	TP					FP					FN				
	Place	People	Org	Role	Total	Place	People	Org	Role	Total	Place	People	Org	Role	Total
NERD-ML	40	47	71	51	209	1	3	30	6	40	4	58	34	35	131
SemiTags	36	57	67	1	161	5	2	7	1	15	8	48	38	85	179
THD	36	12	33	41	122	3	1	55	14	73	8	93	72	45	218
DBpedia-Spotlight	38	70	56	64	228	5	7	93	14	119	6	35	49	22	112
TextRazor	36	57	83	31	207	15	4	79	12	110	8	48	22	55	133
Multi-NER	44	78	100	77	299	**21**	**13**	**157**	**34**	225	**0**	**27**	**5**	**9**	**41**

4.3 Analysis of False Negative Named Entities

We started with a manual inspection of the FN cases in order to understand which are the NEs that the NER tools fail to identify. Typically, by using the Multi-NER approach, it is natural to have high recall values and lower precision values. However, in both Table 3a and b we see that there are many entities of type *person* that are missed (OKE2015 recall - 0.72 and OKE2016 recall - 0.74). When analyzing in detail, we identify three main problems:

– the NER tools have problems in identifying coreferences, or identifying personal and possessive pronouns as named entities

- in $DS2015$, there were 26/27 such cases
- in $DS2016$, there were 83/85 such cases
- there are errors in the ground truth: in $OKE2015$, "One of the them", which is a clear mistake, is considered a correct named entity
- ambiguous combination of type *role* and *people*, *e.g.*, "Bishop Petronius", "Queen Elizabeth II"; "Bishop Petronius" is a *person*, while "Queen Elizabeth II" is a false entity, because "Queen" - *role* and "Elizabeth II" - *people*.

The type *place* seems to be the one that has the lowest number of FNs: in $OKE2016$ all the entities of type *place* were extracted, while in $OKE2015$ only four cases were missed. Here, we identify one main issue, in all four cases the entity is a concatenation of multiple entities of type *place*. Furthermore, in 2/4 cases the ground truth contains errors - the extracted entity span does not match with the given offsets As a general rule, in the $OKE2015$ dataset, the cases "City, Country" were extracted as a single entity of type *place*. However, the annotation guidelines for $OKE2016$ seemed to have changed, since all such cases were considered two different entities of type *place*. Thus, we observe that there is disagreement across the two ground truth datasets.

For the types *organization* and *role*, the general observation is that there is a high disagreement between the single NER tools and they constantly seem to have a high rate of FN for such entities. However, when looking at the Multi-NER approach, we see that overall, only a few cases were missed which means that at least one NER was able to extract the correct entity span. When looking in depth at the entities of type *organization* that were missed, we see two cases:

- in $OKE2016$ the entities missed were actually common entities in 5/5 cases (*e.g.*, "state", "university", "company");
- in $OKE2015$ the entities missed were not common entities, but the GT:
  - contains errors in 2/9 cases (*e.g.*, "Sheffiel", "The Imperial Cancer Research Fund")
  - contains non-English named-entities in 1/9 cases
  - contained combinations of named and common entities in 4/9 cases (*e.g.*, "Boston Brahmin family", "Geiger's staff")

Since the entities of type *role* are common entities, the main issue of the NER tools is the fact that they extract other span alternatives instead of the one in the ground truth. Furthermore, in $OKE2015$ we had a French entity, while in $OKE2016$, in 5/9 cases the entities were highly ambiguous, such as "membership", "originators". Looking further in the FN cases, we see that there are many ambiguities. For example, in $OKE2015$, we have the word phrase "Italian Jewish", where "Italian" is classified as a *person* and "Jewish" as a *role*. In another example for the same dataset, the word phrase "Hungarian Jews" is classified alone as an *organization*, while, in $OKE2016$ we find the word phrase "Jewish mother", where "Jewish" has no type and "mother" is a *role*. We see such inconsistencies across types as well: "independent school" is an incorrect *organization* type, but "independent contractor" is a correct entity of type *role*.

4.4   Analysis of False Positive Named Entities

We performed a similar manual evaluation on the FPs in order to understand how we can correct the results of the NER tools and improve the precision of the Multi-NER approach. For both datasets the precision of extraction an entity of type *organization* is significantly low (OKE2015 - 0.37, OKE2016 - 0.38). This is due to the large number of FP cases, or in other words, the various alternatives for a single entity. The large majority of entities of type *organization* are combinations of *organization* and *place* (*e.g.*, "University of Rome") or combinations of *people* and *organization* (*e.g.*, "Niels Bohr Institute"). Thus, for each such entity, there are at least 2 more FP alternatives that are extracted.

The next type with many FPs is the type *person*. Here we identify:

– the NER tools usually extract correctly the name of the person, but they also extract partial matches (in OKE2015 we have 86/91 such partial matches, while in OKE2016 we have 11/13 such cases); when checking these cases we observe that the names that contain abbreviations, *e.g.*, "J. Hans D. Jensen", are the most prone to get any possible combination of the names;
– the NER tools extract combinations of *role* and *person*, especially when the *person* is an ethnic group (*e.g.*, "French author", "Canadian citizen");
– in OKE2015 we also find a combination of *place* and *person*, due to the ambiguity of the sentence (*e.g.*, "Turin Rita Levi-Montalcini"), which lacks a comma after the word "Turin".

Similarly, for the FP cases on type *place*, in the majority of the cases we identify partial overlaps with the entity in the ground truth, concatenated or nested locations. Moreover, we find:

– combinations of entities of type *role* and entities of type *place* (1/54 cases in OKE2015 and 3/21 cases in OKE2016)
– combinations of entities of type *organization* and entities of type *place* (5/54 cases in OKE2015 and none in OKE2016)

The type *role* is the most ambiguous, especially because these entities are common entities. The main issue of the NER tools is the precision of extracting such entities. Usually, they tend to extract both the most general word phrase that refers to a *role* (*e.g.*, "professor" instead of "assistant professor"), but also the most specific one (*e.g.*, "first black president" instead of "president").

5   Experimental Setup

The aim of our crowdsourcing experiments is two-fold. On the one hand we want to prove that the crowd is able to correct the mistakes of the NER tools. On the other hand, we want to show that the crowd can identify the ambiguities in the GT, which leads to a better NER pipeline performance and an improved GT.

5.1   Crowdsourcing Experimental Data

Our goal is to decrease both the number of false positive and false negative NEs through gathering a crowd-driven ground truth. To achieve this, we select every entity in the ground truth for which the NER tools provided alternatives. It is important to mention that we do not focus on identifying new entities, but only on correcting the ones that exist. Thus, we have the following two cases:

- *Crowd reduces the number of FP*: For each named entity in the ground truth that has multiple span alternatives we create an entity cluster. We also add the largest span among all the alternatives. For example, 'University of Rome' cluster is composed of: 'University', 'Rome' and 'University of Rome', where all entities have been extracted by at least one NER tool.
- *Crowd reduces the number of FN*: For each named entity in the GT that was not extracted, we create an entity cluster that contains the FN named entity and the alternatives returned by the NERs. Further, we add every other combination of words contained in all the alternatives. This step is necessary because we do not want to introduce bias in the task, *i.e.*, the crowd should see all the possibilities, not only the expected one. For example, the entity 'fellow students' was not extracted by any of the NER tools. Instead, they extracted 'fellow' and 'students'. The entity cluster in this case is composed of: 'fellow students', 'fellow' and 'students'.

5.2   Crowdsourcing Annotation Task

For both cases introduced in Sect. 5.1 we designed the same crowdsourcing task on CrowdFlower[8]. The overview of the task is presented in Fig. 2. The goal of the crowdsourcing task is two-fold: *(i)* identification of valid expressions from a list that refer to a highlighted phrase in yellow (Step 2 in Fig. 2 and *(ii)* selection of the type for each expression in the list, from a predefined set of choices - *place, person, organization, role* and *other* (Step 3 in Fig. 2).

The input for this crowdsourcing task consists of *(i)* a sentence from either $OKE2015$ or $OKE2016$, and *(ii)* a list of expressions that could potentially refer to a named entity. The list of expressions was created using the rules described in Sect. 5.1. In total, we ran 303 such pairs, distributed in 7 crowdsourcing jobs. The settings and the distribution per dataset is shown in Table 4.

5.3   CrowdTruth Metrics

We evaluate the crowdsourced data using the CrowdTruth methodology and metrics [10], by adapting the core CrowdTruth metric, the sentence-relation score [24]. In our case, we measure *(1) crowd sentence-entity score* - the likelihood of a sentence to contain a valid entity expression and *(2) crowd entity-type score* - the likelihood of an expression to refer to the given types. These scores are computed using the cosine similarity measure. To identity the low-quality workers

---

[8] www.crowdflower.com.

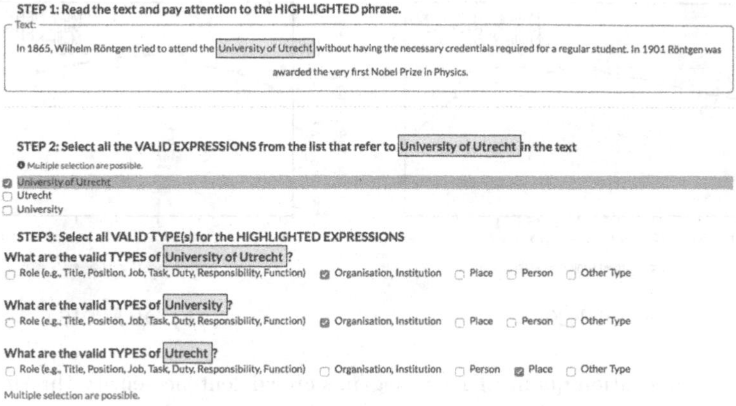

STEP 1: Read the text and pay attention to the HIGHLIGHTED phrase.

Text:

In 1865, Wilhelm Röntgen tried to attend the University of Utrecht without having the necessary credentials required for a regular student. In 1901 Röntgen was awarded the very first Nobel Prize in Physics.

STEP 2: Select all the VALID EXPRESSIONS from the list that refer to University of Utrecht in the text

❶ Multiple selection are possible.

☑ University of Utrecht
☐ Utrecht
☐ University

STEP3: Select all VALID TYPE(s) for the HIGHLIGHTED EXPRESSIONS

What are the valid TYPES of University of Utrecht ?
☐ Role (e.g., Title, Position, Job, Task, Duty, Responsibility, Function)   ☑ Organisation, Institution   ☐ Place   ☐ Person   ☐ Other Type

What are the valid TYPES of University ?
☐ Role (e.g., Title, Position, Job, Task, Duty, Responsibility, Function)   ☑ Organisation, Institution   ☐ Place   ☐ Person   ☐ Other Type

What are the valid TYPES of Utrecht ?
☐ Role (e.g., Title, Position, Job, Task, Duty, Responsibility, Function)   ☐ Organisation, Institution   ☐ Person   ☑ Place   ☐ Other Type

Multiple selection are possible.

Fig. 2. Crowdsourcing annotation task (Color figure online)

Table 4. Experimental Setup for Crowdsourcing Annotations

	Units	Jobs	Judg/Unit	Max Judg/Worker	Worker country	Worker level	Units/Page	Pay/Unit
OKE2015	202	2	15	15	UK, USA, AUS, CAN	3	1	2
OKE2016	101	5						

we apply two CrowdTruth worker metrics [10], the worker-worker agreement and the worker cosine. These measures indicate how much a worker disagrees with the rest of the workers on the units they solved in common and across the entire dataset. Low values for both metrics mean that the workers consistently disagree with the rest of the workers. Their annotations are thus removed.

6 Results

This section presents the crowdsourcing results[9], with focus on analyzing the added value of using the crowd in hybrid Multi-NER pipelines. In short, we gathered 4,545 judgments, from a total of 464 workers. After applying the CrowdTruth metrics, we identified 108 spammers, that contributed to a total of 1,172 low-quality annotations, which were removed from the final data.

We plotted in Figs. 3a and b the F1-score values at each crowd sentence-entity score, as described in Sect. 5.3. When compared with the ground truth, we see that for each crowd sentence-entity score the crowd enhanced Multi-NER (Multi-NER+Crowd) performs much better than the Multi-NER approach. On the $OKE2015$ dataset the crowd performs the best at the crowd entity-score threshold of 0.6 with a F1-score of 0.832, while on $OKE2016$ the crowd has the best performance at a threshold of 0.5, with an F1-score of 0.848. This means

---

[9] http://data.crowdtruth.org/crowdsourcing-ne-goldstandards/.

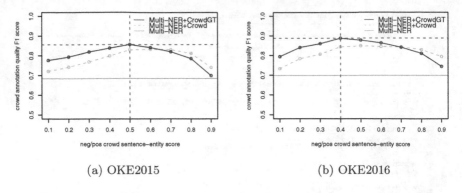

(a) OKE2015   (b) OKE2016

Fig. 3. Annotation quality F1 per neg/pos crowd sentence-entity threshold

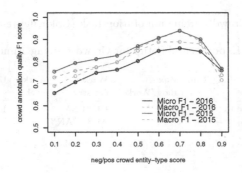

Fig. 4. Annotation quality F1 per neg/pos crowd entity-type threshold

that the crowd can correctly reduce the number of FPs. The difference is also statistically significant for both datasets. Using McNemar's test we get a p-value equal to $6.999e^{-07}$ for $OKE2015$ and p-value 0.01234 for $OKE2016$.

From these graphs, it is natural to assume that the crowd diversity in opinion is indeed not an indication of noise, but signal. In the analysis performed in Sect. 4 we observed that many entities in the ground truth are ambiguous and could have multiple interpretations. Thus, we performed a manual evaluation of the entities in the ground truth and allowed for a richer diversity. When the entities were ambiguous, *"professor"* vs. *"assistant professor"*, *"Bishop Petronius"* vs. *"Bishop"* vs. *"Petronius"*, we included all the possible alternatives. In Fig. 3 this evaluation is indicated by *Multi-NER+CrowdGT*, which stands for enhanced Multi-NER through crowd-driven ground truth gathering. Here we observe that we get even a higher performance ($OKE2015$ - F1 of 0.85 and $OKE2016$ - F1 of 0.88). For both datasets, we see that in this case the best performance threshold is consistently a fraction lower than the one when the crowd is compared with the experts.

We also evaluated the performance of the crowd on the entity types. For this evaluation we considered only the entities in the ground truth that have been used in the crowdsourcing tasks (227 entities in $OKE2015$ and 109 entities in $OKE2016$). Because we deal with multiple classes, in Fig. 4 we plotted the macro F1-score and the micro F1-score based on the crowd weights, *i.e.*, based on the crowd entity-type score. Overall, the crowd is able to capture the correct entity type, as at each threshold all the F1 scores are higher than 0.65, with a maximum performance 0.93 for $OKE2015$ and 0.85 for $OKE2016$.

7   Discussion

Our first hypothesis was that a *Multi-NER approach performs better than a single NER*. As expected, when combining the output of multiple NER tools we increase the number of TP and decrease the number of FN. This observation is in agreement with the fact that in general, single NERs are trained with different data and through different approaches, *i.e.*, entities missed by one NER can be returned by another NER. Furthermore, the conventional belief when dealing with diversity is that *the more instances we have in agreement, the better*. To address this issue and prove the contrary, we follow and apply the CrowdTruth approach, *i.e.*, disagreement is not noise, but signal. In Figs. 1a and b we can see that taking only the entities that have been extracted by many NER tools achieves a lower performance than most of the single NER. In contrast, the more disagreement we allow, the better our Multi-NER performs, which shows that a Multi-NER approach, overall, performs better than any single NER. Interestingly, although NERD-ML seems to overall outperform our approach, when leveraging the NERs diversity, at a 0.4 sentence-entity score threshold, we observe a statistically significant improvement for our method on both datasets.

The NER performance is influenced by the rigidity and the ambiguity of the GT, which can be proved by looking at the FN and FP cases. First, the annotation guidelines of the GT, do not seem to align with the GT used by the NER tools: *(1)* personal and especially possessive pronouns are not considered named entities by NER, in contrast to our GT; *(2)* the GT is inconsistent for the same dataset and across datasets; *(3)* the GT contains ambiguities that are fostered for difficult types such as *role*; *(4)* the GT contains errors. The NER tools tend to extract multiple span alternatives for an entity, while the GT does not allow for multiple perspectives on the entity span. We observe this cuts off meaningful data. Furthermore, many challenge submissions (Sect. 3.1) were off-the-shelf tools, GT-agnostic. The tools performance did not exceed an F1-score of 0.65, which is quite low given that we deal with well-formed English Wikipedia sentences. We argue that the GT ambiguity also impacts their performance.

Overall, the crowd improved the performance of the NER tools. In Fig. 3 it is interesting to see that the best performing threshold for Multi-NER+Crowd is not only a pick, but it is an interval of thresholds (in $OKE2015$ - between 0.5 and 0.7, while in $OKE2016$ - between 0.4 and 0.7). Furthermore, we see that the lower end of the intervals is correlated with the best performing threshold for the crowd-driven ground truth (Multi-NER+CrowdGT). We believe this is an indicator of the fact that entities in that interval are more prone to be ambiguous. Thus, allowing for diversity provides better ground truth. For the type analysis it is interesting to see that the micro F1 and macro F1 differ on each dataset. This behavior is due to the highly unbalanced number of entities in each class for $OKE2016$, where we have 62 entities of type *organization* and only 7 of type *person*. Since in the case of micro averaging larger classes dominate smaller classes, for $OKE2016$ we should consider the macro F1 score as a better indicator. However, for $OKE2015$ the classes are more balanced, so we can give them the same weight, thus, the micro F1 is a better indicator.

8   Conclusion

In this paper we proposed a hybrid Multi-NER crowd-driven approach for improved NER performance. Following the CrowdTruth methodology - *disagreement is not noise but signal*, we showed the added value of leveraging the machines and crowd diversity in a 3-step approach. First, our Multi-NER approach, by considering the data ambiguity, has a significantly higher coverage of entities than Single-NER tools when compared to given ground truth. Furthermore, when leveraging the NERs diversity, we show a significant improvement over state-of-the-art Single-NER on both datasets. Second, through data inspection of the ground truth and the factors that answer for the increased number of false positive and false negative entities, we observed that the NER performance is highly dependent on the ambiguity and inconsistency of such ground truth datasets. Third, our evaluation has shown that the crowd, *by harnessing the inter-annotator disagreement*, is able to correct the mistakes of the NER tools by reducing the total number of false positive cases. Furthermore, the crowd-driven ground truth gathering, that harnesses diversity, perspectives and granularities, proves to be a more reliable way of creating a ground truth when dealing with the natural language ambiguity and the overall task ambiguity.

Although the current performance of the hybrid Multi-NER crowd-driven approach reaches high values, in future work we can focus on reducing the false negative cases related to personal and possessive pronouns. Furthermore, we can optimize the crowdsourcing approach in terms of time and cost by only validating and correcting the named entities with low confidence of being correct.

References

1. Gangemi, A.: A comparison of knowledge extraction tools for the semantic web. In: Cimiano, P., Corcho, O., Presutti, V., Hollink, L., Rudolph, S. (eds.) ESWC 2013. LNCS, vol. 7882, pp. 351–366. Springer, Heidelberg (2013). doi:10.1007/978-3-642-38288-8_24
2. Rizzo, G., van Erp, M., Troncy, R.: Benchmarking the extraction and disambiguation of named entities on the semantic web. In: LREC, pp. 4593–4600 (2014)
3. Derczynski, L., Maynard, D., Rizzo, G., van Erp, M., Gorrell, G., Troncy, R., Petrak, J., Bontcheva, K.: Analysis of named entity recognition and linking for tweets. Inf. Process. Manage. **51**(2), 32–49 (2015)
4. Bayerl, P.S., Paul, K.I.: What determines inter-coder agreement in manual annotations? A meta-analytic investigation. Comput. Linguist. **37**(4), 699–725 (2011)

5. Aroyo, L., Welty, C.: Truth is a lie: CrowdTruth and 7 myths about human computation. AI Mag. **36**(1), 15–24 (2015)
6. Demartini, G., Difallah, D.E., Cudré-Mauroux, P.: ZenCrowd: leveraging probabilistic reasoning and crowdsourcing techniques for large-scale entity linking. In: Proceedings of the 21st International Conference on WWW, pp. 469–478. ACM (2012)
7. Finin, T., Murnane, W., Karandikar, A., Keller, N., Martineau, J., Dredze, M.: Annotating named entities in twitter data with crowdsourcing. In: Proceedings of the NAACL HLT 2010 Workshop on Creating Speech and Language Data with Amazon's Mechanical Turk, pp. 80–88. ACL (2010)
8. Bu, Q., Simperl, E., Zerr, S., Li, Y.: Using microtasks to crowdsource DBpedia entity classification: a study in workflow design. Semant. Web J. (2016)
9. Feyisetan, O., Luczak-Roesch, M., Simperl, E., Tinati, R., Shadbolt, N.: Towards hybrid NER: a study of content and crowdsourcing-related performance factors. In: Gandon, F., Sabou, M., Sack, H., d'Amato, C., Cudré-Mauroux, P., Zimmermann, A. (eds.) ESWC 2015. LNCS, vol. 9088, pp. 525–540. Springer, Cham (2015). doi:10.1007/978-3-319-18818-8_32
10. Inel, O., et al.: CrowdTruth: machine-human computation framework for harnessing disagreement in gathering annotated data. In: Mika, P., et al. (eds.) ISWC 2014. LNCS, vol. 8797, pp. 486–504. Springer, Cham (2014). doi:10.1007/978-3-319-11915-1_31
11. Van Erp, M., Rizzo, G., Troncy, R.: Learning with the web: spotting named entities on the intersection of nerd and machine learning. In: # MSM, pp. 27–30 (2013)
12. Rizzo, G., Troncy, R.: NERD: a framework for unifying named entity recognition and disambiguation extraction tools. In: Proceedings of the Demonstrations at the 13th Conference of the European Chapter of the ACL, pp. 73–76. ACL (2012)
13. Plu, J., Rizzo, G., Troncy, R.: A hybrid approach for entity recognition and linking. In: Gandon, F., Cabrio, E., Stankovic, M., Zimmermann, A. (eds.) SemWebEval 2015. CCIS, vol. 548, pp. 28–39. Springer, Cham (2015). doi:10.1007/978-3-319-25518-7_3
14. Consoli, S., Recupero, D.R.: Using FRED for named entity resolution, linking and typing for knowledge base population. In: Gandon, F., Cabrio, E., Stankovic, M., Zimmermann, A. (eds.) SemWebEval 2015. CCIS, vol. 548, pp. 40–50. Springer, Cham (2015). doi:10.1007/978-3-319-25518-7_4
15. Röder, M., Usbeck, R., Speck, R., Ngomo, A.-C.N.: CETUS – a baseline approach to type extraction. In: Gandon, F., Cabrio, E., Stankovic, M., Zimmermann, A. (eds.) SemWebEval 2015. CCIS, vol. 548, pp. 16–27. Springer, Cham (2015). doi:10.1007/978-3-319-25518-7_2
16. Plu, J., Rizzo, G., Troncy, R.: Enhancing entity linking by combining NER models. In: Sack, H., Dietze, S., Tordai, A., Lange, C. (eds.) SemWebEval 2016. CCIS, vol. 641, pp. 17–32. Springer, Cham (2016). doi:10.1007/978-3-319-46565-4_2
17. Chabchoub, M., Gagnon, M., Zouaq, A.: Collective disambiguation and semantic annotation for entity linking and typing. In: Sack, H., Dietze, S., Tordai, A., Lange, C. (eds.) SemWebEval 2016. CCIS, vol. 641, pp. 33–47. Springer, Cham (2016). doi:10.1007/978-3-319-46565-4_3
18. Dumitrache, A., Aroyo, L., Welty, C.: Achieving expert-level annotation quality with CrowdTruth (2015)
19. Snow, R., O'Connor, B., Jurafsky, D., Ng, A.Y.: Cheap and fast–but is it good? Evaluating non-expert annotations for natural language tasks. In: Proceedings of EMNLP, pp. 254–263. Association for Computational Linguistics (2008)

20. Caselli, T., Sprugnoli, R., Inel, O.: Temporal information annotation: crowd vs. experts. In: LREC (2016)
21. Inel, O., Caselli, T., Aroyo, L.: Crowdsourcing salient information from news and tweets. In: LREC, pp. 3959–3966 (2016)
22. Fromreide, H., Hovy, D., Søgaard, A.: Crowdsourcing and annotating ner for twitter #drift. In: LREC, pp. 2544–2547 (2014)
23. Nowak, S., Rüger, S.: How reliable are annotations via crowdsourcing: a study about inter-annotator agreement for multi-label image annotation. In: Proceedings of the International Conference on Multimedia Information Retrieval. ACM (2010)
24. Aroyo, L., Welty, C.: The three sides of CrowdTruth. J. Hum. Comput. 1, 31–34 (2014)
25. Chen, L., Ortona, S., Orsi, G., Benedikt, M.: Aggregating semantic annotators. Proc. VLDB Endowment 6(13), 1486–1497 (2013)
26. Kozareva, Z., Ferrández, Ó., Montoyo, A., Muñoz, R., Suárez, A., Gómez, J.: Combining data-driven systems for improving named entity recognition. Data Knowl. Eng. 61(3), 449–466 (2007)
27. Hellmann, S., Lehmann, J., Auer, S., Brümmer, M.: Integrating NLP using linked data. In: Alani, H., et al. (eds.) ISWC 2013. LNCS, vol. 8219, pp. 98–113. Springer, Heidelberg (2013). doi:10.1007/978-3-642-41338-4_7
28. Sabou, M., Bontcheva, K., Derczynski, L., Scharl, A.: Corpus annotation through crowdsourcing: towards best practice guidelines. In: LREC, pp. 859–866 (2014)
29. Voyer, R., Nygaard, V., Fitzgerald, W., Copperman, H.: A hybrid model for annotating named entity training corpora. In: Proceedings of LAW IV. ACL (2010)

All that Glitters Is Not Gold – Rule-Based Curation of Reference Datasets for Named Entity Recognition and Entity Linking

Kunal Jha[1], Michael Röder[1(✉)], and Axel-Cyrille Ngonga Ngomo[1,2]

[1] AKSW Research Group, University of Leipzig,
Augustusplatz 10, 04103 Leipzig, Germany
kunal.jha@uni-bonn.de, roeder@informatik.uni-leipzig.de
[2] Data Science Group,
University of Paderborn, Pohlweg 51, 33098 Paderborn, Germany
ngonga@upb.de

Abstract. The evaluation of Named Entity Recognition as well as Entity Linking systems is mostly based on manually created gold standards. However, the current gold standards have three main drawbacks. First, they do not share a common set of rules pertaining to what is to be marked and linked as an entity. Moreover, most of the gold standards have not been checked by other researchers after they were published. Hence, they commonly contain mistakes. Finally, many gold standards lack actuality as in most cases the reference knowledge bases used to link entities are refined over time while the gold standards are typically not updated to the newest version of the reference knowledge base. In this work, we analyze existing gold standards and derive a set of rules for annotating documents for named entity recognition and entity linking. We derive EAGLET, a tool that supports the semi-automatic checking of a gold standard based on these rules. A manual evaluation of EAGLET's results shows that it achieves an accuracy of up to 88% when detecting errors. We apply EAGLET to 13 English gold standards and detect 38,453 errors. An evaluation of 10 tools on a subset of these datasets shows a performance difference of up to 10% micro F-measure on average.

Keywords: Entity recognition · Entity linking · Benchmarks

1 Introduction

The number of information extraction systems has grown significantly over the past few years. This is partly due to the growing need to bridge the text-based document Web and the RDF[1]-based Web of Data. In particular, NER (Named Entity Recognition) frameworks aim to locate named entities in natural language documents while Entity Linking (EL) applications link the recognised entities to a given knowledge base (KB). NER and EL tools are commonly evaluated

---

[1] Resource Description Framework, https://www.w3.org/RDF/.

© Springer International Publishing AG 2017
E. Blomqvist et al. (Eds.): ESWC 2017, Part I, LNCS 10249, pp. 305–320, 2017.
DOI: 10.1007/978-3-319-58068-5_19

using manually created gold standards (e.g., [13]), which are partly embedded in benchmarking frameworks (e.g., [1,20]). While these gold standards have clearly spurred the development of ever better NER and EL systems, they have three main drawbacks: (1) They do not share a common set of rules pertaining to what is to be marked and linked as an entity. (2) Moreover, most of the gold standards have not been checked by other researchers after they have been published and hence commonly contain mistakes. (3) Finally, while in most cases the KB used to link the entities has been refined over time, the gold standards are typically not updated to the newest version of the KB.

We address this drawback of current NER/EL benchmarks through the following contributions: (1) We present a study of existing benchmarks that proposes a unified set of rules for creating NER/EL gold standards. (2) We present a taxonomy of common errors that can be found in the available gold standards that violate the rules. (3) We propose and evaluate EAGLET—a semi-automatic gold standard checking tool that is based on a fully automatic error detection pipeline. (4) We derive improved versions of 3 NER/EL benchmark subsets and quantify the effect of erroneous benchmarks on 10 NER/EL systems.

The rest of this paper is structured as follows. In the subsequent section, we give a brief overview of existing NER/EL gold standards. In Sect. 3, we define a set of rules for the annotation process and identify common annotation errors. EAGLET is described in Sect. 4 and evaluated in Sect. 5 along with state-of-the-art NER/EL tools on improved benchmarks. We conclude the paper with Sect. 6.

2   Related Work

While a large number of publications on new gold standards for the NER/EL tasks are available, only a few describe the process which led to their creation. In the following, we present a non-exhaustive list of English NER/EL benchmarks. *ACE2004* [17] was created using a subset of the ACE co-reference data set which was originally annotated with entities of the types person, organization, facility, location, geo-political entity, vehicle and weapon [3]. The annotations of the subset were linked to Wikipedia articles by Amazon Mechanical Turk workers with an inter-rater agreement of 85% [17]. *AIDA/CoNLL* [6] was created by annotating proper nouns in Reuters newswire articles. People, groups, artifacts and events were linked to the YAGO2 KB if a corresponding entity existed. *AQUAINT* [15] was created based on news articles. The documents were annotated automatically and checked manually. *DBpedia Spotlight*'s [13] evaluation dataset contains 60 natural language sentences from ten different documents with 249 annotated DBpedia entities overall. *IITB* [9] was created based on Web documents gathered from different domains. The authors explicitly state that emerging entities (EEs), i.e., entities that can be found in the text but are not present in the KB [7], should be annotated. *KORE50* [5] is a subset of the larger AIDA dataset. The selection of the KORE 50 dataset followed the objective to be difficult for disambiguation tasks. It contains a large number of first names referring to persons, whose identity needs to be deduced from the given context. However, the authors do not offer a list of the types of entities that have been annotated. *Microposts2014* [19] was created using a set of anonymized twitter messages. Entities have been extracted using the NERD-Framework and linked to DBpedia articles manually by raters. After that, two experts double checked the ratings and managed conflicts. The dataset is separated into two parts—a training and a test dataset. *MSNBC* [2] was created based on news articles. An automatic NER and EL approach has been applied to generate the annotations following which have been checked manually. *OKE* [16] datasets have been created for the Open Knowledge Extraction Challenge 2015. 196 sentences have been annotated manually marking people, organizations, roles and locations.

Recently, van Erp et al. [21] analyzed gold standards and concluded, that the available gold standards are diverse regarding several decisions that their creators have made during the creation process. However, their analysis focused primarily on the entities that have been marked and their characteristics instead of the correctness of gold standard annotations. Ehrmann et al. [4] presented a systematic overview of written and spoken natural language processing resources that can be used for named entity tasks like EL or NER. They pointed out that the quality of these resources is difficult to assess since many gold standards do not have a detailed documentation of the annotation process. In 2015, Ling et al. [11] presented a modular approach for the EL task which is motivated by the same observation as Van Erp et al., i.e., that a common understanding of the task is missing and several different interpretations are possible. The decisions that are made based on these interpretations have a huge impact on the design of a system and the gold standard that is used to benchmark this solution. Regarding the EL task, they list the following 5 major points for discussion:

(P1) It is not defined whether only *named* entities or all resources in the given KB should be linked.

(P2) It is not defined which entity should be chosen if more than one are plausible. The authors motivate this with the example of reoccurring events and different iterations of the same institution, e.g., the different United State Congresses. While these entities can be defined as not distinct to ease the problem, the authors argue that a statement like *"Joe Biden is the Senate President in the 113th United States Congress"* [11] can lead to wrong information if a system extracts Joe Biden as the President of all United States Congresses. On the other hand, they raise the problem that it might not be possible to formulate a statement about an event that will take place in the future since it might not be available in the KB.

(P3) A similar problem is metonymy, i.e., an entity is called not by its own name but by another associated name. A common metonym for a government, e.g., the government of the United States, is the capital in which it is located, e.g., Washington. The authors write that linking to the capital entity as well as to the government entity is possible.

(P4) There is no common set of entity types shared across different gold standards. For example, in some datasets, events are linked as entities while in other datasets, they are not.

(P5) Following the authors, it is not clear whether annotations can overlap. In their example of an U.S. city which is followed by its state— *"Portland, Oregon"*—they argue to annotate the city, the state and both words together since all three markings make sense.

Rehm [18] defined a lifecycle for language resources. Our work can be used in the evaluation and quality control phase during the development of EL/NER gold standards to semi-automatically check the created corpus. Additionally, it can be used after the publication of the gold standards for its maintenance, i.e., to keep the gold standard up to date with new versions of the used KB.

3   Formal Annotation Framework

The creation of a NER/EL gold standard is a difficult task because human annotators commonly have different interpretations of this task as shown by [17]. It is, therefore, important to define a generic set of rules for annotating named entities in natural language text which leaves little if no room for interpretation. An advantage of having such rules is that they can be used to check gold standards automatically. The goal of this section is to present exactly such a set of rules derived from existing benchmarks. Based on the related work described in Sect. 2 we summarize assumptions that we can build upon. Thereafter, we define a set of rules for the preparation of a gold standard followed by a list of errors that we observed in existing gold standards.

3.1   Assumptions

We rely on the following assumptions:

(A1) A single sentence does not need to have a linear structure. However, since state-of-the-art annotation systems do only annotate consecutive words, the gold standards should contain only annotations that can be expressed in this way. The word group *"Barack and Michelle Obama"* contains two persons. To annotate the first person, only the first name of Barack Obama can be annotated and linked to its entity. This assumption has the drawback that in the example *"Mr. and Mrs. Obama"* the word *"Mr."* would have to be linked to Barack Obama.

(A2) The annotation should cover as many consecutive words as possible to represent the entity as precisely as possible. In the word group *"legendary cryptanalyst Alan Turing"* all these words should be part of a single annotation linked to the resource representing Alan Turing. However, this assumption should not be used to annotate whole clauses which will be described as Long Description Error later on.

(A3) Each annotation should be linked to the most precise resource of the KB that is represented by the annotation or it should get a synthetically generated URI if this entity is an EE. Hence, in the example of point P2 described in Sect. 2, *"113th United States Congress"* has to be linked to a resource that represents exactly this 113th congress—not to the resource of the United States Congress in general.

(A4) The annotated string should point to a specific entity. Indirect meanings of a string should not be considered. This assumption is important to make sure that a human annotator does not start to think laterally.

(A5) The decision pertaining to which resources of a knowledge base can be used as entities for linking relies on a given set of entity types T_A. Only those entities that have at least one of the given types should be used for annotation.

3.2   Rule Set

Based on the aforementioned assumptions, we define a set of rules for marking the annotation of entities.

1. Consider each dataset D to be a set of documents and each document d to be an ordered set of words, $d = \{w_1, ..., w_n\}$.

2. Regard every word $w_i \in d$ as a sequence of characters or digits starting either at the beginning of the document or after a white space character and ending either at the end of the document or before a white space or punctuation character.

3. The annotation process relies on the set of entities $E = \{e | \tau(e) \cap T_A \neq \emptyset\}$ where τ is a function that returns the set of types T_e of the entity e and T_A is a given set of types that should be annotated in the corpus. It should be noted that E might contain more entities than the given KB K and that $E \backslash K$ is the set of EEs that can be found in the documents.

4. An annotation $a \in A$ is defined as $a = (S_a, u_a)$, where
   (a) S_a is a maximal sequence of consecutive words, such that
       $S_a = (w_i, w_{i+1}, w_{i+2}, ...)$ and
   (b) u_a is a URI that is used to link the annotated sequence to an entity
       $e = \delta(u_a)$, where δ is the dereferencing function returning the entity that can be identified with the given URI and e is
       i. the most precise entity possible
       ii. that represents a as described in A3.

5. The annotation function $\rho(d, K \cup E, T_A) = A$ creates a set of annotations
   $A = \{a_1, a_2, ..., a_n\}$ that meet the following requirements
   (a) $\delta(u_{a_i}) \in E$,
   (b) $\forall a_i, a_j \in A \ (S_{a_i}, S_{a_j} \subset d) \wedge (S_{a_i} \cap S_{a_j} = \emptyset)$ and
   (c) A has to be complete, i.e., it has to contain all valid annotations that can be found in d.

3.3   Comparison with Related Work

In this section, we compare our rules with the related work—especially with the points raised by Ling et al. [11] described in Sect. 2. Rules 1 and 2 define the structure of a document and the words inside a document. Combined with rule 4a, the possible positions of annotations are defined and the starting or ending of an annotation within a word is prohibited.

Rule 3 solves several issues that are raised in the related work. It answers **P1** by raising the requirement of a predefined entity type set on which the annotation process is based. A definition of the term *named entity* is not needed anymore and the exhaustive linking using all resources of the KB is only a special case in which the set of entity types comprises all types contained in the KB. It also solves **P4** by transforming the need of a common set of entity types that was bound to the unclear term *named entity* into a parameter of the annotation process.

Rule 4b defines the linking step, i.e., the assignment of a URI to an annotated part of the text. With defining e as the most precise entity, the problem of the metonymy described as **P3** is solved, since it becomes clear that *"Washington"* has to be linked to the U.S. government if it is used as its metonym. Note that the last part of the rule *"[...] that represents a directly"* does not object the linking of metonyms but prohibits the linkage of long descriptions which are described in the following section. It also prohibits the linkage of pronouns which aligns with our argumentation that pronouns should not be annotated since this would imply a NER/EL system to include a pronominal coreference resolution—an own, separated field of research that has lead to several solutions for this problem, e.g., the work of Lee et al. [10].

Together with the possible linking of EEs defined in rule 3, Rule 4b solves **P2** as well. In cases in which a statement has to distinguish reoccurring events and different iterations of the same organization, these single events or organizations have to be linked to the most precise entity, i.e., one certain event or iteration. The argument, that events in the future could cause a problem is not valid since based on our rule set, this event would be handled as EE.

Rule 5 defines the annotation function that is based on the other 4 rules. Rule 5b defines annotations as non-overlapping which answers the question raised in **P5**. According to rule 4b, an annotation already contains the most precise link this particular part of the text could have. Adding additional annotations can lead to several problems. First, it would lead to a much larger amount of annotations without adding more information that couldn't be retrieved from the most precise entity, e.g., the fact that `dbr:Portland, _Oregon` is located in `dbr:Oregon`.[2] If this additional information is needed, it should be retrieved using available linked data technologies. Second, it can lead to an unnecessary shift of the focus, since the topic of the example is neither `dbr:Oregon` nor `dbr:Portland` but `dbr:Portland, _Oregon`.

---

[2] Throughout the paper, the prefix `dbr:` stands for http://dbpedia.org/resource/.

3.4 Observations

Having defined these generic rules, we evaluated the human annotated gold standards based on the aforementioned rules and assumptions. The evaluation unveiled various anomalies within the gold standards that we classified into the following categories.

Long Description Error (LDE). The first kind of error stands for annotations of sequences of words which might describe the entity they are linked to but do not contain a surface form of the entity (hence violating rule 4(b)ii).

For example, in Document 2 of Fig. 1, *"a team that won Supporters' Shield in 2014"* is linked to `dbr:Seattle_Sounders_FC` but the marked text is neither equivalent to the surface form of entity nor directly describes the entity.

Positioning Error (PE). The next kind of error lies in marking a portion of a word in a sequence of words as an entity. Given that the rule 4a states that an annotation is only allowed to mark complete words, these errors violate rule 4a and the definition of words in rule 2. In Document 2 of Fig. 1 for example, the *"foot"* in *"football"* is marked as an entity, hence violating the basic definition of the word.

Overlapping Error (OE). The third kind of error involves the presence of two or more annotations that share at least one word, thus violating the rule 5b. In Document 1, *"MLS Cup"* and *"Cup Playoffs"* have been marked over common part of the text *"Cup"*.

Combined Marking (CM). This is a non-trivial tier of errors wherein consecutive word sequences are marked as separate entities while the word sequences, if combined, can be annotated to a more specific entity. These errors are a direct

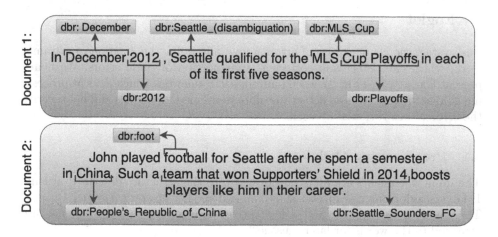

Fig. 1. Example documents.

violation of rule 4(b)i. In Document 1, *"December"* and *"2012"* are two separate consecutive entities which when combined together, *"December 2012"* are more apt in the context, i.e., link to the most precise resource.

URI Error. This error category comprises errors that violate rule 4b and can be separated into the following sub categories.

1. *Outdated URI (OU).* In this category, the entity is linked to an outdated resource which no longer exists in any KB. In Document 2, *"China"* is linked to dbr:People's_Republic_of_China which no longer exists in the KB but instead has to be updated to dbr:China.
2. *Disambiguation URI (DU).* This type of errors involves linking an entity to a non-precise resource page (disambiguation page) instead of a single resource. In Document 1, the entity *Seattle* is annotated with the URI dbr:Seattle_disambiguation, which is a disambiguation page that points to the City dbr:Seattle and the team dbr:Seattle_Sounders_FC. In this case, the team is the correct resource and should also be chosen as annotation.
3. *Invalid URI (IU).* This error category comprises annotations with no valid URI, e.g., an empty URI.

Inconsistent Marking (IM). This category comprises entities that were marked in at least one of the documents but whose occurrence in other documents of the same dataset is not marked as such. For example, the entity *Seattle* was marked in Document 1 but is left out in Document 2.

Missing Entity. The final categorisations of anomalies is a further extension of EM error. This comprises the presence of entities which satisfy the type conditions of the gold standard but were not been marked. This tier of error falls under the dataset completion and violates Rule 5c.

4 Eaglet

The systematic classification of errors above allows for the creation of a framework, which can detect and correct a large portion of these errors. We hence present EAGLET (see Fig. 2), a semi-automatic framework which aims at processing gold standards so as to detect the aforementioned anomalies and help rectify the errors, once reviewed by users.

4.1 Preprocessing Module

The input documents are first transformed into the structure described in Rule 1, i.e., each document is tokenized into an ordered set of words $d = \{w_1, ..., w_n\}$. Thereafter, a POS-tagger and a lemmatizer are applied and the lemmas are attached to the words for later reusage.[3]

---

[3] We used the Stanford CoreNLP suite [12].

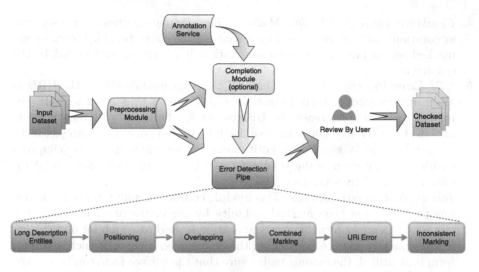

Fig. 2. Eaglet's overview

4.2 Completion Module

The completion module is an optional component. It uses publicly available annotation services to derive a list of entity annotations that are missing in the original dataset. These additional annotations support the work of a user that wants to make sure that the dataset is complete as defined in Rule 5c. However, since state-of-the-art annotation systems are not perfect [20], this module is based on a majority vote, i.e., the majority of the annotation systems have to contain an annotation inside their result list before it is added to the document.

For this module, we relied on the open-source project GERBIL that enables the usage of up to 13 different annotation systems [14].

4.3 Error Detection Pipeline

The error detection pipeline is the primary component of the tool. It tries to identify as many errors as possible in an automatic way based on the rules defined above. Every error type is handled by an own independent module enabling a particular configuration of the pipeline. Annotations that are identified as faulty are marked.

1. *Long Description Detection Module*: This module checks for the Long Description Error by searching for a relative clause inside an annotation.
2. *Wrong Positioning Detection Module*: This module searches for Positioning Errors by searching for mismatches between the start and ending positions of single annotations and the start and ends of words (see Rule 2).
3. *Overlapping Entity Detection Module*: This module checks for entity markings within each document whose positions are intersecting.

4. *Combined Tagging Detection Module*: This module searches for consecutive annotations that are separated by a white space character. Such entities are marked and a larger, combined annotation is generated and added to the document.

5. *URI Error Detection Module*: The URI checking module checks the URIs of all entities regarding their format. If a URI points to a reference KB, the module tries to dereference the URI to check whether (a) the entity exists and (b) the URI does not point to a disambiguation page. For example, if the given KB is the Wikipedia[4] or entities can be directly mapped to Wikipedia entities the module uses the Wikipedia API to determine whether the URI is outdated and derives the new URI.

6. *Inconsistent Marking Module*: This module collects all annotations in the corpus that have not been marked as faulty by one of the other modules. The lemmatized surface form of every annotation is used to search for occurrences of the entity inside the documents that have been missed. If such a surface form is identified, the module makes sure that the surface form can be marked following Rule 2 and that no annotation intersects with the identified occurrence before inserting a new annotation. Since these newly added annotations might be incorrect, e.g., because a URI that is linked to a word in one document does not need to fit to the same word in a different document, they are marked as added by the pipeline and should be checked by the user in the review module.

4.4 Review Module

The list of markings computed by the modules above is sent to the review module allowing the user to review the proposed changes in the dataset. The user interface of our tool allows every user to check each of the documents in the gold standards manually. Users can accept, modify or reject the suggestions of the tool as well as add new entities that have been missed by the completion module.

If a user adds a new entity annotation to a document, it is added to the completion module that processes the remaining documents again, searching for this new entity. This reprocessing aims at reducing the amount of entities the user has to add manually.

5 Evaluation

Our evaluation had three goals: First, we wanted to quantify the number of errors found in existing reference datasets.Secondly, we also wanted to know the accuracy with which EAGLET can detect errors. Finally, we aimed to quantify how much these errors in datasets influences the observed performance of NER/EL tools. We hence evaluated our approach within three different experiments.

---

[4] http://wikipedia.org.

5.1   Experiment I

In our first experiment, we ran EAGLET on the 13 datasets available in the GERBIL evaluation platform at the time of writing. The results are presented in Table 1 as a percentage of the total number of annotations (except for EM) found in each of the reference datasets. Our results show that errors of type PE, CM and URI errors occur often (e.g., up to 36% of CM errors in the IITB dataset) in all the datasets while the numbers for LDE are comparatively lower. OE were found only in DBpediaSpotlight, MSNBC, N3-Reuters-128, OKE2015 Task1 and IITB. We present absolute figures in case of Inconsistent Markings as it, unlike the other errors, involves adding entities to the list of existing annotations. Up to 9904 IM errors are found in a single dataset (IITB).

Table 1. Dataset features and amount of errors. (Abbreviations: $|D|$ = number of documents, $|A|$ = number of annotations, LDE = Long Description Error, PE = Positioning Error, OE = Overlapping Error, CM = Combined Marking, OU = Outdated URI, DU = Disambiguation URI, IU = Invalid URI, IM = Inconsistent Marking (in absolute numbers).)

Systems	Size		Percentage							IM
	$|D|$	$|A|$	LDE	PE	OE	CM	OU	DU	IU	
ACE2004	57	306	0.0	0.3	0.0	0.0	4.6	1.0	23.9	466
AQUAINT	1,393	34,929	<0.1	<0.1	0.0	4.0	2.0	0.2	12.2	6,357
AIDA/CoNLL-Compl	50	727	0.0	0.0	0.0	8.4	10.3	1.4	5.8	586
DBpediaSpotlight	58	330	0.3	3.9	3.6	20.0	6.7	0.3	0.0	11
IITB	104	18,308	<0.1	1.8	0.3	36.0	4.5	7.7	<0.1	9,904
KORE50	50	144	0.0	3.4	0.0	0.0	11.1	0.0	0.0	3
Microposts2014-Test	1,055	1,256	0.2	2.1	0.0	5.8	3.2	0.3	0.4	698
Microposts2014-Train	2,340	3,822	0.2	2.3	0.0	6.6	2.8	<0.1	0.3	2,614
MSNBC	20	755	0.0	2.5	0.5	1.1	16.7	0.9	12.8	70
N3-RSS-500	500	1,000	0.0	0.1	0.0	0.0	2.4	0.0	0.1	193
N3-Reuters-128	128	880	0.3	0.8	0.2	1.6	4.1	1.5	0.9	111
OKE2015 Task1 eval	101	664	0.0	0.0	0.0	10.5	0.5	0.5	0.0	37
OKE2015 Task1 g.s.s	96	338	0.0	0.0	1.2	6.8	2.4	0.0	0.0	52

Table 2. Results of the manual evaluation and the interrater agreement per task in brackets.

Dataset (subset)	Accuracy	Missed entities
ACE2004	0.88 (0.89)	391 (0.81)
AIDA/CoNLL	0.80 (0.93)	71 (0.78)
OKE2015	0.79 (0.98)	14 (0.90)

5.2 Experiment II

To evaluate the accuracy of EAGLET, we analysed a subset of the results of Experiment I manually. As pointed out in Sect. 2, only 4 datasets—ACE2004, AIDA/CoNLL and both OKE2015—come with a definition of the set of entity types that have been used for the annotation process. We randomly chose 25 documents from the ACE dataset, 25 documents from the AIDA/CoNLL dataset and 30 documents from the OKE evaluation dataset. Two researchers checked these documents independently, i.e., they evaluated the errors identified by the error detection pipeline. If at least one of them marked the pipelines decision for an annotation as wrong, the annotation was counted as error. Additionally, the two human annotators searched for entities that should have been marked according to the given type set but have been missed by the original gold standard creators and the error detection pipeline. Table 2 shows the accuracy of the error detection pipeline, the number of missed entities and the inter-rater agreement as F1-measure [8].

The automatic checking of the error detection pipeline was able to classify 79–88% of the annotations correctly. Especially the identification of URI errors worked well with an accuracy of 94%. The performance of the Combined Tagging Detection Module showed some minor flaws. For example, the name of the reporter of a news article directly followed by a city name, e.g., "Steve Pagani VIENNA" (AIDA/CoNLL dataset), was marked as two annotations that should be merged. The module should also be extended to deal with locations that are followed by the state in which they are located, e.g., "Grosse Pointe Park, Mich". These annotations should be merged to fit the rule set and represent the entity, e.g., dbr:Grosse_Pointe_Park,_Michigan.

An important insight revealed by our evaluation is the large number of missed entities in current gold standards—especially for the ACE2004 dataset. The checked subset of the gold standard contained 190 annotations. The Inconsistent Marking Module added 14 correct annotations while the reviewers identified 195 additional annotations. 6 of the 25 documents did not contain any annotations at all in the original gold standard. *Not all that glitters is gold* and our results unveil that the ACE2004 gold standard is not really fit to be used for evaluating NER and EL systems.

We used the annotations added by the reviewers to evaluate the completion module. The module used the ten annotation systems listed in Fig. 4. An annotation was counted as suggested if at least 5 systems marked it. It generated suggestions for 74%, 92% and 57% of the missing entities of the ACE, AIDA and OKE subsets, respectively. Note that these entities would have been considered mistakes as they did not exist in the reference data, clearly pointing towards the need for improved benchmarks.

5.3 Experiment III

The last experiment aimed to quantify the influence of the gold standard quality on the evaluation results of annotation systems. We used GERBIL to benchmark

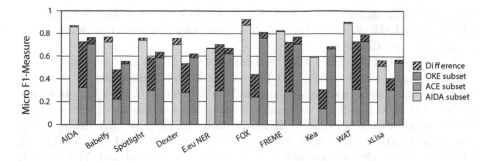

Fig. 3. NER Benchmark result differences of annotation systems on the original and corrected datasets.

10 annotation systems based on two versions of each of the three dataset subsets selected in Experiment II. The first version contained the original annotations while the second version was created based on the manual corrections of the output of the error detection pipeline. The annotation systems were tested using an A2KB (annotation to knowledge base) setting [20], i.e., the annotation systems received plain text, searched for named entities (NER) and linked them to the KB of the dataset (EL). Figure 4 shows the F1-scores for the original datasets as well as the difference to the F1-scores for the corrected datasets.[5] Nearly all annotation systems[6] achieved a higher F1-score on the corrected subsets when compared with the original subsets. On average, the systems' F1-score increased by 16.4% for the ACE, 2.3% for the AIDA and was 1.5% higher for the OKE subset. The high influence of the gold standard quality on the benchmarking results can perhaps be seen most clearly in the ACE subset. While the *xLisa*

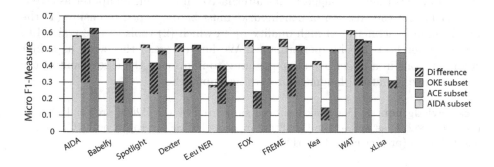

Fig. 4. NER and EL Benchmark result differences of annotation systems on the original and corrected datasets.

---

[5] The complete result table can be found at http://w3id.org/gerbil/experiment?id=201609290008.

[6] The F1-scores of Entityclassifier.eu NER and xLisa for the corrected OKE subset were 1.7% and 0.2% points lower than for the original subset.

annotator has a higher score than *DBpedia Spotlight*, *Dexter*, *Entityclassifier.eu NER* and *FREME NER* on the original subset, its performance is clearly lower on the corrected datasets.

To exclude the possibility that the results of the A2KB task were merely due to the EL step, we also computed the results of the frameworks on the NER substask. Our results (see Fig. 3) show that the corrections have a high influence on the NER task as well. On average, the annotator performance increased with the correction of the ACE and OKE subsets by 29.3% and 4.4%, respectively. The highest enhancement with 43.7% and 6.4% was achieved by the *FREME NER* annotator. The average difference between the original and the corrected AIDA subset was 0%. While the performance of *Dexter* and *FOX* increased by 5.4% and 5.1% the F1-score of *xLisa* and *Babelfy* decreased by 4.8% and 4.6%.

6   Conclusion

We derived a simple set of rules from common practice for benchmark creation. These rules were encoded into the benchmark curation tool EAGLET. A manual evaluation of EAGLET's results suggests it is a reliable tool for improving the quality of gold standards and thus improving the correctness of evaluation results for NER/EL tools. Within our evaluation of existing benchmarks, we were able to automatically detect a significant amount of errors in a large number of corpora. The evaluation of the performance of systems on these datasets and the variation in their performance clearly underlines the importance of having gold standards which really achieve gold standard quality, i.e., which are free of errors. While we have noticed a move towards benchmarking platforms for NER and EL over the last years [1,20], our results suggest the need for a move towards automatic benchmark checking frameworks, the first of which we provide herewith. However, they also suggest that alternative (if possible computer-assisted) approaches for the creation of benchmarks must be developed to ensure (1) the provision of benchmarks of high quality upon which (2) tools can be trained to achieve their best-possible performance. We hence regard this work as a first stepping stone in a larger agenda pertaining to improving the assessment of the performance of natural language processing approaches.

Acknowledgments. This work has been supported by the H2020 project HOBBIT (GA no. 688227) as well as the EuroStars projects DIESEL (project no. 01QE1512C) and QAMEL (project no. 01QE1549C).

References

1. Cornolti, M., Ferragina, P., Ciaramita, M.: A framework for benchmarking entity-annotation systems. In: Proceedings of the 22nd International Conference on World Wide Web (WWW 2013), pp. 249–260, New York, NY, USA. ACM (2013)
2. Cucerzan, S.: Large-scale named entity disambiguation based on wikipedia data. In: EMNLP-CoNLL, pp. 708–716 (2007)

3. Doddington, G., Mitchell, A., Przybocki, M., Ramshaw, L., Strassel, S., Weischedel, R.: Automatic content extraction (ACE) program - task definitions and performance measures. In: Proceedings of the 4th International Conference on Language Resources and Evaluation (2004)
4. Ehrmann, M., Nouvel, D., Rosset, S.: Named entity resources - overview and outlook. In: Calzolari, N., Choukri, K., Declerck, T., Goggi, S., Grobelnik, M., Maegaard, B., Mariani, J., Mazo, H., Moreno, A., Odijk, J., Piperidis, S., (eds.) Proceedings of the Tenth International Conference on Language Resources and Evaluation (LREC 2016), Paris, France. European Language Resources Association (ELRA), May 2016
5. Hoffart, J., Seufert, S., Nguyen, D.B., Theobald, M., Weikum, G.: KORE: keyphrase overlap relatedness for entity disambiguation. In: Proceedings of CIKM (2012)
6. Hoffart, J., Yosef, M.A., Bordino, I., Fürstenau, H., Pinkal, M., Spaniol, M., Taneva, B., Thater, S., Wiegand, M., Weikum, G.: Robust disambiguation of named entities in text. In: Proceedings of EMNLP 2011, 27–31, pp. 782–792, Stroudsburg, PA. ACL, July 2011
7. Hoffart, J., Altun, Y., Weikum, G.: Discovering emerging entities with ambiguous names. In: Proceedings of the 23rd WWW, pp. 385–396. ACM (2014)
8. Rothschild, S.: Agreement, the f-measure, and reliability in information retrieval. J. Am. Med. Inf. Assoc. **12**(3), 296–298 (2005)
9. Kulkarni, S., Singh, A., Ramakrishnan, G., Chakrabarti, S.: Collective annotation of wikipedia entities in web text. In: Proceedings of the 15th ACM SIGKDD, pp. 457–466. ACM (2009)
10. Lee, H., Peirsman, Y., Chang, A., Chambers, N., Surdeanu, M., Jurafsky, D.: Stanford's multi-pass sieve coreference resolution system at the CoNLL-2011 shared task. In: Conference on Natural Language Learning (CoNLL) Shared Task (2011)
11. Ling, X., Singh, S., Weld, D.S.: Design challenges for entity linking. Trans. Assoc. Comput. Linguist. **3**, 315–328 (2015)
12. Manning, C.D., Surdeanu, M., Bauer, J., Finkel, J., Bethard, S.J., McClosky, D.: The Stanford CoreNLP natural language processing toolkit. In: Association for Computational Linguistics (ACL) System Demonstrations, pp. 55–60 (2014)
13. Mendes, P.N., Jakob, M., García-Silva, A., Bizer, C.: Dbpedia spotlight: shedding light on the web of documents. In: Proceedings of the 7th International Conference on Semantic Systems, pp. 1–8. ACM (2011)
14. Michael, R., Usbeck, R., Ngomo, A.-C.N.: Techreport for GERBIL 1.2.2 - V1. Technical report, Leipzig University (2016)
15. Milne, D., Witten, I.H.: Learning to link with wikipedia. In: 17th ACM CIKM, pp. 509–518 (2008)
16. Nuzzolese, A.G., Gentile, A.L., Presutti, V., Gangemi, A., Garigliotti, D., Navigli, R.: Open knowledge extraction challenge. In: Gandon, F., Cabrio, E., Stankovic, M., Zimmermann, A. (eds.) SemWebEval 2015. CCIS, vol. 548, pp. 3–15. Springer, Cham (2015). doi:10.1007/978-3-319-25518-7_1
17. Ratinov, L., Roth, D., Downey, D., Anderson, M.: Local and global algorithms for disambiguation to wikipedia. In: Proceedings of the 49th Annual Meeting of the Association for Computational Linguistics: Human Language Technologies, pp. 1375–1384. ACL (2011)

18. Rehm, G.: The language resource life cycle: towards a generic model for creating, maintaining, using and distributing language resources. In: Calzolari, N., Choukri, K., Declerck, T., Goggi, S., Grobelnik, M., Maegaard, B., Mariani, J., Mazo, H., Moreno, A., Odijk, J., Piperidis, S., (eds.) Proceedings of the Tenth International Conference on Language Resources and Evaluation (LREC 2016), Paris, France, May 2016. European Language Resources Association (ELRA) (2016)

19. Rowe, M., Stankovic, M., Dadzie, A.-S., (eds.): Making Sense of Microposts (#Microposts2014) In: Proceedings of 4th Workshop on Making Sense of Microposts (#Microposts2014): Big Things Come in Small Packages, Seoul, Korea, 7 April 2014

20. Usbeck, R., Röder, M., Ngomo, A.-C.N., Baron, C., Both, A., Brümmer, M., Ceccarelli, D., Cornolti, M., Cherix, D., Eickmann, B., Ferragina, P., Lemke, C., Moro, A., Navigli, R., Piccinno, F., Rizzo, G., Sack, H., Speck, R., Troncy, R., Waitelonis, J., Wesemann, L.: GERBIL - general entity annotation benchmark framework. In: 24th WWW Conference (2015)

21. van Erp, M., Mendes, P., Paulheim, H., Ilievski, F., Plu, J., Rizzo, G., Waitelonis, J.: Evaluating entity linking: an analysis of current benchmark datasets and a roadmap for doing a better job. In: LREC 2016 (2016)

Semantic Annotation of Data Processing Pipelines in Scientific Publications

Sepideh Mesbah[✉], Kyriakos Fragkeskos, Christoph Lofi, Alessandro Bozzon, and Geert-Jan Houben

Delft University of Technology, Mekelweg 4, 2628 CD Delft, The Netherlands
{s.mesbah,k.fragkeskos,c.lofi,a.bozzon,g.j.p.m.houben}@tudelft.nl

Abstract. Data processing pipelines are a core object of interest for data scientist and practitioners operating in a variety of data-related application domains. To effectively capitalise on the experience gained in the creation and adoption of such pipelines, the need arises for mechanisms able to capture knowledge about datasets of interest, data processing methods designed to achieve a given goal, and the performance achieved when applying such methods to the considered datasets. However, due to its distributed and often unstructured nature, this knowledge is not easily accessible. In this paper, we use (scientific) publications as source of knowledge about Data Processing Pipelines. We describe a method designed to classify sentences according to the nature of the contained information (i.e. scientific objective, dataset, method, software, result), and to extract relevant named entities. The extracted information is then semantically annotated and published as linked data in open knowledge repositories according to the DMS ontology for data processing metadata. To demonstrate the effectiveness and performance of our approach, we present the results of a quantitative and qualitative analysis performed on four different conference series.

1 Introduction

Data is now at the centre of almost all fields of technology and science. Data processing workflows (or "pipelines") facilitate the creation, integration, enrichment, and analysis (at scale) of heterogeneous data, thus often opening the field for before unseen innovation. It comes with little surprise that the scientific community is devoting an increasing amount of attention to the design and testing of data processing pipelines, and to their application and validation to big, and open, data sources.

In scientific publications, scientists and practitioners *share* and *seek* information about the properties and limitations of (1) data sources; and (2) of data processing methods (e.g. algorithms) and their implementations. For instance, a researcher in the field of urban planning could be interested in *discovering state of the art methods for point of interest recommendation (e.g. matrix factorisation) that have been applied to geo-located social media data (e.g. Twitter) with good accuracy results.*

© Springer International Publishing AG 2017
E. Blomqvist et al. (Eds.): ESWC 2017, Part I, LNCS 10249, pp. 321–336, 2017.
DOI: 10.1007/978-3-319-58068-5_20

A system able to answer the query above requires access to a structured representation of the knowledge contained in one or more scientific publication repositories. For instance, it should be possible to *access* and *relate* information about: (1) the objective of a given scientific work; (2) the datasets employed in the work; (3) the methods (i.e. algorithms) and tools (e.g. software) developed or used to process such datasets; and (4) the obtained results.

Our vision is to offer support for semantically rich queries focusing on different aspects of data processing pipelines (e.g. methods, datasets, goals). The availability of a semantically rich, interlinked, and machine readable descriptions (metadata) of such knowledge could provide great benefits in terms of retrieval quality, but also for analysing and understanding trends and developments.

Manually inspecting and annotating papers for metadata creation is a nontrivial and time-consuming activity that clearly does not scale with the increasing amount of published work. Alas, scientific publications are also difficult to process in an automated fashion. They are characterised by structural, linguistic, and semantic features that are different from non-scientific publications (e.g. blogs). In this context, general-purpose text mining and semantic annotation techniques might not be suitable analysis and tools. As a consequence, there is a clear need for methodologies and tools for the extraction and semantic representation of scientific knowledge. Recent work focused on methods devoted to the automatic creation of semantic annotations for text snippets, with respect to either structural [8,10,11], argumentative [12,14], or functional [4,6,7] components of a scientific work. However, to the best of our knowledge, there has been no work yet focusing on extracting metadata focusing on properties of data processing pipelines. Therefore, in this paper, we provide the following contributions:

- A novel approach for the classification of text related to data processing pipelines from scientific publications, and for the extraction of named entities. The approach combines distant supervision learning on rhetorical mentions with named entity recognition and disambiguation.

Our system automatically classifies sentences and named entities into five categories (objectives, datasets, methods, software, results). Sentence classification attains an average accuracy of 0.80 and average F-score 0.59.

- A quantitative and qualitative evaluation of the implementation of our approach, performed on a corpus of 3,926 papers published in 4 different conference series in the domain of Semantic Web (ESWC), Social Media Analytics (ICWSM), Web (WWW), and Databases (VLDB).

We provide evidence of the amount and quality of information on data processing pipelines that could be extracted, and we show examples of information needs that can now be satisfied thanks to the availability of a richer semantic annotation of publications' text.

– The annotations resulting from the evaluation are published in an RDF repository, available for query.[1] We employ the DMS [17] ontology to encode properties related to the objectives, datasets, methods, software, and results described in a scientific publication, and then represent them as RDF graphs.

The remainder of the paper is organised as follows: Sect. 3 introduces the DMS ontology; Sect. 4 describes the data processing pipelines knowledge extraction workflow; Sect. 5 reports the results of the evaluations; Sect. 2 describes related work. Finally, Sect. 6 presents our conclusions.

2   Related Work

In the last few years there has been a growing interest in the open and linked publication of metadata related to scientific publications. There are now several ontologies devoted to the description of scholarly information (e.g. SWRC,[2] BIBO,[3] DMS [17]). The Semantic Dog Food [2] and the RKBExplorer [3] are examples of projects devoted to the publication of "shallow" meta data about conferences, papers, presentations, people, and research areas. A large portion of such shallow metadata is already explicitly given by the authors as part of the final document, such as references, author names, keywords, etc. Still, the extraction of that metadata from a layouted document is complex, requiring specialized methods [19] being able to cope with the large variety of layouts or styles used in scientific publication. In contrast, "deep" metadata as for example the topic, objectives, or results of a research publication pose a greater challenge as such information is encoded in the text itself. The manual creation of such metadata related to scientific publications is a tedious and time-consuming activity. Semi-automatic or automatic metadata extraction techniques are viable solutions that enable the creation of large-scale and up-to-date metadata repositories. Common approaches focus on the extraction of relevant entities from the text of publications by means of ruled-based [11,14], machine learning [8], or hybrid (combination of rule based and machine learning) [6,7] techniques.

These approaches share a common assumption: as the number of publications dramatically increases, approaches that exclusively rely on dictionary-based pattern matching (possibly based on pre-existing knowledge bases) are of limited effectiveness. Rhetorical entities (REs) detection [9] is a class of solutions that aims at allowing the identification of relevant entities in scientific publications by analysing and categorising spans of text (e.g. sentences, sections) that contain information related to a given structural [8,10,11] (e.g. Abstract, Introduction, Contributions, etc.), argumentative [12,14] (e.g. Background, Objective, Conclusion, Related Work and Future Work), or functional (e.g. datasets [4], algorithms [6], software [7]) classification.

---

[1] Companion website: http://www.wis.ewi.tudelft.nl/eswc2017.

[2] http://ontoware.org/swrc/.

[3] http://bibliontology.com.

In contrast to existing literature, our work focuses on rhetorical mentions that relate to the description (Objective), implementation (Dataset, Method, Software), and evaluation (Result) of data processing pipelines. Thanks to a distant supervision approach and a simple feature model (bags-of-words), our method does not require prior knowledge about relevant entities [4] or grammatical and part-of-speech characteristics of rhetorical entities [6]. In addition, while in previous work [10,11] only one or few sections of the paper (e.g. abstract, introduction) are the target of rhetorical sentences classification, we make no assumption about the location of relevant information. This adds additional classification noise, due to the uncontrolled context of training sentences: it is more likely for a "Result" section to describe experimental results than for a "Related Work" section, where the likelihood of misclassification is higher [9].

3   The DMS Ontology

The DMS (Dataset, Method, Software) ontology [17] is designed to support the description and encoding of relevant properties of data processing pipelines, while capitalising on established ontologies. DMS has been created in accordance to the *Methondology* guidelines. It has been implemented using OWL 2 DL, and it consists of 10 classes and 30 properties. DMS captures five main concepts, namely *objectives*, *datasets*, *methods*, *software*, and *results*.

In the following, we refer to this initial ontology as DMS-Core. We provide an overview of the five aforementioned core concepts in Fig. 1 (in order to keep compatibility with existing ontologies, for some concepts, we adopt slightly different naming conventions within the ontology and in this text, i.e., *dataset* is encoded as *disco:DataFile* in DMS). Data processing pipelines are composed of one or more methods (*deo:Methods*), and are typically designed and evaluated in the context of a scientific experiment (*dms:Experiment*) described in a publication (*dms:Publication*). An experiment applies data processing methods, implemented by software (*ontosoft:Software* [13]), to one or more datasets (*disco:DataFile*) in order to achieve a given objective (*dms:Objective*), yielding one or more results (*deo:Results*). In each experiment, different implementations or configurations of a method (*dms:MethodImplementation*) or software (*dms:softwareconfiguration*) can be used. However, in this work, we only focus on the core concepts ignoring configurations and implementations.

Our main contribution in this paper is a methodology for the automatic extraction of metadata in accordance with the five core concepts of DMS: objective, dataset, method, software, and result. We reach this goal by labeling each of the sentences in a publication when it contains a *rhetorical mention* of one of the five DMS concepts. To capture knowledge on the properties and results of this extraction process, we introduce an auxiliary module DMS-Rhetorical (Fig. 1) extending DMS-Core as discussed in the following. DMS-rhetorical allows to link any *dms:CorePipelineConcept* (i.e. the supertype of *objective*, *dataset*, *method*, *software*, and *result*) to an extracted rhetorical mention.

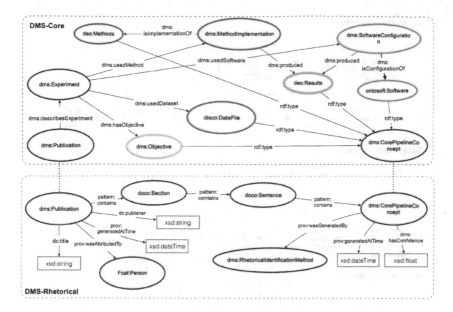

Fig. 1. DMS-Core ontology and the DMS-Rhetorical extension.

This link includes relevant provenance information such as the source of that mention (e.g. the sentence and section within a publication), but also metadata related to the extraction process, such as the classifier used to associate a sentence to a given DMS concept, and the related classification confidence.

We reuse the DoCo [1] ontology for encoding the information on sections and sentences. For each publication, we keep its general metadata including *id, title, authors, year of publication*, and *publisher*. The publication contains (*pattern:contains*) sections and each section of the paper contains several sentences. We store the text of the sentence using the *doco:Sentence* class and link the sentence *pattern:contains* to its *dms:CorePipelineConcept*.

4   DPP Knowledge Extraction Workflow

This section presents the knowledge extraction workflow designed to identify and annotate information referring to data processing pipelines (DPP) along the lines of the main classes of the DMS ontology (i.e. datasets, methods, software, results, and objectives). Our whole approach is summarized in Fig. 2. First, we identify rhetorical mentions of a DMS main class. In this work, for the sake of simplicity, rhetorical mentions are sought at sentence level. Future works will introduce dynamic boundaries, to capture the exact extent of a mention. Then, we extract named entities from the rhetorical mentions. These entities are filtered and, when applicable, linked to pre-existing knowledge bases, creating the final knowledge repository.

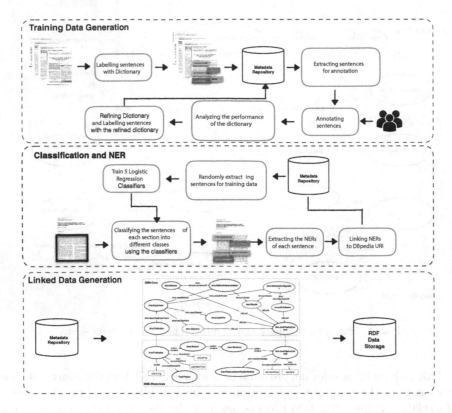

Fig. 2. Data processing pipeline knowledge extraction workflow.

The identification of rhetorical mentions is obtained through a workflow inspired by distant supervision [20], a training methodology for machine learning algorithms that relies on very large, but noisy, training sets. The training sets are generated by means of a simpler classifier, which could rely, for instance, on a mix of expert-provided dictionaries and rules, refined with manual annotations. Intuitively, the training noisiness could be cancelled out by the huge size of the semi-manually generated training data. This method requires significantly less manual effort, while at the same time retaining the performance of supervised classifiers. Furthermore, this approach is more easily adapted to different application domains and changing language norms and conventions.

4.1   Training Data Generation

Data Preparation. Scientific publications, typically available in PDF, are processed using one of the best-state-of-art extraction engines, GeneRation Of BIbliographic Data (GROBID) [18,19]. GROBID extracts a structured full-text representation as Text Encoding Initiative(TEI)-encoded documents, thus providing easy and reliable access paragraphs and sentences.

Dictionary-Based Sentence Annotation. Our goal is to classify each sentence of a given publication with respect to the five main classes of the DMS Ontology (datasets, methods, software, results, and objectives), based on the presence of rhetorical mentions that are related to such classes. Sentence classification could be obtained by means of a traditional supervised machine learning approach, assuming the presence of a large enough training set of sentence-level annotations. In our previous work [17], we manually created a small set of high-quality sentence-level annotations, relying on expert feedback. However, the annotation of a single publication took around 30–60 min per annotator, showing that this approach was not sufficiently scalable. We therefore opted for a workflow inspired by *distant supervision*. All sentences in our corpus were automatically labeled using a lower-quality and noisy dictionary-based classifier and simple heuristic rules, which are created using the following two-steps approach:

- **Reuse of generic scientific rhetorical phrases**: We relied on manually curated and published dictionaries of phrases and words found in [15,16] as an initial starting point to build our own dictionary. Both papers are writing guides giving advise on how to write an academic text based on best practices and commonly used phrases. [16] covers common phrases for introducing different sections in academic literature, e.g. the abstract, problem statement, methodology, or result discussion. [15] presents an extensive manual corpus study on different parts of scientific argumentation, and gives suggestion for accepted and often used phrases split by different disciplines and publication types.
- **Manual refinement and adaptation to the DMS domain**: The set of dictionary words based on [15,16] did not focus specifically on rhetorical mentions of data processing pipelines (even though classes like "result discussion" are quite related). Therefore, we manually refined those dictionaries and adapted them specifically to our 5 DMS classes. This refinement is based on the careful inspection of 20 papers selected from four Web- and data- related conferences series (ESWC, VLDB, ICWSM, and WWW).

The outcome of these two steps is a more class-specific set of dictionaries. For example the rhetorical phrases *"we collected"* and *"we crawled"* indicate a rhetorical mention of the *dataset* class. We used the dictionary to label sentences of 10 publications randomly selected from the four conferences series, to manually check the performance of the dictionary. For instance, we observed that the word *"data"* alone in a sentence is not a good indicator for being related to *dataset*. However if the word *"data"* co-occurs with *"from"*, a relationship with *dataset* is more likely. Several iterations of this manual refinement process lead to the final dictionary used for the following steps. Some example phrases are shown in Table 1.[4] Note that rhetorical mentions used in our refined dictionary are in fact skip n-grams, i.e. we do not expect the terms of each skip n-gram to be adjacent in a sentence (e.g. the rhetorical mention "the aim of this study" stripped of stop words becomes the skip n-gram "aim study").

---

[4] The dictionaries are available at http://www.wis.ewi.tudelft.nl/eswc2017.

Table 1. Excerpt of dictionary of phrases used for classifying sentences

Objective	*this research, this article, aim study, aim article, purpose paper, we aim, we investigate*
Dataset	*dataset, datasource, data source, collected from, database, collect data, retrieve data*
Method	*we present, we develop, we conduct, we propose, methodologies, method, technique*
Sofware	*tool, obtained using, collected using, extracted using, software*
Result	*we find, shows, show, shown, showed, we found, figure, table, we observe, we compare*

Test and Training Data Generation. We created reliable test and training datasets for both training and benchmarking machine learning classifier as follows. By using the phrases dictionary described in the previous subsection, we label all sentences of all research papers collected with appropriate class labels. Most sentences will not receive a label (as they do not contain any rhetorical mentions), but some may obtain multiple labels. This is for instance common for sentences found in an abstract, which often contain information on *datasets*, but also on *methods*, or even *results*. Then, we randomly select a balanced set of sentences with rhetorical mentions of all five classes, and manually inspect the assigned labels. We reclassify them using expert feedback from several annotators, if the pattern-based classifier assigned incorrect labels. Using this approach, we can create a reliable manually annotated and balanced test dataset quicker and cheaper compared to annotating whole publications or random sentences, as the pattern-classifier usually delivers good candidate sentences. Furthermore, this approach allows us to further refine and improve the dictionary by incorporating the expert feedback, allowing us to cheaply re-annotate the whole corpus using the dictionary with higher accuracy compared to the initial classifier.

We assessed the performance of both the dictionary-based classifier and our annotators to decide on the number of manual annotations needed for a reliable test set. We randomly selected 100 sentences from each of the five classes (i.e. 500 in total). Two expert annotators manually checked the assigned labels (a task which was perceived easier by the annotators than applying labels to a random unlabeled sentence). The inter-annotator agreement using the Cohen's kappa measure averaged over all classes was .58 (the Cohen's kappa measures of the individual classes are *objective*: .71, *dataset*: .68, *software*: .37, *result*: .61, and *method*: .53).

4.2 Classification and NER

Machine-Learning-Based Rhetorical Detection. As a second part of our distant supervision workflow, we now train a simple binary Logistic regression classifier for each of the classes using simple TF-IDF features for each sentence.

This simple implementation serves as a proof of concept of our overall approach, and can of course be replaced by more sophisticated features and classifiers in future work.

As a test set, we use the 500 sentences (100 per class) manually labeled with their DMS class by our expert annotators. We associated a single label (some sentences can have multiple labels) to each sentence, decided by a simple majority vote. In order to generate the training data for each class, we randomly selected 5000 positive examples from the sentences labeled with that class by the dictionary-based classifier. We also randomly select 5000 negative examples from sentences which are not labeled with that class by the dictionary classifiers. Sentences from the test set were excluded from the pool of candidate training sentences.

Named Entity Extraction, Linking, and Filtering. In the last step of our method, we extract named entities from the sentences that are classified as related to one of the five main DMS classes, filtering out those entities that are most likely not referring to one of the DMS classes, and retaining the others as an extracted entity of the class matching the sentence label.

Named entity extraction has been performed using the TextRazor API[5]. TextRazor returns the detected entities, possibly decorated with links to the DBpedia or Freebase knowledge bases. As we get all named entities of a sentence, the result list contains many entities which are not specifically related to any of the five classes (e.g. entities like "software", "database"). To filter many of these entities, and after a manual inspection, we opted for a simple filtering heuristic. Named entities are assumed to be not relevant if they come from "common" English language (like software, database), while relevant entities are terms referring to domain-specific terms or specific acronyms (like SVM, GROBID, DMS, Twitter data). The heuristic is implemented as look-up function of each term in *Wordnet*.[6] Named entities that can be found in WordNet are removed. As WordNet is focusing on general English language, only domain-specific terms remain. We present the results of the analysis performed on the quality of the remaining named entities in Sect. 5.

4.3   Linked Data Generation

As a final step, we build a knowledge repository based on the DMS-Core and DMS-Rhetorical ontology (outlined in Sect. 3). The repository is populated with classified sentences, and with the lists of entities for each DMS main class, with links to the sentence where each single entity has been detected. Sentences are linked to the containing publications.

---

[5] http://www.textrazor.com/.
[6] http://wordnet.princeton.edu/.

Listing 1.1 shows an example of a part of an output RDF. The relationships shown in the RDF snippet are from the domain-specific DMS ontology for describing data-processing research. They have not been extracted automatically, as the scope of this work is not on the automatic extraction of relationships between entities.

```
1   PREFIX doco: <http://purl.org/spar/doco>
2   PREFIX prov: <http://www.w3.org/ns/prov#>
3   PREFIX disco: <http://rdf-vocabulary.ddialliance.org/discovery#>
4   PREFIX dms: <https://github.com/mesbahs/DMS/blob/master/dms.owl#>
5   PREFIX rdf: <http://www.w3.org/1999/02/22-rdf-syntax-ns#>
6   PREFIX pattern: <http://www.essepuntato.it/2008/12/pattern>
7   [a dms:Publication;
8   dms:describesExperiment   dms:Ncdec5e68ed864a3a24].
9   dms:Ncdec5e68ed864a3a24   a   dms:Experiment;
10      dms:usedDataset [ a    disco:dataFile ;
11                        rdf:type      dms:Ncdec5e68ed864a ;
12                        prov:value    "Billion Triple Challenge (BTC)"].
13  dms:Ncdec5e68ed864a   a    dms:CorePipelineConcept;
14      pattern:isContainedBy  doco:Ncdec5e68edghgf99.
15  doco:Ncdec5e68edghgf99 a doco:Sentence;
16      prov:value "In our experiments we used real data that were taken from the Billion
                    Triple Challenge (BTC) dataset.";
17      pattern:isContainedBy doco:Ncdec5ehfdjk67.
18  doco:Ncdec5ehfdjk67 a doco:Section;
19      prov:value "Introduction".
```

Listing 1.1. Example of output RDF: A paper describes an experiment which uses a dataset called (BTC). (BTC) is a CorePipelineConcept linked to sentence of the paper.

5   Evaluation

In this section, we analyse the performance of our metadata extraction pipeline in both a quantitative and qualitative fashion. We focused on four major conference series from different communities with notable scientific contributions to data processing pipelines (Table 2): the European Semantic Web Conference (ESWC), International Conference On Web and Social Media (ICWSM), International Conference on Very Large Databases (VLDB), and the International World Wide Web Conference (WWW). We further present the results of both the dictionary-based and logistic regression-based sentence classifiers on the manually annotated test data. Finally, we analyse and discuss the quality of the entities extracted from the classified sentences.

5.1   Dataset

Table 2 summarises the properties of the experimental dataset, including its size, the number of rhetorical mentions extracted for each class (as decided by the regression-based classifier), and the number of unfiltered unique named entities extracted from the rhetorical mentions taken from scientific publications of a particular conference series. The table shows that methods are the most frequent encountered class, followed by datasets. Table 3 summarises statistics on extracted entities as described in the previous section per class (including

Table 2. Quantitative analysis of the rhetorical sentences and named entities extracted from four conference series. Legend: PAP (papers), SNT (sentences), OBJ (objective), DST (dataset), MET (method), SWT (software), RES (results)

Conf.	Size		Rhetorical sentences					Unique named entities				
	#PAP	#SNT	#OBJ	#DST	#MET	#SWT	#RES	#OBJ	#DST	#MET	#SWT	#RES
ESWC	620	129760	12725	13528	26337	9614	22245	4197	4910	6987	4557	6416
ICWSM	793	52094	6096	4277	8936	1830	13848	2830	2241	3658	1538	4499
VLDB	1492	396457	26953	49855	68336	11919	84662	7301	12052	13920	5741	15959
WWW	1021	253401	23378	19783	49331	10293	58212	6616	6499	10793	5164	11869

Table 3. Number of Named Entities after filtering using the Wordnet.

Conf.	Distinct NER with URI					Distinct NER no URI				
	#OBJ	#DST	#MET	#SWT	#RES	#OBJ	#DST	#MET	#SWT	#RES
ESWC	1157	1206	1779	1200	1454	1874	2427	3497	2193	3219
ICWSM	727	555	944	443	1027	1110	900	1588	519	1974
VLDB	1528	2313	2516	1365	2395	3800	6963	8393	2804	10288
WWW	1990	1630	2904	1613	2860	2742	3153	5382	2148	6247

Table 4. Top-5 most frequent methods applied to IMDB dataset.

ESWC	ICWSM	VLDB	WWW
Semantic Web	LDA	Tuple	Web Page
Sem-CF	Classifier_I	XML	Login
User Modeling	SetLock	Query Plan	Faceted Search
Recommender System	Hashtag	XsKetch	Recommender System
FactBox	Future tense	LS-B	Source Rank

filtering and pruning entities using a Wordnet look-up). Furthermore, we report how many of those entities could be linked to Wikipedia by the TextRazor API (columns *with URI*), thus distinguishing well-known entities (e.g. Facebook, Greedy algorithm) from the newly presented or less popular entities (e.g. SIFT Netnews, RW ModMax. columns *no URI*).

Qualitative Analysis. In this section, we showcase how our approach can be used to fulfill a hypothetical information need of a data scientist, namely: *Which methods are commonly applied to a given data set?*

As an example, we use the popular IMDB dataset of movies and actors, and manually inspect the list of top-6 most frequent methods applied to that dataset in publications grouped by their conference series. The results are shown in Table 4, hinting at the different interests conference venues have for that dataset: ignoring the false positives (like "Web Page" or "XML" - we further discuss false positives later in this section), VLDB as a database-centric conference covers methods like XsKetch (summarisers for improving query plans in XML databases) or LSB-Trees for better query plans for nearest-neighbour queries, using the IMDB dataset as a large real-life dataset for evaluation database queries; ICWSM with a focus on Social Media research features LDA topic detection and generic classification to analyse IMDB reviews, while ESWC and WWW are interested in recommendations and user modelling.

5.2   Analysis of Rhetorical Classifiers

In the following, we present the results of both the *dictionary-based* and *logistic regression-based* classifiers on the manually annotated test set, summarised in Table 5, relying on commonly used measurements for accuracy, precision, recall, and F-Score. It can be observed that using logistic regression increases the recall for most classes, while having a slightly negative impact on the precision, showing that this approach can indeed generalise from the manually provided dictionaries to a certain extent.

We believe that better performance can be achieved by employing more sophisticated features and classifiers. Furthermore, the performance gains of the logistic regression classifier come for "free" as we only invested time and effort to train the dictionary-based classifier. The best results are achieved for the *Method* class with F-score = 0.71. We manually inspected the sentences labeled as *Software* and *Dataset* to understand reasons for the comparatively low performance of those classes. To certain extend, this can be attributed to the ambiguity of some n-grams in the dictionary. For example, the word *tool* appearing in different sentences can result to misleading labels: e.g., "extraction tool Poka" is about software, but "current end-user tools" is a general sentence not specifically about a software. Similarly confusion can be observed for the word *dataset* for the *Dataset* class. For instance, "twitter dataset" and "using a dataset of about 2.3 million images from Flickr" are labeled correctly, but "quadruple q and a dataset d" is labeled incorrectly. Thus, we conclude that many terms used in *Software* and *Dataset* are too generic (e.g. dataset, tool, database) leading to higher recall, but having a negative impact on precision, demanding more refined rules in our future work.

Table 5. Estimated Accuracy, Precision, Recall and F-score on manually annotated sentences for Dictionary and Logistic regression based classification

Classes	Dictionary based				Logistic regression based			
	Accuracy	Precision	Recall	F-Score	Accuracy	Precision	Recall	F-Score
Objective	0.85	0.49	0.81	0.61	0.84	0.49	0.81	0.61
Dataset	0.84	0.46	0.68	0.55	0.80	0.41	0.81	0.54
Method	0.76	0.79	0.61	0.69	0.76	0.76	0.67	0.71
Software	0.83	0.39	0.52	0.45	0.84	0.34	0.72	0.46
Result	0.84	0.60	0.68	0.63	0.81	0.53	0.71	0.60

5.3   Quality of Extracted Entities

We studied the performance of the Named Entity (NE) extraction modules of our method by means of a mixed quantitative and qualitative analysis. We calculated the Inverse Document Frequency (IDF) of each named entity NE_i extracted from the corpus. IDF is a measure of informativeness, calculated as $IDF(NE_i) = log\frac{|Sentences|}{|NE_i|}$, that is, the logarithmically scaled inverse fraction of the number of sentences in the corpus and the number of sentences containing NE_i. Figure 3 depicts the distribution of IDF values for each NE in the dataset.

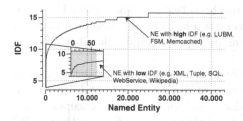
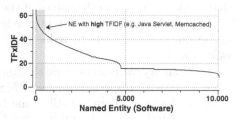

Fig. 3. Distribution of IDF values of extracted named entities.

Fig. 4. Distribution of TFIDF values for NEs contained in *software* sentences.

Only a handful of named entities (about 100) feature a low IDF values (indicating that they are likely not fitting their assigned class well), while a large amount of entities (more than 60%) have relatively high informativeness. But, what is the quality of such entities? Are they useful in the characterization of class-specific sentences? To answer these questions, we first calculated a class-specific TFxIDF value for each named entity NE_i in the dataset as $TFIDF(NE_i, C_j) = (1 + log(|NE_{i,j}|)) \times IDF_{NE_i}$, where $|NE_{i,j}|$ is the raw frequency of a named entity NE_i within the sentences classified as relate to the class C_j. Then, for each class, we ranked named entities in decreasing order of $TFIDF(NE_i, C_j)$, and manually analyzed the first 100 entities.

Figure 4 shows an example distribution of TFIDF values. We excluded from this analysis the *objective* class, as objectives are usually not represented well by a single named entity, but instead require a more elaborate verbal description (which is usually fittingly provided by a rhetorical mention).

Table 6 shows examples of relevant named entities for each considered class. In terms of retrieval precision, we can observe promising results. NEs contained in *method* and *software* sentences feature a precision of 72% and 64%, respectively. On the other hand, NEs contained in *dataset* and *results* sentences resulted in a precision of 23% and 22%. In both cases, however, the returned entities are still relevant and related to the class: False positives in *dataset* sentences are mainly due to terms that are clearly related to data (e.g. Fuzzy set, Data model, Relational Algebra), but not specifically referring to actual datasets. Likewise, false positives in *results* sentences are mainly due to the presence of acronyms

Table 6. Examples of representative Named Entities in different classes

Dataset	Method	Software	Result
MovieLens	Collaborative Filtering	Java Servlet	Expected Value
Enron	Dynamic Programming	Portlet	Standard Deviation
IMDb	Active Learning	PHP	Precision and Recall
YAGO	Support Vector Machine	Memcached	P-value
DBPedia	Language Model	DOM API	MRR

that could be linked to the names of the methods tested in the paper. This type of error can be attributed the sentence-level granularity of our rhetorical mention detection, and can likely be reduced by including a boundary classifier into our workflow.

In summary, we can conclude that our approach is indeed suitable for extracting entities with respect to the five DMS classes in a meaningful and descriptive fashion. However, there are still some false positives of related concepts which cannot easily be recognized using simple statistic means, and which thus invite further deeper semantic filtering in future works.

6 Conclusion

In this paper, we presented the design and evaluation of knowledge extraction workflow aimed at extracting semantically rich metadata from scientific publications. The workflows specialises on the extraction of information related to data processing pipelines, with a focus on rhetorical mentions related to datasets, methods, software, objectives, and results. The extracted information is collected and published as a RDF knowledge base according to the DMS (Data Method Software) ontology, which was specifically designed to enable the description and linking of information related to data processing pipelines. The generated metatada allows researchers and practitioners to access and discover valuable information related to the properties and limitation of data sources and data processing pipelines, based on current literature.

Differently from previous work, our workflow relies on a lightweight distant supervision approach, which features lower training costs (compared to traditional supervised learning) and acceptable performance. These properties make the approach suitable for reuse in additional knowledge domains related to scientific publication. We show that, despite its simple design, it is possible to achieve high precision and recall for all classes. From these classified sentences, we extracted (rather noisy) named entities, which we subsequently filtered and ranked, to select entities which promise high descriptive power for their class.

While promising, the obtained results suggest ample space for future improvements. For instance, it will be interesting to investigate the performance of more complex machine learning classifiers working on richer feature sets (e.g., word-embeddings, POS-tags, parse trees, etc.). Furthermore, for labelling rhetorical mentions, our current granularity is on sentence level. This introduces some additional confusion when extracting named entities in cases that a sentence has multiple labels, or only parts of a sentence refer to a rhetorical mention while others do not. This limitation could be remedied by additionally training boundary classifiers, which can narrow down rhetorical mentions more precisely. Furthermore, we employ sample filtering of entities based on statistics. This could be improved by further utilising semantic information from open knowledge bases.

Finally, we will address the application of our approach to real-life use cases. For instance, applications in the domain of digital libraries seem promising, allowing for both more meaningful queries to find relevant publications, and also allowing for analytic capabilities to track and visualise trends and changes in research fields over time.

References

1. Alexandru, C., Peroni, S., Pettifer, S., Shotton, D., Vitali, F.: The document components ontology (DoCO). Semant. Web **7**(2), 167–181 (2016)
2. Möller, K., Heath, T., Handschuh, S., Domingue, J.: Recipes for semantic web dog food — The ESWC and ISWC metadata projects. In: Aberer, K., et al. (eds.) ASWC/ISWC -2007. LNCS, vol. 4825, pp. 802–815. Springer, Heidelberg (2007). doi:10.1007/978-3-540-76298-0_58
3. Glaser, H., Millard, I.: Knowledge-enabled research support: RKBExplorer.com. In: Proceedings of Web Science, Athens, Greece (2009)
4. Ghavimi, B., Mayr, P., Vahdati, S., Lange, C.: Identifying and improving dataset references in social sciences full texts. arXiv preprint arXiv:1603.01774 (2016)
5. O'Seaghdha, D., Teufel, S.: Unsupervised learning of rhetorical structure with untopic models. In: Proceedings of the 25th International Conference on Computational Linguistics (COLING 2014) (2014)
6. Tuarob, S., et al.: AlgorithmSeer: a system for extracting and searching for algorithms in scholarly big data. IEEE Trans. Big Data **2**(1), 3–17 (2016)
7. Osborne, F., Ribaupierre, H., Motta, E.: TechMiner: extracting technologies from academic publications. In: Blomqvist, E., Ciancarini, P., Poggi, F., Vitali, F. (eds.) EKAW 2016. LNCS (LNAI), vol. 10024, pp. 463–479. Springer, Cham (2016). doi:10.1007/978-3-319-49004-5_30
8. Khodra, M.L., et al.: Information extraction from scientific paper using rhetorical classifier. In: International Conference on Electrical Engineering and Informatics (ICEEI) (2011)
9. Helen, A., Purwarianti, A., Widyantoro, D.H.: Rhetorical sentences classification based on section class and title of paper for experimental technical papers. J. ICT Res. Appl. **9**(3), 288–310 (2015)
10. Burns, G.A., Dasigi, P., de Waard, A., Hovy, E.H.: Automated detection of discourse segment and experimental types from the text of cancer pathway results sections. Database. J. Biol. Databases Curation (2016)
11. Sateli, B., Witte, R.: What's in this paper? Combining rhetorical entities with linked open data for semantic literature querying. In: Proceedings of the 24th International Conference on World Wide Web. ACM (2015)
12. Liakata, M., Teufel, S., Siddharthan, A., Batchelor, C.R.: Corpora for the conceptualisation and zoning of scientific papers. In: LREC (2010)

13. Gil, Y., Ratnakar, V., Garijo, D.: Ontosoft: capturing scientific software metadata. In: International Conference on Knowledge Capture, p. 32. ACM (2015)
14. Groza, T.: Using typed dependencies to study and recognise conceptualisation zones in biomedical literature. PloS One **8**(11), e79570 (2013)
15. Dorgeloh, H., Wanner, A.: Formulaic argumentation in scientific discourse. In: Corrigan, R., Moravcsik, E.A., Ouli, H., Wheatley, K.M. (eds.) Formulaic Language, vol. 2, pp. 523–544. John Benjamins, Amsterdam (2009)
16. English for Writing Research Papers Useful Phrases. http://www.springer.com/cda/content/document/cda_downloaddocument/Free+Download+-+Useful+Phrases.pdf?SGWID=0-0-45-1543172-p177775190
17. Mesbah, S., Bozzon, A., Lofi, C., Houben, G.-J.: Describing data processing pipelines in scientific publications for big data injection. In: WSDM Workshop on Scholary Web Mining (SWM), Cambridge, UK (2017)
18. Lopez, P.: GROBID: combining automatic bibliographic data recognition and term extraction for scholarship publications. In: Agosti, M., Borbinha, J., Kapidakis, S., Papatheodorou, C., Tsakonas, G. (eds.) ECDL 2009. LNCS, vol. 5714, pp. 473–474. Springer, Heidelberg (2009). doi:10.1007/978-3-642-04346-8_62
19. Lipinski, M., Yao, K., Breitinger, C., Beel, J., Gipp, B.: Evaluation of header metadata extraction approaches and tools for scientific PDF documents. In: JCDL, Indianapolis, USA (2013)
20. Mintz, M., Bills, S., Snow, R., Jurafsky, D.: Distant supervision for relation extraction without labeled data. In: International Joint Conference on Natural Language Processing of the AFNLP, Singapore (2009)

Combining Word and Entity Embeddings
for Entity Linking

Jose G. Moreno[1](✉), Romaric Besançon[2], Romain Beaumont[3], Eva D'hondt[3],
Anne-Laure Ligozat[3,4], Sophie Rosset[3], Xavier Tannier[3,5], and Brigitte Grau[3,4]

[1] Université Paul Sabatier, IRIT, 118 Route de Narbonne,
31062 Toulouse, France
jose.moreno@irit.fr
[2] CEA, LIST, Vision and Content Engineering Laboratory,
91191 Gif-sur-Yvette, France
romaric.besancon@cea.fr
[3] LIMSI, CNRS, Université Paris-Saclay, 91405 Orsay, France
{romain.beaumont,eva.dhondt,anne-laure.ligozat,sophie.rosset,
xavier.tannier,brigitte.grau}@limsi.fr
[4] ENSIIE, Évry, France
[5] Univ. Paris-Sud, Orsay, France

Abstract. The correct identification of the link between an entity mention in a text and a known entity in a large knowledge base is important in information retrieval or information extraction. The general approach for this task is to generate, for a given mention, a set of candidate entities from the base and, in a second step, determine which is the best one. This paper proposes a novel method for the second step which is based on the joint learning of embeddings for the words in the text and the entities in the knowledge base. By learning these embeddings in the same space we arrive at a more conceptually grounded model that can be used for candidate selection based on the surrounding context. The relative improvement of this approach is experimentally validated on a recent benchmark corpus from the TAC-EDL 2015 evaluation campaign.

Keywords: Entity Linking · Linked data · Natural language processing and information retrieval

1   Introduction

In this paper, we investigate a new approach to candidate selection in the context of the Entity Disambiguation (or Entity Linking) task. This task consists of connecting an entity mention that has been identified in a text to one of the known entities in a knowledge base [16,25], in order to provide a unique normalization of the mention. Entity Linking sometimes figures as part of a more general framework which globally disambiguates all the concepts in a document with respect to a knowledge base (KB), whether they are named entities or nominal expressions (e.g. Wikify [17] or Babelfy [21]).

© Springer International Publishing AG 2017
E. Blomqvist et al. (Eds.): ESWC 2017, Part I, LNCS 10249, pp. 337–352, 2017.
DOI: 10.1007/978-3-319-58068-5_21

An Entity Disambiguation system usually consists of three main steps [11]. First, it analyzes an input (a text) to identify "entity mentions" that need to be linked to the knowledge base; then, for each mention, the system generates several candidate entities from the knowledge base; finally, it selects the best entity among the candidates. One of the main challenges is the extremely large number of entities present in the knowledge base, and consequently their disambiguation, given that a same mention can refer to different entities, and the correct reference can only be deduced from its surrounding context in a text. Consider the following example: "As soon as he landed at the Bangkok airport, Koirala saw Modi's tweets on the quake, Nepal's Minister ...". The mention "Koirala" is fairly ambiguous, it could refer to "Manisha Koirala", a Nepalese actrice, the "Koirale family", a dominating family in Nepalese politics, "Saradha Koirala", a poet of Nepalese descent or "Sushil Koirala", the Nepalese Prime Minister between 2014 and 2015. In this setting even the context word 'Nepal' will not be of much use, and a disambiguation module must use the information contained within the context to its fullest to arrive at a correct disambiguation. An (accurate) Entity Linking system will map the forms "Koirala" to Wikipedia entity "Sushil Koirala", "Modi" to "Narendra Modi" and "Nepal" to the country entity.

The main contribution of this paper focuses on the last step in this process, i.e. 'candidate selection'. Most of the current approaches to this problem are 'word-based' and consider the words as atomic units when using information on the mention and its context to select the best candidate. We propose a more semantically-oriented approach which is based on the joint learning of word and entity embeddings in the same embedding space. The advantages of learning these representations simultaneously are threefold: (1) The resulting word embeddings are more conceptually grounded as their context (during training) may contain concept vectors which supersede surface variations; (2) Entity embeddings are learned over a large text corpus and attain higher frequencies in training than embeddings that are learned directly over knowledge bases; (3) Since the representations are learned in the same space, we can use a simple similarity measure to calculate the distance between an entity mention and its context (words), and its possible entry (entity) in the KB. In this paper, we focus our efforts on entities that exist in Wikipedia as this is one of the few publicly-available, general purpose corpora that contains both a large amount of text, and is annotated with common entities.

In this paper, we present the following contributions:

- Our EAT model which jointly learns word and entity embeddings (Sect. 3).
- A global Entity Linking pipeline, integrating this EAT model (Sect. 4); Note that this model can be integrated as a feature to any kind of supervised approach for Entity Linking.
- An evaluation of this approach using the TAC 2015 "Entity Discovery and Linking" challenge (Sect. 5). Our result for the task ($P(all) = 0.742$) outperforms a non-EAT baseline and achieves comparable results against the top participants in the challenge.

2   Related Work

Entity Linking approaches are often distinguished by the degree of supervision they require, namely into unsupervised, supervised or semi-supervised methods. The unsupervised methods, proposed for instance in [6,9], usually rely only on some similarity measure between the mention in the text and the entities in the KB. They have the advantage of being simple and easy to implement. However, they also have low performance compared to the supervised methods as was shown in past evaluation campaigns [5]. These supervised methods generally rely on binary classifiers [14,26] or ranking models [4,24] for entity disambiguation.

A lot of recent work in this domain has focused on the similarity calculation between the pieces of text (mentions) and entities. A system such as Babelfy [21] manages to connect structured knowledge like WordNet [19], Babelnet [22], and YAGO2 [10], through the use of random walks with restart over the semantic networks which jointly represent the whole elements. As a result, a signature is obtained for each element and used to calculate similarities between the different sources, i.e., similarities of elements from WordNet against elements from YAGO2. However, recently word representation techniques have shown surprisingly good results for many NLP tasks [18]. Word embeddings are unsupervised strategies based on observation of text regularities in huge text collections to learn reduced vectors which perform better than the traditional count-based representations [1]. The use of embeddings for existing semantic networks has previously been studied by [2]. Representing knowledge information with embeddings consists in transforming each piece of information of the KB – usually represented by triples $(head, relationship, tail)$ – into low dimensional vectors. This transformation is obtained by the optimization of a function that gives high scores when the triples are present in the KB, and low scores otherwise. Based on the work of [2,27] defines a three components function in charge of the optimization of the word embeddings, the knowledge embeddings and their alignment. This technique manages to mix the knowledge and text, which results in a unique representation space for words, entities and relations. These works are interested in the task of knowledge base completion and only few of them are directly related to the task of Entity Linking [8,23,28].

An extension of [27] has been developed in parallel by [8,28]. They applied their respective model to several Entity Linking collections. However, like for [27], these works do not directly use the context of an entity mention to built the vector representation but use an alignment function to achieve some matching between the mention and the entity. [23] prefers to use the document representation of [13] to jointly represent Wikipedia pages and words. In both cases the joint space of entities and words is used to calculate similarities between them. [29] and their later work [30] integrated entity embeddings within their Entity Linking system. However, these entity embeddings were learned over concatenated documents with only sequences of entities (where entities are ordered as they are found in annotated documents or by following short KB paths) followed by the text content to align the word and entity representations.

3   Combining Word and Entity Embeddings

Learning representations of words and entities simultaneously is an interesting approach to the problem of Entity Linking, as it allows for an even larger degree of normalization than the regular grouping of word embeddings (i.e. of words that have similar or related meanings) in the vector space already provides. While previous approaches indirectly addressed the problem by learning separate representations and aligning them [8,28] or by concatenating entity-only with text-only sequences [29,30], we opt to learn them simultaneously. In this section, we present a model that is able to combine word and entity embeddings by only using their context. As a consequence, our model can be considered as an extension of the initial words embedding model [18] or its variation [13].

3.1   Definitions

A corpus is a sequence of words w or anchor texts a, where the words belong to the vocabulary V and an anchor text $a = (w, e)$ is a tuple consisting of a word $w \in (V \cup \emptyset)$, and an entity $e \in \xi$, where ξ is the collection of entities in the KB. In all cases, the bold letters correspond to the respective vectors e, $w \in \mathbb{R}^d$, where $\mathbb{R}^d$ is called the combined space and d defines the number of dimensions.

3.2   Extended Anchor Text

To obtain w and e in the same space, we introduce the concept of Extended Anchor Text (EAT). Through this, we combine entity information with its anchor text, and consequently introduce it into the corpus. To obtain EATs, the mention of an anchor text a_i is redefined as $a'_{ij} = (w'_i, e_j)$ where $w'_i = w_i$ if w_i is not empty, otherwise w'_i is equal to the set of words present in e_j. For an illustration of the decomposition into anchor text and entity, see Fig. 1. We redefine a corpus like a sequence of EATs a', so the full vocabulary is defined by $\mathbb{F} = \{V \cup \xi\}$, that is, the set of embeddings to be learned now contains both words (including mentions) and entities from the knowledge base.

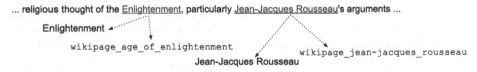

Fig. 1. Illustration of mention-entity decomposition in the EAT model

3.3   The EAT Model

The objective of our model is to find a representation for each element in $\mathbb{F}$ based on the surrounding words/EATs in a sentence. Similarly to [18], we define the probability between two elements in the corpus as in Eq. 1.

$$p(c_o|c_i) = \sum_{f_o \in c_o} \sum_{f_i \in c_i} \frac{exp(\boldsymbol{f_o}^T \boldsymbol{f_i})}{\sum_{j=1}^{|\mathbb{F}|} exp(\boldsymbol{f_j}^T \boldsymbol{f_i})} \tag{1}$$

where the elements in $\mathbb{F}$ are identified as f that can represent either a word or an entity, and the words or EATs in a corpus are identified as c. Note that if c_o and c_i are words, the Eq. 1 becomes the softmax function into two words and the double sum disappears. The optimization process consists of maximizing the average log probability defined in Eq. 1 over a corpus composed by EATs.

The implementation of the EAT models does not imply big changes in actual versions of the original model [18] such as can be found in Word2Vec[1]. As Eq. 1 is equivalent to the originally proposed by [18] when c is a word, we just need to adapt it for the case when c is an EAT. The adaptation consists in the expansion of c into their possible combinations but keeping the context static, e.g., the context is the same for the word and the entity. Similarly, if c is part of the context, it is the context that must be expanded. Figure 2 shows the expansion that occurs when the word vector and entity vector that are linked to the entity mention 'Enlightenment' are trained during the training phase.

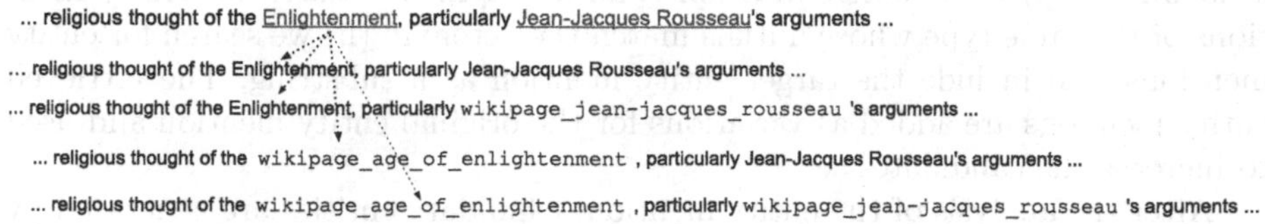

... religious thought of the <u>Enlightenment</u>, particularly <u>Jean-Jacques Rousseau</u>'s arguments ...

... religious thought of the Enlightenment, particularly Jean-Jacques Rousseau's arguments ...

... religious thought of the Enlightenment, particularly wikipage_jean-jacques_rousseau 's arguments ...

... religious thought of the wikipage_age_of_enlightenment , particularly Jean-Jacques Rousseau's arguments ...

... religious thought of the wikipage_age_of_enlightenment , particularly wikipage_jean-jacques_rousseau 's arguments ...

Fig. 2. Illustration of the expanded training contexts for the different embedding types in the EAT model. Please note that only entity mentions are decomposed, and consequently have multiple training moments, i.e. one pass for the anchor text and a separate pass for the entity embedding.

The main advantages of our model are:

- actual methods and embeddings based on the original method proposed by [18] are directly usable within our model including the skip-gram as well as the CBOW configuration;
- anchor texts in publicly available corpus are taken into account within the model allowing us to represent words and entities in a unique space;
- vectors for words and entities are learned using their context instead of indirect relations between them (like the alignment strategy used by [8,28]), which is more similar to distant supervision techniques such as [20].

---

[1] Several open source implementations are available online. We have used Gensim and Hyperwords, available at https://radimrehurek.com/gensim/models/word2vec.html and https://bitbucket.org/omerlevy/hyperwords respectively.

4   Entity Linking System

Our Entity Linking system relies on a standard architecture [11] composed of two main steps: for a given entity mention and its textual context, a first module generates possible candidate entities for the linking and a second one takes as input the different candidate entities and selects the best one.

4.1   Generation of Candidate Entities

The generation of the candidate entities relies on the analysis of the entity mention and its textual context. In this study, we focus mainly on the disambiguation of entities, not their recognition. Therefore, we consider that the offsets of the entity mentions to disambiguate are given as input to the system. A complementary recognition step of entity mentions in the text is nonetheless carried out, in order to associate a type (Person, Location, Organization) with the entity mentions[2] and define their context in terms of surrounding entities (we consider only the explicitly named entity mentions and we ignore the nominal and pronominal mentions). As the entity mention is used to retrieve candidates from the KB, two expansion heuristics are proposed to include variations of the target entity mention. Both of them can be considered as simple co-reference approaches within a document: (i) if the entity mention is an acronym, we search for entity mentions of the same type whose initials match the acronym (ii) we search for entity mentions who include the target entity mention as a substring. The retrieved entity mentions are added as variations for the original entity mention and used to increase the candidates set.

After the analysis of the entity mention, candidate entities are generated by comparing the entity mention (and its variations) to the names of the entities of the KB [7]. We use the following strategies:

- Equality between the forms of the entity mention and an entity in the KB;
- Equality between the form of the entity mention and a variation (alias or translation) of an entity in the KB;
- Inclusion of the form of the entity mention in the name or one of the forms of the variations of an entity in the KB;
- String similarity between the form of the entity mention and a variation of an entity in the KB. We use the Levenshtein distance, which is well suited to overcome the spelling errors and name variations. In the experiments, we considered an entity in the KB as a candidate entity if its form or any of its variations have a distance with the form of the entity mention ≤ 2. For better efficiency, we exploited a BK-tree structure [3] for this selection.
- Information Retrieval model: an information retrieval model is used to index all the known forms of the entities in the KB as documents. We can then select all close variants, weighted by their tf-idf, to find suitable candidates. Lucene was used as search engine.

The candidate entities are also filtered in order to keep only entities that have at least one of the expected entity types (e.g. Person, Location, Organization).

---

[2] Named entities were recognized using MITIE https://github.com/mit-nlp/MITIE.

4.2 Selection of the Best Candidate Entity

The objective of this step is to find the correct candidate entity from the set of generated candidate entities. To this purpose, a classifier is trained to recognize the best entity among the entity candidates, using training data on disambiguated entity mentions. More precisely, each candidate entity is associated with a set of features:

- a set of features associated with the strategy that was used for the generation of this candidate entity: binary features for the simple matching strategies, as well as the value of the similarity score for the Information Retrieval model;
- two similarity scores accounting for a global context obtained by comparing the textual context of the entity mention with a textual description of the entities in the KB;
- one score accounting for global popularity of the entity obtained by counting the number of inlinks present in Wikipedia and applying a log normalization;
- four similarity scores are added based on the EAT embeddings, that account for a narrower context. As this context is made of few words, using embeddings allows to overcome the problem of lexical gaps.

Textual Similarity Scores. For an entity mention q and a possible entity candidate e from the KB, we consider three vectors representing three textual contexts: the document in which q appears, noted $d(q)$, the Wikipedia page associated with e, noted $w(e)$ and a text combining the set of entities that are in relation with e in the KB, noted $r(e)$. Each text is represented by a vector using a standard Vector Space Model, with a *tf-idf* weighting scheme, $d(q) = (d_1, \ldots, d_n)$ with $d_i = tf(t_i, d) \times idf(t_i)$, where $tf(t_i, d)$ is a function of the frequency of the term t_i in the document d and $idf(t_i)$ is an inverse function of the document frequency of the term in the collection. All representations are built in a common vector space, constructed from a full Wikipedia dump (the *idf* scores are therefore computed on this complete collection of documents). The scores are then the cosine similarities between the context vector of the entity mention and each of the context vectors of the entity from the KB:

$$sim_d(q, e) = cos(d(q), w(e)) = \frac{\sum_i d_i . w_i}{||d(q)||.||w(e)||}$$

$$sim_r(q, e) = cos(d(q), r(e)) = \frac{\sum_i d_i . r_i}{||d(q)||.||r(e)||}$$

Similarity Scores Based on EAT Embeddings. From the document $d(q)$, the paragraph $p(q)$ where the mention q occurs is extracted. Using the offsets provided in the data set, we extract the previous, current and next sentence to where the mention was found to built $p(q)$. Then, the average value of cosine similarities between each word from paragraph and the entity is calculated (EAT_1).

The cosine similarity is calculated between the average vector from the paragraph and the entity (EAT_2). The average of the top-k (EAT_3) similarities is used as feature[3]. Finally, the cosine similarity between the entity mention and the entity is added (EAT_4). Equations for the four features are defined below.

$$EAT_1(e, p(q)) = \frac{\sum_{w_i \in p(q)} cos(e, w_i)}{||p(q)||} \qquad EAT_2(e, p(q)) = cos(e, \frac{\sum_{w_i \in p(q)} w_i}{||p(q)||})$$

$$EAT_3(e, p(q)) = \frac{\sum\limits_{i=1\ldots k} argmax_{w_i \in p(q)} cos(e, w_i)}{k} \qquad EAT_4(e, w_m) = cos(e, w_m)$$

where $argmax_{w_i} cos(e, w_i)$ returns the i-th most similar word, in terms of cosine similarity, to $p(q)$.

Classifier Trained for Candidate Selection. We then train a binary classifier that associates the given set of features with a decision whether the candidate entity is the correct one for the entity mention. Using the training data, we generate the candidate entities from the entity mentions. The positive examples for the training are then formed by the (entity mention, candidate) pairs that correspond to the expected link in the reference. The negative examples are pairs with wrong candidates generated for the entity mentions. Since the number of candidates generated for each mention may be very high (between 1 and 460 in our experiments), the positive and negative classes are very imbalanced. We deal with this problem by using undersampling, limiting the number of negative examples to be 10 times the number of positive examples[4]. Each decision of the classifier is then weighted by the probability estimate of the classifier and the candidate entity with the highest probability is selected as the final disambiguated entity. In the standard entity disambiguation task, the system must also be capable of determining when an entity mention does not link to any entity in the KB (these are referred as NIL entities). In our approach, this occurs if no candidate is generated or if all candidates are rejected by the classifier.

Due to the particular nature of the feature vectors which combines dimensions of a very different nature such as binary features versus floats, we tested several classifiers. Models such as Adaboost, Random Forests, Decision Trees and SVM models (linear and RBF kernels) were tested. Combining Adaboost with Decision Trees as base estimator turned out to be the best classifier on the training data (using the non-EATs features and a 10-fold cross validation schema). Further results are obtained using this classifier.

---

[3] In our experiments k was fixed to 3.

[4] We tested several values for this ratio between positive/negative sample on the training data and kept the value that achieved the best result.

5   Experiments and Results

5.1   Learning the Embeddings

In order to apply the EAT model, a collection of documents where entities are referenced within the text is needed. This is the case for Wikipedia[5] where each page is considered as an entity. The anchor texts were slightly modified to indicate the part that corresponds to a word and the part that corresponds to an entity. A mapping to DBpedia is constructed for the entity and a prefix allows us to identify the entities. Next section presents our preliminary results of the implementation used for our model, based on *Gensim* or *Hyperwords*.

5.2   Evaluation of the Embeddings

In our first experiments we want to evaluate the quality of the learned embeddings by testing on the well-known analogy data set [18]. The analogy task goes as follows: Given a pair of words between which a relation holds, e.g. 'Paris' - 'France', predict the empty slot for a new pair such as 'Rome' - <?>. (Spoiler: It's 'Italy'). The original analogy data set consists of syntactic and semantic word pairs. Our experiments were focused on the semantic relations. As our intention is to evaluate the quality of the obtained vectors for words and entities, each example was mapped to their string equivalent entities. This process was possible only for four of the five original semantic relations due the missing entities for the family relations in Wikipedia. Note that the remaining four semantic relations deal only with locations or currencies.

Early experiments were performed with the *Hyperwords* tool used to implement the EAT model. First, we used the suggested configuration by the authors of this tool [15] (skip-gram configuration, *negative_sampling* = 5, *window* = 2, *vectors* = *words* + *context*). Results fairly approximate the values previously reported by [15] and outperform values reported by [18]. The EAT model performs slightly worse than the results obtained by the original *Hyperwords* implementation. The smallest difference is up to 1.2% when the addition function is used (61.9%) and the largest is up to 2.8% when the multiplication function is used (67.6%)[6]. This difference between the basic model and the EAT version is due to the additional points (the extra entities) that are represented in the space. A similar situation was observed during our experiments, e.g., lower performances are obtained when the vocabulary size is increased by the modification of the threshold frequency (frequency values under the threshold are filtered out of the training corpus). As mentioned by [15], correct parametrization is a core element to achieve top performances. Indeed, when a high value is used as frequency threshold the results are competitive compared with the state of the art for the task of analogy (see column *EAT-hyperwords* in Table 1), but many entities are missing. The high number of entities filtered out by the threshold highly

---

[5] We used the data dump available in June 2016.

[6] More details about the addition and multiplication functions can be found in [15].

Table 1. Accuracy by semantic subgroup in the analogy task for words/entities using a high value (EAT-*hyperwords*) or low value (EAT-*Gensim*) as frequency threshold.

Subgroup	EAT-*hyperwords*			EAT-*Gensim*	
	words	entities	entity→word	words	entities
capital-com-countries	95.7%	63.0%	87.5%	75.7%	77.5%
capital-world	77.0%	37.3%	81.3%	49.7%	80.0%
currency	8.2%	0.0%	5.2%	0.0%	0.0%
city-in-state	72.3%	25.8%	62.6%	31.7%	89.8%

impacts the performance obtained with the entities. On the other hand, when the threshold frequency is set to a lower value, many entities are represented but the word analogy performance decays, e.g., when parameters are relaxed the performance for words tends to decrease, but more Wikipedia pages are represented by the model. This situation impacts the results when only the entities are used.

Results of individual subgroups are shown in Table 1 for our model using the EAT-*hyperwords* or EAT-*Gensim* implementations. The *words* column reports the results obtained using only words and *entities* column reports the results for their equivalent entity name. Finally, we have reported experiments in which *word* and *entities* are combined in the *entity→words* column. In the later case, we have replaced by their respective word when the entity was not part of the vocabulary. The results clearly outperform the entity-only column and start to approach the strong words-only based results. Indeed, the *entity→words* results outperform the words-only results for the *capital-world* subgroup. In column EAT-*Gensim*, it is reported the results of our EAT-*Gensim* implementation with the relaxed parameters (frequency threshold equal to 30 to words and to 0 for entities). Results for entities clearly improve those for words in subgroups *capital-world* and *city-in-state*, but no significant changes are observed in subgroups *capital-common-countries* or *currency*[7]. Indeed, overall results using semantic and syntactic groups for only-words are clearly less performant (40.31%) than our EAT-*hyperwords* implementation (61.9%). However, EAT-*Gensim* is preferred because more entities are represented despite the fact that EAT-*Gensim* has a worse performance when compared with EAT-*hyperwords*.

Further experiments are performed with the EAT relaxed parameters version, e.g., our EAT implementation based on *Gensim* in order to have the maximum number of words and entities represented in the joint space.

5.3   Dataset and Evaluation Measures for Entity Linking

To validate our approaches on the Entity Linking task, we use the benchmark from the EDL (Entity Discovery and Linking) task of the TAC 2015 evaluation campaign. We only consider the monolingual English Diagnostic Task, where

---

[7] Results for family are not calculated due the missing entities in Wikipedia.

the entity mentions in the query texts are already given as input, since our main focus in this work is on the linking and not the detection of the entity mentions. Table 2 shows the main features of the used data set: the number of documents, the number of entity mentions to disambiguate (the goal of the task is to disambiguate all the entity mentions present in the considered documents), and the number of entity mentions that do not have a corresponding entity in the knowledge base (mentions NIL). The knowledge base used in this campaign is built from Freebase [12]. The whole Freebase snapshot contains $43M$ entities but a filter was applied to remove some entity types that were not relevant to the campaign (such as music, book, medicine and film), which reduced it to $8M$ entities. Among them, only 3712852 (46%) have an associated content in Wikipedia and can thus be associated with an embedding representation. In order to improve the candidate generation process, we also enriched the knowledge base with new entity expressions automatically extracted from Wikipedia: more precisely, we added all links from disambiguation pages and redirection pages as possible forms of the entities.

Table 2. Description of the dataset used in the evaluation process

	TAC 2015 training	TAC 2015 testing
Nb. docs.	168	167
Nb. mentions	12175	13587
Nb. mentions NIL	3215	3379

For the evaluation scores, we used the standard precision/recall/f-score measures on the correct identification of the KB entity and its type when it exists (*link*), on the correct identification of a NIL mention (*nil*) or the combined score for both cases (*all*). Compared to the official evaluation measures from the campaign, we do not consider the evaluation of the clustering of the NIL mentions referring to the same entity. These measures correspond to the *strong_typed_link_match, strong_typed_nil_match* and *strong_typed_all_match* measures from the TAC EDL campaign. Formally, if we note, for an entity mention e, the KB entity e_r associated with e if in the reference, the KB entity e_t associated with it by our system and $N(x)$ the number of entity mentions that verify x, then these measures are defined by:

$$P(nil) = \frac{N(e_t = \text{NIL} \wedge e_r = \text{NIL})}{N(e_t = \text{NIL})} \quad R(nil) = \frac{N(e_t = \text{NIL} \wedge e_r = \text{NIL})}{N(e_r = \text{NIL})}$$

$$P(link) = \frac{N(e_t = e_r \wedge e_t \neq \text{NIL})}{N(e_t \neq \text{NIL})} \quad R(link) = \frac{N(e_t = e_r \wedge e_t \neq \text{NIL})}{N(e_r \neq \text{NIL})}$$

$$P(all) = \frac{N(e_t = e_r)}{N(e_t)}$$

Note that, for the *all* measure, precision, recall and f-score are equal, provided that the system gave an answer for all the entity mentions ($N(e_t) = N(e_r)$).

5.4   Evaluation of Candidate Entity Generation

In this section we discuss the results of the candidate generation. Table 3 presents some statistics on the candidate generation. We denote C the set of candidates, C_{NIL} the set of mentions for which no candidate is proposed, C_{AVG} the average number of candidates per query and Recall(C) the candidate recall, defined by the percentage of non-NIL queries for which the excepted KB entity is present in the set of candidate entities.

Table 3. Statistics on candidate generation.

	$\|C\|$	$\|C_{NIL}\|$	C_{AVG}	Recall(C)
All candidates				
Training	6843513	781	562.1	95.60%
Test	8339648	499	613.8	94.19%
Entity type filtering				
Training	3179795	952	261.2	92.43%
Test	3810382	626	280.4	90.36%
Lucene+Null simil filtering				
Training	1723470	952	141.6	90.27%
Test	1921577	625	141.4	87.95%

Firstly, when considering all strategies for candidate generation, without any filtering, we achieve a high candidate recall, with 95% of the non-NIL entity mentions found among the candidates. We also note that this leads to a large number of candidate entities per mention. When applying a filtering on the entity types, i.e. we keep only the KB entities for which we can derive one of the expected entity types (PER, LOC, ORG, GPE, FAC), we reduce by more than half the number of candidate entities, which give a sounder base for the classifier: even if the recall is decreased (around 90%), the Entity Linking score is improved.

An analysis of the generated candidates also showed that the candidates returned only by Lucene and the candidates for which the similarity scores are both null were not often the right ones: once again, removing these entities before learning the classifier leads to better results, with a global linking f-score of 72.8%, for a candidate recall around 88%. This last strategy is one used in the following results. Further work and analysis on this candidate generation step is needed, in order to determine more sophisticated filtering strategies, that will allow to keep the good candidates without generating too much noise through spurious candidates.

5.5   Entity Linking Results

In Table 4 we present the results obtained for the global Entity Linking task, using the different features proposed in this paper. The *baseline* result is obtained using only the features from the candidate generation and the cosine similarity scores on the textual context. The other results are obtained when adding each of the scores computed with the embeddings with the EAT model. The last column uses all the scores combined. Since our Entity Linking model relies on methods that have some random elements (negative example selection for undersampling and internal sampling from the Adaboost classifier), the results presented are average results on 10 runs.

Table 4. Entity linking results with the EAT model.

	Baseline	$+EAT_1$	$+EAT_2$	$+EAT_3$	$+EAT_4$	$+EAT_{1/2/3/4}$
P(nil)	0.598	0.604	0.608	0.605	0.605	**0.606**
R(nil)	0.815	0.830	0.825	0.828	0.830	**0.838**
F(nil)	0.690	0.699	0.700	0.700	0.700	**0.704**
P(link)	0.796	0.806	0.800	0.804	0.806	**0.814**
R(link)	0.699	0.706	0.706	0.706	0.707	**0.710**
F(link)	0.745	0.752	0.750	0.752	0.753	**0.759**
P(all)	0.728	0.737	0.735	0.737	0.737	**0.742**

These results show a significant improvement of the scores when using the embeddings of the EAT model, over the baseline. We also note that the improvement obtained with each individual EAT feature is comparable and the combined features give the best results.

When compared to the results from the participants to the TAC-EDL 2015 campaign [12], the best F-score result for the linking task was 0.737, on the *strong_typed_all_match* measure. We therefore achieve with this model better results than the state of the art[8]. A close examination of the results shows some examples of the improvements obtained by using a narrow semantic context through the EAT model. For example, in the ambiguous cases where a person is only referred to using his last name, our model is consistently better in selecting the correct entity. In the phrase "As soon as he landed at the Bangkok airport, Koirala saw Modi's tweets on the quake, Nepal's Minister for Foreign Affairs Mahendra Bahadur Pandey said on Tuesday." the mention "Modi" is correctly identified as "Navendra Modi" instead of "Modi Naturals" an oil processing company based in India. Similar performances were observed for the entity type location. For example, the EAT model correctly identified "Montrouge", as the French town near Paris, instead of the french actor Louis (Émile) Hesnard known as "Montrouge" (who was born in Paris) for the sentence "The other loose guy

---

[8] Our evaluation does not take into account the nominal mentions of entities.

who killed a cop in montrouge seems to have done the same. And there are report of two other armed men running around in Paris. It's kind of a chaos here."

6　Conclusion

In this paper we presented a model capable of jointly representing word and entities into a unique space, the EAT model. The main advantage of our model is the capability of representing entities as well as word embeddings in context during training. Our model – based on anchor texts – accurately represents the entities in the jointly learned embedding space, even better that the words because entities (Wikipedia pages) are used in contexts which clearly represent their meanings. Indeed, this is the main advantage of our model, the direct use of contexts for the construction of the entities embeddings skipping an extra alignment task previously used by [8,27,28] or corpus concatenation [29,30].

We showed that the EAT model can be integrated into a standard entity linking architecture without any extra effort. Four different features have been proposed to encode similarities between the context and the candidates. For evaluation, we have used a recent and competitive entity linking dataset of the TAC-EDL 2015 campaign. The results show that individual EAT features as well as their combination helps to improve classical similarity metrics. Our final result for the task $(P(all) = 0.742)$ outperforms the baseline and hypothetically achieves the first position in the mentioned evaluation campaign.

Acknowledgements. This work was supported by the French National Agency for Research under the grant PULSAR-FUI-18 (PUrchasing Low Signals and Adaptive Recommendation).

References

1. Baroni, M., Dinu, G., Kruszewski, G.: Don't count, predict! a systematic comparison of context-counting vs. context-predicting semantic vectors. In: Proceedings of the 52nd Annual Meeting of the ACL, pp. 238–247, June 2014
2. Bordes, A., Usunier, N., Garcia-Duran, A., Weston, J., Yakhnenko, O.: Translating embeddings for modeling multi-relational data. Adv. Neural Inf. Process. Syst. **26**, 2787–2795 (2013)
3. Burkhard, W.A., Keller, R.M.: Some approaches to best-match file searching. Commun. ACM **16**(4), 230–236 (1973)
4. Cao, Z., Tao, Q., Tie-Yan, L., Ming-Feng, T., Hang, L.: Learning to rank: from pairwise approach to listwise approach. In: 24th International Conference on Machine Learning (ICML 2007), Corvalis, Oregon, USA, pp. 129–136 (2007)
5. Cassidy, T., Chen, Z., Artiles, J., Ji, H., Deng, H., Ratinov, L.A., Zheng, J., Han, J., Roth, D.: CUNY-UIUC-SRI TAC-KBP2011 entity linking system description. In: Text Analysis Conference (TAC 2011) (2011)
6. Cucerzan, S.: Large-scale named entity disambiguation based on Wikipedia data. In: 2007 Joint Conference on EMNLP-CoNLL, pp. 708–716 (2007)

7. Dredze, M., McNamee, P., Rao, D., Gerber, A., Finin, T.: Entity disambiguation for knowledge base population. In: 23rd International Conference on Computational Linguistics (COLING 2010), Beijing, China, pp. 277–285 (2010)
8. Fang, W., Zhang, J., Wang, D., Chen, Z., Li, M.: Entity disambiguation by knowledge and text jointly embedding. In: CoNLL 2016, p. 260 (2016)
9. Han, X., Zhao, J.: NLPR_KBP in TAC 2009 KBP track: a two-stage method to entity linking. In: Text Analysis Conference (TAC 2009) (2009)
10. Hoffart, J., Suchanek, F., Berberich, K., Weikum, G.: YAGO2: a spatially and temporally enhanced knowledge base from Wikipedia. Artif. Intell. **194**, 28–61 (2013)
11. Ji, H., Nothman, J., Hachey, B.: Overview of TAC-KBP2014 entity discovery and linking tasks. In: Text Analysis Conference (TAC 2014) (2014)
12. Ji, H., Nothman, J., Hachey, B., Florian, R.: Overview of TAC-KBP2015 tri-lingual entity discovery and linking. In: Text Analysis Conference (TAC 2015) (2015)
13. Le, Q.V., Mikolov, T.: Distributed representations of sentences and documents. In: Proceedings of the 31st ICML, pp. 1188–1196 (2014)
14. Lehmann, J., Monahan, S., Nezda, L., Jung, A., Shi, Y.: LCC approaches to knowledge base population at TAC 2010. In: Text Analysis Conference (2010)
15. Levy, O., Goldberg, Y., Dagan, I.: Improving distributional similarity with lessons learned from word embeddings. Trans. Assoc. Comput. Linguist. **3**, 211–225 (2015)
16. Ling, X., Singh, S., Weld, D.: Design challenges for entity linking. Trans. Assoc. Comput. Linguist. (TACL) **3**, 315–328 (2015)
17. Mihalcea, R., Csomai, A.: Wikify! linking documents to encyclopedic knowledge. In: Proceedings of the Sixteenth ACM Conference on Conference on Information and Knowledge Management, pp. 233–242. ACM, Lisbon (2007)
18. Mikolov, T., Sutskever, I., Chen, K., Corrado, G.S., Dean, J.: Distributed representations of words and phrases and their compositionality. Adv. Neural Inf. Process. Syst. **26**, 3111–3119 (2013)
19. Miller, G.A.: WordNet: a lexical database for English. Commun. ACM **38**(11), 39–41 (1995)
20. Mintz, M., Bills, S., Snow, R., Jurafsky, D.: Distant supervision for relation extraction without labeled data. In: Proceedings of the Joint Conference of the 47th Annual Meeting of the ACL and the 4th IJCNLP, ACL 2009, pp. 1003–1011 (2009)
21. Moro, A., Raganato, A., Navigli, R.: Entity linking meets word sense disambiguation: a unified approach. Trans. Assoc. Comput. Linguist. (TACL) **2**, 231–244 (2014)
22. Navigli, R., Ponzetto, S.P.: BabelNet: the automatic construction, evaluation and application of a wide-coverage multilingual semantic network. Artif. Intell. **193**, 217–250 (2012)
23. Pappu, A., Blanco, R., Mehdad, Y., Stent, A., Thadani, K.: Lightweight multilingual entity extraction and linking. In: Proceedings of the Tenth ACM International Conference on Web Search and Data Mining, WSDM 2017, pp. 365–374. ACM (2017)
24. Shen, W., Jianyong, W., Ping, L., Min, W.: LINDEN: linking named entities with knowledge base via semantic knowledge. In: Proceedings of the 21st International Conference on World Wide Web (WWW 2012), Lyon, France, pp. 449–458 (2012)
25. Shen, W., Wang, J., Han, J.: Entity linking with a knowledge base: issues, techniques, and solutions. Trans. Knowl. Data Eng. **27**, 443–460 (2015)
26. Varma, V., Bharath, V., Kovelamudi, S., Bysani, P., Santosh, G.S.K., Kiran Kumar, N., Reddy, K., Kumar, K., Maganti, N.: IIT Hyderabad at TAC 2009. In: Text Analysis Conference (TAC 2009) (2009)

27. Wang, Z., Zhang, J., Feng, J., Chen, Z.: Knowledge graph and text jointly embedding. In: The 2014 Conference on Empirical Methods on Natural Language Processing. ACL - Association for Computational Linguistics, October 2014
28. Yamada, I., Shindo, H., Takeda, H., Takefuji, Y.: Joint learning of the embedding of words and entities for named entity disambiguation. In: Proceedings of the 20th SIGNLL CoNLL, pp. 250–259 (2016)
29. Zwicklbauer, S., Seifert, C., Granitzer, M.: DoSeR - a knowledge-base-agnostic framework for entity disambiguation using semantic embeddings. In: Sack, H., Blomqvist, E., d'Aquin, M., Ghidini, C., Ponzetto, S.P., Lange, C. (eds.) ESWC 2016. LNCS, vol. 9678, pp. 182–198. Springer, Cham (2016). doi:10.1007/978-3-319-34129-3_12
30. Zwicklbauer, S., Seifert, C., Granitzer, M.: Robust and collective entity disambiguation through semantic embeddings. In: 39th International ACM Conference on Research and Development in Information Retrieval (SIGIR), pp. 425–434 (2016)

Beyond Time: Dynamic Context-Aware Entity Recommendation

Nam Khanh Tran[✉], Tuan Tran, and Claudia Niederée

L3S Research Center, Leibniz Universität Hannover, Hanover, Germany
{ntran,ttran,niederee}@L3S.de

Abstract. Entities and their relatedness are useful information in various tasks such as entity disambiguation, entity recommendation or search. In many cases, entity relatedness is highly affected by dynamic contexts, which can be reflected in the outcome of different applications. However, the role of context is largely unexplored in existing entity relatedness measures. In this paper, we introduce the notion of contextual entity relatedness, and show its usefulness in the new yet important problem of context-aware entity recommendation. We propose a novel method of computing the contextual relatedness with integrated time and topic models. By exploiting an entity graph and enriching it with an entity embedding method, we show that our proposed relatedness can effectively recommend entities, taking contexts into account. We conduct large-scale experiments on a real-world data set, and the results show considerable improvements of our solution over the states of the art.

Keywords: Contextual entity relatedness · Entity recommendation

1 Introduction

Entities are characterized not only by their intrinsic properties, but also by the manifold relationships between them. Quantifying these entity relationships, which is the idea of entity relatedness [6,10,13], is crucial in several tasks such as entity disambiguation [3,7], contextualization of search results, and improved content analysis [14].

Relationships between entities are not always static. While some relationships are robust and static, e.g. the relationship between a country and its cities, others change frequently, driven by dynamic contexts. In these contexts, time is just one dimension, and alone not sufficient to adequately structure the entity relationship texture. This is illustrated for the entity Brad Pitt in Fig. 1. While time is sufficient to structure the realm of his private relationships, there are other groups of related entities with overlapping timelines, such as the persons he co-acted with in films, which relate to other contexts of his life. Such more fine granular, contextual understanding of the entity relationship texture can be used to refine methods such as entity disambiguation and entity recommendation.

In this paper, we introduce the novel notion of *contextual entity relatedness*, with time and topic as two main ingredients, and show its usefulness in a

© Springer International Publishing AG 2017
E. Blomqvist et al. (Eds.): ESWC 2017, Part I, LNCS 10249, pp. 353–368, 2017.
DOI: 10.1007/978-3-319-58068-5_22

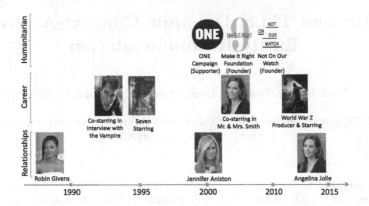

Fig. 1. Related entities with Brad Pitt in different topics and time periods

new yet important problem: Context-aware entity recommendation. We propose to estimate the contextual relatedness using both entity graph extracted from knowledge sources such as Wikipedia, and also to exploit annotated text data using entity embedding methods. Furthermore, while existing work adds temporal aspects into entity relationships [17,20], we go a step beyond by incorporating topic and proposing to enrich the relationships to form a novel *contextual entity graph*: Each entity relation is enriched with the time span and topics indicating *when* and *under which circumstances* it exists.

The main contributions of this paper are: (1) We introduce the idea of a contextual relatedness of entities (2) we define the problem of context-aware entity recommendation for validating the usefulness of contextual relatedness (3) we propose a novel method for tackling the defined problem based on a statistically sound probabilistic model incorporating temporal and topical context via embedding methods, and (4) we evaluate the context-aware recommendation method with large-scale experiments on a real-world data set. The results of the evaluation show the usefulness of contextual entity relatedness as well as the effectiveness of our recommendation method. compared to other approaches.

2   Background and Problem Definition

2.1   Preliminaries

In this work, we use a very general notion of an entity as "a thing with distinct and independent existence" and assume that each entity has a canonical name and is equipped with a unique identifier. Typically, knowledge sources such as Wikipedia or Freebase are used as reference points for identification.

There are relations between entities. These are represented in different ways such as in the form of hyperlinks in Wikipedia or by a fact in an ontological knowledge base asserting a statement between two entities. Entities and their relationships can be captured in an **entity graph**, where the nodes are entities the edges represent relationships between entities.

An entity can be referred to in a text document (e.g. a news article) in the form of an *entity mention*. In our work, we assume that an **annotated corpus** is given, i.e., an annotated text dataset with well disambiguated entities.[1] Such an annotated corpus can be used to create and enrich the entity graph.

We are interested in the relatedness between entities, which is the association of one entity to another. Such a relatedness is often measured by a normalized score indicating the strength of the association. In our work, these scores depend upon the context and we speak of **contextual relatedness**. For ensuring a wide applicability, we use a simple yet flexible model of context, constituted by two dimensions: Time and Topic. We formalize this concept as follows.

Context. *A context c is a tuple* (t, s)*, where t is a time interval* $[t_b, t_e]$ *and s is a topic describing the circumstance of the relationship.*

Our notion of time is a sequence of discrete time units in a specific granularity, e.g. a day. Time points or ranges of other granularities will be mapped to an interval of this granularity. For example, "2016" is converted to [2016-01-01, 2016-12-31]. For the topic s, we use a textual representation. It can be a single word such as *"movies"*, *"wars"*, or a phrase indicating an information interest such as *"scenes in the thriller movie SEVEN"*.

It is important to note that our contextual relatedness is an asymmetric measure, i.e. given a context c, the relatedness of an entity e_2 to an entity e_1 is different from that of e_1 to e_2. For example, in the context (2016, "medals"), 2016_Summer_Olympics is likely to be the highest related entity for Eri_Tosaka, the Japanese female wrestler[2] who won her first Olympics gold medals in Rio. The reversed direction is not true, as there are many winners for the total 306 sets of medals in the games.

2.2   Problem Definition

In this work, we aim to study the usefulness of context in entity relatedness. We do this by undertaking a specific recommendation task, namely *context-aware entity recommendation*. In this task, context reflects a user intent or preference in exploring an entity, and contextual relatedness can be used to guide the exploration. Accordingly, by validating the performance of the recommendation task, the effectiveness of contextual relatedness can be evaluated. More specifically, the input of the recommender system is an entity, which the user wants to explore (e.g., Brad_Pitt), and a context consisting of the aspect she is interested in (e.g., (1995-2015, "awards")); the goal is to find the most related entities given the entity and the context of interest. We give the formal definition as follows:

Context-aware Recommendation: *Given an entity* e_q*, a context of interest* c_q*, an entity graph G, and an annotated corpus D containing annotated and disambiguated entity mentions, find the top-k entities that have the highest relatedness to* e_q *given the context* c_q *(contextual relatedness).*

---

[1] Such collections are increasingly available thanks to the advancement in information extraction research. One example is Freebase annotated KBA dataset: http://trec-kba.org/data/fakba1/.

[2] https://en.wikipedia.org/wiki/Eri_Tosaka.

The query (e_q, c_q) is called an entity-context query. The context-aware entity recommendation problem has some assumptions regarding the query setting. First, query entities can have free text representations, but a *text-to-entity* mapping to resolve the canonical entity name is employed. Such a mapping can be the result of using an entity linking system (e.g., [3]). Second, there is also a map from the textual context representation to the time and topic component, for instance "Black Friday 2016 ads" to ([2016-11-25, 2016-11-25], "ads"). Third, in the absence of time or topic, they will be replaced by some default place holders. For time, we define two special values b_t and e_t to refer to the earliest and latest days represented in the corpus. For topic, we replace missing values by the token "$*$" to indicate an arbitrary topic.

3 Approach

This section gives an overview of our method. In essence, we use a probabilistic model to tackle the recommendation task. To estimate the model, we incorporate different graph enrichment methods. These two components are described below.

3.1 Probabilistic Model

We formalize the context-aware entity recommendation task as estimating the probability $P(e|e_q, c_q)$ of each entity e given a entity-context query (e_q, c_q). The estimation score can be used to output the ranked list of entities. Based on Bayes' theorem, the probability can be rewritten as follows:

$$P(e|e_q, c_q) = \frac{P(e, e_q, c_q)}{P(e_q, c_q)} \propto P(e, e_q, c_q) \tag{1}$$

where the denominator $P(e_q, c_q)$ can be ignored as it does not change the ranking. The joint probability $P(e, e_q, c_q)$ can be rewritten as:

$$\begin{aligned} P(e, e_q, c_q) = P(e, e_q, t_q, s_q) = \quad & P(e_q)P(t_q|e_q)P(e|e_q, t_q)P(s_q|e, e_q, t_q) \\ \overset{\text{rank}}{=} \quad & P(e|e_q, t_q)P(s_q|e, e_q, t_q) \end{aligned} \tag{2}$$

In (2), we drop $P(e_q)$ and $P(t_q|e_q)$ as they do not influence the ranking. The main problem is then to estimate the two components: $P(e|e_q, t_q)$ (*temporal relatedness* model), and $P(s_q|e, e_q, t_q)$ (the *topical relatedness* model).

3.2 Candidate Entity Identification

The entity graph can be very large, e.g. millions of entities and tens of millions of relationships, thus it is costly to estimate $P(e, e_q, c_q)$ for all entities in the graph. To improve the efficiency, we employ a candidate selection process to identify the promising candidates. Given the query (e_q, c_q), we extract all entities directly connected to e_q. Other methods can be used in this step; for example entities that co-occur with the target entity in an annotated corpus can be considered as candidate entities. However, in practice, we observe that this strategy covers sufficiently large amount of entities we need to consider.

3.3   Graph Enrichment

To facilitate the estimation methods for Eq. 2 (see Sect. 4 for more details), we propose to enrich the entity graph, i.e. is to equip all entities as well as their relationships with rich information from the knowledge sources and the annotated corpus. This enrichment extends the entity graph into a **contextual entity graph**, where both nodes and edges are contextualized. We describe the enrichment methods below.

Entity Relationship Enrichment. First, we describe how we enrich the graph edges, i.e. the entity relationships. From the annotated corpus, we extract the set of bounded *text snippets* (e.g. a sentence or paragraph)[3], in which one or multiple entity mentions to the entities can be found. Then, for each edge (e_i, e_j), we construct the set of all text snippets annotating both entities e_i and e_j. For each text snippet, we employ a temporal pattern extraction method to extract the time values, and map them to day granularity, or put a placeholder if no values are found. For each successfully constructed time t, we create a context $c = (t, s)$, where s refers to the textual representation of the snippet. As a result, for each edge (e_i, e_j), we have a set of *relation contexts*, denoted by $C(e_i, e_j)$.

Entity Embedding. To enrich the graph node, i.e. the entity, we propose to learn a continuous vector representation of the entities in the entity graph using a neural network. Our method, *entity embedding*, maps entities to vectors of real numbers so that entities appearing in similar contexts are mapped to vectors close in cosine distance. The vectors can be estimated in a completely unsupervised way by exploiting the distributional semantics hypothesis. Here we extend the Skip-Gram model [9]. The Skip-Gram aims to predict context words given a target word in a sliding window. In our case, we aim to predict context words given a target entity. We train the entities and the words *simultaneously* from the annotated text collection D, using text snippets as the window contexts. Specifically, given a context as a text sequence in which the target entity e appears, i.e., $W = \{w_1, ..., w_M\}$ where w_i might be either an entity or a word, the objective of the model is to maximize the average log probability

$$\mathcal{L}(W) = \frac{1}{M} \sum_{i=1}^{M} \log P(w_i|e) \qquad (3)$$

in which the prediction probability is defined by using a softmax function

$$P(w_i|e) = \frac{\exp(\vec{w_i} \cdot \vec{e})}{\sum_{w \in W} \exp(\vec{w} \cdot \vec{e})} \qquad (4)$$

where $\vec{w}$ and $\vec{e}$ denote the vector representation of w and e respectively. The training example is shown in Fig. 2. The relatedness between two entities e and e_q is then defined as the cosine similarity between their vector representations. In the experiment, we show that the embedding method complements to standard relatedness metrics and help to improve the performance in estimating both models of the contextual relatedness (Eq. 2).

---

[3] In our experiments, we limit to sentences level.

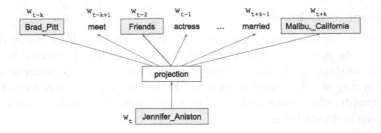

Fig. 2. The training example for the Jennifer Aniston entity

4   Model Parameter Estimation

Our probabilistic model is parameterized by two relatedness models $P(e|e_q, t_q)$ and $P(s_q|e, e_q, t_q)$. In this section, we present in details the estimation of these models based on the contextual entity graph.

4.1   Temporal Relatedness Model

The distribution $P(e|e_q, t_q)$ models the entity relatedness between e and e_q w.r.t t_q. To estimate $P(e|e_q, t_q)$, we take into account both static and dynamic entity relatedness as

$$P(e|e_q, t_q) = \lambda \frac{R_s(e, e_q)}{\sum_{e'} R_s(e', e_q)} + (1 - \lambda) \frac{R_d(e, e_q, t_q)}{\sum_{e'} R_d(e', e_q, t_q)} \tag{5}$$

where $R_s(e, e_q)$ measures the static relatedness between e and e_q, $R_d(e, e_q, t_q)$ measures the dynamic relatedness between e and e_q w.r.t t_q, and λ is a parameter.

Static Relatedness. To measure the static relatedness between entities e and e_q, i.e. $R_s(e, e_q)$, we use the widely adopted method introduced by Milne-Witten et al. using the Wikipedia links [10], and has been effective in various tasks. The Milne-Witten relatedness is measured as:

$$R_s^{MW}(e, e_q) = \frac{\log(\max(|E|, |E_q|)) - \log(|E \cap E_q|)}{\log|V| - \log(\min(|E|, |E_q|))} \tag{6}$$

where E and E_q are the sets of entities that links to e and e_q respectively and V is the set of all entities.

In addition to Milne-Witten, we include the entity embeddings (Sect. 3.3) and define an embedding-based static relatedness measure as the cosine similarity between two corresponding entity vectors:

$$R_s^{Emb}(e, e_q) = \frac{\vec{e} \cdot \vec{e_q}}{\|\vec{e}\|\|\vec{e_q}\|} \tag{7}$$

The two static relatedness measures can be combined in linear fashion to provide the final estimation: $R_s(e, e_q) = R_s^{MW}(e, e_q) + R_s^{Emb}(e, e_q)$.

Dynamic Relatedness. To measure the dynamic relatedness $R_d(e, e_q, t_q)$, we first associate an activation function that captures the importance of an entity e as a function of time: $\alpha_e : T \to \mathbb{R}$. This function can be estimated by analyzing the edit history of Wikipedia, in which the more edits take place for an article in a certain time interval, the higher the value of activation function. Other kinds of estimators are to analyze longitudinal corpora such as news archives. In this work, our estimation is based on Wikipedia page view statistics. The normalized value of the activation function of an entity α_e is estimated as follows:

$$A_e(t) = \frac{\alpha_e(t) - \mu_{\alpha_e}}{\sigma_{\alpha_e}} \text{ with } \mu_{\alpha_e} = \mathbb{E}[\alpha_e] \text{ and } \sigma_{\alpha_e} = \sqrt{\mathbb{E}[(\alpha_e - \mu_{\alpha_e})^2]} \quad (8)$$

where μ_{α_e} and σ_{α_e} are the mean value and standard deviation of the activation function α_e. To assess whether two entities are temporally related, we compare their activity functions. It happens that many entities exhibit very marked peaks of activity at certain points. These peaks are highly representative for an entity. Therefore, we estimate the dynamic relatedness between entities by measuring a form of temporal peak coherence

$$R_d(e, e_q, t_q) = \sum_{t=t_{q_b}}^{t_{q_e}} \max(\min(A_e(t), A_{e_q}(t)) - \theta, 0) \quad (9)$$

where $t_q = [t_{q_b}, t_{q_e}]$ is the time interval of interest and θ is a threshold parameter that is set as 2.5 here to avoid over-interpreting low and noisy values.

4.2 Topical Relatedness Model

The probability $P(s_q|e, e_q, t_q)$ models the likelihood of observing the text snippet s_q in the relationship between entities e and e_q in the time of interest t_q.

For each context $c_i = (t_i, s_i) \in C(e, e_q)$, let $Sim(s_q, c_i, t_q)$ be the similarity between the text snippet s_q and the context c_i w.r.t the time t_q. The likelihood of observing s_q in the relationship between e and e_q w.r.t t_q is estimated as:

$$P(s_q|e, e_q, t_q) = \frac{1}{|C(e, e_q)|} \sum_{c_i \in C(e, e_q)} Sim(s_q, c_i, t_q) \quad (10)$$

Here we assume the context c_i gives less contribution to the overall relevance of the relation w.r.t the time t_q if its time t_i is distant from t_q, then $Sim(s_q, c_i, t_q)$ is estimated as

$$Sim(s_q, c_i, t_q) = \begin{cases} CS(s_q, s_i) \ e^{-\beta|t_q - t_i|}, & \text{if } CS(s_q, s_i) \geq \xi \\ 0, & \text{otherwise} \end{cases} \quad (11)$$

where ξ is a fixed parameter, β is the decay parameter, $|t_q - t_i|$ is the distance between two time intervals t_q and t_i that is calculated by the distance between their middle points. The component $CS(s_q, s_i)$ measures the similarity between two text snippets s_q and s_i. We employ two different methods to estimate $CS(s_q, s_i)$, described below.

Language Model. In this method (called *LM-based*), we represent the relation (e, e_q) by a language model, i.e. the distribution over terms taken from text snippets between two entities in the entity graph. Then by assuming the independence between terms in the snippet s_q, we obtain the following estimation

$$CS(s_q, s_i) = \prod_{w \in s_q} P(w|\theta_{s_i})^{n(w, s_q)} \tag{12}$$

where $n(w, s_q)$ is the number of times the term w occurs in s_q, $P(w|\theta_{s_i})$ is the probability of term w within the language model of the snippet s_i which is estimated with Dirichlet smoothing as follows

$$P(w|\theta_{s_i}) = \frac{n(w, s_i) + \mu \cdot P(w)}{\sum_{w'} n(w', s_i) + \mu} \tag{13}$$

where $n(w, s_i)$ is the frequency of w in s_i, $P(w)$ is the collection language model, and μ is the Dirichlet smoothing parameter.

Embedding Model. The second method is an adaptation of the Word Mover's Distance (WMD) method proposed in [8]. First, we remove all stop words and keep only content words in the text snippets. Then, we define the similarity between two text snippets s_q and s_i using a relaxed version of WMD, where each word in s_q (and s_i) is mapped to its most similar word in s_i (and s_q):

$$CS(s_q, s_i) \propto \frac{1}{2} \left(\frac{\sum_{w \in s_q} \sum_{w' \in s_i} \mathbf{T}_{ww'} \cos(\vec{w}, \vec{w'})}{|s_q|} + \frac{\sum_{w \in s_i} \sum_{w' \in s_q} \mathbf{T}_{ww'} \cos(\vec{w}, \vec{w'})}{|s_i|} \right) \tag{14}$$

where $|s_q|$ and $|s_i|$ are the number words in the text snippets s_q and s_i respectively, $\mathbf{T}_{ww'} = 1$ if $w' = \text{argmax}_{w'} \cos(\vec{w}, \vec{w'})$ or 0 otherwise, $\cos(\vec{w}, \vec{w'})$ is cosine similarity between two vectors. The vector $\vec{w}$ and $\vec{w'}$ are the vector embeddings of the words w and w', respectively learned from the Entity Embedding method described in Sect. 3.3. We denote this as the *WMD-based* method.

5   Experiment Setup

5.1   Entity Graph Construction

The entity graph we use in the context-aware entity recommendation task is derived from Freebase [4] and Wikipedia.[4] More specifically, we extract

---

[4] English Wikipedia dump version dated March 4, 2015.

Wikipedia articles that overlap with Freebase topics, resulting in $3,866,179$ distinct entities, each corresponding to one article. To extract the entity activities for the dynamic temporal relatedness model, we use Wikipedia page view counts[5] in the time frame $01/01/2012$ to $05/31/2016$.

We use the text contents of the articles as the annotated corpus D. Note that due to Wikipedia editing guidelines, an article often ignores the subsequent annotations of an entity in the text, if the entity is already annotated before. For example, within the Wikipedia article of entity Brad_Pitt, *Angelina Jolie* is mentioned 32 times but only 5 of these mentions are annotated. Hence, we employ a machine learning method [11] to identify more entity mentions. In average, 12 new entity mentions were added to each Wikipedia article.

To extract text snippets for the graph enrichment, we cleaned and parsed the sentences from the contents, resulting in 108 millions sentences in total. We use Stanford Temporal Tagger[6] to extract temporal patterns from these annotated sentences. For the edges of the entity graphs, we establish the undirected edge (e_1, e_2) if the corresponding Wikipedia article of e_1 or e_2 (after adding new mentions using [11]) contains a hyperlink to the article of the other.

5.2   Automated Queries Construction

We use the recently published Wikipedia clickstream dataset [18] from February 2015 and structural information from Wikipedia for constructing entity-context queries and the Ground Truth.

The clickstream dataset contains about 22 million (referrer, resource) pairs and their respective request count extracted from the request logs of the main namespace of the English Wikipedia. The referrers can be categorized in internal and external traffic; in this work, we only focus on request pairs stemming from internal Wikipedia traffic, i.e., referring page and requested resource are both Wikipedia pages from the main namespace.

Wikipedia articles are collaboratively and iteratively organised in sections and paragraphs, such that each section is concerned with particular aspects or contexts of the entity profile [5]. Each entity mentions within these sections are therefore highly relevant to the source entity in the respective context.

Based on these observations, we propose an automated entity-context query construction using the following heuristics: (i) For each pair of source and target entities, we first extract the section heading where the target entity is mentioned in the source page (ii) The source entity is then used as query entity and the extracted heading is used as context to create a entity-context query; here we filter out noisy headings such as *"further reading"*, *"see also"*. (iii) We only keep queries for which at least 5 entities are clicked in the clickstream dataset.

To construct the query time, we use the publication time of clickstream dataset, which is February 2015, and convert it to [2015-02-01,2015-02-28].

---

[5] https://dumps.wikimedia.org/other/pagecounts-ez/.
[6] http://nlp.stanford.edu/software/sutime.shtml.

Table 1. Example of entity-context queries and related entities with the number of clicks extracted from the clickstream dataset

Entity	Context	Related entities
Brad Pitt	Humanitarian and political causes	University of Missouri (101), John Kerry (80), Barack Obama (26)...
Brad Pitt	Career	Fury (2014 film) (1772), Mr. & Mrs. Smith (2005 film) (973), Legends of the Fall (893)...
Brad Pitt	Personal life	Angelina Jolie (16564), Jennifer Aniston (11306), Gwyneth Paltrow (3383)...
Brad Pitt	In the media	Supercouple (798), People (magazine) (126)...

Table 1 presents example queries created for the entity Brad_Pitt. In total, we have 219, 844 entity-context queries. To accommodate the impact of time in the queries, we define the *ratio of views*, denoted by r, which is the ratio between the number of times the entity was clicked in February 2015 and in January 2015. The intuition is that if r is very high, the corresponding query entities and topics might have some underlying information interests emerging in February 2015 (for instance, the release of a new movie, etc.). We divide our query set into 4 subsets based on different value ranges of r (Table 2).

Table 2. The different set of queries Q_r with varying ratios of interest

Query set	$Q_{r>0}$	$Q_{r>1}$	$Q_{r>5}$	$Q_{r>10}$
Number of queries	219,844	69,489	1,263	493

Ground Truth. For each query in the query set, we establish the ground truth through the click information available in the clickstream dataset. Existing work suggests that the Wikipedia viewing behaviour can be used as a good proxy of entity relevance to current user interest [12,15]. Transferring this idea to navigational traffic within Wikipedia networks (as they are reflected in the click streams), we can consider an increased navigation between two entities as a signal for the importance of the relationship between the corresponding source and the target entities.

Thus, given an entity-context query, the larger number of clicks a candidate entity gets, the higher related the entity is. Based on this, for each query we take the most clicked entity as the relevant entity, and measure how good recommendation approaches rank the entity using MRR metric. In addition, we extract the top-5 clicked entities for each query to measure the recall. We publish our code and data to encourage future similar research.[7]

---

[7] http://www.l3s.de/~ntran/dycer.html.

Evaluation Metrics. To measure the performance of different approaches, we use two evaluation metrics. The first metric is *mean reciprocal rank* (**MRR**) which is computed as

$$MRR = \frac{1}{|Q_{test}|} \sum_{i=1}^{|Q_{test}|} \frac{1}{rank(e_{q_i})} \tag{15}$$

where $|Q_{test}|$ is the number of queries, and $rank(e_{q_i})$ represents the rank of the ground truth entity e_{q_i} in the results for the query q_i. Notice that a larger MRR indicates better performance.

We also use **recall** at rank k ($R@k$) as another evaluation metric. $R@k$ is measured as the ratio of the retrieved and relevant entities up to rank k over the total number of relevant results. The larger $R@k$ indicates better performance.

5.3 Baselines

We implemented several baselines to compare to our methods on the task. The first group of baselines are static methods using an ad hoc ranking function without considering the given context. We consider the baselines that only use Milne-Witten or entity embeddings-based relatedness, and the combination. We denote these static methods as **Static**$_{mw}$, **Static**$_{emb}$, and **Static**$_{mw\&emb}$.

The second group of baselines are time-aware methods which are similar to our probabilistic model but without taking into account the search topic s_q. We reimplemented the approach proposed by [20] and extended it by combining the entity embedding and link based similarities to integrate into the model. We denote these time-aware methods as **Temp**$_{mw}$ [20] and **Temp**$_{mw\&emb}$.

Finally, we denote our methods as **Dycer**$_{lm}$ and **Dycer**$_{wmd}$ where **Dycer**$_{lm}$ uses the LM-based method and **Dycer**$_{wmd}$ uses the WMD-based method for estimating the similarity between text snippets.

Parameter Settings. We empirically set the similarity threshold ξ to 0.35, and the decay parameter β to 0.5. The Dirichlet smoothing parameter is fixed to 2000, and the parameter λ is set to 0.3 by default and will be discussed in detail in the experiments.

6   Results and Discussion

Figure 3 presents a detailed comparison between the MRR for the different methods. The proposed methods outperform the baselines on all query sets. In addition, when increasing the ratio of views r, our method progressively improves, with its highest score $MRR = 0.282$ on the query set $Q_{r>10}$. In contrast, the performance of the static methods is not changed much and around $MRR = 0.145$. This conforms the effectiveness of our model in capturing the dynamic contexts. Even without context, our relatedness model ($Static_{mw\&emb}$) already performs better compared to the $Static_{mw}$ and $Static_{emb}$ methods. Interestingly, the time-aware methods gain comparable, even worse results compared

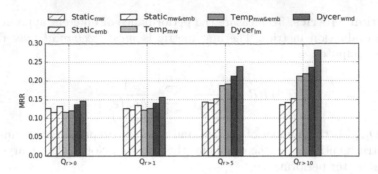

Fig. 3. Performance of the different approaches on the different query sets

to the static methods on the query sets $Q_{r>0}$ and $Q_{r>1}$, however they obtain significantly better MRR scores on the query sets $Q_{r>5}$ and $Q_{r>10}$. This can be explained by the fact that the entities in $Q_{r>5}$ and $Q_{r>10}$ are more sensitive to time because of high user interests. Furthermore, the adapted implementation $Temp_{mw\&emb}$ outperforms the original method $Temp_{mw}$, which again indicates the effectiveness of the combination of the embedding-based and link-based methods. The best overall performing approach is the WMD-based method $Dycer_{wmd}$. The method performs better than the LM-based method $Dycer_{lm}$, which is due to the fact that the WMD-based method takes into account the semantic meaning of words using word embeddings for the textual similarity estimation, while the LM-based method purely uses the surface form of words.

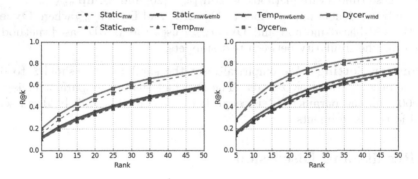

Fig. 4. $R@k$ for the different entity recommendation approaches under comparison. (*Left*) All queries $Q_{r>0}$. (*Right*) Queries with high ratios $Q_{r>5}$

Next, we analyse the recall at rank k ($R@k$) as quality criteria. The results of $R@k$ with varying k for different methods are shown in Fig. 4. We compute the performance of methods on the different query sets $Q_{r>0}$ and $Q_{r>5}$. Figure 4 shows that the proposed methods outperform the baselines on both sets of queries. On the first query set $Q_{r>0}$ the WMD-based method $Dycer_{wmd}$ gains

7.9%, 10.5%, 15.2%, and 14.3% improvements compared to the static method $Static_{mw\&emb}$, and 9.3%, 13.0%, 16.6% and 17.6% improvements compared to the time-aware method $Temp_{mw\&emb}$ when the rank k is 5, 10, 20, and 30 respectively. On the query set $Q_{r>5}$, it even obtains much better improvements. In addition, similar to our findings for MRR, the time-aware methods achieve comparable results compared to the static methods overall, but perform considerably better on the query set with the high ratio of views $Q_{r>5}$.

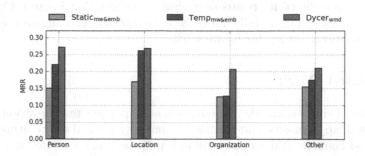

Fig. 5. MRR of relevant entity for different query entity types in $Q_{r>5}$ and for different approaches (note, we show the results for the best method in each group)

In addition, we also compare the performance in different query types, as for each type, users often have different intents and expectations. Figure 5 shows the comparison in terms of MRR for four groups of high-level types. It can be seen that the performance differences vary quite noticeably in different type groups. Nevertheless, in all cases, the highest result is achieved by the $Dycer_{wmd}$ approach. Interestingly, for the type "*Person*" and "*Location*" the $Temp_{mw\&wmd}$ approach gains large improvements compared to the static method. One possible explanation for this is that the "*Person*" and "*Location*" entities usually involve in events which highly relate to time. Consequently, taking time into account helps improving the performance. In the case of "*Organization*", the time-aware method does not show any improvement compared to the $Static_{mw\&emb}$ method whereas the WMD-base method still obtains a huge improvement. It demonstrates the usefulness of contextual information for the task.

Table 3 shows the impact of λ on the performance of the time-aware and the proposed method using the query set $Q_{r>5}$. The $\lambda = 0.3$ yields the best results on average using both methods, which is then used as in our experiments.

Table 3. MRR of relevant entity using the query set $Q_{r>5}$ for different λ (with the best results in bold)

Method	$\lambda = 0$	$\lambda = .1$	$\lambda = .2$	$\lambda = .3$	$\lambda = .4$	$\lambda = .5$	$\lambda = .6$	$\lambda = .7$	$\lambda = .8$	$\lambda = .9$	$\lambda = 1$
$Temp_{mw\&emb}$	0.1700	0.1894	0.1902	**0.1914**	0.1913	0.1913	0.1903	0.1906	0.1888	0.1898	0.1511
$Dycer_{wmd}$	0.2153	0.2371	**0.2372**	**0.2372**	0.2359	0.2366	0.2365	0.2360	0.2361	0.2363	0.1828

Discussions. While we use Wikipedia for building the model in the experiments, the proposed approach can also use other knowledge bases (e.g. Freebase) to construct the entity graph, and any text collections (e.g. news archives, web archives) can also be used to enrich the entity graph. Our choice of using Wikipedia is driven by the availability of rich and high-quality meta-data in the collection, which enables us to focus on the effectiveness of the models. In addition, we focus on frequent entities in our experiments, however the proposed method leverages both link structures and the textual representation from the document collection to estimate the entity relatedness; thus we believe that it can achieve good performance with the long-tail entities, as been shown in existing approaches [6]. These evaluations are left for future work.

7   Related Work

Estimation of entity semantic relatedness is an important task in various semantic and NLP applications, and has been extensively studied in literature [10,13]. Strube and Ponzetto [13] proposed using Wikipedia link structures and the hierarchy of Wikipedia categories to provide a light-weight related estimation. Milne and Witten [10] followed a similar approach, and carefully designed the relatedness measure based on Wikipedia incoming links, inspired by the Google distance metric. These methods are close to our work in the sense that we also combine various similarity measures, but do so in an advanced probabilistic model, taking into account context information. Hence, while the aforementioned works are static, our proposed measure is context-aware and dynamic to time.

One main issue with relatedness measures based on link structures is that they perform poorly for long-tail entities with little or no connections. Hoffart et al. [6] (KORE) addressed this issue by extracting key phrases from surrounding texts of entity mentions, and incorporate the overlaps of such key phrases between two entities. In our work, we also use the text surrounding of entity mentions. However, in contrast to KORE that uses these texts to enrich the entities, we use the texts to enrich the relations between entities, and in this regard, can contextualize the relatedness directly. In addition, KORE is still a static quantity, while our measure is fully dynamic to time and context.

Several approaches have been proposed to add temporal dimension to entity semantic relationships [16,17]. Wang et al. [17] extracted temporal information for entities with focus on infobox, categories and events. Tuan et al. [16] also extracted information from infobox and categories, but defined a comprehensive model comprising time, location and topic. However, these studies are limited to predefined types of relations, and cannot be easily extended to address the semantic relatedness. Recently, Zhang et al. [20] incorporated various correlation metrics to complement the semantic relatedness, proposed a new metric that is sensitive to time. We extend this work, but incorporate time and topic in an consistent context model, and also introduce the entity embedding method.

From the application perspective, entity recommendation is one of the directed applications of entity semantic relatedness. It assumes the input entities encode some user activities or information needs, and suggests a list of entities, normally ordered, that are most relevant. Blanco et al. [2] introduced Spark that links a user search query to an entity in a knowledge base and suggests a ranked list of related entities for further exploration. Similarly, [1,19] proposed personalized entity recommendation which uses several features extracted from user click logs. Our work is distinguished from this work in that we take into account context as an additional information need, not just input entities.

8 Conclusions

In this work, we have presented the concept of contextual entity relatedness and proposed effective methods for using it in the task of entity recommendation. We have shown the usefulness of the contextual information for this task as well as the effectiveness of our method.

Our work leaves ample space for further investigations in contextual entity relatedness. One research direction is the investigation of further more fine granular context dimensions. In addition, we have planned to look into more depth the evolution of entity relationship and to exploit contextual information for improving performance of other tasks such as entity disambiguation.

Acknowledgments. This work was partially funded by the German Federal Ministry of Education and Research (BMBF) for the project eLabour (01UG1512C).

References

1. Bi, B., Ma, H., Hsu, B.J.P., Chu, W., Wang, K., Cho, J.: Learning to recommend related entities to search users. In: WSDM (2015)
2. Blanco, R., Cambazoglu, B.B., Mika, P., Torzec, N.: Entity recommendations in web search. In: Alani, H., Kagal, L., Fokoue, A., Groth, P., Biemann, C., Parreira, J.X., Aroyo, L., Noy, N., Welty, C., Janowicz, K. (eds.) ISWC 2013. LNCS, vol. 8219, pp. 33–48. Springer, Heidelberg (2013). doi:10.1007/978-3-642-41338-4_3
3. Blanco, R., Ottaviano, G., Meij, E.: Fast and space-efficient entity linking in queries. In: WSDM (2015)
4. Bollacker, K., Evans, C., Paritosh, P., Sturge, T., Taylor, J.: Freebase: a collaboratively created graph database for structuring human knowledge. In: SIGMOD (2008)
5. Fetahu, B., Markert, K., Anand, A.: Automated news suggestions for populating wikipedia entity pages. In: CIKM (2015)
6. Hoffart, J., Seufert, S., Nguyen, D.B., Theobald, M., Weikum, G.: KORE: keyphrase overlap relatedness for entity disambiguation. In: CIKM (2012)
7. Hoffart, J., Yosef, M.A., Bordino, I., Fürstenau, H., Pinkal, M., Spaniol, M., Taneva, B., Thater, S., Weikum, G.: Robust disambiguation of named entities in text. In: EMNLP (2011)
8. Kusner, M.J., Sun, Y., Kolkin, N.I., Weinberger, K.Q.: From word embeddings to document distances. In: ICML (2015)
9. Mikolov, T., Chen, K., Corrado, G., Dean, J.: Efficient estimation of word representations in vector space. In: ICLR (2013)

10. Milne, D., Witten, I.H.: An effective, low-cost measure of semantic relatedness obtained from wikipedia links. In: AAAI (2008)
11. Noraset, T., Bhagavatula, C., Downey, D.: Adding high-precision links to wikipedia. In: EMNLP (2014)
12. Ratkiewicz, J., Flammini, A., Menczer, F.: Traffic in social media i: paths through information networks. In: SocialCom (2010)
13. Strube, M., Ponzetto, S.P.: Wikirelate! computing semantic relatedness using wikipedia. In: AAAI (2006)
14. Tran, N.K., Ceroni, A., Kanhabua, N., Niederée, C.: Supporting interpretations of forgotten stories by time-aware re-contextualization. In: WSDM (2015)
15. Tran, T.A., Niederee, C., Kanhabua, N., Gadiraju, U., Anand, A.: Balancing novelty and salience: adaptive learning to rank entities for timeline summarization of high-impact events. In: CIKM (2015)
16. Tuan, T.A., Elbassuoni, S., Preda, N., Weikum, G.: CATE: context-aware timeline for entity illustration. In: WWW (2011)
17. Wang, Y., Zhu, M., Qu, L., Spaniol, M., Weikum, G.: Timely YAGO: harvesting, querying, and visualizing temporal knowledge from wikipedia. In: EDBT (2010)
18. Wulczyn, E., Taraborelli, D.: Wikipedia clickstream (2015)
19. Yu, X., Ma, H., Hsu, B.J.P., Han, J.: On building entity recommender systems using user click log and freebase knowledge. In: WSDM (2014)
20. Zhang, L., Rettinger, A., Zhang, J.: A probabilistic model for time-aware entity recommendation. In: Groth, P., Simperl, E., Gray, A., Sabou, M., Krötzsch, M., Lecue, F., Flöck, F., Gil, Y. (eds.) ISWC 2016. LNCS, vol. 9981, pp. 598–614. Springer, Cham (2016). doi:10.1007/978-3-319-46523-4_36

Vocabularies, Schemas, and Ontologies Track

Patterns for Heterogeneous TBox Mappings to Bridge Different Modelling Decisions

Pablo Rubén Fillottrani[1,2] and C. Maria Keet[3(✉)]

[1] Departamento de Ciencias e Ingeniería de la Computación,
Universidad Nacional del Sur, Bahía Blanca, Argentina
prf@cs.uns.edu.ar
[2] Comisión de Investigaciones Científicas,
Buenos Aires, Provincia de Buenos Aires, Argentina
[3] Department of Computer Science,
University of Cape Town, Cape Town, South Africa
mkeet@cs.uct.ac.za

Abstract. Correspondence patterns have been proposed as templates of commonly used alignments between heterogeneous elements in ontologies, although design tools are currently not equipped with handling these definition alignments nor pattern alignments. We aim to address this by, first, formalising the notion of design pattern; secondly, defining typical modelling choice patterns and their alignments; and finally, proposing algorithms for integrating automatic pattern detection into existing ontology design tools. This gave rise to six formalised pattern alignments and two efficient local search and pattern matching algorithms to propose possible pattern alignments to the modeller.

1 Introduction

Ontology developers face choices on how to represent the subject domain knowledge during the ontology authoring stage. In some instances, there really is a right and a wrong way of representing it—e.g., not confusing subsumption with parthood—but this is not always the case. For instance, whether Marriage should be represented as a class or as an object property. Some OWL files may also lean more toward being a logic-based conceptual data model for a single application, rather than have its knowledge represented such that many applications may avail of it, such as with ontologies for ontology-based data access applications [2] versus foundational ontologies such as DOLCE [16], respectively. In the context of ontology alignment, such different modelling decisions result in modeller-intended alignments that cannot be found automatically with current alignment tools [19,23] and may have a mismatch logically despite a deemed 'sufficient semantic equivalence' conceptually. For instance, ontology O_1 may have Employee ⊑ Person, which is a typical subsumption in conceptual modelling, versus Person ⊑ PhysicalObject, Employee ⊑ Role, and Employee ⊑ ∃inheresIn.Person in ontology O_2 that is inspired by foundational ontology modelling practices, and where PhysicalObject ⊑ ¬Role is also asserted. Aligning Employee and Person

© Springer International Publishing AG 2017
E. Blomqvist et al. (Eds.): ESWC 2017, Part I, LNCS 10249, pp. 371–386, 2017.
DOI: 10.1007/978-3-319-58068-5_23

across these two ontologies with a simple 1:1 equivalence axiom will then result in an unsatisfiable Employee in O_1. This and similar situations raise the question of how to link or integrate O_1 with O_2 when there is no single vocabulary element to match with in *both* ontologies, or when intuitively the knowledge represented is deemed sufficiently the same with respect to the subject domain knowledge, but an alignment will result in an inconsistency nevertheless.

This question can be recast as one of somehow having to manage and resolve heterogeneous modelling *patterns*, rather than the customary pairwise alignments of single elements [3,19]. This requires the capability to align patterns *as a whole*, and thus also involve different types of elements that will have to be bridged, such as a class and an object property, which has been noted widely as an open issue in ontology matching [19]. Ontology Design Patterns (ODPs) have a type of pattern called correspondence pattern, with a few correspondences proposed in [20–22] and several submissions are included in the ODP catalogue[1]. They are mostly definition alignments, however, i.e., with a class on the left-hand side of the equivalence and a pattern (the definition of the class) on the right-hand side (automated only recently [7]), rather than patterns on both sides. Ghidini et al. [8] focus on logic and reasoning complexity and cover only a small subset of possible correspondences with single heterogeneous elements. In short, heterogeneous alignments between modelling patterns is a manual, individual, process at present.

We aim to address this mapping impasse by identifying common modelling patterns, their rationale, and providing a formalisation of the corresponding ODPs, inclusive of a formalisation of an ontology pattern itself. This resulted in six complex alignments between modelling patterns. Secondly, we show that *automated finding* of the patterns and *checking correctness* of a possible pattern-based alignment is indeed possible. Algorithmically, finding such patterns can be reduced to localised pattern matching in the ontology. We present two such algorithms for one of the pattern alignments to demonstrate that feasibility; the algorithms for the other patterns follow the same design. They are being implemented in the ICOM tool [4] that already caters for inter-ontology assertions between single elements in different OWL ontologies and reason over them.

The remainder of this paper is structured as follows. Section 2 describes related work. Sections 3 and 4 present the main theoretical contribution of the paper with the formal representation of a design pattern and the pattern alignments. Section 5 describes the optimised algorithms, which is followed by a discussion (Sect. 6) and conclusions (Sect. 7).

2   Related Works

The work most closely related to ours is that of [20–22], who present mostly definition alignments as correspondence patterns within the context of ontology design patterns. That is, there is one named concept in ontology O_1 that matches with an axiom in O_2, such as PinotageWine in a simple wine ontology O_1

---

[1] http://www.ontologydesignpatterns.org.

being aligned to Wine ⊓ ∃grapeVariety.Pinotage from a more detailed *wyn ontologie* O_2. Based on the alignment naming, Scharffe et al.'s (no longer available) library [22] also includes a few homogeneous mappings that involve more than one element, such as a range restriction. Somewhat related to this are entity transformations within the same ontology [25], which covers single-entity swaps and one pattern similar to one in [21]. Such correspondences were, at the time, a substantial extension to the 1:1 mappings common in ontology matching [3,19] and independent of later automated alignment efforts [19]. Definition alignments and automatic finding thereof now has a logic-based tool to assist with this [7].

Other recent efforts in automated ontology matching has commenced considering aligning domain ontologies to a foundational ontology [23], which brings afore a separate set of issues. One is that such matching involves mostly subsumption rather than equivalence [23]. The domain ontology↔foundational ontology alignment issues regarding modelling choices, such as a process as a class (e.g., Running) or an object property (*Op*) (e.g., runs) and those mentioned in the introduction of this paper, were not addressed. Logic-based approaches for heterogenous alignments were investigated in [8] that focussed on the class ↔ *Op* mappings (DL concept, role), such as the marriage example mentioned in the introduction, and *Op* ↔ *Dp* (DL role, attribute) mappings, using bridge rules. These two patterns are a subset of the range needed for heterogeneous alignments and the logic is a subset of OWL ($\mathcal{ALCQI}_b$). The Distributed Ontology Language [17] takes another approach by defining a framework by which to link and extend logical theories. This is useful for the case where the ontologies are represented in different logics, but it does not consider patterns yet.

Conceptual model alignment patterns have been proposed by [5,14], which resemble some of the cases in ontologies, such as their class↔relationship and the relationship↔attribute, and ORM Value Type↔UML Attribute. They used a rule-based approach to inter-model links in general and more applied with ATL in particular, which is popular in model-driven engineering but not immediately usable for ontologies, and they avail of a particular unifying metamodel to mediate between the conceptual models, which is different for OWL.

In sum, while some reusable ideas have been proposed, logic-based linking of heterogeneous patterns between ontologies it yet to be solved.

3   Formal Representation of Patterns and Alignments

We need a way to formally describe a pattern before being able to declare links between patterns. Such precision is useful for determining a logical or a 'subject domain semantics' equivalence or subsumption or another motivation for aligning the patterns. To the best of our knowledge, there is no formal definition of an ODP yet, therefore we introduce this first before the actual alignment patterns. Note that the focus is on the ontology and logic aspects; for a specification of ODP documentation and metadata, the reader is referred to [12].

Definition 1 (Language of pattern instantiation). *OWL Ontology O with language specification adhering to the W3C standard [18], which has classes $C \in V_C$, object properties $OP \in V_{OP}$, data properties $D \in V_D$, data types $DT \in V_{DT}$ of the permitted XML schema types, axiom components ('language features') $X \in V_X$, and such that $Ax \in V_{Ax}$ are the axioms.*

Of course, this could be another ontology language, but it is easier for the current presentation to cast it in the context of OWL. The 'axiom components' in Definition 1 include features such as functionality on an object property "≤ 1", transitivity, 'at least one' "$\exists$", and cardinality that can be used according to the syntax of the language.

The pattern itself is a meta-level specification, alike UML's stereotyping. In addition, because some patterns will refer to categories from a foundational ontology, yet which are also included in the ontology, we acknowledge their status here with respect to the context of a pattern.

Definition 2 (Language for patterns: Vocabulary $\mathcal{V}$). *The meta-level (second order) elements (or* stereotypes*) for patterns are:*

- *class $C \in V_C$ as $\mathcal{C}$ in the pattern;*
- *object property $OP \in V_{OP}$ as $\mathcal{R}$ in the pattern;*
- *data property $D \in V_D$ as $\mathcal{D}$ in the pattern;*
- *data type $DT \in V_{DT}$ as $\mathcal{DT}$ in the pattern;*
- *reserved set of entities from a foundational ontology, as $\mathcal{F}$ in the pattern;*

where added subscripts i with $1 \leq i \leq n$ may be different elements. Two elements in the vocabulary are called homogeneous *iff they belong to the same type, i.e., they are both classes, or both object properties, and so on. Elements can be used in axioms $Ax \in V_{Ax}$ that consists of axiom components $x \in V_X$ in the pattern such that the type of axioms are those supported in the ontology language in which the instance of the pattern is represented.*

Subsumption and equivalence axiom components relate *homogeneous* elements; functionality and at least one axiom components relate *heterogeneous* elements from the vocabulary. The set of "reserved entities" in $\mathcal{F}$ depends on which foundational ontology is used; e.g., for DOLCE, this set includes, among others, *Perdurant, Endurant*, and *qt* and for BFO, *Independent Continuant* and *Quality*.

An ontology pattern P can now be defined as follows.

Definition 3 (Ontology Pattern P). *An ontology pattern P consists of more than one element from vocabulary $\mathcal{V}$ which relate through at least one axiom component from V_X. Its specification contains the:*

- *pattern name;*
- *pattern elements from $\mathcal{V}$;*
- *pattern axiom component(s) from V_X;*
- *pattern's full formalisation.*

For instance, a simple *named class subsumption* pattern with axiom component $\sqsubseteq$ (in V_X) and two named classes is formalised as $C_1 \sqsubseteq C_2$, which can be instantiated with classes, e.g., Human, Animal $\in V_C$ as, in DL notation, Human $\sqsubseteq$ Animal. A slightly more comprehensive one, the *basic all-some* pattern, then has as specification: *pattern name*: *basic all-some*, with *pattern elements*: C_1, C_2, $\mathcal{R}$, the *pattern axiom component(s)*: $\sqsubseteq$, $\exists$, and the *pattern's full formalisation*: $C_1 \sqsubseteq \exists \mathcal{R}.C_2$. An instantiation of this pattern is, e.g., Professor $\sqsubseteq$ ∃teaches.Course.

Now that we have a precise notion of patterns and instantiations, we can proceed to define mappings between elements in different ontologies in two steps: the component of common homogeneous *mappings* and the homogeneous with heterogeneous *alignments* that are typically needed to align patterns as a whole. The former is defined in Definition 4, which is a shorthand version of the lengthy specification in [3], and the latter in Definition 5.

Definition 4 (Homogeneous mapping). *Let O, O' be two ontologies with vocabularies V, V'. A mapping is a subsumption or equivalence axiom relating two homogeneous elements, one in V and the other in V'.*

For example, to relate the class Teacher in O to Instructor in O' with the mapping Instructor $\sqsubseteq$ Teacher. Mappings provide bindings between homogeneous elements in different ontologies, constituting the basis for ontology pattern alignment.

Definition 5 (Ontology Pattern Alignment, OPA). *An ontology pattern alignment OPA consists of two ontology patterns, P and P', such that its signature Σ is a subset of the signature of the respective ontologies O and O', i.e., $\Sigma(P) \subseteq \Sigma(O)$ and $\Sigma(P') \subseteq \Sigma(O')$, and alignment axioms*

- *alignment pattern name;*
- *pattern elements;*
- *alignment patterns' context, consisting of:*
  - *O's pattern P*
  - *O''s pattern P'*
- *alignment pattern axiom component(s) from V_X;*
- *pattern alignment's formalisation, composed of:*
  - *a (possibly empty) set of mappings between homogeneous elements in P and P'*
  - *a set of axioms made from components in V_X connecting heterogeneous elements in P and P'*

An OPA thus relates different patterns in separate ontologies, based on previous knowledge of class or Op alignments.

Note that here, and in the remainder of the paper, a pattern must neither be inconsistent nor have unsatisfiable classes or Ops. That is, a pattern is well-formed, and verified to be so before any alignment of patterns will occur.

4 Aligning Alternate Modelling Patterns

In presenting the patterns, we first introduce those motivated by 'conceptual' or 'subject domain semantics-motivated' mappings that have sufficient semantic approximation and are typical for differences in modelling decisions between domain ontologies and foundational ontologies (Sect. 4.1). Subsequently, we analyse, update, and formalise those presented elsewhere, which are mostly of the definitorial type rather than true pattern mappings (Sect. 4.2). In interest of space, we omit from the presentation the "pattern axiom component(s)", for they they are principally relevant for the language fragments aspects, which is not the current scope and they can be seen from the full formalisation anyway.

4.1 Matching Modelling Patterns

Five pattern alignments are introduced, of which common examples are shown in Fig. 1. While they may look different, some are slight variants of the same pattern, so, in the interest of space, we formalise the principal cases only.

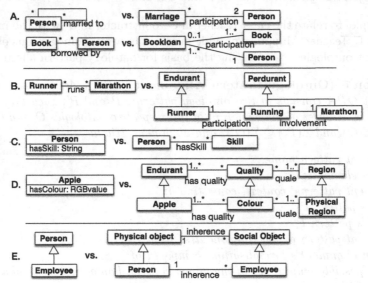

Fig. 1. Examples of the modelling patterns. A: generic class vs *Op*; B: Perdurant class vs *Op*; C: class vs. *Dp*; D: qualities vs *Dp* s; E: example on representing roles.

A: Class vs. Object Property (*Op*) is the more generic, unconstrained version of Fig. 1-B ('Perdurant class vs. *Op*'; see below),which is called *granularity mismatch* in [6]. The typical example is Marriage. It may well be that most practical cases turn out to be of that pattern because typically verbs are used as names of *Op* s, which, in turn, are typically reified as dolce:Perdurant or bfo:Occurrent.

- *alignment pattern name*: class-OP
- *pattern elements*: $C_1, C_2, \mathcal{R}_1$ from O, $C'_3, C'_4, C'_5, \mathcal{R}'_2, \mathcal{R}'_3$ from O'
- *alignment pattern' contexts*:
  - *pattern P in O*: $\exists \mathcal{R}_1.C_2 \sqsubseteq C_1$ and $\exists \mathcal{R}_1^-.C_1 \sqsubseteq C_2$;
  - *pattern P' in O'*: $\exists \mathcal{R}'_2.C'_4 \sqsubseteq C'_3$, $\exists \mathcal{R}'^-_2.C'_3 \sqsubseteq C'_4$, $\exists \mathcal{R}'_3.C'_5 \sqsubseteq C'_3$, $\exists \mathcal{R}'^-_3.C'_3 \sqsubseteq C'_5$, $C'_3 \sqsubseteq (\exists \mathcal{R}'_2)$, and $C'_3 \sqsubseteq (\exists \mathcal{R}'_3)$.
- *pattern's full formalisation*:
  - *homogeneous mappings*: between C_1 and C'_4 and between C_2 and C'_5, which may be subsumption or equivalence relations.
  - *heterogeneous alignments*: $\exists \mathcal{R}_1 \sqsubseteq C'_3$, $\exists \mathcal{R}_1^- \sqsubseteq C'_3$, $C'_3 \sqsubseteq \exists \mathcal{R}_1 \sqcap \exists \mathcal{R}_1^- \sqcap (\leq 1 \mathcal{R}_1) \sqcap (\leq 1 \mathcal{R}_1^-)$.

Figure 1-A includes two instantiations of this pattern. In the first one, $C_1 \equiv C_2$ for Person, $C'_4 \equiv C'_5$ for Person as well, and the pattern associates the role married to ($\mathcal{R}_1$) with the class Marriage (instantiating C'_3). There must be an equivalence or a subsumption mapping between the Person classes in both ontologies. Note that due to cardinality constraints in P' that are not in P, the alignment *de facto* generates a new property that is a sub property of $\mathcal{R}_1$.

B: Perdurant Class vs. Op. This mismatch tends to occur more often when aligning a domain ontology to a foundational ontology (FO). FOs typically have a class Perdurant (Occurrent) for objects unfolding in time (cf. endurants that are wholly present at each time instant they exist), which is the branch in the TBox to represent entities such as Running and Performance. They relate to endurants, like a runner participates in a running (event) like a marathon. Domain ontologies may use a modelling approach more typical of conceptual modelling, where there would not be a 'running', but an *Op* runs between a class runner and the class marathon, and an *Op* performs etc. That is, the choice between the verb or its reification. For the FO, there are thus three classes (runner, running, and marathon) with two *Op* s, whereas in the domain ontology, there are two classes (runner and marathon) and one *Op*. The formalisation of this pattern is similar to case A, but $\mathcal{F}$ is non-empty such that $C'_3 \sqsubseteq PD$ (DOLCE's Perdurant, or an equivalent [13]) and at least C_1, C_2, C'_4 and C'_5 in the pattern are subsumed by ED (Endurant) or its equivalent.

C: Class vs Data Property (Dp). A typical example that also persists in conceptual modelling, is whether 'skill' should be modelled as a class Skill with subclasses or as a *Dp* hasSkill with data values. From an ontological viewpoint, the former is better, as the latter diminishes the chances of interoperability and usability of ontologies, but it does occur. Note that in the pattern alignment, we take *Dp* s to be specialisations of *Op* s, which are recast as a subsumption between two unnamed classes.

- *alignment pattern name*: class-DP
- *pattern elements*: $\mathcal{C}_1, \mathcal{D}_1, \mathcal{DT}_1$ from O, $\mathcal{C}'_2, \mathcal{C}'_3, \mathcal{R}'_1$ from O'.
- *alignment pattern contexts*:
    - *pattern P in O*: $\exists \mathcal{D}_1 \sqsubseteq \mathcal{C}_1$ and $\exists \mathcal{D}_1.\mathcal{DT}_1 \sqsubseteq \mathcal{C}_1$;
    - *pattern P' in O'*: $\exists \mathcal{R}'_1 \sqsubseteq \mathcal{C}'_2$, and $\exists \mathcal{R}'^{-}_1 \sqsubseteq \mathcal{C}'_3$.
- *pattern's full formalisation*:
    - *homogeneous mappings*: $\mathcal{C}_1 \sqsubseteq \mathcal{C}'_2$;
    - *heterogeneous alignments*: $\exists \mathcal{D}_1.\mathcal{DT}_1 \sqsubseteq \exists \mathcal{R}'_1.\mathcal{C}'_3$

Op vs Dp. This is a consequence of the 'class vs data property' choice, for the Dp turns into a class, which therewith forces the introduction of an Op. The details of Op vs. Dp are a fragment of the previous one, and is therefore omitted.

D: Qualities vs. Dps. This distinction is another case of FO modelling vs the more widely-known option from conceptual modelling practices. Instead of a Dp such as hasColour and hasHeight that are known in philosophy as *attributions*, one can turn that into a class of type Quality (or similar) and then have two properties, one from the class of type Endurant to the Quality and one from the Quality to a Region for the value space. This is a modification of case C, for it entails 'class vs. data property' as the attribute is turned into a class, yet it is not just any class but specifically Quality from a FO (i.e., it is in $\mathcal{F}$), with associated Op s and a subclass of Abstract $\in \mathcal{F}$ for the category of value regions, and associated constraints.

- *alignment pattern name*: quality-DP
- *pattern elements*: $\mathcal{C}_1, \mathcal{D}_1, \mathcal{DT}_1$ from O, $\mathcal{C}'_2, \mathcal{C}'_3, \mathcal{C}'_4, \mathcal{C}'_5, \mathcal{C}'_6, \mathcal{C}'_7, \mathcal{R}'_1, \mathcal{R}'_2$ from O' where $\mathcal{C}'_3, \mathcal{C}'_5, \mathcal{C}'_7, \mathcal{R}'_1, \mathcal{R}'_2 \in \mathcal{F}$.
- *alignment pattern contexts*:
    - *pattern P in O*: $\exists \mathcal{D}_1.\mathcal{DT}_1 \sqsubseteq \mathcal{C}_1$;
    - *pattern P' in O'*: $\mathcal{C}'_2 \sqsubseteq \mathcal{C}'_3$, $\mathcal{C}'_4 \sqsubseteq \mathcal{C}'_5$, $\mathcal{C}'_6 \sqsubseteq \mathcal{C}'_7$ $\exists \mathcal{R}'_1.\mathcal{C}'_5 \sqsubseteq \mathcal{C}'_3$, $\exists \mathcal{R}'^{-}_1.\mathcal{C}'_3 \sqsubseteq \mathcal{C}'_5$, $\exists \mathcal{R}'_2.\mathcal{C}'_7 \sqsubseteq \mathcal{C}'_5$, $\exists \mathcal{R}'^{-}_1.\mathcal{C}'_5 \sqsubseteq \mathcal{C}'_7$, $\mathcal{C}'_5 \sqsubseteq \exists \mathcal{R}'^{-}_1.\mathcal{C}'_3$, $\mathcal{C}'_5 \sqsubseteq \exists \mathcal{R}'_2.\mathcal{C}'_7$, and these are inherited down the hierarchy for $\mathcal{C}'_2$, $\mathcal{C}'_4$, and $\mathcal{C}'_6$.
- *pattern's full formalisation*: (in addition to the "alignment pattern contexts")
    - *homogeneous mappings*: $\mathcal{C}_1 \sqsubseteq \mathcal{C}'_2$;
    - *heterogeneous alignments*: $\exists \mathcal{D}_1.\mathcal{DT}_1 \sqsubseteq \exists \mathcal{R}'_1.(\exists \mathcal{R}'_2.\mathcal{C}'_6)$

If $\mathcal{F}$ draws from DOLCE's vocabulary, then $\mathcal{C}'_3 \equiv ED$ (Endurant), $\mathcal{C}'_5 \equiv Q$ (Quality), $\mathcal{C}'_7 \equiv R$ (Region), $\mathcal{R}'_1 \equiv qt$ ('has quality'), and $\mathcal{R}'_2 \equiv ql$ ('has quale'); see [13] for their mappings to BFO and GFO.

E: Representing Roles. The differences for this case can be traced back to both conceptual modelling practices and OntoClean [9] that conflict with FO guidelines. For instance, one can represent Employee $\sqsubseteq$ Person in an ontology, or, more generally in OntoClean terminology, that an anti-rigid property (like

Employee) is subsumed by a rigid one (like Person). FOs put those anti-rigid properties—they being the roles played by rigid properties (Independent Continuant in BFO or Physical Object in DOLCE)—in another branch in the taxonomy, typically as subclasses of, e.g., Role in BFO or Social Object in DOLCE. The two are then related through an Op generally known as *inherence*, where the role inheres in the physical object or the physical object is the bearer of the role.

- *alignment pattern name*: subs-Role-inherence
- *pattern elements*: C_1, C_2 from O, $C'_3, C'_4, C'_5, C'_6, R'_1$ from O' and $C'_4, C'_6, R'_1 \in \mathcal{F}$.
- *alignment pattern contexts*:
  - *pattern P in O*: $C_1 \sqsubseteq C_2$;
  - *pattern P' in O'*: $C'_3 \sqsubseteq C'_4$, $C'_5 \sqsubseteq C'_6$, $\exists R'_1.C'_6 \sqsubseteq C'_4$, $\exists R'_1.C'_5 \sqsubseteq C'_3$, $\exists R'^-_1.C'_4 \sqsubseteq C'_6$, $\exists R'^-_1.C'_3 \sqsubseteq C'_5$, $C'_6 \sqsubseteq = 1\,R'_1.C'_4$, $C'_5 \sqsubseteq = 1\,R'_1.C'_3$.
- *pattern's full formalisation*:
  - *homogeneous mappings*: $C_1 \equiv C'_5$, $C_2 \equiv C'_3$;
  - *heterogeneous alignments*: the subsumption relation in O aligns with $\mathcal{R}_1$, which is not expressible in DL or OWL.

If $\mathcal{F}$ draws from DOLCE, then $C'_4 \equiv POB$ (Physical object), $C'_6 \equiv SOB$ (Social object), and $\mathcal{R}'_1 \equiv OD$ (one-sided constant dependence) or $\mathcal{R}'_1 \equiv OGD$ (one-sided generic constant dependence); see [13] for their mappings to BFO and GFO.

4.2 Assessment and Formalisation of Other Correspondence Patterns

As noted, correspondence patterns have been proposed before in [20–22] and there are several submissions in the ODP catalogue. The analysis of these CPs brings afore the difference between patterns that are *patterns* and those that are essentially *definition mappings*. That is, the former has more than one vocabulary element on the left-hand side of the mapping (recall Definition 3 of an ODP) *and* more than one vocabulary element on the right-hand side of the inclusion or equivalence, whereas the latter has *one* vocabulary element on the left-hand side and more than one on the right-hand side (or vv.). We first provide the analysis of the proposed CPs and refine and formalise the the patterns afterward.

Ritze et al. [20] surveyed multiple sources, and proposed four patterns based on that, three of which are of the definition-mapping type:

CAT: Class Attribute Type: $C_1 \equiv \exists R_2.C_2$, where R can be an Op or Dp, with as example: PositiveReviewedPaper $\equiv$ Paper $\sqcap \exists$hasEvaluation.positive [20]. By that example, R is a Dp, not an Op, and the right-hand side in the pattern misses a parent class. This would need to be separated into a CAT for Op s and CAT for Dp s, so as to recognise the pattern properly in alignment tools. The omission of the named class on the rhs seems to have been unintended, for [20] note it is the same as the CAT in the ODP catalogue (see below), which does include the class on the rhs, as does its 'inverse', below.

CAT$^-$: $C_1 \equiv C_2 \sqcap \exists R^-_2.C_3$; e.g., Researcher $\equiv$ Person $\sqcap \exists$researchedBy$^-$.$\top$ [20].

CAV: Class Attribute Value, with nominals: $C_1 \equiv \exists R_2.\{...\}$, with as example submittedPaper $\equiv$ submission.{true}, where 'true' is discussed as if it were a Boolean but is represented as one of the values, like the one for passing the course as hasExamScore.{A, B, C, D} [20]. However, with nominals/one-of, which is a class, CAV turns out to be a variant of CAT.

PC: $R_1 \equiv S_2 \circ S_3$, where S_3 is a Dp [20]. However, OWL2 does not permit a chain combining Op s and a Dp, and if S_3 were to be an Op, then the ontologies+mapping assertion goes beyond OWL 2 DL, for true role composition is undecidable. This can be corrected by asserting subsumption rather than equivalence, which brings us back to our pattern D.

Scharffe and Fensel [21,22] claim to have a library of 35 patters, but the URL in the paper is broken, therefore, we assess only those described in [21,22], three of which are definitional. Those patterns all have observed instances, such as when trying to align the wine ontology with the ontologie du vin, DOAP with OSSV, and FOAF with itself, whose examples are omitted here for brevity.

CAT: Class Attribute Type: $C_1 \equiv C_2 \sqcap \exists R_2.C_3$ [21], as intended by [20].

CRD: $C_1 \sqsubseteq \exists R_2.(\exists S_2.datatype) \equiv D_3 \sqsubseteq \exists R_3.datatype$, so S_2 and R_3 are Dp s [21], and noting that the 'Property-Relation Correspondence' in [22] is a fragment of CRD. This is a sort of specialisation of our pattern D, as a Dp is a kind of Op.

UI: Union and intersection patterns: they were not formalised in [21,22], but are obvious, being: $C_1 \equiv C_2 \sqcup C_3$ and $C_1 \sqcap C_2 \equiv C_3$.

The other correspondence patterns listed in [22] include homogeneous mappings of one or more entities, such as 'Equivalent Attribute' and domain/range axioms, whereas a 'Class to instance' mapping constitutes a modelling error in either O or O' and is therefore not considered here.

The ODP catalogue (see footnote 1) has two types of correspondence patterns: reengineering and alignment patterns. The former can be seen as a 'swapping' of elements within the same ontology for syntax refactoring, whereas the latter are matchings between two ontologies. There are 13 submitted alignment patterns. They include the aforementioned Class Union (UI), CAT, CAV, Class by path attribute value (alike the PC of [20]), Class correspondence by relation domain with an unsupported AttributeOccurenceCondition, (named) Class Equivalence, (named) Class Subsumption, disjointness between (named) classes, and three Vocabulary Alignment Patterns (VAP) that have the same formalisation in the catalogue, amounting to an Op subsumption where the subsumed one has a domain and range axiom whose classes are aligned as well.

This results in the following two heterogeneous alignments inspired by [21]'s CRD and the catalogue's VAP:

Class-role-attribute pattern

- *alignment pattern name*: *class-OP-DP*
- *pattern elements*: $\mathcal{C}_1$, $\mathcal{R}_1$, $\mathcal{D}_1$, $\mathcal{DT}_1$ from O, and $\mathcal{C}'_2$, $\mathcal{D}'_2$, $\mathcal{DT}'_2$ from O'.
- *alignment pattern' contexts*:
  - *pattern*   *P*   *in*   *O*: $\mathcal{C}_1 \sqsubseteq \exists \mathcal{R}_1.(\exists \mathcal{D}_1.\mathcal{DT}_1)$;
  - *pattern P' in O'*: $\mathcal{C}'_2 \sqsubseteq \exists \mathcal{D}'_2.\mathcal{DT}'_2$.
- *pattern's full formalisation*:
  - *homogeneous mappings*: $\mathcal{C}_1 \equiv \mathcal{C}'_2$, $\mathcal{DT}_1 \equiv \mathcal{DT}'_2$;
  - *heterogeneous alignments*: $\exists \mathcal{D}'_2 \sqsubseteq \exists \mathcal{R}_1.(\exists \mathcal{D}_1.\mathcal{DT}_1)$.

Vocabulary alignment pattern

- *alignment pattern name*: *OP-subs*
- *pattern elements*: $\mathcal{C}_1$, $\mathcal{R}_1$, $\mathcal{C}_2$ from O, and $\mathcal{C}'_3$, $\mathcal{R}'_2$, and $\mathcal{C}'_4$ from O'.
- *alignment pattern' contexts*:
  - *pattern P in O*: $\mathcal{C}_1$, $\mathcal{R}_1$, $\mathcal{C}_2$ (i.e., three independent entities);
  - *pattern P' in O'*: $\exists \mathcal{R}'_2.\top \sqsubseteq \mathcal{C}'_3$, $\exists \mathcal{R}'^-_2.\top \sqsubseteq \mathcal{C}'_4$.
- *pattern's full formalisation*:
  - *homogeneous mappings*: $\mathcal{C}_1 \sqsubseteq \mathcal{C}'_3$, $\mathcal{C}_2 \sqsubseteq \mathcal{C}'_4$, $\mathcal{R}_1 \sqsubseteq \mathcal{R}'_2$;
  - *heterogeneous alignments*: $\exists \mathcal{R}_1.\top \sqsubseteq \mathcal{C}_1$, $\exists \mathcal{R}^-_1.\top \sqsubseteq \mathcal{C}_2$.

The formalisation of the six OPAs, and thus 12 ODPs, presented in this section constitute typical cases of modelling decisions. Although this might still be shown to be not exhaustive, it does comprise a systematic approach for extension with other complex alignments.

5   Alignment Pattern Search and Checking Algorithms

In order to show viability of our proposal, we introduce in this section two algorithms for automated finding of, and checking correctness for, for handling pattern A from Sect. 4.1. These algorithms can be incorporated in an ontology development tool based on a DL/OWL reasoner, evidencing patterns are practically relevant in ontology engineering (lest the pattern becomes an abstract constraint relegated to a handwaiving manual check at best). We assume the tool already has the capability of simultaneously handling two or more ontologies, and also can represent inter-ontology homogeneous mappings between classes and between Op s. ICOM [4] is an example of such a tool.

Inter-ontology mappings can be explicitly added by the user, or implicitly detected by the tool assisted by the reasoner. Both algorithms are described using calls to OWLink [15] services in order to make them independent of the actual DL/OWL reasoner used in the tool.

Algorithm 1 searches for all possible pattern matching instantiations in any pair of ontologies such that the homogeneous mappings have already been analysed. The search is done based on reasoner services and proposes to the user to select which of the instantiations found are meaningful for the ontology integration process. It uses OWLink's IsOPSatisfiable for checking Op satisfiability and GetSubClasses with the 'direct' flag activated in order to get the immediate descendants of a class expression, and it needs to be run each time before accepting a suggestion made by the user. Algorithm 1's running time is proportional to $m^2 r^3 c$, where m is the number of detected homogeneous mappings, r the number of Op s and c the number of classes in the integration scenario.

Algorithm 1. Alignment pattern **A** Search

Precondition: All mappings between ontologies O and O' already found. Mappings
are of the form $CrelC'$ with $C \in \mathcal{V}_C$, $C' \in \mathcal{V}'_C$ and rel $\in \{\sqsubseteq, \sqsupseteq, \equiv\}$.

1: **function** ALIGNMENT PATTERN **A** CANDIDATES( )
2:     $A \leftarrow \emptyset$ ▷ A is the set of candidate alignments
3:     **for** each pair of mappings $C_1\mathrm{rel}_1C'_4$ and $C_2\mathrm{rel}_2C'_5$, and relations $R_1 \in \mathcal{V}_{OP}$ and
    $R'_2, R'_3 \in \mathcal{V}'_{OP}$ **do**
4:         **for** each ClassSynset S in GetSubClasses($O', \exists R'_2 \sqcap \exists R'_3$,direct=true) **do**
5:             $C'_3 \leftarrow$ a representative class in S
6:             **if** (IsOPSatisfiable(O', $(\exists R'_2.C'_4 \sqsubseteq C'_3) \sqcap (\exists R'^-_2.C'_3 \sqsubseteq C'_4)$) and
7:                 IsOPSatisfiable(O', $(\exists R'_3.C'_5 \sqsubseteq C'_3) \sqcap (\exists R'^-_3.C'_3 \sqsubseteq C'_5)$) and
8:                 IsOPSatisfiable(O, $(\exists R_1.C_2 \sqsubseteq C_1) \sqcap (\exists R^-_1.C_1 \sqsubseteq C_2)$)) **then**
9:                 Add $\langle R_1 \in O, C'_3, R'_2, R'_3 \in O' \rangle$ to A
10:            **end if**
11:        **end for**
12:    **end for**
13:    **return** A
14: **end function**

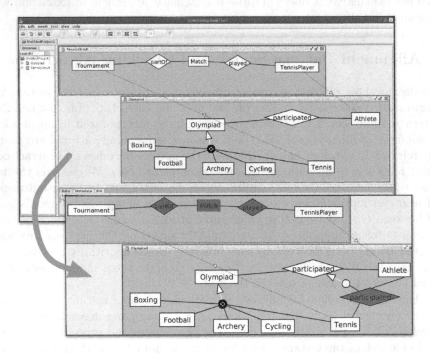

Fig. 2. ICOM integration project with mappings (top) and pattern alignment suggestion (inset).

This is acceptable for being on-demand available when the user considers the homogeneous mappings stable.

To illustrate the algorithm with an example, suppose we are integrating two pre-existing ontologies, represented as different ICOM schemas within the same grand project as shown in Fig. 2. The **TennisCircuit** ontology (top one) contains knowledge about tennis matches and the **Olympiad** ontology (second one) describes athletes who participated in Olympic competitions (setting aside whether they are 'good'). ICOM's homogeneous mappings are shown with thin cross-ontology lines, being TennisPlayer ⊑ Athlete and Tennis ⊑ Tournament, which may have been explicitly added by the user, detected by deduction from the tool, or (when integrated) suggested by an ontology alignment technique [1]. After assessing the mappings, the user may ask the tool to find possible alignment patterns. It runs Algorithm 1, and in line 5 it finds class Match such that it is connected to partOf and played. Then it asks the reasoner to check if these two *Op* s are related to the already mapped classes Tournament and TennisPlayer from **TennisCircuit** (lines 6–7), and finds participated that connects the mapped classes Tennis and Athlete from **Olympiad** in line 8. Given everything is satisfiable, it sets a new candidate alignment connecting a subproperty of participated in **Olympiad** with the concept Match in **TennisCircuit**, and presents it to the user in a different colour as depicted in Fig. 2 (inset). Observe that the matched *Op* in **Olympiad** is not directly defined over the mapped class, but instead over a class higher in the hierarchy. Therefore, the suggested alignment is based on the restriction of the original *Op*. The role of the reasoner in the algorithm is essential in order to resolve which *Op* is applicable for the alignment.

Algorithm 2 checks whether a proposed pattern A instantiation follows the formalised pattern properties. This would admit the alignment pattern to be included in the integration process. The application scenario for this algorithm is that the designer suspects the pattern is relevant for the integration, and has already set the alignments in the project. Then she asks tool to verify all conditions for the alignment pattern are met. For this algorithm there is no precondition for its execution, and it simply communicates with the reasoner checking satisfiability of the pattern formalisation within both ontologies.

Algorithms for the other patterns can be developed analogously, and are not shown due to space limitations. If the pattern references a FO, as in pattern B, D, and E, this fact can be used to shorten the algorithm execution time by restricting the search of candidates to only the descendants of the FO's elements in the pattern. Pattern C can be detected by exhaustively analysing each *Dp* involved in a given mapped class. Pattern D can be handled by similar algorithms as Algorithms 1 and 2 by replacing the *Op* in the first ontology (*O*) with a *Dp*, and then the search shortened with the FO optimisation.

6 Discussion

The patterns and their alignments proposed in the previous sections allow for both specifying and finding more complex alignment than the customary 1:1

Algorithm 2. Alignment pattern **A** Check

1: **function** ALIGNMENT PATTERN **A** CHECK($R_1 \in \mathcal{V}_{OP}$, $C_3' \in \mathcal{V}_C'$, $R_2'.R_3' \in \mathcal{V}_{OP}'$)
2:     **for** each ClassSynSet S_1 in GetSubClasses(O, $\exists R_1$, direct=true) **do**
3:         **for** each ClassSynSet S_2 in GetSubClasses(O, $\exists R_1^-$, direct=true) **do**
4:             $C_1, C_2 \leftarrow$ a representative class in S_1, S_2
5:             **if** not IsOPSatisfiable(O, $(\exists R_1.C_2 \sqsubseteq C_1) \sqcap (\exists R_1^-.C_1 \sqsubseteq C_2)$ **then**
6:                 **return** *false*
7:             **end if**
8:         **end for**
9:     **end for**
10:     **for** each ClassSynSet S_3 in GetSubClasses(O', $\exists R_2'^-$, direct=true) **do**
11:         **for** each ClassSynSet S_4 in GetSubClasses(O', $\exists R_3'^-$, direct=true) **do**
12:             $C_4', C_5' \leftarrow$ a representative class in S_3, S_4
13:             **if** not IsOPSatisfiable(O', $(\exists R_2'.C_4' \sqsubseteq C_3') \sqcap (\exists R_3'.C_5' \sqsubseteq C_3') \sqcap (\exists R_2'^-.C_3' \sqsubseteq C_4') \sqcap (\exists R_3'^-.C.3' \sqsubseteq C_5')$ **then**
14:                 **return** *false*
15:             **end if**
16:         **end for**
17:     **end for**
18:     **return** *true*
19: **end function**

---

mappings and definition mappings. Reconsidering the ODP formalisation of Sect. 3, this bears similarity to the framework in [21], for it takes into account a certain 'layering' of components, from the actual pattern instantiations in an ontology represented in a particular language and their mapping to another one (called "ground correspondences" in [21]), a "correspondence level" where the correspondences between the patterns are represented in an abstract format cf. the ground correspondence in the application, and the abstraction of that as a correspondence pattern. Having patterns on both sides of the (equivalence or subsumption) mapping bridge essentially extends these layers to the more complex scenarios of pattern alignments considered in this paper. Further, the algorithms have been designed such that one can do a local search compared to brute force exhaustive searches across the whole ontology. This speeds up the alignment process.

The pattern alignments do not come 'for free' computationally, however. Some issues and consequences of aligning patterns were already mentioned in passing: one cannot assert that some relation R in ontology O is the same as $S \circ T$ in ontology O', for $R \equiv S \circ T$ is undecidable, and thus outside OWL 2 DL. Other combinations may also result in undecidability whilst the separate ontologies are within OWL 2 DL, especially due to the regularity constraint on role hierarchies [10]. Our alignments do remain within OWL 2 DL, but it does not guarantee to remain within the O's or O''s fragments; e.g., pattern alignment A adds a conjunction.

While the scope here is complex alignments of ontologies, one also could use the results obtained as a basis to *replace* patterns in one single ontology, i.e., moving toward a *refactoring* of a single ontology as a preparatory step for alignment. For instance, to prepare a domain ontology for linking to a foundational ontology, which thereby can substantially extend the entity transformations of PatOMat [25]. It is expected to be useful also for the process of transforming a foundational ontology inspired domain ontology into an 'application ontology' for ontologically well-founded conceptual models, whose ideas were proposed in [11,24], but that yet still lack a formalisation and automation.

7 Conclusions

The paper introduced a first formalisation of an ontology design pattern, as a prerequisite for complex alignments between different common modelling patterns. Twelve patterns for six pattern alignments were motivated from a modelling viewpoint and formalised, augmented with two alignments inspired by definition alignments. They are supported by two efficient local search and pattern matching algorithms that can propose possible pattern alignments to the modeller. We are currently implementing the algorithms in ICOM. As future work, the proposed pattern alignments may be recast as *refactoring* patterns within a same ontology that may be of use in test-driven development of ontologies or facilitate the automation of the link to ontology-driven conceptual modelling.

References

1. Ardjani, F., Bouchiha, D., Malki, M.: Ontology-alignment techniques: survey and analysis. Int. J. Modern Edu. Comput. Sci. **7**(11), 67 (2015)
2. Calvanese, D., Liuzzo, P., Mosca, A., Remesal, J., Rezk, M., Rull, G.: Ontology-based data integration in EPNet: production and distribution of food during the Roman Empire. Eng. Appl. AI **51**, 212–229 (2016)
3. Euzenat, J., Shvaiko, P.: Ontology Matching. Springer, Heidelberg (2007)
4. Fillottrani, P.R., Franconi, E., Tessaris, S.: The ICOM 3.0 intelligent conceptual modelling tool and methodology. Semant. Web J. **3**(3), 293–306 (2012)
5. Fillottrani, P.R., Keet, C.M.: Conceptual model interoperability: a metamodel-driven approach. In: Bikakis, A., Fodor, P., Roman, D. (eds.) RuleML 2014. LNCS, vol. 8620, pp. 52–66. Springer, Cham (2014). doi:10.1007/978-3-319-09870-8_4
6. Gangemi, A.: Ontology design patterns for semantic web content. In: Gil, Y., Motta, E., Benjamins, V.R., Musen, M.A. (eds.) ISWC 2005. LNCS, vol. 3729, pp. 262–276. Springer, Heidelberg (2005). doi:10.1007/11574620_21
7. Geleta, D., Payne, T.R., Tamma, V.: An investigation of definability in ontology alignment. In: Blomqvist, E., Ciancarini, P., Poggi, F., Vitali, F. (eds.) EKAW 2016. LNCS (LNAI), vol. 10024, pp. 255–271. Springer, Cham (2016). doi:10.1007/978-3-319-49004-5_17
8. Ghidini, C., Serafini, L., Tessaris, S.: Complexity of reasoning with expressive ontology mappings. In: Eschenbach, C., et al. (eds.) Proceedings of FOIS 2008. Frontiers in Artificial Intelligence and Applications, vol. 183, pp. 151–163. IOS Press (2008)
9. Guarino, N., Welty, C.: An overview of OntoClean. In: Staab, S., Studer, R. (eds.) Handbook on Ontologies, pp. 151–159. Springer, Heidelberg (2004)

10. Horrocks, I., Kutz, O., Sattler, U.: The even more irresistible $\mathcal{SROIQ}$. In: Proceedings of KR-2006, pp. 452–457 (2006)
11. Jarrar, M., Demy, J., Meersman, R.: On using conceptual data modeling for ontology engineering. J. Data Semant. **1**(1), 185–207 (2003)
12. Karima, N., Hammar, K., Hitzler, P.: How to document ontology design patterns. In: Proceedings of 7th WS on Ontology Patterns (WOP 2016), Kobe, Japan, 18 October 2016
13. Khan, Z.C., Maria Keet, C.: Foundational ontology mediation in ROMULUS. In: Fred, A., Dietz, J.L.G., Liu, K., Filipe, J. (eds.) IC3K 2013. CCIS, vol. 454, pp. 132–152. Springer, Heidelberg (2015). doi:10.1007/978-3-662-46549-3_9
14. Khan, Z.C., Keet, C.M., Fillottrani, P.R., Cenci, K.: Experimentally motivated transformations for intermodel links between conceptual models. In: Pokorný, J., Ivanović, M., Thalheim, B., Šaloun, P. (eds.) ADBIS 2016. LNCS, vol. 9809, pp. 104–118. Springer, Cham (2016). doi:10.1007/978-3-319-44039-2_8
15. Liebig, T., Luther, M., Noppens, O., Wessel, M.: OWLlink. Semant. Web J. **2**(1), 23–32 (2011)
16. Masolo, C., Borgo, S., Gangemi, A., Guarino, N., Oltramari, A.: Ontology library. WonderWeb Deliverable D18 (ver. 1.0, 31-12-2003) (2003)
17. Mossakowski, T., Kutz, O., Codescu, M., Lange, C.: The distributed ontology, modeling and specification language. In: Proceedings of 7th International WS Modular Ontologies (WoMo 2013). CEUR-WS, vol. 1081, Corunna, Spain, 15 September 2013
18. Motik, B., Patel-Schneider, P.F., Parsia, B.: OWL 2 web ontology language structural specification and functional-style syntax. W3c Recommendation, W3C, 27 October 2009. http://www.w3.org/TR/owl2-syntax/
19. Otero-Cerdeira, L., Rodríguez-Martínez, F.J., Gómez-Rodríguez, A.: Ontology matching: a literature review. Expert Syst. Appl. **42**, 949–971 (2015)
20. Ritze, D., Meilicke, C., Svab-Zamazal, O., Stuckenschmidt, H.: A pattern-based ontology matching approach for detecting complex correspondences. In: Shvaiko, P., et al. (ed.) Ontology Matching OM-2009 (2009)
21. Scharffe, F., Fensel, D.: Correspondence patterns for ontology alignment. In: Gangemi, A., Euzenat, J. (eds.) EKAW 2008. LNCS (LNAI), vol. 5268, pp. 83–92. Springer, Heidelberg (2008). doi:10.1007/978-3-540-87696-0_10
22. Scharffe, F., Zamazal, O., Fensel, D.: Ontology alignment design patterns. Knowl. Inf. Syst. **40**, 1–28 (2014)
23. Schmidt, D., Trojahn, C., Vieira, R.: Analysing top-level and domain ontology alignments from matching systems. In: Ontology Matching OM-2016, Kobe, Japan, 18 October 2016
24. Sugumaran, V., Storey, V.C.: The role of domain ontologies in database design: an ontology management and conceptual modeling environment. ACM TODS **31**(3), 1064–1094 (2006)
25. Zamazal, O., Svatek, V.: PatOMat - versatile framework for pattern-based ontology transformation. Comput. Inform. **34**, 305–336 (2015)

Exploring Importance Measures
for Summarizing RDF/S KBs

Alexandros Pappas[1], Georgia Troullinou[2], Giannis Roussakis[2],
Haridimos Kondylakis[2(✉)], and Dimitris Plexousakis[2]

[1] Computer Science Department, University of Crete, Heraklion, Greece
apappas@csd.uoc.gr
[2] Institute for Computer Science, FORTH, Heraklion, Greece
{troulin,rousakis,kondylak,dp}@ics.forth.gr

Abstract. Given the explosive growth in the size and the complexity of
the Data Web, there is now more than ever, an increasing need to develop
methods and tools in order to facilitate the understanding and exploration of RDF/S Knowledge Bases (KBs). To this direction, summarization approaches try to produce an abridged version of the original data
source, highlighting the most representative concepts. Central questions
to summarization are: how to identify the most important nodes and then
how to link them in order to produce a valid sub-schema graph. In this
paper, we try to answer the first question by revisiting six well-known
measures from graph theory and adapting them for RDF/S KBs. Then,
we proceed further to model the problem of linking those nodes as a graph
Steiner-Tree problem (GSTP) employing approximations and heuristics
to speed up the execution of the respective algorithms. The performed
experiments show the added value of our approach since (a) our adaptations outperform current state of the art measures for selecting the most
important nodes and (b) the constructed summary has a better quality
in terms of the additional nodes introduced to the generated summary.

Keywords: Semantic summaries · Schema summary · RDF/S Knowledge Bases · Graph theory

1 Introduction

The recent explosion of the Data Web and the associated Linked Open Data
(LOD) initiative have led to an enormous amount of widely available RDF
datasets. These datasets often have extremely complex schemas which are difficult to comprehend, limiting the exploration and the exploitation potential of
the information they contain. In addition, a user, in order to formulate queries,
has to examine carefully the entire schema in order to identify the interesting
elements of the schema and the data. As a result, there is now, more than ever,
an increasing need to develop methods and tools in order to facilitate the quick
understanding and exploration of these data sources [11].

© Springer International Publishing AG 2017
E. Blomqvist et al. (Eds.): ESWC 2017, Part I, LNCS 10249, pp. 387–403, 2017.
DOI: 10.1007/978-3-319-58068-5_24

To this direction several works try to provide overviews on the ontologies [15,20,25] maintaining however the most important ontology elements. Such an overview can also be provided by means of an ontology summary. Ontology summarization [29] is defined as the process of distilling knowledge from an ontology in order to produce an abridged version. While summaries are useful, creating a good summary is a non-trivial task. A summary should be concise, yet it needs to convey enough information to enable a decent understanding of the original schema. Moreover, the summarization should be coherent and provide an extensive coverage of the entire ontology.

In this paper, we focus on RDF/S Knowledge Bases (KBs) and explore efficient and effective methods to automatically create high-quality summaries. The goal is to construct better summaries in terms of selecting the most important part of the schema as end-users perceive importance. We view an RDF/S KB as two distinct and interconnected graphs, i.e. the schema and the instance graph. As such, a summary constitutes a valid sub-schema graph containing the most important nodes, summarizing the instances as well. Central questions to the process of summarization is how to identify the most important nodes and then how to link those nodes to produce a valid sub-schema graph. For answering the first question various importance measures have been proposed trying to provide real-valued measures for ranking the nodes of a graph. In this paper, we adapt, for RDF/S KBs, six well-known importance measures from graph theory, covering a wide range of alternatives for identifying importance. Then we try to answer the second question by modelling the problem of selecting a valid sub-schema graph as a Steiner-Tree problem which we resolve using approximations with heuristics. More specifically our contributions are the following:

- We explore the *Degree*, the *Betweeness*, the *Bridging Centrality*, the *Harmonic Centrality*, the *Radiality* and the *Ego Centrality* measures adapting them for RDF/S KBs to consider instance information as well. Our experiments show that the adapted versions of these importance measures greatly outperform other proposed measures in the domain.
- Besides identifying the most important nodes, we try next to identify the proper paths connecting those nodes. We achieve this by modelling the problem as a graph Steiner-Tree Problem trying to minimize the total number of the additional nodes introduced when constructing the summary sub-graph. Since the problem is NP-complete and the exact algorithms proposed require significant execution time, we proceed further to explore three approximations, the SDIST, the CHINS and the HEUM trying to optimize either the insertion of a single component or the connection of the components using their shortest paths. On top of these approximations we implement an improvement procedure using heuristics the I-MST, ensuring that all leaves are terminal nodes.
- Finally, we perform a detailed two-stage experimental evaluation using two versions of the DBpedia. In the first stage we compare the applicability of the adapted measures for identifying the nodes' importance comparing our adaptations with the most frequent nodes queried in the corresponding query logs. We identify that overall our adaptation of Betweeness outperforms all

other important measures. In the second stage, we evaluate the quality the of the selected sub-graphs in terms of additional nodes introduced showing that CHINS performs better without too much overhead in the execution time.

To the best of our knowledge, this is the first time that these six diverse importance measures are adapted and compared for RDF/S summarization purposes. In addition although other recent works focus on using the maximum cost spanning tree [24,25] for linking the selected nodes, this is the first time the problem of summarization is formulated as a Steiner-Tree problem using approximations for the fast identification of the corresponding summaries with many benefits as we shall show in the sequel.

The rest of the paper is organized as follows: Sect. 2 introduces the formal framework of our solution and Sect. 3 describes the various measures for estimating importance. Then in Sect. 4 we present the algorithms for selecting the proper subgraphs, whereas Sect. 5 presents our evaluation. Section 6 presents related work and finally Sect. 7 concludes this paper and presents directions for future work.

2   Preliminaries

In this paper, we focus on RDF/S KBs, as RDF is among the widely-used standards for publishing and representing data on the Web. Representation of RDF data is based on three disjoint and infinite sets of *resources*, namely: URIs ($\mathcal{U}$), literals ($\mathcal{L}$) and blank nodes ($\mathcal{B}$). We impose typing on resources, so we consider 3 disjoint sets of resources: classes ($\mathbf{C} \subseteq \mathcal{U} \cup \mathcal{B}$), properties ($\mathbf{P} \subseteq \mathcal{U}$), and individuals ($\mathbf{I} \subseteq \mathcal{U} \cup \mathcal{B}$). The set $\mathbf{C}$ includes all classes, including RDFS classes and XML datatypes (e.g., xsd:string, xsd:integer). The set $\mathbf{P}$ includes all properties, except rdf:type, which connects individuals with the classes they are instantiated under. The set $\mathbf{I}$ includes all individuals, but not literals. In addition, we should note that our approach adopts the unique name assumption, i.e. that resources that are identified by different URIs are different.

In this work, we separate between the schema and instances of an RDF/S KB, represented in separate graphs (G_S, G_I, respectively). The schema graph contains all classes and the properties they are associated with; note that multiple domains/ranges per property are allowed, by having the property URI be a label on the edge (via a labelling function λ) rather than the edge itself. The instance graph contains all individuals, and the instantiations of schema properties; the labelling function λ applies here as well for the same reasons. Finally, the two graphs are related via the τ_c function, which determines which class(es) each individual is instantiated under. Formally:

Definition 1 (RDF/S KB). *An RDF/S KB is a tuple $V = \langle G_S, G_I, \lambda, \tau_c \rangle$, where:*

– G_S is a labelled directed graph $G_S = (V_S, E_S)$ such that V_S, E_S are the nodes and edges of G_S, respectively, and $V_S \subseteq \mathbf{C} \cup \mathcal{L}$.

- G_I is a labelled directed graph $G_I = (V_I, E_I)$ such that V_I, E_I are the nodes and edges of G_I, respectively, and $V_I \subseteq \mathbf{I} \cup \mathcal{L}$.
- A labelling function $\lambda : E_S \cup E_I \mapsto 2^{\mathbf{P}}$ determines the property URI that each edge corresponds to (properties with multiple domains/ranges may appear in more than one edge).
- A function $\tau_c : \mathbf{I} \mapsto 2^{\mathbf{C}}$ associating each individual with the classes that it is instantiated under.

For simplicity, we forego extra requirements related to RDFS inference (subsumption, instantiation) and validity (e.g., that the source and target of property instances should be instantiated under the property's domain/range, respectively), because these are not relevant for our results below and would significantly complicate our definitions.

In the following, we will write $p(v_1, v_2)$ to denote an edge e in G_S (where $v_1, v_2 \in V_S$) or G_I (where $v_1, v_2 \in V_I$) from node v_1 to node v_2 such that $\lambda(e) = p$. In addition, a path from a schema node v_s to v_i, denoted by $path(v_s, v_i)$, is the finite sequence of edges, which connect a sequence of nodes, starting from the node v_s and ending in the node v_i. The length of a path, denoted by $d_{path(v_s, v_i)}$, is the number of the edges that exist in that path whereas $d(v_s, v_i)$ is the number of the edges that exist in the shortest path linking v_s and v_i. Finally, having a schema/instance graph G_S/G_I, the closure of G_S/G_I, denoted by $Cl(G_S)/Cl(G_I)$, contains all triples that can be inferred from G_S/G_I using inference. Since in our algorithms we use the closure of the corresponding schema/instance graphs, from now on when we use G_S/G_I we will mean $Cl(G_S)/Cl(G_I)$ for reasons of simplicity unless stated otherwise. This is to ensure that the result will be the same, independent of the number of inferences applied in the input.

Now as an example, consider the DBpedia 3.8 shown in Fig. 1(a). Obviously, it is really difficult to examine all the nodes in order to understand the schema. However, focusing only on the schema summary, shown in Fig. 1(b), allows the user to get a quick overview on the contents of the ontology, identifying and linking the most important nodes. We have to note, that our approach handles OWL ontologies as well, considering however only the RDF/S fragment of these ontologies.

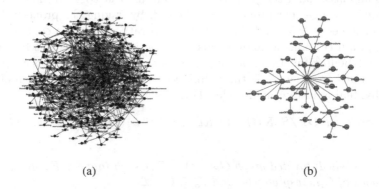

(a)   (b)

Fig. 1. The DBpedia 3.8 schema graph (a) and a corresponding schema summary (b).

3 Importance Measures

Schema summarization aims to highlight the most representative concepts of a schema, preserving important information and reducing the size and the complexity of the schema [17]. Despite the significance of the problem, there is still no universally accepted measurement on the importance of nodes in an RDF/S graph. In this section, we describe six alternative measures that have been proposed for capturing importance in directed graphs. Then we will show how we adapt those measures to consider instance information as well for summarizing RDF/S KBs. We selected the *Betweeness*, the *Bridging Centrality*, the *Degree*, the *Harmonic Centrality*, the *Radiality* and the *Ego Centrality* as they constitute the state of the art geometric measures for generic graphs [2]. We do not compare with spectral measures (HITS, PageRank etc.) because they are based on external factors and spectral properties. The complexities of all aforementioned measures is shown in Table 1.

Table 1. The complexities of the examined importance measures.

Measure	Complexity
Degree (DE)	$O(V_S + E_S)$
Betweeness (BE)	$O(V_S \cdot (V_S \cdot E_S))$
Bridging Centrality (BC)	$O(V_S \cdot (V_S \cdot E_S))$
Harmonic Centrality (HC)	$O(V_S \cdot (V_S + E_S))$
Radiality (RA)	$O(V_S \cdot (V_S + E_S))$
Ego Centrality (EC)	$O(V_S + E_S)$

– The simplest importance measure for a graph is the Degree, that is defined as the number of edges incident to a node.

Definition 2 *(Degree)*. *Let $G_S = (V_S, E_S)$ be an RDF/S schema graph with V_S nodes and E_S edges. The Degree of a node $v \in V_S$ is defined as follows:*

$$DE(v) = deg(v) \tag{1}$$

where $deg(v)$ is the number of edges incident to the node.

– The Betweenness measure is equal to the number of the shortest paths from all nodes to all others that pass through that node. Calculating the betweenness for all nodes in a graph requires the computation of the shortest paths between all nodes.

Definition 3 *(Betweenness)*. *Let $G_S = (V_S, E_S)$ be an RDF/S schema graph. The Betweenness of a node $v \in V_S$ is defined as follows:*

$$BE(v) = \sum_{s \neq v \neq t} \frac{\sigma_{st}(v)}{\sigma_{st}} \tag{2}$$

where σ_{st} is the total number of shortest paths from node s to node t and $\sigma_{st}(v)$ is the number of those paths that pass through v.

- The Bridging Centrality tries to identify the information flow and the topological locality of a node in a network. It is widely used for clustering or in order to identify the most critical points interrupting the information flow for network protection and robustness improvement purposes. A node with high Bridging Centrality is a node connecting densely connected components in a graph. The bridging centrality of a node is the product of the betweenness centrality and the bridging coefficient, which measures the global and local features of a node respectively.

Definition 4 (Bridging Centrality). *Let* $G_S = (V_S, E_S)$ *be an RDF/S schema graph. The bridging centrality of a node* $v \in V_S$ *is defined as follows:*

$$BC(v) = B_C(v) \cdot BE(v) \tag{3}$$

where $B_C(v)$ is the bridging coefficient of a node which determines how well the node is located between high degree nodes and $BE(v)$ is the betweenness centrality. The bridging coefficient of a node v is defined:

$$B_C(v) = \frac{DE(v)^{-1}}{\sum_{i \in N(v) \cdot \frac{1}{DE(i)}}} \tag{4}$$

where $DE(v)$ is the degree of node v, and $N(v)$ is the set of it's neighbors.

- The Harmonic Centrality was initially defined for undirected graphs by Rochat [19] in 2009 and later for directed graphs by Boldi and Vigna [2]. It is a modification of the *Closeness* [2], replacing the average distance with the harmonic mean of all distances, requiring again the computation of the shortest paths between all nodes.

Definition 5 (Harmonic Centrality). *Let* $G_S = (V_S, E_S)$ *be an RDF/S schema graph. The Harmonic Centrality of a node* $v \in V_S$ *is defined as follows:*

$$HC(v) = \frac{1}{\sum_{u \neq v} d(u, v)} \tag{5}$$

- The Radiality was first proposed by Valente and Foreman [26], to provide information on how close a node is to all other nodes in a graph (i.e. the integration measure of a node to a graph). In order to compute the diameter of a graph we need to compute the shortest paths between all nodes.

Definition 6 (Radiality). *Let* $G_S = (V_S, E_S)$ *be an RDF/S schema graph. The Radiality of a node* $v \in V_S$ *defined as:*

$$RA(v) = \frac{1}{\sum_{u \neq v} (\Delta_G - (1/d(u, v)))} \tag{6}$$

where Δ_{G_S} is the Diameter of graph G_s.

– The Ego Centrality (EC) was first introduced in the iManageCancer[1] project. For a node v, EC is the induced subgraph of G, which contains v, its neighbors, and all the edges between them, trying to identify how important a node is to his neighborhood.

Definition 7 (Ego Centrality). *Let $G_S = (V_S, E_S)$ be an RDF/S schema graph. The Ego Centrality of a node $v \in G_S$ is defined as follows:*

$$EC(v) = \sum_{i=1}^{i=n^{in}} W_i * e.ego_i + \sum_{i=1}^{i=n^{out}} W_i * e.ego_i \tag{7}$$

where:

$$W_i = \sum_{i=1}^{i=n^{in}} 1/v_i^{out} + \sum_{i=1}^{i=n^{out}} 1/v_i^{in} \tag{8}$$

and $e.ego = 1/v_i^{out}$, v_i the adjacent node of a node v using the incoming edge e and $e.ego = 1/v_i^{in}$, v_i the adjacent node of a node v using the outgoing edge e.

3.1 Adapted Importance Measures

In order to take into consideration the instances of each class, we adapt the aforementioned importance measures. To achieve that we first normalize each importance measure IM_i on a scale of 0 to 1:

$$normal(IM_i(v)) = \frac{IM_i(v) - \min(IM_i(g))}{\max(IM_i(g)) - \min(IM_i(g))} \tag{9}$$

Where i one of the DE, BE, BC, HC, RA, EC. $IM_i(v)$ is the importance value of a node v in the schema graph g, $min(IM_i(g))$ is the minimum and $max(IM_i(g))$ is the maximum importance value in the graph. Similarly, we normalize the number of instances (InstV) that belong to a schema node. As such, the *adapted importance measure* (AIM) of each node is the sum of the normalized values of the importance measures and the instances.

$$AIM_i(v) = normal(IM_i(v)) + normal(InstV(v)) \tag{10}$$

4 Construction of the RDF/S Summary Schema Graph

Using the aforementioned AIM we select the top-k important nodes of a directed schema graph (also known as *terminals* in graph theory). Then we to focus on the paths that link those nodes, trying to produce a valid sub-schema graph. We have to note that in the stage of constructing the final RDF/S summary schema graph we are not interested in the direction of the edges since we only want to get a connected schema sub-graph.

---

[1] http://imanagecancer.eu/.

The latest approaches in the area [25] identify a maximum cost spanning tree (MST) in the graph and then link the most important nodes using paths from the selected maximum-cost spanning tree. The main idea there is to select the paths that maximize the total weight of the selected sub-graph. However, the main problems with this approach is that although the MST identifies the paths with the maximum weight in the whole graph, the paths selected out of the MST might not maximize the weight of the selected summary (remember that MST connects all nodes in the schema graph whereas summaries only select the most important nodes to be connected using paths from the MST). A second problem there is that many additional nodes are introduced in the result, since there is only one path to be selected between two nodes and in this path many other not important nodes might appear as well.

A different idea that we explore in this paper is to model the problem of linking the most important nodes as a variation of the well-known graph Steiner-Tree problem (GSTP) exploiting the optimal solutions there proposed by Hakimi [6] and Levin [12] independently. The problem is an NP-hard [8] problem and remains NP-complete if all edge weights are equal.

Definition 8 *(The Graph Steiner-Tree problem (GSTP)). Given an undirected graph $G = (V, E)$, with edge weights $w : E \to \mathbb{R}^+$ and a node set of terminals $S \subseteq V$, find a minimum-weight tree $T \in G$ such that $S \subseteq V_t$ and $E_t \subseteq E$.*

In our case, we consider as G the G_S ignoring as well the direction in the edges. As such the objective now is not to increase the total weight of the summary graph but to introduce as much as possible less number of additional nodes. This is due to the fact that introducing a lot of additional nodes shifts the focus of the summary and decreases summary's quality.

4.1 Algorithms, Approximation and Heuristics

There had been various exact algorithms for the GSTP. Hakimi [6] proposed the first brute force algorithm that enumerates all minimum spanning trees of sub-networks of G included by super-sets of terminals that runs in $O(2^{V-t} \cdot V^2 + V^3)$. The first dynamic programming algorithms were proposed independently by Dreyfus & Wagner [4] and by Levin [12]. The former runs in $O(3^t \cdot V + 2^t \cdot V^2 + V^3)$ whereas the latter in $O(3^t \cdot V + 2^t \cdot V^2 + t^2 \cdot V)$ and they are based on the optimal decomposition property by creating two node sets, removing one node at each step and solving the GSTP by connecting each set. Levin's method uses a recursive optimization approach that pre-computes the possible sub-trees. Since all aforementioned algorithms have an exponential running time, various approximations such as [16, 18, 27] have been proposed in order to find good solutions for large networks. A central theme in these approximations, is the use of some principles known from the two classic algorithms for solving the minimum spanning tree problem, Prim's and Kruskal's [27]. We will use the following top-three well-known and good performing methods SDISTG, CHINS and HEUM [27]. These approximations have a worst case bound of 2, i.e., $Z_T/Z_{opt} \leq 2 \cdot (1 - l/|Q|)$, where Z_T and Z_{opt} denote the objective function values of a feasible solution and an optimal solution respectively, Q the set of terminals and l a constant [1].

SDISTG (Shortest distance graph)

1. Construct a complete graph G' for the node set Q (set of terminal nodes) with each edge having the weight of a shortest path between the corresponding nodes in G.
2. Construct a minimum spanning tree T' of G'.
3. Replace each edge of the tree T' by its corresponding shortest path in G.

CHINS (Cheapest insertion)

1. Start with a partial solution $T = (w, 0)$ consisting of a single terminal node w.
2. While T does not contain all terminal nodes do
   find the nearest nodes $u* \in V_t$ and $p*$ being a terminal node not in V_t.

HEUM (Heuristic measure)

1. Start with a partial solution $T = (Q, 0)$ consisting of Q singleton components (terminal nodes).
2. While T is not connected do
   choose a node u using a heuristic function F and unite the two components of T which are nearest to u by combining them with u via shortest paths (the nodes and edges of these paths are added to T).
   Up to now the most promising way is to choose F according to:
   $\min_{i \leq t \leq \sigma} \left\{ \frac{1}{t} \cdot \sum_{i=0}^{t} d(u, T_i) \right\}$ where $T_0, \ldots, T_\sigma$ are the components of T such that $d(u, T_i) \leq d(u, T_i) \forall i, j \in \sigma, i < j$.

Besides these approximations, many heuristics can be employed to improve even more the corresponding algorithms. The most promising ones are the I-MST+P and the TRAFO [27]. I-MST+P is a pruning routine that ensures that all leaves are terminal nodes whereas TRAFO transforms a feasible solution to another one trying to overcome the deficiency of bad local optima by allowing the temporary deterioration of the actual solutions. In this paper we use only the I-MST+P since TRAFO requires considerable more time to run and the improvements are insignificant - due to the sparsity of the examined ontologies.

I-MST+P (Improvement procedure with MST+P)

1. Let $T = (V_t, E_t)$ be a feasible solution of the GSTP. The subgraph of G induced by V_t will be defined as G_t.
2. Construct a minimum spanning tree $T = (V_t', E_t')$ of G_t.
3. While there exists a leaf of T' being a terminal do
   delete that leaf and its incident edge.

Table 2. Worst-case complexities for linking the most important nodes in a graph.

Algorithm	Weighted graph	Un-weighted graph
MST	$O(E \cdot logV)$	$O(|V + E|)$
SDISTG	$O(Q \cdot |VlogV|)$	$O(Q \cdot |V + E|)$
CHINS	$O(Q \cdot |VlogV|)$	$O(Q \cdot |V + E|)$
HEUM	$O(V \cdot |VlogV|)$	$O(V \cdot |V + E|)$

Complexities. Using a Breadth-first search requires $O(V + E)$ time (where E is $O(V)$) to find shortest paths from a source vertex v to all other vertices in the graph. All-pairs shortest paths for unweighted undirected graphs can be computed in $O(V \cdot (V + E))$ time by running the BFS algorithm for each node of the graph. The complexity of SDISTG, CHINS and HEUM differs, due to the usage of different heuristics. Table 2 provides the worst case complexities of those algorithms for weighted/un-weighted graphs.

5   Evaluation

To evaluate our approach, we used two versions of the DBpedia[2]. DBpedia 3.8 is consisted of 359 classes, 1323 properties and more that 2.3M instances, whereas DBpedia 3.9 is consisted of 552 classes, 1805 properties and more than 3.3M instances. Those two versions offer two interesting use-cases for exploration.

To identify the most important nodes of those two versions we do not rely on a limited amount of domain experts with subjective opinions as past approaches do [20,23,25]. Instead, we exploit the query logs from the corresponding DBpedia endpoints trying to identify the schema nodes that are more frequently queried. For DBpedia 3.8 we were able to get access to more than 50K queries whereas for 3.9 we were able to get access to more than 110K queries.

For each examined version, we considered the corresponding query log trying to identify the most important classes. We assess as the most important, the ones that have higher *frequency of appearance* in the queries. A class appears within a query either directly or indirectly. Directly when the said class appears within a triple pattern of the query and undirectly when (a) the said class is the type of an instance or the domain/range of a property that appear in a triple pattern of the query.

In addition, we compare our approach with *relevance*, another measure recently published, combining both syntactic and semantic information, shown to outperform past approaches in the area [23,25].

5.1   Spearman's Rank Correlation Coefficient

Initially we tried to understand the statistical dependence between the ranking of all nodes using the aforementioned measures. To do this we used the Spearman's

---

[2] http://wiki.dbpedia.org/.

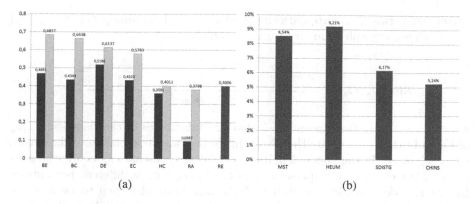

Fig. 2. (a) Spearman's rank correlation for the adapted (yellow) and the non-adapted (blue) importance measures with the frequency ranking and (b) the percentage of additional nodes introduced (Color figure online)

rank correlation coefficient [21], a nonparametric measure of rank correlation. It assesses how well the relationship (measures the strength and direction of association) between two variables can be described using a monotonic function.

Spearman correlation indicates the direction of the association between two variables X, Y. It can vary between -1 and 1, where 1 is total positive correlation, 0 is no correlation, and -1 is total negative correlation. If Y tends to decrease when X increases, the Spearman correlation coefficient is negative. A Spearman correlation of zero indicates that there is no tendency for Y to either increase or decrease when X increases. When X and Y are perfectly monotonically related, the Spearman correlation coefficient becomes 1. The results of our experiments are shown in Fig. 2(a).

As shown, our adapted importance measures show a higher dependence to the frequency ranking than the pure structural ones wit. In addition we can see that measures like the AIM_{BE}, the AIM_{BC} and the AIM_{DE} show a really high correlation with the frequency ranking. Finally, we can see that all adapted measures - except Radiality (AIM_{RA})- show a better correlation than Relevance.

5.2   The Similarity Measure

Next, we would like to evaluate the measures identified in Sect. 3 for their quality with respect to identifying the nodes' importance. Measures like precision, recall and F-measure, used by the previous works [15,17,20] are limited in exhibiting the added value of a summarization system because of the "disagreement due to synonymy" [3] meaning that they fail to identify closeness with the ideal result when the results are not exactly the same with the reference ones. On the other hand, content-based metrics compute the similarity between two summaries in a more reliable way [29]. To this direction, we use the *similarity measure*, first defined in [23], denoted by $Sim(G_S, G_R)$, in order to define the level of agreement between an automatically produced graph summary $G_S = (V_S, E_S)$ and a reference graph summary $G_R = (V_R, E_R)$:

$$Sim(G_S, G_R) = \frac{|V_S \cap V_R| + a \cdot \sum_{i=k}^{p} \frac{1}{d_{p(c_i, c_i')}} + b \cdot \sum_{i=m}^{n} \frac{1}{d_{p(c_i, c_i')}}}{|V_R|}$$

where $c_k, ..., c_p$ are the classes in V_R that are sub-classes of the classes $c_k', ..., c_p'$ of V_S and that $c_m, ..., c_n$ are the classes in V_R that are superclasses of the classes $c_m', ..., c_n'$ of V_S. In the above definition a and b are constants assessing the existence of sub-classes and super-classes of G_S in G_R with a different percentage. In [25] the ideal weights for RDF/S KBs have been identified to be $a = 0.6$ and $b = 0.3$ which we use in this paper as well, giving more weight to the super-classes. The idea behind that is that the super-classes, since they generalize their sub-classes, are assessed to have a higher weight than the sub-classes, which limit the information that can be retrieved. Consequently, the effectiveness of a summarization system is calculated by the average number of the similarity values between the summaries produced by the system and the set of the corresponding experts' summaries. The results of our experiments are shown in Fig. 3 and present the average similarity values for generating summaries from 1% to 50% of the corresponding schema graph size. As shown again our adapted measures (in yellow) outperform the pure structural ones (in blue) in all cases. In addition, all measures but AIM_{BA} outperform Relevance showing again the high value of our adaptations. When comparing between the ontology versions we can observe that although AIM_{BE} is the clear winner in all cases, the second best in DBpedia 3.8 is the AIM_{BC} whereas in DBpedia 3.9 is the AIM_{DE}. To interpret these results we shall consider that 193 more classes were added in DBPedia 3.9 introducing only a small number of new edges. This results in a reduction of 37% of the density and an increase of the diameter from 9 to 13. As such, only a few number of nodes have more than one out-going edge and the degree performs better in this case as it captures more effectively the importance of more sparse graphs.

5.3   Additional Nodes Introduced

Next we would like to identify the overhead imposed by the algorithms for linking the most important nodes in terms of the additional nodes that are introduced. The average number of additional nodes introduced per algorithm is shown in Fig. 2(b). We can observe that MST used by our previous work introduces on average 8.5% of additional nodes, whereas CHINS only 4.7% additional nodes. For example, for DBpedia 3.9 this corresponds to 19 additional nodes using MST over CHINS when requesting a summary of 10% of the nodes. This is reasonable, since the Steiner-Tree approximations have the objective of minimizing the additional nodes introduced in the selected subgraph confirmed by our experiments.

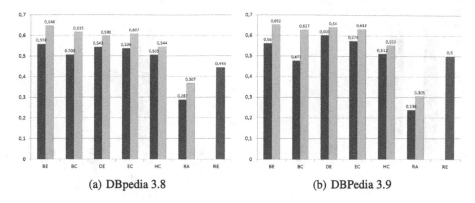

(a) DBpedia 3.8 (b) DBPedia 3.9

Fig. 3. Comparing the average similarity of the adapted (in yellow) and the non-adapted (in blue) importance measures in (a) DBpedia 3.8 and (b) DBpedia 3.9 for a summary of 1–50%. (Color figure online)

5.4 Execution Time

Finally, to test the efficiency of our system, we measured the average time of 50 executions in order to produce the corresponding summaries of the two KBs. The experiments run on a Intel(R) Xeon(R) CPU E5-2630 running at 2.30 GHz with 64 GB memory running Ubuntu 12.04 LTS.

The mean execution times for identifying the most important nodes and constructing the corresponding summaries are shown in Fig. 4. As we can observe, the execution times of the various measures can be divided into three categories. The measures that need to compute the shortest paths of all pairs, the measures that need to iterate only the nodes and the edges of the graph and the measures that need to execute queries on external databases or combine complex measures. The Betweenness, the Bridging Centrality, the Harmonic Centrality and the Radiality belong to the first category since they assign weights by calculating the shortest paths between all pairs of nodes. As such they have similar execution times. The Betweenness differentiates from the rest since the set of all shortest paths should be computed for each pair of nodes. The Bridging Centrality uses the Betweenness and as such it takes almost the same time. In the second category we find the Degree and the Ego Centrality. The Degree needs only to iterate over all edges of the graph and "submit" the weight to each node. The Ego Centrality needs one more iteration over all nodes and edges of the graph. Relevance is not included in these graphs since it requires significantly more time (two orders of magnitude larger) using an external triple store to be able to handle mass amounts of data /while the aforementioned algorithms load everything in memory.

For linking the most important nodes, the complexity the SDISTG and CHINS approximation algorithms show that there is a linear function relationship between their execution time and the input data size (the number of the nodes and the edges). HEUM is the only one that has a quadratic time, and this

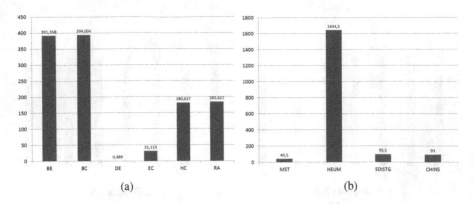

Fig. 4. The average execution times of the importance measures (msec) (a) and the algorithms for linking those nodes (msec) (b)

is due to the fact that it has to construct the shortest paths of all pairs. As such SDISTG and CHINS have a better execution time as the number of terminal nodes is small and HEUM the worst execution time, which grows linearly to the number of nodes. MST is slightly faster than other algorithms because it depends on the size of graph (nodes and edges), in contrast to CHINS and SDISTG that are highly dependent on the number of the terminals and the shortest paths between them.

6   Related Work

The latest years summarization approaches for linked data are constantly gaining ground. For example, a wide variety of research works [5,9,10,14] focus on extracting statistics and producing visual summaries of linked datasets, try to create mainly instance summaries, exploiting the instances' semantic associations [7,13,22] or focus on peer-to-peer systems [15]. However, our system differs from the above in terms of both goals and techniques.

More closely related works are Peroni et al. [20] and Wu et al. [28]. The former [20] try to automatically identify the key concepts in an ontology, combining cognitive principles, lexical and topological measurements such as the density and the coverage, whereas in the latter the authors [28] use similar algorithms to identify the most important concepts and relations in an iterative manner. However, both of these works focus only on returning the most important nodes and not on returning an entire graph summary. Zhang et al. [29] uses measures such as the degree-centrality, the betweenness and the eigenvector centrality to identify not the most important nodes but the most important RDF sentences. In Queiroz-Sousa et al. [17] the authors try to combine user preferences with the degree centrality and the closeness to calculate the importance of a node and then they use an algorithm to find paths that include the most important nodes in the final graph. However the corresponding algorithm prioritizes direct neighbors ignoring that the selection of other paths that could maximize the total importance of the selected summary. Finally Troulinou et al. [23,25] employee relevance for identifying the most important nodes and then they try to connect those nodes by generating and pruning appropriately the maximum cost spanning tree. However, many additional nodes might be introduced and the selected summary does not guarantee to maximize the total importance of the selected sub-graph.

7   Discussion and Conclusion

In this paper, we try to provide answers to the two main questions in constructing RDF/S summaries: how to identify the most important nodes and how to link the selected nodes to create the final summary. To this direction, we adapt six diverse measures for identifying node's importance and we implement three graph Steiner-Tree approximations in order to link those nodes.

To evaluate our approach we do not rely on subjective reference summaries generated by a limited amount of domain experts but instead we exploit the query logs from the DBpedia endpoints. The performed evaluation shows that the adapted measures perform significant better that the pure structural ones for RDF/S KBs. In addition all but the adapted version of Radiality outperform past approaches in the area. The adaptation of Betweeness is the winner in all cases. In addition, we show that the Steiner-Tree approximation algorithms introduce less additional nodes to the result schema graph. CHINS seems to be the best choice in terms of the quality of the generated summary offering an optimal trade-off between quality and execution time.

As future work, an interesting topic would be to extend our evaluation to spectral properties as well or to focus on how to combine the various measures in order to achieve the best results according to the specific characteristics of the input ontologies. Finally, another interesting topic would be to extend our approach to handle more constructs from OWL ontologies such as class restrictions, disjointness and equivalences.

Acknowledgments. This work was partially supported by the EU projects iManage-Cancer and CloudSocket under the contracts H2020-643529, H2020-644690.

References

1. Du, D.-Z., Smith, J.M., Rubinstein, J.H. (eds.): Advances in Steiner Trees. Kluwer Academic Publishers, Dordrecht (2000)
2. Boldi, P., Vigna, S.: Axioms for centrality. Internet Math. **10**(3–4), 222–262 (2014)
3. Donaway, R.L., Drummey, K.W., Mather, L.A.: A comparison of rankings produced by summarization evaluation measures. In: NAACL-ANLP Workshop, pp. 69–78 (2000)
4. Dreyfus, S.E., Wagner, R.A.: The steiner problem in graphs. Networks **1**(3), 195–207 (1971)

5. Dudáš, M., Svátek, V., Mynarz, J.: Dataset summary visualization with LODSight. In: Gandon, F., Guéret, C., Villata, S., Breslin, J., Faron-Zucker, C., Zimmermann, A. (eds.) ESWC 2015. LNCS, vol. 9341, pp. 36–40. Springer, Cham (2015). doi:10.1007/978-3-319-25639-9_7

6. Hakimi, S.L.: Steiner's problem in graphs and its implications. Networks 1(2), 113–133 (1971)

7. Jiang, X., Zhang, X., Gao, F., Pu, C., Wang, P.: Graph compression strategies for instance-focused semantic mining. In: Qi, G., Tang, J., Du, J., Pan, J.Z., Yu, Y. (eds.) CSWS 2013. CCIS, vol. 406, pp. 50–61. Springer, Heidelberg (2013). doi:10.1007/978-3-642-54025-7_5

8. Karp, R.M.: Reducibility among combinatorial problems. In: Jünger, M., Liebling, T.M., Naddef, D., Nemhauser, G.L., Pulleyblank, W.R., Reinelt, G., Rinaldi, G., Wolsey, L.A. (eds.) 50 Years of Integer Programming 1958–2008 - From the Early Years to the State-of-the-Art, pp. 219–241. Springer, Heidelberg (2010)

9. Khatchadourian, S., Consens, M.P.: Explod: summary-based exploration of interlinking and RDF usage in the linked open data cloud. In: ESWC, pp. 272–287 (2010)

10. Khatchadourian, S., Consens, M.P.: Exploring RDF usage and interlinking in the linked open data cloud using explod. In: LDOW (2010)

11. Kondylakis, H., Plexousakis, D.: Ontology evolution: assisting query migration. In: Atzeni, P., Cheung, D., Ram, S. (eds.) ER 2012. LNCS, vol. 7532, pp. 331–344. Springer, Heidelberg (2012). doi:10.1007/978-3-642-34002-4_26

12. Levin, A.Y.: Algorithm for the shortest connection of a group of graph vertices. Sov. Math. Dokl. 12, 1477–1481 (1971)

13. Navlakha, S., Rastogi, R., Shrivastava, N.: Graph summarization with bounded error. In: ACM SIGMOD, pp. 419–432. ACM (2008)

14. Palmonari, M., Rula, A., Porrini, R., Maurino, A., Spahiu, B., Ferme, V.: ABSTAT: linked data summaries with ABstraction and STATistics. In: Gandon, F., Guéret, C., Villata, S., Breslin, J., Faron-Zucker, C., Zimmermann, A. (eds.) ESWC 2015. LNCS, vol. 9341, pp. 128–132. Springer, Cham (2015). doi:10.1007/978-3-319-25639-9_25

15. Pires, C.E., Sousa, P., Kedad, Z., Salgado, A.C.: Summarizing ontology-based schemas in pdms. In: ICDEW, pp. 239–244 (2010)

16. Plesnik, J.: Worst-case relative performances of heuristics for the steiner problem in graphs (1991)

17. Queiroz-Sousa, P.O., Salgado, A.C., Pires, C.E.: A method for building personalized ontology summaries. J. Inf. Data Manage. 4(3), 236 (2013)

18. Rayward-Smith, V.J., Clare, A.: On finding steiner vertices. Networks 16(3), 283–294 (1986)

19. Rochat, Y.: Closeness centrality extended to unconnected graphs: the harmonic centrality index. In: Applications of Social Network Analysis (ASNA) (2009)

20. Peroni, S., Motta, E., d'Aquin, M.: Identifying key concepts in an ontology, through the integration of cognitive principles with statistical and topological measures. In: Domingue, J., Anutariya, C. (eds.) ASWC 2008. LNCS, vol. 5367, pp. 242–256. Springer, Heidelberg (2008). doi:10.1007/978-3-540-89704-0_17

21. Spearman, C.: The proof and measurement of association between two things. Am. J. Psychol. 15(1), 72–101 (1904)

22. Tian, Y., Hankins, R.A., Patel, J.M.: Efficient aggregation for graph summarization. In: ACM SIGMOD, pp. 567–580. ACM (2008)

23. Troullinou, G., Kondylakis, H., Daskalaki, E., Plexousakis, D.: RDF digest: efficient summarization of RDF/S KBs. In: Gandon, F., Sabou, M., Sack, H., d'Amato, C., Cudré-Mauroux, P., Zimmermann, A. (eds.) ESWC 2015. LNCS, vol. 9088, pp. 119–134. Springer, Cham (2015). doi:10.1007/978-3-319-18818-8_8
24. Troullinou, G., Kondylakis, H., Daskalaki, E., Plexousakis, D.: RDF digest: ontology exploration using summaries. In: ISWC (2015)
25. Troullinou, G., Kondylakis, H., Daskalaki, E., Plexousakis, D.: Ontology understanding without tears: the summarization approach. Semant. Web J. (2017). IOS press
26. Valente, T.W., Foreman, R.K.: Integration and radiality: measuring the extent of an individual's connectedness and reachability in a network. Soc. Netw. **20**(1), 89–105 (1998)
27. Voß, S.: Steiner's problem in graphs: heuristic methods. Discrete Appl. Math. **40**(1), 45–72 (1992)
28. Wu, G., Li, J., Feng, L., Wang, K.: Identifying potentially important concepts and relations in an ontology. In: Sheth, A., Staab, S., Dean, M., Paolucci, M., Maynard, D., Finin, T., Thirunarayan, K. (eds.) ISWC 2008. LNCS, vol. 5318, pp. 33–49. Springer, Heidelberg (2008). doi:10.1007/978-3-540-88564-1_3
29. Zhang, X., Cheng, G., Qu, Y.: Ontology summarization based on RDF sentence graph. In: WWW, pp. 707–716 (2007)

Data-Driven Joint Debugging of the DBpedia Mappings and Ontology

Towards Addressing the Causes Instead of the Symptoms of Data Quality in DBpedia

Heiko Paulheim[✉]

Data and Web Science Group, University of Mannheim, Mannheim, Germany
`heiko@informatik.uni-mannheim.de`

Abstract. DBpedia is a large-scale, cross-domain knowledge graph extracted from Wikipedia. For the extraction, crowd-sourced mappings from Wikipedia infoboxes to the DBpedia ontology are utilized. In this process, different problems may arise: users may create wrong and/or inconsistent mappings, use the ontology in an unforeseen way, or change the ontology without considering all possible consequences. In this paper, we present a data-driven approach to discover problems in mappings as well as in the ontology and its usage in a joint, data-driven process. We show both quantitative and qualitative results about the problems identified, and derive proposals for altering mappings and refactoring the DBpedia ontology.

Keywords: Knowledge graph construction · Knowledge graph debugging · Ontology debugging · Data quality · Data-driven approaches · DBpedia

1 Introduction

Knowledge graphs on the Web are a backbone of many information systems that require access to structured knowledge, be it domain-specific or domain-independent [17]. The idea of feeding intelligent systems and agents with general, formalized knowledge of the world dates back to classic Artificial Intelligence research in the 1980s [21]. More recently, with the advent of Linked Open Data [3] sources like DBpedia [14] or YAGO [24], and by Google's announcement of the Google Knowledge Graph in 2012[1], representations of general world knowledge as graphs have drawn a lot of attention again.

In the Linked Open Data cloud, the DBpedia knowledge graph has become a central hub and widely used resource [22]. DBpedia is created from Wikipedia by harvesting information from infoboxes in Wikipedia pages. Those are mapped to an ontology in a community effort. Using those mappings, an A-box for the

---

[1] http://googleblog.blogspot.co.uk/2012/05/introducing-knowledge-graph-things-not.html.

© Springer International Publishing AG 2017
E. Blomqvist et al. (Eds.): ESWC 2017, Part I, LNCS 10249, pp. 404–418, 2017.
DOI: 10.1007/978-3-319-58068-5_25

ontology T-box is automatically extracted using the DBpedia extraction framework [14]. While this approach allows for a reasonable coverage, there are various steps where errors can be introduced: for example, users may create wrong mappings or use the ontology in an unforeseen way. Those errors may lead to wrong extractions, which may limit the utility of the knowledge graph.

To address those shortcomings, various methods for *knowledge graph refinement* have been proposed. In many cases, those methods are developed by researchers outside the organizations or communities which *create* the knowledge graphs. They rather take the extracted DBpedia A-box and try to increase its quality by various means [17]. However, most of those approaches which target error correction focus on identifying single wrong assertions in the knowledge graph as a post-processing step to the construction. This means that the outcome is usually a long list of problematic assertions, which, depending on how reliable the approach is and how defensive the removal of assertions from the graph should be, need to be checked manually, which is a time-consuming process. Furthermore, upon a regeneration of the knowledge graph (e.g., with an updated set of heuristics and/or input corpus), the process needs to be run again. As new DBpedia releases are usually created on a bi-yearly basis[2], this means that the output of those approaches cannot be easily reused, and hence is discarded after six months in most of the cases.[3]

Therefore, a more sustainable way of handling data quality in DBpedia is required. In this paper, we propose a data-driven process which enriches the DBpedia knowledge graph with provenance information that allows for tracking the mapping which was responsible for creating an assertion in the DBpedia knowledge graph, or the ontology assertion that caused a statement to be inconsistent. By automatically identifying clusters inconsistent statements and corresponding mapping assertions, we are able to rank and pinpoint wrong mappings, as well as issues in the DBpedia ontology and its usage. By this, we can identify mappings to be changed, as well as proposals for refactoring the DBpedia ontology. Thereby, we step from identifying the *symptoms* of data quality in DBpedia to addressing their *causes*.

The rest of this paper is structured as follows. Section 2 discusses related work. In Sect. 3, we sketch our approach, followed by a quantitative and qualitative analysis of the findings in Sect. 4. We close with a summary and an outlook on future work.

2   Related Work

In this paper, we target the identification of *systematic errors* in the construction of the large-scale knowledge graph DBpedia.

There is a larger body of work which targets at finding errors in web knowledge graphs such as DBpedia. The approaches vary both with respect to the

---

[2] http://wiki.dbpedia.org/why-is-dbpedia-so-important.

[3] A continuously updated knowledge graph, like DBpedia Live [10], generates a whole new set of challenges, which are out of scope of this paper.

methods employed as well as to the targeted type of assertions – i.e., identifying wrong type assertions, relational assertions, literals, etc. Methods found in the literature range from statistical methods [18] and outlier detection [6,16,27] to using external sources of knowledge, such as web search engines [13]. In addition, crowdsourcing [1] and games with a purpose [26] have been proposed as non-automatic means for identifying errors in knowledge bases. While such approaches can lead to a high precision, their main problem lies in scalability to larger knowledge bases [18].

Since many (but not all) wrong statements in a knowledge graph may surface as an inconsistency w.r.t. the underlying ontology, a few approaches rely on the use of reasoning given the knowledge graph's ontology for detecting inconsistencies. However, the DBpedia ontology – as many schemas used for providing Linked Open Data – is not very expressive, in particular with respect to the presence of disjointness axioms. Thus, there is a natural limitation for reasoning-based approaches. Hence, such approaches are often combined with ontology learning as a preprocessing step to enrich the ontology at hand [12,15,25], or exploit upper level ontologies [11,19,23]. Like the work presented in this paper, the latter two try to identify root causes: the former use T-box level reasoning to identify unsatisfiable concepts, the latter performs clustering on the reasoner's outcome to identify clusters of similar inconsistencies, which can often be attributed to a common root cause.

Reasoning on large-scale knowledge graphs, however, is a resource-intensive problem [20]. An approximation to the problem has thus been proposed in [4], in which schema-level reasoning on the DBpedia mappings and the DBpedia ontology is used to find inconsistent mapping assertions. Thus, the computational problem of dealing with a massive A-box is circumvented, while the T-box used in the reasoner is by several orders of magnitude smaller. In contrast to the work presented in this paper, this works at much faster runtimes, but cannot detect certain types of defects (e.g., the range of object properties is not respected).

3   Approach

Wrong assertions in a knowledge graph like DBpedia often surface as an inconsistency with the underlying ontology[4]. Therefore, in our approach, we first determine whether a single relation assertion is consistent with the subject's and object's types. Furthermore, we identify the DBpedia mappings that are responsible for the assertion at hand, and group the inconsistencies by the mapping. For each mapping, we can then compute scores which determine how frequently the mapping is involved in inconsistent statements, and hence, identify mappings which should be inspected by an expert. The inspection often reveals that either the mapping as such or the definition of the ontology concept it maps to are problematic.

---

[4] However, not all wrong assertions lead to inconsistencies, and not all inconsistencies are due to wrong assertions.

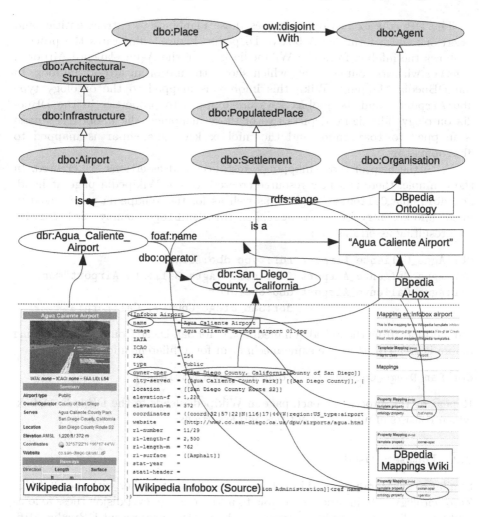

Fig. 1. Example of the Extraction from DBpedia. Infoboxes in Wikipedia (lower left) are mapped to the DBpedia ontology in the DBpedia Mappings Wiki (lower right). With the help of those mappings, entities and assertions are extracted, the DBpedia A-box (middle), which use the DBpedia ontology (top).

3.1 Preliminaries

DBpedia is extracted from Wikipedia infoboxes. In the DBpedia Mappings Wiki[5], infoboxes are mapped to classes in the DBpedia ontology, and infobox keys are mapped to properties in the DBpedia ontology. These mappings are created in a community-driven process.

---

[5] http://mappings.dbpedia.org.

The DBpedia extraction code uses those mappings to extract entities and assertions which form the A-box of DBpedia. Figure 1 illustrates the process. It shows the infobox from the Wikipedia page of the Agua Caliente Airport[6], together with its source code[7], which shows the use of an airport infobox. In the DBpedia Mappings Wiki, this infobox is mapped to the ontology type dbo:Airport[8], and its infobox keys are mapped to properties in the DBpedia ontology. The figure depicts two of those mappings: the infobox key name is mapped to foaf:name, and the infobox key owner-oper is mapped to dbo:operator.

With the help of those mappings, instances and assertions are created. In the example, there is a new resource created for the Wikipedia page at hand, i.e., dbr:Agua_Caliente_Airport, as well as for the Wikipedia page linked to in the infobox entry for owner-oper, i.e., dbr:San_Diego_County,_California. The resulting assertions are:

```
dbr:Agua_Caliente_Airport rdf:type dbo:Airport.
dbr:Agua_Caliente_Airport foaf:name "Agua Caliente Airport"@en.
dbr:Agua_Caliente_Airport dbo:operator
                          dbr:San_Diego_County,_California.
```

Furthermore, from the linked Wikipedia page for San Diego County, the following axiom is created, using the same mechanism for a different infobox:

```
dbr:San_Diego_County,_California a dbo:Settlement.
```

Repeating this process for each page in Wikipedia leads to the DBpedia A-box, which consists of millions of entities and assertions.

3.2   Datasets Used

Similar to our work proposed in [19], we use DBpedia together with the DOLCE-Zero ontology [7,8] in order to be able to add more top level disjointness axioms and hence discover more inconsistencies. We use the most recent DBpedia 2016-04 release with the corresponding ontology.

However, the mapping from the DBpedia 2016-04 ontology to DOLCE has an issue with the mappings which have originally been defined as rdfs:subclassOf relations, but been changed to owl:equivalentClass mappings in the current release.[9] This leads to certain problems, as this example illustrates:

---

[6] https://en.wikipedia.org/wiki/Agua_Caliente_Airport, retrieved on December 6th, 2016.

[7] https://en.wikipedia.org/w/index.php?title=Agua_Caliente_Airport\&action=edit, retrieved on December 6th, 2016.

[8] Throughout this paper, we use the following namespace conventions: dbo=http://dbpedia.org/ontology/, dbr=http://dbpedia.org/resource/, foaf=http://xmlns.com/foaf/0.1/, rdf=http://www.w3.org/1999/02/22-rdf-syntax-ns, rdfs=http://www.w3.org/2000/01/rdf-schema.

[9] See https://sourceforge.net/p/dbpedia/mailman/message/35452137/.

dbo:GovernmentType owl:equivalentClass dul:Concept.
dbo:MusicGenre owl:equivalentClass dul:Concept.

From this, a reasoner would conclude that dbo:GovernmentType owl:equiva-
lentClass dbo:MusicGenre also holds. The class dbo:MusicGenre thereby
becomes unsatisfiable. In order to work around these issues, we have replaced
all owl:equivalentClass mappings between the DBpedia and DOLCE-Zero
ontology with rdfs:subclassOf.[10]

3.3   Identifying and Grouping Inconsistencies

An inconsistency arises if an assertion (or a set of assertions) is incompatible with the ontology that defines constraints on those assertions.
In the example in Fig. 1, the range of the relation dbo:operator is
restricted to the class dbo:Organisation. However, the object of the relation at hand is dbr:San_Diego_ County,_California, which is an instance of
dbr:Settlement.

From the ontology, we can see that dbo:Agent (which is a superclass of
dbo:Organisation) and dbo:Place (which is a superclass of dbo:Settlement)
are disjoint. Hence, the relation assertion, together with the object's type, forms
an inconsistency.

For detecting such inconsistencies, we create minimal A-boxes, consisting of
a relation assertion and its subject's and object's types. Those are then loaded
into the HermiT reasoner [9], together with the ontologies. The reasoner computes whether the A-box is consistent or not. Furthermore, the proof tree the
reasoner provides contains all statements that together form the inconsistency.[11]
In the example in Fig. 1, those consist of the object's type assertion, the relation assertion, the range statement, as well as chain of subclass statements for
dbo:Organisation and dbo:Settlement up to dbo:Agent and dbo:Place, and
the disjointness axiom between the latter two.

Since we are interested in identifying wrong mapping assertions as well as
problems within the ontology, we identify the mapping statements that contribute to an inconsistency. However, in DBpedia, there is no provenance information which tracks which mapping element was used to create which assertion.
Hence, we first need to reconstruct that provenance information. For this reconstruction, we use two sources:

1. The set of templates which are used on subject's and object's original Wikipedia pages. That information is provided with the DBpedia release.
2. The mappings from the DBpedia Mappings Wiki translated to RDF using
   the RML vocabulary [4,5].

---

[10] Note that this change only weakens the original assertions, thus, it cannot introduce
   additional inconsistencies.
[11] There can be multiple proof trees for the same inconsistency. However, in our approach, we only pick one at random.

```
1    Given an assertion a (S r O)
2    Get T(S) and T(O) // templates used in the assertion's subject's and object's Wikipedia page
3    MapR(r) = Mapping elements in T(S) that assign a relation r
4    For all asserted types T of S
5            MapT(S) = Mapping elements in T(S) that assign a type from T
6    For all asserted types of T of O
7            MapT(O) = Mapping elements in T(O) that assign a type from T
8
9    Use reasoner to compute consistency of a
10   If a is inconsistent
11           If explanation contains a
12                   Mark all entries in MapR(r)
13           If explanation contains subject type assertion with type T
14                   Mark all entries in MapT(S)
15           If explanation conatins object type assertion with type T
16                   Mark all entries in MapT(O)
17
18   For all marked mapping elements
19           increase inconsistency counter for element
20   For all non−marked mapping elements
21           increase consistency counter for element
```

Fig. 2. Pseudocode for identifying mapping elements contributing to inconsistencies

In the first step, we identify those templates which are used on the subject's and object's original Wikipedia page. In our running example, that is `Infobox airport` for the subject, and `Infobox settlement`, `Authority control`, and `See also` for the object.

In a second step, we retrieve all mapping elements for those templates. From that subset, we identify those which are used to assert the types and the relation at hand. In our example, this would be the mapping elements that assert the class `dbo:Airport` for `Infobox airport` (1), the relation `dbo:organization` for the infobox key `owner-oper` (2), and the type `dbo:Settlement` for the `Infobox settlement` (3).

When this identification is done, we run the reasoner to compute the inconsistency, and, if the statement is inconsistent, the explanation. Using that explanation, we mark all the mapping elements that are responsible for a statement used in the explanation. With those marks, we maintain two counters for each mapping element m: how often it was generating an assertion involved in an inconsistency (inconsistency counter i_m), and how often it was not (consistency counter c_m). Figure 2 illustrates this approach.

In our example, i_m is increased for (2) and (3), i.e., the mapping elements responsible for the relation assertion and the object type assertion, while c_m is increased for (1), i.e., the mapping element responsible for the subject type assertion, is increased, since the type assertion for the subject was not involved in the inconsistency.

3.4   Scoring Inconsistencies

Given the consistency and inconsistency counts for each mapping element, we can compute a number of scores for each mapping element. Given the hypothesis that a mapping element m is problematic, and the two counts for the statements produced by m that were involved in explanations for inconsistencies (i_m) and those that were not (c_m), we first use two metrics borrowed from association rule mining, i.e., support and confidence [2]:

$$s(m) := \frac{i_m}{N} \tag{1}$$

$$c(m) := \frac{i_m}{i_m + c_m} \tag{2}$$

Here, N is the total number of statements in the knowledge base. As we are looking only at relation assertions, N is roughly 17.6 million. In turn, this means that while the confidence can easily grow towards 1 (e.g., if a wrong mapping element produces mostly wrong statements leading to inconsistencies), this is not true for support: even a mapping element leading to 100,000 inconsistent cases (which is a mapping element which we would want to achieve a high score) would have a support value of only 0.005. That makes it difficult to compute a common score from the two, since, although they both theoretically produce a score in the $[0; 1]$ interval, the actual scores practically come in different scales. Hence, to overcome this problem, we propose to use *logarithmic support*, defined as

$$logs(m) := \frac{log(i_m + 1)}{log(N + 1)} \tag{3}$$

In contrast to standard support, logarithmic support works in orders of magnitude, i.e., a log support of 0.5 means that the order of magnitude of the counted incoherences affects half the order of magnitude of all the axioms. Thus, a mapping element leading to 100,000 inconsistent cases, as discussed above, would achieve a fairly high logarithmic support score of 0.69.

Finally, to assign ratings to the mapping elements and pick the most promising cases for inspection by an expert, we want to consider mapping elements which achieve both a high (logarithmic) support, i.e., that lead to a significant number of inconsistencies, and that have a high confidence, i.e., those inconsistencies are not mere noise. Hence, we use the harmonic mean of logarithmic support and confidence as a final rating score for statements:

$$score(m) := \frac{2 \cdot logs(m) \cdot c(m)}{logs(m) + c(m)} \tag{4}$$

For our experiments, we computed mean (μ) and standard deviation (σ) of s, logs, and c. We observe $\mu_s = 0.0002$, $\sigma_s = 0.003$, $\mu_{logs} = 0.179$, $\sigma_{logs} = 0.139$, $\mu_c = 0.114$, and $\sigma_c = 0.260$. This shows that the distribution of logs and c actually resemble one another, and they can thus be safely combined using the harmonic mean.

Using that rating function, we can group the output of the identified inconsistencies by mapping elements, assign a score to each mapping element, and inspect the high scoring elements for an identification of common problems in the construction process of DBpedia.

Note that in our running example, we had flagged two mapping elements: one mapping the type dbo:Settlement to Infobox settlement, and one mapping the infobox key owner-oper to dbo:operator. The property mapping has a confidence of 0.153 and a logarithmic support of 0.377, leading to a score of 0.218. In contrast, the type mapping for the object has only a confidence of 0.0004, at a similar logarithmic support of 0.344, leading to a score of 0.0008. Thus, in this case, the likelihood that the statements extracted from the property mapping are involved in an inconsistency is much higher than the likelihood of the object type mapping. Hence, given a suitable lower bound for the overall score, we would examine the case only once. Here, an expert would diagnose that the range assertion of dbo:operator is not compatible with a larger fraction of the assertions using the property.

4   Findings

In this section, we report about the number of problems identified, as well as present typical problems and proposed solutions.

4.1   Quantitative Results

In total, there are 63,981 mapping elements in the snapshot of the RML translation of the DBpedia Mappings Wiki we used in our experiments.[12] Out of those, 3,454 are identified to produce a statement which is in one of the A-boxes we inspect.[13] We identified a total number 1,117 (i.e., 32.3%) of mapping statements involved in at least one inconsistency.

Figure 3 depicts the distribution of the mapping elements that produce at least one statement involved in an inconsistency. We can observe that there is a larger number of mapping elements with confidence 1, i.e., all the statements they produce are involved in inconsistencies. Furthermore, there is a larger cluster of mapping elements with a confidence below 0.05, i.e., we can assume that the inconsistencies are produced by other effects than the mapping element or the ontology (e.g., wrong statements in Wikipedia, incorrect extraction of URIs from hashed URLs, etc. – see, e.g., [19] for a discussion).

We mainly inspected those mapping elements which receive a high overall score as defined in (4), i.e., those that are depicted in the top right corner of the scatter plot. From those, we derived a set of typical problems.

---

[12] http://rml.io/data/DBpediaAll.rml.nt, retrieved on November 2nd, 2016.

[13] This is indeed an interesting discrepancy. A larger number of mapping elements produces literal-valued statements, which are out of scope of our inspection. Furthermore, there might be outdated mappings for infoboxes no longer in use.

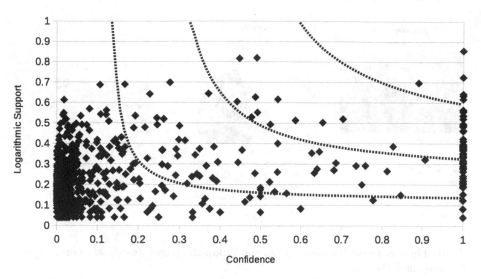

Fig. 3. Distribution of flagged mapping elements according to logarithmic support and confidence. The three dashed lines depict the isometrics of the overall score of 0.25, 0.5, and 0.75.

4.2 Mapping Errors

Pure erroneous mappings are scarce. In most cases, there is a mapping to a property which seems correct, but the property has a different meaning defined in the ontology, as can be seen by inspecting the domain and range of the property. When inspecting the distributions of the actual usage of the property, however, it becomes evident its intended semantics is often different from how it is understood and used by the crowd contribution to the DBpedia Mappings Wiki.

Mapping to Wrong Property. In the simplest case, an infobox key is mapped to a wrong property in the DBpedia ontology, where a correct one actually exists.

One example for a wrong mapping is the `branch` key in the `Military unit` infobox, which is mapped to `dbo:militaryBranch`. However, that property expects a person in the subject position. The correct property would be `dbo:commandStructure`. This affects 12,172 assertions in DBpedia (i.e., 31% of all assertions of `dbo:militaryBranch`).

Another interesting example is the mapping of depictions in various infoboxes (e.g., music covers) to `dbo:picture`. The extraction code does not generate a link to the media object as such, but tries to parse the image file name into a DBpedia resource. Thus, the 3,354 assertions using that property are mostly inconsistent, covering mostly places (2,021, or 64.5%) and persons (465, or 23.0%). This is one of the rare cases where a mapping should not be altered, but discarded altogether. Some instantiations of that problem are the statements

Fig. 4. Three example infoboxes (from the Wikipedia pages *Justify My Love*, *Brixton Academy*, and *Cannes*)

dbo:Brixton_Academy dbo:picture dbo:Brixton.
dbo:Justify_My_Love dbo:picture dbo:Madonna_(entertainer).

The corresponding two infobox snippets are depicted in Fig. 4 (left and middle).

A similar example is the mapping of the property dbo:mayor. For a majority of instances of that property (436, or 47.3%), the corresponding party is extracted as an object of the property, while a considerably smaller fraction (234, or 25.4%) are actually persons. A corresponding infobox is depicted on the right hand side in Fig. 4.

Missing Properties. Similarly to mappings to wrong properties, some infobox keys are mapped to an inadequate ontology property, but a correct one cannot be found in the ontology. Here, the ontology has to be enhanced.

The property dbo:president is designed to assert that a person is the president of an organization (e.g., a state or a company). However, this is only the minority of the cases (2,600 assertions, or 23.7%). The vast majority of the subjects is of type dbo:Person (8,354, or 76.2%). Here, the property is used to assert that a person (e.g., a minister) was serving *for* a certain president. Here, the introduction of a new property (e.g., dbo:servedFor) should be considered.

Another example is the mapping of the infobox key instruments in the Infobox music genre to dbo:instrument, whose domain is dbo:Artist (i.e., the property is intended to relate a artist to the instrument (s)he plays). This affects 707 assertions in total. Again, the introduction of a new property (e.g., dbo:characteristicInstrument) should be considered.

Further examples include the use of dbo:species for fictional characters in movies or books, while it is intended (and most widely used) for defining biological taxonomies, or `dbo:director`, which is meant to be the director of a movie, but also used to denote, e.g., the director of a festival or other event.

4.3   Problems in the Ontology

While some errors can be tracked down to individual mappings, others keep recurring for many infobox types. Hence, their root is usually not the mappings as such, but problematic definitions in the ontology. One example is the `dbo:operator` property, which has been used as a running example above. While having `Organisation` as its defined range, out of 13,425 objects of `dbo:operator`, 9,529 are of that type, while 1,092 have populated places (i.e., cities, counties, etc.) being their operator. This holds for many classes, such as airports, libraries, or stadiums. In that case, the range of `dbo:operator` should either be broadened, or cities, counties etc. should be a subclass of both `dbo:Place` and `dbo:Agent` in order to allow for the observed polysemy (cities, at least as they are used in DBpedia, are both a geographic and a social object).

A similar case is the range of properties like `dbo:architect`, `dbo:designer`, `dbo:engineer`, etc. Those often define a `dbo:Person` in the range, but in many cases, there are companies in the object position: `dbo:architect` has 1,480 (8.6%) organizations and 8,538 (49.5%) persons, designer has 806 organizations (7.8%) and 5,298 (50,4%) persons, and `dbo:engineer` has even a majority 153 (58.4%) organizations and 32 (12.2%) persons. Here, the range of those properties should be broadened to `dbo:Agent` for covering both persons and organizations.

A large-scale problematic mapping is the mapping of the property `dbo:team`, which is mainly used for subjects of type `dbo:CareerStation` (351,580 out of 1,287,645, or 27.3%)[14], but also for `dbo:Person` (160,452 out of 1,287,645, or 12.5%). The two are clearly not compatible, and since the corresponding super property in the DOLCE-Zero (`dul:isSettingFor`) expects a `dul:Situation` in the subject position, the latter cases lead to inconsistencies. Here, a more uniform usage of the property and hence, a more consistent modeling of athletes and the team(s) they belong to over time, should be enforced.

Another large-scale issue in the ontology is the use of bands (i.e., `dbo:Band`) and musical artists (i.e., `dbo:MusicalArtist`) together with properties that expect either one or the other (e.g., `dbo:associatedBand`). However, the corresponding objects are more or less equally distributed across both classes. In fact, the corresponding infobox keys, which do not distinguish between associated musical artists and bands, are currently mapped to *both* `dbo:associatedBand` and `dbo:associatedMusicalArtist`. Here, we propose the refactoring into a common property `dbo:associatedMusicAct`, whose range is the union of `dbo:Band` and `dbo:MusicalArtist`.

4.4   Problems with the Mapping to DOLCE-Zero

The mapping to DOLCE-Zero provides additional disjointness axioms which help discovering more inconsistencies. However, in a few cases, a property is

---

[14] A large majority of the subjects is not typed at all.

used for a certain purpose in the majority of cases, which is incompatible with the formalization DOLCE.

One such example is the property `dbo:commander`, although not defining an explicit domain in the DBpedia ontology, is a subproperty of `dul:coparticipatesWith`, which has the domain `dul:Object` by inference using the DOLCE-Zero ontology, which is disjoint with `dul:Event`. However, the majority (i.e., 11,831 out of 12,841, or 92.1%) of all subjects using this property are `dbo:MilitaryConflicts`, a subclass of `dul:Event`. Those are mainly created by a mapping in the `Military conflict` infobox.

5   Conclusion and Outlook

In this paper, we have introduced a data-driven approach targeted at increasing the data quality in DBpedia. The proposed approach searches for inconsistencies with the underlying ontology, also leveraging the linked top level ontology DOLCE-Zero, and tries to identify common root causes of those inconsistencies.

With our approach, we were able to identify quite a few problems of different kinds. We found pure mapping errors, hints for missing properties, as well as problematic domain and range restrictions in the DBpedia ontology.

To leverage the results, we have started fixing the problematic mappings, where appropriate, so that the next version of DBpedia will not contain the problematic assertions anymore. In cases where changes to the ontology are required, be it domain/range changes or the introduction of new properties, discussion threads have been started, since some of the changes may have long-reaching consequences and should thus be considered carefully.

For the moment, an expert has to review the problems identified in our approach, and manually classify them as wrong mappings, problems in the ontology, etc. In future work, we aim at developing a set of data-driven heuristics which perform these classifications automatically. As a long term high level vision, we foresee the creation of a knowledge graph validator which, provided with a knowledge graph and statement-level provenance information (i.e., what were the mechanisms that led to the inclusion of a particular assertion in the knowledge graph), issues fine-grained qualified suggestions on how to revise the creation process for increasing the knowledge graph's data quality. Tooling-wise, this could be implemented in an automatic reporting system, or even in the mappings Wiki for issuing live warnings upon editing time.

The work presented in this paper is a first step in this direction. While most of the approaches for increasing the data quality in knowledge graphs discussed in the literature so far are pure post-processing approaches which aim at the identification, elimination, or correction of problematic statements, the work presented in this paper is one of the rare examples that go one step beyond by identifying the root causes and fixing the cause of the data quality problem instead of remedying the symptoms.

Acknowledgements. The author would like to thank the numerous people involved in the DBpedia project for their past, ongoing, and future efforts, as well as the authors of [4] for providing the DBpedia mappings in RML.

References

1. Acosta, M., Zaveri, A., Simperl, E., Kontokostas, D., Auer, S., Lehmann, J.: Crowdsourcing linked data quality assessment. In: Alani, H., et al. (eds.) ISWC 2013. LNCS, vol. 8219, pp. 260–276. Springer, Heidelberg (2013). doi:10.1007/978-3-642-41338-4_17
2. Agrawal, R., Srikant, R., et al.: Fast algorithms for mining association rules. In: Proceedings of the 20th International Conference on Very Large Data Bases, VLDB, vol. 1215, pp. 487–499 (1994)
3. Bizer, C., Heath, T., Berners-Lee, T.: Linked data - the story so far. Int. J. Semant. Web Inf. Syst. **5**(3), 1–22 (2009)
4. Dimou, A., Kontokostas, D., Freudenberg, M., Verborgh, R., Lehmann, J., Mannens, E., Hellmann, S.: DBpedia mappings quality assessment. In: International Semantic Web Conference - Posters and Demonstrations (2016)
5. Dimou, A., Vander Sande, M., Colpaert, P., Verborgh, R., Mannens, E., Van de Walle, R.: RML: a generic language for integrated RDF mappings of heterogeneous data. In: LDOW (2014)
6. Fleischhacker, D., Paulheim, H., Bryl, V., Völker, J., Bizer, C.: Detecting errors in numerical linked data using cross-checked outlier detection. In: Mika, P., et al. (eds.) ISWC 2014. LNCS, vol. 8796, pp. 357–372. Springer, Cham (2014). doi:10.1007/978-3-319-11964-9_23
7. Gangemi, A., Guarino, N., Masolo, C., Oltramari, A.: Sweetening WordNet with DOLCE. AI Mag. **24**(3), 13–24 (2003)
8. Gangemi, A., Mika, P.: Understanding the semantic web through descriptions and situations. In: Meersman, R., Tari, Z., Schmidt, D.C. (eds.) OTM 2003. LNCS, vol. 2888, pp. 689–706. Springer, Heidelberg (2003). doi:10.1007/978-3-540-39964-3_44
9. Glimm, B., Horrocks, I., Motik, B., Stoilos, G., Wang, Z.: Hermit: an owl 2 reasoner. J. Autom. Reasoning **53**(3), 245–269 (2014)
10. Hellmann, S., Stadler, C., Lehmann, J., Auer, S.: DBpedia live extraction. In: Meersman, R., Dillon, T., Herrero, P. (eds.) OTM 2009. LNCS, vol. 5871, pp. 1209–1223. Springer, Heidelberg (2009). doi:10.1007/978-3-642-05151-7_33
11. Jain, P., Hitzler, P., Yeh, P.Z., Verma, K., Sheth, A.P.: Linked data is merely more data. In: AAAI Spring Symposium: Linked Data Meets Artificial Intelligence, vol. 11 (2010)
12. Lehmann, J., Bühmann, L.: ORE - a tool for repairing and enriching knowledge bases. In: Patel-Schneider, P.F., Pan, Y., Hitzler, P., Mika, P., Zhang, L., Pan, J.Z., Horrocks, I., Glimm, B. (eds.) ISWC 2010. LNCS, vol. 6497, pp. 177–193. Springer, Heidelberg (2010). doi:10.1007/978-3-642-17749-1_12
13. Lehmann, J., Gerber, D., Morsey, M., Ngonga Ngomo, A.-C.: DeFacto - deep fact validation. In: Cudré-Mauroux, P., et al. (eds.) ISWC 2012. LNCS, vol. 7649, pp. 312–327. Springer, Heidelberg (2012). doi:10.1007/978-3-642-35176-1_20
14. Lehmann, J., Isele, R., Jakob, M., Jentzsch, A., Kontokostas, D., Mendes, P.N., Hellmann, S., Morsey, M., van Kleef, P., Auer, S., Bizer, C.: DBpedia - a large-scale, multilingual knowledge base extracted from Wikipedia. Semant. Web J. **6**(2), 167–195 (2015)

15. Ma, Y., Gao, H., Wu, T., Qi, G.: Learning disjointness axioms with association rule mining and its application to inconsistency detection of linked data. In: Zhao, D., Du, J., Wang, H., Wang, P., Ji, D., Pan, J.Z. (eds.) CSWS 2014. CCIS, vol. 480, pp. 29–41. Springer, Heidelberg (2014). doi:10.1007/978-3-662-45495-4_3

16. Paulheim, H.: Identifying wrong links between datasets by multi-dimensional outlier detection. In: International Workshop on Debugging Ontologies and Ontology Mappings, vol. 1162, pp. 27–38. CEUR Workshop Proceedings (2014)

17. Paulheim, H.: Knowledge graph refinement: a survey of approaches and evaluation methods. Semant. Web **8**(3), 489–508 (2017)

18. Paulheim, H., Bizer, C.: Improving the quality of linked data using statistical distributions. Int. J. Semant. Web Inf. Syst. (IJSWIS) **10**(2), 63–86 (2014)

19. Paulheim, H., Gangemi, A.: Serving DBpedia with DOLCE – more than just adding a cherry on top. In: Arenas, M., et al. (eds.) ISWC 2015. LNCS, vol. 9366, pp. 180–196. Springer, Cham (2015). doi:10.1007/978-3-319-25007-6_11

20. Paulheim, H., Stuckenschmidt, H.: Fast approximate A-box consistency checking using machine learning. In: Sack, H., Blomqvist, E., d'Aquin, M., Ghidini, C., Ponzetto, S.P., Lange, C. (eds.) ESWC 2016. LNCS, vol. 9678, pp. 135–150. Springer, Cham (2016). doi:10.1007/978-3-319-34129-3_9

21. Russell, S., Norvig, P.: Artificial Intelligence: A Modern Approach. Pearson, London (1995)

22. Schmachtenberg, M., Bizer, C., Paulheim, H.: Adoption of the linked data best practices in different topical domains. In: Mika, P., et al. (eds.) ISWC 2014. LNCS, vol. 8796, pp. 245–260. Springer, Cham (2014). doi:10.1007/978-3-319-11964-9_16

23. Sheng, Z., Wang, X., Shi, H., Feng, Z.: Checking and handling inconsistency of DBpedia. In: Wang, F.L., Lei, J., Gong, Z., Luo, X. (eds.) WISM 2012. LNCS, vol. 7529, pp. 480–488. Springer, Heidelberg (2012). doi:10.1007/978-3-642-33469-6_60

24. Suchanek, F.M., Kasneci, G., Weikum, G.: YAGO: a core of semantic knowledge unifying WordNet and Wikipedia. In: 16th International Conference on World Wide Web, pp. 697–706. ACM, New York (2007)

25. Töpper, G., Knuth, M., Sack, H.: DBpedia ontology enrichment for inconsistency detection. In: Proceedings of the 8th International Conference on Semantic Systems, pp. 33–40. ACM, New York (2012)

26. Waitelonis, J., Ludwig, N., Knuth, M., Sack, H.: WhoKnows? - evaluating linked data heuristics with a quiz that cleans up DBpedia. Int. J. Interact. Technol. Smart Educ. **8**(4), 236–248 (2011)

27. Wienand, D., Paulheim, H.: Detecting incorrect numerical data in DBpedia. In: Presutti, V., d'Amato, C., Gandon, F., d'Aquin, M., Staab, S., Tordai, A. (eds.) ESWC 2014. LNCS, vol. 8465, pp. 504–518. Springer, Cham (2014). doi:10.1007/978-3-319-07443-6_34

Rule-Based OWL Modeling with ROWLTab Protégé Plugin

Md. Kamruzzaman Sarker[1(✉)], Adila Krisnadhi[1,2], David Carral[3],
and Pascal Hitzler[1]

[1] Data Semantics (DaSe) Laboratory, Wright State University, Dayton, OH, USA
sarker.3@wright.edu
[2] Faculty of Computer Science, Universitas Indonesia, Depok, Indonesia
[3] Center for Advancing Electronics Dresden (cfaed), TU Dresden, Dresden, Germany

Abstract. It has been argued that it is much easier to convey logical statements using rules rather than OWL (or description logic (DL))
axioms. Based on recent theoretical developments on transformations
between rules and DLs, we have developed ROWLTab, a Protégé plugin
that allows users to enter OWL axioms by way of rules; the plugin then
automatically converts these rules into OWL 2 DL axioms if possible,
and prompts the user in case such a conversion is not possible without
weakening the semantics of the rule. In this paper, we present ROWLTab,
together with a user evaluation of its effectiveness compared to entering
axioms using the standard Protégé interface. Our evaluation shows that
modeling with ROWLTab is much quicker than the standard interface,
while at the same time, also less prone to errors for hard modeling tasks.

1 Introduction

About a decade ago, not long after description logics [1] had been chosen
as the basis for the then-forthcoming W3C Recommendation for the Web
Ontology Language OWL [6], a rather aggressively voiced discussion as to
whether a rule-based paradigm might have been a better choice emerged in
the Semantic Web community [7,19,23]. On the one hand, this eventually led to
a new W3C Recommendation on the Rule Interchange Format RIF [9], based on
the rules paradigm, while an alternative approach which layered rules on top of
the existing OWL standard, known as SWRL [8], remained a mere W3C member
submission. However, SWRL has proven significantly more popular than RIF.
To see this, it may suffice to compare the Google Scholar citation numbers for
SWRL – over 2500 since 2004 – and RIF – just over 50 since 2009.

At the same time, researchers kept investigating more elaborate ways to
bridge between the two paradigms [10,12,13,16,20], and in particular how to
convert rules into OWL [2,11,14,15]. These results regarding conversion now
make it possible to express axioms first as rules, and only then to convert them
into OWL. We have consistently used this approach to model complex OWL
axioms throughout the last few years, as rules are, arguably, easier to understand
and produce than OWL (or description logic) axioms in whichever syntax.

© Springer International Publishing AG 2017
E. Blomqvist et al. (Eds.): ESWC 2017, Part I, LNCS 10249, pp. 419–433, 2017.
DOI: 10.1007/978-3-319-58068-5_26

Consider the sentence: "If a person has a parent who is female, then this parent is a mother", which we consider to be of medium difficulty in terms of modeling it in OWL. As a first-order logic rule, this can be expressed as

$$\text{Person}(x) \wedge \text{hasParent}(x, y) \wedge \text{Female}(y) \rightarrow \text{Mother}(y).$$

This can be expressed in OWL using description logic syntax as follows:

$$\text{Female} \sqcap \exists \text{hasParent}^-.\text{Person} \sqsubseteq \text{Mother}$$

Based on anecdotal evidence, many people find it easier to come up with the rule than directly with the OWL axiom. Following this lead, we have produced a Protégé [18] plugin, called ROWLTab, which accepts rules as input, and adds them as OWL axioms to a given ontology, provided the rule is expressible by an equivalent set of such axioms. In case the rule is not readily convertible, the user is prompted and asked whether the rule shall be saved as SWRL rule.

In order to assess the usefulness of the ROWLTab, we have furthermore conducted a user experiment in which we compare the ROWLTab interface for adding axioms to the standard Protégé interface. Our hypotheses for the user evaluation were that given complex relationships expressed as natural language sentences as above, users will be quicker to add them to an ontology using the ROWLTab than with the standard Protégé interface, and that they will also make less mistakes in doing so. The first hypothesis has been fully confirmed by our experiment, the second has been partially confirmed.

The rest of the paper will be structured as follows. In Sect. 2 we explain in more detail the rule-to-OWL conversion algorithm used. In Sect. 3 we present the ROWLTab Protégé plugin. In Sect. 4 we present our user evaluation and results, and in Sect. 5 we conclude. More information about the plugin can be found at http://daselab.org/content/modeling-owl-rules. A preliminary report on the plugin, without evaluation and with much fewer details, was presented as a software demonstration at the ISWC2016 conference [21].

2   SWRL Rules to OWL Axioms Transformation

In this section we introduce theoretical notions employed across the paper. Note that, due to space constraints, some of the definitions below are simplified and may not exactly correspond with existing definitions from different sources.

Let $\mathbf{C}$, $\mathbf{R}$, $\mathbf{I}$ and $\mathbf{V}$ be pairwise disjoint, countably infinite sets of *classes*, *properties*, *individuals* and *variables*, respectively, where $\top, \bot \in \mathbf{C}$, the *universal property* $U \in \mathbf{R}$ (i.e., `owl:topObjectProperty`) and, for every $R \in \mathbf{R}$, $R^- \in \mathbf{R}$ and $R^{--} = R$. A *class expression* is an element of the grammar $\mathbf{E} ::= (\mathbf{E} \sqcap \mathbf{E}) \mid \exists R.\mathbf{E} \mid \exists R.\text{Self} \mid C \mid \{a\}$ where $C \in \mathbf{C}$, $R \in \mathbf{R}$ and $a \in \mathbf{I}$. Furthermore, let $\mathbf{T} = \mathbf{I} \cup \mathbf{V}$ be the set of *terms*. An *atom* is a formula of the form $C(t)$ or $R(t, u)$ where $C \in \mathbf{E}$, $R \in \mathbf{R}$ and $t, u \in \mathbf{T}$. For the remainder of the paper, we identify pairs of atoms of the form $R(t, u)$ and $R^-(u, t)$. Furthermore, we identify a conjunction of formulas with the set containing all the formulas in the conjunction and vice-versa.

An *axiom* is a formula of the form $C \sqsubseteq D$ or $R_1 \circ \ldots \circ R_n \sqsubseteq R$ with $C, D \in \mathbf{E}$ and $R_{(i)} \in \mathbf{R}$. A *rule* is a first-order logic formula of the form $\forall \boldsymbol{x}(\beta(\boldsymbol{x}) \to \eta(\boldsymbol{z}))$ with β and η are conjunctions of atoms, $\boldsymbol{x}$ and $\boldsymbol{z}$ are non-empty sets of terms where $\boldsymbol{z} \subseteq \boldsymbol{x}$. As customary, we often omit the universal quantifier from rules. Axioms and rules are also referred to as *logical formulas*. Axioms as defined above essentially correspond to OWL 2 EL axioms [4] plus inverse property expression, while rules correspond to SWRL rules minus (in)equality and built-in atoms.

Consider some terms t and u and a conjunction of atoms β. We say that t *and u are directly connected in* β if both terms occur in the same atom in β. We say t *and u are connected in* β if there is some sequence of terms $t_1, \ldots, t_n$ with $t_1 = t$, $t_n = u$, and t_{i-1} and t_i are directly connected in β for every $i = 2, \ldots, n$.

The notions of *interpretation* and of an interpretation *entailing an axiom* follow the standard definitions for description logics [5]. For rules ρ of the form $\beta \to \eta$ we say that an interpretation $\mathcal{I}$ entails ρ if, for every substitution σ we have that $\mathcal{I}, \sigma \models \beta$ implies $\mathcal{I}, \sigma \models \eta$, i.e., the semantics of rules follows the standard semantics of first-order predicate logic. We say that two sets of logical formulas $\mathcal{S}$ and $\mathcal{S}'$ are *equivalent* if and only if every interpretation $\mathcal{I}$ that entails $\mathcal{S}$ also entails $\mathcal{S}'$ and vice-versa. Furthermore, we say that $\mathcal{S}'$ is a *conservative extension* of $\mathcal{S}$ if and only if (i) every interpretation that entails $\mathcal{S}'$ also entails $\mathcal{S}$ and (ii) every interpretation that entails $\mathcal{S}$ and is only defined for the symbols in $\mathcal{S}$ can be extended to an interpretation entailing $\mathcal{S}'$ by adding suitable interpretations for additional signature symbols. It is well-known that a set of logical formulas can be replaced by another set without affecting the outcome of reasoning tasks if the latter set is a conservative extension of the former.

We now formally discuss the transformation of rules into axioms. We do not include a comprehensive description of this transformation, which was introduced in [15], but only present a simplified version in an attempt to make this publication more self-contained. Specifically, our presentation makes the following assumptions about rules, all without loss of generality.

1. Rules do not contain constants. Note that, an atom of the form $R(a, b)$ (resp. $A(a)$) with $a, b \in \mathbf{I}$ in the body of a rule may be replaced by an equivalent atom $\exists U.(\{a\} \sqcap \exists R.\{b\})(x)$ (resp. $\exists U.(\{a\} \sqcap A)(x)$) where x is any arbitrarily chosen variable occurring in the body of the rule. Furthermore, atoms of the form $R(x, a)$ with $x \in \mathbf{V}$ and $a \in \mathbf{I}$ occurring in the body may be replaced by $\exists R.\{a\}(x)$. Similar transformations may be applied to the head of a rule in order to remove all occurrences of constants.
2. The head of a rule is of the form $C(x)$ or $S(x, y)$ with $C \in \mathbf{E}$, $S \in \mathbf{R}$ and $x, y \in \mathbf{V}$. Note that, a rule of the form $\beta \to \eta$ is equivalent to a set of rules $\{\beta \to \eta_1, \ldots, \beta \to \eta_n\}$ provided that $\eta = \eta_1 \cup \ldots \cup \eta_n$.
3. All of the variables in the body of a rule are connected. If two variables x and y are not connected in the body of a rule, we may simply add the atom $U(x, y)$ to the body of the rule resulting in a semantically equivalent rule.

The preprocessing implied by the aforementioned assumptions has been implemented in the ROWLTab plugin and thus, such constraints need not be considered by the end users. Moreover, such preprocessing is carefully implemented in an attempt to minimize the number of necessary modifications for a rule to satisfy assumptions (1–3). We now proceed with the definition of our translation.

Definition 1. *Given some rule $\rho = \beta \to \eta$, let $\delta(\beta \to \eta)$ be the rule that results from exhaustively applying transformations (1–3) where (1) and (2) should be applied with higher priority than (3).*

1. *Replace every atom of the form $R(x,x)$ in β or in η with $\exists R.\mathsf{Self}(x)$.*
2. *Replace every maximal subset of the form $\{C_1(y)\ldots,C_n(y)\} \subseteq \beta$ with the atom $C_1 \sqcap \ldots \sqcap C_n(y)$.*
3. *For every variable y not occurring in η that occurs in exactly one binary atom in β of the form $R(z,y)$, do the following:*
   - *If there is some atom of the form $C(y) \in \beta$, then replace the atoms $R(z,y)$ and $C(y)$ in β with the atom $\exists R.C(z)$.*
   - *Otherwise, replace the atom $R(z,y)$ in β with $\exists R.\top(z)$.*

Example 1. Consider the rule $\rho = \mathrm{Person}(x) \wedge \mathrm{hasParent}(x,y) \wedge \mathrm{Female}(y) \to \mathrm{Mother}(y)$. Then, the transformation presented in the previous definition would sequentially produce the following sequence of rules

$$(\exists \mathrm{hasParent}^-.\mathrm{Person})(y) \wedge \mathrm{Female}(y) \to \mathrm{Mother}(y)$$

$$(\exists \mathrm{hasParent}^-.\mathrm{Person} \sqcap \mathrm{Female})(y) \to \mathrm{Mother}(y)$$

Rule $\delta(\rho)$ from the previous example can be directly transformed into an axiom as indicated in the following lemma.

Lemma 1. *Consider some rule ρ. If $\delta(\rho)$ is of the form $C(x) \to D(x)$, then ρ is equivalent to the axiom $C \sqsubseteq D$.*

Proof. Let υ and υ' be some rules such that υ' results by applying some of the transformations (1–3) introduced in Definition 1 to υ. Note that, by definition, we can conclude equivalency between υ and υ'. Thus, we can show via induction that ρ is equivalent to $\delta(\rho)$. Furthermore, if $\delta(\beta \to \rho)$ is of the form $C(x) \to D(x)$, then, by the definition of the semantics of rules and axioms, $C \sqsubseteq D$ is equivalent to $\delta(\beta \to \rho)$. Since the equivalence relation is transitive, we can conclude that ρ is equivalent to $C \sqsubseteq D$.

As indicated by Lemma 1, rule ρ from Example 1 is equivalent to the axiom $\exists \mathrm{hasParent}^-.\mathrm{Person} \sqcap \mathrm{Female} \sqsubseteq \mathrm{Mother}$.

Lemma 2. *Consider some rule ρ. If the rule $\delta(\rho)$ is of the form $\bigwedge_{i=2}^{n}(C_i(x_{i-1})\wedge R_i(x_{i-1},x_i)) \wedge C_n(x_n) \to S(x_1,x_n)$, then the set of axioms $\{C_i \sqsubseteq \exists R_{C_i}.\mathsf{Self} \mid i = 1,\ldots,n\} \cup \{R_{C_1} \circ R_1 \circ \ldots \circ R_{C_{n-1}} \circ R_n \circ R_{C_n} \sqsubseteq S\}$ where all R_{C_i} are fresh properties unique for every class C_i is a conservative extension of the rule ρ.*

Proof. As shown in proof of Lemma 1, rules ρ and $\delta(\rho)$ are indeed equivalent. Thus, the lemma follows from the fact that the set of rules presented in the statement of the lemma is a conservative extension of $\delta(\rho)$.

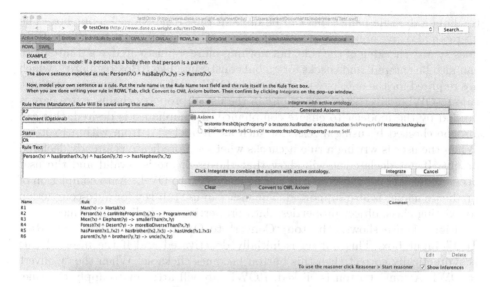

Fig. 1. The ROWLTab interface with generated axioms.

We finalize the section with some brief comments about the presented transformation. First, the transformation is sound, but has not been proven to be complete. That is, there may be some rules which are actually expressible as axioms, but cannot be handled by our translation algorithm. Moreover, the transformation of a given rule may in some cases produce axioms that violate some of the global syntactic restrictions of OWL 2 DL, such as regularity restriction on property inclusion/hierarchy, when added to an ontology.

3   Plugin Description and Features

Figure 1 shows the user interface of the ROWLTab plugin with generated axioms from a rule and also shows previously saved rules in the bottom part of the user interface. As seen in the figure, the plugin consists of two tabs: ROWL and SWRL. The latter is really the SWRLTab input interface, while the former is our implementation of rule-to-OWL conversion functionality. We have in fact reused the source code of the SWRLTab, kept its functionality intact and added extra functionality by means of the ROWL tab. The upper part of the interface is for rule insertion and the bottom part is for rule modification. At the top a modeling example is also shown.

A user can enter a rule in the "Rule Text" input box using the standard SWRL syntax, e.g.

```
Person(?x) ^ hasChild(?x,?y) ^ Female(?y) -> hasDaughter(?x,?y)
```

Every rule needs a distinct name to be eligible to be saved for later modification. A suggested rule name is automatically generated but the user also has the option to change the rule name; this can be done in the "Rule Name" input box. The user can also give annotations to a rule in the "Comment" input box. The plugin does syntax checking of the rule, but nothing more sophisticated, e.g. tautologies can also be entered, and checks for global constraints like RBox regularity, which are required for the ontology to stay within OWL 2 DL, are not performed. Since rule-to-OWL conversion often results in the use of property chains, extra care is needed by the modeler to ensure compliance, if compliance is desired; this can also be checked by using a reasoner such as HermiT [17] from within Protégé. When the user is writing a rule it checks whether a predicate is already declared or not. If not declared it will show that the predicate is invalid and the user needs to declare it before the rule can be converted to OWL. Auto completion of predicate names is also supported. The user can use tab-key for auto-completion to existing class, object properties, data properties and individual names.

Figure 1 also shows a button "Convert to OWL Axiom" below the "Rule Text" input box. This button is initially deactivated and if the inserted rule is syntactically correct then this button becomes clickable. When the "Convert to OWL Axiom" button is clicked, ROWLTab will attempt to apply the rule-to-OWL transformation described in the previous section to the given rule. If successful, a pop-up will appear displaying one or more OWL axioms resulting from the transformation, presented in Manchester syntax. These axioms can then be integrated into the active ontology by clicking the "Integrate" button of the pop-up interface. If the given rule cannot be transformed into OWL axioms, ROWLTab will prompt the user if (s)he still want to insert the rule into the ontology as a SWRL rule. If the user agrees, ROWLTab will switch to its SWRL tab and proceed in the same way as adding a rule via the original SWRLTab.

As described in Sect. 2, the translation of a rule into OWL axioms may sometimes require the introduction of fresh object properties, which will be automatically created by the plugin when necessary. The namespace for these fresh object properties is taken from the default namespace. To create a unique fresh object property, the plugin counts the number of existing fresh object properties in the active ontology (including imports) then increments the counter by 1 and creates the new object property with the incremented counter as part of its identifier.

Once the axioms generated from a rule are added to the ontology, the rule will be saved with the ontology for later modification; the rule is in fact added as an annotation to every OWL axiom generated by the rule. Figure 1 shows saved rules displayed on the bottom left of the ROWLTab plugin. A user can modify or delete rules at any time. If a rule is modified or deleted, the axioms generated by that rule will be affected. That means if a rule is deleted then the axioms generated by the rule will also be deleted. To edit a rule which was previously used to generate axioms the user needs to select the rule first and then click the "Edit" button at the bottom right part of the interface. The rule will then appear in the "Rule Text" input box for modification. The user can also double click on the rule to edit that rule. To delete a rule, the user needs to select that rule and then click the "Delete" button on the bottom right of the interface.

A feature of ROWLTab not found in the SWRLTab is the possibility to automatically add declarations for classes and properties if the inserted rule contains classes or properties not yet defined in the ontology. For example, in the rule above, the original SWRLTab requires that hasChild and hasDaughter be already defined as object properties, and Person and Female as classes in the ontology. This means that the user does not need to first exit the plugin and declare the classes and properties outside the ROWLTab.

Another feature of the ROWLTab plugin is that it actually works as a super-set of SWRLTab plugin. So if a user need to work with the SWRLTab plugin, the user does not need to install the SWRLTab plugin separately, as the ROWLTab contains a full instance of the SWRLTab plugin. If the ROWLTab plugin is installed the user only need to switch the tab from ROWL to SWRL to get the full SWRLTab functionality. This also creates a limitation that when a new version of SWRLTab is available the developer of ROWLTab has to embed the newer version of SWRLTab explicitly.

To manage the source code and to be able to modify the source code efficiently in the future we have separated the view module from the control module. The view module consist of the user interface and the control module implements the rule-to-OWL transformation. Besides those two modules we have a separate listener module, which acts as a bridge between the view and controller module. We have used Maven as our build system to easily manage the dependency of various APIs. This plugin is open source and the source code is available at the DaseLab website (http://dase.cs.wright.edu/content/modeling-owl-rules).

4   Evaluation

For the evaluation of the ROWLTab plugin, we conducted a user evaluation to answer the following three questions:

- Is writing OWL axioms into Protégé via the ROWLTab plugin quicker than writing them directly through the standard Protégé interface?
- Is writing OWL axioms into Protégé via ROWLTab plugin less error-prone than writing them directly through the standard Protégé interface?
- Do users view modeling OWL axioms via ROWLTab plugin to be an easier task than directly through the standard Protégé interface?

The evaluation was conducted by asking the participants to model a set of natural language sentences as rules using the ROWLTab or as OWL axioms using the standard Protégé interface. We recorded time and number of keyboard and mouse clicks required for each question (see Sect. 4.1) and also recorded the responses which we subsequently assessed for correctness (see Sect. 4.2). Finally, the participants answered a brief questionnaire (see Sect. 4.3).

Before describing the experiment in detail, we would like to encourage the readers to do it themselves; it should take less than an hour, the software can be obtained from the ROWLTab website already indicated.

For the experiment we recruited 12 volunteers from among the graduate students at Wright State University. Our sole selection criterion was that the participants had at least some basic knowledge of OWL, and had at least minimal exposure to Protégé. All participants were then given a half-hour briefing in which we explained, by means of examples, how to model natural language sentences with and without the ROWLTab in Protégé.

Each participant was given the same twelve natural language sentences to model. The sentences are listed in Table 1 where group A consists of sentence 1 to 6 and group B consists of sentence 7 to 12. As indicated in the table, each group contains two easy, two medium, and two hard sentences to model. Each participant modeled one of the sets of sentences using the ROWLTab and the other group without using the ROWLTab, and we randomly assigned whether the participant will model Group A using the ROWLTab or Group B using the ROWLTab. In order to minimize learning effects which may come from different sentences, we made sure that for each sentence in Group A there is a very similar sentence in Group B, and vice-versa: Each sentence number n in Group A corresponds to sentence number $n+6$ in Group B. We furthermore randomized whether the participant will first model using ROWLTab, and then without the ROWLTab, or vice-versa, also to control for a possible learning effect during the course of the experiment. There was no time limit for the modeling; participants were informed that it should usually take no longer than an hour to model all twelve sentences. Participants were also informed that they cannot go back to earlier sentences during the course of the experiment.

Table 1. Evaluation Questions

Group A	Group B	Difficulty
1. Every father is a parent 2. Every university is an educational institution	7. Every parent is a human 8. Every educational institution is an organization	easy
3. If a person has a mother then that mother is a parent 4. Any educational institution that awards a medical degree is a medical school	9. If a person has a parent who is female, then this parent is a mother 10. Any university that is funded by a state government is a public university	medium
5. If a person's brother has a son, then that son is the first person's nephew 6. All forests are more biodiverse than any desert	11. If a person has a female child, then that person would have that female child as her daughter 12. All teenagers are younger than all twens	hard

Our categorization into easy, medium, and hard sentences was done as follows: Easy sentences expressed simple subclass relationships. Medium sentences required the use of property restrictions to model them in OWL; the medium sentence 9 was discussed in Sect. 1. Hard sentences could only be expressed using two or three OWL axioms, together with a technique called *rolification* [11,14]. For example, sentence 5 when expressed as a rule becomes

$$\text{Person}(x) \wedge \text{hasBrother}(x, y) \wedge \text{hasSon}(y, z) \rightarrow \text{hasNephew}(x, z).$$

In order to express this sentence as OWL axioms, one first has to *rolify* the class Person by adding the axiom Person $\sqsubseteq \exists R_{\text{Person}}.\text{Self}$, where R_{Person} is a fresh property name, and to then add the property chain axiom

$$R_{\text{Person}} \circ \text{hasBrother} \circ \text{hasSon} \sqsubseteq \text{hasNephew}.$$

We informed all the participants regarding the total number of easy, medium, and hard sentences the participants would face. With each sentence, we also displayed the suitable class and property-names which had been pre-defined by us, i.e. the participants did not have to declare them in Protégé, and directed participants to use the displayed class and property names to the maximum extent possible. For example, the pre-defined classes and properties for sentence 9 were Person, hasParent, Female, Mother, while for sentence 5 they were Person, hasBrother, hasSon, hasNephew. An exception to this is when modeling the hard sentences (5, 6, 11, and 12) via standard Protégé. Those sentences contain class names that need to be rolified, which necessitates one to declare one or more fresh object properties. In this case, we informed the participants that the hard sentences may require them to declare additional object properties without disclosing that this is due to rolification.

4.1 Time Used for Modeling

Our hypothesis was that, on medium and hard sentences, participants would be able to model quicker with the ROWLTab than without it. Cumulated data is given in Table 2.

Table 2. Average and standard deviation of time (in seconds), number of clicks (keyboard and mouse), and correctness score per difficulty category of sentences.

Sentence Category	Time (in secs)		# clicks		Correctness	
	Protégé	ROWL	Protégé	ROWL	Protégé	ROWL
	avg/std	avg/std	avg/std	avg/std	avg/std	avg/std
Easy	79/41	47/9	44/38	59/19	2.9/0.3	2.9/0.3
Medium	312/181	116/61	216/131	141/91	2.2/0.5	2.5/0.8
Hard	346/218	160/66	351/318	228/168	0.9/0.7	2.5/0.7

For the statistical analysis, our null hypothesis was that there is no difference between the time taken with ROWLTab versus Protégé. Since each participant had modeled sentences from each difficulty class, we could perform a paired (two-tailed) t-test – note that assuming normal distributions appears to be perfectly reasonable for this data. For the medium sentences the null hypothesis was rejected with $p \approx 0.002 < 0.01$. For the hard questions the null hypothesis was rejected with $p \approx 0.020 < 0.05$. Both results are statistically significant with $p < 0.05$, thus confirming our hypotheses.

Interestingly, if we run the same t-test also on the easy sentences, the same null hypothesis is also rejected with $p \approx 0.019 < 0.05$. We will reflect on this further below.

In order to aid us in interpreting the results, we also recorded the number of clicks (keyboard plus mouse) required for modeling each sentence; cumulative data is provided also in Table 2. If we run the number of clicks through the same t-test as before, the null hypothesis being that there is no difference between the two interfaces. The corresponding p-values are 0.092 (easy), 0.030 (medium) and 0.173 (hard), i.e. we have $p < 0.05$ only for the medium sentences. The click analysis may provide us with a partial answer to the better performance of the ROWLTab regarding time used: Fewer clicks may in this case simply translate into less time required. However, this observation does not explain the data for easy and hard questions. We will return to this discussion at the end of Sect. 4.2, after we have looked at answer correctness.

4.2 Correctness of Modeled Axioms

Our hypothesis was that for medium and hard questions, participants would provide more correct answers with the ROWLTab than without it. To see if we can confirm this hypothesis, we verify the correctness of the axioms in the OWL files obtained from the participants. The correct set of axioms for each modeling question is given in Table 3. Since the sentences are short and the resulting OWL axioms are relatively simple, the verification was done manually. Also, for modeling tasks where the participants were asked to model the sentence via ROWLTab, we check the correctness of the rules by examining the OWL axioms obtained after translation, which are annotated with information regarding the actual rule input given to ROWLTab. We then assign a score of 0, 1, 2, or 3 to each answer from the participants as follows:

- Modeling a sentence via ROWLTab, the score is:
    - 3 if the participant's rule is fully correct (equivalent to the answer key),
    - 2 if the participant's mistakes are only in the incorrect use of variables (wrong placement, missing/spurious variables), i.e., the rule still employs the correct predicates in the rule body and head and no spurious predicates are used,
    - 1 if there's a missing predicate in the participant's rule or spurious predicates are used that makes the rule not equivalent to the correct answer,
    - 0 if the participant provides no answer.
- Modeling a sentence via the standard Protégé interface, the score is:
    - 3 if the participant's OWL axioms are fully correct,

- 2 if the participant's OWL axioms employ the correct set of class and property names, but there is a mistake in the use of logical constructs,
- 1 if there is a missing or spurious class names or property names, or even missing some necessary OWL axioms,
- 0 if the participant provides no answer.

The average and standard deviation of the correctness score for easy, medium, and hard questions can be found in Table 2. Here, our null hypothesis was that there is no difference in correctness of the answers given with ROWLTab versus Protégé. For the same reasons as before, we thus performed a paired (two-tailed) t-test, the null hypothesis being that there be no difference whether using ROWLTab or not. For the medium sentences we obtained $p \approx 0.18 > 0.05$, so the null hypothesis could not be rejected. But for the hard questions the null hypothesis was rejected with $p \approx 0.0001 < 0.01$. The latter result is statistically significant with $p < 0.01$. If we run the same t-test also on the easy sentences, the same null hypothesis cannot be rejected; we in fact obtain $p \approx 1.0000$.

Table 3. Answers to Evaluation Questions – Rui and Axi are answers for question i in the form of rule and OWL axioms, resp. where $R_1, \ldots, R_7$ are fresh (object) properties generated due to rolification

Ru1: Father$(x) \rightarrow$ Parent(x) Ax1: Father $\sqsubseteq$ Parent
Ru2: University$(x) \rightarrow$ EducationalInstitution(x) Ax2: University $\sqsubseteq$ EducationalInstitution
Ru3: Person$(x) \wedge$ hasMother$(x,y) \rightarrow$ Parent(y) Ax3: $\exists$hasMother$^-$.Person $\sqsubseteq$ Parent
Ru4: $\dfrac{\text{EducationalInstitution}(x) \wedge \text{awards}(x,y) \wedge \text{MedicalDegree}(y)}{\rightarrow \text{MedicalSchool}(x)}$ Ax4: EducationalInstitution $\sqcap \exists$awards.MedicalDegree $\sqsubseteq$ MedicalSchool
Ru5: Person$(x) \wedge$ hasBrother$(x,y) \wedge$ hasSon$(y,z) \rightarrow$ hasNephew(x,z) Ax5: Person $\sqsubseteq \exists R_1$.Self, $R_1 \circ$ hasBrother $\circ$ hasSon $\sqsubseteq$ hasNephew
Ru6: Forest$(x) \wedge$ Desert$(y) \rightarrow$ moreBiodiverseThan(x,y) Ax6: Forest $\sqsubseteq \exists R_2$.Self,   Desert $\sqsubseteq \exists R_3$.Self, $R_2 \circ U \circ R_3 \sqsubseteq$ moreBiodiverseThan
Ru7: Parent$(x) \rightarrow$ Human(x) Ax7: Parent $\sqsubseteq$ Human
Ru8: EducationalInstitution$(x) \rightarrow$ Organization(x) Ax8: EducationalInstitution $\sqsubseteq$ Organization
Ru9: Person$(x) \wedge$ hasParent$(x,y) \wedge$ Female$(y) \rightarrow$ Mother(x) Ax9: Person $\sqcap \exists$hasParent.Female $\sqsubseteq$ Mother
Ru10: University$(x) \wedge$ fundedBy$(x,y) \wedge$ StateGovernment$(y) \rightarrow$ PublicUniversity(x) Ax10: University $\sqcap \exists$fundedBy.StateGovernment $\sqsubseteq$ PublicUniversity
Ru11: Person$(x) \wedge$ hasChild$(x,y) \wedge$ Female$(y) \rightarrow$ hasDaughter(x,y) Ax11: Person $\sqsubseteq \exists R_4$.Self,   Female $\sqsubseteq \exists R_5$.Self, $R_4 \circ$ hasChild $\circ R_5 \sqsubseteq$ hasDaughter
Ru12: Teenager$(x) \wedge$ Twen$(y) \rightarrow$ youngerThan(x,y) Ax12: Teenager $\sqsubseteq \exists R_6$.Self,   Twen $\sqsubseteq \exists R_7$.Self, $R_6 \circ U \circ R_7 \sqsubseteq$ youngerThan

We thus confirm our hypothesis that ROWLTab helps users in modeling hard sentences correctly; however we could not confirm this for medium sentences on this population of participants. It could be hypothesized that the participants were sufficiently familiar with Protégé to perform well on medium difficulty, thus use of the ROWLTab only had an effect on time used, as shown in Sect. 4.1.

At the same time, the correctness analysis also sheds further light on the hard questions: While participants used less time for these on the ROWLTab, they did not use significantly fewer clicks; however answer correctness was much higher on the ROWLTab. This seems to indicate that the additional time using Protégé was spent thinking (and indeed, rather unsuccessfully) about the problem, while this additional thinking was not required when using the ROWLTab.

4.3   Participant Survey

We finally used a questionnaire with four questions to assess the subjective value which the use of the ROWLTab had to the participants. For this, we asked all participants to indicate to what extent they agree with each of the following statements.

1. ROWLTab is a useful tool to help with ontology modeling.
2. Modeling rules with ROWLTab was easier for me than modeling without it.
3. Given some practice, I think I will find modeling rules with the ROWLTab easier than modeling without it.
4. The ROWLTab is better for ontology modeling than the SWRLTab.

Participants were asked to click, on screen whether they agree with each statement, on a scale from -3 (strongly disagree) to +3 (strongly agree). It turns out that participants agreed highly with all three statements:

Question number	Mean	Standard deviation
1 (ROWL is a useful tool.)	2.83	0.39
2 (ROWL makes modeling easier.)	3.00	0.00
3 (Modeling with ROWL easier with some practice.)	2.75	0.45
4 (ROWLTab better than SWRLTab)	1.75	1.22

In assessing these responses, we need to be aware that the pool of participants came from the investigators' institution, and many of them were either associated with the investigators' lab or had attended classes by one of the investigators. Hence the scores should be interpreted with caution. Nevertheless, the scores for the first three questions indicate strong agreement with the usefulness of the ROWLTab.

Regarding the fourth question, it should be noted that our briefing did not include a briefing on the SWRLTab. As discussed in Sect. 3, the user interaction of the ROWLTab is very similar to that of the SWRLTab, so the only

Table 4. Summary of evaluation results. Entries indicate whether the difference between using the ROWLTab and not using it were statistically significant.

Category	Time	Clicks	Correctness
Easy	Significant ($p < 0.05$)	Not significant	Not significant
Medium	Significant ($p < 0.01$)	Significant ($p < 0.05$)	Not significant
Hard	Significant ($p < 0.05$)	Not significant	Significant ($p < 0.01$)

substantial difference would be in the fact that the ROWLTab produces OWL axioms, while the SWRLTab produces SWRL axioms with a different semantics. We do not know to what extent the participants were aware of this difference. A quarter of the participants answered this question with "0" (neutral). Results of the experiment can be found at http://dase.cs.wright.edu/content/rowl and raw result can be found at https://github.com/md-k-sarker/ROWLPluginEvaluation/tree/master/results.

5   Conclusions and Further Work

We have presented the Protégé ROWLTab plugin for rule-based OWL modeling in Protégé, and its underlying algorithms. We have furthermore reported on a user evaluation for assessing the improvements arising from the use of ROWLTab.

The evaluation results are summarized in Table 4: We have a significant time improvement in all three categories (it was hypothesized by us only for medium and hard sentences). In the medium category, where answer correctness was not significantly different, the ROWLTab required significantly less clicks. In the hard category, the difference in answer correctness was also significant.

The evaluation results are rather encouraging, and we also already received direct feedback from users that the ROWLTab is considered very useful. But while basic functionality is already in place, we already see further improvements that can be made to the plugin:

- When rolification is used for the transformation of a rule to OWL, the ROWLTab currently invents an artificial property name for the fresh object property. It may be helpful to more directly support a renaming of these properties, or to come up with a standard naming scheme for properties arising out of rolification. Note, however, that it is not sufficient to have one fresh property for each defined atomic class, as in some cases complex classes need to be rolified [11].
- The translation of rules into OWL often leads to the use of property chains, which may result in a non-regular property hierarchy, thus violating a global syntactic restriction of OWL 2 DL. While standard tools such as reasoners, which can be called from within Protégé, can detect this issue, it may be helpful to catch this earlier, e.g. directly at the time when a rule is translated.
- Currently, if a rule is input which cannot be translated to OWL, it is simply saved as a SWRL rule, i.e., with a significantly modified (and, in a sense, restricted) semantics. However, through the use of so-called *nominal schemas* [14] it is possible to recover more of the first-order semantics of the input rules, and it has even been shown that the use of such nominal schemas can lead to performance improvements of reasoners compared to SWRL [22].

More substantial possible future work would carry the ROWLTab theme beyond the basic rule paradigm currently supported:

- The rule syntax could be extended to allow for capturing OWL features which cannot be expressed by means of the basic rules currently supported. In particular, these would be right-hand side (head) disjunctions and existentials as well as cardinality restrictions, as well as left-hand side universal quantifiers. It would even be conceivable to add additional shortcut notation, e.g. for witnessed universals [3], or for nominal schemas [14].
- The development of a full-blown rule syntax for all of OWL 2 DL would then also make it possible to perform all ontology modeling using rules, i.e., to establish an interface where the user would get a pure rules view on the ontology, if desired.

We are looking forward to feedback by ontology modelers on the route which we should take with the plugin in the future.

Acknowledgements. This work was supported by the National Science Foundation under award 1017225 *III: Small: TROn – Tractable Reasoning with Ontologies* and the German Research Foundation (DFG) within the Cluster of Excellence "Center for Advancing Electronics Dresden" (cfaed). We would also like to thank Tanvi Banerjee and Derek Doran for some advise on statistics.

References

1. Baader, F., et al. (eds.): The Description Logic Handbook: Theory, Implementation, and Applications, 2nd edn. Cambridge University Press, Cambridge (2010)
2. Carral Martínez, D., Hitzler, P.: Extending description logic rules. In: Simperl, E., Cimiano, P., Polleres, A., Corcho, O., Presutti, V. (eds.) ESWC 2012. LNCS, vol. 7295, pp. 345–359. Springer, Heidelberg (2012). doi:10.1007/978-3-642-30284-8_30
3. Carral, D., Krisnadhi, A., Rudolph, S., Hitzler, P.: All but not nothing: left-hand side universals for tractable OWL profiles. In: Keet, C.M., Tamma, V.A.M. (eds.) Proceedings of the 11th International Workshop on OWL: Experiences and Directions (OWLED 2014). CEUR Workshop Proceedings, vol. 1265, Riva del Garda, Italy, 17–18 October 2014, pp. 97–108. CEUR-WS.org (2014)
4. Cuenca Grau, B., Motik, B., Wu, Z., Fokoue, A., Lutz, C.: OWL 2 Web Ontology Language Profiles, 2nd edn. W3C Recommendation, 11 December 2012. http://www.w3.org/TR/owl2-profiles/
5. Hitzler, P., Krötzsch, M., Rudolph, S.: Foundations of Semantic Web Technologies. CRC Press, Chapman & Hall (2010)
6. Hitzler, P., et al. (eds.): OWL 2 Web Ontology Language Primer, 2nd edn. W3C Recommendation, 11 December 2012. http://www.w3.org/TR/owl2-primer/
7. Horrocks, I., Parsia, B., Patel-Schneider, P., Hendler, J.: Semantic web architecture: stack or two towers? In: Fages, F., Soliman, S. (eds.) PPSWR 2005. LNCS, vol. 3703, pp. 37–41. Springer, Heidelberg (2005). doi:10.1007/11552222_4
8. Horrocks, I., et al.: SWRL: A Semantic Web Rule Language Combining OWL and RuleML. W3C Member Submission, 21 May 2004. http://www.w3.org/Submission/SWRL/
9. Kifer, M., Boley, H. (eds.): RIF Overview, 2nd Edn. W3C Working Group Note, 5 February 2013. https://www.w3.org/TR/rif-overview/

10. Knorr, M., Hitzler, P., Maier, F.: Reconciling OWL and non-monotonic rules for the semantic web. In: Raedt, L.D., et al. (eds.) 20th European Conference on Artificial Intelligence, Montpellier, ECAI 2012, France, 27–31 August 2012, pp. 474–479. IOS Press (2012)

11. Krisnadhi, A., Maier, F., Hitzler, P.: OWL and rules. In: Polleres, A., d'Amato, C., Arenas, M., Handschuh, S., Kroner, P., Ossowski, S., Patel-Schneider, P. (eds.) Reasoning Web 2011. LNCS, vol. 6848, pp. 382–415. Springer, Heidelberg (2011). doi:10.1007/978-3-642-23032-5_7

12. Krötzsch, M.: Description Logic Rules, Studies on the Semantic Web, vol. 8. IOS Press, Amsterdam (2010)

13. Krötzsch, M., Hitzler, P., Vrandecic, D., Sintek, M.: How to reason with OWL in a logic programming system. In: Eiter, T., et al. (eds.) Proceedings of the Second International Conference on Rules and Rule Markup Languages for the Semantic Web, RuleML 2006, pp. 17–26. IEEE Computer Society, Athens, Georgia (2006)

14. Krötzsch, M., Maier, F., Krisnadhi, A., Hitzler, P.: A better uncle for OWL: nominal schemas for integrating rules and ontologies. In: Srinivasan, S., et al. (eds.) Proceedings of the 20th International Conference on World Wide Web, WWW 2011, Hyderabad, India, 28 March–1 April, pp. 645–654. ACM (2011)

15. Krötzsch, M., Rudolph, S., Hitzler, P.: Description logic rules. In: Ghallab, M., et al. (eds.) Proceeding of the 18th European Conference on Artificial Intelligence, Patras, Greece, 21–25 July, vol. 178, pp. 80–84. IOS Press, Amsterdam (2008)

16. Krötzsch, M., Rudolph, S., Hitzler, P.: ELP: tractable rules for OWL 2. In: Sheth, A., Staab, S., Dean, M., Paolucci, M., Maynard, D., Finin, T., Thirunarayan, K. (eds.) ISWC 2008. LNCS, vol. 5318, pp. 649–664. Springer, Heidelberg (2008). doi:10.1007/978-3-540-88564-1_41

17. Motik, B., Shearer, R., Horrocks, I.: Hypertableau reasoning for description logics. J. Artif. Intell. Res. **36**, 165–228 (2009)

18. Musen, M.A.: The protégé project: a look back and a look forward. AI Matters **1**(4), 4–12 (2015)

19. Patel-Schneider, P.F., Horrocks, I.: A comparison of two modelling paradigms in the semantic web. In: Proceedings of the Fifteenth International World Wide Web Conference (WWW 2006), pp. 3–12. ACM (2006)

20. Rudolph, S., Krötzsch, M., Hitzler, P., Sintek, M., Vrandecic, D.: Efficient OWL reasoning with logic programs – evaluations. In: Marchiori, M., Pan, J.Z., Marie, C.S. (eds.) RR 2007. LNCS, vol. 4524, pp. 370–373. Springer, Heidelberg (2007). doi:10.1007/978-3-540-72982-2_34

21. Sarker, M.K., Carral, D., Krisnadhi, A.A., Hitzler, P.: Modeling OWL with rules: the ROWL protege plugin. In: Kawamura, T., Paulheim, H. (eds.) Proceedings of the ISWC 2016 Posters & Demonstrations Track. CEUR Workshop Proceedings, vol. 1690, Kobe, Japan, 19 October, CEUR-WS.org (2016)

22. Steigmiller, A., Glimm, B., Liebig, T.: Reasoning with nominal schemas through absorption. J. Autom. Reasoning **53**(4), 351–405 (2014)

23. W3C Workshop on Rule Languages for Interoperability, 27–28 April 2005, Washington, DC, USA. W3C (2005). https://www.w3.org/2004/12/rules-ws/accepted

Chaudron: Extending DBpedia
with Measurement

Julien Subercaze[✉]

Univ Lyon, UJM-Saint-Etienne, CNRS, Laboratoire Hubert Curien,
UMR 5516, 42023 Saint-Etienne, France
`julien.subercaze@univ-st-etienne.fr`

Abstract. Wikipedia is the largest collaborative encyclopedia and is used as the source for DBpedia, a central dataset of the LOD cloud. Wikipedia contains numerous numerical measures on the entities it describes, as per the general character of the data it encompasses. The DBpedia Information Extraction Framework transforms semi-structured data from Wikipedia into structured RDF. However this extraction framework offers a limited support to handle measurement in Wikipedia.

In this paper, we describe the automated process that enables the creation of the Chaudron dataset. We propose an alternative extraction to the traditional mapping creation from Wikipedia dump, by also using the rendered HTML to avoid the template transclusion issue.

This dataset extends DBpedia with more than 3.9 million triples and 949.000 measurements on every domain covered by DBpedia. We define a multi-level approach powered by a formal grammar that proves very robust on the extraction of measurement. An extensive evaluation against DBpedia and Wikidata shows that our approach largely surpasses its competitors for measurement extraction on Wikipedia Infoboxes. Chaudron exhibits a F1-score of .89 while DBpedia and Wikidata respectively reach 0.38 and 0.10 on this extraction task.

Keywords: Wikipedia · Extraction · DBpedia · Measurement · RDF · Formal grammar

1   Introduction

Wikipedia is a free content internet encyclopedia, currently the largest encyclopedia with more than five million articles in its English version; overall 38 milllion articles in over 250 languages. Wikipedia is currently in the top ten of the most viewed websites in the world, with, on average 18 billion page views and nearly 500 million unique visitors per month.

As a central knowledge repository on the Web, Wikipedia serves as a seed for the creation of knowledge bases such as Google's Knowledge Graph or Wolfram Alpha. The Intelligence in Wikipedia project [23] used Wikipedia to derive an ontology [26] and fostered open information extraction [27]. The linked open data movement has long understood the central role of Wikipedia in the web of

© Springer International Publishing AG 2017
E. Blomqvist et al. (Eds.): ESWC 2017, Part I, LNCS 10249, pp. 434–448, 2017.
DOI: 10.1007/978-3-319-58068-5_27

knowledge. The DBpedia effort is, since 2007, an open source project that aims at translating entries from Wikipedia into RDF and making it publicly available [2]. Another extraction of Wikipedia into RDF was made in the Yago project [19] using a heuristic based approach [6]. After a tentative to create a semantic backend for Wikipedia [7], the same authors created a new Wikimedia project, entitled Wikidata [21,22], that aims at structuring Wikipedia data in order make them machine readable.

Triples in the DBpedia dataset are mainly extracted from Wikipedia dumps through the DBpedia Information Extraction Framework (DIEF for the rest of the paper). This framework uses manually defined mappings to transform Wikipedia semi-structured content – called Infoboxes – into structured content. These mappings encompass, among others, the structuration of physical measurements intro RDF triples. A measurement is the assignment of a number to a property of an object or an event. For instance, the fact that Lionel Messi is 1.70 m tall is written as follows in DBpedia:

```
1  @prefix dbr: <http://dbpedia.org/resource/> .
2  @prefix dbo: <http://dbpedia.org/ontology/> .
3  @prefix dbd: <http://dbpedia.org/datatype/> .
4
5  dbr:Lionel_Messi dbo:Person/Height  "170"^^dbd:centimeter .
6  dbr:Lionel_Messi dbo:Height  "1.70"^^xsd:double .
```

The DIEF, while providing one the most valuable asset to the LOD cloud [18], is not perfect: 40% of the property occurrences are unmapped, leaving room for focused automated mappings. Some mappings do not include units [9,24] or some are incorrect[1] and transcluded Infoboxes are not properly managed, especially in chemistry[2] [12].

In this paper, we tackle the issue of structuring measures from Wikipedia Infoboxes into RDF, where the property is a physical quantity. We take into account the above mentioned issues and design a novel approach that aims for robustness. The contributions of this paper are threefold. *First*, we extract and make public Chaudron, a dataset of over 900.000 measurements of the utmost practical interest that complements DBpedia. *Second*, to create this dataset, we devised a novel robust approach based on a formal grammar for units detection. *Third*, we show that processing rendered HTML is a viable approach for extracting triples from Wikipedia.

The paper is organized as follows: Sect. 2 presents measurements and how measurements are managed in Wikipedia's Infoboxes. Section 3 describes Chaudron's extraction techniques. Section 4 presents the Chaudron dataset and its characteristics. Section 5 describes the dataset availability and its potential

---

[1] The Bowatenna Dam has a `dbp:plantCapacity` of 40 W instead of 40 MW.

[2] See for instance DBpedia resources corresponding to chemical elements, compounds and drugs, e.g. `Iron` and `Nicotine`.

applications. The quality evaluation of the dataset is presented in Sect. 6. Section 7 provides the concluding remarks.

2   Measurement

Measurement is, the assignment of a number to a characteristic or event, which can be compared with other objects or events [13]. To facilitate comparisons in various fields, measurement systems have been developed. Although large progress have been made towards unification, there still exists various measurement systems used to describe similar physical dimensions. The metrication – the process of converting to the metric system of units of measurement, also known as the International System of Units (ISU) – began in France after the revolution and spread across the globe. Nowadays, except the metric system, the US customary units and the burmese units of measurement are heavily used in the United Sates and Burma respectively. Due to the large volume of trade between the United States and Canada, the US customary is also in use in some limited contexts in Canada, mainly agriculture and engineering. Although United Kingdom officially uses the metric system, imperial units is widespread among the public; commonwealth countries are currently in similar situations. Furthermore, non standard units are also widely used in some special contexts or areas. Computer scientists, for instance, are common users of the *rack unit* – one rack (U) being 44.45 mm – that is used to measure rack-mountable computer equipment such as servers or network elements. In conclusion, there exists a large body of custom units, thus they should be catered or converted accordingly.

This diversity of units system, each of them used by large number of people on earth, is reflected in Wikipedia's articles. In its english version, height, length and the weight are usually displayed in both ISU and US customary units. Non-standard units also appear in Wikipedia articles. This leads to the first challenge in this study, i.e., to identify the set of units that are possibly used in Wikipedia Infoboxes. We address this issue in Sect. 3.2. We first start with how measurement are described within these Infoboxes.

2.1   Measurement in Wikipedia's Infoboxes

Wikipedia contains a large number of measures, most of them are stored in Infoboxes. An Infobox is a Wikipedia template that is used to represent a summary of the article, as stated in Wikipedia's documentation. It presents a summary of the most relevant facts in form of key-value pairs that are displayed as a table. Units are mainly from the three most common systems of units (ISU, US customary & Imperial) and may include other customary systems. In some rare cases, other unit systems may be used. These cases are sufficiently rare to be ignored. As noted by the authors of DBpedia [10], most of the editors do not follow the recommendations and good practices for formatting. As a consequence, the units chosen for display are not necessarily standards and may belong either to the three main systems, as well as isolated units of measure that are specialized in some domains. We investigated how measurements are stored and displayed in Wikipedia's Infoboxes. From our experience, they are pigeonholed in three categories:

Conversion. In this case, units are displayed through the template {{Convert}}. This template allows to display a measurement with two units. It is mainly used to display measurement in both metric and Imperial or US customary systems. The main syntax is the following: {{Convert | val | $unit_{source}$ | $unit_{dest}$}}. For instance the following expression {{Convert|1|lb|kg}} gives "1 pound (0.45 kg)". The {{Convert}} template supports a closed list of units suitable for conversion.

Formatted display. The formatted display is used when one does not require a unit conversion. The {{Val}} template formats a measure with units to a readable form. It is also used to display uncertainty in the measurement. The main syntax is the following: {{Val|number|ul=unit code}}. Similarly to the {{Convert}} template, {{Val}} supports a predefined list of units[3], however it also supports arbitrary units that may be customised by the editor.

Free riding. When none of the above described templates are used, editors use custom formatting to describe the measurements. This is the less formatted case of all three. Whereas common practice lead to output that are formatted similarly to {{Val}}, edge cases happen where the output does not resemble to any given format.

Some Infoboxes automatically format certain values, i.e., when an editor fills up the template instance with a key-value pair, the template will process the value with one of the two above described templates or formats it along a predefined pattern. Figure 1 depicts this process: the height is processed by the {{Convert}} templates. Other Infoboxes do not support automatic formatting, this is either due to the impracticality of using a single prefix – range of values maybe too large – or due to the laziness of the Infobox template editors.

3   Measures Extraction

To extract measures from Infobox, we are facing the following problem: we must determine whether a key/value pair is of the form: <physicalQuantity/ numericalValue Unit>. Our approach is divided in three parts. In the first part, we filter the key/value pairs to discard irrelevant entries. This process is described in Sect. 3.1. In the second part, we apply patterns to the values of key-value pairs for the single unit case. This pattern matching approach requires a list of units to be matched. We detail in Sect. 3.2 how we obtained such a list of units. In the third part, we develop a novel technique based on formal grammar to match complex units: this robust technique is presented in Sect. 3.3.

---

[3] https://en.wikipedia.org/wiki/Template:Val/list.

1) Definition

TEMPLATES CALLS THE
CONVERT MODULE TO
CONVERT (FT IN) TO M

```
{{infobox [...]

| label9    = Listed height

|   data9   = {{#if:{{{height ft|}}}
              |{{convert|{{{height ft|0|}}}|ft|
              {{#if:{{{height in|}}}|{{{height in}}}|0}}
              |in|m|2|abbr=on|order={{{height order|}}}}}}
              {{{height footnote|}}}}
```

2) Instance

TEMPLATE   INFOBOX
BASKETBALL BIOGRAPHY
FILLED WITH THE VALUES
OF KOBE BRYANT

```
{{Infobox basketball biography
| name      = Kobe Bryant
| image     = Kobe Bryant 2014.jpg
| caption   = Bryant in 2014
| position  = [[Shooting guard]]<!-
| height ft = 6
| height in = 6
| weight lb = 212
| league    =
| team      =
| number    = 8, 24
```

3) Display

FORMATTED TEXT IS DISPLAYED
WITH BOTH U.S. CUSTOMARY AND
SI UNITS.

Personal information	
Born	August 23, 1978 (age 38)
	Philadelphia, Pennsylvania
Nationality	American
Listed height	6 ft 6 in (1.98 m)[a]
Listed weight	212 lb (96 kg)
Career information	
High school	Lower Merion
	(Ardmore, Pennsylvania)
NBA draft	1996 / Round: 1 / Pick: 13th overall
	Selected by the Charlotte Hornets
Playing career	1996–2016
Position	Shooting guard
Number	8, 24
Career history	
1996–2016	Los Angeles Lakers

Fig. 1. Example of Infobox that uses the {{Convert}} template to format the height of a person. Here the basketball player Kobe Bryant.

3.1 Infobox Parsing and Filtering

As stated in the previous section, measures are often displayed by the mean of templates. These latter may be combined with other templates, making the extraction from the markup text a complex procedure. The DBpedia Information Extraction Framework extracts triples directly from the markup. There also exist very convenient tools to programmatically access Wikipedia raw data, as well as its edit history [3].

For our process, such an approach based on wiki markup is not suitable. Templates, such as {{Convert}} or {{Val}}, are complex pieces of code from different languages (Lua, PHP) and are subject to changes. Infoboxes may be aliased (e.g. {{chembox}},{{drugbox}}), or transcluded[4] as for instance chemical elements whose Infoboxes are templates from a template[5]. Therefore trying to reproduce the code stack from Mediawiki[6] in another language is overly complex. This is why the DIEF is unable to extract triples from the Infoboxes of chemical elements. To give an insight into this complexity, the Infobox settlement, one of the most used Infobox in Wikipedia, transcludes more than 50 different templates and more than 40 modules. Moreover, these modules and templates may also themselves transclude other modules and templates.

We chose to parse HTML generated code using a local copy of Wikipedia. This method, while less computationally performant, is more robust. To reduce the performance overhead, we first filter the content from the Wikipedia dump using a custom MWDumper[7] filter. Thus, for a page to pass the filter, its Infobox must either contains digits and units from the list (see Sect. 3.2) or the template of this Infobox must contains units or refer the templates for conversion or formatted display. The filter only retains the Infobox text and discards the rest of the page. This allows us to drastically reduce the size and the number of entries on our local Wikipedia copy, i.e., from 16 million entries in the original dump to 1.4 million entries. The size of our local database dump is 2 GB against 50 GB for its original counterpart. Consequently, this speeds up the overall process as compared to processing the whole set of Wikipedia articles.

While processing HTML from the local Wikipedia pages, a second filter is applied on each key/value pair. This filter is activated before the pair is sent to the units extractor. To avoid useless computation, we verify that the value of the key/value pair contains digits, but not only digits: ensuring that a unit is potentially present after the digits. Once this filter is successfully passed, we try to match the value against the following pattern: numericalValue Unit, where the list of all unit is thereafter defined.

3.2   List of Units

In order to apply the pattern matching approach, we need a list of units used in Wikipedia. For this purpose, we extracted the pages contained in the category Units of measurement and its subcategories. These pages are not necessarily only units and outliers have to be discarded. To ensure that we retain only units, we determine if the first sentence of the page validates certain textual patterns

---

[4] https://www.mediawiki.org/wiki/Transclusion.

[5] See for instance the element Iron (https://en.wikipedia.org/wiki/Iron) and its Infobox (https://en.wikipedia.org/w/index.php?title=Template:Infobox_iron &action=edit) which is itself an instance of the Infobox element (https://en. wikipedia.org/wiki/Template:Infobox_element).

[6] Mediawiki is the software that powers Wikipedia.

[7] https://www.mediawiki.org/wiki/Manual:MWDumper.

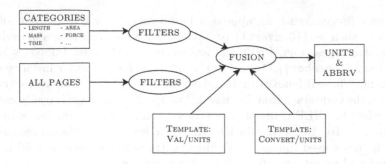

Fig. 2. Extracting the list of units from Wikipedia

(e.g. the first sentence contains "is a unit of") or contains a link to unit pages (Unit of measurement, SI units, Derived SI units, etc.). Not every unit is linked to its respective category, we therefore apply more restrictive patterns on every page to extract a second list of units.

In Sect. 2, we described the usage of the {{Convert}} and {{Val}} templates. Each of these templates has an associated list of units, that we also process to extract our third and fourth lists. Finally the four lists are merged, and in total, 464 atomic units and their abbreviations are extracted. When available, we also retain the URL of the Wikipedia article that describes the unit. Figure 2 summarizes this process and depicts the fusion process from the different sources.

This list of units allows us to apply a simple pattern matching technique to extract measurements. This technique consists of detecting numerical values followed by one of the unit from the list. This approach covers simple cases where the measurement can be expressed using a single physical dimension like height or weight. However it falls short when the quantity measured is described by a combination of several units. For example the molar mass expressed in $g \cdot mol^{-1}$ cannot be extracted using this technique.

3.3   Extraction with Formal Grammar

The above described approach only covers simple cases but falls short on multiple complex cases, i.e., when the dimension of the physical quantities are given using a combination of units. For instance the molar heat capacity of a substance is given in $kJ/(mol \cdot K)$. The formula that describes the unit of such a measure follows, by its very nature, a formal grammar. Formal grammars precisely define the structure of valid sentences in a language. Formal grammars are one of the common underlying tools of formal languages such as computer programming languages, but are not expressive enough to describe natural language where ambiguity arise. However there exists some approach to describe natural language using formal grammar such as the Attempto Controlled English [5] that can be used for knowledge representation purposes [4].

Herein we develop a formal grammar that validates whether or not a string is a valid formula. A valid a formula is made up of a single unit or of an arithmetic

$$\langle\text{UNIT}\rangle \models m \mid g \mid K \mid A \mid J \mid F \mid V \mid \ldots$$
$$\langle\text{PREFIX}\rangle \models Y \mid Z \mid E \mid P \mid T \mid G \mid M \mid \ldots$$
$$\langle\text{OP}\rangle \models \cdot \mid /$$
$$\langle\text{SIGN}\rangle \models + \mid -$$
$$\langle\text{NUM}\rangle \models (1\ldots9)+$$
$$\langle\text{EXP}\rangle \models \langle\text{EXPT}\rangle\ (\langle\text{OP}\rangle\ \langle\text{EXPT}\rangle)*$$
$$\langle\text{EXPT}\rangle \models \langle\text{PREFIX}\rangle?\ \langle\text{UNIT}\rangle\ (\langle\text{POW}\rangle)?\ \mid \text{"("}\ \langle\text{EXP}\rangle\ \text{")"}$$
$$\langle\text{POW}\rangle \models \text{"*"}(\langle\text{SIGN}\rangle)?\langle\text{NUM}\rangle$$

Fig. 3. Simplified BNF grammar of the unit formula checker. *, + and ? are the standard Kleene operators. The exponentiation is denoted "*", the multiplication · and the division /.

expression of valid prefix and units couple. The prefixes are the 26 valid prefixes from the ISU. The abbreviations of the seven core units from the ISU as from its derived units[8] are incorporated in the list. Since Wikipedia makes a very large use of conversion to display measurement in both SI and imperial, it is sufficient to extract the data.

We defined valid formula to be standard arithmetic formula including parentheses, along with the operators of multiplication and division between (prefix/units) and exponentiation on the units. We give a simplified version of the grammar in Fig. 3. This version omits the complicated details required for units that are not made of a single terminal symbol. For example the mole which is abbreviated `mol` is one of these cases, since its first letter `m` collides with the letter used for meter. This requires some special treatment in the grammar than do not carry a particular interest to be precisely described here. The complete grammar is available on Chaudron's website[9].

An input formula is described as valid, if it can syntactically be parsed by the grammar. An example is given in Fig. 4 for the ISU definition of the Volt.

3.4 Extraction and Discussion

The extraction process is run against the rendered HTML pages obtained from the local Wikipedia copy. The HTML is filtered using CSS selectors in order to retain only the Infobox. Afterwards, the Infobox – which is concretely a `table` in HTML – is parsed to extract keys and values. Values, if not matched by the simple pattern matching described above, are analyzed by the parser obtained from our grammar. If a value is matched by the parser, the formula is parsed to extract the unit and to determine whether the numerical value represents an interval or not. The representation of intervals is discussed in the next section.

As we stated in Scct. 2, we aim at identifying the largest possible set of physical measurement in Wikipedia Infoboxes. This excludes any measurement that

---

[8] https://en.wikipedia.org/wiki/International_System_of_Units#Derived_units.
[9] http://w3id.org/chaudron/.

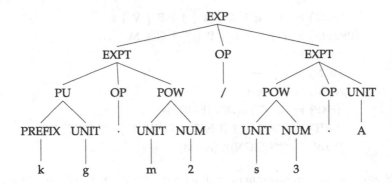

Fig. 4. Simplified parse tree for the SI equivalent of the Volt : $(\text{kg} \cdot \text{m}^2)/(\text{s}^3 \cdot \text{A})$ using the grammar defined in Fig. 3. Note that the grammar also parses the equivalent formulation: $\text{kg} \cdot \text{m}^2 \cdot \text{s}^{-3} \cdot \text{A}^{-1}$.

uses non standard units. Figure 5 depicts such an example. Fields highlighted in green are candidates to be extracted since they contain standard units. The field highlighted in red contains a measurement using a non-standard unit *rounds per minute*, that in fact represents a frequency. These measurements could be interesting to extract but they raise some issues. First of all, being non standard, these units do not connect to any existing system and are of limited use to their very context. Second, the notation of these units is also non-standard, i.e. it does not follow a rigorous notation. For instance *rounds per minute* appears in some articles under the abbreviation *rpm*; this, under a different context, means *rotation per minute* which is also a frequency, but of a very different type of events. Therefore these non-standard units – that do not measure physical quantities but are characteristics of some entities – are edge cases that are difficult to handle.

4 Dataset

The dataset that we obtain by extracting measures from the Infoboxes in the English Wikipedia contains over 900.000 measurements represented in 3.9 million triples. These measures extend existing knowledge for more than 450,000 DBpedia resources. More than 3.000 properties are measured with more than 600 units. Tables 1 and 2 present the most encountered units and properties in the dataset.

Representation. Representing units of measures and measurement has been largely discussed within the community and several approaches have been proposed [8,11,14,15,17]. We refer the reader to [16] for an overview and interesting discussions. The most common issue with these approaches is that they come with closed lists (often large, but closed) of supported units that do not contain the whole set of units of our dataset. Our approach is an open formalism that resembles those used to represent measures using reification to describe the various properties of the measure. For instance, to describe the fact that the

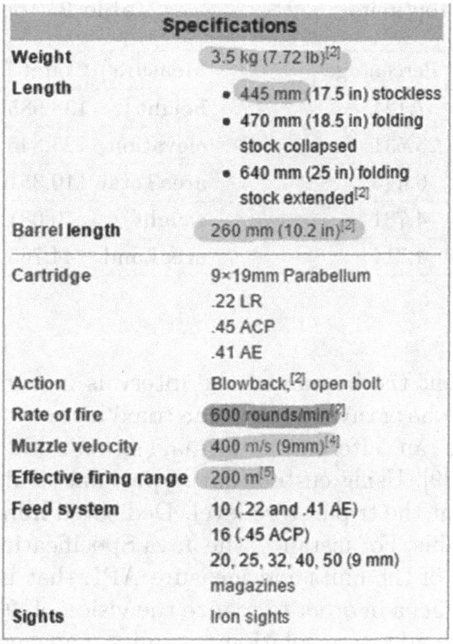

Specifications	
Weight	3.5 kg (7.72 lb)[2]
Length	• 445 mm (17.5 in) stockless
	• 470 mm (18.5 in) folding stock collapsed
	• 640 mm (25 in) folding stock extended[2]
Barrel length	260 mm (10.2 in)[2]
Cartridge	9×19mm Parabellum
	.22 LR
	.45 ACP
	.41 AE
Action	Blowback,[2] open bolt
Rate of fire	600 rounds/min[2]
Muzzle velocity	400 m/s (9mm)[4]
Effective firing range	200 m[5]
Feed system	10 (.22 and .41 AE)
	16 (.45 ACP)
	20, 25, 32, 40, 50 (9 mm) magazines
Sights	Iron sights

Fig. 5. Example of Infobox that contains standard and non standard units

Three Gorges Dam has a Nameplate Capacity of 22,500 Megawatts, we obtain the following representation:

```
1  @prefix dbr: <http://DBpedia.org/resource/> .
2  @prefix cha: <http://w3id.org/chaudron/ontology/> .
3
4  dbr:Three_Gorges_Dam
5    cha:measure [
6      cha:value ''22,500''^^xsd:decimal;
7      cha:unit ''MW''^^xsd:string;
8      cha:physicalProperty ''NameplateCapacity''^^xsd:string;
9      cha:DBpediaResource dbr:Nameplate_Capacity
10   ] .
```

We are able to link the physical quantity measured to the DBpedia resource using cha:DBpediaResource dbr:Nameplate_Capacity. This is possible since the key in the key/value pair in the Infobox links to the Wikipedia page Nameplate Capacity.

Our extraction framework also supports the extraction of interval of values. Intervals are commonly encountered in Wikipedia to describe measure that are variable under the context (dimension of soccer field) or could not be precisely

Table 1. Top units

Units	Count	Percentage
m	352,506	37.124
km^2	242,423	25.531
kg	61,202	6.445
mm	45,401	4.781
km	44,763	4.714

Table 2. Top measures

Measure	Count	Percentage
height	138,985	14.637
elevation	125,345	13.201
areaTotal	110,351	11.622
weight	70,081	7.381
areaLand	44,765	4.714

measured. To represent the bounds of the intervals in our dataset, we use the following properties: `cha:minValue` and `cha:maxValue`.

We also propose an alternative formalism suitable to process custom datatypes, based on [9]. Using custom datatypes, one could also integrate conversion and calculus at the triple store level. Dedicated libraries to process units are now widely available. For instance, the Java Specification Request 363 led to an implementation[10] of the units and measure API, that is a perfect candidate to be integrated with Jena in order to realize the vision of [9]. To our opinion this would provide a more adequate solution for units management than extensible SPARQL queries with Javascript [25].

5   Availability and Applications

The Chaudron dataset is fully integrated into the Linked Open Data Web. The resources that are qualified with measurements refer to Wikipedia articles and therefore could be directly binded to DBpedia resources. As a linked dataset, our goal is to ensure the best availability and description of Chaudron to ensure its wide adoption. We follow the current practices [1] to describe and publish our dataset.

License Chaudron is available under the terms of the Attribution-ShareAlike 3.0 license at the following persistent address: http://www.w3id.org/chaudron/. The dataset is registered at Datahub[11] and described using VoID[12]. We also provide a public SPARQL endpoint, that is available from Chaudron's homepage.

Applications. The Chaudron dataset extends DBpedia in a way that allows new applications to be developed as well as to enhance existing ones. Since measurement concern universal values, the triples do not only extend the english version of DBpedia but the complete set of localized DBpedia versions. Chaudron also fills the gap with Chemistry [12] data and leads the way to the development of new applications in this field. DBpedia has been successfully used as the primary knowledge source for Question answering Systems, especially within

---

[10] https://github.com/unitsofmeasurement/.
[11] https://datahub.io/dataset/chaudron.
[12] http://w3id.org/chaudron/voID.ttl.

the Question Answering over Linked Data workshop [20]. With the integration of Chaudron, existing Q&A systems will see their performance increased, since Chaudron provides them with triples from the utmost practical interest. Notably, Question Answering systems would be able to answer a new class of queries, that is, queries comparing entities by an attribute, including top-k queries. "What is the heaviest element between Oxygen and Gallium ?", or "What are the ten fastest roller coasters in the world ?" are examples of such questions. More complex questions such as "In which country, the biggest dam of the 19th century was built?", combine knowledge from both DBpedia and Chaudron.

6   Evaluation

To assess the quality of the dataset, we conducted a thorough evaluation to determine the precision and recall of our approach and compare with other datasets extracted from Wikipedia. We compare our extraction with the one of DBpedia [2], and Wikidata [22]. We previously discussed the construction of DBpedia; Wikidata follows a similar process as Wikipedia: the approach is to crowdsource the data acquisition.

In order to evaluate the quality of the dataset, we manually annotated over 300 statements from Infoboxes over 40 Wikipedia articles of different nature with the objective to maximize the coverage of categories: our evaluation includes vehicles, cities, persons, weapons, celestial objects, regions, buildings, drugs, rockets and others types of articles containing measurement of different nature. These measurements encompass height, weight, force, diameter, boiling point, torque, volume. The goal of this manual evaluation is to cover many types of categories and measurements. The detailed evaluation is open and can be consulted online[13].

Table 3. Evaluation of the measurement extraction task for Chaudron, DBpedia and Wikidata.

	Chaudron	DBpedia	Wikidata
Precision	**.976** $\left(\frac{248}{254}\right)$	.555 $\left(\frac{50}{90}\right)$	.941 $\left(\frac{16}{17}\right)$
Recall	**.817** $\left(\frac{248}{311}\right)$	.289 $\left(\frac{90}{311}\right)$	.055 $\left(\frac{17}{311}\right)$
F1	**.889**	.381	.103

The evaluation was conducted as follows: articles categories (buildings, drugs, weapons, ...) were identified and among them 40 articles were chosen. For each article, the fields in the Infobox have been manually identified. For each dataset,

---

[13] https://docs.google.com/spreadsheets/d/1yKFU1MMakEsKF08b3jMNBe4m91SRV XlsG6WXtNTPzlw/edit?usp=sharing.

we manually identified one of the following situations: the measurement – including its units – is correctly extracted; the measurement is missing; the measurement is incorrectly extracted. For DBpedia, the latter case is very common. That is, measurement is partially extracted, mostly only the numerical value, but the units are missing. Units are sometimes present in the relationship between the subject and the numerical values (height, weight, length) for instance, but in the majority of the cases this crucial information is missing, making the measurement valueless. As stated in the definition given in Sect. 2, measurement are meant to be compared and thus required units. Without units, a numerical value assigned to a unitless property does not constitute a measurement.

The results of this evaluation are presented in Table 3. Wikidata, with its crowdsourced approach offers a high precision, but the crowdsourced approach shows its limit on the recall. Out of the 311 measures in this evaluation, Wikidata contains only 17, out of which 16 are correct. We believe that this result to the lack of incentive of people to translate measurement into Wikidata. DBpedia uses also a manual approach, that defines translation for given Infobox templates. Therefore the recall is much higher than Wikidata (.289 vs. 0.055) but it indicates that there is room for improvement in the manually defined templates. The main issue with DBpedia and measurement is how units are handled. While some properties contain the unit as the `rdf:label`, most of properties do not include this information. This is the main reason of the low precision (.55) exhibited by DBpedia on this task. Our approach outperforms its competitors on both precision (.976) and recall (.817), naturally F1-score follows. The formal grammar approach proves very robust on the extraction task. The issues encountered are mainly due to the lack of support for measurement precision (using symbol ±) and to multiple values in the same cell. For instance, when an attribute gets different value depending on context: the Tesla S has different electric range depending on the model. This remains an open issue that could be fixed in the next version of Chaudron.

An important remark is that this evaluation has been conducted to cover the largest body of domains containing measurement. However data is not uniformly distributed as shown in the Infobox usage list[14]. For instance, the template `Infobox:settlement` reaches by far the first place. More precisely, at the time of writing, out the 3.7 millions use of Infoboxes in Wikipedia articles, the top 10 Infoboxes cover more than the half of use in articles. The top 100 covers 84 percents of the total usage. Would we have conducted an evaluation that follows the distribution, we would have had mainly settlements, taxobox and person articles. However, such an evaluation would leave aside complete domains that are of the utmost practical interest.

7   Conclusion and Future Work

We presented Chaudron, to our knowledge, the first dataset of measurement that extends DBpedia with more than 949K measurements. We described the different

---

[14] https://en.wikipedia.org/wiki/Wikipedia:List_of_infoboxes.

techniques used to filter and structure measures from Wikipedia Infoboxes into RDF, including pattern matching and formal grammar. The evaluation shows that our automated approach proves very robust on this extraction task and largely outperforms DBpedia and Wikidata. Our representation of measurement, in its n-ary version, offers the opportunity to support custom datatype processing using recent research results [9].

A side result of this work, nevertheless interesting for other practitioners is to demonstrate the soundness of extracting Wikipedia data from the rendered HTML instead of using Wiki markup. Future work will include the extraction of measurement from Wikipedia's plain text and will benefit from the Chaudron data for validation. Since Wikipedia's text contains way more information than the Infoboxes, we therefore hope to lift a larger body of measurements.

References

1. Alexander, K., Cyganiak, R., Hausenblas, M., Zhao, J.: Describing linked datasets. In: LDOW (2009)
2. Auer, S., Bizer, C., Kobilarov, G., Lehmann, J., Cyganiak, R., Ives, Z.: DBpedia: a nucleus for a web of open data. In: Aberer, K., et al. (eds.) ASWC/ISWC 2007. LNCS, vol. 4825, pp. 722–735. Springer, Heidelberg (2007). doi:10.1007/978-3-540-76298-0_52
3. Ferschke, O., Zesch, T., Gurevych, I., Wikipedia revision toolkit: efficiently accessing Wikipedia's edit history. In: Proceedings of the 49th Annual Meeting of the Association for Computational Linguistics, pp. 97–102 (2011)
4. Fuchs, N.E., Kaljurand, K., Kuhn, T.: Attempto controlled english for knowledge representation. In: Baroglio, C., Bonatti, P.A., Małuszyński, J., Marchiori, M., Polleres, A., Schaffert, S. (eds.) Reasoning Web. LNCS, vol. 5224, pp. 104–124. Springer, Heidelberg (2008). doi:10.1007/978-3-540-85658-0_3
5. Fuchs, N.E., Schwertel, U., Schwitter, R.: Attempto controlled english — not just another logic specification language. In: Flener, P. (ed.) LOPSTR 1998. LNCS, vol. 1559, pp. 1–20. Springer, Heidelberg (1999). doi:10.1007/3-540-48958-4_1
6. Kasneci, G., Ramanath, M., Suchanek, F., Weikum, G.: The yago-naga approach to knowledge discovery. ACM SIGMOD Rec. **37**(4), 41–47 (2009)
7. Krötzsch, M., Vrandečić, D., Völkel, M.: Semantic MediaWiki. In: Cruz, I., Decker, S., Allemang, D., Preist, C., Schwabe, D., Mika, P., Uschold, M., Aroyo, L.M. (eds.) ISWC 2006. LNCS, vol. 4273, pp. 935–942. Springer, Heidelberg (2006). doi:10.1007/11926078_68
8. Leal, D., Schröder, A.: RDF vocabulary for physical properties, quantities and units. Technical report, ScadaOn-Web (2002). http://www.s-ten.eu/scadaonweb/NOTE-units/2002-08-05/NOTE-units.html
9. Lefrançois, M., Zimmermann, A.: Supporting arbitrary custom datatypes in RDF and SPARQL. In: Sack, H., Blomqvist, E., d'Aquin, M., Ghidini, C., Ponzetto, S.P., Lange, C. (eds.) ESWC 2016. LNCS, vol. 9678, pp. 371–386. Springer, Cham (2016). doi:10.1007/978-3-319-34129-3_23
10. Lehmann, J., Isele, R., Jakob, M., Jentzsch, A., Kontokostas, D., Mendes, P.N., Hellmann, S., Morsey, M., van Kleef, P., Auer, S., et al.: Dbpedia-a large-scale, multilingual knowledge base extracted from Wikipedia. Semant. Web **6**(2), 167–195 (2015)

11. Masolo, C., Borgo, S., Gangemi, A., Guarino, N., Oltramari, A.: Wonderweb deliverable d18, ontology library (final). ICT project, 33052 (2003)
12. Murray-Rust, P.: Chemistry for everyone. Nature **451**(7179), 648–651 (2008)
13. Pedhazur, E.J., Schmelkin, L.P.: Measurement, Design, Analysis: An Integrated Approach. Psychology Press, New York (2013)
14. Pinto, H.S. Martins, J.: Revising and extending the units of measure "subontology". In: Proceedings of IJCAI's Workshop on IEEE Standard Upper Ontology, Seattle, WA. Citeseer (2001)
15. Probst, F.: Observations, measurements and semantic reference spaces. Appl. Ontol. **3**(1–2), 63–89 (2008)
16. Rijgersberg, H., van Assem, M., Top, J.: Ontology of units of measure and related concepts. Semant. Web **4**(1), 3–13 (2013)
17. Rijgersberg, H., Wigham, M., Top, J.L.: How semantics can improve engineering processes: a case of units of measure and quantities. Adv. Eng. Inform. **25**(2), 276–287 (2011)
18. Schmachtenberg, M., Bizer, C., Paulheim, H.: Adoption of the linked data best practices in different topical domains. In: Mika, P., et al. (eds.) ISWC 2014. LNCS, vol. 8796, pp. 245–260. Springer, Cham (2014). doi:10.1007/978-3-319-11964-9_16
19. Suchanek, F.M., Kasneci, G., Weikum, G.: Yago: a core of semantic knowledge. In: Proceedings of the 16th International Conference on World Wide Web, pp. 697–706. ACM (2007)
20. Unger, C., Forascu, C., Lopez, V., Ngomo, A.-C.N., Cabrio, E., Cimiano, P., Walter, S.: Question answering over linked data (QALD-5). In: Working Notes of CLEF (2015)
21. Vrandečić, D.: Wikidata: a new platform for collaborative data collection. In: Proceedings of the 21st International Conference on World Wide Web, pp. 1063–1064. ACM (2012)
22. Vrandečić, D., Krötzsch, M.: Wikidata: a free collaborative knowledgebase. Commun. ACM **57**(10), 78–85 (2014)
23. Weld, D.S., Wu, F., Adar, E., Amershi, S., Fogarty, J., Hoffmann, R., Patel, K., Skinner, M.: Intelligence in Wikipedia. In AAAI, vol. 8, pp. 1609–1614 (2008)
24. Wienand, D., Paulheim, H.: Detecting incorrect numerical data in DBpedia. In: Presutti, V., d'Amato, C., Gandon, F., d'Aquin, M., Staab, S., Tordai, A. (eds.) ESWC 2014. LNCS, vol. 8465, pp. 504–518. Springer, Cham (2014). doi:10.1007/978-3-319-07443-6_34
25. Williams, G.: Extensible SPARQL functions with embedded javascript. In: Auer, S., Bizer, C., Heath, T., Grimnes, G.A. (eds.) Proceedings of the ESWC 2007 Workshop on Scripting for the Semantic Web, SFSW, Innsbruck, Austria. CEUR Workshop Proceedings, vol. 248. CEUR-WS.org, 30 May 2007
26. Wu, F., Weld, D.S.: Autonomously semantifying Wikipedia. In: Proceedings of the Sixteenth ACM Conference on Information and Knowledge Management, pp. 41–50. ACM (2007)
27. Wu, F., Weld, D.S.: Open information extraction using Wikipedia. In: Proceedings of the 48th Annual Meeting of the Association for Computational Linguistics, pp. 118–127. Association for Computational Linguistics (2010)

SM4MQ: A Semantic Model
for Multidimensional Queries

Jovan Varga[1]([⊠]), Ekaterina Dobrokhotova[1], Oscar Romero[1],
Torben Bach Pedersen[2], and Christian Thomsen[2]

[1] Universitat Politècnica de Catalunya, BarcelonaTech, Barcelona, Spain
{jvarga,oromero}@essi.upc.edu, ekaterina.dobrokhotova@est.fib.upc.edu
[2] Aalborg Universitet, Aalborg, Denmark
{tbp,chr}@cs.aau.dk

Abstract. On-Line Analytical Processing (OLAP) is a data analysis
approach to support decision-making. On top of that, Exploratory OLAP
is a novel initiative for the convergence of OLAP and the Semantic Web
(SW) that enables the use of OLAP techniques on SW data. Moreover,
OLAP approaches exploit different metadata artifacts (e.g., queries) to
assist users with the analysis. However, modeling and sharing of most of
these artifacts are typically overlooked. Thus, in this paper we focus on
the query metadata artifact in the Exploratory OLAP context and propose an RDF-based vocabulary for its representation, sharing, and reuse
on the SW. As OLAP is based on the underlying multidimensional (MD)
data model we denote such queries as MD queries and define *SM4MQ*: A
Semantic Model for Multidimensional Queries. Furthermore, we propose
a method to automate the exploitation of queries by means of SPARQL.
We apply the method to a use case of transforming queries from *SM4MQ*
to a vector representation. For the use case, we developed the prototype
and performed an evaluation that shows how our approach can significantly ease and support user assistance such as query recommendation.

Keywords: Semantic Web · Vocabulary · OLAP · Query modeling

1 Introduction

On-Line Analytical Processing (OLAP) is a well-established approach for data
analysis to support decision-making [14]. Due to its wide acceptance and successful use by non-technical users, novel tendencies endorse broadening of its use
from solutions working with in-house data sources to analysis considering external and non-controlled data. A vision of such settings is presented as Exploratory
OLAP [1] promoting the convergence of OLAP and the Semantic Web (SW).
The SW provides a technology stack for publishing and sharing of data with their
semantics and many public institutions, such as Eurostat, already use it to make
their data publicly available. The Resource Description Framework (RDF) [7] is
the backbone of the SW representing data as directed triples that form a graph

© Springer International Publishing AG 2017
E. Blomqvist et al. (Eds.): ESWC 2017, Part I, LNCS 10249, pp. 449–464, 2017.
DOI: 10.1007/978-3-319-58068-5_28

where each triple has its semantics defined. Querying of RDF data is supported by SPARQL [17], the standard query language for RDF.

To facilitate data analysis, OLAP systems typically exploit different metadata artifacts (e.g., queries) to assist the user with analysis. However, although extensively used, little attention is devoted to these metadata artifacts [23]. This originates from traditional settings where very few (meta)data are open and/or shared. Thus, [23] proposes the Analytical Metadata (AM) framework, which defines AM artifacts such as schema and queries that are used for user assistance in settings such as Exploratory OLAP. In this context, analysis should be collaborative and therefore these metadata artifacts need to be open and shared. Thus, SW technologies are good candidates to model and capture these artifacts.

A first step for (meta)data sharing among different systems is to agree about (meta)data representation, i.e., modeling. As RDF uses a triple representation that is generic, the structure of specific (meta)data models is defined via RDF vocabularies providing semantics to interpret the (meta)data. Thus, the AM artifacts are modeled in [22] proposing SM4AM: a semantic metamodel for AM. Due to the heterogeneity of systems, the metamodel abstraction level is used to capture the common semantics and organization of AM. Then, metadata models of specific systems are defined at the model level instantiating one or more AM artifacts. For instance, the schema artifact for Exploratory OLAP can be represented using the QB4OLAP vocabulary to conform data to a multidimensional (MD) data model for OLAP on the SW [24]. QB4OLAP further enables running of MD queries to perform OLAP on the SW [25]. However, the representation of these queries to support their sharing, reuse, and more extensive exploitation on the SW is yet missing. Thus, in the present paper we propose a model for MD queries and explain how it supports sharing, reuse, and can also be used to facilitate metadata processing, e.g., for user assistance exploitations such as query recommendations. In particular, the contributions of this paper are:

- We propose *SM4MQ*: A Semantic Model for MD Queries formalized as an RDF-based vocabulary of typical OLAP operations (see [21]). The model captures the semantics of common OLAP operations at the conceptual level and supports their sharing and reuse via the SW.
- We define a method to automate the exploitation of *SM4MQ* queries by means of SPARQL. The method is exemplified on a use case to transform a query from *SM4MQ* to a vector representation. The use case shows an example of generating vectors (forming a matrix) as analysis-ready data structures that can be used by existing recommender systems [2] and approaches to query recommendations (e.g., [6]).
- We developed a prototype and used a set of MD queries to evaluate our approach for the chosen use case. The evaluation shows that *SM4MQ* significantly eases and supports the automation of query exploitation by means of SPARQL.

The paper is organized as follows. The next section explains the preliminaries of our approach. Then, Sect. 3 proposes the MD query model. Section 4 presents the proposed method and the related use case. Section 5 discusses the use case evaluation results. Finally, Sect. 6 discusses the related work and Sect. 7 concludes the paper.

2   Background

In this section, we introduce the necessary preliminaries and a running example used throughout the paper. First, we explain the MD model and the most popular OLAP operations. Then, we discuss the use of SW and QB4OLAP for MD models. The formalization of QB4OLAP concepts and OLAP operations can be found in [10] and in the present paper we provide the necessary intuition for understanding the proposed query model. The running example is incrementally introduced in each of the subsections.

2.1   Multidimensional Model and OLAP Operations

The MD model organizes data in terms of *facts*, i.e., data being analyzed, and *dimensions*, i.e., analytical perspectives [14]. Dimensions consist of *levels* representing different data granularities that are hierarchically organized into dimension *hierarchies*. Levels can have *attributes* that further describe them. Facts contain *measures* that are typically numerical values being analyzed. Data conforming to an MD schema are referred to as a *data cube* that is being navigated (e.g., data granularity is changed) via OLAP operations. For instance, Fig. 1 illustrates an MD schema created for the European Union asylum applicants data set available in a Linked Data version of the Eurostat data[1]. In the data set, the number of asylum applications as a measure, can be analyzed according to the age, sex, type of application (Asyl_app), destination country (Geo), country of origin (Citizenship), and month of application (RefPeriod) levels of related dimensions. Moreover, the data can be aggregated from months to quarters and likewise to years, from country of origin to continent, and from destination country to continent or government type as additional levels in related dimensions.

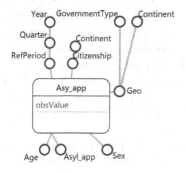

Fig. 1. Asylum data set schema (in DFM Notation [12])

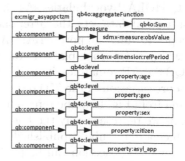

Fig. 2. Asylum data set QB4OLAP schema representation

---

[1] http://eurostat.linked-statistics.org/.

To navigate a data cube, OLAP operations are used and different OLAP algebras have been proposed [18]. In the present paper, we consider the set of OLAP operations used in [9,10] that are defined at the conceptual level as discussed in [8]. The considered OLAP operations are described in the following.

The *ROLL-UP* operation aggregates data from a finer granularity level to a coarser granularity level in a dimension hierarchy of a data cube. For instance, in case of the schema in Fig. 1 data can be aggregated from the month level to the year level for the RefPeriod dimension. Similarly, the *DRILL-DOWN* operation as its inverse disaggregates data from a coarser granularity level to a finer granularity level in a dimension hierarchy of a data cube. Furthermore, the *DICE* operation takes a data cube and applies a boolean condition expressed over a level (attribute) and/or measure value over it. For instance, for the schema in Fig. 1 a user may be interested in number of asylum applications only for the years 2009 and 2010. Finally, the *SLICE* operation removes a dimension or a measure from a data cube. For instance, a user may not be interested in the age of the applicants and thus remove the related dimensions.

2.2   The Semantic Web Technologies

As mentioned in the introduction, the SW technologies provide means for flexible (meta)data representation and sharing. RDF, which is the SW backbone, represents data in terms of directed subject - predicate - object triples that comprise an RDF graph where subjects and objects are nodes and predicates are edges. Subject and predicate are represented with IRIs, i.e., unique resource identifiers on the SW, while objects can either be IRIs or literal values. Furthermore, RDF supports representation of the data semantics via the rdf:type property. In the context of sharing, the Linked Data initiative [13] strongly motivates interlinking of RDF data on the SW to support identification of related/similar/same concepts. Finally, the RDF data can be queried with SPARQL [17], the standardized query language for RDF, which supports their systematic exploration.

To support the publishing of MD data and their OLAP analysis directly on the SW, two RDF vocabularies were proposed, namely the RDF Data Cube (QB) and QB4OLAP vocabularies. As the former vocabulary was primarily designed for statistical data sets, the latter one was proposed to extend QB with necessary concepts to fully support OLAP. A detailed discussion on this is presented in [24] where it is also explained how existing QB data sets can be enriched with the QB4OLAP semantics. Thus, in the present paper we consider QB4OLAP for the representation of the MD data on the SW and Fig. 2 illustrates how the MD schema from Fig. 1 can be represented with QB4OLAP. Note that for simplicity reasons we represent only the finest granularity levels.

Once a data cube is published using QB4OLAP, the OLAP operations from the previous subsection can be performed. However, a metadata model for representing these queries is yet missing. Such a model can be created by instantiating the SM4AM metamodel (see [21,22] for the initial and extended SM4AM versions, respectively). The metamodel represents the query AM artifact with several meta classes that we explain next. First, the sm4am:UAList element is a complex element that combines atomic elements that include data exploration actions (i.e., sm4am:DataExplorationAction with its sm4am:ManipulationAction subclass). Thus, the metamodel elements can be instantiated as an MD query that combines OLAP operations and we present the details in the next section.

3   A Semantic Model for Multidimensional Queries

In this section, we define *SM4MQ* as an RDF-based vocabulary to represent the introduced OLAP operations. The model is created by instantiating the related SM4AM metamodel elements. It is built around the QB4OLAP model for representing an MD schema and explained with examples related to the running example schema.

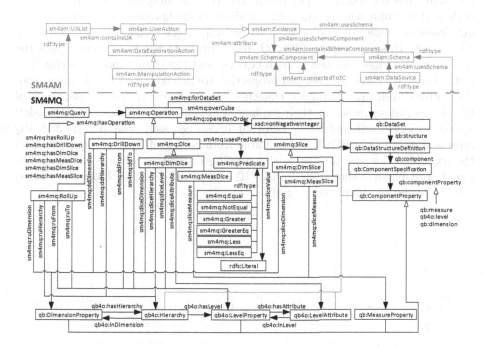

Fig. 3. A semantic model for multidimensional queries

3.1   MD Query Model

Figure 3 illustrates the complete *SM4MQ* query model. Furthermore, the figure also shows how *SM4MQ* relates to the SM4AM metamodel concepts and reuses concepts from the QB, QB4OLAP, and RDFS vocabularies. The central concept of the model is sm4mq:Query representing a query. A query can be related to a simply ordered set of OLAP operations (i.e., subclasses of sm4mq:Operation) via subproperties of sm4mq:hasOperation. OLAP operations are organized in a simply ordered set as each of them directly relates to the query and they are mutually ordered, e.g., the order of ROLL-UP and DRILL-DOWN operations is relevant to determine the final granularity of the data cube. Each operation relates to a data cube schema (i.e., qb:DataStructureDefinition) via sm4mq:over-Cube and to a data set (i.e., qb:DataSet) to which data belong to via sm4mq:-forDataSet. This way, operations belonging to a single query can operate over different schemata and data sets (inspired by the federated queries mechanism in SPARQL). In the next subsections, we explain each of the OLAP operations. Note that we follow the formalization given in [9] and we also enrich the model with additional information needed to facilitate sharing.

3.2   ROLL-UP and DRILL-DOWN Operations

Following the definition in [9], ROLL-UP (i.e., sm4mq:RollUp) is represented with a data cube schema (i.e., qb:DataStructureDefinition), a dimension (i.e., qb:DimensionProperty), and a level to roll-up to (i.e., qb4o:LevelProperty). In addition to these concepts, *SM4MQ* also represents the level from which the roll-up is performed, its order in the query, and the dimension hierarchy (i.e., qb4o:Hierarchy) used. The related properties are illustrated in Fig. 3. In general, the roll-up from and hierarchy concepts can be inferred from a sequence of OLAP operations, however their explicit representations makes the ROLL-UP operation model self-contained such that it can be easily shared. The *SM4MQ* representation of the ROLL-UP example from Sect. 2 of aggregating data from the month to the year level over the running example schema is illustrated in Fig. 4. The example shows the ROLL-UP instance and also includes the related *SM4MQ* concepts (depicted in gray).

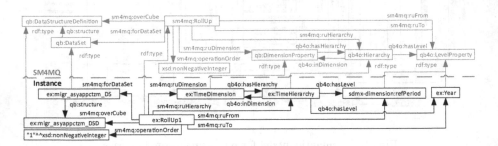

Fig. 4. ROLL-UP instance example

Following the definition in [9], DRILL-DOWN (i.e., sm4mq:DrillDown) is represented with a data cube schema (i.e., qb:DataStructureDefinition), a dimension (i.e., qb:DimensionProperty), and a level to drill-down to (i.e., qb4o:LevelProperty). In addition to these concepts, *SM4MQ* also represents the level from which the drill-down is performed, its order in the query, and the dimension hierarchy (i.e., qb4o:Hierarchy) used for the same reasons as in the case of ROLL-UP. An example of DRILL-DOWN is analogous to the ROLL-UP one and we omit it for space reasons.

3.3   DICE Operation

Following the definition in [9], DICE (i.e., sm4mq:Dice) is represented with a data cube schema (i.e., qb:DataStructureDefinition) and a boolean condition (i.e., sm4mq:Predicate) over a dimension (i.e., qb:DimensionProperty) or a measure (i.e., qb:MeasureProperty). We represent these two cases separately for the atomicity of operations that also facilitates sharing. Thus, in *SM4MQ* we create two subclasses for DICE, sm4mq:DimDice as DICE applied over a dimension and sm4mq:MeasDice as DICE applied over a measure. The former one relates to a dimension, hierarchy, level, and optionally level attribute, while the latter relates to a measure. For both cases we define the order and consider a set of relational predicates that includes equals to (i.e., sm4mq:Equal), not equals to (i.e., sm4mq:NotEqual), greater than (i.e., sm4mq:Greater), greater than or equal (i.e., sm4mq:GreaterEq), less than (i.e., sm4mq:Less), and less than or equal (i.e., sm4mq:LessEq). Each specific relational operator is an instance of the sm4mq:-Predicate class (similarly to the case of aggregate functions in QB4OLAP), and it is related to rdfs:Literal used for the representation of the concrete values. The *SM4MQ* representation of the DICE example from Sect. 2 where a user is interested in number of asylum applications only for the years 2009 and 2010 for the running example schema is illustrated in Fig. 5. The example shows the DICE instance and includes the related *SM4MQ* concepts (depicted in gray).

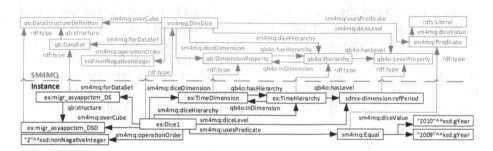

Fig. 5. DICE instance example

3.4   SLICE Operation

Following the definition in [9], SLICE (i.e., sm4mq:Slice) is represented with a data cube schema (i.e., qb:DataStructureDefinition) and a dimension (i.e., qb:-DimensionProperty) or a measure (i.e., qb:MeasureProperty). Again, we represent these two cases separately for the atomicity of operations that also facilitates sharing. Thus, in *SM4MQ* we create two subclasses for SLICE, sm4mq:DimSlice as SLICE applied over a dimension and sm4mq:MeasSlice as SLICE applied over a measure. For both cases we also define the order. The *SM4MQ* representation of the SLICE example from Sect. 2 where a user is not interested in the age of the applicants and thus removes the related dimensions for the running example schema (see Fig. 1) is illustrated in Fig. 6. The example shows the SLICE instance and also includes the related *SM4MQ* concepts (depicted in gray).

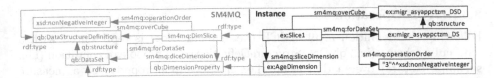

Fig. 6. SLICE instance example

4 Exploiting SM4MQ

In this section, we discuss the benefits of having a semantic model for MD queries. First, we discuss why modeling and capturing of the MD query semantics is essential for their exploitation. We then propose a method on how *SM4MQ* semantics can serve to automate the exploitation of *SM4MQ* queries by means of SPARQL. We also present a use case that shows how can the method be used to define the query transformations from *SM4MQ* to a vector representation.

4.1 Modeling and Semantics

We first explain the challenges behind the current state of the art to represent queries and explain how *SM4MQ* overcomes them. MD queries in existing approaches are typically stored in query logs [23]. From logs, queries are parsed to extract semantics needed for further processing. The parsing is dependent on the particular technology used (e.g., SQL) where different patterns need to be applied to identify OLAP operations introduced in Sect. 2. Once identified, the OLAP operations are represented with internal data structures and any processing of the queries is directly dependent on these internals. In the context of Exploratory OLAP [1] and next generation BI systems [23], this situation leads to several challenges:

1. *Repetitive model designing of MD queries* – Instead of considering query metadata as a first-class citizen and conforming them to a dedicated model, repetitive efforts are invested into designing ad-hoc query representations for each system.
2. *Repetitive adjustments for exploitation* – The use of hard-coded and ad-hoc query models hinders the use of existing algorithms for their exploitation. As the internals on query representation are typically not available, existing algorithms need to be adjusted to each system-specific query model.
3. *Burdensome query sharing* – Overlooking of query modeling obstructs query reuse among different systems. This becomes especially relevant, considering Exploratory OLAP and public data sets on the SW where not only data but also queries can be shared among the users. Moreover, once modeled, queries can be made publicly available so that users can exploit them for different purposes.
4. *The need for IT people support* – Working with internal query representation requires technical skills that are not characteristic of OLAP end-users.

Thus, preparing these queries (e.g., extracting the relevant semantics) requires the support of IT people. In the OLAP context, the data preparation by means of a correct ETL process may take up to 80% of the entire DW project as reported by Gartner [20]; illustrating the enormous efforts even from trained professionals.

The *SM4MQ* model for MD queries captures the MD semantics at the conceptual abstraction level using the SW technologies. Thus, it overcomes the previous challenges in the following way:

1. *SM4MQ* is a model of MD queries covering the OLAP operations specified in Sect. 2 that are commonly accepted and used in OLAP systems. Furthermore, the use of RDF makes it flexible to be extended with additional operations. Moreover, it can be linked with other RDF-based models using Linked Data principles (see Sect. 2).
2. Being an RDF-based model, *SM4MQ* provides a common semantics which exploitation algorithms can query in a standardized way via SPARQL.
3. The conforming of queries to *SM4MQ* directly supports their sharing via Linked Data principles, i.e., publishing on the SW in the RDF format. This way, different systems can make their queries available and reusable.
4. The semantics of MD queries captured at the conceptual abstraction level is understandable even for non-technical OLAP end-users [10,24]. It can automatically be transformed into different data structures (e.g., vectors) via SPARQL and thereby benefit from algorithms working over the related structures (e.g., computing cosine similarity between vectors).

Once MD queries are represented with *SM4MQ*, we can use off-the-shelf tools (i.e., triple stores) to store and query them in a standardized way via SPARQL.

4.2   Automating SM4MQ Exploitation

Instead of extracting and parsing queries like in typical settings, *SM4MQ* supports automation of query exploitations as their RDF representation can be directly retrieved via SPARQL. Using this benefit, we propose a method that automates the transformation of queries from *SM4MQ* to other query representations. We also provide a use case where we exemplify our claims by transforming *SM4MQ* queries into analytical vectors that can be used to perform advanced analysis such as query comparison and undertake recommendations. Vector-based representations have been typically used to compute similarities (e.g., the cosine similarity) that have been widely used in recommender systems [2]. Hence, several of the state of the art query recommendation approaches such as [6] use vectors to represent queries and compute similarities (see [3]). Next, we explain the method (sub)tasks and exemplify each task on our use case.

Task 1. Choosing the analytical structure is the initial task where the target data structure needs to be chosen. The analytical structure then directly determines the following tasks that define its *analytical features* and populate the analytical structure. An analytical feature is a set of one or more model elements (i.e., nodes or edges) representing a model characteristic, e.g., an operation used in a query, that is relevant for the desired analysis such as comparison of queries.

Use Case. We focus on a vector representation as an analytical structure.

Task 2. Defining analytical features identifies the elements of a given model (e.g., *SM4MQ*) that should be considered as analytical features. As an analytical feature can involve one or more model elements, it can either be *atomic*, i.e., consisting of one model element, or *composite* including two or more model elements. We next explain the two subtasks of Task 2.

Task 2.1. Selecting model element(s) to form analytical features is the task where one or more model elements are selected to form an atomic or composite feature, respectively. It is performed for each analytical feature. Here, a single or several connected elements of the model are selected to be considered as an atomic or a composite feature, respectively.

Use Case. Analytical features will be a part of the vector representation, i.e., they will be represented with vector elements. As the whole vector will represent a query, sm4mq:Query is not an analytical feature. Instead, the analytical features to be defined are different OLAP operations, i.e., the subclasses of sm4mq:Operation, with their successor nodes (e.g., schema elements such as level). Thus, each OLAP operation is a composite analytical feature. Considering that all the queries are run over the same data cube, the data set and schema elements are omitted, as well as the operation order.

Task 2.2. Defining the level of detail for each analytical feature relates to the model elements forming analytical features. For each analytical feature, the level of detail needs to be defined for each of the model elements forming that feature. There are two possible levels of detail. One is the *basic level of detail*, meaning that the model element in an analytical feature should be considered without its instances. The other option is the *comprehensive level of detail* where a model element in an analytical feature should be considered together with its instances.

Use Case. In a vector representation, this means that a model element with the base level of detail takes a single vector element, e.g., if an operation is used or not. Accordingly, in case of a model element with the comprehensive level of detail, there is a vector element for each instance of the model element, e.g., one vector element for each possible dimension that can be used in an operation. This way, a vector length is defined by vector elements needed for the model elements in each analytical feature.

As an example, we next define the level of detail in case of the ROLL-UP operation as an analytical feature. Here, there should be a vector element indicating if there is any ROLL-UP in the query (i.e., the basic level of detail) and for each dimension used in one or more ROLL-UPs there should be a sequence of vector elements (due to the comprehensive level of detail) including: an element identifying the dimension, an element for each dimension hierarchy, an element for each possible from-level, and an element for each possible to-level. Note that the aggregate function used in ROLL-UP is defined by the data structure definition and thus the same for all queries. For example, Table 1 illustrates a part

Table 1. A roll-up piece of vector instance

ROLL-UP (Operation)	D1 (Dimension)	H1 (Hierarchy)	RefPeriod (From-level)	Quarter (From-level)	Year (From-level)	RefPeriod (To-level)	Quarter (To-level)	Year (To-level)	D2 (Dimension)	H2 (Hierarchy)	Citizenship (From-level)	Continent (From-level)	Citizenship (To-level)	Continent (To-level)	...	D6 (Dimension)	H7 (Hierarchy)	Age (From-level)	Age (To-level)	DRILL-DOWN (Operation)	...
1	1	1	1	0	0	0	0	1	0	0	0	0	0	0	...	0	0	0	0	0	...

of the vector instance for the running example schema related to the ROLL-UP operation from Fig. 4. The non-zero values shown from the first vector element at the left specify that it is a ROLL-UP operation, over the dimension D1 and hierarchy H1, from the RefPeriod level to the Year level.

Task 3. Populating the analytical structure focuses on taking a query instance in *SM4MQ* and populating the chosen analysis-ready data structures. This task depends on the analytical feature definition from the previous task and can be automated with SPARQL query templates. It consists of the two subtasks that we explain in the sequel.

Use Case. This task refers to the population of all vector elements.

Task 3.1. Retrieving model instances is the task where template SPARQL queries are defined to automatically retrieve the model element instances related to the analytical features. There are two types of templates. The first type retrieves all instances of the model elements that are nodes and belong to one or more analytical features. The related SPARQL query is shown in Query 1. Note that variables between two '?' are parameters that should be replaced with the related IRIs, e.g., node IRIs in the previous case. This way, all instances of model elements related to either atomic or composite analytical features are retrieved.

Query 1. *Retrieve Model Element Instances*

```
1  SELECT DISTINCT ?i
2  WHERE {
3    i? rdf:type ?nodeIRI? . }
```

Query 2. *Retrieve Roll-Ups*

```
1  SELECT DISTINCT ?r ?d ?h ?fromL ?toL
2  WHERE {
3    ?q? rdf:type sm4mq:Query ;
4      sm4mq:hasRollUp ?r .
5    ?r rdf:type sm4mq:RollUp ;
6      sm4mq:ruDimension ?d ;
7      sm4mq:ruHierarchy ?h ;
8      sm4mq:ruFrom ?fromL ;
9      sm4mq:ruTo ?toL . }
```

The other template type retrieves the instances of graph patterns that include both nodes and edges related to composite analytical features. This way, all model instances that are used in composite analytical features are retrieved. For instance, Query 2 retrieves the model elements related to ROLL-UP (as a composite analytical feature) of a particular query, where the ?q? parameter is the query IRI.

Use Case. Benefiting from the *SM4MQ* model, this task can be automated with Algorithm 1. The algorithm takes a metadata graph containing the *SM4MQ* queries and QB4OLAP schema for a data set and returns the matrix populated with vectors of all the queries. For simplicity of explanation we consider that the graph contains metadata for a single data set. In lines 2 and 3, the algorithm first retrieves queries and schema triples and this can be automatically performed with SPARQL queries based on the *SM4MQ* and QB4OLAP semantics. Then, line 4 initializes the matrix, i.e., defines the number of columns, based on the schema (see above for the vector instance structure). The rest of the algorithm belongs to the following task and is explained in the sequel.

***Task 3.2.* Computing values for each analytical feature** is the final task that takes the model instances previously retrieved and processes them according to the defined analytical features to populated the chosen analytical structure. This processing can be simple and generate values 1 or 0 to show if a model element instance has been used or not in an analytical feature, or be a customized function.

Use Case. To populate vector elements, we apply a simple computation and consider that the values of vector elements are 1 or 0 (meaning either the presence or absence of that element). Using this logic, lines 5 to 13 of Algorithm 1 populate the matrix as follows. For each query, a vector is created again based on the schema (see line 6), populated according to the OLAP operations used in the query (see lines 7 to 12), and added to the matrix (see line 13). The nested functions in lines 7 to 12 run SPARQL queries and return specific OLAP operations for an MD query. An example of a SPARQL query for $q.getRollUps()$ used in line 7 is illustrated in Query 2. Finally, line 14 returns the resulting matrix. Thus, benefiting from the *SM4MQ* query modeling and semantics, the population of the matrix can be completely automated using Algorithm 1. The populated matrix can be used for computing similarities (e.g., cosine similarity) between vectors and other exploitations. In the next section, we present a prototype for our use case and discuss on practical benefits of our approach.

Algorithm 1. Populate the matrix of queries

```
Input: graph;                          // metadata graph with queries and schema
Output: matrix;                        // matrix representing queries
1  begin
2      queries = graph.getQueries();
3      schema = graph.getSchema();
4      matrix.init(schema);
5      foreach q ∈ queries do
6          vector.init(schema);
7          vector.addRollUps(q.getRollUps());
8          vector.addDrillDowns(q.getDrillDowns());
9          vector.addDimDices(q.getDimDices());
10         vector.addMeasDices(q.getMeasDices());
11         vector.addDimSlices(q.getDimSlices());
12         vector.addMeasSlices(q.getMeasSlices());
13         matrix.addVector(vector);
14     return matrix;
```

---

5   Use Case Evaluation

To evaluate our approach for the specified use case, we implemented a prototype and used a set of 15 MD queries related to the running example data set (see [21] for more details). The prototype includes the MD querying module (MDM) for the query generation and the to-Vector Transformation and Comparison module (VTC). MDM provides a GUI that enables a user to generate MD queries. It automatically generates *SM4MQ* query metadata as well as OLAP queries in Cube Query Language (CQL) [10] used by QB4OLAP explorer[2] for querying of QB4OLAP data cubes. The *SM4MQ* queries are stored in the SPARQL end-point. Then, VTC retrieves the (*SM4MQ*) query and (QB4OLAP) schema meta-data, transforms queries into vectors using Algorithm 1, and compares queries using the cosine similarity. Assuming the existence of a single data set and its metadata on the endpoint, VTC takes the endpoint address as a parameter and based on the *SM4MQ* and QB4OLAP semantics automatically retrieves the needed metadata. For space reasons, more implementation and evaluation details can be found in the technical report (see [21]) and next we focus on the key evaluation aspects.

One evaluation aspect of our approach for the selected use case is the size of the vector space, i.e., number of vector elements. As discussed in [2], for content-based recommender systems the pre-processing phase includes the definition of vector features. In the case of the running example data set, the vector defined by VTC is of size 180 and follows the structure (i.e., elements ordering) as defined in Sect. 4. Thus, without *SM4MQ* and VTC, this task needs to be performed *manually* entailing the tedious analysis of CQL algebra and exploration of data set schema. Once having the vector structure, the other evaluation aspect considers the population of the vector structures. Benefiting from *SM4MQ*, VTC automatically generates SPARQL queries to populate the vector for each query. Without *SM4MQ* and VTC, this task entails creation of a parser for CQL queries that populates the manually defined vector structures.

Finally, we show the degree of automation achieved in this process with the number of automatically triggered SPARQL queries by VTC. First, for the running example data set, VTC automatically triggered 32 SPARQL queries to retrieve its schema. The number of SPARQL queries in this context depends on the schema structure, e.g., how many dimensions or hierarchies exist. Furthermore, in the context of queries, we used a set of 15 *SM4MQ* queries and VTC automatically triggered 91 SPARQL queries related to *SM4MQ* queries including one SPARQL query to retrieve all IRIs of *SM4MQ* queries and 6 SPARQL queries for each *SM4MQ* query (one for each OLAP operation in *SM4MQ*). The execution of all the SPARQL queries took 2 s in average. Otherwise, if a user runs each SPARQL query manually using prepared templates and takes only 2 s/query in average, it would take 246 s. Thus, in this case VTC enables speed up of at least 100 times just for this task. Finally, VTC calculated similarity between the queries and the results indeed reflected similarities from OLAP and

---

[2] https://www.fing.edu.uy/inco/grupos/csi/apps/qb4olap/queries.

business perspectives. Thus, we confirmed that our approach also significantly eases the exploitation of MD queries represented with $SM4MQ$.

6   Related Work

Typically, traditional approaches consider MD queries in terms of OLAP algebras, i.e., formalization of possible actions over a data cube. A thorough overview of such approaches is given in [18] where authors argue that an OLAP algebra needs to be closed (i.e., each operation produces a data cube), minimal (i.e., no operation can be expressed in terms of other operations), and complete (i.e., covering all the relevant operation). Thus, the algebras mainly focus on the correctness and satisfiability of queries in terms of underlying MD schemata. Furthermore, considering [18] as a backbone state of the art, the authors of [16] propose an approach for representing MD queries at the conceptual abstraction level by defining them as OCL constraints over UML schema representing data cube. Such a solution is a step towards interoperability of different platform specific models. However, the focus still remains on the validity of queries. On the other hand, our approach focuses on representing queries as metadata to support their sharing and reuse by means of SW technologies. As discussed in Sect. 2, $SM4MQ$ is based on the OLAP algebra proposed in [8,9].

Furthermore, different OLAP query recommendation approaches typically store queries in logs and focus on their processing. According to a comprehensive overview found in [3], query in this context are represented with their syntactic (SQL) representation, resulting data, as vectors of features, sets of query fragments, or as graphs. The vector representation is used in several approaches, as for example in [6]. Moreover, some other approaches like [5] again use algebraic query representation. However, query representations are typically system specific that again hinders their sharing and reuse.

The modeling and representing of queries on the SW is just in its infancy. Just recently, in [19] the authors proposed an RDF-based model for representing SPARQL queries. They have used their model to represent a portion of queries over DBpedia and other public SPARQL endpoints with the following suggested use cases: generation of benchmarks, query feature analysis, query cashing, usability analysis, and meta-querying. Another interesting use case that can be added to the previous ones is the user assistance (e.g., query recommendations). Moreover, [11,15] propose vocabularies to represent SPARQL and SQL queries in RDF, respectively. Thus, there is a movement towards opening not only data but also metadata such as queries so that they could be explored with SPARQL. However, although needed for the context of Exploratory OLAP [1], an MD query model is still missing. By now, most of the efforts have been devoted to the schema modeling with vocabularies such as QB4OLAP (see [24]). Thus, the present paper proposes an RDF-based model to support sharing and reuse of MD queries on the SW that as well facilitates their exploitation.

7 Conclusion and Future Work

We have proposed *SM4MQ*, a Semantic Model for Multidimensional Queries. Using RDF and MD semantics, *SM4MQ* is a step towards Exploratory OLAP and sharing and reuse of MD queries on the SW. We also proposed a method to automate the transformation of the SM4MQ queries into other analysis-ready representations via SPARQL. The method is exemplified with the use case of transforming *SM4MQ* into a vector representation. To evaluate our approach, we have developed a prototype implementing the method for the vector use case. The evaluation showed that *SM4MQ* supports automation of transformation tasks that would otherwise require significant manual efforts (e.g., at least 100 times more just for SPARQL queries). In our future work, we plan to work on the exploitation side of the queries, e.g., develop richer transformations to support advanced user support techniques such as [4]. We also plan to apply our method to other use cases and support analytical feature definition via high-level GUIs.

Acknowledgments. This research has been funded by the European Commission through the Erasmus Mundus Joint Doctorate IT4BI-DC and it has been partially supported by the Secretaria d'Universitats i Recerca de la Generalitat de Catalunya, the grant number 2014 SGR-1534.

References

1. Abelló, A., et al.: Using semantic web technologies for exploratory OLAP: a survey. IEEE Trans. Knowl. Data Eng. **27**(2), 571–588 (2015)
2. Aggarwal, C.C.: Recommender Systems - The Textbook. Springer, New York (2016)
3. Aligon, J., et al.: Similarity measures for OLAP sessions. Knowl. Inf. Syst. **39**(2), 463–489 (2014)
4. Aligon, J., et al.: A collaborative filtering approach for recommending OLAP sessions. Decis. Support Syst. **69**, 20–30 (2015)
5. Aufaure, M.-A., Kuchmann-Beauger, N., Marcel, P., Rizzi, S., Vanrompay, Y.: Predicting your next OLAP query based on recent analytical sessions. In: Bellatreche, L., Mohania, M.K. (eds.) DaWaK 2013. LNCS, vol. 8057, pp. 134–145. Springer, Heidelberg (2013). doi:10.1007/978-3-642-40131-2_12
6. Chatzopoulou, G., et al.: The QueRIE system for personalized query recommendations. IEEE Data Eng. Bull. **34**(2), 55–60 (2011)
7. Cyganiak, R., et al.: Resource description framework (RDF): concepts and abstract syntax (2014). http://www.w3.org/TR/2014/REC-rdf11-concepts-20140225/
8. de Aguiar Ciferri, C.D., et al.: Cube algebra: a generic user-centric model and query language for OLAP cubes. IJDWM **9**(2), 39–65 (2013)
9. Etcheverry, L., et al.: Modeling and querying data cubes on the semantic web. CoRR, abs/1512.06080 (2015)
10. Etcheverry, L., Vaisman, A.A.: Querying semantic web data cubes. In: AMW (2016)
11. Follenfant, C., Corby, O.: SQL abstract syntax trees vocabulary (2014). http://ns.inria.fr/ast/sql/index.html

12. Golfarelli, M., et al.: The dimensional fact model: a conceptual model for data warehouses. Int. J. Coop. Inf. Syst. **7**(2–3), 215–247 (1998)
13. Heath, T., Bizer, C., Data, L.: Evolving the Web into a Global Data Space. Morgan & Claypool Publishers, Seattle (2011)
14. Jensen, C.S., et al.: Multidimensional Databases and Data Warehousing. Synthesis Lectures on Data Management. Morgan & Claypool Publishers, Seattle (2010)
15. Knublauch, H.: SPIN - SPARQL syntax (2013). http://www.spinrdf.org/sp.html
16. Pardillo, J., et al.: Extending OCL for OLAP querying on conceptual multidimensional models of data warehouses. Inf. Sci. **180**(5), 584–601 (2010)
17. Prud'hommeaux, E., Seaborne, A.: SPARQL 1.1 Query Language for RDF (2011)
18. Romero, O., Abelló, A.: On the need of a reference algebra for OLAP. In: Song, I.Y., Eder, J., Nguyen, T.M. (eds.) DaWaK 2007. LNCS, vol. 4654, pp. 99–110. Springer, Heidelberg (2007). doi:10.1007/978-3-540-74553-2_10
19. Saleem, M., Ali, M.I., Hogan, A., Mehmood, Q., Ngomo, A.-C.N.: LSQ: the linked SPARQL queries dataset. In: Arenas, M., et al. (eds.) ISWC 2015. LNCS, vol. 9367, pp. 261–269. Springer, Cham (2015). doi:10.1007/978-3-319-25010-6_15
20. Strange, K.: ETL was the key to this data warehouse's success. Gartner Research, CS-15-3143, 3 (2002)
21. Varga, J.: SM4MQ materials (2016). http://www.essi.upc.edu/~jvarga/sm4mq-page.html
22. Varga, J., et al.: SM4AM: a semantic metamodel for analytical metadata. In: DOLAP, pp. 57–66 (2014)
23. Varga, J., Romero, O., Pedersen, T.B., Thomsen, C.: Towards next generation BI systems: the analytical metadata challenge. In: Bellatreche, L., Mohania, M.K. (eds.) DaWaK 2014. LNCS, vol. 8646, pp. 89–101. Springer, Cham (2014). doi:10.1007/978-3-319-10160-6_9
24. Varga, J., et al.: Dimensional enrichment of statistical linked open data. J. Web Sem. **40**, 22–51 (2016)
25. Varga, J., et al.: QB2OLAP: enabling OLAP on statistical linked open data. In: ICDE, pp. 1346–1349 (2016)

Using Insights from Psychology and Language to Improve How People Reason with Description Logics

Paul Warren[✉], Paul Mulholland, Trevor Collins, and Enrico Motta

Knowledge Media Institute, The Open University, Milton Keynes,
Buckinghamshire MK7 6AA, UK
paul.warren@cantab.net,
{paul.mulholland,trevor.collins,enrico.motta}@open.ac.uk

Abstract. Inspired by insights from theories of human reasoning and language, we propose additions to the Manchester OWL Syntax to improve comprehensibility. These additions cover: functional and inverse functional properties, negated conjunction, the definition of exceptions, and existential and universal restrictions. By means of an empirical study, we demonstrate the effectiveness of a number of these additions, in particular: the use of *solely* to clarify the uniqueness of the object in a functional property; the replacement of *and* with *intersection* in conjunction, which was particularly beneficial in negated conjunction; the use of *except* as a substitute for *and not*; and the replacement of *some* with *including* and *only* with *noneOrOnly*, which helped in certain situations to clarify the nature of these restrictions.

Keywords: Description logics · Psychology of reasoning · Philosophy of language · Empirical studies

1 Introduction

The motivation for the research reported here is to mitigate the difficulties which occur when an ontologist reasons using DLs, in particular Manchester OWL Syntax (MOS). The ontologist needs to be able to perform such reasoning to understand the consequences of an ontology design, and also to understand why an entailment leads to a particular inference when debugging an ontology. In this context, the ontologist might be a computer scientist with a relatively deep knowledge of logic or a domain expert with less training in formal logic.

Previous studies by the authors have investigated the difficulties people experience in comprehending and reasoning with Description Logics (DLs). Warren et al. (2014) studied the difficulties experienced with DL features drawn from commonly used patterns, using a simplified version of the Manchester OWL Syntax (MOS). They identified particular difficulties with negated conjunction and functional properties. A subsequent study (Warren et al. 2015) investigated these features in more detail, as well as looking at the effect of combining negation and a restriction and also nested restrictions. As a result, Warren et al. (2015) made some recommendations regarding training. Both these studies sought to explain participants' difficulties in terms of theories of human

© Springer International Publishing AG 2017
E. Blomqvist et al. (Eds.): ESWC 2017, Part I, LNCS 10249, pp. 465–481, 2017.
DOI: 10.1007/978-3-319-58068-5_29

reasoning. Based on these studies, the authors have proposed some syntactic additions to MOS to mitigate the difficulties identified. The study reported here investigates the effect of these additions to MOS.

Section 2 describes related research on the difficulties experienced with DLs. Section 3 then describes the theories of human reasoning and language employed by the authors. Section 4 reviews the findings of the authors' two previous studies and their interpretation in terms of the theories discussed in Sect. 3. Section 5 provides an overview of this study and makes some general observations. Section 6 investigates the use of an additional keyword, *solely*, to clarify the nature of functional and inverse functional properties. Section 7 is concerned with Boolean concept constructors. Specifically it investigates the effect of replacing *and* and *or* with *intersection* and *union*, the use of *except* as a substitute for *and not*, and also the use of prefix notation for conjunction and disjunction. Sections 8 and 9 investigate the use of *including* and *noneOrOnly* in place of *some* and *only*, with the intention of clarifying the nature of these restrictions. These two sections also investigate the effect of replacing the keyword *some* with *any* where the associated property is preceded by a negation. Finally, Sect. 10 draws some conclusions and proposes some future work.

2   Related Work

It has long been recognized that the original formal notation of DLs posed problems for those who were not logicians. This was the motivation for the Manchester OWL Syntax (Horridge et al. 2006). However, based on their experience of teaching DLs, Rector et al. (2004) observed that, even with a notation more akin to natural language, DLs still posed problems of comprehensibility. Indeed, Rector et al. (2004) point out that the use of natural language can create ambiguities, observing that *and* and *or* in everyday use do not always correspond to their meanings in logic.

There has been some empirical work investigating both the comprehensibility of DLs and the facility of human reasoning with them. Horridge et al. (2011) have investigated the difficulties experienced by users trying to understand how subsets of an ontology justified particular entailments, as is necessary when debugging an ontology. Participants were presented with a justification and an entailment, expressed in a formal notation, and were asked whether the entailment followed from the justification. The empirical results were related to an ad-hoc model of cognitive complexity with 12 parameters, e.g. number of axiom types, number of class constructors and maximum depth of class expressions in the justification.

Nguyen et al. (2012) were concerned with automatically creating proof trees, composed from deduction rules. When choosing between alternative possibilities they needed a measure of the comprehensibility of each of 51 deduction rules. They created English equivalents of these deduction rules, using 'nonsense' words to avoid any effect of domain knowledge. Study participants, drawn from a crowdsourcing service, were required to confirm or reject the validity of the various deduction rules, thereby determining a comprehensibility rating for each rule. The intention was that these ratings could be used by an algorithm to create a proof tree, optimized for comprehensibility.

None of this work made any use of psychological theory. Our work goes beyond this previous work by looking more precisely at the difficulties experienced with specific OWL constructs, and interpreting those difficulties in terms of psychological theories.

3   Human Reasoning and Human Language

There has been considerable research into how people reason. Two early opposing approaches were the rule-based and model-based approaches. The former assumes that 'naïve users', i.e. people not trained in logic, use rules similar to that of the logician (Rips 1983). The model-based view, as argued by Johnson-Laird (2010), assumes that people create mental models of a given situation. In this view, any putative deduction is tested against the various models. If the deduction is true in every model, then it is a valid conclusion.

The mental model theory can be used to explain the mistakes that people make in reasoning. According to the theory, mistakes frequently arise when a situation requires more than one mental model. It may happen that some of these models are never formed, or get forgotten under situations of cognitive stress. For example, conjunction, exclusive disjunction and inclusive disjunction are represented by one, two and three mental models respectively[1]. Johnson-Laird et al. (1992) have confirmed that inclusive disjunction gives rise to more errors than exclusive disjunction. Khemlani et al. (2012) also demonstrated that people make more errors when reasoning about inclusive disjunction than conjunction.

Relational complexity (RC) theory complements these approaches by providing a measure of the complexity of a reasoning step. Complexity is defined "as a function … of the number of variables that can be related in a single cognitive representation" (Halford and Andrews 2004). As an example Halford et al. (2004) note that reasoning with transitivity has an RC of 3. A transitive relation, e.g. 'greater than', is binary since it relates two individuals. However, integrating two instantiations of a transitive relation in a deductive step requires concurrent attention to three individuals and hence has an RC of 3. Proponents of the theory argue that the likelihood of error in any chain of reasoning is determined by the maximum RC of the individual steps. In this study, RC theory is used to provide a measure of difficulty and enable comparison between different reasoning steps.

Besides theories of reasoning, studies of language offer useful insights. Of particular value is the concept of *implicature*, developed by Grice (1975) to describe a conclusion to which a speaker or writer leads an audience, but which is not a strictly logical implication of what has been said or written, e.g. "some of the students are industrious", which leads the reader to assume that not all the students are industrious. The existence of implicatures is of particular relevance in considering mental model theory. In certain situations language may lead people to form an incomplete set of mental models. This is discussed further in Sect. 8.

---

[1] conj: A and B; excl disj: not A and B, A and not B; incl disj: A and B, not A and B, A and not B.

4 Previous Studies

In a previous study (study 1), Warren et al. (2014) identified a difficulty with functional object properties. A question requiring reasoning about a functional object property was only answered correctly by 50% of the participants, i.e. 6 out of 12. Since participants were presented with a binary choice between valid and non-valid, this is exactly equivalent to chance. The question required a reasoning step of RC 4 and it was not clear whether the difficulty was because of the complexity of this step or was a specific problem with functionality. In a subsequent study (study 2), Warren et al. (2015) compared reasoning steps of equal complexity using functional and transitive properties, which suggested that there was a problem specific to functional properties. This topic is returned to in Sect. 6.

In addition, study 1 identified a difficulty with negated conjunction. Only 25% of participants (3 out of 12) correctly answered a question with negated conjunction, compared with 92% (11 out of 12) who correctly answered a question with negated disjunction. The difference in ability to handle negated conjunction and negated disjunction has also been observed by Khemlani et al. (2012), who interpret this in terms of the mental model theory. Negated conjunction requires three mental models (*not A and not B; not A and B; A and not B*), whereas negated disjunction requires only one model (*not A and not B*). This topic is returned to in Sect. 7.

Study 2 also identified problems with understanding universal and existential restrictions, using the MOS keywords *only* and *some*. These difficulties were interpreted in terms of mental model theory. Table 1 shows the mental models corresponding to the two restrictions. In each case there are two models, the first of which is the more obvious and the second of which may be overlooked. This is well known for the case of *only* (Rector et al. 2004) but was found also to be the case in certain situations for *some*. This topic is returned to in Sects. 8 and 9.

Table 1. Mental models for universal and existential restrictions

	Universal restriction	Existential restriction
MOS	P only X	P some X
Mental models	P x	P x
	P ⊥	P x P ¬x

5 Overview of Current Study

The principal objective of the current study was to determine whether certain additions to MOS would lead to improved human performance with the features found difficult in the previous two studies. Consequently, the majority of questions were isomorphic to questions in the previous studies, in particular study 2.

Each question followed the same pattern as in the previous studies. A set of statements were provided, plus a putative conclusion; participants had to indicate whether the conclusion was valid or not valid. The language features used in this and the previous studies were chosen because of their relatively common use; see the discussion in

Warren et al. (2014). Performance was measured by accuracy of response and time taken to respond. MediaLab[2] was used to collect the responses and measure response time. The assumption is that response time can be regarded as a proxy for difficulty. All statistical analysis was undertaken using the R statistical package (R Core Team 2014).

The study used a simplified form of MOS, with additions as explained below. At the beginning of the study each participant was given a handout that explained the syntax used. Participants retained this handout during the study and could refer to it at any time.

The study comprised four sections, each containing eight questions. There were two variants of the study, referred to as variant 1 and variant 2, permitting some of the questions to be different in the two variants. There were 30 participants, 15 for each variant, drawn from the authors' own university, another U.K. university and an industrial research laboratory. At the beginning of the study participants were asked about their knowledge of formal logic, and of OWL or other DL formalisms. The breakdown for knowledge of formal logic was: none 3%; little 23%; some 47%; expert 27%. The breakdown for knowledge of OWL or other DL formalisms was: none 13%; little 47%; some 27%; expert 13%. Thus, the majority of the participants had a reasonable knowledge of logic but fewer had much knowledge of DL; in the latter respect the majority represented occasional users of ontologies.

Examination of the distribution of the response times revealed a positive skew, indicating a considerable deviation from normality. This phenomenon has been reported elsewhere (Blake et al. 2012; Warren et al. 2015). Further analysis suggested that the logarithmic transformation of time, selected from Tukey's ladder of powers (Scott 2012), resulted in a distribution closer to the normal. Since ANOVA and the t-test require approximately normal populations, this transformation has been applied prior to all such tests on time data reported in this paper[3].

To prevent any bias due to question position, the order of the sections and of the questions within each section were randomized, using a randomization feature provided within MediaLab.

In reporting all statistical results, the convention adopted is the usual one of taking $p < 0.05$ as representing significance. For significant results, p is reported as being <0.05, <0.01, <0.001 etc. Non-significant results are identified either by explicitly stating $p \geq 0.05$ or using the abbreviation n.s. (not significant).

6   Functional and Inverse Functional Object Properties

The motivation for this part of the study came from a comparison in study 2 between functional object properties and transitive object properties. The comparison concluded that, under conditions of equal relational complexity, the proportion of correct responses to the transitive questions was not significantly different from that for the functional questions (Fisher's Exact Test), but the latter took significantly

---

[2] Provided by Empirisoft: http://www.empirisoft.com/.

[3] An alternative would have been to use a non-parametric test. Hopkins et al. (2009) note that a transformation to reduce skewness followed by a parametric test provides greater statistical power at small sample sizes than does a non-parametric test.

longer (t(87.484) = 2.2376, p < 0.05), implying that the participants were finding the functional questions harder. A possible explanation for this is confusion between functionality and inverse functionality, i.e. whether it is the subject or object of the property that is unique. In this study the keyword *solely* was added after the property name and before the object to indicate that it is the object which is unique. This keyword was chosen because of its anticipated power to convey uniqueness. A possibly more natural choice, *only*, was rejected because of its existing use in MOS.

This leads to the following hypothesis:

H1 The introduction of the additional keyword *solely* between a functional property and its object will improve participant performance.

Previous studies did not investigate inverse functional properties. However, it is possible that similar difficulties will arise as with functional properties. This study investigates the effect of introducing the keyword *solely* before the subject of an inverse functional property, to indicate that in this case it is the subject which is unique. This leads to a second hypothesis:

H2 The introduction of the additional keyword *solely* before the subject of an inverse functional property will improve participant performance.

These two hypotheses are investigated in the following two subsections.

6.1   Functional Object Properties – Comparison with Study 2; Hypothesis H1

In this study six questions were created isomorphic to the six functional questions in study 2, with the additional keyword, *solely*, being used as described above. Table 2 shows the six questions. For brevity F is used to represent the property. In practice, this and study 2 used the object property *has_nearest_neighbour*. Note that consecutive questions share the same axioms; but have different putative conclusions. Three reasoning steps are required to arrive at each of the valid conclusions. The table shows the RC of each step. For example, question 1 starts by using the first two axioms in an inference of RC 3 to deduce that *s* and *t* are identical. A step of RC 2 replaces *s* with *t* in axiom 3 and, finally, another inference of RC 3 concludes that *v* and *w* are identical.

Table 2. Functional object property questions

	Axioms (F = *has_nearest_neighbour*)	Putative conclusion	Validity	Relational complexity
1	r F solely s; r F solely t;	v sameAs w	Valid	3, 2, 3
2	s F solely v; t F solely w	r sameAs t	Not valid	
3	r F solely s; r F solely t; v F solely s;	v DifferentFrom w	Valid	3, 2, 4
4	w F solely x; t DifferentFrom x	r DifferentFrom v	Not valid	
5	r F solely s; t F solely v;	w DifferentFrom x	Valid	4, 2, 4
6	s DifferentFrom v; w F solely r; x F solely z; t SameAs z	r DifferentFrom x	Not valid	

Table 3 shows the percentage of correct responses and the mean time to respond for each question in the two studies. The data is also provided aggregated over all six questions and over the three valid questions and the three non-valid questions. For four of the six questions the percentage of correct responses was greater for the current study than for study 2; these included the two questions with the worst performance in study 2. However a Fisher's Exact test indicated no significant difference between the two studies ($p \geq 0.05$). This was also the case when the comparison was limited to the valid questions and to the non-valid questions.

Table 3. Functional object property questions: accuracy and response times

	Study 2 – without *solely*		**Current study** – with *solely*	
	%age correct N = 28	Mean time (SD) - secs; N = 24	%age correct N = 30	Mean time (SD) - secs; N = 30
1	75%	52 (36)	83%	39 (31)
2	96%	61 (46)	83%	50 (29)
3	61%	84 (67)	70%	58 (27)
4	79%	92 (66)	83%	78 (49)
5	43%	109 (79)	63%	73 (37)
6	71%	96 (47)	70%	90 (46)
All six questions	71%	83 (61)	76%	65 (41)
Valid questions	60%	81 (67)	72%	57 (35)
Non-valid questions	82%	83 (55)	79%	73 (45)

For each question the mean response time was less for the current study. A two-factor ANOVA indicated that response time varied significantly between the studies ($F(1, 320) = 7.559$, $p < 0.01$) and between the valid and non-valid questions ($F(1, 320) = 4.928$, $p < 0.05$), with no significant interaction ($F(1, 320) = 1.761$, n.s.). A subsequent Tukey Honest Significant Difference (HSD) analysis revealed a significant difference in response time between the two studies for the valid questions ($p < 0.05$) but not for the non-valid.

In summary, the study partially supports hypothesis H1. The introduction of *solely* significantly reduces response time for the valid questions, although it has no significant effect on the non-valid questions, nor on the accuracy of responses.

6.2   Inverse Functional Object Properties; Hypothesis H2

As neither of the previous studies included inverse functional properties, comparison of any syntactic changes could not be made between studies. Consequently, participants in this study were given two questions employing inverse functional properties. In variant 1, the questions used our simplified version of MOS. In variant 2, the keyword *solely* was included before the subject to indicate its uniqueness. Table 4 shows the format of the questions in the two variants. For brevity, I is used to represent the property. In the study *is_nearest_neighbour_of* was used. Note that questions 7 and 8 were created from questions 1 and 5 in Table 2 by restating the axioms using *is_nearest_neighbour_of* rather than its inverse *has_nearest_neighbour* and by interchanging the individual names.

Table 4. Inverse functional object property questions

Axioms (I = *is_nearest_neighbour_of*)	Putative conclusion	Validity	Relational complexity
Variant 1			
7	r I s; t I s; v I r; w I t	v SameAs w	valid	3, 2, 3
8	r I s; t I v; r DifferentFrom t; s I w; x I z; v SameAs x	w DifferentFrom z	valid	4, 2, 4
Variant 2			
7	solely r I s; solely t I s; solely v I r; solely w I t	v SameAs w	valid	3, 2, 3
8	solely r I s; solely t I v; r DifferentFrom t; solely s I w; solely x I z; v SameAs x	w DifferentFrom z	valid	4, 2, 4

Table 5 shows the results of the study. Considering the two questions aggregated, there was no significant difference in accuracy of response between the two variants (Fisher's Exact Test). In addition, a two-way ANOVA showed that there was a significant difference in response time between the questions ($F(1, 56) = 26.836, p < 0.00001$), reflecting their difference in complexity, but not between the two variants ($F(1, 56) = 0.640$, n.s.). There was no interaction effect (($F1, 56) = 0.382$, n.s.).

Table 5. Inverse functional object property questions: accuracy and response times

	Variant 1 – without *solely*; N = 15		**Variant 2** – with *solely*; N = 15	
	% corr	Mean time (SD) - secs	% corr	Mean time (SD) - secs
Question 7	73%	38 (18)	87%	48 (23)
Question 8	73%	105 (92)	73%	90 (43)
Both questions	73%	72 (74)	80%	69 (40)

Thus, the study offered no support for hypothesis H2. In the case of inverse functional properties, the use of the keyword *solely* has no significant effect on performance. It may be that *solely* was taken to refer to the whole <subject, predicate, object> triple without making clear the uniqueness of the subject rather than object. Other keywords and other choices of position might improve performance. One could, for example, experiment with the use of *alone* after the subject, e.g. *s alone has_nearest_neighbour r*.

7   Boolean Concept Constructors

The eight questions in this part of the study were designed to test out whether amendments to MOS could reduce the difficulties experienced with negated conjunction. Study 1 noted that negated conjunction was significantly harder than negated disjunction, as has also been observed by Khemlani et al. (2012). At the same time, it is known that *and* and *or* in everyday language are used ambiguously, e.g. see Mendonça et al. (1998). It was thought that performance might be improved by the use of unambiguous terminology for conjunction and, for consistency, disjunction. In particular, the use of *intersection* rather than *and* may avoid the implicature, based on normal usage, that *and* represents union. This leads to the hypothesis:

H3 The use of the keyword *intersection* in place of *and* will improve performance for negated conjunction.

In some syntaxes, OWL already uses the terms *intersection* and *union* as part of prefix operators. A relevant question is how such a prefix notation compares with the infix used in MOS. The hypothesis was proposed:

H4 There will be a difference in performance between the prefix notation *IntersectionOf()* and *UnionOf()*, and the infix notation *intersection* and *union*.

Study 2 included some questions which contained the consecutive keywords *and not*. The original motivation came from the discussion of exceptions in Rector (2003), where exceptions were defined using *and not*. For this study, it was thought that *except*

Table 6. Boolean concept constructor questions as used in the study

	Axioms	Putative conclusion	Validity
Study 1			
1	Entity DisjointUnionOf Event, Abstract, Quality, Object; A Type Entity; A Type not (Event and Quality);	A Type (Abstract or Object)	Not valid
2	Entity DisjointUnionOf Event, Abstract, Quality, Object; A Type Entity; A Type not (Event or Quality);	A Type (Abstract or Object)	Valid
Study 2			
3	Z EquivalentTo (TOP_CLASS and not A and not B); TOP_CLASS DisjointUnionOf A, B, C	Z EquivalentTo C	Valid
4	Z EquivalentTo (TOP_CLASS and not (A or B)); TOP_CLASS DisjointUnionOf A, B, C	Z EquivalentTo C	Valid
5	Z EquivalentTo (TOP_CLASS and not (A and not A_1)); TOP_CLASS DisjointUnionOf A, B; A DisjointUnionOf A_1, A_2	Z EquivalentTo (B or A_1)	Valid
6	As for question 5	Z EquivalentTo B	Not valid
7	Z EquivalentTo (TOP_CLASS and not (A and not (A_1 and not A_1_X))); TOP_CLASS DisjointUnionOf A, B; A DisjointUnionOf A_1, A_2; A_1 DisjointUnionOf A_1_X, A_1_Y	Z EquivalentTo (B or A_1_Y)	Valid
8	As for question 7	Z EquivalentTo A_1_Y	Not valid

might be more intuitively understandable, including within nested exceptions, i.e. *and not (… and not …)*, which give rise to negated conjunction. This leads to:

H5 The use of *except* in place of *and not* will improve performance.

Table 6 shows the original questions from the two previous studies, used to generate the questions in this study.

In the case of study 1, the two questions originally made use of an ontology pattern. In the table, the essence of the questions has been extracted out. In this study, variant 1 consists of questions isomorphic to the eight questions shown, with *and* replaced by *intersection*, *or* replaced by *union*, and in the case of questions 3 to 8, *and not* replaced by *except*. Variant 2 consists of questions isomorphic to the questions shown, with *and* replaced by the prefix form *IntersectionOf()*, and *or* replaced by the prefix form *UnionOf()*. As a result, H3 can be tested by comparing performance on variant 1 of question 1 with performance on the analogous question in study 1. Question 2 was included to determine that the change in terminology did not have a deleterious effect on negated disjunction. H4 can be tested by comparing the two variants of this study, using questions 1 and 2. Finally, H5 can be tested by comparing the two variants of the study, using questions 3 to 8. Table 7 shows the relevant data.

Table 7. Boolean concept constructor questions: accuracy and response times

	%age correct	Mean times (SD) - secs	%age correct	Mean time (SD) - secs	%age correct	Mean time (SD) - secs
	Study 1 *and, or, not*		**Current study variant 1** *intersection, union, not*		**Current study variant 2** *IntersectionOf, UnionOf, not*	
	N = 12	N = 12	N = 15	N = 15	N = 15	N = 15
1	25%	75 (48)	80%	53 (41)	67%	47 (16)
2	92%	44 (19)	87%	42 (28)	100%	39 (25)
	Study 2 *and, or, not*		**Current study variant 1** *intersection, union, except*		**Current study variant 2** *IntersectionOf, UnionOf, not*	
	N = 28	N = 24	N = 15	N = 15	N = 15	N = 15
3	82%	39 (26)	100%	39 (25)	100%	47 (30)
4	86%	43 (29)	100%	35 (26)	93%	46 (35)
5	61%	96 (56)	53%	61 (37)	53%	65 (38)
6	64%	105 (78)	100%	44 (25)	73%	82 (57)
7	54%	90 (48)	60%	97 (75)	40%	156 (126)
8	68%	94 (47)	80%	88 (60)	60%	93 (51)
Q1 and 2	58%	60 (39)	83%	47 (35)	83%	43 (21)
Q3 to 8	69%	78 (56)	82%	61 (50)	70%	82 (74)
Q1 to 8	68%	75 (54)	82%	57 (47)	73%	72 (67)

7.1 Negated Conjunction and Disjunction; Hypotheses H3

Table 7 shows that question 1 was answered much more accurately in variant 1 of this study, with the use of *intersection*, than in study 1 which used *and* (p < 0.01, Fisher's Exact Test). However, there was no significant difference in response times (t(23.988) = 1.6031, n.s.). Thus, hypothesis H3 is supported with regard to accuracy, but not response time. For question 2, with negated disjunction, there was no significant difference in accuracy between variant 1 and study 1 (p ≥ 0.05, Fisher's Exact Test); nor was there a significant difference in response time (t(24.433) = 0.5372, n.s.). Thus, the change in notation had no effect on negated disjunction.

7.2 Use of Prefix Notation; Hypothesis H4

The two variants of this study enable a comparison of infix and prefix notation. However, if all questions were used this comparison would be confounded with the effect of using *except* in variant 1. For a more controlled comparison, only questions 1 and 2 are used. Taking these two questions together, the percentage of correct responses is the same for both variants. Moreover, there was no significant difference in response time between the two variants (t(56.443) = 0.21532, n.s.). In summary, there is no evidence of a difference in performance between infix and prefix notation.

7.3 Use of *except* in Place of *and not*; Hypothesis H5

Questions 3 to 8 in variant 1 were intended to test the effect of replacing *and not* with *except*. For technical reasons concerned with the fact that the order of questions in study 2 was not fully randomized, it is not possible to compare study 2 with variant 1 of this study. Instead the two variants of this study are compared on the assumption, supported by the evidence of the previous subsection, that the use of infix and prefix notation makes no significant difference. For questions 3 to 8 aggregated, there was no significant difference in accuracy (p ≥ 0.05, Fisher's Exact Test). However, there was a significant difference in response time (t(176.22) = 2.3962, p < 0.05), with variant 1, using *except*, having the lower response time. Thus, hypothesis H5 is not supported in respect of accuracy but is supported in respect of time.

8 Negation and Restriction

As already noted, difficulties with the universal and existential restrictions may in part be due to a failure to form both the required mental models. This part of the study investigated whether the replacement of *only* with *noneOrOnly* and *some* with *including* would improve performance with these restrictions. *noneOrOnly* was intended to draw attention to the fact that, e.g. the class *has_child noneOrOnly MALE* includes those individuals who have no children at all, i.e. to avoid the implicature that *only MALE* suggests the existence of some male child, and thereby help participants to form both the necessary mental models, as shown in Table 1. *including* was intended to draw attention to the fact that, e.g. the class *has_child including MALE* may contain individuals who have a female child in addition to a male one, i.e. again to help participants form both the mental models. The hypothesis to be investigated is:

H6 The use of *noneOrOnly* in place of *only* and *including* in place of *some* will lead to improved participant performance.

In addition, it was thought that the use of *not ... some* might read unnaturally, and that *not ... any* corresponds to more normal English usage; this led to the hypothesis:

H7 The use of *any* to indicate the existential restriction, when the corresponding property is preceded by a negation, will lead to improved performance.

The original questions, as in study 2, are shown in Table 8. Each question is comprised of two axioms. For the first axiom there are four variants; for the second axiom two variants. All questions have the same putative conclusion. In Table 8, two different typefaces are used for the first axioms to indicate semantic equivalence. The questions can be grouped into four semantically equivalent pairs: {1, 4}; {2, 3}; {5, 8}; {6, 7}. For this study these questions were modified by the replacements described above. For six of the questions there was no difference between the two variants. However, in their original form questions 3 and 7 contain the class description *not (has_child some MALE)*. In variant 1 *some* was replaced with *including* as in the other questions. In variant 2 *any* was used, i.e. *not (has_child any MALE)*. Table 9 shows the accuracy and response times for the questions in study 2 and the current study. For the latter, separate data are shown for the two variants when the questions differ.

Table 8. Questions employing negation and restrictions; form as in study 2. N.B. the putative conclusion in each case was *X DisjointWith Y*.

	First axiom	Second axiom	Validity
1	*X SubClassOf has_child some (not MALE)*	Y SubClassOf has_child only MALE	Valid
2	X SubClassOf has_child only (not MALE)		Not valid
3	X SubClassOf not (has_child some MALE)		Not valid
4	*X SubClassOf not (has_child only MALE)*		Valid
5	*X SubClassOf has_child some (not MALE)*	Y SubClassOf has_child some MALE	Not valid
6	X SubClassOf has_child only (not MALE)		Valid
7	X SubClassOf not (has_child some MALE)		Valid
8	*X SubClassOf not (has_child only MALE)*		Not valid

Table 9. Negation and restriction questions: accuracy and response times.

	Study 2 *only, some*		Current study: overall; N = 30		Current study: variant 1; N = 15		Current study: variant 2; N = 15	
			noneOrOnly, including				*not … any*	
	%age correct N = 28	Mean time (SD) - secs N = 24	%age correct	Mean time (SD) - secs	%age correct	Mean time (SD) - secs	%age correct	Mean time (SD) - secs
1	61%	52 (39)	80%	42 (33)				
2	50%	33 (18)	73%	29 (20)				
3	68%	45 (22)	70%	69 (127)	67%	55 (49)	73%	84 (175)
4	75%	43 (25)	90%	41 (32)				
5	64%	41 (30)	70%	30 (21)				
6	50%	44 (40)	70%	33 (33)				
7	79%	43 (37)	80%	29 (16)	73%	26 (14)	87%	33 (18)
8	68%	60 (37)	67%	38 (24)				
Exc Q3 and 7	61%	45 (33)	75%	35 (28)				
Q3 and 7	73%	44 (30)	75%	49 (92)	70%	40 (38)	80%	58 (125)

8.1 *noneOrOnly* and *including*; Hypothesis H6

To avoid the confounding effect of the introduction of *not … any* in variant 2 for questions 3 and 7, an analysis was conducted based on the six questions excluding questions 3 and 7. Table 9 shows the mean results for these six questions. The use of *noneOrOnly* and *including* led to a significant increase in accuracy (Fisher's Exact Test, p < 0.01) and reduction in response time (t(284.57) = 2.7897, p < 0.01), supporting H6. This suggests that the new keywords do support the creation of the two mental models necessary for each of the restrictions. It is also noteworthy that there is an appreciable increase in accuracy for the two questions which were answered worst in study 2, i.e. questions 2 and 6. The former requires the second model for the universal restriction.

8.2 *not … any*; Hypothesis H7

Questions 3 and 7 provide an opportunity to investigate the effect of using *not … any* in variant 2. There was no support for H7, i.e. no significant difference between the variants, in accuracy (p ≥ 0.05, Fisher's Exact Test) or time (t(57.832) = 0.79496, n.s.).

9 Nested Restrictions

Study 2 included eight questions making use of nested restrictions, as shown in Table 10; Table 11 shows the associated data. Analogous questions in variant 1 of this study were created by replacing *only* with *noneOrOnly* and *some* with *including*, enabling a further investigation of hypothesis H6. For variant 2, questions 5 to 8 were as for variant 1, except that in the final axiom, *not ... some* in study 2 has been replaced by *not ... any*, enabling a further investigation of hypothesis H7

Table 10. Questions employing nested restriction; form as in study 2. N.B. the putative conclusion in each case was *a Type (not X)*.

	First axiom(s)	Final axiom	Validity
1	X SubClassOf (has_child some (has_child some FEMALE))	a Has_child b; b Type has_child some (not FEMALE)	Not valid
2	X SubClassOf has_child some Y; Y EquivalentTo has_child only FEMALE		Not valid
3	X SubClassOf (has_child only (has_child some FEMALE))		Not valid
4	X SubClassOf has_child only Y; Y EquivalentTo has_child only FEMALE		Valid
5	X SubClassOf has_child some Y; Y EquivalentTo has_child some FEMALE	a Has_child b; b Type (not (has_child some FEMALE))	Not valid
6	X SubClassOf (has_child some (only FEMALE))		Not valid
7	X SubClassOf has_child only Y; Y EquivalentTo has_child some FEMALE		Valid
8	X SubClassOf (has_child only (has_child only FEMALE))		Not valid

Table 11. Nested restriction questions: accuracy and response times

	Study 2		Current study: var 1		Current study: var 2	
	%age correct	Mean time (SD) 0 secs	%age correct	Mean time (SD) - secs	%age correct	Mean time (SD) - secs
	N = 28	N = 24	N = 15	N = 15	N = 15	N = 15
	only, some		*noneOrOnly, including*			
1	71%	69 (45)	80%	47 (30)	60%	73 (59)
2	57%	79 (53)	40%	65 (20)	67%	68 (41)
3	71%	63 (43)	60%	54 (31)	53%	79 (46)
4	57%	63 (39)	53%	100 (87)	47%	66 (49)
					not ... any	
5	54%	88 (62)	40%	64 (30)	67%	74 (67)
6	64%	73 (45)	73%	85 (82)	73%	99 (90)
7	71%	80 (36)	53%	64 (39)	47%	97 (102)
8	50%	55 (30)	60%	83 (35)	53%	63 (37)
Mean for all questions	62%	71 (45)	58%	70 (51)		
Mean for Q5 to 8	60%	74 (46)	57%	74 (51)	60%	83 (77)

9.1 *noneOrOnly* and *including*; Hypothesis H6

For the eight questions aggregated there was no significant difference between study 2 and variant 1, neither in accuracy ($p \geq 0.05$, Fisher's Exact Test) nor in response time ($t(288.79) = 0.40724$, n.s.). Thus, unlike the questions discussed in Subsect. 8.1, these questions offered no support for hypothesis H6.

9.2 *not … any*; Hypothesis H7

For questions 5 to 8, there was no significant difference in accuracy ($p \geq 0.05$, Fisher's Exact Test) between the two variants of this study, nor in response time ($t(106.86) = 0.19408$, n.s.), i.e. as with Subsect. 8.2, there was no support for H7.

10 Conclusions and Future Work

Table 12 summarizes the findings from the study. Based on the empirical evidence, the following extensions to MOS can be recommended: *intersection* in place of *and*, avoiding the associated ambiguity; *except* as an alternative to *and not*, providing a natural way to think about exceptions; *solely* with functional properties, identifying that the object of the property is unique. Whilst *noneOronly* and *including* improve reasoning in some cases, more research is needed to investigate under what circumstances these keywords are beneficial, and whether alternatives might be preferable. Research is also needed to improve performance with inverse functional properties.

Table 12. Summary of findings

Hypothesis (results section shown in brackets)	Accuracy	Response time
H1 – *solely* with functional properties (6.1)	No advantage	Advantage for valid questions
H2 – *solely* with inverse funct. properties (6.2)	No advantage	
H3 – *intersection* in place of *and* (7.1)	Advantage	No advantage
H4 – prefix versus infix notation (7.2)	No difference	
H5 – *except* in place of *and not* (7.3)	No advantage	Advantage
H6 – *noneOrOnly* and *including* (8.1, 9.1)	Advantage with single restriction but not with nested restrictions	
H7 – *not … any* (8.2, 9.2)	No advantage	

   This work has shown how theory can be used to guide language development. An understanding of the mental models associated with logical constructs can help choose keywords which emphasize all the models, as was done here with the universal and existential restrictions. An understanding of the implicatures present in natural language can help avoid the use of words which, seemingly user friendly, are ambiguous or even misleading, as is the case for *and*. However, when ambiguities and misleading implicatures are avoided, the use of natural language can aid human reasoning, as was shown with *except*. More generally, it is proposed that theories of reasoning and language will be able to provide support in the development of a range of computer languages.

Acknowledgements. The authors would like to thank all the study participants, and in particular Dr. Gem Stapleton, of Brighton University, and Dr. John Davies, of BT Research, for facilitating experimental sessions at their respective institutions.

References

Blake, A., Stapleton, G., Rodgers, P., Cheek, L., Howse, J.: Does the orientation of an Euler diagram affect user comprehension? In: DMS, pp. 185–190 (2012)

Grice, H.P.: Logic and conversation. In: Cole, P., Morgan, J.L. (eds.) Syntax and Semantics: Speech Acts, vol. 3, pp. 41–58. New York, Academic Press (1975)

Halford, G.S., Andrews, G.: The development of deductive reasoning: how important is complexity? Think. Reason. **10**(2), 123–145 (2004)

Hopkins, W., Marshall, S., Batterham, A., Hanin, J.: Progressive statistics for studies in sports medicine and exercise science. Med. Sci. Sports Exerc. **41**(1), 3 (2009)

Horridge, M., Bail, S., Parsia, B., Sattler, U.: The cognitive complexity of OWL justifications. In: Aroyo, L., Welty, C., Alani, H., Taylor, J., Bernstein, A., Kagal, L., Noy, N., Blomqvist, E. (eds.) ISWC 2011. LNCS, vol. 7031, pp. 241–256. Springer, Heidelberg (2011). doi: 10.1007/978-3-642-25073-6_16

Horridge, M., Drummond, N., Goodwin, J., Rector, A., Stevens, R., Wang, H.H.: The manchester owl syntax. In: OWL: Experiences and Directions (2006)

Johnson-Laird, P.N.: Against logical form. Psychologica Belgica **50**(3), 193–221 (2010)

Johnson-Laird, P.N., Byrne, R.M., Schaeken, W.: Propositional reasoning by model. Psychol. Rev. **99**(3), 418 (1992)

Khemlani, S., Orenes, I., Johnson-Laird, P.N.: Negating compound sentences. Naval Research Lab, Washington DC, Navy Center for Applied Research in Artificial Intelligence. http:// mindmodeling.org/cogsci2012/papers/0110/paper0110.pdf

Mendonça, E.A., Cimino, J.J., Campbell, K.E., Spackman, K.A.: Reproducibility of interpreting 'and' and 'or' in terminology systems. In: Proceedings of the AMIA Symposium, p. 790. American Medical Informatics Association (1998)

Nguyen, T.A.T., Power, R., Piwek, P., Williams, S.: Measuring the understandability of deduction rules for OWL. In: Presented at the First International Workshop on Debugging Ontologies and Ontology Mappings, Galway, Ireland (2012)

R Core Team (2014): R: A language and environment for statistical computing. R Foundation for Statistical Computing, Vienna, Austria (2013). ISBN 3-900051-07-0

Rector, A. et al.: OWL pizzas: practical experience of teaching OWL-DL: common errors and common patterns. In: Motta, E., Shadbolt, N.R., Stutt, A., Gibbins, N. (eds.) Engineering Knowledge in the Age of the Semantic Web, EKAW 2004. LNCS, vol. 3257, pp. 63–81. Springer, Berlin (2004)

Rector, A.L.: Defaults, context, and knowledge: alternatives for OWL-indexed knowledge bases. In: Pacific Symposium on Biocomputing, pp. 226–237 (2003)

Rips, L.J.: Cognitive processes in propositional reasoning. Psychol. Rev. **90**(1), 38 (1983)

Scott, D.: Tukey's ladder of powers. Rice University (2012). http://onlinestatbook.com/2/ transformations/tukey.html

Warren, P., Mulholland, P., Collins, T., Motta, E.: Making sense of description logics. In: Proceedings of the 11th International Conference on Semantic Systems, pp. 49–56. ACM (2015)

Warren, P., Mulholland, P., Collins, T., Motta, E.: The usability of description logics. In: Presutti, V., d'Amato, C., Gandon, F., d'Aquin, M., Staab, S., Tordai, A. (eds.) ESWC 2014. LNCS, vol. 8465, pp. 550–564. Springer, Cham (2014). doi:10.1007/978-3-319-07443-6_37

Reasoning Track

Updating Wikipedia via DBpedia Mappings and SPARQL

Albin Ahmeti[1]([✉]), Javier D. Fernández[1,2], Axel Polleres[1,2], and Vadim Savenkov[1]

[1] Vienna University of Economics and Business, Vienna, Austria
{albin.ahmeti,javier.fernandez,axel.polleres,vadim.savenkov}@wu.ac.at
[2] Complexity Science Hub Vienna, Vienna, Austria

Abstract. DBpedia crystallized most of the concepts of the Semantic Web using simple mappings to convert Wikipedia articles (i.e., infoboxes and tables) to RDF data. This "semantic view" of wiki content has rapidly become the focal point of the Linked Open Data cloud, but its impact on the original Wikipedia source is limited. In particular, little attention has been paid to the benefits that the semantic infrastructure can bring to maintain the wiki content, for instance to ensure that the effects of a wiki edit are consistent across infoboxes. In this paper, we present an approach to allow ontology-based updates of wiki content. Starting from DBpedia-like mappings converting infoboxes to a fragment of `OWL 2 RL` ontology, we discuss various issues associated with translating SPARQL updates on top of semantic data to the underlying Wiki content. On the one hand, we provide a formalization of DBpedia as an Ontology-Based Data Management framework and study its computational properties. On the other hand, we provide a novel approach to the inherently intractable update translation problem, leveraging the pre-existent data for disambiguating updates.

1 Introduction

DBpedia [24] is a community effort that has created the most important cross-domain dataset in RDF [7] in the focal point of the Linked Open Data (LOD) cloud [3]. At its core is a set of declarative mappings extracting data from Wikipedia *infoboxes* and tables into RDF. However, DBpedia makes knowledge machine readable only, rather than also *machine writable*. This not only restricts the possibilities of automatic curation of the DBpedia data that could be semi-automatically propagated back to Wikipedia, but also prevents maintainers from evaluating the impact of their edits on the consistency of knowledge; indeed, previous work confirms that there are such inconsistencies discoverable in DBpedia [6,11] arising most likely from inconsistent content in Wikipedia itself with respect to the mappings and the DBpedia ontology. Excluding the DBpedia taxonomy from the editing cycle is thus a — as we will show, unnecessary — drawback, but rather can be turned into an advantage for helping editors to create and maintain consistent content inside infoboxes, which we aim to address.

© Springer International Publishing AG 2017
E. Blomqvist et al. (Eds.): ESWC 2017, Part I, LNCS 10249, pp. 485–501, 2017.
DOI: 10.1007/978-3-319-58068-5_30

To this end, in this paper we want to make a case for DBpedia as a practical, real-world benchmark for Ontology-Based Data Management (OBDM) [25]. Although based on fairly restricted mappings—which we cast as a variant of so-called nested tuple-generating dependencies (tgds) herein—and minimalistic TBox language, accommodating DBpedia updates is intricate from different perspectives. The challenges are both conceptual (what is an adequate semantics for DBpedia SPARQL updates?) and practical, when having to cope with high ambiguity of update resolutions. While general updates in OBDM remain largely infeasible, we still arrive at reasons to believe, that for certain use cases within DBpedia updates, reasonable and practically usable conflict resolution policies could be defined; we present the first serious attempt with DBpedia as a potential benchmark use case in this area.

Pushing towards the vision of a "Read/Write" Semantic Web,[1] the unifying capabilities of SPARQL extend beyond the mere querying of heterogeneous data. Indeed, the standardization of update functionality introduced in SPARQL1.1 renders SPARQL as a strong candidate for the role of web data manipulation language. For a concrete motivation example consider Listing 1, where a simple SPARQL Update request would reflect a recent merger of French administrative regions: for each settlement belonging to either Upper or Lower Normandy, we set the corresponding administrative attribution property to be just Normandy. In our scenario, the user should have means to write this update in SPARQL and let it be reflected in the underlying Wikipedia data.

Despite clear motivation, updates in the information integration setting abound with all sorts of challenges, starting from obvious data security concerns, to performance, data quality issues and, last but not least the technical issues of side effects and lack of unique semantics, demonstrated already in the classical scenarios such as database views and deductive databases [4,9]. Although based on a very special join-free mapping language, the DBpedia setting is no different in this respect. With a high-quality curated data source at the backend, we set our goal not at ultimate transparency and automatic translation of updates, but rather at maximally support users in choosing the most economic and *typical* way of accommodating an update while maintaining (or at least, not degrading) consistency and not losing information inadvertently. As for DBpedia, if such RDF frontend systems have their own taxonomy (TBox) with also class and property disjointness assertions as well as functionality of properties, updates can result in inconsistencies with the data already present. In particular, we make the following contributions in this paper:

- we formalize the actual ontology language used by DBpedia as an OWL 2 RL fragment, and DBpedia mappings as a variant of so-called nested tuple-generating dependencies (tgds); based on this formalization
- we propose a semantics of OBDA updates for DBpedia and its Wikipedia mappings
- we discuss how such updates can be practically accommodated by suitable conflict resolution policies: the number of consistent revisions are in the worst

---

[1] cf. https://www.w3.org/community/rww/.

case exponential in the size of the mappings and the TBox, so we investigate policies for choosing the "most reasonable" ones, e.g. following existing patterns in the data, that is choosing most popular fields in the present data to be filled upon inserts.

Note that, since neither the SPARQL Update language [14] nor the SPARQL Entailment regimes [15] specification covers the behaviour of updates in the presence of TBox axioms, the choice of semantics in such cases remains up to application designers. In [1,2] we have discussed how SPARQL updates relating to the ABox can be implemented with TBoxes allowing no or limited form of inconsistency (class disjointness), a work we partially build upon herein: as a requirement from this prior work (as a consequence of the common postulates for updates in belief revision), such an update semantics needs to ensure that no mutually inconsistent pairs of triples are inserted in the ABox. In order to achieve this, a policy of conflict resolution between the new and the old knowledge is needed. To this end, in our earlier work [2] we defined *brave, cautious* and *fainthearted* semantics of updates. Brave semantics removes from the knowledge base all facts clashing with the inserted data. Cautious semantics discards entirely an update if it is inconsistent w.r.t. knowledge base, otherwise brave semantics is applied. Fainthearted semantics is in-between the two, amounts to adding an additional filter to the WHERE clause of SPARQL update in order to discard variable bindings which make inserted facts contradict prior knowledge. In the present work, we stick to these three basic cases, extending them to the OWL fragment used by DBpedia. However, since our goal is to accommodate updates as Wiki infobox revisions for which no batch update language exists, we restrict our considerations to *grounded updates* (u^+, u^-) of triples over URIs and literals that are to be inserted or, respectively, deleted (instead of considering the whole general SPARQL Update language).[2]

The rest of the paper is organized as follows. Section 2 provides our formalization on the DBpedia ontology and mapping language, defining the translation of Wiki updates to DBpedia updates and their (local) consistency. Section 3 outlines the main sources of worst-case complexity for automatic update translation that cannot be mitigated by syntactic restrictions of the mapping language. Section 4 discusses our pragmatic approach to OBDM in the DBpedia setting including our specific update conflict resolution strategies for DBpedia. Section 5 gives an overview of related work, and finally Sect. 6 provides concluding remarks.

Listing 1. SPARQL 1.1. Update that merges two regions in France.

```
DELETE { ?X :region :Upper_Normandy . ?Y :region :Lower_Normandy .}
INSERT { ?X :region :Normandy . ?Y :region :Normandy}
WHERE { {?X :region :Upper_Normandy} UNION {?Y :region :Lower_Normandy} }
```

---

[2] We emphasize though that such an extension is a fairly straightforward extension of the discussions in [2], since general SPARQL Updates can be viewed as templates which are instantiated into exactly such sets of INSERTed and DELETed triples.

2   The DBpedia OBDM Setting

We define the declarative **WikiDBpedia framework** (WDF) $\mathcal{F}$ as a triple $(\mathbf{W}, \mathcal{M}, \mathcal{T})$ where $\mathbf{W}$ is a relational schema encoding the infoboxes, $\mathcal{M}$ is a set of rules transforming it into RDF triples (the DBpedia ABox), and $\mathcal{T}$ is a TBox. The rules in $\mathcal{M}$ are given by a custom-designed declarative DBpedia mapping language [19]. This language can be captured by the language of *nested tuple generating dependences* (nested tgds) [12,22], enhanced with negation in the rule bodies and interpreted functions for arithmetics, date, string and geocoordinate processing.

A *WDF instance* of a WDF $(\mathbf{W}, \mathcal{M}, \mathcal{T})$ is an infobox instance I satisfying $\mathbf{W}$. We now specify the language used to formalize the TBox $\mathcal{T}$, the tgds language of $\mathcal{M}$ and the infobox schema $\mathbf{W}$.

Table 1. Description of DBpedia (English) mappings.

Type of mappings	Declared	Description
Template	958	Map Wiki templates to DBpedia classes
Property	19,972	Map Wiki template properties to DBpedia properties
IntermediateNode	107	Generate a blank node with a URI
Conditional	31	Depend on template properties and their values
Calculate	23	Compute a function over two properties
Date	106	Mappings that generate a starting and ending date

DBpedia ontology language. DBpedia uses a fragment of OWL 2 RL profile, which we call DBP. It includes the RDF keywords `subClassOf` (which we abbreviate as `sc`), `subPropertyOf` (`sp`), `domain` (`dom`) and `range` (`rng`), `disjointWith` (`dw`), `propertyDisjointWith` (`pdw`), `inversePropertyOf` (`inv`) as well as `functionalProperty` (`func`). At present, functional properties in DBpedia are limited to data properties, and inverse functional roles are not used.

Many concepts in the actual DBpedia are copied from external ontologies like Yago [30] and UMBEL[3]. All DBpedia resources also instantiate the concepts in DBpedia ontology, with the namespace http://dbpedia.org/ontology, to which we refer as DBP. They can be listed by the following SPARQL query:

```
SELECT DISTINCT ?x WHERE {{?x a owl:Class}
    UNION {?x a owl:ObjectProperty}
    UNION {?x a owl:DatatypeProperty}
    FILTER(strstarts(str(?x), "http://dbpedia.org/"))}
```

As of December 2016, this query retrieves 758 concepts, 1104 object and 1756 datatype properties for the English Live DBpedia[4]. Herewith, we only consider the facts from this core vocabulary set instantiated with the set of DBpedia

---

[3] http://techwiki.umbel.org/index.php/UMBEL_Vocabulary.

[4] http://live.dbpedia.org/sparql.

mappings $\mathcal{M}$, and not the imported assertions from the external ontologies. We denote this vocabulary by $\mathbf{T}$ and, analogously to the infobox part of the system, call it "schema".

Infobox schema W. Each Wiki page is identified by a URI which translates to a subject IRI in DBpedia. A page can contain several *infoboxes* of distinct *types*. We model this semistructured data store using a relational schema $\mathbf{W}$ with two ternary relations $W_i = $ UTI and $W_d = $ IPV, attribute I storing infobox identifiers, U page URI, T infobox type, and P and V being respectively property names and values. That is, unlike the real Wiki where infoboxes may belong to different pages or be separate tables of distinct types, we use an auxiliary surrogate key I to horizontally partition the single key-value store W_d. Our schema $\mathbf{W}$ assumes key constraints UT $\rightarrow$ I, IP $\rightarrow$ V and the inclusion dependency $W_d[\mathsf{I}] \subseteq W_i[\mathsf{I}]$ which we encode as the set of rules $\mathcal{W}$:

$$\mathcal{W} = \{\forall i \forall p \forall v \big(W_d(i,p,v) \rightarrow \exists u \exists t \; W_i(u,t,i)\big),$$
$$\forall u \forall t \forall i_1 \forall i_2 \big(W_i(u,t,i_1) \wedge W_i(u,t,i_2) \wedge i_1 \neq i_2 \rightarrow \bot\big),$$
$$\forall i \forall p \forall v_1 \forall v_2 \big(W_d(i,p,v_1) \wedge W_d(i,p,v_2) \wedge v_1 \neq v_2 \rightarrow \bot\big)\}.$$

Mapping constraints $\mathcal{M}$. The specification [19] distinguishes several types of DBpedia mappings summarized in Table 1 along with their figures in the English DBpedia. All these mappings can be represented as *nested tgds* [12,22] extended with *negation and constraints* in the antecedents for capturing the conditional mappings and interpreted functions in the conclusions of implications, in the case of calculated mappings handling, e.g., dates or geo-coordinates. A crucial limitation of the mapping language (which we call *DBpedia tgds*) is the *impossibility of comparisons between infobox property values*. Infobox type W_i.T and property names W_d.P must be specified explicitly.

Example 1. Figure 1(a) shows a conditional mapping transferring the information about clerics from French wiki pages with an infobox *Prélat catholique* (d). Under these conditions, the except shown in Fig. 1(c) as an instance over the schema $\mathbf{W}$ gives rise to the triples depicted in Fig. 1(b). A tgd formalizing a French DBpedia mapping for clergy is given below:

$\forall U \forall I \big(W_i(U, \text{'fr:Prélat catholique'}, I) \rightarrow$
  $\big(W_d(I, \text{'titre'}, \text{'Pape'}) \rightarrow$
    $\exists Y \big(\mathsf{Pope}(U) \wedge \mathsf{occupation}(U,Y) \wedge \mathsf{PersonFunction}(Y)$
      $\wedge \, \mathsf{title}(Y, \text{'Pape'})\big) \quad // \text{ "Intermediate node mapping"}$
      $\wedge \ldots$
      $\wedge \, \forall X(W_d(I, \text{'prédécesseur pape'}, X) \rightarrow \mathsf{predecessor}(Y,X))\big)$
  $\ldots$
  $\wedge \, (W_d(I, \text{'titre'}, \text{'Prêtre'}) \rightarrow \mathsf{Priest}(U))$
  $// \text{ The "otherwise" branch:}$
  $\wedge \, (\neg W_d(I, \text{'titre'}, \text{'Pape'}) \wedge \ldots \wedge \neg W_d(I, \text{'titre'}, \text{'Prêtre'}) \rightarrow \mathsf{Cleric}(U))$
  $\wedge \, \forall X(W_d(I, \text{'nom'}, X) \rightarrow \mathsf{foaf:name}(U,X))$
  $\ldots$
  $\wedge \, \forall X(W_d(I, \text{'nom naissance'}, X) \rightarrow \mathsf{birthName}(U,X))))$

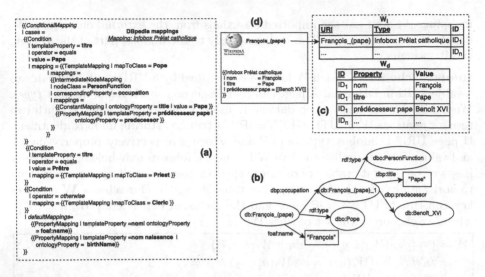

Fig. 1. (a) DBpedia mappings (b) the RDF graph, and the Infobox as an instance of the schema **W** (c) and in the native format (d).

The specification stipulates that conditions are evaluated in the natural order, and thus every next condition has to include the negation of all preceding conditions. In our case, this is only illustrated by the last, default ("otherwise") case, since the conditions are mutually exclusive. Note also that no universally quantified variable besides the page URI U and the technical infobox identifier I) — i.e., no X variable representing an infobox property — can occur more than once on the left-hand side of an implication, due to the lack of support for comparisons between infobox properties.

One further particularity of the chase with tgds is the handling of existentially quantified variables that represent so-called "intermediate nodes" (e.g., Y in Example 1). A usual approach is to instantiate such variables by null values, which could become blank nodes on the RDF storage side. The strategy currently followed by DBpedia is different: instead of blank nodes, the chase produces fresh IRIs. By appending an incremented number to the Wiki page address it avoids clashes with existing page URIs. We name it *constant inventing chase*.

Updates. We consider updates that can be specified on both the infobox and the DBpedia sides. Since DBpedia is a materialized extension constructed based on the contents of infoboxes, persistent modifications must be represented as infobox updates. We consider updates based on ground facts to be inserted or deleted, each update being limited to exactly one schema, the infobox **W** or DBpedia **T**.

Definition 1. *Let* **S** *be a schema and* J *an instance of* **S**. *An* update u *of* J *is a pair* (u^-, u^+) *of sets of ground atoms over* **S** *in which* u^+ *signifies facts to be inserted to* I *and* u^- *facts to be removed from* I. *Deletions are applied prior to insertions.*

Since WDF includes the mapping and TBox rules, special care is needed to make update effective and enforce or maintain the consistency of the affected WDF instance apply a minimal necessary modifications. Our formalization is close to the usual definition of formula based belief revision operators. A WDF instance I is identified with a conjunctive formula over $\mathbf{W}$ closed under the integrity constraints $\mathcal{W}$ of the infobox schema. The notation $u(I)$ is understood as $(I \setminus u^-) \cup u^+$ where $I \setminus u^-$ denotes the removal of all conjuncts occurring in u^- from I, and $I \cup u^+$ is the same as the conjunction $I \wedge u^+$.

We define a partial order $\preceq$ relation between updates as follows $u \preceq e$ iff $u^- \subseteq e^-$ and $u^+ \subseteq e^+$. One can as well consider other, e.g. cardinality based, partial orders.

Definition 2. *Let $\mathcal{F}$ be a WDF $(\mathbf{W}, \mathcal{M}, \mathcal{T})$, I be an $\mathcal{F}$-instance and let u be an update over $\mathcal{F}$. The consistency-oblivious semantics $\{\!\{u\}\!\}$ of u is the set of smallest (w.r.t. $\preceq$) updates $[u]$ over the infobox schema $\mathbf{W}$ such that the conditions $[u](I) \cup \mathcal{W} \cup \mathcal{M} \cup \mathcal{T} \not\models u^-$, $[u](I) \cup \mathcal{W} \cup \mathcal{M} \cup \mathcal{T} \models u^+$ and $I \cup \mathcal{W} \not\models \perp$ hold.*

The former two conditions ensure the effectiveness of the update, that is, that all desired insertions and deletions are performed. The conformance with $\mathbf{W}$ ensures that the update can be accommodated in the physical infobox storage model, which the constraints $\mathcal{W}$ simulate. The following definition of the semantics $\{\!\{[u]\}\!\}$ restricts the semantics $\{\!\{u\}\!\}$ in order to ensure that the DBpedia instance can be used under entailment w.r.t. $\mathcal{T}$, denoted as closure $cl(I, \mathcal{M})$. Note that both semantics $\{\!\{u\}\!\}$, $\{\!\{[u]\}\!\}$ depend on $\preceq$, $\mathcal{F}$ and on I—which is not explicit in our notation for the sake of readability.

Definition 3. *Let $\mathcal{F}$ be a WDF $(\mathbf{W}, \mathcal{M}, \mathcal{T})$, I be an $\mathcal{F}$-instance and let u be an update over $\mathcal{F}$. The consistency-aware semantics $\{\!\{[u]\}\!\}$ of u is the set of smallest (w.r.t. $\preceq$) updates $[u]$ such that $[u] \in \{\!\{u\}\!\}$ and $[u](I) \cup \mathcal{W} \cup \mathcal{M} \cup \mathcal{T} \not\models \perp$.*

3   Challenges of DBpedia OBDM

We consider the EXISTENCE OF SOLUTIONS problem and show that it is in general intractable even for the consistency-oblivious semantics.

Problem EXSOL-OBL. Parameter: WDF $\mathcal{F} = (\mathbf{W}, \mathcal{M}, \mathcal{T})$. Input: $\mathcal{F}$-instance I, update u. Test if $\{\!\{u\}\!\} \neq \emptyset$.

Proposition 1. EXSOL-OBL *is NP-complete.*

Proof (Sketch). Consider a DBpedia update u, and the WDF instance I. For the membership in NP, observe that enforcing the constraints in $\mathcal{M}$ and in $\mathcal{T}$ (e.g., via chase) terminates in polynomial time for every fixed WDF $\mathcal{F}$, which gives a bound on the size of the infobox instance witnessing $\{\!\{u\}\!\} \neq \emptyset$ for an instance I. For each condition in the mapping $\mathcal{M}$ (limited to comparing a single infobox value with a fixed constant), we can define a canonical way of satisfying it, and thus defining *canonical witnesses*, whose size and active domain is determined by u, I and $\mathcal{F}$. As a result, the test comes down to guessing a canonical witness and checking it by the chase with constraints, that u^+ is inserted and u^- deleted, which is feasible in poly time for the constraints in DBP.

For the hardness, consider the following reduction from the 3-COLORABILITY problem. Let I be empty and let the set of atoms A that the DBpedia update $u = (\emptyset, A)$ inserts represents an undirected graph $G = (V, E)$ of degree at most 4 (for which 3COL is intractable [13]). A represents the vertices V as IRIs and each edge $(x, y) \in E$ for the IRIs x, y is represented by a collection of 8 atoms of the form $\mathsf{a}(x, y)$, $\mathsf{a}(y, x)$, $\mathsf{b}(x, y)$, $\mathsf{b}(y, x)$, $\mathsf{c}(x, y)$, $\mathsf{c}(y, x)$, $\mathsf{d}(x, y)$, and $\mathsf{d}(y, x)$, for which the assertions $\mathsf{a} = \mathsf{a}^{-1}$, $\mathsf{b} = \mathsf{b}^{-1}$, $\mathsf{c} = \mathsf{c}^{-1}$ and $\mathsf{d} = \mathsf{d}^{-1}$ are defined in $\mathcal{T}$.

Each infobox encodes a single vertex of the graph, together with all its adjacent vertices (at most four direct neighbors). Together these 1-neighborhoods cover the graph. The encoding ensures that the only way to obtain the regular DBpedia representation of the graph, with exactly eight property assertions for each pair of vertices, is only possible if every vertex is assigned the same color in each infobox. This is achieved by distributing the a, b, c and d between each pair of adjacent nodes depending on the node color. The rules for that are given in Fig. 2(c). For instance, an edge between a red I and a green vertex II is composed from the properties a(I,II) b(I,II) whose creation is triggered by the infobox of page I, and the other two properties b, c are created by chasing the infobox I: c(II,I), d(II,I). Due to symmetry, this results in the eight property assertions.

The excerpt of the mapping for the neighborhood types 'r_ggb', 'b_rgg', 'g_rb' illustrated by a graph in Fig. 2(b) is shown below.

$\forall U \forall I \, (W_i(U, \text{'vertex'}, I) \rightarrow$
$\quad (W_d(I, \text{'n-type'}, \text{'r\_ggb'}) \rightarrow (\mathsf{Node}(U) \wedge \forall X \, (W_d(I, \text{'n1'}, X) \rightarrow \mathsf{a}(U, X) \wedge \mathsf{b}(U, X))$
$\quad\quad \wedge \forall X \, (W_d(I, \text{'n2'}, X) \rightarrow \mathsf{a}(U, X) \wedge \mathsf{b}(U, X)) \wedge \forall X \, (W_d(I, \text{'n3'}, X) \rightarrow \mathsf{a}(U, X) \wedge \mathsf{c}(U, X)))$
$\quad \wedge \, (W_d(I, \text{'n-type'}, \text{'b\_rgg'}) \rightarrow (\mathsf{Node}(U) \wedge \forall X \, (W_d(I, \text{'n1'}, X) \rightarrow \mathsf{b}(U, X) \wedge \mathsf{d}(U, X))$
$\quad\quad \wedge \forall X \, (W_d(I, \text{'n2'}, X) \rightarrow \mathsf{b}(U, X) \wedge \mathsf{b}(U, X)) \wedge \forall X \, (W_d(I, \text{'n3'}, X) \rightarrow \mathsf{b}(U, X) \wedge \mathsf{c}(U, X)))$
$\quad \wedge \, (W_d(I, \text{'n-type'}, \text{'g\_rb'}) \rightarrow (\mathsf{Node}(U) \wedge \forall X \, (W_d(I, \text{'n1'}, X) \rightarrow \mathsf{c}(U, X) \wedge \mathsf{d}(U, X))$
$\quad\quad \wedge \forall X \, (W_d(I, \text{'n2'}, X) \rightarrow \mathsf{a}(U, X) \wedge \mathsf{d}(U, X)))$
$\wedge \ldots$ etc for other 1-neighborhood types $\ldots)$ $\qquad\qquad\qquad\qquad\qquad\qquad\qquad\quad$ □

If we bring the TBox and infobox schema constraints along with non-monotonicity of mapping rules into the picture, the potential challenges of accommodating updates start piling up quickly. An interplay of the following features of the framework can make update translation unwieldy: (i) inconsistencies due to the TBox assertions, namely the class and role disjointness and functional properties; (ii) many-to-many relationships between infobox and ontology properties defined by the mappings, and (iii) infobox schema constraints.

Example 2 **Deletions due to infobox constraints.** Consider the update u_1 inserting an alternative foaf:name value for an existing cleric (cf. the mapping in Example 1). The infobox key IP → V would deprecate this, since there is only

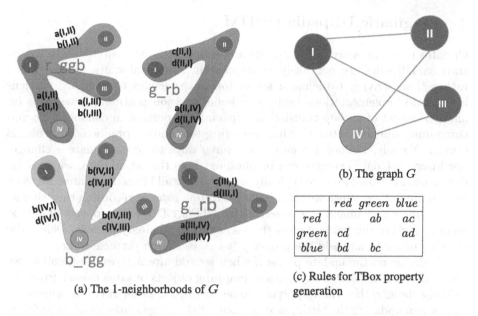

(a) The 1-neighborhoods of G

(b) The graph G

	red	green	blue
red		ab	ac
green	cd		ad
blue	bd	bc	

(c) Rules for TBox property generation

Fig. 2. Concepts of the proof of Proposition 1

one infobox property matching foaf:name. Therefore, all updates in $\{u_1\}$ will extend u_1 with the deletion of the old name.

Insertions and many-to-many property matches. Several Wiki properties are mapped to the same DBpedia property and the insertion cannot be uniquely resolved. E.g., infoboxes of football players in the English Wikipedia have the properties 'full name', 'name' and 'short name' all mapped to 'foaf:name'.

Deletions with conditional mappings. Triples generated by a conditional mapping can be deleted either by removing the corresponding Wiki property or by modifying the Infobox property so that the condition is no longer satisfied. E.g., in Example 1, deleting the triple predecessor (Nicholas_II, Alexander_II) can be done either by unsetting the infobox property 'prédécesseur pape' or the property 'titre' used in the condition.

The above considerations suggest that despite the syntactic restrictions of the DBpedia mappings, the problem of update translation is hard in the worst case. Furthermore, numerous translations of an ABox update often exist (exponentially many in the size of the mapping: e.g., each n-to-m property match increases the total size of possible translations by the factor of mn). Due to the interplay between the mapping conditions and TBox axioms a complete solution of the OBDM problem, presenting and explaining to the user *all possible ways* of accommodating an arbitrary update is not practical. Our pragmatic approach to the problem is described next.

4   Pragmatic DBpedia OBDM

Updates in the presence of constraints and mappings over a curated data source such as DBpedia are not likely to happen in a fully atomatic mode. Thus, rather than striving to define a set of formal principles to compare particular update implementations (akin, e.g. belief revision postulates) we focus on another aspect of update translation, especially important in collaborative and community-oriented settings, where adhering to standard practices and rules is crucial. Namely, we look for *most customary* ways to accommodate a change. For insertions, data evidence can be obtained from the actual data, whereas for deletions, additional logs are typically required. For all kinds of updates, we use a special kind of log, which we call *update resolution pattern*, recording the "shape" of each update command (e.g. *inserting a* birthPlace DBpedia property of a Pope instance, where the Infobox property 'lieu de naissance' is alredy present. Delete the existing property and add the property 'lieu naissance' *with the new value*).

To decide on the update pattern, when several alternatives are possible, we try to derive most customary ways of mapping objects of same classes from the existing data, rather than applying some principled belief revision semantics. E.g., when updating the birth place, we look at the usage statistics of the Infobox properties 'lieu naissance' and 'lieu de naissance' and choose the one used most often. If most infoboxes have both, we will not delete the already existing property but just add a second one. This way, we might resolve a DBpedia's foaf:name as two infobox properties (e.g., 'name' and 'full name') at once if most existing records of a given type follow this pattern, even if it would contradict the minimal change principle which typically governs belief revision.

A translation procedure we discuss next proceeds essentially on the best effort basis, exploring the most likely update accommodations and facilitating reuse of standard practices through update resolution patterns. It takes a SPARQL update and transforms it into a set of Infobox updates for the user to apply and save as an update reolution pattern. The source code of our system is openly available[5].

4.1   Update Translation Steps

From the very beginning, we turn our SPARQL update into a set of ground atoms, which are then grouped by subject (corresponding to the Wikipedia page). The idea of our update translation procedure is to create or to re-use existing update patterns for each grouped update extracted from the user input. A user update request related to a particular Wikipedia page (DBpedia entries grouped by a commond subject) becomes a core pattern, which gives rise to a number of possible translations as a wiki insert.

For each translation, the mapping and the TBox constraints are applied, in order to see which further atoms have to be added and if there are inconsistencies

---

[5] https://github.com/aahmeti/DBpedia-SUE, a screencast is available at https://goo.gl/BQhDYf.

with the pre-existing facts. All such inconsistencies are removed, resulting in a further update, giving rise to an update resolution pattern nested within the root one, and the translation process procedes recursively.

Pruning is essential in this process, since resolution patterns can sprout actively (e.g., some DBpedia properties are mapped to tens of Wikipedia ones). Potentially non-terminating, with the current DBpedia mappings inconsistencies can typically be resolved within the scope of one or two subjects (Wikipedia pages), and thus pattern trees resulting from this process are not deep. The reason is that functionality is currently only used for data properties, and only very few properties are declared disjoint.

4.2 Update Resolution Policies

Given the large number of possible translations of an update, potentially resulting in different clash patterns, an update can be translated in various ways, from which the user must select one. The crucial issue here is that the number of choices can be too large even for a very simple update, and that updates can cause side effects outlined in the previous section.

Here, we consider *update resolution policies* aimed at reducing the number of options for the user in the specific case of n-to-1 alternatives to insert. We currently consider two different alternatives in accordance with some concise principles, namely *infobox-frequency-first* and *similar-subject-first*.

We exemplify the application of such policies looking at the ambiguities in the top 10 most used Infoboxes[6]. In particular, we find and inspect the ambiguities in 'Settlement', 'Taxobox', 'Person', 'Football biography' and 'Film'. For the sake of clarity, we show a selection of the most representative ambiguities in Table 2, while other ambiguities in the infoboxes follow the same patterns. For instance, all 'name', 'fullname' and 'player name' in a 'football biography' infobox map to a foaf:name property. Table 2 also reports the number of subjects (i.e., wikipedia pages) of each infobox type, converted from the English Wikipedia.

Infobox-frequency-first. This policy considers that, for an insertion in a subject with an infobox W, resulting in a n-to-1 alternatives, we infer that the most likely accommodation would be the most frequent property in all the subjects with such infobox W, among all the alternatives not fulfilled in the subject we are currently updating. Statistics on frequent properties can be computed seamless, concurrently to the DBpedia conversion. Overall, this approximation could help users to inspect frequent properties for the update, so that rare or infrequent properties can be quickly discarded. In contrast, the approach may fail to guess the concrete purpose or real users, who may choose to accommodate different alternatives.

Figure 3 evaluates the distribution of frequencies of the Wikipedia properties involved in $n-1$ mappings from Table 2, considering all the subjects in the infobox (series *Infobox-frequency-first*). Results show that the application

---

[6] http://mappings.dbpedia.org/server/statistics/en/.

Table 2. Examples of n-to-1 alternatives in DBpedia (English) mappings.

Infobox	Subjects	Ambiguous $n-1$ Mapping	
		Wikipedia prop.	dbpedia prop.
Settlement	369,024	*area_total_km2*	dbp:areaTotal
		area_total_sq_mi	
		area_total	
		TotalArea_sq_mi	
Taxobox	293,715	*species*	dbp:species
		subspecies	
		variety	
		species_group	
		species_subgroup	
		species_complex	
Person	168,372	*website*	foaf:homepage
		homepage	
Football biography	128,602	*name*	foaf:name
		fullname	
		playername	
Film	106,254	*screenplay*	dbp:writer
		writer	

of this policy can certainly filter out infrequent property candidates, but it may require further elaboration for a more informed recommendation, specially in those cases in which the property is not extensively used in the infoboxes. For instance, all properties with no or marginal presence can be discarded, such as 'area_total' and 'TotalArea_sq_mi' in 'Settlement' (Fig. 3(a)), 'variety', 'species_group', 'species_subgroup' and 'species_complex' in 'Taxobox' (Fig. 3(b)), 'homepage' in 'Person' and 'playername' in 'Football biography' (Fig. 3(d)). In turn, some properties are much more represented than others, and shall be the first ranked suggestion when inserting an ambiguous mapping. This is the case of most of the infoboxes, such as the frequent 'area_total_km2' property in 'Settlement', 'species' in 'Taxobox', 'website' in 'Person', and 'writer' in 'Film'. In contrast, only one case, 'Football biography', showed two properties that are almost equally distributed, with 'name' slightly more used than 'fullname'.

Similar-subject-first. The objective of this strategy is to refine the previous *Infobox-frequency-first* policy by delimiting a set of similar subjects for which the frequent properties are inspected. The reason of this strategy is that most of the properties in infoboxes are optional, so that different Wikipedia resources can, and often are, described with different levels of detail. Thus, finding "similar" subjects could effectively recommend more frequent patterns. For finding similar entities, we focus for the moment on a simple approach on sampling m subjects described with the same target infobox W and described with the same DBpedia property as the update u.

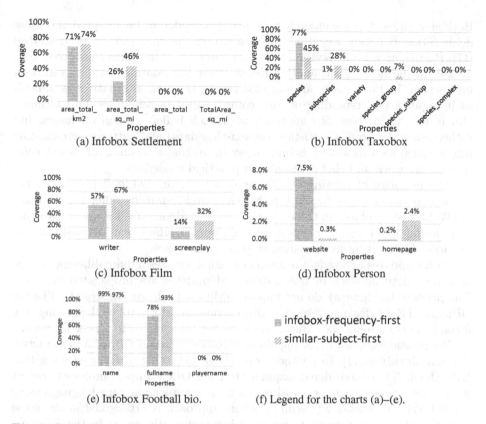

Fig. 3. Statistics obtained by *infobox-frequency-first* and *similar-subject-first* policies on four different infoboxes.

Figure 3 evaluates the distribution of property frequencies in such scenario (series *Similar-subject-first*), sampling $m = 1,000$ subjects of each infobox described the DBpedia property to be inserted (dbp:areaTotal, dbp:species, foaf:homepage, foaf:name or dbp:writer respectively). Results show that this policy allows the system to perform more informed decisions. For instance, in the 'Person' use case (Fig. 3(c)), the 'homepage' property cannot be discarded (as suggested by the *Infobox-frequency-first* approach), given that a particular type of persons are more frequently associated with homepage instead of websites (e.g., those who are not related to a company). Similarly, in 'Taxobox' (Fig. 3(b)), some particular species also include 'subspecies' and 'subsepecies_group', hence they should be included and ranked as potential accommodations for the user query.

5   Related Work

The problem of knowledge base update and belief revision has been extensively studied in the literature for various description logics, cf. e.g. [5,8,10,16,21].

Both semantics strongly enforcing the new knowledge to be accepted and those deliberating between accepting and discarding the change have been studied [17]. Particularly, belief revision with user interaction has been considered, e.g., in [26]. In the same spirit, another recent study [32] considers repairs with user interaction. In both cases, an informed choice between alternatives is difficult as it requires understanding of such complex update semantics. Our ultimate aim is not to compare our approach with such belief revision operators, but rather using/developing statistics (pre-existing data) and patterns (pre-existing interactions) as a means for helping users in making a meaningful choice, complementing work on belief revision with practical guidelines.

The majority of existing OBDM approaches (e.g., [23,28,29]) consider the problem of query answering only rather than updates, using different fragments of OWL. The emphasis in those approaches is in algorithms for query rewriting considering one-to-many, many-to-many mappings, where queries consist of also variables (without an instantiation step as in our case).

As for updates and tgds, the approach [20] addresses a quite different setting of a peer data network in which data and updates are propagated via tgds. The peers in the network do not impose additional schema constraints (like the DBpedia TBox), features like class disjointness are not part of the setting, the focus is on combining the external data with local updates in a peer network.

We mentioned work reporting about inconsistencies in DBpedia in the introduction already [6,11]. In another work about detecting inconsistencies within DBpedia [27] have considered mappings to the DOLCE upper ontology to detect even more inconsistencies, operating in a more complex ontology language using a full OWL DL reasoner (HermiT). Their approach is orthogonal in the sense that they focus on detecting and resolving systematic errors in the DBpedia ontology itself, rather than automatically fixing the assertions, leave alone the data in Wikipedia itself. Nonetheless, it would be an interesting future direction to combine these two orthogonal approaches.

It is also worth mentioning work in the domain of applying statistical methods for disambiguating updates, e.g., [31], namely for enriching the TBox based on the data, which is actually not our scope, as we do not modify the TBox here.

Recently Wikipedia partially shifted to another, structured datasource than infoboxes, namely, Wikidata. We note that the model of Wikidata is different to DBpedia; different possible representations in plain RDF or Relational models have been recently suggested/discussed [18]. Our approach could potentially help in bridging between the two, which we leave to future work.

6   Conclusion

Little attention has been paid to the benefits that the semantic infrastructure can bring to maintain the wiki content, for instance to ensure that the effects of a wiki edit are consistent across infoboxes. In this paper, we present first insights to allow ontology-based updates of wiki content.

Various worst-case scenarios of update translation, especially those exhibiting the intractability of update handling, can be hardly realized in the current

DBpedia version (mappings and ontology). From the practical point of view, the following aspects of OBDM appear crucial for the DBpedia case. Firstly, it is the *inherent ambiguity of update translation*; mappings often create a many-to-one or many-to-many relationships between infobox and DBpedia properties. Second, concisely presenting a large number of options to the user is a challenge, hence an *automatic selection of most likely update translations* is likely required. Finally, being a curated system, Wiki also requires curated updates. Thus, *splitting a SPARQL update into small independent pieces* to be verified by the Wiki maintainers is needed as well. Note that human intervention is often unavoidable, since calculating mappings involve non-invertible functions.

The main distinguishing characteristic of our approach is the DBpedia OBDM setting, and the focus on update accommodation strategies which are simple, comprehensible for the user and can draw from pre-existing meta-knowledge, such as already existing mapping patterns resp. usage frequencies of certain infobox fields, to decide update ambiguities upon similar, prototypical objects in the underlying data, estimating probabilities of alternative update translations. Our goal in this work was twofold: on the one hand, to understand and to formalize the DBpedia setting from the OBDM perspective, and on the other hand, to explore more pragmatic approaches to OBDM. To the best of our knowledge, it is the first attempt to study DBpedia mappings from the formal point of view. We found out that although the worst-case complexity of OBDM can be prohibitively high (even with low expressivity ontology and mapping languages), the real data, mappings and ontology found in DBpedia do not necessarily hit this full potential complexity; indeed, we conclude that the study and development of best-effort pragmatic approaches — some of which we have explored — is worthwhile.

Our early practical experiments with a DBpedia-based OBDM prototype shows that high worst case complexity of update translation can have little to do with actual challenges of OBDM for curated data. Rather, simple and comprehensible update resolution policies, reliable methods of confidence estimation and the ability to automatically learn and use best practices should be considered.

Acknowledgements. The work of Fernández was supported by the Austrian Science Fund (FWF): M1720-G11. Ahmeti was supported by the Vienna Science and Technology Fund (WWTF), project SEE: ICT12-15. Savenkov was supported by the Austrian Research Promotion Agency (FFG), project 855407.

References

1. Ahmeti, A., Calvanese, D., Polleres, A.: Updating RDFS ABoxes and TBoxes in SPARQL. In: Mika, P., et al. (eds.) ISWC 2014. LNCS, vol. 8796, pp. 441–456. Springer, Cham (2014). doi:10.1007/978-3-319-11964-9_28
2. Ahmeti, A., Calvanese, D., Polleres, A., Savenkov, V.: Handling inconsistencies due to class disjointness in SPARQL updates. In: Sack, H., Blomqvist, E., d'Aquin, M., Ghidini, C., Ponzetto, S.P., Lange, C. (eds.) ESWC 2016. LNCS, vol. 9678, pp. 387–404. Springer, Cham (2016). doi:10.1007/978-3-319-34129-3_24

3. Auer, S., Bizer, C., Kobilarov, G., Lehmann, J., Cyganiak, R., Ives, Z.: DBpedia: a nucleus for a web of open data. In: Aberer, K., et al. (eds.) ASWC/ISWC - 2007. LNCS, vol. 4825, pp. 722–735. Springer, Heidelberg (2007). doi:10.1007/978-3-540-76298-0_52
4. Bancilhon, F., Spyratos, N.: Update semantics of relational views. ACM Trans. Database Syst. **6**(4), 557–575 (1981)
5. Benferhat, S., Bouraoui, Z., Papini, O., Würbel, E.: A prioritized assertional-based revision for dl-lite knowledge bases. In: Fermé, E., Leite, J. (eds.) JELIA 2014. LNCS (LNAI), vol. 8761, pp. 442–456. Springer, Cham (2014). doi:10.1007/978-3-319-11558-0_31
6. Bischof, S., Krötzsch, M., Polleres, A., Rudolph, S.: Schema-agnostic query rewriting in SPARQL 1.1. In: Mika, P., et al. (eds.) ISWC 2014. LNCS, vol. 8796, pp. 584–600. Springer, Cham (2014). doi:10.1007/978-3-319-11964-9_37
7. Brickley, D., Guha, R. (eds.): RDF Vocabulary Description Language 1.0: RDF Schema. W3C Recommendation (2004)
8. Calvanese, D., Kharlamov, E., Nutt, W., Zheleznyakov, D.: Evolution of *DL–Lite* knowledge bases. In: Patel-Schneider, P.F., Pan, Y., Hitzler, P., Mika, P., Zhang, L., Pan, J.Z., Horrocks, I., Glimm, B. (eds.) ISWC 2010. LNCS, vol. 6496, pp. 112–128. Springer, Heidelberg (2010). doi:10.1007/978-3-642-17746-0_8
9. Cong, G., Fan, W., Geerts, F., Li, J., Luo, J.: On the complexity of view update analysis and its application to annotation propagation. IEEE TKDE **24**(3), 506–519 (2012)
10. De Giacomo, G., Lenzerini, M., Poggi, A., Rosati, R.: On instance-level update and erasure in description logic ontologies. J. Log. Comput. **19**(5), 745–770 (2009)
11. Dimou, A., Kontokostas, D., Freudenberg, M., Verborgh, R., Lehmann, J., Mannens, E., Hellmann, S., Walle, R.: Assessing and refining mappingsto RDF to improve dataset quality. In: Arenas, M., et al. (eds.) ISWC 2015. LNCS, vol. 9367, pp. 133–149. Springer, Cham (2015). doi:10.1007/978-3-319-25010-6_8
12. Fuxman, A., Hernández, M.A., Ho, C.T.H., Miller, R.J., Papotti, P., Popa, L.: Nested mappings: schema mapping reloaded. In: Proceedings of the VLDB, pp. 67–78 (2006)
13. Garey, M., Johnson, D., Stockmeyer, L.: Some simplified np-complete graph problems. Theoret. Comput. Sci. **1**(3), 237–267 (1976)
14. Gearon, P., Passant, A., Polleres, A.: SPARQL 1.1 update. W3C Recommendation (2013)
15. Glimm, B., Ogbuji, C.: SPARQL 1.1 entailment regimes. W3C Recommendation (2013)
16. Grau, B.C., Jiménez-Ruiz, E., Kharlamov, E., Zheleznyakov, D.: Ontology evolution under semantic constraints. In: Proceedings of the (KR 2012), pp. 137–147. AAAI Press (2012)
17. Hansson, S.O.: A survey of non-prioritized belief revision. Erkenntnis **50**(2–3), 413–427 (1999)
18. Hernández, D., Hogan, A., Riveros, C., Rojas, C., Zerega, E.: Querying Wikidata: comparing SPARQL, relational and graph databases. In: Groth, P., Simperl, E., Gray, A., Sabou, M., Krötzsch, M., Lecue, F., Flöck, F., Gil, Y. (eds.) ISWC 2016. LNCS, vol. 9982, pp. 88–103. Springer, Cham (2016). doi:10.1007/978-3-319-46547-0_10
19. Jentzsch, A., Sahnwaldt, C., Isele, R., Bizer, C.: Dbpedia mapping language. Technical report (2010). http://mappings.dbpedia.org
20. Karvounarakis, G., Green, T.J., Ives, Z.G., Tannen, V.: Collaborative data sharing via update exchange and provenance. ACM ToDS **38**(3), 19:1–19:42 (2013)

21. Kharlamov, E., Zheleznyakov, D., Calvanese, D.: Capturing model-based ontology evolution at the instance level: the case of DL-Lite. J. Comput. Syst. Sci. **79**(6), 835–872 (2013)
22. Kolaitis, P.G., Pichler, R., Sallinger, E., Savenkov, V.: Nested dependencies: structure and reasoning. In: Proceedings of the PODS 2014, pp. 176–187 (2014)
23. Kontchakov, R., Rezk, M., Rodríguez-Muro, M., Xiao, G., Zakharyaschev, M.: Answering SPARQL queries over databases under OWL 2 QL entailment regime. In: Mika, P., et al. (eds.) ISWC 2014. LNCS, vol. 8796, pp. 552–567. Springer, Cham (2014). doi:10.1007/978-3-319-11964-9_35
24. Lehmann, J., Isele, R., Jakob, M., Jentzsch, A., Kontokostas, D., Mendes, P.N., Hellmann, S., Morsey, M., van Kleef, P., Auer, S., et al.: Dbpedia-a large-scale, multilingual knowledge base extracted from Wikipedia. SWJ **6**(2), 167–195 (2015)
25. Lenzerini, M.: Ontology-based data management. In: Proceedings of the 20th ACM International Conference on Information and Knowledge Management, CIKM 2011, pp. 5–6. ACM (2011)
26. Nikitina, N., Rudolph, S., Glimm, B.: Interactive ontology revision. JWS **12–13**, 118–130 (2012)
27. Paulheim, H., Gangemi, A.: Serving DBpedia with DOLCE – more than just adding a cherry on top. In: Arenas, M., et al. (eds.) ISWC 2015. LNCS, vol. 9366, pp. 180–196. Springer, Cham (2015). doi:10.1007/978-3-319-25007-6_11
28. Priyatna, F., Corcho, O., Sequeda, J.: Formalisation and experiences of r2rml-based sparql to sql query translation using morph. In: Proceedings of the WWW 2014, pp. 479–490. ACM (2014)
29. Rodríguez-Muro, M., Rezk, M.: Efficient SPARQL-to-SQL with R2RML mappings. JWS **33**(1), 141–169 (2015)
30. Suchanek, F.M., Kasneci, G., Weikum, G.: Yago: a core of semantic knowledge. In: Proceedings of the WWW 2007, pp. 697–706. ACM (2007)
31. Töpper, G., Knuth, M., Sack, H.: Dbpedia ontology enrichment for inconsistency detection. In: Proceedings of the I-SEMANTICS 2012, pp. 33–40. ACM (2012)
32. Bienvenu, M., Bourgaux, C., Goasdoué, F.: Query-driven repairing of inconsistent DL-Lite knowledge bases. In: Proceedings of the IJCAI 2016, pp. 957–964. IJCAI/AAAI Press (2016)

Learning Commonalities in RDF

Sara El Hassad, François Goasdoué[(⊠)], and Hélène Jaudoin

IRISA, Univ. Rennes 1, Lannion, France
{sara.el-hassad,fg,helene.jaudoin}@irisa.fr

Abstract. Finding the commonalities between descriptions of data or knowledge is a foundational reasoning problem of Machine Learning introduced in the 70's, which amounts to computing a *least general generalization (lgg)* of such descriptions. It has also started receiving consideration in Knowledge Representation from the 90's, and recently in the Semantic Web field. We revisit this problem in the popular Resource Description Framework (RDF) of W3C, where descriptions are RDF graphs, i.e., a mix of data *and* knowledge. Notably, and in contrast to the literature, our solution to this problem holds for the *entire* RDF standard, i.e., we do not restrict RDF graphs in any way (neither their structure nor their semantics based on RDF entailment, i.e., inference) and, further, our algorithms can compute lggs of *small-to-huge* RDF graphs.

Keywords: RDF · RDFS · RDF entailment · Least general generalization

1 Introduction

Finding the commonalities between descriptions of data or knowledge is a foundational reasoning problem of Machine Learning, which was formalized in the early 70's as computing a *least general generalization* (lgg) of such descriptions [25,26]. Since the early 90's, this problem has also received consideration in the Knowledge Representation field, where least general generalizations were rebaptized *least common subsumers* [12], in Description Logics, e.g., [9,12,19,33] and in Conceptual Graphs [11].

In this paper, we revisit this old reasoning problem, from both the theoretical and the algorithmic viewpoints, in the *Resource Description Framework (RDF)*: the prominent Semantic Web data model by W3C. In this setting, the problem amounts to computing the lggs of RDF graphs, i.e., a mix of data *and* knowledge, the semantics of which is defined through RDF entailment, i.e., inference using entailment rules from the RDF standard.

To the best of our knowledge, the only proposal in that direction is the recent work [13,14], which brings a limited solution to the problem. It allows finding the commonalities between *single entities* extracted from RDF graphs (e.g., users in a social network), *ignoring* RDF entailment. In contrast, we further aim at considering the problem in all its generality, i.e., finding the commonalities between *general* RDF graphs, hence modeling *multiple interrelated entities* (e.g., social networks of users), accurately w.r.t. their standard semantics.

© Springer International Publishing AG 2017
E. Blomqvist et al. (Eds.): ESWC 2017, Part I, LNCS 10249, pp. 502–517, 2017.
DOI: 10.1007/978-3-319-58068-5_31

More precisely, we bring the following contributions:

1. We define and study the problem of computing an lgg of RDF graphs in the *entire RDF standard*: we do not restrict RDF graphs in any way, i.e., neither their structure nor their semantics defined upon RDF entailment.
2. We provide three algorithms for our solution to this problem, which allow computing lggs of *small-to-huge general RDF graphs* (i.e., that fit either in memory, in data management systems or in MapReduce clusters) w.r.t. *any set of entailment rules* from the RDF standard.

The paper is organized as follows. First, we introduce the RDF data model in Sect. 2. Then, in Sect. 3, we define and study the problem of computing an lgg of RDF graphs, for which we provide algorithms in Sect. 4. Finally, we discuss related work and conclude in Sect. 5.

Proofs of our technical results are available in the online research report [16].

2   The Resource Description Framework (RDF)

RDF Graphs. The RDF data model allows specifying *RDF graphs*. An RDF graph is a set of *triples* of the form (s, p, o). A triple states that its *subject* s has the *property* p, the value of which is the *object* o. Triples are built using three pairwise disjoint sets: a set $\mathcal{U}$ of *uniform resources identifiers (URIs)*, a set $\mathcal{L}$ of *literals* (constants), and a set $\mathcal{B}$ of *blank nodes* allowing to support *incomplete information*. Blank nodes are identifiers for missing values in an RDF graph (unknown URIs or literals). *Well-formed triples*, as per the RDF specification [31], belong to $(\mathcal{U} \cup \mathcal{B}) \times \mathcal{U} \times (\mathcal{U} \cup \mathcal{L} \cup \mathcal{B})$; we only consider such triples hereafter.

Notations. We use s, p, o in triples as placeholders. We note $\mathtt{Val}(\mathcal{G})$ the set of *values* occurring in an RDF graph $\mathcal{G}$, i.e., the URIs, literals and blank nodes; we note $\mathtt{Bl}(\mathcal{G})$ the set of blank nodes occurring in $\mathcal{G}$. A blank node is written b possibly with a subscript, and a literal is a string between quotes. For instance, the triples $(b, \mathrm{hasTitle}, \text{"LGG in RDF"})$ and $(b, \mathrm{hasContactAuthor}, b_1)$ state that *something (b) entitled "LGG in RDF" has somebody (b_1) as contact author.*

A triple models an assertion, either for a *class* (unary relation) or for a *property* (binary relation). Table 1 (top) shows the use of triples to state such assertions. The RDF standard [31] provides built-in classes and properties, as URIs within the rdf and rdfs pre-defined namespaces, e.g., rdf:type which can be used to state that the above b is a conference paper with the triple $(b, \mathrm{rdf:type}, \mathrm{ConfPaper})$.

Adding Ontological Knowledge to RDF Graphs. An essential feature of RDF is the possibility to enhance the descriptions in RDF graphs by declaring *ontological constraints* between the classes and properties they use. This is achieved with *RDF Schema (RDFS)* statements, which are triples using particular built-in properties. Table 1 (bottom) lists the allowed constraints and the triples to state them; *domain* and *range* denote respectively the first and second attribute of every property. For example, the triple (ConfPaper, rdfs:subClassOf, Publication) states that *conference papers are publications*, the triple (hasContactAuthor, rdfs:subPropertyOf, hasAuthor) states that *having a contact author is having an author*, the triple (hasAuthor, rdfs:domain, Publication) states that *only publications may have authors*, and the triple (hasAuthor, rdfs:range, Researcher) states that *only researchers may be authors of something*.

Notations. For conciseness, we use the following shorthands for built-in properties: τ for rdf:type, $\preceq_{sc}$ for rdfs:subClassOf, $\preceq_{sp}$ for rdfs:subPropertyOf, $\hookleftarrow_d$ for rdfs:domain, and $\hookrightarrow_r$ for rdfs:range.

Figure 1 displays the usual representation of the RDF graph $\mathcal{G}$ made of the seven above-mentioned triples, which are called the *explicit triples* of $\mathcal{G}$. A triple (s, p, o) corresponds to the p-labeled directed edge from the s node to the o node. Explicit triples are shown as solid edges, while the *implicit ones*, which are derived using ontological constraints (see below), are shown as dashed edges.

Importantly, it is worth noticing the deductive nature of ontological constraints, which begets implicit triples within an RDF graph. For instance, in Fig. 1, the constraint (hasContactAuthor, $\preceq_{sp}$, hasAuthor) together with the triple $(b, \text{hasContactAuthor}, b_1)$ implies the implicit triple $(b, \text{hasAuthor}, b_1)$, which, further, with the constraint (hasAuthor, $\hookrightarrow_r$, Researcher) yields another implicit triple $(b_1, \tau, \text{Researcher})$.

Deriving the Implicit Triples of an RDF Graph. The RDF standard defines a set of *entailment rules* in order to derive automatically *all* the triples that are implicit to an RDF graph. Table 2 shows the strict subset of these rules that we will use to illustrate important notions as well as our contributions in the next sections; importantly, our contributions hold for the whole set of entailment rules of the RDF standard, and any subset of thereof. The rules in Table 2 concern the derivation of implicit triples using ontological constraints (i.e., *RDFS statements*). They encode the *propagation* of assertions through constraints (`rdfs2`, `rdfs3`, `rdfs7`, `rdfs9`), the *transitivity* of the $\preceq_{sp}$ and $\preceq_{sc}$ constraints (`rdfs5`, `rdfs11`), the *complementation* of domains or ranges through $\preceq_{sc}$ (`ext1`, `ext2`), and the *inheritance* of domains and of ranges through $\preceq_{sp}$ (`ext3`, `ext4`).

Table 1. RDF & RDFS statements.

RDF statement	Triple
Class assertion	$(s, \text{rdf:type}, o)$
Property assertion	(s, p, o) with $p \neq \text{rdf:type}$

RDFS statement	Triple
Subclass	$(s, \text{rdfs:subClassOf}, o)$
Subproperty	$(s, \text{rdfs:subPropertyOf}, o)$
Domain typing	$(s, \text{rdfs:domain}, o)$
Range typing	$(s, \text{rdfs:range}, o)$

Table 2. Sample RDF entailment rules.

Rule [32]	Entailment rule
`rdfs2`	$(p, \hookleftarrow_d, o), (s_1, p, o_1) \rightarrow (s_1, \tau, o)$
`rdfs3`	$(p, \hookrightarrow_r, o), (s_1, p, o_1) \rightarrow (o_1, \tau, o)$
`rdfs5`	$(p_1, \preceq_{sp}, p_2), (p_2, \preceq_{sp}, p_3) \rightarrow (p_1, \preceq_{sp}, p_3)$
`rdfs7`	$(p_1, \preceq_{sp}, p_2), (s, p_1, o) \rightarrow (s, p_2, o)$
`rdfs9`	$(s, \preceq_{sc}, o), (s_1, \tau, s) \rightarrow (s_1, \tau, o)$
`rdfs11`	$(s, \preceq_{sc}, o), (o, \preceq_{sc}, o_1) \rightarrow (s, \preceq_{sc}, o_1)$
`ext1`	$(p, \hookleftarrow_d, o), (o, \preceq_{sc}, o_1) \rightarrow (p, \hookleftarrow_d, o_1)$
`ext2`	$(p, \hookrightarrow_r, o), (o, \preceq_{sc}, o_1) \rightarrow (p, \hookrightarrow_r, o_1)$
`ext3`	$(p, \preceq_{sp}, p_1), (p_1, \hookleftarrow_d, o) \rightarrow (p, \hookleftarrow_d, o)$
`ext4`	$(p, \preceq_{sp}, p_1), (p_1, \hookrightarrow_r, o) \rightarrow (p, \hookrightarrow_r, o)$

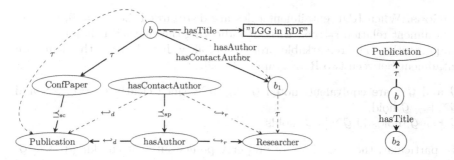

Fig. 1. Sample RDF graph $\mathcal{G}$.   **Fig. 2.** Sample RDF graph $\mathcal{G}'$.

The *saturation (a.k.a. closure)* of an RDF graph $\mathcal{G}$ w.r.t. a set $\mathcal{R}$ of RDF entailment rules, is the RDF graph $\mathcal{G}^\infty$ obtained by adding to $\mathcal{G}$ all the implicit triples that can be derived from $\mathcal{G}$ using $\mathcal{R}$. Roughly speaking, the saturation $\mathcal{G}^\infty$ *materializes* the semantics of $\mathcal{G}$. It corresponds to the fixpoint obtained by repeatedly applying the rules in $\mathcal{R}$ to $\mathcal{G}$ in a forward-chaining fashion. In RDF, the saturation is *always* finite and unique (up to blank node renaming), and does not contain implicit triples [31,32].

The saturation of the RDF graph $\mathcal{G}$ shown in Fig. 1 corresponds to the RDF graph $\mathcal{G}^\infty$ in which all the $\mathcal{G}$ implicit triples have been made explicit. It is worth noting how, starting from $\mathcal{G}$, applying RDF entailment rules *mechanizes* the construction of $\mathcal{G}^\infty$. For instance, recall the reasoning sketched above for deriving the triple $(b_1, \tau, \text{Researcher})$. This is automated by the following sequence of applications of RDF entailment rules: $(\text{hasContactAuthor}, \preceq_{\text{sp}}, \text{hasAuthor})$ and $(b, \text{hasContactAuthor}, b_1)$ trigger `rdfs7` that adds $(b, \text{hasAuthor}, b_1)$ to the RDF graph. In turn, this new triple together with $(\text{hasAuthor}, \hookrightarrow_r, \text{Researcher})$ triggers `rdfs3` that adds $(b_1, \tau, \text{Researcher})$.

Comparing RDF Graphs. The RDF standard defines a generalization/specialization relationship between two RDF graphs, called *entailment between graphs*. Roughly speaking, an RDF graph $\mathcal{G}$ is more specific than another RDF graph $\mathcal{G}'$, or equivalently $\mathcal{G}'$ is more general than $\mathcal{G}$, whenever there is an embedding of $\mathcal{G}'$ into the *saturation* of $\mathcal{G}$, i.e., the complete set of triples that $\mathcal{G}$ models.

More formally, given any subset $\mathcal{R}$ of RDF entailment rules, an RDF graph $\mathcal{G}$ *entails* an RDF graph $\mathcal{G}'$, denoted $\mathcal{G} \models_\mathcal{R} \mathcal{G}'$, iff there exists an homomorphism ϕ from $\text{Bl}(\mathcal{G}')$ to $\text{Val}(\mathcal{G}^\infty)$ such that $[\mathcal{G}']_\phi \subseteq \mathcal{G}^\infty$, where $[\mathcal{G}']_\phi$ is the RDF graph obtained from $\mathcal{G}'$ by replacing every blank node b by its image $\phi(b)$.

Figure 2 shows an RDF graph $\mathcal{G}'$ entailed by the RDF graph $\mathcal{G}$ in Fig. 1 w.r.t. the RDF entailment rules displayed in Table 2. In particular, $\mathcal{G} \models_\mathcal{R} \mathcal{G}'$ holds for the homomorphism ϕ such that: $\phi(b) = b$ and $\phi(b_2) = $ "LGG in RDF". By contrast, when $\mathcal{R}$ is empty, this is not the case (i.e., $\mathcal{G} \not\models_\mathcal{R} \mathcal{G}'$), as the dashed edges in $\mathcal{G}$ are not materialized by saturation, hence the $\mathcal{G}'$ triple $(b, \tau, \text{Publication})$ cannot have an image in $\mathcal{G}$ through some homomorphism.

Notations. When RDF entailment rules are disregarded, i.e., $\mathcal{R} = \emptyset$, we note the entailment relation $\models$ (i.e., without indicating the rule set at hand).

Importantly, some remarkable properties follow directly from the definition of entailment between two RDF graphs [31,32]:

1. $\mathcal{G}$ and $\mathcal{G}^\infty$ are equivalent, noted $\mathcal{G} \equiv_\mathcal{R} \mathcal{G}^\infty$, since clearly $\mathcal{G} \models_\mathcal{R} \mathcal{G}^\infty$ and $\mathcal{G}^\infty \models_\mathcal{R} \mathcal{G}$ hold,
2. $\mathcal{G} \models_\mathcal{R} \mathcal{G}'$ holds iff $\mathcal{G}^\infty \models \mathcal{G}'$ holds.

In particular, the second above property points out that checking $\mathcal{G} \models_\mathcal{R} \mathcal{G}'$ can be done in two steps: a reasoning step that computes the saturation $\mathcal{G}^\infty$ of $\mathcal{G}$, followed by a standard graph homomorphism step that checks if $\mathcal{G}^\infty \models \mathcal{G}'$ holds.

3 Finding Commonalities Between RDF Graphs

In Sect. 3.1, we define the largest set of commonalities between RDF graphs as a particular RDF graph representing their *least general generalization* (lgg for short). Then, we devise a technique for computing such an lgg in Sect. 3.2.

3.1 Defining the lgg of RDF Graphs

A *least general generalization* of n descriptions $d_1, \ldots, d_n$ is a most specific description d generalizing every $d_{1 \leq i \leq n}$ for some generalization/specialization relation between descriptions [25,26]. In RDF, we use RDF graphs as descriptions and entailment between RDF graphs as relation for generalization/specialization:

Definition 1 (lgg of RDF graphs). *Let $\mathcal{G}_1, \ldots, \mathcal{G}_n$ be RDF graphs and $\mathcal{R}$ a set of RDF entailment rules.*

- *A generalization of $\mathcal{G}_1, \ldots, \mathcal{G}_n$ is an RDF graph $\mathcal{G}_g$ such that $\mathcal{G}_i \models_\mathcal{R} \mathcal{G}_g$ holds for $1 \leq i \leq n$.*
- *A least general generalization (lgg) of $\mathcal{G}_1, \ldots, \mathcal{G}_n$ is a generalization $\mathcal{G}_{\text{lgg}}$ of $\mathcal{G}_1, \ldots, \mathcal{G}_n$ such that for any other generalization $\mathcal{G}_g$ of $\mathcal{G}_1, \ldots, \mathcal{G}_n$, $\mathcal{G}_{\text{lgg}} \models_\mathcal{R} \mathcal{G}_g$ holds.*

Importantly, in the RDF setting, the following holds:

Theorem 1. *An lgg of RDF graphs always exists; it is* unique *up to entailment.*

Intuitively, we can always construct a (possibly empty) RDF graph that is the lgg of RDF graphs, in particular *the cover graph of RDF graphs* devised in the next Sect. 3.2. Further, an lgg is unique up to entailment (since $\mathcal{G}_{\text{lgg}} \models_\mathcal{R} \mathcal{G}_g$ holds for any $\mathcal{G}_g$ in Definition 1): if it were that many lggs exist, pairwise incomparable w.r.t. entailment, then their merge[1] would be a single strictly more specific lgg, a contradiction.

---

[1] The merge of RDF graphs is their union *after renaming* their blank nodes, so that these RDF graphs do not join on such values which are *local* to them (Sect. 2).

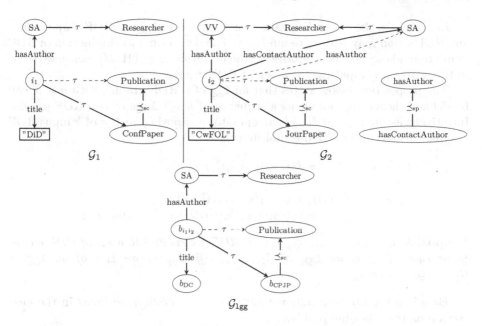

Fig. 3. Sample RDF graphs $\mathcal{G}_1$, $\mathcal{G}_2$ and $\mathcal{G}_{lgg}$, with $\mathcal{G}_{lgg}$ the minimal lgg of $\mathcal{G}_1$ and $\mathcal{G}_2$; their implicit triples (i.e., derived by the rules in Table 2) are shown as dashed edges.

Figure 3 displays two RDF graphs $\mathcal{G}_1$ and $\mathcal{G}_2$, as well as their minimal lgg (with lowest number of triples) when we consider the RDF entailment rules shown in Table 2: $\mathcal{G}_{lgg}$. $\mathcal{G}_1$ describes a conference paper i_1 with title "Disaggregations in Databases" and author Serge Abiteboul, who is a researcher; also conference papers are publications. $\mathcal{G}_2$ describes a journal paper i_2 with title "Computing with First-Order Logic", contact author Serge Abiteboul and author Victor Vianu, who are researchers; moreover, journal papers are publications and having a contact author is having an author. $\mathcal{G}_{lgg}$ states that their common information comprises the existence of a resource ($b_{i_1 i_2}$) having some type ($b_{C(onf)P(aper)J(our)P(aper)}$), which is a particular case of publication, with some title ($b_{D(iD)C(wFOL)}$) and author Serge Abiteboul, who is a researcher.

Though unique up to entailment (i.e., semantically unique), an lgg may have many syntactical forms due to *redundant* triples. Such triples can be either explicit ones that could have been left implicit if the set of RDF entailment rules at hand allows deriving them from the remaining triples (e.g., materializing the only $\mathcal{G}_{lgg}$ implicit triple in Fig. 3 would make it redundant if we consider the entailment rules in Table 2) or triples generalizing others without needing RDF entailment rules, i.e., w.r.t. $\models_\emptyset$ (e.g., adding the triple $(b, \text{hasAuthor}, b')$ to $\mathcal{G}_{lgg}$ in Fig. 3 would be redundant w.r.t. $(b_{i_1 i_2}, \text{hasAuthor}, SA)$). Also, an lgg may have *several minimal* syntactical variants obtained by pruning out redundant triples. For example, think of a minimal lgg comprising the triples $(A, \preceq_{sc}, B)$, $(B, \preceq_{sc}, A)$ and (b, τ, A), i.e., there exists an instance of the class A, which is equivalent to class B. Clearly, an equivalent and minimal variant of this lgg is the RDF graph comprising the triples $(A, \preceq_{sc}, B)$, $(B, \preceq_{sc}, A)$ and (b, τ, B).

Importantly, the above discussion is <u>not specific</u> to lggs of RDF graphs, since any RDF graph may feature redundancy. The detection and elimination of RDF graph redundancy has been studied in the literature, e.g., [21,24], hence we focus in this work on computing *some* lgg of RDF graphs.

The proposition below states that an lgg of n RDF graphs, with $n \geq 3$, can be defined (hence computed) as a sequence of $n - 1$ lggs of *two* RDF graphs. Intuitively, assuming that $\ell_{k \geq 2}$ is an operator computing an lgg of k input RDF graphs, the next proposition establishes that:

$$\ell_3(\mathcal{G}_1, \mathcal{G}_2, \mathcal{G}_3) \equiv_{\mathcal{R}} \ell_2(\ell_2(\mathcal{G}_1, \mathcal{G}_2), \mathcal{G}_3)$$
$$\cdots \qquad \cdots$$
$$\ell_n(\mathcal{G}_1, \ldots, \mathcal{G}_n) \equiv_{\mathcal{R}} \ell_2(\ell_{n-1}(\mathcal{G}_1, \ldots, \mathcal{G}_{n-1}), \mathcal{G}_n)$$
$$\equiv_{\mathcal{R}} \ell_2(\ell_2(\cdots \ell_2(\ell_2(\mathcal{G}_1, \mathcal{G}_2), \mathcal{G}_3) \cdots, \mathcal{G}_{n-1}), \mathcal{G}_n)$$

Proposition 1. *Let $\mathcal{G}_1, \ldots, \mathcal{G}_{n \geq 3}$ be n RDF graphs and $\mathcal{R}$ a set of RDF entailment rules. $\mathcal{G}_{1\text{gg}}$ is an lgg of $\mathcal{G}_1, \ldots, \mathcal{G}_n$ iff $\mathcal{G}_{1\text{gg}}$ is an lgg of an lgg of $\mathcal{G}_1, \ldots, \mathcal{G}_{n-1}$ and $\mathcal{G}_n$.*

Based on the above result, without loss of generality, we focus in the next section on the following problem:

Problem 1. Given *two* RDF graphs $\mathcal{G}_1, \mathcal{G}_2$ and a set $\mathcal{R}$ of RDF entailment rules, we want to compute *some* lgg of $\mathcal{G}_1$ and $\mathcal{G}_2$.

3.2   Computing an lgg of RDF Graphs

We first devise the *cover graph* of two RDF graphs $\mathcal{G}_1$ and $\mathcal{G}_2$ (to be defined shortly, Definition 2 below), which is central to our technique for computing an lgg of $\mathcal{G}_1$ and $\mathcal{G}_2$. We indeed show (Theorem 2) that this particular RDF graph corresponds to an lgg of $\mathcal{G}_1$ and $\mathcal{G}_2$ when considering their explicit triples *only*, i.e., ignoring RDF entailment rules. Then, we show the main result of this section (Theorem 3): an lgg of $\mathcal{G}_1$ and $\mathcal{G}_2$, for *any* set $\mathcal{R}$ of RDF entailment rules, is the cover graph of their saturations w.r.t. $\mathcal{R}$. We also provide the worst-case size of cover graph-based lggs, as well as the worst-case time to compute them.

Definition 2 (Cover graph). *The* cover graph *$\mathcal{G}$ of two RDF graphs $\mathcal{G}_1$ and $\mathcal{G}_2$ is the RDF graph, which may be empty, such that for every property p in both $\mathcal{G}_1$ and $\mathcal{G}_2$:*

$$(t_1, p, t_2) \in \mathcal{G}_1 \text{ and } (t_3, p, t_4) \in \mathcal{G}_2 \text{ iff } (t_5, p, t_6) \in \mathcal{G}$$

with $t_5 = t_1$ if $t_1 = t_3$ and $t_1 \in \mathcal{U} \cup \mathcal{L}$, else t_5 is the blank node $b_{t_1 t_3}$, and, similarly $t_6 = t_2$ if $t_2 = t_4$ and $t_2 \in \mathcal{U} \cup \mathcal{L}$, else t_6 is the blank node $b_{t_2 t_4}$.

The cover graph is a *generalization* of $\mathcal{G}_1$ and $\mathcal{G}_2$ (first item in Definition 1) as each of its triple (t_5, p, t_6) is a *least general anti-unifier* of a triple (t_1, p, t_2) from $\mathcal{G}_1$ and a triple (t_3, p, t_4) from $\mathcal{G}_2$. The notion of *least general anti-unifier* [25,26, 29] is dual to the well-known notion of *most general unifier* [28,29]. Observe that $\mathcal{G}$'s triples result from anti-unifications of $\mathcal{G}_1$ and $\mathcal{G}_2$ triples with *same* property

URI. Indeed, anti-unifying triples of the form (s_1, p, o_1) and (s_2, p', o_2), with $p \neq p'$, would lead to a non-well-formed triples of the form $(s, b_{pp'}, o)$ (recall that property values *must* be URIs in RDF graphs), where $b_{pp'}$ is the blank node required to generalize the distinct values p and p'.

Further, the cover graph is an lgg for the explicit triples in $\mathcal{G}_1$ and those in $\mathcal{G}_2$ (second item in Definition 1) since, intuitively, we capture their *common structures* by consistently naming, across all the anti-unifications begetting $\mathcal{G}$, the blank nodes used to generalize pairs of distinct subject values or of object values: each time the distinct values t from $\mathcal{G}_1$ and t' from $\mathcal{G}_2$ are generalized by a blank node while anti-unifying two triples, it is *always* by the <u>same</u> blank node $b_{tt'}$ in $\mathcal{G}$. This way, we establish *joins* between $\mathcal{G}$ triples, which reflect the common join structure on t within $\mathcal{G}_1$ and on t' within $\mathcal{G}_2$. For instance in Fig. 3, the *explicit* triples $(i_1, \tau, \text{ConfPaper}), (\text{ConfPaper}, \preceq_{sc}, \text{Publication}), (i_1, \text{title}, \text{"DiD"})$ in $\mathcal{G}_1$, and $(i_2, \tau, \text{JourPaper}), (\text{JourPaper}, \preceq_{sc}, \text{Publication}), (i_2, \text{title}, \text{"CwFOL"})$ in $\mathcal{G}_2$, lead to the triples $(b_{i_1 i_2}, \tau, b_{\text{CPJP}}), (b_{\text{CPJP}}, \preceq_{sc}, \text{Publication}), (b_{i_1 i_2}, \text{title}, b_{\text{DC}})$ in the cover graph of $\mathcal{G}_1$ and $\mathcal{G}_2$ shown in Fig. 4 (top). The first above-mentioned $\mathcal{G}$ triple results from anti-unifying i_1 and i_2 into $b_{i_1 i_2}$, and, ConfPaper and JourPaper into b_{CPJP}. The second results from anti-unifying *again* ConfPaper and JourPaper into b_{CPJP}, and, Publication and Publication into Publication (as a constant is its own least general generalization). Finally, the third results from anti-unifying *again* i_1 and i_2 into $b_{i_1 i_2}$, and, "DiD" and "CwFOL" into b_{DC}. By reusing consistently the same blank node name $b_{i_1 i_2}$ for each anti-unification of the constants i_1 and i_2 (resp. b_{CPJP} for ConfPaper and JourPaper)), the cover graph triples join on $b_{i_1 i_2}$ (resp. b_{CPJP}) in order to reflect that, in $\mathcal{G}_1$ and in $\mathcal{G}_2$, there exists a particular case of publication (i_1 in $\mathcal{G}_1$ and i_2 in $\mathcal{G}_2$) with some title ("DiD" in $\mathcal{G}_1$ and "CwFOL" in $\mathcal{G}_2$).

The next theorem formalizes the above discussion by stating that the cover graph of two RDF graphs is an lgg of them, *just in case of an empty set of RDF entailment rules.*

Theorem 2. *The* cover graph $\mathcal{G}$ *of the RDF graphs* $\mathcal{G}_1$ *and* $\mathcal{G}_2$ *exists and is an* lgg *of them for the empty set* $\mathcal{R}$ *of RDF entailment rules (i.e.,* $\mathcal{R} = \emptyset$*).*

We provide below worst-case bounds for the time to compute a cover graph and for its size; these bounds are met when *all the triples of the two input graphs use the same property URI* (i.e., every pair of $\mathcal{G}_1$ and $\mathcal{G}_2$ triples begets a $\mathcal{G}$ triple).

Proposition 2. *The cover graph of two RDF graphs* $\mathcal{G}_1$ *and* $\mathcal{G}_2$ *can be computed in* $O(|\mathcal{G}_1| \times |\mathcal{G}_2|)$; *its size is bounded by* $|\mathcal{G}_1| \times |\mathcal{G}_2|$.

The main theorem below generalizes Theorem 2 in order to take into account <u>any set</u> of entailment rules from the RDF standard. It states that it is sufficient to compute the cover graph of the saturations of the input RDF graphs, instead of the input RDF graphs themselves.

Theorem 3. *Let* $\mathcal{G}_1$ *and* $\mathcal{G}_2$ *be two RDF graphs, and* $\mathcal{R}$ *a set of RDF entailment rules. The* cover graph $\mathcal{G}$ *of* $\mathcal{G}_1^\infty$ *and* $\mathcal{G}_2^\infty$ *exists and is an* lgg *of* $\mathcal{G}_1$ *and* $\mathcal{G}_2$.

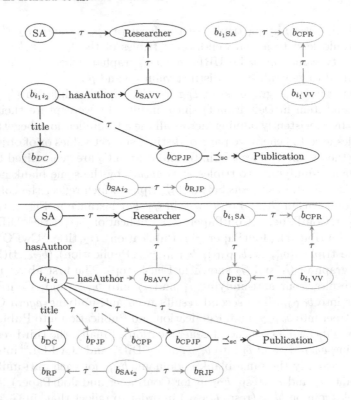

Fig. 4. Cover graphs of $\mathcal{G}_1$ and $\mathcal{G}_2$ in Fig. 3 (top) and of their saturations w.r.t. the entailment rules in Table 2 (bottom). Triples shown in gray are redundant w.r.t. those shown in black: they are part of the graph, while implicit to the triples shown in black.

As an immediate consequence of the above results, we get the following worst-case bounds for the time to compute a cover graph-based lgg of two RDF graphs $\mathcal{G}_1$ and $\mathcal{G}_2$, and for its size. Here, we assume given the saturation $\mathcal{G}_1^\infty$ and $\mathcal{G}_2^\infty$, as the times to compute them and their sizes depend on the RDF entailment rules at hand.

Corollary 1. *An* lgg *of two RDF graphs* $\mathcal{G}_1$ *and* $\mathcal{G}_2$ *can be computed in* $O(|\mathcal{G}_1^\infty| \times |\mathcal{G}_2^\infty|)$ *and its size is bounded by* $|\mathcal{G}_1^\infty| \times |\mathcal{G}_2^\infty|$.

Remark that computing naively the cover graph-based lgg of n RDF graphs of size M based on Proposition 1 may lead to an lgg of size M^n, in the unlikely worst-case where all the triples of all the RDF graphs use the same property URI. However, removing the redundant triples from the intermediate and final cover graph-based lggs limits their size to at most M.

Figure 4 (bottom) displays an lgg of the RDF graphs $\mathcal{G}_1$ and $\mathcal{G}_2$ in Fig. 3 w.r.t. the entailment rules shown in Table 2. In contrast to Fig. 4 (top), which shows an lgg of the same RDF graphs when RDF entailment rules are ignored, we further learn that Serge Abiteboul is an author of some particular publication (i_1 in $\mathcal{G}_1$ and i_2 in $\mathcal{G}_2$). Moreover, removing the redundant triples (the gray ones) yields precisely the lgg $\mathcal{G}_{\text{lgg}}$ of $\mathcal{G}_1$ and $\mathcal{G}_2$ shown in Fig. 3.

4   Algorithms

We provide algorithms to compute lggs of RDF graphs based on the results obtained in the preceding Section. In Sect. 4.1, we present an algorithm for computing the least general anti-unifiers of triples. Then, in Sect. 4.2, we give three algorithms to compute a cover graph-based lgg of RDF graphs, which allow handling RDF graphs of increasing size, i.e., when the input and output RDF graphs fit in memory, in data management systems or in MapReduce clusters. Also, to choose between these algorithms, we show how the exact size of a cover graph-based lgg they produce can be calculated, *without computing this* lgg.

4.1   Least General Anti-unifier of Triples

Algorithm 1, called lgau, computes a least general anti-unifier (t_1^T, t_2^T, t_3^T) of two triples (t_1, t_2, t_3) and (t'_1, t'_2, t'_3). This is achieved by setting the ith value t_i^T of the output triple to the least general generalization of the values found at the ith positions of the two input triples: t_i and t'_i. Recall that a pair of a same constant is generalized by this constant itself, otherwise the generalization leads to a blank node (Sect. 3.2). Crucially, such a blank node uses the consistent naming scheme devised in Sect. 3.2, which allows us preserving the common structure of input RDF graphs across the anti-unifications of their triples.

Algorithm 1. Least general anti-unification: lgau

In: triples $T_1 = (t_1, t_2, t_3)$ and $T_2 = (t'_1, t'_2, t'_3)$
Out: least general anti-unification T of T_1 and T_2

1: **for** $i = 1$ **to** 3 **do**   ▷ *for each pair of T_1 and T_2 i^{th} values*
2:     **if** $t_i = t'_i$ and $t_i \in \mathcal{U} \cup \mathcal{L}$ **then**
3:         $t_i^T \leftarrow t_i$ ▷ *generalization of a same constant by itself*
4:     **else**
5:         $t_i^T \leftarrow b_{t_i t'_i}$ ▷ *otherwise generalization by a blank node*
6: **return** (t_1^T, t_2^T, t_3^T)

4.2   lgg of RDF Graphs

Following Definition 2, Algorithm 2, called lgg4g, computes the cover graph $\mathcal{G}$ of two input RDF graphs $\mathcal{G}_1$ and $\mathcal{G}_2$: $\mathcal{G}$ comprises the least general anti-unifier of every pair of $\mathcal{G}_1$ and $\mathcal{G}_2$ triples with *same* property. Therefore, given two RDF graphs $\mathcal{G}_1$ and $\mathcal{G}_2$, a call lgg4g$(\mathcal{G}_1, \mathcal{G}_2)$ produces the cover graph-based lgg of $\mathcal{G}_1$ and $\mathcal{G}_2$ ignoring RDF entailment (Theorem 2), while a call lgg4g$(\mathcal{G}_1^\infty, \mathcal{G}_2^\infty)$ produces the cover graph-based lgg of $\mathcal{G}_1$ and $\mathcal{G}_2$ taking into account the set of RDF entailment rules at hand (Theorem 3). In the latter case, the input RDF graphs can be saturated using standard algorithms implemented in RDF reasoners or data management systems, like Jena [3] and Virtuoso [8].

Importantly, lgg4g assumes that the input RDF graphs, as well as their output cover graph, fit in memory. Checking whether this is the case for the input RDF graphs under consideration can be done as follows.

Algorithm 2. Cover graph of two RDF graphs: lgg4g

In: RDF graphs $\mathcal{G}_1$ and $\mathcal{G}_2$
Out: $\mathcal{G}$ is the cover graph of $\mathcal{G}_1$ and $\mathcal{G}_2$
1: $\mathcal{G} \leftarrow \emptyset$
2: **for all** $T_1 = (\mathsf{s}_1, \mathsf{p}_1, \mathsf{o}_1) \in \mathcal{G}_1$ **do**
3:     **for all** $T_2 = (\mathsf{s}_2, \mathsf{p}_2, \mathsf{o}_2) \in \mathcal{G}_2$ with $\mathsf{p}_1 = \mathsf{p}_2$ **do**
4:         $\mathcal{G} \leftarrow \mathcal{G} \cup \{\mathtt{lgau}(T_1, T_2)\}$ ▷ *add to $\mathcal{G}$ the least general anti-unifier of T_1 and T_2*
5: **return** $\mathcal{G}$

The size of the input RDF graphs $\mathcal{G}_1$ and $\mathcal{G}_2$ can be computed with the following SPARQL queries counting how many triples each of them holds: SELECT count(*) as ?size FROM $\mathcal{G}_i$ with $i \in [1, 2]$. Recall that the worst-case size of the output cover graph is $|\mathcal{G}| = |\mathcal{G}_1| \times |\mathcal{G}_2|$ in the unlikely case where *all* the $\mathcal{G}_1$ and $\mathcal{G}_2$ triples use the same property (Proposition 2 and Corollary 1).

The precise size of the output cover graph $\mathcal{G}$ can be computed, *without* computing $\mathcal{G}$, with SPARQL queries. First, we calculate for each input RDF graph $\mathcal{G}_i$, with $i \in [1, 2]$, how many triples it holds per distinct property p:

$$S_{\mathcal{G}_i} = \{(p, n_i) \mid |\{(\mathsf{s}, p, \mathsf{o}) \in \mathcal{G}_i\}| = n_i\}$$

This can be computed with the SPARQL query: SELECT ?p count(*) as ?n_i FROM $\mathcal{G}_i$ WHERE $\{(?s, ?p, ?o)\}$ GROUP BY ?p. Then, since every $\mathcal{G}_1$ triple with property p anti-unifies with every $\mathcal{G}_2$ triple with same property p in order to beget $\mathcal{G}$, the size of $\mathcal{G}$ is: $|\mathcal{G}| = \sum_{(p,n_1) \in S_{\mathcal{G}_1}, (p,n_2) \in S_{\mathcal{G}_2}} n_1 \times n_2$.

This can be computed with the SPARQL query: SELECT SUM(?n_1*?n_2) as ?size WHERE $\{\{S_{\mathcal{G}_1}\}\{S_{\mathcal{G}_2}\}\}$ with $S_{\mathcal{G}_1}$ and $S_{\mathcal{G}_2}$ denoting the above SPARQL queries computing these two sets, which join on their common answer variable ?p.

When the input RDF graphs or their output cover graph cannot fit in memory, we propose variants of lgg4g that either assume that RDF graphs are stored in data management systems (DMSs, in short) or in a MapReduce cluster.

Handling Large RDF Graphs Using DMSs. Algorithm 3, called lgg4g-dms, is an adaptation of lgg4g, which assumes that the input RDF graphs (already saturated if needed) and their cover graph are all stored in one or several DMSs. It further assumes that the system(s) storing the input RDF graphs $\mathcal{G}_1$ and $\mathcal{G}_2$ feature(s) the well-known database mechanism of *cursor* [17, 27]. This is for instance the case for RDF graphs stored in relational servers like DB2 [1], MySQL [4], Oracle [5] and PostgreSQL [6], or in RDF servers like Jena-TDB [3] and Virtuoso [8]. Roughly speaking, a cursor is a pointer or iterator on tuples held in a DMS (e.g., stored as relation or computed as the results to a query) that can be used to access these tuples. In particular, a cursor can be used by an application to iteratively traverse all the tuples by fetching n of them at a time.

lgg4g-dms uses cursors to proceed similarly to lgg4g (remark that lines 5–7 in Algorithm 3 are almost the same as lines 2–4 in Algorithm 2) on pairs of n-triples subsets of $\mathcal{G}_1$ and of $\mathcal{G}_2$, instead of on the whole RDF graphs themselves. It follows that the worst-case number of triples kept in memory by lgg4g-dms is $M = (2 \times n) + 1$ at line 7 (i.e., n for B_1, n for B_2, and the anti-unifier triple output by lgau), with:

Algorithm 3. Cover graph of two RDF graphs: lgg4g-dms

In: cursor c_1 on RDF graph $\mathcal{G}_1$, cursor c_2 on RDF graph $\mathcal{G}_2$, data access path d to an empty RDF graph $\mathcal{G}$, integer n

Out: $\mathcal{G}$ is the cover graph of $\mathcal{G}_1$ and $\mathcal{G}_2$

1: c_1.init()   ▷ c_1 at beginning of $\mathcal{G}_1$ triples set
2: **while** $B_1 = c_1$.next(n) **do**   ▷ fetch the next block B_1 of n $\mathcal{G}_1$ triples
3:      c_2.init()   ▷ c_2 at beginning of $\mathcal{G}_2$ triples set
4:      **while** $B_2 = c_2$.next(n) **do**   ▷ fetch the next block B_2 of n $\mathcal{G}_2$ triples
5:          **for all** $T_1 = (s_1, p_1, o_1) \in B_1$ **do**
6:              **for all** $T_2 = (s_2, p_2, o_2) \in B_2$ with $p_1 = p_2$ **do**
7:                  d.insert(lgau(T_1, T_2))

$$3 \leq M \leq |\mathcal{G}_1| + |\mathcal{G}_2| + 1$$

The above lower bound is met for n set to 1, while the upper one is met for n set to $\max(|\mathcal{G}_1|, |\mathcal{G}_2|)$. Importantly, lgg4g-dms allows *choosing* the value of n in order to reflect the memory devoted to handling triples. For instance, if one wants to use 4 GB of RAM for triples, assuming that any triple fits in less one 1 KB (this value is much less when using dictionary encoding [22], i.e., when triples values are mapped to integers), the value of n can be set to 2M.

This clearly contrasts with the worst-case number of triples kept in memory by lgg4g: $M = |\mathcal{G}_1| + |\mathcal{G}_2| + |\mathcal{G}|$ at line 5, with:

$$|\mathcal{G}_1| + |\mathcal{G}_2| \leq M \leq |\mathcal{G}_1| + |\mathcal{G}_2| + (|\mathcal{G}_1| \times |\mathcal{G}_2|)$$

The above lower bound is met when $\mathcal{G}_1$ and $\mathcal{G}_2$ have no property in common in their triples (i.e., $|\mathcal{G}| = 0$), while the upper one is met in the unlikely case where $\mathcal{G}_1$ and $\mathcal{G}_2$ use a same property in all their triples (i.e., $|\mathcal{G}| = |\mathcal{G}_1| \times |\mathcal{G}_2|$).

Handling Huge RDF Graphs Using MapReduce. Algorithm 4, called lgg4g-mr, is a MapReduce (MR) variant of lgg4g. MR is a popular massively parallel programming framework [15], implemented by large-scale data processing systems, like Hadoop [2] and Spark [7], which orchestrate clusters of compute nodes.

Algorithm 4. Cover graph of two RDF graphs: lgg4g-mr

In: file G_1 for RDF graph $\mathcal{G}_1$, file G_2 for RDF graph $\mathcal{G}_2$

Out: $\mathcal{G}$ is the cover graph of $\mathcal{G}_1$ and $\mathcal{G}_2$, stored in G-* files

Map(key: file G_i, value: triple $T_i = (s_i, p_i, o_i)$)

1: emit($\langle p_i, (G_i, T_i) \rangle$)

Reduce(key: p, values: set $\mathcal{V}$ of values emitted for key p)

1: $f \leftarrow$ open(G-p)
2: **for all** $(G_1, T_1 = (s_1, p, o_1)) \in \mathcal{V}$ **do**
3:      **for all** $(G_2, T_2 = (s_2, p, o_2)) \in \mathcal{V}$ **do**
4:          f.write(lgau(T_1, T_2))
5: close(f)

A MR program is organized in successive *jobs*, each of which comprises a *Map task* followed by a *Reduce task*. The Map task consists in reading some input data from the distributed file system[2] of the cluster, so as to partition the data into $\langle k, v \rangle$ key-value pairs. Importantly, an MR engine transparently processes the Map task by running *Mapper processes* in parallel on cluster nodes, each process taking care of partitioning a portion of the input data by applying a Map(key: file, value: data unit) function on every data unit of a given input file. Key-value pairs thus produced are shuffled across the network, so that *all* pairs with *same* key $\langle k, v_1 \rangle \cdots \langle k, v_n \rangle$ are shipped to a same compute node. The Reduce task consists in running *Reducer processes* in parallel, for every distinct key k received by every compute node. Each process takes care of the set $\mathcal{V}$ of values $\{v_1, \ldots, v_n\}$ emitted with key k, by applying a Reduce(key: k, values: $\mathcal{V}$) function, and writing its results in a file. The result of an MR job comprises the data, stored in a distributed fashion, in all the files output by Reducers.

In lgg4g-mr, the Map function applies to every (s_i, p_i, o_i) triple of the input RDF graph $\mathcal{G}_i$ stored in file G_i, and produces the corresponding key-value pair $\langle p_i, (G_i, (s_i, p_i, o_i)) \rangle$, for $i \in [1, 2]$. Hence, all the $\mathcal{G}_1$ and $\mathcal{G}_2$ triples with a same key/property p are shipped to the same cluster node. Then, similarly to lgg4g at lines 2–4, the Reduce functions process, on each node, the set $\mathcal{V}$ of values emitted for every received key p. The least general anti-unifier triples obtained at line 4 are stored in the output file G-p. At the end of the MR job, the lgg $\mathcal{G}$ of $\mathcal{G}_1$ and $\mathcal{G}_2$ is stored in the G-$*$ files of the distributed file system, where $*$ denotes any key/property p.

A Map function holds at most a single $\mathcal{G}_1$ or $\mathcal{G}_2$ triple in memory. In contrast, the worst-case number of triples handled by a Reduce function for a given key p is: $M = |\mathcal{G}_1| + |\mathcal{G}_2| + 1$ at line 4. This upper bound is met in the unlikely case where $\mathcal{G}_1$ and $\mathcal{G}_2$ use the same property p in all their triples. Similarly to lgg4g-dms, this upper bound can set to $M = (2 \times n) + 1$, with $3 \leq M \leq |\mathcal{G}_1| + |\mathcal{G}_2| + 1$, by first splitting the input RDF graphs in k_i files of n $\mathcal{G}_i$ triples (files $G_i^1, \ldots, G_i^{k_i}$), and then by processing every pair of such files with an MR job (i.e., with $k_1 \times k_2$ jobs), instead of a single MR job for the entire two input RDF graphs.

Finally, to take into account RDF entailment, input RDF graphs can be saturated before being stored in the MR cluster using standard (centralized) techniques, or within the MR cluster using MR-based saturation techniques [30]. Also, it is worth noting that RDF graphs, hence lggs of them, stored in an MR cluster can be queried with MR-based SPARQL engines [18,23].

5   Related Work and Conclusion

We revisited the Machine Learning problem of computing a *least general generalization (lgg) of some descriptions* in the setting of RDF; it was introduced to generalize First Order Logic clauses w.r.t. θ-subsumption [25,26], a non-standard specialization/generalization relation widely used in Machine Learning.

---

[2] We assume w.l.o.g. that input and output data of an MR job is stored on disk, like in Hadoop, while it can also reside in in-memory shared data structures, like in Spark.

This problem has also been investigated in Knowledge Representation, for formalisms whose expressivity overlaps with our RDF setting, notably Description Logics (DLs) [9,19,33] and Conceptual Graphs (CGs) [11]. Finally, recently, this problem has started receiving attention in the Semantic Web field [13,14,20].

In DLs, computing an lgg of concepts (formulae) has been studied for $\mathcal{EL}$ and extensions thereof [9,19,33]. The $\mathcal{EL}$ setting translates into *particular tree-shaped* RDF graphs, which may feature RDFS subclass and domain constraints, and for which RDF entailment is limited to the use of these two constraints only[3]. In these equivalent RDF and $\mathcal{EL}$ fragments, the $\mathcal{EL}$ technique that computes an lgg of $\mathcal{EL}$ concepts, which is an $\mathcal{EL}$ concept, provides *only a (non least general) generalization* of their corresponding tree-shaped RDF graphs w.r.t. the problem we study: the (minimal) cover graph-based lgg of tree-shaped RDF graphs is clearly a forest-shaped RDF graph in general. In CGs, the so-called *simple CGs with unary and binary relations* correspond to *particular* RDF graphs (e.g., a property URI in a triple cannot be the subject or object of another triple, a class - URI or blank node - in a τ triple cannot be the subject of another τ triple nor the subject or object of another non-τ triple, etc.), which may feature the four RDFS constraints, and for which RDF entailment is limited to the use of these RDFS constraints only [10]. In these equivalent RDF and CG fragments, we may interchangeably compute lggs with the CG technique in [11] or ours.

In RDF, computing an lgg has been studied for *particular* RDF graphs, called *r-graphs, ignoring RDF entailment* [13,14]. An r-graph is an *extracted subgraph* of an RDF graph $\mathcal{G}$, *rooted* in the $\mathcal{G}$ value r and comprising the $\mathcal{G}$ triples *reachable from r through directed* paths of length at most n. Such a rooted and directed r-graph can be defined recursively as $\mathcal{S}(\mathcal{G}, r, n)$, with:

$$\mathcal{S}(\mathcal{G}, v, 0) = \emptyset \mid \mathcal{S}(\mathcal{G}, v, n) = \bigcup_{(v,p,v')\in\mathcal{G}} \{(v, p, v')\} \cup \mathcal{S}(\mathcal{G}, v', n-1) \cup \mathcal{S}(\mathcal{G}, p, n-1)$$

Intuitively, this purely structural definition of r-graph attempts carrying $\mathcal{G}$'s knowledge about r. lggs of r-graphs allow finding the commonalities between *single root entities*, while with general RDF graphs we further allow finding the commonalities between *sets of multiple interrelated entities*. The technique for computing an lgg of two r-graphs exploits their rooted and directed structure: it starts from their respective root and traverses them simultaneously considering triples reachable through directed paths of increasing size, while incrementally constructing an r-graph lgg. In contrast, the general RDF graphs we consider are unstructured; our technique blindly traverses the input RDF graphs to anti-unify their triples with same property, and captures their common structure across these anti-unifications thanks to the consistent naming scheme we devised

---

[3] An $\mathcal{EL}$ concept C recursively translates into the RDF graph rooted in the blank node b_r returned by the call $\mathcal{G}(C, b_r)$, with: $\mathcal{G}(\top, b) = \emptyset$ for the universal $\mathcal{EL}$ concept $\top$, $\mathcal{G}(A, b) = \{(b, \tau, A)\}$ for an atomic $\mathcal{EL}$ concept A, $\mathcal{G}(\exists r.C, b) = \{(b, r, b')\} \cup \mathcal{G}(C, b')$, with b' a fresh blank node, for an $\mathcal{EL}$ existential restriction $\exists r.C$, and $\mathcal{G}(C_1 \sqcap C_2, b) = \mathcal{G}(C_1, b) \cup \mathcal{G}(C_2, b)$ for an $\mathcal{EL}$ conjunction $C_1 \sqcap C_2$; the $\mathcal{EL}$ constraints $A_1 \sqsubseteq A_2$ and $\exists r.\top \sqsubseteq A$ correspond to $(A_1, \preceq_{sc}, A_2)$ and $(r, \hookleftarrow_d, A)$ resp.

for the blank nodes they generate. The r-graph technique that computes an lgg of r-graphs, which is an r-graph, gives *only a (non least general) generalization* of them w.r.t. the problem we study: the (minimal) cover graph-based lgg of r-graphs is clearly a general RDF graph.

In SPARQL, computing an lgg has been considered for *unary tree-shaped conjunctive queries (UTCQ)* [20]; a UTCQ lgg is computed by a simultaneous root-to-leaves traversal of the input queries. UTCQs are tree-shaped RDF graphs, when variables are viewed as blank nodes, for which RDF entailment is ignored [13]. The UTCQ technique that computes an lgg of UTCQs, which is a UTCQ, yields *only a (non least general) generalization* of their corresponding tree-shaped RDF graphs w.r.t. the problem we study: the (minimal) cover graph-based lgg of tree-shaped RDF graphs is clearly a forest-shaped RDF graph.

Our work significantly extends the state of the art on computing lggs of RDF graphs by considering the *entire* RDF standard of W3C. Crucially, we neither restrict RDF graphs nor RDF entailment in any way, while related works consider *particular* RDF graphs and, further, *ignore* RDF entailment, hence do not accurately capture the semantics of RDF graphs. Also, we provide a set of algorithms that allows computing lggs of small-to-huge general RDF graphs.

As future work, we want to study heuristics in order to efficiently prune out as much as possible redundant triples, while computing lggs. Indeed, as for instance Fig. 4 shows, our cover graph technique does produce redundant triples. This would allow having more compact lggs, and reducing the a posteriori elimination effort of redundant triples using standard technique from the literature.

References

1. DB2. www.ibm.com/analytics/us/en/technology/db2
2. Hadoop. hadoop.apache.org
3. Jena. jena.apache.org
4. MySQL. www.mysql.com
5. Oracle. www.oracle.com/database
6. PostgreSQL. www.postgresql.org
7. Spark. spark.apache.org
8. Virtuoso. virtuoso.openlinksw.com
9. Baader, F., Sertkaya, B., Turhan, A.Y.: Computing the least common subsumer w.r.t. a background terminology. J. Appl. Logic 5(3), 392–420 (2007)
10. Baget, J., Croitoru, M., Gutierrez, A., Leclère, M., Mugnier, M.: Translations between RDF(S) and conceptual graphs. In: ICCS (2010)
11. Chein, M., Mugnier, M.: Graph-Based Knowledge Representation - Computational Foundations of Conceptual Graphs. Springer, London (2009)
12. Cohen, W.W., Borgida, A., Hirsh, H.: Computing least common subsumers in description logics. In: AAAI (1992)
13. Colucci, S., Donini, F., Giannini, S., Sciascio, E.D.: Defining and computing least common subsumers in RDF. J. Web Semant. 39, 62–80 (2016)
14. Colucci, S., Donini, F.M., Sciascio, E.D.: Common subsumers in RDF. In: AI*IA (2013)

15. Dean, J., Ghemawat, S.: MapReduce: simplified data processing on large clusters. In: OSDI (2004)
16. El Hassad, S., Goasdoué, F., Jaudoin, H.: Learning commonalities in RDF and SPARQL (research report) (2016). https://hal.inria.fr/hal-01386237
17. Garcia-Molina, H., Ullman, J.D., Widom, J.: Database Systems - The Complete Book. Pearson Education, Harlow (2009)
18. Goasdoué, F., Kaoudi, Z., Manolescu, I., Quiané-Ruiz, J., Zampetakis, S.: Cliquesquare: flat plans for massively parallel RDF queries. In: ICDE (2015)
19. Küsters, R.: Non-standard Inferences in Description Logics. LNCS, vol. 2100. Springer, Heidelberg (2001)
20. Lehmann, J., Bühmann, L.: AutoSPARQL: let users query your knowledge base. In: Antoniou, G., Grobelnik, M., Simperl, E., Parsia, B., Plexousakis, D., Leenheer, P., Pan, J. (eds.) ESWC 2011. LNCS, vol. 6643, pp. 63–79. Springer, Heidelberg (2011). doi:10.1007/978-3-642-21034-1_5
21. Meier, M.: Towards rule-based minimization of RDF graphs under constraints. In: Calvanese, D., Lausen, G. (eds.) RR 2008. LNCS, vol. 5341, pp. 89–103. Springer, Heidelberg (2008). doi:10.1007/978-3-540-88737-9_8
22. Neumann, T., Weikum, G.: The RDF-3X engine for scalable management of RDF data. VLDB J. **19**(1), 91–113 (2010)
23. Papailiou, N., Tsoumakos, D., Konstantinou, I., Karras, P., Koziris, N.: H2rdf+: an efficient data management system for big RDF graphs. In: SIGMOD (2014)
24. Pichler, R., Polleres, A., Skritek, S., Woltran, S.: Complexity of redundancy detection on RDF graphs in the presence of rules, constraints, and queries. Semant. Web **4**(4), 351–393 (2013)
25. Plotkin, G.D.: A note on inductive generalization. Mach. Intell. **5**, 153–163 (1970)
26. Plotkin, G.D.: A further note on inductive generalization. Mach. Intell. **6**, 101–124 (1971)
27. Ramakrishnan, R., Gehrke, J.: Database Management Systems. McGraw-Hill, New York (2003)
28. Robinson, J.A.: A machine-oriented logic based on the resolution principle. J. ACM **12**(1), 23–41 (1965)
29. Robinson, J.A., Voronkov, A. (eds.): Handbook of Automated Reasoning. Elsevier and MIT Press, Weidenbach (2001)
30. Urbani, J., Kotoulas, S., Maassen, J., van Harmelen, F., Bal, H.E.: WebPIE: a web-scale parallel inference engine using MapReduce. J. Web Semant. **10**, 59–75 (2012)
31. Resource description framework 1.1. https://www.w3.org/TR/rdf11-concepts
32. RDF 1.1 semantics. https://www.w3.org/TR/rdf11-mt/
33. Zarrieß, B., Turhan, A.: Most specific generalizations w.r.t. general EL-TBoxes. In: IJCAI (2013)

Lean Kernels in Description Logics

Rafael Peñaloza[1]($\boxtimes$), Carlos Mencía[2], Alexey Ignatiev[3],
and Joao Marques-Silva[3]

[1] Free University of Bozen-Bolzano, Bolzano, Italy
penaloza@inf.unibz.it
[2] University of Oviedo, Gijón, Spain
cmencia@gmail.com
[3] University of Lisbon, Lisbon, Portugal
{aignatiev,jpms}@ciencias.ulisboa.pt

Abstract. Lean kernels (LKs) are an effective optimization for deriving
the causes of unsatisfiability of a propositional formula. Interestingly,
no analogous notion exists for explaining consequences of description
logic (DL) ontologies. We introduce LKs for DLs using a general notion
of consequence-based methods, and provide an algorithm for computing them which incurs in only a linear time overhead. As an example,
we instantiate our framework to the DL $\mathcal{ALC}$. We prove formally and
empirically that LKs provide a tighter approximation of the set of relevant axioms for a consequence than syntactic locality-based modules.

1 Introduction

Description logics (DLs) [6] are logic-based knowledge representation formalisms
characterized by having an intuitive syntax and formal, well-understood semantics. These logics have been successfully used for representing the terminological
knowledge of several domains, and are the logical formalism behind OWL 2, the
standard ontology language for the Semantic Web. Along with the availability of
better editors, this had led to the creation of larger ontologies; indeed, observing
ontologies with tens of thousands of axioms is increasingly common.

Depending on the reasoning task of interest, not all axioms in an ontology
may be relevant at any given time. To improve the efficiency, or even guarantee
the feasibility of a task over large ontologies it is thus fundamental to focus only
on a subset of pertinent axioms, usually called a *module* [12,14]. For example, in
axiom pinpointing, where the task is to identify all the minimal subsets of axioms
that entail a given consequence (called *MinAs* or *justifications*), only a small fraction of the ontology is relevant [34,37]. Similarly, error-tolerant and probabilistic
reasoning can usually be restricted to a subset of relevant axioms [26,31]. Since
the performance of reasoning methods depends on the size of the input ontology,
a useful optimization consists in computing first a small module containing all
the MinAs. Ideally, this module would contain exactly the union of all MinAs;
however, computing this set is computationally expensive [30]. Thus, different
approximations that are easy to compute have been proposed.

© Springer International Publishing AG 2017
E. Blomqvist et al. (Eds.): ESWC 2017, Part I, LNCS 10249, pp. 518–533, 2017.
DOI: 10.1007/978-3-319-58068-5_32

The notion of a MinA in DLs is conceptually closely related to that of a MUS in propositional satisfiability [10,20,24,27]. Given a propositional formula in CNF, a MUS is a minimal subset of clauses that is still unsatisfiable. As in DLs, computing the union of all MUSes is computationally expensive, even for Horn formulas.[1] In this context, the lean kernel has been proposed as a tight and easier-to-compute overapproximation of the union of all MUSes [20–22]. As such, it is an effective way to improve MUS enumeration. Briefly, the lean kernel (LK) is the set of all clauses that are used in some resolution proof for unsatisfiability. Recent work has shown that the LK can be obtained by solving maximum satisfiability [25] or by finding a minimal correction subset [28], thus requiring at most a logarithmic number of calls to a witness-producing NP oracle (e.g. a SAT solver).

Interestingly, the analogous of the lean kernel has never been studied in the context of DLs or, to the best of our knowledge, any other ontology language. Perhaps one reason for this is that the notion of LK depends on a specific derivation procedure (i.e., resolution). In this paper, we introduce lean kernels for description logics. To keep our approach as general as possible, we do not focus on a specific DL or reasoning algorithm, but rather base our definitions on abstract *consequence-based methods*, of which many instances exist in the literature. We then present an algorithm for computing the LKs of all consequences derivable through these methods with only a linear time overhead. As an example of our general methods, we focus on the consequence-based algorithm for $\mathcal{ALC}$, which generalizes the well-known completion method for the light-weight DL $\mathcal{EL}^+$. We compare the LKs in this setting with locality-based modules, which have been used for optimizing axiom pinpointing in DLs, and show formally that LKs are in general strictly smaller than those modules.

Through an empirical analysis, we show that the lean kernel is typically smaller (in some cases much smaller) than locality-based modules. More precisely, we compute the LKs and the locality-based modules for all atomic subsumption relations derivable from well-known large ontologies written in $\mathcal{EL}^+$. In these instances, the size of the LK is typically less than 1% of the size of the original ontology, and in many cases one-tenth or less of the locality-based modules. Moreover, the time required to compute these LKs is small, and the set obtained often coincides with the union of all MinAs. These results show that lean kernel computation is an effective approximation of the union of all MinAs, which can be used for solving other related reasoning problems.

2   Preliminaries

Description logics (DLs) [6] are a family of knowledge representation formalisms that have been successfully used to handle the knowledge of many application domains. They are also the logical formalism underlying the standard Web Ontology Language (OWL 2). As prototypical examples, we briefly introduce $\mathcal{ALC}$ [32],

---

[1] Finding the union of MUSes is at least as hard as testing MUS membership, which is Σ_2^p-complete for arbitrary CNF formulas [23, Theorem 4].

$$\begin{aligned}
\top^{\mathcal{I}} &:= \Delta^{\mathcal{I}} \\
\bot^{\mathcal{I}} &:= \emptyset \\
\neg C^{\mathcal{I}} &:= \Delta^{\mathcal{I}} \setminus C^{\mathcal{I}} \\
(C \sqcap D)^{\mathcal{I}} &:= C^{\mathcal{I}} \cap D^{\mathcal{I}} \\
(C \sqcup D)^{\mathcal{I}} &:= C^{\mathcal{I}} \cup D^{\mathcal{I}} \\
(\exists r.C)^{\mathcal{I}} &:= \{d \in \Delta^{\mathcal{I}} \mid \exists e \in C^{\mathcal{I}}.(d,e) \in r^{\mathcal{I}}\} \\
(\forall r.C)^{\mathcal{I}} &:= \{d \in \Delta^{\mathcal{I}} \mid \forall e.(d,e) \in r^{\mathcal{I}} \Rightarrow e \in C^{\mathcal{I}}\}
\end{aligned}$$

Fig. 1. Interpretation of complex concepts

the smallest propositionally closed DL, and $\mathcal{EL}^+$ [5], the logic underlying the OWL 2 EL profile.[2]

Let N_C and N_R be two disjoint sets of *concept-* and *role-names*, respectively. $\mathcal{ALC}$ *concepts* are constructed via the grammar rule

$$C ::= A \mid \top \mid \bot \mid \neg C \mid C \sqcap C \mid C \sqcup C \mid \exists r.C \mid \forall r.C, \tag{1}$$

where $A \in N_C$ and $r \in N_R$. $\mathcal{EL}$ *concepts* are obtained from the rule (1) by disallowing the constructors $\bot$ (bottom), $\neg$ (negation), $\sqcup$ (disjunction), and $\forall$ (value restrictions). Knowledge is represented by a *TBox*. An $\mathcal{ALC}$ *TBox* is a finite set of *general concept inclusions* (GCIs) $C \sqsubseteq D$ with C, D $\mathcal{ALC}$ concepts. An $\mathcal{EL}^+$ *TBox* is a finite set of GCIs formed by $\mathcal{EL}$ concepts, and *role inclusions* (RIs) $r_1 \circ \cdots \circ r_n \sqsubseteq s$, $n \geq 1$, with $r_i, s \in N_R$. We use the term *axiom* to denote both GCIs and RIs. Given an axiom $\alpha = x \sqsubseteq y$, we denote by $\mathsf{siglhs}(\alpha)$ and $\mathsf{sigrhs}(\alpha)$ the set of all symbols from N_C and N_R appearing in x and y, respectively, and $\mathsf{sig}(\alpha) = \mathsf{siglhs}(\alpha) \cup \mathsf{sigrhs}(\alpha)$.

The semantics of DLs is based on *interpretations* of the form $\mathcal{I} = (\Delta^{\mathcal{I}}, \cdot^{\mathcal{I}})$ where $\Delta^{\mathcal{I}}$ is a non-empty *domain* and $\cdot^{\mathcal{I}}$ maps every $A \in N_C$ to a subset $A^{\mathcal{I}} \subseteq \Delta^{\mathcal{I}}$ and every $r \in N_R$ to a binary relation $r^{\mathcal{I}} \subseteq \Delta^{\mathcal{I}} \times \Delta^{\mathcal{I}}$. This function is extended to arbitrary concepts as shown in Fig. 1. The interpretation $\mathcal{I}$ *satisfies* the GCI $C \sqsubseteq D$ if $C^{\mathcal{I}} \subseteq D^{\mathcal{I}}$; it *satisfies* the RI $r_1 \circ \cdots r_n \sqsubseteq s$ if $r_1^{\mathcal{I}} \circ \cdots r_n^{\mathcal{I}} \subseteq s^{\mathcal{I}}$. $\mathcal{I}$ is a *model* of the TBox $\mathcal{T}$ if it satisfies all axioms in $\mathcal{T}$.

One of the main reasoning problems in DLs is *subsumption* between concepts; that is, to decide whether every model of a TBox $\mathcal{T}$ also satisfies the GCI $C \sqsubseteq D$ (denoted by $C \sqsubseteq_{\mathcal{T}} D$). Without loss of generality, we focus only on *atomic subsumption*, where C and D are restricted to be concept names. It is often important to determine, in addition, the axioms that are responsible for a subsumption to follow.

Definition 1 (MinA). *Let $\mathcal{T}$ be a TBox and $A, B \in N_C$. A MinA for $A \sqsubseteq B$ w.r.t. $\mathcal{T}$ is a subset $\mathcal{M} \subseteq \mathcal{T}$ s.t. $A \sqsubseteq_{\mathcal{M}} B$ and for every $\mathcal{N} \subsetneq \mathcal{M}$, $A \not\sqsubseteq_{\mathcal{N}} B$.*

Example 2. Consider the TBox $\mathcal{T}_{\mathsf{exa}} := \{\mathsf{ax}_1, \ldots \mathsf{ax}_6\}$, with

$$\begin{array}{lll}
\mathsf{ax}_1 = A \sqsubseteq B & \mathsf{ax}_2 = A \sqsubseteq \exists r.A & \mathsf{ax}_3 = \exists r.B \sqsubseteq B \\
\mathsf{ax}_4 = B \sqsubseteq C & \mathsf{ax}_5 = A \sqsubseteq \exists s.C & \mathsf{ax}_6 = \exists s.A \sqsubseteq B
\end{array}$$

---

[2] https://www.w3.org/TR/owl2-overview/.

Then $A \sqsubseteq_{\mathcal{T}_{exa}} C$. Moreover, $\{ax_1, ax_4\}$ is the only MinA for $A \sqsubseteq C$ w.r.t. $\mathcal{T}_{exa}$.

The importance of the computation of MinAs, also known as *axiom pinpointing*, for ontology debugging and repair is well-documented [8,16]. Other applications of this task are error-tolerant [26] and context-based reasoning [7]; and probabilistic reasoning under distribution [31] and Bayesian semantics [11], to name just a few recent examples.

One fundamental step for handling large ontologies is to extract a small subset of axioms (or *module*) that preserves the relevant properties of the original TBox. In the case of axiom pinpointing and its associated reasoning tasks, such a module should contain all the MinAs [38].

Definition 3 (MinA-preserving module). *Let $\mathcal{T}$ be a TBox and $A, B \in N_C$. A subset $S \subseteq \mathcal{T}$ is a MinA-preserving module for $A \sqsubseteq B$ if for every MinA $\mathcal{M}$ for $A \sqsubseteq B$ w.r.t. $\mathcal{T}$ it holds that $\mathcal{M} \subseteq S$.*

To improve the efficiency of reasoning, one would start with the smallest possible MinA-preserving module, and extract all the MinAs from this set. Clearly, the smallest MinA-preserving module is formed by the union of all MinAs. However, computing this union is known to be hard, even for restricted sublogics of $\mathcal{EL}$ [30]. Thus, other approaches, like reachability- and locality-based modules [13,36], have been suggested to compute a small module more efficiently.

To improve readability, we introduce syntactic locality modules only for TBoxes that are in *normal form*; that is, where all the GCIs are of the form

$$A_1 \sqcap \cdots \sqcap A_n \sqsubseteq B, \quad A \sqsubseteq B_1 \sqcup \cdots \sqcup B_n, \quad \exists r.A \sqsubseteq B, \quad A \sqsubseteq \exists r.A, \quad A \sqsubseteq \forall r.B \quad (2)$$

with $n \geq 0$, $A_i, A \in N_C$, and $B_i, B \in N_C$. As usual, we identify the empty conjunction with $\top$ and the empty disjunction with $\bot$. Every TBox can be transformed to normal form preserving all relevant subsumption relations in polynomial time [5,35]. Moreover, modules obtained from a normalized TBox can be easily mapped to modules of the original TBox preserving the same properties [3,4].

Definition 4 (locality-based module). *Let $\mathcal{T}$ be a TBox in normal form, and Σ a signature. An axiom $\alpha \in \mathcal{T}$ is $\bot$-local w.r.t. Σ if $\mathsf{siglhs}(\alpha) \not\subseteq \Sigma$; it is $\top$-local w.r.t. Σ if $\mathsf{sigrhs}(\alpha) \not\subseteq \Sigma$. Locality is extended to sets of axioms in the obvious way.*

Let $A, B \in N_C$ and $x \in \{\bot, \top\}$. The x-module for $\mathcal{T}$ w.r.t. $A \sqsubseteq B$, denoted $\mathcal{M}^x_{A,B}$, is the smallest subset $\mathcal{M} \subseteq \mathcal{T}$ s.t. $\mathcal{T} \setminus \mathcal{M}$ is x-local w.r.t. $\{A, B\} \cup \mathsf{sig}(\mathcal{M})$. The $\bot\top^*$-module for $\mathcal{T}$ w.r.t. $A \sqsubseteq B$ is the fixpoint reached from iteratively extracting the $\bot$- and $\top$-modules for $\mathcal{T}$ w.r.t. $A \sqsubseteq B$.

Example 5. Consider again the TBox $\mathcal{T}_{exa}$ from Example 2, which is already in normal form. Clearly, ax_1, ax_3, and ax_5 are not $\bot$-local w.r.t. $\{A\}$. Moreover, ax_3, ax_4, and ax_6 are not $\bot$-local w.r.t. $\mathsf{sig}(\{ax_1, ax_3, ax_5\})$. Hence, $\mathcal{M}^\bot_{A,C} = \mathcal{T}_{exa}$. Similarly, $\mathcal{M}^\top_{A,C} = \mathcal{T}_{exa}$ and thus $\mathcal{M}^{\bot\top^*}_{A,C} = \mathcal{T}_{exa}$.

In the following, the term *locality-based module* (LBM) refers to any of the three kinds of modules defined above. LBMs are MinA-preserving modules that can be computed in polynomial time. It has been shown, through various empirical studies, that these modules are typically small for realistic ontologies, in particular for $\mathcal{EL}^+$ [4,37]. In the next section we consider a new notion of module that has been previously considered in the context of propositional logic.

3 Lean Kernels

Intuitively, the lean kernel for a consequence c—e.g. a subsumption relation—is the set of all axioms that can appear in some proof for c. In general, the notion of a proof depends not only on the logic, but also on the decision method used. We now define lean kernels based on a general notion of consequence-based methods.

Abstracting from particularities, a *consequence-based method* is an algorithm that works on a set $\mathcal{A}$ of *assertions*, and uses rules to extend this set. The algorithm has two phases. First, the *normalization* phase transforms all the axioms into a suitable normal form. The *saturation* phase initializes the set $\mathcal{A}$ and extends it through rule applications. A *rule* is of the form $(\mathcal{B}_0, \mathcal{S}) \rightarrow \mathcal{B}_1$, where $\mathcal{B}_0, \mathcal{B}_1$ are finite sets of assertions, and $\mathcal{S}$ is a finite set of axioms in normal form. This rule is *applicable* to a set of axioms $\mathcal{T}$ and a set of assertions $\mathcal{A}$ if $\mathcal{B}_0 \subseteq \mathcal{A}, \mathcal{S} \subseteq \mathcal{T}$, and $\mathcal{B}_1 \not\subseteq \mathcal{A}$. Its *application* extends $\mathcal{A}$ to $\mathcal{A} \cup \mathcal{B}_1$. $\mathcal{A}$ is *saturated* if no rule is applicable to it. The method *terminates* if $\mathcal{A}$ is saturated after finitely many rule applications. After termination, the consequences of $\mathcal{T}$ can be read directly from $\mathcal{A}$; that is, to decide whether a consequence c follows from $\mathcal{T}$ it suffices to verify whether an assertion from the distinguished set check(c) appears in $\mathcal{A}$. The set check(c) contains the assertions that suffice for deciding that the consequence c holds. Given a rule $R = (\mathcal{B}_0, \mathcal{S}) \rightarrow \mathcal{B}_1$, we will use $pre(R)$, $ax(R)$ and $res(R)$ to denote the sets $\mathcal{B}_0$ of premises, $\mathcal{S}$ of axioms that trigger R, and $\mathcal{B}_1$ of assertions resulting of its applicability, respectively.

A simple example of a consequence-based method is the algorithm for reasoning with $\mathcal{ALC}$ TBoxes presented in [35]. Before describing this algorithm we introduce some necessary notation. A *literal* is either a concept name or a negated concept name. In the following, H, K denote (possibly empty) conjunctions of literals, and M, N are (possibly empty) disjunctions of concept names. For simplicity, we will often treat these conjunctions and disjunctions as sets.

The consequence-based algorithm for $\mathcal{ALC}$ works on assertions of the form (H, M) and (H, N, r, K). Intuitively, these assertions express $H \sqsubseteq_\mathcal{T} M$ and $H \sqsubseteq_\mathcal{T} N \sqcup \exists r.K$, respectively. The normalization phase transforms all GCIs to be of the form (2) introduced before. The saturation phase initializes $\mathcal{A}$ to contain the assertions (H, A) for all concept names A and all conjuctions of literals H from the normalized TBox, such that $A \in H$. The rules applied during saturation are depicted in the upper part of Table 1. After termination, for every two concept names A, B, it holds that $A \sqsubseteq_\mathcal{T} B$ iff $(A, B) \in \mathcal{A}$ or $(A, \bot) \in \mathcal{A}$. Thus, in this case, if the desired consequence c is the subsumption $A \sqsubseteq B$, then check$(c) = \{(A, B)\}$.

Table 1. $\mathcal{ALC}$ and $\mathcal{EL}^+$ consequence-based algorithm rules $(\mathcal{B}_0, \mathcal{S}) \rightarrow \mathcal{B}_1$

$\mathcal{B}_0$	$\mathcal{S}$	$\mathcal{B}_1$
$(H \sqcap \neg A, N \sqcup A)$	$\emptyset$	(H, N)
$(H, N_1 \sqcup A_1), \ldots, (H, N_n \sqcup A_n)$	$A_1 \sqcap \cdots \sqcap A_n \sqsubseteq B$	$(H, \bigsqcup_{i=1}^n N_i \sqcup B)$
$(H, N \sqcup A)$	$A \sqsubseteq \exists r.B$	(H, N, r, B)
$(H, M, r, K), (K, N \sqcup A)$	$\exists r.A \sqsubseteq B$	$(H, M \sqcup B, r, K \sqcap \neg A)$
$(H, M, r, K), (K, \bot)$	$\emptyset$	(H, M)
$(H, M, r, K), (H, N \sqcup A)$	$A \sqsubseteq \forall r.B$	$(H, M \sqcup N, r, K \sqcap B)$
$(A_0, \emptyset, r_1, A_1), \ldots (A_{n-1}, \emptyset, r_n, A_n)$	$r_1 \circ \cdots \circ r_n \sqsubseteq s$	$(A_0, \emptyset, s, A_n)$

We emphasize that this is only one of many consequence-based algorithms available. The completion-based algorithm for $\mathcal{EL}^+$ [5] is obtained by restricting the assertions to be of the form (A, B) and $(A, \emptyset, r, B)$ with $A, B \in N_C \cup \{\top\}$ and $r \in N_R$, and adding the rule in the last row of Table 1. Other examples include LTUR approach for Horn clauses [29], and methods for more expressive and Horn DLs [9,18,19]. For the rest of this section, we consider an arbitrary, but fixed, consequence-based method, that is sound and complete for deciding consequences from a set of axioms.

For the following definition, we need to weaken the notion of applicability of a rule. The rule $(\mathcal{B}_0, \mathcal{S}) \rightarrow \mathcal{B}_1$ is *weakly applicable* to $\mathcal{T}$ and $\mathcal{A}$ if $\mathcal{B}_0 \subseteq \mathcal{A}$ and $\mathcal{S} \subseteq \mathcal{T}$. In other words, the last condition of applicability is ignored.

Definition 6 (proof). *A proof for a consequence c is a finite sequence of rules $\mathcal{P} = (R_1, \ldots, R_n)$ such that: (i) for all $i, 1 \leq i \leq n$, R_i is weakly applicable after $R_1, \ldots, R_{i-1}$ have been applied, (ii) check$(c) \cap res(R_n) \neq \emptyset$, and (iii) for every $i, 1 \leq i < n$, there is a non-initial assertion $b \in res(R_i)$ and a $j > i$ where $b \in pre(R_j)$. Pf(c) denotes the set of all proofs for c.*

Given a proof $\mathcal{P} = (R_1, \ldots, R_n)$, we denote by $\mathcal{T}_\mathcal{P}$ the set of all axioms appearing in $\mathcal{P}$; that is, $\mathcal{T}_\mathcal{P} := \bigcup_{i=1}^n ax(R_i)$. Using this notation, it is now possible to define a general notion of the lean kernel, which corresponds to the set of all axioms appearing in at least one proof for the consequence.

Definition 7 (lean kernel). *The* lean kernel *for a consequence c is the set* $LK(c) := \bigcup_{\mathcal{P} \in \mathsf{Pf}(c)} \mathcal{T}_\mathcal{P}$.

Notice that if there is a proof $\mathcal{P}$ for a consequence c, then the subset of axioms $\mathcal{T}_\mathcal{P}$ already entails c. Since the consequence-based algorithm for $\mathcal{ALC}$ is sound and complete for deciding subsumptions entailed by a TBox, it follows that the lean kernel is a MinA-preserving module. In fact, this is true for any consequence-based method $\mathcal{C}$.

Theorem 8. *Let $\mathcal{T}$ be a set of axioms, c a consequence, and $\mathcal{C}$ a sound and complete consequence-based method. Then $LK(c)$ is a MinA-preserving module for $\mathcal{T} \models c$.*

Table 2. A proof for $A \sqsubseteq C$ in $\mathcal{T}_{\mathsf{exa}}$.

	$\mathcal{B}_0$	$\mathcal{S}$	$\mathcal{B}_1$
R_1	(A, A)	$A \sqsubseteq B$	(A, B)
R_2	(A, A)	$A \sqsubseteq \exists r.A$	$(A, \emptyset, r, A)$
R_3	$(A, \emptyset, r, A), (A, B)$	$\exists r.B \sqsubseteq B$	$(A, B, r, A \sqcap \neg A)$
R_4	$(A, B, r, A \sqcap \neg A), (A \sqcap \neg A, \bot)$	$\emptyset$	(A, B)
R_5	(A, B)	$B \sqsubseteq C$	(A, C)

Proof. Let $\mathcal{M}$ be a MinA for c w.r.t. $\mathcal{T}$. Then, there is a sequence of rule applications that uses only the axioms in $\mathcal{M}$ and eventually adds an assertion from check(c) to $\mathcal{A}$. This sequence can be minimized by iteratively removing all superfluous rule applications, thus yielding a proof $\mathcal{P}$. If $\mathcal{T}_\mathcal{P} \subsetneq \mathcal{M}$, then $\mathcal{M}$ cannot be a MinA. Thus, we get that $\mathcal{M} \subseteq LK(c)$. □

Example 9. In our running example, there are two proofs for $A \sqsubseteq_{\mathcal{T}_{\mathsf{exa}}} C$ (modulo reordering of the rules) w.r.t. the consequence-based algorithm: one that uses the axioms $\mathsf{ax}_1, \mathsf{ax}_4$, and another one that uses ax_1–ax_4 as shown in Table 2. Thus $LK(A \sqsubseteq_{\mathcal{T}_{\mathsf{exa}}} C) = \{\mathsf{ax}_1, \ldots, \mathsf{ax}_4\}$.

Notice that the lean kernel from this example contains some axioms that do not belong to any MinA; specifically, ax_2 and ax_3 are not fundamental for deriving this consequence. On the other hand, this LK is a strict subset of the $\bot\mathcal{T}^*$-module for the same consequence (see Example 5). As we show next, this property holds in general for $\mathcal{ALC}$ TBoxes in normal form.

Theorem 10. *Let $\mathcal{T}$ be an $\mathcal{ALC}$ TBox and $A, B \in N_C$. Then, w.r.t. the completion algorithm, $LK(A \sqsubseteq_\mathcal{T} B) \subseteq \mathcal{M}_{A,B}^{\bot\mathcal{T}^*}$.*

Proof. To obtain this result, it suffices to show that $LK(A \sqsubseteq_\mathcal{T} B) \subseteq \mathcal{M}_{A,B}^\bot$ and $LK(A \sqsubseteq_\mathcal{T} B) \subseteq \mathcal{M}_{A,B}^\top$ hold. Assume that $LK(A \sqsubseteq_\mathcal{T} B) \not\subseteq \mathcal{M}_{A,B}^\bot$. Then there exists a proof $\mathcal{P} \in \mathsf{Pf}(c)$ such that $\mathcal{T}_\mathcal{P} \not\subseteq \mathcal{M}_{A,B}^\bot$; let α be the first axiom appearing in $\mathcal{P}$ such that $\alpha \notin \mathcal{M}_{A,B}^\bot$. Then α is $\bot$-local w.r.t. $\Sigma := \mathsf{sig}(\mathcal{M}_{A,B}^\bot) \cup \{A, B\}$; i.e., $\mathsf{siglhs}(\alpha) \not\subseteq \Sigma$. By construction, for a rule to be weakly applicable to the axiom α, $\mathsf{siglhs}(\alpha)$ must have been already derived. (This connection between axioms and rules is a property of this specific algorithm.) Thus, α cannot be $\bot$-local. An analogous but dual argument can be used to show that $LK(A \sqsubseteq_\mathcal{T} B) \subseteq \mathcal{M}_{A,B}^\top$ also holds. □

If we consider assertions and axioms as propositional variables, the rules in a consequence-based method can be seen as implications. In particular, they can be seen as (generalized) Horn clauses. This insight was exploited in [33] to encode the execution of the $\mathcal{EL}^+$ completion algorithm in a Horn formula. In [34], the notion of *COI module* was introduced based on this encoding. As formally defined, the COI module for a given consequence is in general a superset of its

LK, as it considers also all possible derivations of initial assertions, which can be seen as tautologies, and are disregarded by our definition of proof. However, this notion can be adapted to correspond to the LKs defined here.

4   Computing Lean Kernels

We now describe a method for computing LKs based on modifying consequence-based algorithms to keep track of the relevant axioms used in the derivation of the consequences. To achieve this, we first provide a unique label to each axiom in $\mathcal{T}$, which will be used to identify it. At the normalization phase, every normalized axiom α obtained is labeled with the set $\mathsf{lab}(\alpha)$ of the original axioms that produce it. To label all the derived assertions with the set of the relevant axioms that generate them, we modify the rule applicability condition, as well as the result of applying it.

First, all assertions a obtained at initialization are labeled with the empty set $\mathsf{lab}(a) = \emptyset$. The rule $R = (\mathcal{B}_0, \mathcal{S}) \to \mathcal{B}_1$ is *LK-applicable* to $\mathcal{T}$ and $\mathcal{A}$ if $\mathcal{B}_0 \subseteq \mathcal{A}$, $\mathcal{S} \subseteq \mathcal{T}$, and there exists some non-initial $b \in \mathcal{B}_1$ such that $b \notin \mathcal{A}$ or $\mathsf{lab}(R) := \bigcup_{a \in \mathcal{B}_0} \mathsf{lab}(a) \cup \bigcup_{\alpha \in \mathcal{S}} \mathsf{lab}(\alpha) \nsubseteq \mathsf{lab}(b)$. Its *application* extends $\mathcal{A}$ to $\mathcal{A} \cup \mathcal{B}_1$, sets $\mathsf{lab}(b) = \mathsf{lab}(R)$ for all new assertions b, and modifies the label of all previously existing assertions b from $\mathcal{B}_1$ to $\mathsf{lab}(b) \cup \mathsf{lab}(R)$. $\mathcal{A}$ is *LK-saturated* if no rule is LK-applicable to it. The LK of the consequence c is obtained as the union of the labels of all assertions in $\mathsf{check}(c)$. Given a consequence-based algorithm, we call the variant using these applicability conditions its *LK extension*.

Example 11. An example execution of the LK extension of the $\mathcal{ALC}$ algorithm over $\mathcal{T}_{\mathsf{exa}}$ appears in Table 3. Each step shows the preconditions for a rule application, along with the assertion added and its label. For simplicity, we removed all steps that derive obvious tautologies. In step 4, the assertion (A, B) is not added again; rather its label is extended to the set $\{\mathsf{ax}_1, \mathsf{ax}_2, \mathsf{ax}_3\}$. Notice, moreover, that the rule at step 4 would not be applicable in the original algorithm, but becomes LK-applicable, as more axioms are detected as potentially relevant.

It is easy to see that the LK extension of a consequence-based algorithm has essentially the same asymptotic run-time behavior as the original algorithm.

Table 3. Execution of the LK extension over $\mathcal{T}_{\mathsf{exa}}$.

step	$\mathcal{B}_0$	$\mathcal{S}$	$\mathcal{B}_1$	label
1	(A, A)	$A \sqsubseteq B$	(A, B)	$\{\mathsf{ax}_1\}$
2	(A, A)	$A \sqsubseteq \exists r.A$	$(A, \emptyset, r, A)$	$\{\mathsf{ax}_2\}$
3	$(A, \emptyset, r, A), (A, B)$	$\exists r.B \sqsubseteq B$	$(A, B, r, A \sqcap \neg A)$	$\{\mathsf{ax}_1, \mathsf{ax}_2, \mathsf{ax}_3\}$
4	$(A, B, r, A \sqcap \neg A), (A \sqcap \neg A, \bot)$	$\emptyset$	(A, B)	$\{\mathsf{ax}_1, \mathsf{ax}_2, \mathsf{ax}_3\}$
5	(A, B)	$B \sqsubseteq C$	(A, C)	$\{\mathsf{ax}_1, \ldots, \mathsf{ax}_4\}$
6	(A, A)	$A \sqsubseteq \exists s.C$	$(A, \emptyset, s, C)$	$\{\mathsf{ax}_5\}$

Indeed, the extension generates the same set of assertions. The main difference is that, while the original algorithm generates each assertion only once, its LK extension may update its label several times. Notice, however, that each update strictly extends the set of axioms in the label. Thus, each label can be updated at most $|\mathcal{T}|$ times, and the run-time of the LK extension is increased by a linear factor. Another important feature of this extension is that the resulting labels do not depend on the order in which the rules are applied.

Theorem 12. *If the LK extension of a consequence-based method is executed until LK saturation, then* lab(check(c)) *is the lean kernel for c.*

Proof. Let $\mathcal{P}$ be a proof for c. Then $\mathcal{P}$ is a sequence of weakly applicable rules. Consider the LK application of the same sequence. If one rule $(\mathcal{B}_0, \mathcal{S}) \rightarrow \mathcal{B}_1$ is not LK-applicable, it means that its application would not change the label of the assertions in $\mathcal{B}_1$. Condition (iii) in Definition 6 guarantees that all axioms in $\mathcal{T}_{\mathcal{P}}$ appear in the labels of the elements of check(c) obtained after the execution of this sequence; otherwise, the rule applications in which the missing axioms appear could be removed from the sequence. Hence, $\mathcal{T}_{\mathcal{P}} \subseteq$ lab(check(c)).

Conversely, for every axiom $\alpha \in$ lab(check(c)) there exists a sequence of rule LK-applications that eventually adds the axiom α to this label. Since every LK-applicable rule is also weakly applicable, such a sequence can be trivially transformed into a proof. Thus, lab(check(c)) $\subseteq LK(c)$. □

Notice that the runtime behaviour of the LK extension is governed by the underlying decision procedure. For instance, our approach would compute the LK of all atomic subsumption relations following from an $\mathcal{ALC}$ TBox in exponential time. This is in contrast to LBM and other MinA-preserving modules (see e.g. [4]), which can be computed very efficiently, at the cost of losing soundness and potentially including many superfluous axioms.

Example 13. Let $\mathcal{T}'_{\mathsf{exa}} := \mathcal{T}_{\mathsf{exa}} \cup \{\exists s.A \sqsubseteq B_i, B_i \sqsubseteq C \mid i \in I\}$, where I is an arbitrarily large set of indices and $\mathcal{T}_{\mathsf{exa}}$ is the TBox from Example 2. Then the $\bot\mathsf{T}^*$-module for $\mathcal{T}'_{\mathsf{exa}}$ w.r.t. $A \sqsubseteq C$ is $\mathcal{T}'_{\mathsf{exa}}$, while $LK(A \sqsubseteq_{\mathcal{T}'_{\mathsf{exa}}} C) = \{\mathsf{ax}_1, \dots, \mathsf{ax}_4\}$.

Moreover, the TBox $\mathcal{T}''_{\mathsf{exa}} := \mathcal{T}'_{\mathsf{exa}} \setminus \{\mathsf{ax}_4\}$ does not entail $A \sqsubseteq C$, and hence $LK(A \sqsubseteq_{\mathcal{T}''_{\mathsf{exa}}} C) = \emptyset$. However, the $\bot\mathsf{T}^*$-module for $\mathcal{T}''_{\mathsf{exa}}$ w.r.t. $A \sqsubseteq C$ is the set $\{\mathsf{ax}_1\} \cup \{\exists s.A \sqsubseteq B_i, B_i \sqsubseteq C \mid i \in I\}$.

This example shows that LBMs may not provide much information about the entailments or their axiomatic causes. As we see in the following section, this phenomenon can be observed in realistic TBoxes used in practice.

5   Experiments

We performed an experimental study aimed at assessing the sizes of the LKs in practice, and comparing them with the $\bot$-, T- and $\bot\mathsf{T}^*$-modules (see Definition 4). A prototype for computing the different modules was implemented in C++ for the DL $\mathcal{EL}^+$ and several experiments were run on a Linux cluster

Table 4. Ontologies considered in the experiments

$\mathcal{T}$	# GCIs	# RIs	$|\text{class}(\mathcal{T})|$
GENE	20465	1	164743
NCI	46800	0	252519
NOT-GALEN	3937	442	27980
FULL-GALEN	35530	1014	453674
SNOMED-CT	307692	12	5333580

(2 GHz, 128 GB) on different well-known bio-medical ontologies. These ontologies are: GENE, NCI, NOT-GALEN, FULL-GALEN and SNOMED-CT (v. 2009). Table 4 shows the number of GCIs, RIs and atomic subsumption relations in these ontologies. For each ontology, except SNOMED, the different modules were computed for all atomic subsumption relations entailed by the ontology. For SNOMED, both the LK and $\bot$-modules were computed for all atomic subsumption relations. Computing all $\top$-modules was infeasible; hence we considered the 240 concept names subsuming the largest number of concepts. These yield the $\top$-modules for around 2.7 million subsumption relations, for which also the $\bot\top^*$-modules were computed.

Figure 2 compares the sizes of the LKs against the size of locality-based modules, shown as a percentage. Theorem 10 guarantees that the size of the lean kernel never exceeds the size of the $\bot$, $\top$ and $\bot\top^*$-modules. A result of 10^2 means that the size of the lean kernel matches the size of the other module, with smaller values indicating more significant size reductions. Notice that in most cases the LK is smaller than both $\bot$- and $\top$-modules, often achieving a significant reduction. The difference is more significant w.r.t. $\top$-modules, where the LK is usually at least two orders of magnitude smaller. Compared to $\bot$-modules, LKs are usually smaller in a factor or 2 or more. Despite their large size, $\top$-modules appear to be useful as shown in the sizes of $\bot\top^*$-modules. Recall that $\bot\top^*$-modules are the fixpoint from the iterative application of $\bot$- and $\top$-modules. It is thus guaranteed that $\bot\top^*$-modules will not be greater than those modules. Noticeably, for GENE $\bot\top^*$-modules are in general equal to the LK; only in 4.5% of the cases is the LK strictly smaller. For NCI, $\bot\top^*$-modules match the LKs for all subsumption

Table 5. Module sizes

$\mathcal{T}$	$\bot$			$\top$			$\bot\top^*$			LK		
	Min	Avg	Max	Min	Avg	Max	Min	Avg	Max	Min	Avg	Max
GENE	2	18.33	68	1	2012.04	8786	1	5.70	66	1	5.45	52
NCI	1	36.70	398	1	2431.16	11572	1	7.08	85	1	7.08	85
NOT-GALEN	1	88.06	495	1	3433.56	3796	1	23.47	300	1	13.78	80
FULL-GALEN	1	9727.15	15543	1	34244.03	35733	1	8940.22	14586	1	68.90	416
SNOMED-CT	1	51.71	264	3269	216213.80	307704	1	41.46	264	1	40.19	220

Table 6. Running times (in seconds). For LKs, the time refers to the total time for computing all LKs; for LBMs the time refers to each subsumption relation.

$\mathcal{T}$	LKs	$\bot$			$\top$			$\bot\top^*$		
		Min	Avg	Max	Min	Avg	Max	Min	Avg	Max
GENE	1.98	0.00	0.01	0.04	0.00	0.01	0.04	0.00	0.01	0.06
NCI	3.79	0.00	0.01	0.10	0.01	0.01	0.09	0.01	0.02	0.13
NOT-GALEN	4.22	0.00	0.01	0.02	0.00	0.02	0.04	0.00	0.01	0.03
FULL-GALEN	461.03	0.00	0.36	1.12	0.01	1.60	2.90	0.02	2.46	8.06
SNOMED-CT	11200.53	0.11	0.65	5.34	0.15	2081.77	5209.57	0.26	3.91	28.92

relations. For NOT-GALEN and FULL-GALEN, LKs are significantly smaller than $\bot\top^*$-modules, especially in the latter case. Regarding SNOMED, the reduction w.r.t. $\bot$-modules is slightly smaller than in other cases: LKs are usually 75% of the $\bot$-modules. $\top$-modules are in general quite large, and $\bot\top^*$-modules improve over $\bot$-modules, getting close to the LKs in most cases.

These results are confirmed in Table 5, which shows, for each ontology the minimum, average and maximum size of each module over the selected 2.7 million instances of SNOMED and all the atomic subsumptions of the other ontologies. Observe that $\top$-modules usually represent a large fraction of the ontology, while $\bot$-modules, and especially $\bot\top^*$-modules, are quite small for GENE, NCI, and SNOMED representing in general less than one percent of the ontology. For NOT-GALEN and FULL-GALEN, $\bot$- and $\bot\top^*$-modules are not so small; in the latter case representing around half of the ontology in most cases. Noticeably, LK modules are in general a small fraction of the ontologies, representing less than 1% of it in most cases.

For the five ontologies we computed the 99.9% confidence interval for the mean size of the LKs and $\bot\top^*$-modules. In all cases, excepting only NCI, these intervals do not overlap. Thus, we can conclude that the difference in size between LKs and $\bot\top^*$-modules is highly statistically significant ($p < 0.001$). In NCI, this difference does not exist as both kinds of modules coincide.

In order to get a more detailed view on the ability of the different modules to approximate the union of MinAs (UMinAs), we computed UMinAs for all the atomic subsumption relations of GENE and NOT-GALEN using BEACON [1]. The results show that LKs match UMinAs for all subsumption relations in GENE, whereas $\bot\top^*$-modules match UMinAs in 95.49% of the cases, being in average around 3.26% larger than UMinAs. For NOT-GALEN, both the LKs and the $\bot\top^*$-modules equal UMinAs for a similar percentage of the subsumption relations (43.78% and 41.28% respectively), and in average LKs are around 154.41% larger than UMinAs whereas $\bot\top^*$ modules are around 362.06% larger than UMinAs. These experiments confirm earlier results on the accuracy of locality-based modules, and reveal that LKs constitute tighter and more stable approximations.

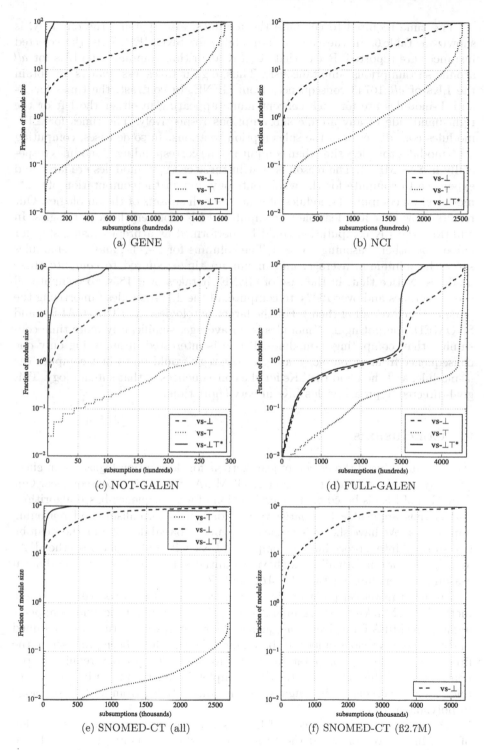

Fig. 2. LK module size w.r.t. locality-based modules (in percentage)

The time required to compute the modules, after parsing the ontology, is shown in Table 6. In the case of ⊤-modules for SNOMED, only the selected instances are reported. Recall that the LK algorithm computes the LKs for *all* atomic subsumptions simultaneously. Thus, e.g., it takes less than 2s. to obtain the LKs of *all* 16743 consequences from GENE. In contrast, the times for ⊥- and ⊤-modules are for each concept name appearing in either the left or the right-hand side of any atomic subsumptions respectively. The times for ⊥⊤*-modules consider each of the subsumption relations. In some cases, computing ⊥⊤*-modules took less time than computing the corresponding ⊤-modules (especially in SNOMED). The reason is that in these cases, ⊤-modules are large and expensive to compute for the whole ontology, but in the computation of ⊥⊤*-modules, we compute ⊤-modules of usually small subsets of the ontologies. Our results show that it is feasible to compute all LKs for very large ontologies. In addition, the LK computation could be performed as a preprocessing step for enhancing other reasoning services. The columns for ⊥-, ⊤-, and ⊥⊤*-modules show the minimum, average, and maximum time required to compute these modules. Notice that, in the case of GENE, it takes over 185s. to compute all the ⊥-modules and over 1647s. to compute all the ⊥⊤*-modules (amortizing the parsing time over all of them). For the larger ontologies, i.e. FULL-GALEN and SNOMED, computing ⊥⊤* modules is, in average, significantly more time consuming than computing ⊥-modules. If one is interested in analyzing only one consequence from an ontology, a better strategy would be to first compute the ⊥-module, and then find the LKs for the consequences of that subontology. This goal-directed approach yields LKs in very short time.

6   Conclusions

We have introduced the notion of lean kernels for description logics, as an effective way to approximate the union of all MinAs for a given consequence. Our definition of LKs is based on a general notion of a consequence-based algorithm, and is thus applicable to a large variety of logical formalisms and reasoning approaches. We have shown how the consequence-based decision method can be transformed into a procedure for computing LKs, with only a linear overhead. As an example for our formalism, we have instantiated the definitions to well-known reasoning algorithms for the DLs $\mathcal{ALC}$ and $\mathcal{EL}^+$.

From a theoretical point of view, we have shown that LKs based on these methods are MinA-preserving modules; i.e., they contain all the axioms appearing in some MinA for a given consequence. More interestingly, they are contained in the different versions of locality-based modules that have been proposed in the literature. While the computation of LKs for an $\mathcal{ALC}$ ontology requires exponential time in the worst case, the $\mathcal{EL}^+$ completion algorithm can be adapted to find the lean kernels of all atomic subsumptions entailed from an $\mathcal{EL}^+$ ontology in polynomial time.

To evaluate the effectiveness of LKs as a means to approximate the union of all MinAs, we computed the LKs and the three variants of locality-based modules for all atomic subsumptions derivable from large bio-medical ontologies that are commonly used as benchmarks for $\mathcal{EL}^+$ systems. Overall, more than 6 million subsumption relations were analyzed. Our experiments show that LKs are often (much) smaller than LBMs, and in most cases only a small fraction of the original ontology. Interestingly, for the Gene Ontology LKs in fact coincide with the union of all MinAs. Moreover, they can be effectively computed. Thus, lean kernel computation can improve the runtime of axiom pinpointing and other related reasoning tasks. In future work we will analyze the practical benefits of computing these modules for those tasks.

It is known that syntactic LBMs are also MinA preserving, and can be computed in polynomial time even for very expressive DLs where reasoning is 2ExpTime-hard [2–4,17]. However, an additional reasoning step is required to verify whether the consequence follows from the ontology. Following our approach, we can compute the LKs of all relevant consequences of an ontology simultaneously, while reasoning. In that sense, LKs are more closely related to *semantic* locality-based modules [13]. It has been observed in [15] that the difference between syntactic and semantic LBMs is not statistically significant. Our results thus suggest that the size difference between semantic LBMs and LKs is also statistically significant, although a full empirical analysis is needed to justify this conclusion.

Acknowledgements. We would like to thank Beatriz Peñaloza for her help on statistical methods. Carlos Mencía is supported by grant TIN2016-79190-R.

References

1. Arif, M.F., Mencía, C., Ignatiev, A., Manthey, N., Peñaloza, R., Marques-Silva, J.: BEACON: An efficient SAT-based tool for debugging EL+ ontologies. In: SAT, pp. 521–530 (2016)
2. Romero, A.A.: Ontology module extraction and applications to ontology classification. Ph.D. thesis, University of Oxford, UK (2015)
3. Romero, A.A., Kaminski, M., Grau, B.C., Horrocks, I.: Ontology module extraction via datalog reasoning. In: AAAI, pp. 1410–1416 (2015)
4. Romero, A.A., Kaminski, M., Cuenca Grau, B., Horrocks, I.: Module extraction in expressive ontology languages via datalog reasoning. JAIR **55**, 499–564 (2016)
5. Baader, F., Brandt, S., Lutz, C.: Pushing the $\mathcal{EL}$ envelope. In: IJCAI, pp. 364–369 (2005)
6. Baader, F., Calvanese, D., McGuinness, D., Nardi, D., Patel-Schneider, P.F. (eds.): The Description Logic Handbook: Theory, Implementation, and Applications. Cambridge University Press, Cambridge (2003)
7. Baader, F., Knechtel, M., Peñaloza, R.: Context-dependent views to axioms and consequences of semantic web ontologies. J. Web Semant. **12–13**, 22–40 (2012)
8. Baader, F., Suntisrivaraporn, B.: Debugging SNOMED CT using axiom pinpointing in the description logic $\mathcal{EL}^+$. In: KR-MED (2008)
9. Bate, A., Motik, B., Grau, B.C., Simancik, F., Horrocks, I.: Extending consequence-based reasoning to SRIQ. In: KR, pp. 187–196 (2016)
10. Belov, A., Lynce, I., Marques-Silva, J.: Towards efficient MUS extraction. AI Commun. **25**(2), 97–116 (2012)

11. Ceylan, İI., Peñaloza, R.: The bayesian description logic $\mathcal{BEL}$. In: Demri, S., Kapur, D., Weidenbach, C. (eds.) IJCAR 2014. LNCS (LNAI), vol. 8562, pp. 480–494. Springer, Cham (2014). doi:10.1007/978-3-319-08587-6_37

12. Grau, B.C., Horrocks, I., Kazakov, Y., Sattler, U.: Just the right amount: Extracting modules from ontologies. In: WWW, pp. 717–726 (2007)

13. Cuenca Grau, B., Horrocks, I., Kazakov, Y., Sattler, U.: Modular reuse of ontologies: Theory and practice. J. Artif. Intell. Res. (JAIR) **31**, 273–318 (2008)

14. Cuenca Grau, B., Horrocks, I., Kazakov, Y., Sattler, U.: Extracting modules from ontologies: A logic-based approach. In: Stuckenschmidt, H., Parent, C., Spaccapietra, S. (eds.) Modular Ontologies. LNCS, vol. 5445, pp. 159–186. Springer, Heidelberg (2009). doi:10.1007/978-3-642-01907-4_8

15. Vescovo, C., Klinov, P., Parsia, B., Sattler, U., Schneider, T., Tsarkov, D.: Empirical study of logic-based modules: Cheap is cheerful. In: Alani, H., Kagal, L., Fokoue, A., Groth, P., Biemann, C., Parreira, J.X., Aroyo, L., Noy, N., Welty, C., Janowicz, K. (eds.) ISWC 2013. LNCS, vol. 8218, pp. 84–100. Springer, Heidelberg (2013). doi:10.1007/978-3-642-41335-3_6

16. Kalyanpur, A., Parsia, B., Horridge, M., Sirin, E.: Finding all justifications of OWL DL entailments. In: Aberer, K., Choi, K.-S., Noy, N., Allemang, D., Lee, K.-I., Nixon, L., Golbeck, J., Mika, P., Maynard, D., Mizoguchi, R., Schreiber, G., Cudré-Mauroux, P. (eds.) ASWC/ISWC -2007. LNCS, vol. 4825, pp. 267–280. Springer, Heidelberg (2007). doi:10.1007/978-3-540-76298-0_20

17. Kaminski, M., Nenov, Y., Grau, B.C.: Datalog rewritability of disjunctive datalog programs and its applications to ontology reasoning. In: AAAI, pp. 1077–1083 (2014)

18. Kazakov, Y.: Consequence-driven reasoning for Horn SHIQ ontologies. In: Boutilier, C. (ed.) IJCAI 2009, pp. 2040–2045 (2009)

19. Kazakov, Y., Krötzsch, M., Simancik, F.: The incredible ELK - from polynomial procedures to efficient reasoning with $\mathcal{EL}$ ontologies. JAR **53**(1), 1–61 (2014)

20. Büning, H.K., Kullmann, O.: Minimal unsatisfiability and autarkies. In: Biere, A., Heule, M., van Maaren, H., Walsh, T. (eds.) Handbook of Satisfiability, Frontiers in Artificial Intelligence and Applications, vol. 185, pp. 339–401. IOS Press (2009)

21. Kullmann, O.: Investigations on autark assignments. Discrete Appl. Math. **107**(1–3), 99–137 (2000)

22. Kullmann, O., Lynce, I., Marques-Silva, J.: Categorisation of clauses in conjunctive normal forms: Minimally unsatisfiable sub-clause-sets and the lean kernel. In: Biere, A., Gomes, C.P. (eds.) SAT 2006. LNCS, vol. 4121, pp. 22–35. Springer, Heidelberg (2006). doi:10.1007/11814948_4

23. Liberatore, P.: Redundancy in logic I: CNF propositional formulae. Artif. Intell. **163**(2), 203–232 (2005)

24. Liffiton, M.H., Previti, A., Malik, A., Marques-Silva, J.: Fast, flexible MUS enumeration. Constraints **21**(2), 223–250 (2016)

25. Liffiton, M., Sakallah, K.: Searching for autarkies to trim unsatisfiable clause sets. In: Kleine Büning, H., Zhao, X. (eds.) SAT 2008. LNCS, vol. 4996, pp. 182–195. Springer, Heidelberg (2008). doi:10.1007/978-3-540-79719-7_18

26. Ludwig, M., Peñaloza, R.: Error-tolerant reasoning in the description logic $\mathcal{EL}$. In: JELIA, pp. 107–121 (2014)

27. Marques-Silva, J., Ignatiev, A., Mencía, C., Peñaloza, R.: Efficient reasoning for inconsistent horn formulae. In: Michael, L., Kakas, A. (eds.) JELIA 2016. LNCS (LNAI), vol. 10021, pp. 336–352. Springer, Cham (2016). doi:10.1007/978-3-319-48758-8_22

28. Marques-Silva, J., Ignatiev, A., Morgado, A., Manquinho, V.M., Lynce, I.: Efficient autarkies. In: ECAI, pp. 603–608 (2014)
29. Minoux, M.: LTUR: A simplified linear-time unit resolution algorithm for horn formulae and computer implementation. Inf. Process. Lett. **29**(1), 1–12 (1988)
30. Peñaloza, R., Sertkaya, B.: On the complexity of axiom pinpointing in the $\mathcal{EL}$ family of description logics. In: KR (2010)
31. Riguzzi, F., Bellodi, E., Lamma, E., Zese, R.: Probabilistic description logics under the distribution semantics. Semant. Web **6**(5), 477–501 (2015)
32. Schmidt-Schauß, M., Smolka, G.: Attributive concept descriptions with complements. Artif. Intell. **48**(1), 1–26 (1991)
33. Sebastiani, R., Vescovi, M.: Axiom pinpointing in lightweight description logics via horn-SAT encoding and conflict analysis. In: Schmidt, R.A. (ed.) CADE 2009. LNCS (LNAI), vol. 5663, pp. 84–99. Springer, Heidelberg (2009). doi:10.1007/978-3-642-02959-2_6
34. Sebastiani, R., Vescovi, M.: Axiom pinpointing in large $\mathcal{EL}^+$ ontologies via SAT and SMT techniques. Technical Report DISI-15-010, DISI, University of Trento, Italy, April 2015. http://disi.unitn.it/rseba/elsat/elsat_techrep.pdf
35. Simancik, F., Kazakov, Y., Horrocks, I.: Consequence-based reasoning beyond horn ontologies. In: IJCAI 2011, pp. 1093–1098 (2011). IJCAI/AAAI
36. Suntisrivaraporn, B.: Module extraction and incremental classification: A pragmatic approach for $\mathcal{EL}^+$ ontologies. In: ESWC, pp. 230–244 (2008)
37. Suntisrivaraporn, B.: Polynomial-Time Reasoning Support for Design and Maintenance of Large-Scale Biomedical Ontologies. Ph.D. thesis, TU Dresden (2009)
38. Suntisrivaraporn, B., Qi, G., Ji, Q., Haase, P.: A modularization-based approach to finding all justifications for OWL DL entailments. In: Domingue, J., Anutariya, C. (eds.) ASWC 2008. LNCS, vol. 5367, pp. 1–15. Springer, Heidelberg (2008). doi:10.1007/978-3-540-89704-0_1

Social Web and Web Science Track

Linked Data Notifications: A Resource-Centric Communication Protocol

Sarven Capadisli[1]([✉]), Amy Guy[2], Christoph Lange[1,3], Sören Auer[1,3], Andrei Sambra[4], and Tim Berners-Lee[4]

[1] University of Bonn, Bonn, Germany
info@csarven.ca, {langec,auer}@cs.uni-bonn.de
[2] School of Informatics, University of Edinburgh, Edinburgh, UK
amy@rhiaro.co.uk
[3] Fraunhofer IAIS, Sankt Augustin, Germany
[4] Decentralized Information Group, CSAIL, MIT, Cambridge, USA
deiu@mit.edu, timbl@w3.org

Abstract. In this article we describe the Linked Data Notifications (LDN) protocol, which is a W3C Candidate Recommendation. Notifications are sent over the Web for a variety of purposes, for example, by social applications. The information contained within a notification is structured arbitrarily, and typically only usable by the application which generated it in the first place. In the spirit of Linked Data, we propose that notifications should be reusable by multiple authorised applications. Through separating the concepts of *senders*, *receivers* and *consumers* of notifications, and leveraging Linked Data principles of shared vocabularies and URIs, LDN provides a building block for decentralised Web applications. This permits end users more freedom to switch between the online tools they use, as well as generating greater value when notifications from different sources can be used in combination. We situate LDN alongside related initiatives, and discuss additional considerations such as security and abuse prevention measures. We evaluate the protocol's effectiveness by analysing multiple, independent implementations, which pass a suite of formal tests and can be demonstrated interoperating with each other. To experience the described features please open this document in your Web browser under its canonical URI: http://csarven.ca/linked-data-notifications.

Keywords: Communications protocol · Decentralisation · Linked Data · Social Web

1 Introduction

Notifications are sent over the Web for a variety of purposes, including social applications: "You have been invited to a graduation party!", "Tim commented on your blog post!", "Liz tagged you in a photo". The notification data may be displayed to a human to acknowledge, or used to trigger some other application-specific process (or both). In a decentralised architecture, notifications can be a

© The Author(s) 2017
E. Blomqvist et al. (Eds.): ESWC 2017, Part I, LNCS 10249, pp. 537–553, 2017.
DOI: 10.1007/978-3-319-58068-5_33

key element for federation of information, and application integration. However in centralised systems which prevail today, this data is structured arbitrarily and typically only usable by the application that generated it in the first place. Current efforts towards *re-decentralising* the Web [1–3] are moving towards architectures in which data storage is decoupled from application logic, freeing end users to switch between applications, or to let multiple applications operate over the same data. So far, notifications are considered to be *ephemeral* resources which may disappear after transport, and thus are excluded from being designed for reuse.

We argue that notification data should not be locked into particular systems. We designed the *Linked Data Notifications (LDN)* protocol to support sharing and reuse of notifications *across* applications, regardless of how they were generated or what their contents are. We describe how the principles of identification, addressability and semantic representation can be applied to notifications on the Web. Specifying LDN as a formal protocol allows independently implemented, heterogeneous applications which generate and use notifications, to seamlessly work together. Thus, LDN supports the decentralisation of the Web as well as encourages the generation and consumption of Linked Data.

We build on existing W3C standards and Linked Data principles. In particular, the storage of notifications is compatible with the Linked Data Platform standard; notifications are identified by HTTP URIs; and notification contents are available as JSON-LD. A key architectural decision is the separation of concerns between *senders*, *receivers*, and *consumers* of notifications. Implementations of the protocol can play one or more of these roles, and interoperate successfully with implementations playing the complementary roles. This means that notifications generated by one application can be reused by a completely different application, accessed via the store where the notification data resides, through shared Linked Data vocabularies. LDN also pushes the decentralised approach further by allowing any *target* resource to advertise its Inbox anywhere on the Web; that is, targets do not need to be coupled to or controlled by a receiver, and can make use of a third-party *Inbox as a service*.

LDN is a W3C Candidate Recommendation via the Social Web Working Group [4]. The first two authors of this article co-edited the specification.

Use cases for decentralised notifications are particularly evident in social networking (status updates, interactions, games); scholarly communication (reviews, citations); and changes of state of resources (datasets, versioning, sensor readings, experimental observations). We describe the requirements which guided the development of the protocol and discuss related work, including current alternative approaches and complementary protocols which can work alongside LDN. We summarise the protocol itself, and specific architectural considerations that were made. We built a test suite which can be used to confirm that implementations conform with the specification, and we describe 17 implementations which interoperate with each other.

As the following terms used throughout this article may be subject to different interpretations by different communities, we provide some definitions here.

By decentralisation, we mean data and applications are loosely coupled, and users are empowered to choose where their data is stored or held. We focus on Web-based decentralisation, where content is transported over HTTP, and resources are identified with URIs. An Inbox is a container or directory (attached to a Web resource) which is used to store and serve a collection of notifications. A notification is a retrievable resource which returns RDF. The contents of notifications are intended to describe a change in state of some other resource, or contain new information for the attention of a user or process, and may be subject to constraints of the Inbox it is contained in.

2   Related Work

Here we review previous and ongoing efforts towards delivering notifications in a decentralised manner. Many systems which make use of notifications operate either in a completely centralised way, or are decentralised only in the sense that different instances of the *same* codebase need to interoperate; we restrict our review to mechanisms which do not expect the notification to be received or used only by the same software or platform which sent it.

The contents of a notification is either: (1) URLs, indicating relations between Web resources, or (2) a 'fat ping' containing a blob of information. Semantic Pingback, Webmention, and Provenance Pingback follow the first form, and are also known as linkbacks, the suite of protocols that essentially allows Web documents to automatically reciprocate hyperlinks. This has the advantage that a verification mechanism can be tightly specified (the URL of the target must appear in the content of the source), but the disadvantage that notifications are only available for use cases involving Web publishing.

Semantic Pingback [2] and Webmention [5] both update the original Pingback [6] mechanism by replacing the XML-RPC transport mechanism by a `x-www-form-urlencoded` request with two parameters (`source` and `target`). Resources which are the target for a notification advertise the respective receiving service or endpoint via a `Link` relation, either in HTTP headers or HTML. Semantic Pingback additionally enables discovery of the Pingback service where target is available as RDF. While the content at source may indicate (in any convention or serialisation format) the type of relation between the source and target URLs, this information about the relation is not transmitted to the receiver's endpoint; only the source and target URLs are sent. As such, there is also no way to distinguish between multiple potential mentions of the target at the source; this is left up to the receiver to interpret. Semantic Pingback does encourage generation of additional semantics about the relation(s) between the source and the target by processing the source as RDF if possible, and also defines specific ways for a receiving server to handle incoming pingback data in order to add the source data to an RDF knowledge base [2]. Beyond verifying that the source contains the URL of the target, Webmention does not specify any further requirements of the receiving server; nor is it expected that "mentions" are retrievable once they have been sent.

A Provenance Pingback endpoint is also advertised via the HTTP Link header; it accepts a list of URIs for provenance records describing uses of the resource [7]. Provenance Pingback does not specify any further behaviour by the receiving server, but the contents at the URIs listed in the notification body must be semantic data.

Other notification mechanisms send more information than just URLs in the notification body; due to each mechanism's focused use case, the payload is restricted to a particular vocabulary.

DSNotify is a centralised service which crawls datasets and observes changes to links with the specific use case of preserving link integrity between Linked Open Data resources. Third-party applications can register with the sending service to receive notifications of changes in the form of a specific XML payload [8]. With the sparqlPuSH service, users may input a SPARQL query, the results of which are the specific updates they are interested in. The query is run periodically by the service, and the results are converted to RSS and Atom feeds, which is sent to a PubSubHubbub hub to which the user can subscribe [9]. The ResourceSync Change Notification specification also sends update notifications via a PuSH hub, this time with an XML payload based on the Sitemap format [10]. Each of these mechanisms is triggered by subscription requests. That is, a user must actively solicit messages from a particular service, rather than having a way for a service to select a notification target and autonomously discover where to send notifications to.

3 Requirements and Design Considerations

In this section we discuss our considerations for a Web notification protocol that conforms to the Linked Data design principles, as well as best practices for applications. We use these considerations to establish both concrete requirements and points of implementation-specific flexibility for the protocol.

3.1 R1 Modularity

To encourage modularity of applications, one should differentiate between different classes of implementation of the protocol. Two parties are involved in the creation of a notification: a *sender*, generating the notification data, and a *receiver*, storing the created resource. We also have the role of a *consumer*, which reads the notification data and repurposes it in some way. A software implementation can of course play two or all three of these roles; the important part is that it need not. A consuming application can read and use notification data without being concerned about ever sending or storing notifications.

3.2 R2 Reusable Notifications

The relationship between the *consumer* and *receiver* roles is key to notifications being reusable. A consumer must be able to autonomously find the location of notifications for or about the particular resource it is interested in. To achieve this we place a requirement on the receiver to expose notifications it has been sent in such away to permit other applications to access them; and specify how any resource can advertise its receiving endpoint for consumers to discover. To promote fair use or remixing of notification contents, applications can incorporate rights and licensing information into the data. Similarly, applications may include additional information on licensing resources that the notification refers to. The presence of this type of information is important for consumers to assess the (re)usability of data.

3.3   R3 Persistence and Retrievability

There is a social expectation and technical arguments for ensuring the persistence of identifiers of Web resources [11]. This is inconsistent with the traditionally ephemeral nature of notifications. Applications may benefit from referring to or reusing notifications if the notifications are known to be available in the long term, or indicate their expected lifespan [12].

A *RESTful architecture* [13] is well suited for persistent notifications, as it involves organisation of atomic resources, their discovery and description, and a lightweight API for the CRUD (create, read, update, and delete) operations [14]. This enforces the notion that notifications are considered resources in their own right, with their own dereferencable URIs.

We need to consider both the needs of software systems and humans when large amounts of notification data are being generated and shared between diverse applications which may be operating without knowledge of each other. To organise and manage large amount of notifications over time, mechanisms should be in place to break representations of collections of notifications into multiple paged responses that may be easier to consume by applications.

Relatedly, receivers may carry out resource management or garbage collection, or permit consumers or other applications to do so. For example, an application to consume messages might let an authenticated and authorised user 'mark as read' by adding a triple to the notification contents.

3.4   R4 Adaptability

Linked Data applications benefit from domain-driven designs; that is, functionality being small and focussed on a particular purpose, rather than generic. We believe a notification protocol should be adaptable for different domains, but that there is no need to create multiple domain-specific notification protocols; the fundamental mechanics are the same.

R4-A: Any resource may be the *target* of a notification. By target, we mean a notification may be addressed *to* the resource, be *about* the resource, or for a sender to otherwise decide that it is appropriate to draw the attention of the resource (or resource owner) to the information in the notification body. As such, any Web resource must be able to advertise an endpoint to which it can receive notifications. Resources can be RDF or non-RDF (such as an image, or CSV dataset), and may be informational (a blog post, a user profile) or non-informational (a person).

R4-B: We do not purport to be able to design a notifications ontology which is appropriate for every domain. Thus we consider the *contents* of a notification to be application specific. From a sender's perspective, we derive two core principles: a notification can contain *any data*; a notification can use *any vocabulary*. From a consumer's perspective, interoperability between different applications occurs through vocabulary reuse, and shared understanding of terms. This is in accordance with Linked Data principles in general. The practical upshot of this is that a calendar application which consumes event invitations using the RDF Calendar vocabulary is likely to completely ignore notifications containing the PROV Ontology, even if it finds them all stored in the same place. For two independent applications operating in the *same* domain, a shared understanding of appropriate vocabulary terms is assumed.

However from a receiver's perspective, exposing itself to receive any blobs of RDF data from unknown senders may be problematic. Thus, R4-C: it should be possible for the receiver to enforce restrictions and accept only notifications that are acceptable according to its own criteria (deemed by e.g., user configuration; domain-specific receivers). This can be used as an anti-spam measure, a security protection, or for attaining application and data integrity.

Rejecting notifications which do not match a specific pattern in their contents, or the *shape* of the data, is one way to filter. For example, if the Inbox owner knows that they will only ever use a consuming application which processes friend requests, they can configure their receiver to filter out anything that does not match the pattern for a friend request, helping their consumer to be more efficient. If the notification constraints are also advertised by the receiving service as structured descriptions, generation and consumption of the notifications can be further automated. Possible specifications for doing so are W3C Shapes Constraint Language (SHACL) [15] or ShEx.

Receivers may wish to filter notifications by verifying the sender, through for example a whitelist or a Web of trust. This requires an authentication mechanism and since different authentication mechanisms are appropriate for different applications, the notification protocol should ideally be usable alongside various methods such as clientside certificates, e.g., WebID+TLS, token-based, e.g., OAuth 2.0, or digital signatures.

As "anyone can say anything about anything" a receiver may choose to resolve any external resources referred to by the notification, and cross-check the notification contents against authoritative sources. This is similar to how Semantic Pingback and Webmention require fetching and parsing of the source URL to verify existence of the target link.

3.5 R5 Subscribing

In general, applications may require that new notifications are pushed to them in real-time, or to request them at appropriate intervals. To take this into account, we expand our definition of senders, receivers and consumers with the following interaction expectations: notifications are *pushed* from senders to receivers; and *pulled* from receivers by consumers.

Thus, an application which offers an endpoint or callback URL to which notifications should be sent directly is a receiver, and an application which fetches notifications from an endpoint on its own schedule is a consumer. Much of the related work *requires* notifications to be explicitly solicited to trigger sending. Since in a decentralised model, receivers may not be aware of possible sources for notifications, our sender-receiver relationship depends on the sender's autonomy to make such decisions by itself. This does not preclude the scenario in which a receiver may wish to solicit notifications from a particular sender, but as there are already subscription mechanisms in wide use on the Web, we do not need to specify it as part of LDN. For example, WebSub (recent W3C evolution of PubSubHubbub), the WebSocket Protocol, or HTTP Web Push.

Given our adoption of Linked Data principles and a RESTful architecture, a further design decision was to ensure minimal compatibility with the Linked Data Platform (LDP) specification [16]. LDP is a RESTful read-write API for RDF resources, which groups related resources together into constructs known as "Containers". Thus, existing LDP servers can be used to store notifications, as new notifications can be created by POSTing RDF to a container.

4 The LDN Protocol

The *Linked Data Notifications (LDN)* protocol describes how servers (receivers) can receive messages pushed to them by applications (senders), as well as how other applications (consumers) may retrieve those messages. Any resource can advertise a receiving endpoint (Inbox) for notification messages. Messages are expressed in RDF, and can contain arbitrary data. It is not dependent on a complete implementation of LDP, but comprises an easy-to-implement subset (Fig. 1).

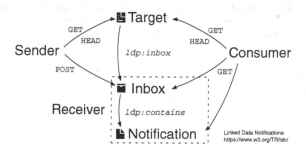

Fig. 1. Overview of Linked Data Notifications

4.1   Sender to Receiver Interactions

The following steps (in order without skipping) describe the interaction between sender and receiver:

(1) A sender is triggered, either by a human or an automatic process, to deliver a notification; (2) The sender chooses a target resource to send notifications to; (3) The sender discovers the location of the target's *Inbox* through the ldp:inbox relation in the HTTP Link header or RDF body of the target resource; (4) The sender creates the body of the notification according to the needs of application; (5) The sender makes a POST to the Inbox URL, containing the body in JSON-LD or in another serialisation acceptable by the server; (6) The receiver optionally applies filtering rules, and sends the appropriate HTTP response code to accept or reject the notification; (7) The receiver exposes the notification data (according to appropriate access control) for use by consumers.

4.2   Consumer to Receiver Interactions

The following steps (in order without skipping) describe the interaction between consumer and receiver:

(1) A consumer selects a target and discovers the location of its Inbox in the same way as the sender; (2) A receiver responds to GET requests made to the Inbox URL with a listing of the URLs of notifications that have previously been accepted, linked to the Inbox with the ldp:contains predicate; (3) The receiver responds to GET requests made to the individual notification URLs with JSON-LD (or optionally other serialisations); (4) Following the retrieval of notification listings or individual notifications, the consumer may perform further processing, combine with some other data, or simply present the results in a suitable human-readable way.

4.3   Example Notifications

For more example notification payloads, see the LDN specification.

```
1  {"@context": { "sioc": "http://rdfs.org/sioc/ns#" }
2    "@id": "",
3    "@type": "sioc:Comment",
4    "sioc:content": "This is a great article!",
5    "sioc:reply_of": { "@id": "http://example.org/article" },
6    "sioc:created_at": { "@value": "2015-12-23T16:44:21Z" } }
```

A notification about a comment created by a user (JSON-LD).

```
1   @prefix as: <https://www.w3.org/ns/activitystreams#> .
2   @prefix cito: <http://purl.org/spar/cito/> .
3   <> a as:Announce
4     as:object <https://linkedresearch.org/resources#r-903b83> ;
5     as:target <http://csarven.ca/dokieli#architecture> .
6   <https://linkedresearch.org/resources#r-903b83>
7     cito:citesAsPotentialReading
8       <http://csarven.ca/linked-data-notifications#protocol> .
```

An announcement of a specific citation relation between two entities (Turtle).

5   Implementations

Here we summarise the 17 LDN implementations we are aware of to date. They are built by 10 different teams or individuals using different tool stacks (5 clientside JavaScript, 3 PHP, 3 NodeJS, 3 Python, 1 Perl, 1 Virtuoso Server Pages, 1 Java) and have submitted implementation reports as part of the W3C standardisation process. We note that any LDP implementation is a conforming LDN receiver; we refer here to the ones we have tested. We discuss the value of these implementations further in the Evaluation section (Table 1).

We highlight social scholarly communication use cases with dokieli, a clientside editor for decentralised scientific article publishing, annotations and social interactions [17]. dokieli uses LDN to send and consume notifications: When a reader comments on a fragment of text in an article, the application discovers the article's Inbox and sends a notification about the annotation. dokieli also consumes notifications from this Inbox to fetch and display the annotation as marginalia (Fig. 2). A reader can share a dokieli-enabled article with their contacts; dokieli discovers each contact's Inbox and sends a notification there (Fig. 3). When editing an article, the author can add a citation. If an Inbox is discovered in the cited article, dokieli sends a notification there to indicate what part of the article was cited by whom and where. dokieli-enabled articles also consume citation notifications to display these metrics for the author and other readers (Fig. 4).

Notifications sent by dokieli can be reused by any consuming applications that recognise the vocabulary terms; similarly, dokieli can consume notifications sent by different applications.

Further social use cases are demonstrated by sloph, a personal publishing and quantified self platform which acts as a node in a decentralised social network. When new content is created on the server, sloph performs discovery on URLs it finds as values of particular properties of the new content, as well as any URLs in the body of the content, and sends notifications accordingly. For instance:

- If a *Like* activity is generated on the server, sloph uses the `object` of the *Like* as the target for a notification. Since dokieli uses the same vocabulary for social interactions (ActivityStreams 2.0 [18]), if the target is a dokieli article, this *Like* will be displayed (Fig. 4).

Table 1. LDN implementations

Implementation	Class*	Description
CarbonLDP	R	Data storage platform (LDP)
dokieli[a]	S, C	Clientside editor and annotator
errol[a]	S	Generic message sending client
Fedora Commons	R	Open source repository platform (LDP)
IndieAnndroid	R	Personal blogging platform
Linked Edit Rules	S	Statistical dataset consistency checker
mayktso[a]	R	Personal data store (LDP)
OnScreen[a]	C	Notifications display client
pyldn	R	Standalone Inbox
RDF-LinkedData-Notifications	R	Standalone Inbox
sloph[a]	S, R	Social publishing & quantified self
Solid Words	S	Foreign language learning app
solid-client	S	Clientside library for LDP
solid-inbox	C	Clientside social message reader
solid-notifications	S, C	Clientside library for LDN
solid-server	R	Personal data storage server (LDP)
Virtuoso+ ODS Briefcase	R, C	Personal data storage server (LDP)

*Conformance classes: S – sender, C – consumer, R – receiver.
[a]Implementations by the authors
Source: https://github.com/w3c/ldn

- If the user publishes a blog post containing a link, which may be semantically annotated to indicate the reason for linking, sloph sends a notification to any Inbox discovered at that link.
- As a receiver, sloph accepts all incoming notifications, but holds for moderation (i.e. places behind access control) any that it cannot automatically verify refer to third-party content published on another domain. If an article written with dokieli publishes a citation of a blog post which advertises a sloph Inbox, sloph will fetch the article and verify whether the relation matches the contents of the notification before exposing the notification for re-use.

Linked Edit Rules and Solid Words are specialised senders. Linked Edit Rules checks the consistency of statistical datasets against structured constraints, and delivers the consistency report as a notification to the user. Solid Words is a clientside game for learning new words in a foreign language; it delivers the player's score for each round to their Inbox. OnScreen is a (crude) generic consumer; as such, it can display notifications sent by both of the aforementioned senders (Fig. 5).

Fig. 2. Video of dokieli Web Annotation   **Fig. 3.** Video of dokieli Share

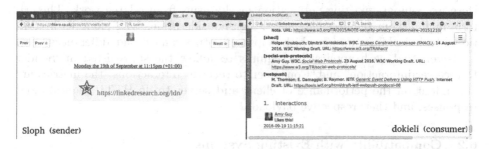

Fig. 4. A *Like* notification created by sloph, displayed by dokieli.

Fig. 5. A: Solid Words (a sender), B: Linked Edit Rules (a sender), C: OnScreen (a consumer) displaying notifications sent by A and B.

6   Analysis and Evaluation

The LDN protocol describes the discovery of a resource's Inbox whence notifications are sent or consumed, and the sending and exposure of those notifications. Here we analyse how well features of LDN achieve the requirements identified previously, and compare this to related work.

We have already examined implementations of the specification and described how they interoperate with each other; this can be further tested by running the test suite: https://linkedresearch.org/ldn/tests/. We can use this towards an evaluation of its feasibility and effectiveness at interoperability. Given the relatively early stage in the standardisation process (LDN entered Candidate Recommendation in 2016-11), the fast adoption of the LDN specification, quantity of the implementations, and their diversity is promising and further shows LDN's feasibility. Furthermore, during the development of the specification issues have been raised or discussed by 28 different people (excluding the authors; 21 outside of the Social Web Working Group, 7 within) and the specification has undergone formal review by internationalisation, accessibility, and security specialists. We also discuss in more depth particular challenges that were raised and resolved as part of this process.

6.1   Comparison Summary

Here we compare existing notification mechanisms from related work. The criteria includes our requirements and design considerations (Rx) along with additional technical information which helps to capture some design differences (Tx).

Given that each application requires to follow the steps listed in "Sender to receiver interaction" and "Consumer to receiver interactions" the metrics are dependent on the performance of client and server to do HTTP requests and responses, and their respective payloads.

6.2   Compatibility with Existing Systems

Per R1 and R4 we have tried to optimise LDN for use as a module of a larger system. The success of this is demonstrated by implementations which use LDN alongside existing protocols according to their specific needs.

The Solid suite of tools, Virtuoso+ODS-Briefcase, and dokieli use Web Access Control along with an authentication mechanism to apply fine grained access controls to restrict who can send notifications, or who can retrieve notifications from the Inbox. sloph demonstrates an Inbox as a Webhooks callback URL, for requesting notifications from APIs which post JSON-based payloads. ActivityPub is a W3C CR for decentralised social media [19]. It uses LDN for delivery of notifications with the ActivityStreams 2.0 (AS2) vocabulary, and specifies additional specialised receiver behaviour; also used by sloph. dokieli uses the Web Annotation Protocol, an LDP-based mechanism for creating new content, which acts as a trigger for notifications to be sent to the Inbox of the annotation target. The Fedora API Specification is in the process of being formalised (as an extension of LDP) by the Fedora community. The repository event stream draws upon the LDN specification, allowing LDN consumers and senders to react asynchronously to repository events.

Any existing LDP implementation can serve as an LDN receiver. Simply advertising any `ldp:Container` as the Inbox for a resource is sufficient. We confirmed this with four LDP servers which were developed independently with different code bases, prior to the LDN specification (CarbonLDP, Fedora Commons, Solid Server, Virtuoso).

LDN has been integrated into existing domain specific systems: dokieli, Fedora Commons, IndieAnndroid, Linked Edit Rules, sloph, solid-client,

Solid Words. Standalone implementations of LDN are also straightforward as a result of this modularity, i.e.: errol, mayktso, onscreen, pyLDN, RDF-LinkedData-Notifications, solid-inbox, solid-notifications (Table 2).

6.3 Optimising Implementation

We have considered tradeoffs between the HTTP operations receivers and publishers are *required* to respond to, and ways in which developers may wish to optimise senders or consumers by reducing outbound requests.

HEAD requests are low cost, and GET requests may be high cost if the body of the resource is large.

Given that an Inbox may be discovered from the HTTP headers of a resource, senders and consumers can optimise by attempting a HEAD request for discovery, and only continuing with a GET request if the HEAD is not successful. On the other hand, senders and consumers may be attempting discovery upon RDF resources which they already intend to parse into their own storage. In this case, there is no need for a HEAD request, as a GET will yield both HTTP Link headers and an RDF body, either of which could include the Inbox triple. This means that resources advertising an Inbox must respond to GET requests (even if only with HTTP headers) and may respond to HEAD requests.

6.4 Data Formats and Content Negotiation

Handling data irrespective of the particular RDF serialisation permits some flexibility, but can be costly to support. We take into account: (a) application interoperability, (b) maintenance of RDF parsers and serialisation libraries, (c) complexity of their inclusion in applications, (d) run-time efficiency.

To address these issues, LDN requires all applications to create and understand the JSON-LD syntax, both for the contents of Inbox as well as for individual notifications. Choosing a single serialisation to *require* is necessary for consistent interoperability, as well as keeping processing requirements or external code dependencies minimal.

JSON-LD is advantageous in being familiar for developers who are used to JSON-based APIs but not RDF [20], and it is compatible with existing JSON libraries or in some cases native programming language data structures.

Optionally, applications may attempt to exchange different RDF serialisations by performing content negotiation (receivers can expose Accept-Post headers for senders, and consumers can send Accept headers to receivers).

6.5 Precision

In placing no constraints on the contained information, LDN enables a sender to be precise and lossless with the data it is transmitting. Approaches which send only URLs rely on the receiver interpreting a third-party resource, which may or may not contain structured markup or be under the control of the sender.

Table 2. Comparison of notification mechanisms

Mechanism	T1	T2	T3	R1	R2	R3	R4-A	R4-B	R4-C^P	R4-C^v	R4-C^o	R5
Semantic Pingback	Linkback	POST	RDF	S R	/	/	Any^r	form urlencoded^k	!	! parse source	Any^r	X
Webmention	Linkback	POST	HTML	S R	-	-	Any^h	form urlencoded^k	!	! parse source	Any^h	X
Provenance Pingback	Linkback	OST	RDF	S R	/	/	/	URI list	/	/	RDF^q	X
DSNotify	Fat ping	POST, PUT	XML, PuSH	S U	/	-	-	XML	/	-	RDF^t	!
sparqlPuSH	Fat ping	POST	XML, SPARQL, PuSH	S U	-	-	-	XML^{ra}	/	-	RDF^t	!
ResourceSync	Fat ping	POST	XML, PuSH	S U	/	-	-	XML^s	/	-	?	!
Linked Data Notifications	Fat ping	POST	JSON-LD	S R C	!	! URI	Any	JSON-LD^j	+ app	+ app	-	O app

T1: Notification type

T2: Delivery method

T3: Dependencies

R1: Modularity (application classes: S Sender, R Receiver, C Consumer, U Subscriber/User)

R2: Reusability

R3: Persistence - required? how?

-: not applicable, out of scope

/: not specified, in scope

X: explicitly disallowed

app: application specific decision

^{h}: HTML recommended

^{j}: Alternate RDF formats can be negotiated

^{k}: source and target key-value pairs is required

^{q}: Provenance records with PROV Ontology

R4-A: Target representation

R4-B: Notification body

R4-C^P: Payload processing required?

R4-C^v: Verification - required? how?

R4-C^o: Requirements for referenced resources?

R5: Subscription

!: required (*MUST*)

+: recommended (*SHOULD*)

O: optional (*MAY*)

PuSH: PubSubHubbub

^{r}: RDF representation recommended

^{ra}: SPARQL results transformed to RSS/Atom

^{s}: Sitemaps

^{t}: Described in an RDF store or dataset

Approaches which offer additional guidance to aid the receiver in interpreting the source document(s) nonetheless still restricts the sender. LDN therefore offers flexibility to senders, increasing the potential uses for the notification mechanism. LDN compensates for increased complexity on the receiver's end by recommending filtering mechanisms, and moving some of the burden of understanding notifications to the consumer role. As such LDN can cover a broader variety of use cases.

6.6 Accommodating Different Targets

Per *R4 Adaptability*, we want LDN to be available for all resources in any publishing context. We consider lowering the bar for publishers of target resources to be a worthwhile trade-off against slightly increased complexity for senders and consumers. This is why we require that senders and consumers must be equipped to discover Inboxes through both HTTP headers and RDF content.

Since binary formats such as images and video cannot contain an RDF relation, the HTTP header is essential for including them. It also allows the inclusion of resources for which it is undesirable or impractical to add individual Inbox relations, such as to elements in a dataset; or circumstances where the developer responsible for the Inbox relation is unable to modify the content. Conversely, non-informational resources (represented with fragment URIs or 303 redirects) are unable to express HTTP headers. Their relation to an Inbox must be expressed in an RDF source. However, if a sender or consumer has a domain-specific requirement to *only* ever target non-informational resources, they are exempt from the requirement of discovery via HTTP headers.

7 Conclusions

In this article we describe LDN, a protocol for decentralised semantic notifications, currently undergoing standardisation at the W3C. Key elements are:

- Notifications as retrievable, reusable entities with their own URIs.
- Distinct conformance classes for senders, receivers, and consumers.
- Deliberately not defining the vocabulary of notification contents to allow for use in a range of different application domains.
- Flexibility of authentication and verification, for the same reason.

We outlined design requirements, describe how LDN meets these, and compare this with related work. We consider LDN to have greater modularity and adaptability to different scenarios, as well as good conformance with Linked Data principles. This specification has potential to have high impact in increasing interoperability between decentralised Linked Data applications in related domains, as well as generating new discoverable content for the LOD Cloud. This is evidenced by 17 diverse implementations which can be shown to interoperate with each other, including generic libraries and datastores, and domain-specific applications. Being on the W3C standards track increases the likelihood of further adoption.

Acknowledgements. The initial work on the Inbox was done by the Solid project at MIT. This material is based on work supported by the Qatar Computing Research Institute (QCRI), and by the DFG project "Opening Scholarly Communication in Social Sciences" (grant agreement AU 340/9-1).

References

1. Mansour, E., Sambra, A., Hawke, S., Zereba, M., Capadisli, S., Ghanem, A., Aboulnaga, A., Berners-Lee, T.: A demonstration of the solid platform for social web applications. In: WWW, Demo (2016)
2. Tramp, S., Frischmuth, P., Ermilov, T., Shekarpour, S., Auer, S.: An architecture of a distributed semantic social network. Semant. Web J. **5**(1), 77–95 (2014)
3. Arndt, N., Junghanns, K., Meissner, R., Frischmuth, F., Radtke, N., Frommhold, M., Martin, M.: Structured feedback. In: WWW, LDOW (2016)
4. Capadisli, S., Guy, A.: Linked data notifications, W3C Candidate Recommendation (2016). https://www.w3.org/TR/ldn/
5. Parecki, A.: Webmention, W3C Proposed Recommendation (2016). https://www.w3.org/TR/webmention/
6. Langridge, S., Hickson, I.: Pingback 1.0 (2002). http://www.hixie.ch/specs/pingback/pingback
7. Klyne, G., Groth, P.: PROV-AQ: Provenance Access and Query, W3C Note (2013). http://www.w3.org/TR/prov-aq/
8. Haslhofer, B., Popitsch, N.: DSNotify - detecting and fixing broken links in linked data sets. In: WWW (2010)
9. Passant, A., Mendes, P.N.: sparqlPuSH: proactive notification of data updates in RDF stores using PubSubHubbub. In: SFSW, vol. 699. CEUR-WS.org (2010)
10. Klein, M., Van de Sompel, H., Warner, S., Klyne, G., Haslhofer, B., Nelson, M., Lagoze, C., Sanderson, R.: ResourceSync framework specification - change notification (2016). http://www.openarchives.org/rs/notification/1.0/notification
11. Berners-Lee, T.: Cool URIs don't change, W3C (1998). https://www.w3.org/Provider/Style/URI.html
12. Archer, P., Loutas, N., Goedertier S., Kourtidis, S.: Study on persistent URIs (2012). http://philarcher.org/diary/2013/uripersistence/
13. Fielding, R.T.: Architectural styles and the design of network-based software architectures. Doctoral dissertation, University of California, Irvine (2000)
14. Page, K.R., De Roure, D.C., Martinez, K.: REST and linked data: a match made for domain driven development? In: WWW, WS-REST (2011)
15. Knublauch, H., Kontokostas, D.: Shapes constraint language, W3C Working Draft (2016)
16. Speicher, S., Arwe, J., Malhotra, A.: Linked data platform, W3C Recommendation (2015). https://www.w3.org/TR/ldp/
17. Capadisli, S., Guy, A., Verborgh, R., Lange, C., Auer, S., Berners-Lee, T.: Decentralised Authoring, Annotations and Notifications for a Read-Write Web with dokieli (2017). http://csarven.ca/dokieli-rww
18. Snell, J., Prodromou, E.: Activity streams 2.0, W3C Candidate Recommendation (2016). https://www.w3.org/TR/activitystreams-core/
19. Webber, C., Tallon, J.: ActivityPub, W3C Candidate Recommendation (2016). https://www.w3.org/TR/activitypub/
20. Sporny, M.: JSON-LD and why i hate the semantic web (2014). http://manu.sporny.org/2014/json-ld-origins-2/

Crowdsourced Affinity: A Matter of Fact or Experience

Chun Lu[1,2(✉)], Milan Stankovic[1,2], Filip Radulovic[1], and Philippe Laublet[2]

[1] Sépage, 27 rue du chemin vert, 75011 Paris, France
{chun,milstan,filip}@sepage.fr
[2] STIH, Université Paris-Sorbonne, 28 rue Serpente, 75006 Paris, France
philippe.laublet@paris-sorbonne.fr

Abstract. User-entity affinity is an essential component of many user-centric information systems such as online advertising, exploratory search, recommender system etc. The affinity is often assessed by analysing the interactions between users and entities within a data space. Among different affinity assessment techniques, content-based ones hypothesize that users have higher affinity with entities similar to the ones with which they had positive interactions in the past. Knowledge graph and folksonomy are respectively the milestones of Semantic Web and Social Web. Despite their shared crowdsourcing trait (not necessarily all knowledge graphs but some major large-scale ones), the encoded data are different in nature and structure. Knowledge graph encodes factual data with a formal ontology. Folksonomy encodes experience data with a loose structure. Many efforts have been made to make sense of folksonomy and to structure the community knowledge inside. Both data spaces allow to compute similarity between entities which can thereafter be used to calculate user-entity affinity. In this paper, we are interested in observing their comparative performance in the affinity assessment task. To this end, we carried out a first experiment within a travel destination recommendation scenario on a gold standard dataset. Our main findings are that knowledge graph helps to assess more accurately the affinity but folksonomy helps to increase the diversity and the novelty. This interesting complementarity motivated us to develop a semantic affinity framework to harvest the benefits of both data spaces. A second experiment with real users showed the utility of the proposed framework and confirmed our findings.

Keywords: Crowdsourcing · Affinity · Similarity · Semantic · Knowledge graph · Folksonomy · Travel · e-tourism · Semantic affinity framework · Diversity · Novelty

1 Introduction

User-entity affinity is the likelihood of a user to be attracted by an entity or to perform an action (click, purchase, like, share) related to an entity. The entity can be book, film, artist etc. User-entity affinity has a big impact from both economic and user experience point of view. It is an essential component of many user-centric information systems such as online advertising systems, exploratory search systems and recommender systems. It is crucial for predicting the click-through rate which is central to the multi-billion-dollar online advertising industry [1]. In exploratory search systems, it is

© Springer International Publishing AG 2017
E. Blomqvist et al. (Eds.): ESWC 2017, Part I, LNCS 10249, pp. 554–570, 2017.
DOI: 10.1007/978-3-319-58068-5_34

leveraged to retrieve interesting entities that might satisfy a user's fuzzy intention [2]. It is proper to recommender systems which are designed to mitigate the information overload by suggesting entities in affinity with the user.

Among different common affinity assessment techniques [3, 4], content-based ones [5] hypothesize that users would have higher affinity with entities that are similar to the ones with which they had positive interactions in the past. The emergence of knowledge graph and folksonomy has boosted this family of techniques by providing a large amount of data about entities [6].

Knowledge graph and folksonomy are respectively the milestones of Semantic Web and Social Web. On the Semantic Web, people contribute to the creation of large public knowledge graphs like DBpedia[1] and Wikidata[2] [7]. On the Social Web, people annotate and categorize entities with freely chosen texts called tags which form the folksonomy. Despite their shared crowdsourcing trait (not necessarily all knowledge graphs but some major large-scale ones above-mentioned), the encoded data are different in nature and structure. Knowledge graph encodes factual data with a formal ontology. Folksonomy encodes experience data with a loose structure. We give a concrete example to illustrate their difference. On DBpedia, the film dbr:Jumanji is linked to facts like dbr:Joe_Johnston by the property dbo:director and to dbr:Robin_Williams by the property dbo:starring. In the folksonomy of users of MovieLens[3], the same film is abundantly tagged with "nostalgic", "not funny", "natural disaster" etc. Even though these folksonomy tags are less formally structured, they reflect the experience that different users had with the film and thus a sort of intersubjectivity which is lacking in factual knowledge graph.

After in-depth study of the literature (Sect. 2), we struggled to find helpful insights about which data space is more effective in affinity-based systems. While both data spaces continue to proliferate on the web (Twitter hashtags, Instagram, Flickr, Mendeley) [8], it is more necessary than ever to shed some light on their comparative performance and contribution to the user-entity affinity assessment.

We conducted a first experiment within a travel destination recommendation scenario (Sect. 3). The findings motivated us to develop a semantic affinity framework which harvests the benefits of both Social Web and Semantic Web (Sect. 4). We used the proposed framework to compute a travel affinity graph which was evaluated in a second experiment with real users (Sect. 5). Section 6 concludes the paper with some advice for future development of affinity-based systems.

2   Related Work

For the past ten years, researchers have been closely studying the relatedness between knowledge graph and folksonomy. They are respectively the milestones of Semantic Web and Social Web. The general idea behind many research efforts is to enhance semantics in the Social Web with the help of Semantic Web technologies [9]. In the case

---

[1] http://wiki.dbpedia.org/about.
[2] https://www.wikidata.org/.
[3] https://movielens.org/.

of folksonomy, semantics are leveraged to (1) guide and control the tagging process (2) make sense of folksonomy. In [10], the authors proposed the MOAT ontology and a collaborative framework to guide tagging system users to specify the meaning of a tag with an existing resource on the Semantic Web. In [11], the author stated that although the semantics is much more implicit in folksonomy, the collective actions of a large number of individuals can still lead to the emergence of semantics. He suggested building lightweight ontologies from folksonomies. In [12], the authors used Semantic Web and natural language processing techniques to automatically classify folksonomy tags to four categories based on the intention: content-based, context-based, subjective, organisational. Some authors tried to ground folksonomy tags semantically by mapping pairs of tags in Del.icio.us to pairs of synsets in WordNet. Then WordNet-based measures for semantic distance are used to derive semantic relations between the mapped tags [13].

Some authors tried to mine user interests from folksonomies and other platforms of the Social Web. [14] observed that users have multiple profiles in various folksonomies and they proposed a method for the automatic consolidation of user profiles across different social platforms. Wikipedia categories was used to represent user interests. A similar method was proposed in [15] for the automatic creation and aggregation of interoperable and multi-domain user interest profile. Instead of using Wikipedia categories, user interests are represented with semantic concepts. The paper [16] also studied the multi-folksonomies problem. The authors found that the overlap between different tag-based profiles from different tagging systems is small and aggregating tag-based profiles lead to significantly more information about individual users, which impacts positively the personalized recommendations. Some recent work [17, 18] focused on one particular social platform: Twitter. They extract semantic concepts from tweets. These concepts are then enriched within the Wikipedia categorization system [17] or within a whole knowledge graph with multi-strategy enrichment [18].

Some authors studied the advantage of using folksonomies in some recommendation scenarios. The authors of [19] argued that although folksonomies provide structures that are formally weak or unmotivated, they are strongly connected with the actual use of the terms in them and the resources they describe. They may provide data about the perceptions of users, which is what counts in the recommendation context. The authors showed the advantage of folksonomies over using keywords in a movie recommendation scenario. Another study [20] in the cultural heritage domain showed that using both static official descriptions of items and user-generated tags about items allows to increase the precision of recommendations than using solely one of them.

User data mined from folksonomies are certainly useful for understanding the user and finding entities in affinity with him/her. But these data are not always easy to acquire for a system which is not built within social platforms. Il requires the system to ask users to log in with their social accounts or to purchase user data from some data management platforms. In this paper, we are not interested in social tagging actions of one particular user but the community knowledge generated by all users' tagging actions.

In [21], a class of applications called collective knowledge systems was proposed and was aimed at unlocking the collective intelligence of the Social Web with knowledge representation and reasoning techniques of the Semantic Web. On the location-based social network Foursquare, the venue similarity partially relies on the "taste" similarity[4]. Their taste map[5] was launched on 2014. It contains more than 10,000 short descriptive tags for restaurants sourced from 55 million tips[6]. Tastes can be as simple as a favourite dish like "soup dumplings" or a vibe like "good for dates". This collective knowledge is useful to characterize different venues and calculate their similarity. In [22], the authors proposed a data structure named "tag genome" which extends a traditional tagging model. It records how strongly each tag applies to each item on a continuous scale. It encodes each item based on its relationship to a common set of tags. As a case study, a tag genome is computed for MovieLens[7]. Since the tag genome is a vector space model, the entity similarity can be calculated with measures like cosine similarity.

In the literature about the semantic similarity in knowledge graph, we can find papers in two general directions: (1) similarity between classes [23] (2) similarity between entities. In this paper, we are interested in the latter direction.

In [24–28], different algorithms were proposed to compute the semantic similarity between entities by exploring the semantic properties. In [24], the author proposed an algorithm named Linked Data Semantic Distance (LDSD) which calculates the dissimilar degree of two resources in a semantic dataset such as DBpedia. In [25], the authors proposed an algorithm which is built on the top of LDSD. The modification consists of incorporating normalizations that use both resources and global appearances of paths. In [26], the authors proposed to use a vector space model to compute the similarity between RDF resources. It calculates a similarity score between two entities at the same property then sum all property-level similarity scores. These papers consider only three types of properties between two entities (1) direct property (2) outbound property pointing to the same entity (3) inbound property pointing from the same entity. The paper [27] proposed the *SPrank* algorithm to compute top-N item recommendations exploiting the information available on Linked Open Data. It consists of exploring paths in a semantic graph to find items that are related to the ones the user is interested in. A supervised learning to rank approach is leveraged to find what paths are most relevant for the recommendation task. A recent approach named RDF2Vec learns latent numerical representations of entities in RDF graphs [28].

Some variants of the spreading activation algorithm were presented in [29–31]. In [29], the algorithm aims to make cross-domain recommendation in DBpedia. [30] is in the same direction and the authors gave a concrete example of recommending places of interest from music artists. In [31], the algorithm is used to support an exploratory search system. The system retrieves similar/related entities to the ones initially entered by the user.

After in-depth study of the literature, we did not find any study comparing how the entity similarity calculated with knowledge graph and folksonomy performs in the user-entity affinity assessment task. We also did not find any approach with clear instructions

---

[4] http://engineering.foursquare.com/2015/12/08/finding-similar-venues-in-foursquare/.

[5] http://engineering.foursquare.com/2014/10/21/exploring-the-foursquare-taste-map/.

[6] Statistics reported in this article: http://www.theverge.com/2014/8/6/5973627/foursquare-8-review-the-ultimate-food-finder.

[7] http://www.movielens.org.

on how to tackle the affinity challenge in modern-day e-commerce systems such as e-tourism systems. This paper represents an initiative towards shedding light on these issues. We hope that the findings of our study can guide future design and development of affinity-based systems.

3 First Experiment: Gold Standard Study

We conducted an experiment within a travel destination recommendation scenario. It is a real and important problem. Web is today one of the most important sources for travel inspiration and purchase. More than 80% of people do travel planning online[8]. However, travelers feel bogged down by the myriad of options[9]. They feel overwhelmed and are obliged to spend a lot of time browsing multiple websites[10] before booking a trip. 68% of travelers begin searching online without having a clear travel destination in mind. Recommender systems can help travelers find more efficiently destinations in affinity with them. There is no consensual definition of travel destination, it can be a village, a city, a region or a country. In this paper, we consider cities as travel destinations. We use both terms "city" and "travel destination" interchangeably. In this section, we present an experiment within the travel destination recommendation scenario. We compare two representative approaches of knowledge graph and folksonomy within a gold standard dataset.

3.1 Experiment Dataset

To the best of our knowledge, there is no publicly available and widely used dataset for the evaluation of travel destination recommender systems. Thus, we use a dataset of our recent previous work [32]. It is constructed from Yahoo! Flickr Creative Commons 100 M (YFCC100 M) dataset[11] [33]. The original dataset contains 100 million of geotagged photos and videos published on Flickr. We processed the original dataset to make it suitable for our use case. Firstly, we took the file "yfcc100m_dataset". We filtered all the lines where latitude and longitude data were missing and where the accuracy level was below 16 (the highest accuracy level in Flickr). In other words, we retained only geotagged photos and videos with the highest geo-location accuracy. Secondly, we mapped each photo/video to a travel attraction entity in a travel knowledge graph constructed during that work. In the travel knowledge graph, travel attraction entities are linked to their city entities. So, once the mapping is done, we know the cities users have been to. We eliminated users who have been to only one city because in our evaluation, we need at least one city as user profile and another city as ground truth. Finally, for each user, we sorted the visited cities in a chronological order by considering the

---

[8] The 2013 Traveler: http://www.thinkwithgoogle.com//research-studies/2013-traveler.html.

[9] Reaching the connected customer: http://info.boxever.com/reaching-the-connected-customer.

[10] Custom Research: Exploring the Traveler's Path to Purchase: https://info.advertising.expedia.com/travelerspathtopurchase.

[11] Yahoo Webscope: http://webscope.sandbox.yahoo.com.

dates photos/videos were taken. After the processing, we know the travel sequence of each user. A travel sequence is a list of cities that a user visited in a chronological order. For example, the travel sequence "dbr:Munich, dbr:Stockholm, dbr:New_York_City" means that the user has visited respectively Munich, Stockholm and New York City. Table 1 shows the statistics about the experiment dataset.

Table 1. Statistics about the experiment dataset

# users	3878
# cities	705
avg# cities per user	5.27

The dataset is published for future benchmarking and reproducibility[12]. In the dataset, some users have the same sequence. We intentionally retained these seemly duplicated records because in a real-world scenario, some sequences are more frequent than others. A system which can produce high quality recommendations based on these sequences should be somehow "rewarded" or on the contrary "penalized".

3.2 Folksonomy Engineering

For the folksonomy part, we crawled data from the website of a collaborative travel platform where users are invited to tag cities after their trips there. They are restricted to use existing tags such as "Kayaking", "Great for wine", "People watching". The crawled dataset contains 234 tags about 26,237 cities in 154 countries. To give readers a clearer idea about the dataset, in Fig. 1, we show the distribution of tags in terms of the number of times they are applied (Applications) and the number of cities on which they are applied (Cities). However, due to the space limit and for a better readability,

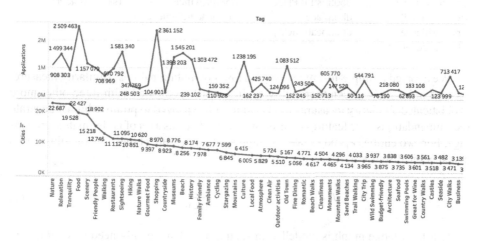

Fig. 1. Distribution of folksonomy tags in the travel domain

---

[12] https://bitbucket.org/sepage/semantic-affinity-framework.

we can only show a part of the distribution. We modeled this folksonomy dataset in a tag genome fashion [22]. The tag relevance score is calculated with the Term Frequency-Inverse Document Frequency scheme. In our case, terms are tags and documents are cities. As in [22], we compute the similarity between cities by calculating the cosine of the angle between their vectors.

3.3 Knowledge Graph Engineering

For the knowledge graph part, we manually selected some inbound and outbound properties shown in Table 2. We put *skos:broader* in brackets because it is not directly linked to cities but indirectly linked via *dct:subject*.

For each of the 705 cities in the evaluation dataset, we ran SPARQL queries with all selected properties. We gave a special treatment to the property *dct:subject*. For each retrieved direct linked category, we also retrieved its parent categories by using *skos:broader*. We put all direct and parent categories together, deduplicated the list and put it under the property *dct:subject*. Then we eliminated nodes which are linked to only one city because they do not contribute to the similarity calculation between two cities. 501365 nodes were initially retrieved. After the cleaning, only 29743 nodes were retained.

Table 2. Selected properties for calculating city similarity in knowledge graph

Inbound		Outbound	
dbo:birthPlace	dbo:broadcastArea	dbo:isPartOf	dbo:part
dbo:location	dbo:nearestCity	dbo:country	dbo:twinTown
dbo:deathPlace	dbo:ground	dbo:timeZone	dbo:saint
dbo:city	dbo:foundationPlace	dbo:Mayor	dbo:district
dbo:capital	dbo:assembly	dbo:region	dct:subject
dbo:hometown	dbo:restingPlace	dbo:province	(skos:broader)
dbo:recordedIn	dbo:place	dbo:leaderName	
dbo:residence	dbo:locationCity		
dbo:headquarter			

We adopted a simple-to-implement and low-computational-cost similarity measure: Jaccard index. It has been thoroughly evaluated and compared with three other more sophisticated similarity measures in [34]. It has been proved to produce highly accurate recommendations. The features of an entity are modeled as a set of nodes in its surroundings. For two entities e_1 and e_2, we do graph walking to collect their surrounding nodes at a specific distance d: $N_d(e_1)$ and $N_d(e_2)$.

$$J(e_1, e_2) = \frac{|N_d(e_1) \cap N_d(e_2)|}{|N_d(e_1) \cup N_d(e_2)|} \tag{1}$$

Our knowledge graph is modelled in such a way that we only need to set d to 1 to get all interesting nodes. With this measure, we can easily consider two interesting graph patterns in Fig. 2. For example, GP1 allows to capture that a person was born (dbo:birthPlace) in a city and resides (dbo:residence) in another city. GP2 allows to capture entities linked to different direct categories which have a common broader category, for example:

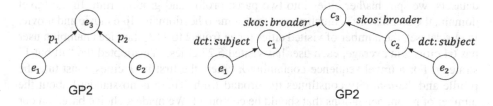

Fig. 2. Examples of two interesting graph patterns exploited by our similarity measure

   dbr:Pushkar → dbc:Hindu_holy_cities → dbc:Holy_cities ← dbc:Bethlehem ← dbr:Bethlehem
   dbr:New_Delhi → dbc:Capitals_in_Asia → dbc:Capitals ← dbc:Capitals_in_Europe ← dbr:Athens
   One-hop category has been proven to be useful in many personalization tasks [18, 26].

3.4 Candidate Approaches

Since our experiment aims to compare the performance of knowledge graph and folksonomy on user-city affinity assessment, we implemented two candidate approaches: FOLK and KG which use respectively the data and techniques presented in Sects. 3.2 and 3.3.

3.5 Common Affinity Prediction Algorithm

To ensure a fair environment for comparing them, we used a common affinity prediction algorithm. This assessment methodology was adopted in a comparative study on knowledge graph and similarity measure choices [34]. Given a user profile *profile(u)* containing a list of cities that the user u has visited in the past, the affinity score of a candidate city c_i is calculated with Eq. 2, which is the sum of pairwise similarity with each city in the user profile divided by the total number of cities in the user profile. The pairwise similarity $Sim(c_i, c_j)$ is calculated by the candidate approaches and feeds the common affinity prediction algorithm. The affinity score is only influenced by the similarity score calculated by the candidate approaches.

$$affinity(u, c_i) = \frac{\sum_{c_j \in profile(u)} Sim(c_i, c_j)}{|profile(u)|} \qquad (2)$$

3.6 Protocol

We use the "all but n" protocol. It is aligned with the common practice of offline experiments in the recommender system community [5, 35, 36]. For each user in the evaluation datasets, we split his/her cities into two parts: profile and ground truth. In the travel domain, the user history is much poorer than some other domains like music and movie. In our dataset, the number of visited cities range from 2 to 112 (the vice champion user has 64 cities), in average, each user has visited 5.27 cities. We adopted the "all but 1" strategy. For a travel sequence containing n cities, the first $n-1$ cities constitute the profile and the n-th city constitutes the ground truth. There is no standard about the number of recommendations that should be computed. We made a choice based on our past research work [37, 38], experience with the clients of Sépage company and practices on several popular travel websites (expedia, tripadvisor, kayak etc.). The number of recommendations depends on different contexts. In a recommendation or an advertising banner, the number is relatively limited, in an inspirational browsing environment, more cities are displayed. For these reasons, we decided to compute top-10, top-20 and top-30 recommendations.

3.7 Quality Dimensions and Metrics

In the recommendation scenario, the user-entity affinity assessment capacity can be most reflected by the accuracy of the recommendations. To measure the accuracy, we used two metrics: Success and Mean Reciprocal Rank (MRR).

$$Success = \frac{\sum_{u \in U} rel_{g,u}}{|U|} \text{ where } rel_{g,u} = \begin{cases} 1, & \text{if ground truth } g \text{ is in top} - N \\ 0, \text{ otherwise} \end{cases} \tag{3}$$

The Success metric (Eq. 3) calculates the number of users for whom the candidate approaches recommend cities that are in affinity with the users divided by the total number of users. It is an alternative metric to classic precision and recall which are not perfectly adapted in our case. Because each user's ground truth contains only one city. The precision and recall of each user are binary values. $1/N$ or 0 for the precision, 1 or 0 for the recall. For this reason, it would be more intuitive to compare the number of users for whom the system can actually recommend the ground truth.

The Mean Reciprocal Rank is calculated as in Eq. 4.

$$MRR = \frac{1}{|U|} \sum_{u \in U} \frac{1}{rank_u} \tag{4}$$

where $rank_u$ is the rank position of the ground truth of the user u. It shows how early the ground truth appears in the recommendation list. A higher MRR reveals a better capacity of a candidate approach to detect affinity cities. It is crucial if we can only recommend a very limited number of cities.

Currently, the recommender system community has a growing interest in generating diverse and novel recommendations, even at the expense of the accuracy. Apart from our main focus, we are also interested in knowing how our approaches perform on these two quality dimensions.

In ESWC 2014 Challenge on Linked Open Data-enabled Book Recommendation [35], the diversity was considered with respect to two properties in DBpedia: *dbo:author* and *dct:subject*. In our case, we considered the diversity with respect to *dbo:country* and *dct:subject*. The Eqs. 5 and 6 measure the intra-list similarity (ILS)

$$ILS_u@N = \sum_{i \in L_u^N} \sum_{j \in L_u^N} \frac{sim(i,j)}{|pairs|} \tag{5}$$

$$ILS@N = \frac{1}{|U|} \sum_{u \in U} ILS_u@N \tag{6}$$

where $sim(i,j)$ is the aggregated similarity score with respect to the two properties. We give equal importance (0.5) to them in this calculation. The higher ILS is, the less diverse the recommendation list is.

We calculate the novelty with respect to the capacity of recommending long-tail cities. Following the power law distribution[13], we consider the 80% less popular cities as long-tail cities. We use the DBpedia pagerank[14] value as the popularity index.

$$Novelty@N = \frac{number\ of\ recommended\ long-tail\ cities}{N * |U|} \tag{7}$$

3.8 Results and Discussions

In Table 3, we show the scores of the two approaches on four metrics when top-10, top-20 and top-30 recommendations are computed. Paired t-tests show that the differences between the two approaches among all metrics in all settings are statistically significant with $p < .01$.

Table 3. Scores of two candidate approaches on four metrics when top-10, top-20 and top-30 recommendations are computed

	Top-10		Top-20		Top-30	
	KG	FOLK	KG	FOLK	KG	FOLK
Success	**0.232**	0.06	**0.33**	0.116	**0.386**	0.166
MRR	**0.047**	0.003	**0.047**	0.003	**0.047**	0.003
ILS	0.257	**0.089**	0.208	**0.072**	0.176	**0.065**
Novelty	0.717	**0.824**	0.722	**0.772**	0.723	**0.755**

We can observe a net advantage of KG over FOLK in terms of success and MRR. Higher scores on success and MRR reflect the capacity of a system to detect cities in high affinity with the user and to give them better rankings. Recommendations produced

---

[13] https://en.wikipedia.org/wiki/Power_law.
[14] http://people.aifb.kit.edu/ath/.

by FOLK are generally more diverse and novel than those produced by KG. Actually, in the folksonomy, we ignore some aspects that are considered by DBpedia such as the geography (*dbo:country, dbo:region*), the people (*dbo:birthPlace, dbo:residence*), the related categories (*dct:subject, skos:broader*). The folksonomy contains travel-related traits like "Luxury Brand Shopping", "Clean Air", "Traditional food". These traits can be shared by different cities in the world and by less popular cities.

The very different performances of the two approaches on different quality dimensions led to us to believe in their complementarity in obtaining a balanced trade-off and in yielding recommendations that are equitably accurate, diverse and novel.

4   Semantic Affinity Framework

Motivated by the findings of our first experiment, we propose a Semantic Affinity Framework which harvests the benefits of both Semantic Web and Social Web. It is designed for user-centric information systems which aim to provide personalised user experience by leveraging the affinity. The framework integrates, aggregates, enriches and cleans entity (here we refer to main objects of the system, e.g., book, film, city) data from knowledge graphs and folksonomies. The Fig. 3 shows the pipeline of the framework.

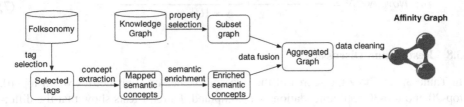

Fig. 3. Pipeline of semantic affinity framework, from folksonomy and knowledge graph to affinity graph

For the folksonomy data, a tag selection rule should be specified. For example, for data which are structured in the tag genome fashion, we can define a threshold above which tags can be considered as relevant for the entity. Then selected tags are mapped to semantic concepts by using concept extractors. For example, the tag "Skyline" can be mapped to "dbr:Skyline" with tools like Babelfy[15] and Dandelion[16]. After this, we can conduct semantic enrichment operations on mapped concepts, such as the 1-hop category enrichment (GP2 in Fig. 2). For example, "dbr:Skyline" can be enriched with "dbc:Skycrappers" and "dbc:Towers". The enriched semantic concepts are fusioned to an aggregated graph together with the subset knowledge graph resulting from a property selection process (like in Sect. 3.3). More precisely, the enriched semantic concepts are linked to the entities that they describe. In the aggregated graph, data from the knowledge graph use their original properties, data from the folksonomy use a common

---

[15] http://babelfy.org/.
[16] http://dandelion.eu/.

"has_characteristic" property. In the current version of the framework, the data cleaning process uses three heuristics: 1. Eliminating nodes linked to only one entity in the graph; 2. Deduplicating nodes which appear multiple times due to the fusion of processed knowledge graph and folksonomy data; 3. Privileging properties from the knowledge graph than "has_characteristic".

Finally, an affinity graph is generated. In addition to the recommendation usage, it provides the possibility to explain the recommendations [24, 39] in a feature style [36]. For example, one can recommend "Ljubljana" because of the feature "dbc:Capitals_in_Europe". Given a user profile containing a list of entities, the affinity graph searches the most common features shared by the entities. Since features are linked to entities explicitly with different properties, we have control on the diversity of the features to display to the user. We developed a diversity function which maximizes the number of properties of the displayed features. The function iterates on the list of features ordered by their occurrences and selects a feature if no other feature sharing the same property has been selected previously. The iteration ends when desired number of features are selected. In case it is not reached after an iteration on the whole list, more iterations are then conducted on the unselected features.

5 Second Experiment: User Study

To assess the usefulness and the efficiency of the Semantic Affinity Framework, we conducted a qualitative user study which is complementary to the quantitative evaluation on a gold-standard dataset. We computed an affinity graph with data from both knowledge graph and folksonomy. Apart from the aspects that were already mentioned in the first experiment, in this user study, we are especially interested in the explanation capacity of different approaches. This aspect is of qualitative nature and can only be assessed by questioning real users. Three approaches are compared: KG, FOLK (which are compared in the first experiment) and AG (which uses the affinity graph). The explanation generation mechanism is described for AG in Sect. 4. KG and FOLK compute explanations within their respective data spaces and follow the same logic with the exception that the diversity function is not applied on FOLK because there are no differentiated semantic properties.

5.1 Protocol and Metrics

We used a well-oiled protocol [40] to simulate a travel planning process. Firstly, participants put themselves into the scenario of looking for the destination for the next trip. They were free to choose to plan for a weekend trip or a long holiday. Secondly, they went to the evaluation interface where they could visualize the 705 cities. To reduce bias towards cities shown at the top of the page, the presentation order of the cities was randomized. Thirdly, they chose several cities that appealed to them at first glance. Fourthly, they submitted their choices and got three sets of top-5 highest scored cities generated by the three candidate approaches accompanied by 5 semantic concepts to explain the recommendations. In Fig. 4, we show an example about how the recommendations and explanations were presented. Finally, participants rated respectively the recommendations and the explanations as a whole in a five-level Likert scale on different quality dimensions. For the recommendations, the same dimensions as in the first experiment were reused: relevance, diversity and novelty. For the explanations, instead of novelty, we opted for interestingness. This dimension measures the capacity of arousing the attention or interest. Participants were guided by the exact meaning of the scale. For example, on the relevance dimension, the scale was: 1 – not relevant, 2 – weakly relevant, 3 – moderately relevant, 4 – relevant, 5 – strongly relevant. We consider 4 and 5 as positive ratings. We use the percentage of positive ratings as our metric.

You submitted:	You might like:	We recommend you:
dbr:Rome	dbc:Clothing	dbr:The_Hague
dbr:Florence	dbr:Food	dbr:Haarlem
dbr:Amsterdam	dbr:David_de_Haen	dbr:Naples
	dbr:Italy	dbr:Milan
	dbr:History	dbr:Turin

Fig. 4. Example of recommendations and explanations generated by the AG approach for a user having submitted dbr:Rome, dbr:Florence and dbr:Amsterdam

5.2 Results and Discussions

37 people participated in our study. They work in different companies located at the "Pépinière 27" in Paris, France at the moment of this study. They have between 25 to 38 years old, 19 males, 18 females.

The results on the recommendations are in line with the results obtained in our first experiment. KG yields clearly better accuracy than FOLK. FOLK has more advantage on diversity and novelty. AG obtains indeed balanced and good scores on all three dimensions (Fig. 5).

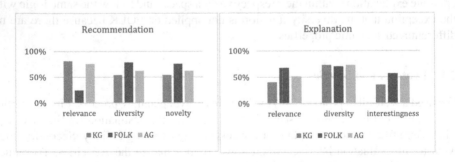

Fig. 5. Percentage of users having given positive ratings to recommendations and explanations

We discuss more about the explanations part which is not covered in the first experiment. The results showed that explanations provided by FOLK were the most appreciated. Our folksonomy dataset was crowdsourced by travelers, it is by nature highly relevant and it covers different travel aspects (food, activity, transport). The explanation capacity of AG was boosted by the inclusion of features from FOLK which allowed it to outperform KG where only knowledge graph features were used. Participants were relatively skeptical about some knowledge graph features. Some users found the features very general, for example, "dbc:Leisure" which comes from the 1-hop category enrichment. A possible solution to this problem is to use the DBpedia category tree [17]. Since we know the level of all categories, we can define a threshold above which categories and its related concepts are too general for the explanation task. Some users found some features difficult to understand such as "dbr: China_Record_Corporation". The user who got this explanation submitted "dbr:Shanghai", "dbr:Shenzhen", "dbr:Beijing". "dbr: China_Record_Corporation" is linked to all these three cities by "dbo:location". Actually we picked "dbo:location" because it allows to capture interesting links, for example it can link two cities via a television series. However, this property is also used by entities having the type "dbo:Company". This problem can be solved with additional engineering efforts, such as blacklisting certain types.

To sum up, knowledge graph allows to better yield entities in high affinity with the users, folksonomy performs better on diversity and novelty, and it also brings high quality explanations. Harvesting both data spaces, the affinity graph results in equitable and competitive performance on multiple quality dimensions in both recommendation and explanation tasks. To make explanations more user-friendly, additional engineering efforts are needed and it can be very helpful to leverage knowledge graphs especially ontology types and DBpedia category tree.

6 Conclusion

In this paper, we are interested in the problem of user-entity affinity assessment which is essential in many user-centric information systems. Among different assessment techniques, content-based ones predict higher affinity scores for entities similar to the ones with which a user had positive interactions in the past. Knowledge graph and folksonomy have boosted the similarity calculation by providing a large amount of data on entities. Despite the shared crowdsourcing trait between knowledge graph (some major large-scale ones e.g. DBpedia and Wikidata) and folksonomy, the encoded data are different in nature and structure. Knowledge graph encodes factual data with a formal ontology. Folksonomy encodes experience data with a loose structure. Existing work has proven their efficiency in separate settings. To the best of our knowledge, this paper is the first work trying to shed some light on their comparative performance in the affinity assessment task. We made a comprehensive state of the art. We have selected the most representative approach of each category for comparison. We conducted two experiments. The first one within a travel destination recommendation scenario on a gold standard dataset has shown a net advantage of knowledge graph in affinity assessment accuracy. However, folksonomy contributes more to two other important quality dimensions which are diversity and novelty. This interesting complementarity motivated us to develop the Semantic Affinity Framework to harvest the benefits of both knowledge graph and folksonomy. The framework integrates, aggregates, enriches and cleans entity data from both spaces, and finally produces an affinity graph. A second experiment with real users confirmed the findings of the first experiment and showed the utility and the efficiency of the proposed semantic affinity framework. In addition to the recommendation task, we evaluated the capacity of explaining the recommendations. The inclusion of folksonomy data in the affinity graph has clearly increased the relevance and interestingness of the explanations. The travel domain within which our two experiments were conducted is the predominant domain of e-commerce. We hope that our findings can guide the design and the development of affinity-based systems in this important domain. We also hope that the ideas and the methodology of this paper can serve as an instigator of further similar comparative studies in other domains.

References

1. McMahan, H.B., Holt, G., Sculley, D., Young, M., Ebner, D., Grady, J., Chikkerur, S.: Ad click prediction: a view from the trenches. In: Proceedings of the 19th ACM SIGKDD International Conference on Knowledge Discovery and Data Mining. pp. 1222–1230. ACM (2013)
2. Waitelonis, J., Sack, H.: Towards exploratory video search using linked data. Multimed. Tools Appl. **59**(2), 645–672 (2012)
3. Krulwich, B.: Lifestyle finder: intelligent user profiling using large-scale demographic data. AI Mag. **18**(2), 37 (1997)
4. Candillier, L., Meyer, F., Boullé, M.: Comparing state-of-the-art collaborative filtering systems. In: Perner, P. (ed.) MLDM 2007. LNCS (LNAI), vol. 4571, pp. 548–562. Springer, Heidelberg (2007). doi:10.1007/978-3-540-73499-4_41
5. Lops, P., De Gemmis, M., Semeraro, G.: Content-based recommender systems: state of the art and trends. In: Ricci, F., Rokach, L., Shapira, B., Kantor, P.B. (eds.) Recommender Systems Handbook, pp. 73–105. Springer, New York (2011)
6. Di Noia, T., Ostuni, V.C.: Recommender systems and linked open data. In: Faber, W., Paschke, A. (eds.) Reasoning Web 2015. LNCS, vol. 9203, pp. 88–113. Springer, Cham (2015). doi:10.1007/978-3-319-21768-0_4
7. Pellissier Tanon, T., Vrandečić, D., Schaffert, S., Steiner, T., Pintscher, L.: From freebase to wikidata: the great migration. In: Proceedings of the 25th International Conference on World Wide Web, pp. 1419–1428 (2016)
8. Schmachtenberg, M., Bizer, C., Paulheim, H.: Adoption of the linked data best practices in different topical domains. In: Mika, P., et al. (eds.) ISWC 2014. LNCS, vol. 8796, pp. 245–260. Springer, Cham (2014). doi:10.1007/978-3-319-11964-9_16
9. Bontcheva, K., Rout, D.: Making sense of social media streams through semantics: a survey. Semant. Web **5**(5), 373–403 (2014)
10. Passant, A., Laublet, P.: Meaning of a tag: a collaborative approach to bridge the gap between tagging and linked data. In: Proceedings of Linked Data on the Web Workshop (2008)
11. Mika, P.: Ontologies are us: a unified model of social networks and semantics. Web Semant. Sci. Serv. Agents World Wide Web **5**(1), 5–15 (2007)
12. Cantador, I., Konstas, I., Jose, J.M.: Categorising social tags to improve folksonomy-based recommendations. Web Seman. Sci. Serv. Agents World Wide Web **9**(1), 1–15 (2011)
13. Cattuto, C., Benz, D., Hotho, A., Stumme, G.: Semantic grounding of tag relatedness in social bookmarking systems. In: Sheth, A., Staab, S., Dean, M., Paolucci, M., Maynard, D., Finin, T., Thirunarayan, K. (eds.) ISWC 2008. LNCS, vol. 5318, pp. 615–631. Springer, Heidelberg (2008). doi:10.1007/978-3-540-88564-1_39

14. Szomszor, M., Alani, H., Cantador, I., O'Hara, K., Shadbolt, N.: Semantic modelling of user interests based on cross-folksonomy analysis. In: Sheth, A., Staab, S., Dean, M., Paolucci, M., Maynard, D., Finin, T., Thirunarayan, K. (eds.) ISWC 2008. LNCS, vol. 5318, pp. 632–648. Springer, Heidelberg (2008). doi:10.1007/978-3-540-88564-1_40

15. Orlandi, F., Breslin, J., Passant, A.: Aggregated, interoperable and multi-domain user profiles for the social web. In: Proceedings of the 8th International Conference on Semantic Systems, pp. 41–48. ACM (2012)

16. Abel, F., Herder, E., Houben, G.J., Henze, N., Krause, D.: Cross-system user modeling and personalization on the social web. User Model. User-Adap. Inter. 23(2–3), 169–209 (2013)

17. Kapanipathi, P., Jain, P., Venkataramani, C., Sheth, A.: User interests identification on twitter using a hierarchical knowledge base. In: Presutti, V., d'Amato, C., Gandon, F., d'Aquin, M., Staab, S., Tordai, A. (eds.) ESWC 2014. LNCS, vol. 8465, pp. 99–113. Springer, Cham (2014). doi:10.1007/978-3-319-07443-6_8

18. Piao, G., Breslin, J.: Exploring dynamics and semantics of user interests for user modeling on Twitter for link recommendations. In: Proceedings of the 12th International Conference on Semantic Systems (SEMANTiCS 2016), Leipzig, Germany (2016)

19. Szomszor, M., Cattuto, C., Alani, H., O'Hara, K., Baldassarri, A., Loreto, V., Servedio, V.D.P.: Folksonomies, the semantic web, and movie recommendation. In: Proceedings of the Workshop on Bridging the Gap between Semantic Web and Web 2.0 at the 4th ESWC (2007)

20. Semeraro, G., Lops, P., De Gemmis, M., Musto, C., Narducci, F.: A folksonomy-based recommender system for personalized access to digital artworks. J. Comput. Cult. Heritage (JOCCH) 5(3), 11 (2012)

21. Gruber, T.: Collective knowledge systems: where the social web meets the semantic web. Web Web Semant. Sci. Serv. Agents World Wide Web 6(1), 4–13 (2008)

22. Vig, J., Sen, S., Riedl, J.: The tag genome: encoding community knowledge to support novel interaction. ACM Trans. Interact. Intell. Syst. 2(3), Article 13 (2102)

23. Zhu, G., Iglesias, C.A.: Computing semantic similarity of concepts in knowledge graphs. IEEE Trans. Knowl. Data Eng. 29(1), 72–85 (2016)

24. Passant, A.: dbrec—music recommendations using DBpedia. In: Patel-Schneider, P.F., et al. (eds.) Proceedings of the 9th International Semantic Web conference. LNCS, vol. 6497, pp. 209–224. Springer, Heidelberg (2010)

25. Piao, G., Breslin, J.: Measuring semantic distance for linked open data-enabled recommender systems. In: Proceedings of the 31st Annual ACM Symposium on Applied Computing. ACM (2016)

26. Di Noia, T., Mirizzi, R., Ostuni, V.C., Romito, D., Zanker, M.: Linked open data to support content-based recommender systems. In: Proceedings of the 8th International Conference on Semantic Systems, pp. 1–8. ACM (2012)

27. Di Noia, T., Ostuni, V.C., Tomeo, P., Di Sciascio, E.: Sprank: semantic path-based ranking for top-n recommendations using linked open data. ACM Trans. Intell. Syst. Technol. (TIST) 8, 9 (2016)

28. Ristoski, P., Paulheim, H.: RDF2Vec: RDF graph embeddings for data mining. In: Groth, P., Simperl, E., Gray, A., Sabou, M., Krötzsch, M., Lecue, F., Flöck, F., Gil, Y. (eds.) ISWC 2016. LNCS, vol. 9981, pp. 498–514. Springer, Cham (2016). doi:10.1007/978-3-319-46523-4_30

29. Heitmann, B.: An open framework for multi-source, cross-domain personalisation with semantic interest graphs. Doctoral dissertation. National University of Ireland, Galway (2014)

30. Kaminskas, M., Fernández-Tobías, I., Ricci, F., Cantador, I.: Knowledge-based identification of music suited for places of interest. Inf. Technol. Tourism 14(1), 73–95 (2014)

31. Marie, N.: Linked data based exploratory search. Doctoral dissertation. Université de Nice Sophia-Antipolis (2014)

32. Lu, C., Laublet, P., Stankovic, M.: Travel attractions recommendation with knowledge graphs. In: Blomqvist, E., Ciancarini, P., Poggi, F., Vitali, F. (eds.) EKAW 2016. LNCS (LNAI), vol. 10024, pp. 416–431. Springer, Cham (2016). doi:10.1007/978-3-319-49004-5_27

33. Thomee, B., Shamma, D.A., Friedland, G., Elizalde, B., Ni, K., Poland, D., Li, L.J.: YFCC100M: the new data in multimedia research. Commun. ACM **59**(2), 64–73 (2016)

34. Nguyen, P.T., Tomeo, P., Noia, T., Sciascio, E.: Content-based recommendations via DBpedia and freebase: a case study in the music domain. In: Arenas, M., Corcho, O., Simperl, E., Strohmaier, M., d'Aquin, M., Srinivas, K., Groth, P., Dumontier, M., Heflin, J., Thirunarayan, K. (eds.) ISWC 2015. LNCS, vol. 9366, pp. 605–621. Springer, Cham (2015). doi: 10.1007/978-3-319-25007-6_35

35. Noia, T., Cantador, I., Ostuni, V.C.: Linked open data-enabled recommender systems: ESWC 2014 challenge on book recommendation. In: Presutti, V., Stankovic, M., Cambria, E., Cantador, I., Iorio, A., Noia, T., Lange, C., Reforgiato Recupero, D., Tordai, A. (eds.) SemWebEval 2014. CCIS, vol. 475, pp. 129–143. Springer, Cham (2014). doi: 10.1007/978-3-319-12024-9_17

36. Bobadilla, J., Ortega, F., Hernando, A., Gutiérrez, A.: Recommender systems survey. Knowl.-Based Syst. **46**, 109–132 (2013)

37. Lu, C., Laublet, P., Stankovic, M.: Ricochet: context and complementarity-aware, ontology-based POIs recommender system. In: Proceedings of SALAD at the 14th Extended Semantic Web Conference, pp. 10–17 (2014)

38. Lu, C., Stankovic, M., Laublet, P.: Leveraging semantic web technologies for more relevant E-tourism behavioral retargeting. In: Proceedings of the 24th International Conference on World Wide Web, pp. 1287–1292. ACM (2015)

39. Vig, J., Sen, S., Riedl, J.: Tagsplanations: explaining recommendations using tags. In: Proceedings of the 14th International Conference on Intelligent User Interfaces, pp. 47–56. ACM (2009)

40. Lu, C., Stankovic, M., Laublet, P.: Desperately searching for travel offers? formulate better queries with some help from linked data. In: Gandon, F., Sabou, M., Sack, H., d'Amato, C., Cudré-Mauroux, P., Zimmermann, A. (eds.) ESWC 2015. LNCS, vol. 9088, pp. 621–636. Springer, Cham (2015). doi:10.1007/978-3-319-18818-8_38

A Semantic Graph-Based Approach for Radicalisation Detection on Social Media

Hassan Saif[1(✉)], Thomas Dickinson[1], Leon Kastler[2], Miriam Fernandez[1], and Harith Alani[1]

[1] Knowledge Media Institute, The Open University, Milton Keynes, UK
{h.saif,thomas.dickinson,m.fernandez,h.alani}@open.ac.uk
[2] University of Koblenz Landau, Mainz, Germany
lkastler@uni-koblenz.de

Abstract. From its start, the so-called Islamic State of Iraq and the Levant (ISIL/ISIS) has been successfully exploiting social media networks, most notoriously Twitter, to promote its propaganda and recruit new members, resulting in thousands of social media users adopting a pro-ISIS stance every year. Automatic identification of pro-ISIS users on social media has, thus, become the centre of interest for various governmental and research organisations. In this paper we propose a semantic graph-based approach for radicalisation detection on Twitter. Unlike previous works, which mainly rely on the lexical representation of the content published by Twitter users, our approach extracts and makes use of the underlying semantics of words exhibited by these users to identify their pro/anti-ISIS stances. Our results show that classifiers trained from semantic features outperform those trained from lexical, sentiment, topic and network features by 7.8% on average F1-measure.

Keywords: Radicalisation detection · Semantics · Feature engineering · Twitter

1 Introduction

Traditionally, the process of radicalisation has occurred directly, person to person. However, in the age of social media platforms and access to the Internet, this process has moved to a virtual sphere where terrorist organisations use 21st century technology to promote their ideology and recruit individuals. Particularly, the so-called Islamic State of Iraq and the Levant (ISIL/ISIS) is one of the leading terrorist organisations on the use of social media to share their propaganda, raise money and radicalise and recruit individuals. According to a 2015 U.S government report[1] this organisation has lured more than 25,000 foreigners to fight in Syria and Iraq, including 4,500 from Europe and North America.

---

[1] https://homeland.house.gov/wp-content/uploads/2015/09/TaskForceFinalReport.pdf.

© Springer International Publishing AG 2017
E. Blomqvist et al. (Eds.): ESWC 2017, Part I, LNCS 10249, pp. 571–587, 2017.
DOI: 10.1007/978-3-319-58068-5_35

Aiming to hinder ISIS recruiting efforts via social media, researchers, governments and organisations are actively working on identifying ISIS-linked or ISIS-supporting social media accounts. A popular example was the campaign launched by the hacker community Anonymous as a response to the Paris attacks,[2] where they claimed taking down more than 20,000 Twitter accounts linked to ISIS. A key criticism received by this initiative was the wrong categorisation as pro-ISIS of multiple Twitter accounts, including the ones of the U.S president Barack Obama, and the one of BBC news.[3] While it is unclear the strategy used by Anonymous to identify these accounts, this incident emphasises the difficulty and sensitivity of the problem at hand.

Current research works that have aimed to analyse radicalisation and pro-ISIS stances of social media users mainly rely on features extracted from the lexical representation of words (e.g., word n-grams, topics, sentiment), or from the online profile of users (e.g.,network features). While effective, these approaches provide limited capabilities to grasp and exploit the conceptualisations involved in content meanings. This involves limitations such as the inability to capture relations between terms (countries *attacking* ISIS vs. countries *attacked* by ISIS), or the weakness to properly capture contextual information by understanding which groups of terms co-occur together and how they relate to one another. The above limitations constitute a problem when trying to discriminate the stance expressed by users in social media. We therefore hypothesise that, by exploiting the latent semantics of words expressed in tweets, we could identify additional pro-ISIS and anti-ISIS signals that will complement and enhance the ones extracted by previous approaches.

Starting from this position, this paper investigates the use of ontologies and knowledge bases to support a graph-based analysis of tweets' content. Entities are extracted from the tweets of users' timelines (e.g. *"ISIS"*, *"Syria"*, *"United Nations"*) and expanded with their corresponding semantic concepts (e.g. *"Jihadist_Group"*, *"Country"*, *"Organisation"*) and relations (*e.g., Military_intervention_against_ISIL, place, Syria*) by using DBpedia[4]. Frequent subgraph mining is applied over the extracted semantic graphs to capture patterns of semantic relations that help discriminating the radicalisation stances of users. These patterns are then used as features (so-called *semantic features* in our work) for detecting the radicalisation stances of users on Twitter.

The effectiveness of semantic features to identify pro-ISIS and anti-ISIS stances is compared against several baseline features, particularly unigram features, sentiment features, topic features and network features. This comparison is performed by creating classifiers, based on the different sets of features, from a training dataset of 1,132 European Twitter users equally divided in pro-ISIS and anti-ISIS. Our results show how classifiers trained with semantic features outperform the baselines by 7.8% on average F1-measure, showing a positive

---

[2] https://en.wikipedia.org/wiki/November_2015_Paris_attacks.

[3] http://www.bbc.co.uk/newsbeat/article/34919781/anonymous-anti-islamic-state-list-features-obama-and-bbc-news.

[4] http://dbpedia.org.

impact on the use of semantic information to identify pro and anti ISIS stances. An additional analysis is performed over the data to identify signals of radicalisation. Our results show that pro-ISIS users' discussions tend to mention entities and relations focused on religion, historical events and ethnicity, such as *"Allah"*, *"Prophet"*, *"(Mohamed, ethnicity, Arab)"*, while anti-ISIS users' discussions tend to focus more around politics, geographical locations, and interventions against ISIS (*e.g., Military_intervention_against_ISIL, place, Syria*). Anti-ISIS users tend to mention the entity ISIS with a higher frequency than pro-ISIS users.

The rest of the paper is structured as follows: Sect. 2 assesses the related work in the areas of radicalisation studies. Section 3 describes our graph-based approach for radicalisation detection. Section 4 describes our experimental setup, including the dataset used for the analysis and the baseline features selected for comparison. Section 5 reports the results of comparing semantic features against the baselines. Section 6 discusses our identified pro- and anti-ISIS signals. Discussions and conclusions of this work are reported in Sects. 7 and 8 respectively.

2 Related Work

Understanding how an individual becomes radicalised online has become a burgeoning, albeit relatively-new, topic of research, and has recently focused on the role that social media plays in radicalisation. Existing works in this space have spanned multiple research domains (social science, psychology, computer science) and have often sought to understand the *pathway* to radicalisation: (i) picking out key signifiers of increasingly radicalised behaviour [2] (e.g. distribution of jihad videos); (ii) defining pathway models of the stages towards radicalisation (e.g. isolation, disillusionment, anger, etc.) [9], and; (iii) the process used by those radicalised to recruit others [4,8,18]. Indeed, in our own prior work [14], by Rowe and Saif, we found that users adopted radicalised rhetoric from other users with whom they shared common interactions and connections - suggesting a potential *nascent* community of influence.

Moving away from more general studies of *how* radicalisation occurs, towards *predicting* who will become radicalised has been the focus of several recent works. For instance, although O'Callaghan et al. [11] did not necessary *label* Twitter users as being pro or anti-ISIS (as we do in this paper), the authors instead clustered Twitter users collected from Twitter lists related to the Syria conflict into high-modularity clusters. The authors subsequently identified a cluster of *'jihadist'* users, which contained those who support ISIS. Inspection of the videos shared by users in that cluster found that videos were often shared from YouTube channels related to ISIS, the Nusra Front, and Aleppo (a key city that has been under ISIS control). In a more direct approach, Berger and Morgan [3] collected 90K ISIS supporters, manually, from Twitter and then induced a machine learning model (it is not clear which) to differentiate between pro and anti-ISIS supporters. Using this approach, the authors found that pro-ISIS supporters could be accurately predicted ($\sim$ 94% accuracy) from their profile descriptions' terms alone: with keywords such as succession, linger, Islamic State, Caliphate

State or In Iraq being key indicators of ISIS-support. Similarly, Magdy et al. [10] were able to accurately (87% F1) differentiate between pro and anti-ISIS users, finding that ISIS supporters talked a lot more about the Arab Spring than ISIS opponents. Users were defined from a collection of Arabic Tweets as pro-ISIS if they used '*Islamic State*' more, and anti-ISIS is they used '*ISIS*' more.

Unlike the above reviewed works, this paper investigates the role of semantics in classifying users as pro or anti-ISIS - as opposed to using terms in users' profile descriptions [3] or terms alone [10]. In carrying out our work, our results empirically validate the utility of semantics (i.e. semantic concepts, entities and relations) over merely unigrams.

3   Semantic Graph-Based Approach for Pro-ISIS Stance Detection

In this section we describe our semantic graph-based approach for detecting pro-ISIS stances on Twitter. As discussed before, the discriminative power of features used for radicalisation detection often relies on the latent semantic interdependencies that exist between certain words in tweets. As such, the proposed approach aims to extract and use such interdependencies and relations to learn patterns of radicalisation.

The proposed semantic graph-based approach breaks down into four main steps, as depicted in Fig. 1: (1) extract named entities and their semantic concepts in tweets, (2) build a semantic graph per user representing the concepts and semantic relations extracted from her posted content, (3) apply frequent sub-graph mining on the semantic graphs to capture patterns of semantic relations that discriminatingly characterise the radicalisation stances of users, and lastly (4) use the extracted patterns as features for radicalisation classifier training. Our approach uses a dataset of 1,132 European Twitter users (together with their timelines) equally divided in pro-ISIS and anti-ISIS. This dataset is further described in Sect. 4.1.

Step 1. Conceptual Semantics Extraction Given a training set, consisting on labelled (pro-ISIS, anti-ISIS) users' timelines, this steps extracts named-entities from the tweets of the users' timelines (e.g. ISIL, Syria,

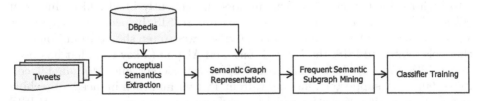

Fig. 1. Pipe of detecting pro-ISIS stances using semantic sub-graph mining-based feature extraction

Table 1. Total number and top 10 frequent entities and their associated semantic concepts extracted from our dataset.

	pro-ISIS		anti-ISIS	
No. of unique entities	32,406		30,206	
No. of unique concepts	35		36	
Top 10 frequent entities & their concepts	Entity	Concept	Entity	Concept
	MSNBC	Company	BBC	Company
	Iraq	Country	UK	Country
	Allah	Person	Kobane	City
	America	Continent	London	City
	Muslim	Person	ISIS	Organisation
	Officer	JobTitle	Syria	Country
	Wounds	Healthcondition	Europe	Continent
	Syria	Country	Iran	Country
	WAPO	PrintMedia	Kurdish	Person
	Israel	Country	Police	Organisation

Al-Baghdadi[5]) and expands them with their corresponding semantic concepts (e.g. Jihadist_Group, Country, Leader). The semantic extraction tool AlchemyAPI[6] is used for this purpose due to its accuracy and high coverage of semantic types and subtypes in comparison with other semantic extraction services [13,15]. Table 1 lists the total number of unique entities and concepts and the top 10 frequent entities and concepts, extracted from our dataset, for both pro-ISIS and anti-ISIS user accounts. Visible differences can be observed within these top 10 entities and concepts.

Step 2. Semantic Graph Representation The second step in our approach aims to extract the sets of semantic relations for every pair of named-entities (e.g., Syria, ISIL) co-occurring together in tweets, and represent these relations as graph structures. For the purpose of our study we extract semantic relations using the approach proposed by Pirro [12] over DBpedia, since DBpedia is a large generic knowledge graph which captures a high variety of relations between terms. To extract the set of relations between two named entities this approach takes as input the identifiers (i.e., URIs) of the source entity e_s, the target entity e_t and an integer value K that determines the maximum path length of the relations between the two named entities. The output is a set of SPARQL queries that enable the retrieval of paths of length at most K connecting e_s and e_t. Note that in order to extract all the paths, all the combinations of ingoing/outgoing edges must be considered. For example, if we were interested in finding paths of length $K <= 2$ connecting $e_s = Syria$ and $e_t = ISIL$ our approach will consider the following set of SPARQL queries:

---

[5] Abu Bakr Al-Baghdadi is the leader of the Islamic State in Iraq and Syria http:// dbpedia.org/page/Abu_Bakr_al-Baghdadi.
[6] http://www.alchemyapi.com/.

```
SELECT * WHERE {:Syria ?p1 :ISIL}
SELECT * WHERE {:ISIL ?p1 :Syria}
SELECT * WHERE {:Syria ?p1 ?n1. ?n1 ?p2 :ISIL}
SELECT * WHERE {:Syria ?p1 ?n1. :ISIL ?p2 ?n1}
SELECT * WHERE {?n1 ?p1 :Syria. :ISIL ?p2 ?n1}
SELECT * WHERE {?n1 ?p1 :Syria. ?n1 ?p2 :ISIL}
```

As it can be observed, the first two queries consider paths of length one. Since a path may exist in two directions, two queries are required. The retrieval of paths of length 2 requires 4 queries. In general, given a value K, to retrieve paths of length K, 2^k queries are required. Figure 2 shows an example of the semantic relations for the entities Syria and ISIL. As can be noted, these two entities are either connected via a direct semantic relation (e.g., $ISIL < headquarters > Syria$) or via linking nodes (e.g., $ISIL < ideology > Pan - Islam < ideology > MuslimsBrotherhood < location > Syria$).

Once we have the entities' pairwise semantic relations (i.e., paths) extracted from the users' timelines, we represent these relations for each user as a directed graph $G = (V, E)$ comprising a set V vertices of co-occurring entities with a set of edges E denoting the semantic relations between these entities.

Step 3. Frequent Patterns Mining In this step we apply frequent pattern mining to the users' semantic graphs which we extracted in step 2. As mentioned earlier, the goal behind this approach is to find patterns of similar semantics among pro- and anti-ISIS users, which can help characterise their stances. To this end, we apply frequent pattern mining to the users' semantic graphs and extract the sub-graphs appearing more than n times, where n is set to 2 in our experiments. This allows us to identify as many sub-graphs as possible, without returning a user's entire graph.

To mine these frequent sub-graphs we use CloseGraph [20], which performs an exhaustive sub-graph search, returning all frequent closed graphs within our dataset. We use the Parallel and Sequential Mining Suites (ParSeMiS[7])

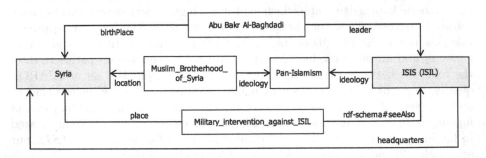

Fig. 2. Example for semantic relations between the entities Syria and ISIS with a path length of = 3

---

[7] https://www2.informatik.uni-erlangen.de/EN/research/zold/ParSeMiS/index.html.

implementation of this algorithm. Applying the aforementioned algorithm to the users' semantic graphs results in 187 unique sub-graphs for pro-ISIS users and 723 unique sub-graphs for anti-ISIS users in total, with the top frequent sub-graph appearing more than 500 times.

Figure 6 depicts three of the top most discriminative semantic sub-graphs mined from pro-ISIS and anti-ISIS users. These sub-graphs differ in the underlying semantics represented by the entities and the semantic relations. While the pro-ISIS sub-graphs (Fig. 6a) denote entities and relations around historical and religious topics, e.g., "Muhammad < religion> Islam", the anti-ISIS sub-graphs (Fig. 6b) denote entities and relations around key military interventions and geographical locations, e.g., " Military_intervention_against_ISIL < place>Syria". Further details on the analysis of these sub-graphs are provided in Sect. 6.

Step 4. Classifier Training This step takes as input a training set $\mathcal{T}^{train} = \{(\mathbf{U}_n; c_n) \in \mathcal{U} \times \mathcal{C} : 1 \leq n \leq N^{train}\}$ of users U in our dataset along with their class labels $\mathcal{C} = \{\text{pro-ISIS, anti-ISIS}\}$). After that, it constructs for each user $U \in \mathbf{U}$ a semantic vector $\mathbf{v}_{us} = (e_1, e_2, ..., e_l, s_1, s_2, ..., s_m, b_1, b_2, ..., b_j)$ as the joined vector of entities $\mathbf{e} = (e_1, e_2, ..., e_l)$, concepts $\mathbf{s} = (s_1, s_2, ..., s_m)$, and $\mathbf{b} = (b_1, b_2, ..., b_j)$ semantic sub-graphs (patterns) extracted from the user's timelines as explained in the previous steps. The generated semantic vectors are then used to train multiple machine learning classifiers. SVM was selected in our experiments as the best performing one. Further details on the creation of this classifier are provided in Sect. 4.3.

4   Experimental Setup

Our proposed approach, as shown in the previous section, extracts frequent patterns of semantics commonly expressed by users of a certain radicalised stance. We assess the extracted patterns by using them as features to train supervised classifiers for user-level radicalisation classification, i.e., classifying users in our dataset according to their stance as pro-ISIS or anti-ISIS. Hence, our experimental setup requires the selection of (i) an annotated dataset of Twitter users (pro-ISIS and anti-ISIS) together with their timelines, (i) baseline features for cross-comparison and (ii) a supervised classification method. These elements are explained in the following subsections.

4.1   Dataset of pro-ISIS and anti-ISIS Twitter Users

Our approach relies on a training dataset of 1, 132 European Twitter users (together with their timelines) collected in our previous work [14]. In this work the pro-ISIS stance of 727 Twitter users was determined based on their sharing of incitement material from known pro-ISIS accounts and on their use of extremist language. By the time of conducting this research, 161 of these Twitter accounts were suspended or changed the privacy to *protected*, preventing us from accessing their profile information. As such, we resorted to remove them

Table 2. Statistics of the Twitter dataset used for evaluation

	pro-ISIS users	anti-ISIS users
Total number of tweets	602,511	1,368,827
Average number of tweets per user	1,065	2,418
Total number of words	3,945,815	9,375,841
Average number of words per user	6,971	16,570

from the original set, resulting in 566 pro-ISIS users in total. To balance our dataset, we added 566 anti-ISIS users, whose stance is determined by the use of anti-ISIS rhetoric. Table 2 shows the total number, and distribution of tweets and words for each user group. As we can observe, both the number of tweets and words for anti-ISIS users are significantly higher than the ones for pro-ISIS users. We refer the reader to the body of our work [14] for more details about the construction and annotation of this dataset.

4.2   Baseline Features

Unigrams Features: Word unigrams are features traditionally used for various classification tasks of tweets data. For example, in the context of a sentiment analysis task, models trained from word unigrams were shown to outperform random classifiers by 20%. [1] We generate the user's unigram vector t_{uunig} as the vector $t_{uunig} = (w_1, w_2, ..., w_m)$ of the words in his timeline. Note that stopwords, non-English words and special characters are removed from the timeline prior to building t_{uunig} in order to reduce its dimensionality.

Sentiment Features: Sentiment features denote the sentiment orientation (positive, negative, neutral) of users in our dataset. The rational behind using these features is that the sentiment conveyed by the users' posts may help discriminating between pro- and anti-ISIS stances. To extract these features for a given user u, we first extracted the sentiment orientation of each tweet in the user's timeline. To this end, we used SentiStrength [17], a lexicon-based sentiment detection method for the social web. To construct the sentiment vector $t_{usentiment}$ for user u, we augment the unigrams feature vector t_{uunig} with the extracted sentiment orientation of tweets as: $t_{usentiment} = (w_1, w_2, ..., w_m, p_{pos}, p_{neg}, p_{neu})$, where p_{pos}, p_{neg} and p_{neu} are the numbers of positive, negative and neutral posts in the user's timeline. Note that, due to the low dimensionality of sentiment features, the sentiment vector for each user is constructed as a combination of ngrams and sentiment attributes.

Topic Features: Topic features denote the latent topics extracted from tweets using the probabilistic generative model, LDA [6]. LDA assumes that a document is a mixture of topics and that each topic is a mixture of probabilities of words that are more likely to co-occur together under the topic. For example the topic "ISIS" is more likely to generate words like "behead" and "terrorism".

Therefore, LDA topics represent groups of words that are contextually related. To extract these latent topics from our dataset we use an implementation of LDA provided by Mallet.[8] The topic feature vector for user u is constructed as $t_{utopic} = (t_1, t_2, ..., t_k)$, where $t_i \in t_{utopic}$ represents a topic extracted from the user's timeline. It is worth noting that LDA requires defining the number of topics to extract before applying it on the data. To this end, we ran LDA with different choices of numbers of topics between 5 and 10,000. We trained a SVM classifier from the features extracted by each of these choices and measured the classification performance in F1-measure. The best performance 82.9% F1 is reached with 5,000 topics, which is the number of topics used in our analysis.

Network Features: Network features refer to the profile information/attributes of Twitter users.[9] This includes: *number of followers, number of followee, number of hashtags, number of mentions* (i.e., @user), *favourites count, status count, profile description,* and *geographic location.* The notion behind using these features for radicalisation detection is that users of a certain radicalisation stance are more likely to interact with other users of the same stance than with users from a different stance, as we discussed in our previous work [14]. The network feature vector for user u is constructed as $t_{unetwork} = (n_1, n_2, ..., n_l)$, where $n_i \in t_{unetwork}$ represents an attribute derived from the user's profile.

4.3 Classification Method

We tested several machine learning classifiers with our semantic features including Naive Bayes, Maximum Entropy, and SVMs with linear, polynomial, sigmoid, and RBF kernels. The classifiers were tested using 10-fold cross validation over 30 runs. Results showed that SVM with RBF kernel produced the highest and most consistent performance in accuracy and F1-measure among all the other classification methods. Section 5 reports performance using this classifier. Note that, to generate these classifiers, we perform a feature selection process on the 4 baseline feature sets, as well as our semantic features, by excluding features with low discrimination power. To this end, we use Information Gain (IG) [7] to compute the discriminative score of features in each feature set and filter out those with low scores (IG $\approx$ 0) from the feature space.[10]

Figure 3 shows the original number of features under each feature set and the impact of feature selection using the IG method on them. Here we notice a reduction rate of 88% for all feature sets on average. The highest reduction rate of 97% is achieved on the semantic features, reducing the number of features from 346,512 (considering entities, concepts and semantic sub-graphs) to 8,429 only. This indicates that pro- and anti-ISIS users share a high degree of terminology and semantics when posting in Twitter.

---

[8] http://mallet.cs.umass.edu/.

[9] https://dev.twitter.com/overview/api/users.

[10] IG measures the decrease in entropy when the feature is given vs. absent [21].

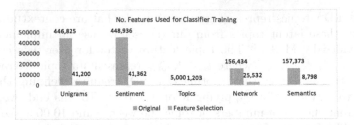

Fig. 3. Number of features used for classification with and without feature selection

5   Evaluation Results

In this section, we report the results obtained from using the proposed semantic features for user-level radicalisation classification. Our baselines of comparison are SVM classifiers trained from the 4 sets of features described in Sect. 4.2. Results in all experiments are computed using 10-fold cross validation over 10 runs of different random splits of the data to test their significance. Statistical significance is done using *Wilcoxon signed-rank test* [16]. Note that all the results in average Precision, Recall and F1-measure reported in this section are statistically significant with $\rho < 0.001$.

Table 3 shows the results of our binary stance classification (pro-ISIS vs. anti-ISIS) using *Unigrams, Sentiment, Topic*, and *Semantic* features after feature selection, applied over the 1,132 users in our dataset. The table reports three sets of precision (P), recall (R), and F1-measure (F1), one for anti-ISIS stance identification, one for pro-ISIS stance identification, and the third shows the averages of the two. The table also reports the total number of features used for classification under each feature set.

According to the results presented in Table 3, the proposed Semantic features outperform the 4 baseline feature sets in all average measures by a large margin. In particular, classifiers trained from Semantic features produce 7.8% higher Recall, 7.7% higher precision, and 7.82% higher F1 than all baselines on average. Network features come next, followed by Unigrams features, with approximately

Table 3. Classification performance of the five feature sets with IG feature selection. The values highlighted in grey correspond to the best results obtained for each feature. Results in average P, R and F1 are statistically significant with $\rho < 0.001$.

	No. of features	anti-ISIS			pro-ISIS			Average		
		P	R	$F1$	P	R	$F1$	P	R	$F1$
UNIGRAMS	41,200	0.814	0.919	0.863	0.907	0.79	0.844	0.86	0.854	0.854
SENTIMENT	41,362	0.814	0.919	0.863	0.907	0.79	0.844	0.86	0.854	0.854
TOPICS	992	0.771	0.943	0.848	0.927	0.719	0.81	0.849	0.831	0.829
NETWORK	25,532	0.897	0.827	0.86	0.839	0.905	0.871	0.868	0.866	0.866
SEMANTICS	8,798	0.994	0.852	0.917	0.87	0.995	0.928	0.932	0.923	0.923

87% and 85% in average F1 respectively. On the other hand, Topic features produce the lowest classification performance with 82.9% in average F1. We also notice that sentiment features, which consist of both, word unigrams and their sentiment (Sect. 4.2), have no impact on the classification performance compared to using Unigrams only.

As for per-stance classification performance, we observe that Unigrams, Sentiment and Topic features produce higher performances on detecting anti-ISIS stance than pro-ISIS stance. For example, the F1 produced by Unigrams when identifying anti-ISIS stance is 2.2% higher than the F1 produced when identifying pro-ISIS stance. This might be due to the imbalanced distribution of words and tweets in both classes. As described in Sect. 4.1, the number word unigrams in anti-ISIS users' timelines is $\approx$ 2.5 the number of those in pro-ISIS users' timelines.

On the other hand, classifiers trained from either Network or Semantic features seem to be more tolerant to the imbalanced distribution of words in our dataset. Specifically, the performance of both, anti-ISIS and pro-ISIS classification becomes more consistent, with $\approx$ 1% difference in F1 only when using Network or Semantic features.

The above results show the effectiveness of using semantic features for radicalisation classification of users on Twitter, substantially improving per-stance, as well as overall, classification performance. It is worth noting that our results here are directly comparable to prior work of Magdy et al. [10]. However, while in this work the authors used unigrams alone to identify pro and anti-ISIS users, our work shows the enhancement obtained by using semantics for this task.

6 Signals of Radicalisation (pro-ISIS vs. anti-ISIS)

In the previous sections we showed how to detect radicalisation stances of users on Twitter and investigated which features help achieving higher performance levels. In this section, we reflect on the semantics expressed by both, pro- and anti-ISIS users, aiming at finding signals of radicalisation or anti-radicalisation in their timelines.

To this purpose we look for possible variations in the types of semantics in both, pro-ISIS and anti-ISIS users and study whether such variations indicate radicalisation or anti-radicalisation stances. We compute the frequency distribution of the entities, concepts and semantic sub-graphs that pro-ISIS and anti-ISIS users adopt in their tweets and we discard entities with zero information gain score (IG $\approx$ 0), since they have no discrimination power for identifying pro- and anti-ISIS stances, as discussed in Sect. 4.3. We also compute the sentiment of each entity by taking the average sentiment of the tweets where the entity is mentioned in. Looking at the entities and concepts used by pro-ISIS users (Fig. 4(a)) we observe that the majority of discriminative entities found within pro-ISIS users' discussions focus on religion, with many positively mentioned entities such as "Allah", "Prophet", and "Khilafah" (Khilafat). On the other hand, the most discriminative entities and concepts found within anti-ISIS

582   H. Saif et al.

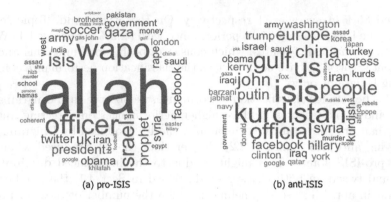

(a) pro-ISIS   (b) anti-ISIS

Fig. 4. Word clouds of the top-50 named-entities published by pro-ISIS and anti-ISIS users, the colour indicates the sentiment attached to the entity - with red being negative, and green being positive. (Color figure online)

users' discussions focus on locations and politics, with entities like "Kurdistan", "Europe" and "Putin". We can also observe several common entities between pro- and anti-ISIS users. Some of these entities are perceived positively by both groups (e.g., "Hillary", "Gulf", "China") although the vast majority are commonly mentioned with negative sentiment (e.g., "Assad", "Syria", "Israel").

Figure 5 shows the per-user distribution of six highly frequent and commonly mentioned entities by pro-ISIS and anti-ISIS users. We can notice that, although these words receive similar sentiment by users in both groups (see Fig. 4), they have different frequency distributions. For example, an entity like "Allah" is used more frequently by pro-ISIS users (mean = 42.64) than anti-ISIS users (mean = 1.28). On the other hand anti-ISIS users tend to use the word "ISIS" more frequently than pro-ISIS users, with a mean frequency of 46.88 for the former and 18.13 for the latter. While the sentiment of "ISIS" is negative for both groups, a manual analysis of the tweets containing "ISIS" reveals that pro-ISIS users generally refer positively to the term, although the context in which it is mention might be negative. For example, the tweet, "If your goal is killing Abu Sayyaf then our goal is killing Obama and the worshipers of the cross. #ISIS", shows support to "ISIS" but expresses negativeness towards Americans and Christians, and therefore, the tweet is categorised as negative. This comes inline with the work of Magdy et al. [10] (see Sect. 2), which reported that "ISIS', when mentioned in a negative context, often indicates anti-ISIS stance whereas it indicates pro-ISIS stance when it is mentioned in a positive context.

Figure 6 shows the three most discriminative sub-graphs for pro-ISIS (a) and anti-ISIS (b) users. As we can see in this figure pro-ISIS sub-graphs are formed of concepts and relations related to religion and ethnicity (Muhammad, Islam, Arabs), to relevant historical events (the invasion of Badr[11]) and to the United

---

[11] https://www.britannica.com/event/Battle-of-Badr.

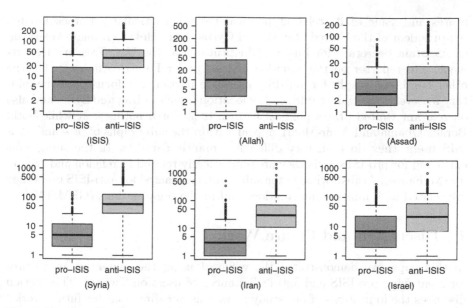

Fig. 5. Per-user distribution of the named-entities: *ISIS, Allah, Assad, Syria, Iran and Israel* for pro-ISIS and anti-ISIS users.

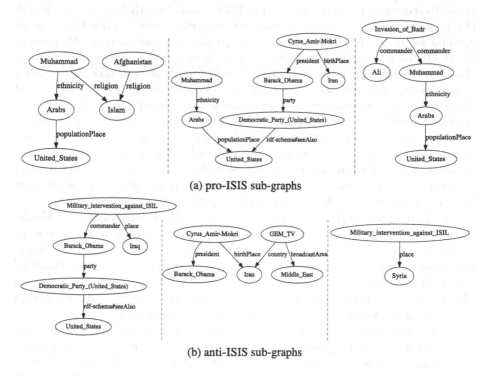

(a) pro-ISIS sub-graphs

(b) anti-ISIS sub-graphs

Fig. 6. Example of 3 of the most discriminative semantic sub-graphs for (a) pro-ISIS and (b) anti-ISIS users in our Twitter dataset

States and some of its relevant political figures, particularly Barack Obama (ex-president of the United States), and Cyrus Amir-Mokri, (Iranian-American, ex-Assistant Secretary for Financial Institutions at the U.S. Treasury Department, and reporter on the situation of Iran under Foreign Affairs[12]). On the other hand, sub-graphs for anti-ISIS users are particularly focused on the military interventions against ISIS, on key locations, such as Iraq and Syria, and also on relevant United States political figures. It is worth noting that while both, Barack Obama and Amir-Mokri, appear within the sub-graphs of pro- and anti-ISIS users, they do it in very different semantic contexts. In the sub-graphs extracted for pro-ISIS users they are semantically related to religion and ethnicity (Muhamad, Arabs), while in the sub-graphs extracted for anti-ISIS users they are related with military interventions and media organisations (GEM_TV).

7   Discussion and Future Work

In this paper we demonstrated the value of using the semantics as features for identifying pro-ISIS and anti-ISIS stances of users on Twitter. This section discusses the limitations of our study as well as our directions for future work.

We experimented with a dataset of $1,132$ Twitter users. These users were annotated as pro-ISIS or anti-ISIS based on their sharing of content from known pro-ISIS accounts and on their use of pro-ISIS or anti-ISIS rhetoric [14]. Although the ratio of pro-ISIS users (727 pro-ISIS accounts were identified from a an original set of 154 K accounts) is similar to the one identified in other works [3], the dataset is relatively sparse, which leads to question the *generality* of the obtained results. Our future work will therefore replicate the process described in this paper with similar datasets. It is also relevant to observe that word distribution in this dataset was found to be skewed towards anti-ISIS users, with approximately 58% more words found in anti-ISIS users' timelines (see Table 2). Although we reduced the impact of such bias by performing feature selection, our future work will study whether such differences in content are similar for other datasets. Since this dataset focuses on users based in Europe, anti-ISIS users may express their views more freely and actively than pro-ISIS users, which may lead to the variation in the amount and diversity of generated content.

While previous works focus on identifying pro-ISIS users mainly in Middle East [3,10], we are interested in studying European radicalisation stances. In addition, as opposed to these works, which include Arabic tweets in their analyses, we only processed English tweets. Note that extracting conceptual semantics for the Arabic language is a challenging task, with little research being done in this regard. As future work, we plan to explore different ways for extracting semantics from Arabic tweets and include them in our analysis.

As described in Sect. 3, our proposed semantic features include entities, concepts and semantic sub-graphs. We consider these three elements for completeness, since semantic sub-graphs might not be found in the timelines of all Twitter users. The generation of these sub-graphs depends on the richness of semantics

---

[12] https://www.foreignaffairs.com/articles/iran/2015-10-20/windfall-iran.

existing within the users' content, the accuracy of AlchemyAPI to extract these semantics, and the coverage of the knowledge base, in this case DBpedia, to capture the semantic relations between the extracted entities.

To extract semantic relations from DBpedia our approach considers a maximum path length of two (see Sect. 3). While a larger maximum path length could increase the likelihood of a semantic relation existing between two entities, as well as the amount of existing semantic relations, higher values of maximum path length come close to the diameter of the DBpedia graph itself, and may lead to an explosion in the number of extracted relationships.[13] Additionally, extracting semantic relations between a high number of entities via a SPARQL endpoint is a high-cost process [12]. Our implementation uses multithreading to enhance the performance (i.e., queries are sent in parallel), and relations are extracted once per Twitter dataset. These relations are stored to be reused for future experiments.

When performing sub-graph mining we need to consider scalability as well as redundancy issues (similar extracted sub-graphs). In our experiments, the number of training instances was limited, but a greater number of users, as well as larger graphs per user, may be difficult to scale. To deal with scalability our plan includes the use of the parallel processing [5]. Regarding redundancy, although our work already filters sub-graphs with low discrimination power based on IG, our plan for minimising redundancy also includes using compression techniques to cluster together sub-graphs with similar information [19].

When performing feature selection (see Sect. 4.3) we observed a reduction rate of 88% for all feature sets on average and 97% on the semantic features. This indicates that pro- and anti-ISIS users share a high degree of terminology and semantics when posting in Twitter. Our analysis (see Sect. 6) has therefore focused on analysing those key entities, concepts and semantic sub-graphs with higher discrimination power, i.e., those that are different among the two groups. Our future work aims to complement this analysis by investigating whether entities associated to particular semantic concepts (e.g., countries, organisations) have more discriminative power than other ones.

Although we use Twitter as a case study in our analysis, our approach is not tied to Twitter data. Room for future work is investigating the applicability and performance of our semantic features on other social media platforms such as Facebook and Instagram.

8   Conclusions

In this paper we proposed the use of the conceptual semantics of words for detecting pro-ISIS and anti-ISIS stances of users on social media. We used Twitter as case study of social media platforms, and investigated: (i) how semantic graphs can be created by extracting entities in tweets, together with their corresponding semantic concepts and relations, (ii) how frequent semantic sub-graphs

---

[13] The effective estimated diameter of DBpedia is 6.5082 edges. See http://konect. uni-koblenz.de/networks/dbpedia-all.

can be mined from these graphs and, (iii) how entities, concepts and semantic sub-graphs can be used as features to train machine learning classifiers for stance detection of Twitter users.

We experimented with our semantic features on a Twitter dataset of 1,132 pro-ISIS and anti-ISIS users and compared the performance of a SVM classifier trained from semantic features against classifiers trained from Unigrams, Sentiment, Topics, and Network features. We also studied the impact of feature selection on the performance of our classifiers and showed that, using the most discriminative semantic features in radicalisation classification improves performance by 7.8% F1 over the average performance of all baselines.

We performed an exploratory analysis on the variations of semantics and sentiment used by pro-ISIS and anti-ISIS users in our dataset and showed that pro-ISIS users tend to discuss about religion, historical events and ethnicity while anti-ISIS users focus more on politics, geographical locations and interventions against ISIS.

Acknowledgment. This work was supported by the EU H2020 projects COMRADES (grant no. 687847) and TRIVALENT (grant no. 740934).

References

1. Agarwal, A., Xie, B., Vovsha, I., Rambow, O., Passonneau, R.: Sentiment analysis of Twitter data. In: Proceedings of ACL 2011 Workshop on Languages in Social Media, pp. 30–38 (2011)
2. Bartlett, J., Miller, C.: The edge of violence: towards telling the difference between violent and non-violent radicalization. Terrorism Polit. Violence **24**(1), 1–21 (2012)
3. Berger, J., Morgan, J.: The ISIS Twitter census: defining and describing the population of ISIS supporters on Twitter. Brookings Proj. US Relat. Islamic World **3**(20), 265–284 (2015)
4. Berger, J.M.: Tailored online interventions: The islamic state's recruitment strategy. Combatting Terrorism Center (2015)
5. Bhuiyan, M.A., Al Hasan, M.: FSM-H: frequent subgraph mining algorithm in hadoop. In: 2014 IEEE International Congress on Big Data, pp. 9–16. IEEE (2014)
6. Blei, D.M., Ng, A.Y., Jordan, M.I.: Latent dirichlet allocation. J. Mach. Learn. Res. **3**, 993–1022 (2003)
7. Forman, G.: An extensive empirical study of feature selection metrics for text classification. J. Mach. Learn. Res. **3**, 1289–1305 (2003)
8. Hall, J.: Canadian foreign fighters and ISIS (2015)
9. King, M., Taylor, D.M.: The radicalization of homegrown Jihadists: a review of theoretical models and social psychological evidence. Terrorism Polit. Violence **23**(4), 602–622 (2011)
10. Magdy, W., Darwish, K., Weber, I.: # failedrevolutions: using Twitter to study the antecedents of ISIS support. First Monday **21**(2), 1481–1492 (2016)
11. O'Callaghan, D., Prucha, N., Greene, D., Conway, M., Carthy, J., Cunningham, P.: Online social media in the Syria conflict: encompassing the extremes and the in-betweens. In: Proceedings of International Conference on Advances in Social Networks Analysis and Mining (ASONAM 2014) (2014)

12. Pirró, G.: Explaining and suggesting relatedness in knowledge graphs. In: Arenas, M., Corcho, O., Simperl, E., Strohmaier, M., d'Aquin, M., Srinivas, K., Groth, P., Dumontier, M., Heflin, J., Thirunarayan, K., Staab, S. (eds.) ISWC 2015. LNCS, vol. 9366, pp. 622–639. Springer, Cham (2015). doi:10.1007/978-3-319-25007-6_36
13. Rizzo, G., Troncy, R.: Nerd: evaluating named entity recognition tools in the web of data. In: Workshop on Web Scale Knowledge Extraction (WEKEX 2011), vol. 21 (2011)
14. Rowe, M., Saif, H.: Mining pro-ISIS radicalisation signals from social media users. In: Proceeedings of the International Conference on Weblogs and Social Media (2016)
15. Saif, H., He, Y., Alani, H.: Semantic sentiment analysis of Twitter. In: Cudré-Mauroux, P., Heflin, J., Sirin, E., Tudorache, T., Euzenat, J., Hauswirth, M., Parreira, J.X., Hendler, J., Schreiber, G., Bernstein, A., Blomqvist, E. (eds.) ISWC 2012. LNCS, vol. 7649, pp. 508–524. Springer, Heidelberg (2012). doi:10.1007/978-3-642-35176-1_32
16. Siegel, S.: Nonparametric statistics for the behavioral sciences (1956)
17. Thelwall, M., Buckley, K., Paltoglou, G.: Sentiment strength detection for the social web. J. Am. Soc. Inf. Sci. Technol. **63**(1), 163–173 (2012)
18. Winter, C.: Documenting the virtual 'caliphate'. Quilliam Foundation (2015)
19. Xin, D., Han, J., Yan, X., Cheng, H.: On compressing frequent patterns. Data Knowl. Eng. **60**(1), 5–29 (2007)
20. Yan, X., Han, J.: Closegraph: mining closed frequent graph patterns. In: Proceedings of the Ninth ACM SIGKDD International Conference on Knowledge Discovery and Data Mining, pp. 286–295. ACM (2003)
21. Yang, Y., Pedersen, J.O.: A comparative study on feature selection in text categorization. In: ICML, vol. 97, pp. 412–420 (1997)

Semantic Web and Transparency Track

Modeling and Querying Greek Legislation Using Semantic Web Technologies

Ilias Chalkidis[1]([✉]), Charalampos Nikolaou[2], Panagiotis Soursos[1], and Manolis Koubarakis[1]

[1] Department of Informatics and Telecommunications,
National and Kapodistrian University of Athens, Athens, Greece
ihalk@di.uoa.gr
[2] Department of Computer Science, University of Oxford, Oxford, UK

Abstract. In this work, we study how legislation can be published as open data using semantic web technologies. We focus on Greek legislation and show how it can be modeled using ontologies expressed in OWL and RDF, and queried using SPARQL. To demonstrate the applicability and usefulness of our approach, we develop a web application, called Nomothesia, which makes Greek legislation easily accessible to the public. Nomothesia offers advanced services for retrieving and querying Greek legislation and is intended for citizens through intuitive presentational views and search interfaces, but also for application developers that would like to consume content through two web services: a SPARQL endpoint and a RESTful API. Opening up legislation in this way is a great leap towards making governments accountable to citizens and increasing transparency.

Keywords: Legislative knowledge representation · Linked Open Data · Public services · CEN MetaLex · European Legislation Identifier

1 Introduction

Recently, there has been an increased interest in making government data open and easily accessible to the public. Technological advances in the area of the semantic web have given rise to the development of the so-called Web of Data, which has given an even stronger push to such efforts. The research area of linked data studies how data that are expressed in RDF will be available on the Web and interconnected with other data with the aim of increasing its value for everybody. Therefore, semantic web and linked data provide a set of technologies for making government data open [16].

An important kind of government data is the data related to legislation. Legislation applies to every aspect of people's living and evolves continuously building a huge network of interlinked legal documents. Therefore, it is important for a government to offer services that make legislation easily accessible to the public aiming at informing them, enabling them to defend their rights, or to use

© Springer International Publishing AG 2017
E. Blomqvist et al. (Eds.): ESWC 2017, Part I, LNCS 10249, pp. 591–606, 2017.
DOI: 10.1007/978-3-319-58068-5_36

legislation as part of their job. Towards this direction, there are already many European Union (EU) countries that have computerized the legislative process by developing platforms for archiving legislation documents and offering on-line access to them.

In Greece, so far, there has been a limited degree of computerization of the legislative process and even the discovery of legislation related to a specific topic can be a hard task. The legislative work of the Greek government has been published since 1907 in the form of a gazette by the National Printing Office (http://www.et.gr/). Legislation is published on a daily basis in that gazette and it is distributed only in a PDF file format. A small step for making Greek legislation more accessible to the public has been made by Diavgei@ (https://diavgeia.gov.gr/), a Greek program introduced in 2010, enforcing transparency over government and public administration by requiring that government and public administration have to upload their decisions on the Web. However, no common file format is enforced nor any structuring of their textual content.

In this work, we are following the footsteps of other successful efforts in Europe and aim at modernizing the way Greek legislation is made public. We envision a new state of affairs in which ordinary citizens have advanced search capabilities at their fingertips on the content of legislation. We also envision that legislation is published in a way so that developers can consume it, and so that it can be also combined with other open data to increase its value for interested people. Currently, there is no other effort in Greece that takes this perspective on legislation and related decisions made by government institutions and administration alike.

Technically, we view legislation as a collection of legal documents with a standard structure. Legal documents may be linked in complex ways. A legal document might refer to another legal document or it may modify the content of other legal documents. As a result, a complex semantic structure arises from legal documents and their interrelationships. Our aim is to develop intelligent services that not only present the textual content of legal documents but are able to answer complex analytics such as "Which are the 5 most frequently modified legal documents during 2008–2013?" or "Who are the 3 past government members that have signed the most legal documents during their service in 2008–2015?". In addition, we would like to be able to interlink legislation with other linked data sources (e.g., the administrative geography of Greece) so that queries like "Which legal documents refer to geographical areas that belong to the region of Macedonia-Thrace and have population more than 100,000?" can be posed.

Contributions. Our contributions towards achieving the aforementioned goals are summarized as follows.

We follow the latest European standards and best practices and develop an OWL ontology, called Nomothesia ontology (Nomothesia means legislation in Greek), for modeling the content of Greek legislation documents. The first benefit of this approach is the formulation of rich queries over the content of Greek legislation. The second one is the representation of legislative modifications which enables tracking of the evolution of a legislative document and compact storage of its content in the form of differences (i.e., *deltas* in the jargon of version control).

We build a tool tailored to the particularities of Greek legislation for populating Nomothesia ontology with all issues of the government gazette during 2006–2015 and linking the resulting dataset with DBpedia and the datasets Greek Administrative Geography and Greek Public Buildings (http://linkedopendata.gr/). Overall, the tool processed 2,676 legal documents and stored approximately 1,85M RDF triples. This is the first time such a substantial part of the legislative corpus of Greece is organized in this manner, linked with external sources, and made publicly available.

We develop a prototype web application, called Nomothesia, that offers advanced presentational views, search, and analytics functionality over Greek legislation. Among others, one may view how legal documents evolve through time in response to amendments, retrieve a piece of legislation in various formats, and search for legislative documents based on their metadata or textual content. Moreover, Nomothesia offers a SPARQL endpoint and a RESTful API which enable the formulation of complex queries, such as the ones presented previously.

Organization. The rest of the paper is organized as follows. Section 2 discusses related work in legislative knowledge representation using semantic web technologies. Section 3 gives background information on the structure and encoding of Greek legislation. Section 4 presents Nomothesia ontology that is developed for the modeling of Greek legislation, the challenges faced in populating it based on a large part of Greek legislation corpus, and discusses interlinking with other publicly available open data. Section 5 gives examples of the RDF representation of two real legal documents according to Nomothesia ontology and demonstrates how the resulting data can be queried using SPARQL. Section 6 presents the Nomothesia web platform. Section 7 presents preliminary user feedback on Nomothesia. Last, Sect. 8 summarizes our contributions and discusses future work.

2 Related Work

Modernization of the way citizens access legislation has been a primary concern of many governments across the world. The development of information systems archiving the content and metadata of legal documents has been a common practice towards making legislation easily accessible to the public [5]. To name a few examples, the MetaLex document server [11] (http://doc.metalex.eu/) offers Dutch national regulations published by the official portal of the Dutch government, while the United Kingdom publishes legislation on its official portal (http://www.legislation.gov.uk/). In the same spirit, [9] presents a service that offers Finnish legislation as linked data. Last, the Publications Office of the EU has developed a central content and metadata repository, called CELLAR, for storing official publications and bibliographic resources produced by the institutions of the EU [8]. The content of CELLAR, which includes EU legislation, is made publicly available by the EUR-lex service (http://eur-lex.europa.eu).

All of the above endeavors adopt semantic web technologies for modeling, querying, and making legislative content easily accessible to the public. The adoption of web standards like XML, RDF, SPARQL as well as common practices for publishing such data as linked data has been a common practice among these efforts in the design of vocabularies and ontologies for legislative documents. One such vocabulary is the XML schema Akoma Ntoso (http://www.akomantoso.org/), which was funded by the United Nations for publishing African parliamentary proceedings and supporting legislative documentation in parliaments [2]. Another one is MetaLex [3], which was originally proposed as an XML vocabulary for the encoding of the structure and content of legislative documents, but updated later on with functionality related to timekeeping and version management [4]. After its adoption by the European Committee for Standardization (CEN) [7], MetaLex evolved to an OWL ontology called CEN MetaLex (in the following just MetaLex). More recently, the European Council introduced the European Legislation Identifier (ELI) [6] as a new common framework that has to be adopted by the national legal publishing systems in order to unify and link national legislation with European legislation. ELI, as a framework, proposes a URI template for the identification of legal resources on the web and it also provides an OWL ontology, which is used for expressing metadata of legal documents and legal events. ELI, like Akomo Ntoso and MetaLex, is not a one-size-fit-all model but it has to be extended to capture the particularities of national legislation systems.

Both Akoma Ntoso and MetaLex have been the keystones for the adoption of relevant practices in the legal domain. Akoma Ntoso has been the vocabulary of choice for the development of the XML schemata LegalDocML [14] and LegalRuleML [1] that aim to offer vocabularies for the modeling and representation of legal documents as well as reasoning capabilities with respect to the normative rules appearing in such documents. MetaLex was extended in the context of EU Project ESTRELLA (http://www.estrellaproject.org/) which developed the Legal Knowledge Interchange Format (LKIF) [12]. LKIF is an ontology for modeling both legal and legislative concepts that mainly express structure, content, and events regarding the process of legislation. In the same spirit, the European Case Law Identifier (ECLI), a sister endeavor of ELI, was introduced recently for modeling case laws [13].

The present work follows in the footsteps of the aforementioned efforts by offering to citizens a prototype system through which they may browse, search, and query Greek legislation. Similar to other efforts, the system is based on semantic web technologies; it offers an OWL ontology that extends MetaLex and ELI and is linked with DBpedia and two Greek linked geospatial datasets.

3   Background on Greek Legislation

Greek legislation is published through different types of documents depending on the government body enacting the legislation. Each piece of legislation has a standardized structure and follows an appropriate encoding. Both of these aspects are discussed in detail below.

3.1   Types and Encoding of Greek Legislation

The encoding of Greek legislation follows the rules set out in the document "Manual Directives for the encoding of legislation", which has been issued by the Greek Central Committee of Encoding Standards and has been legislated in Law 2003/3133. In this work we are considering the encoding of five primary sources (*types*) of Greek legislation: *constitution, presidential decrees, laws, acts of ministerial cabinet,* and *ministerial decisions.* We also consider two secondary sources of Greek legislation: *legislative acts* and *regulatory provisions.* These sources of legislation are materialized in *legal documents* (see Fig. 1 for an example) that adhere to a typical structure organized in a tree hierarchy around the concept of *fragments* (divisions).

PRESIDENTIAL DECREE No. 54

Article 1

Establishment, Title, Responsibilities

1. A Civil Service entitled "Civil Service of Public Constructions for Flood Protection of river Evros' valley and its tributaries"(EYDE EVROY), with responsibilities relating to the special and important work of flood protection of the valley of the rivers: Evros, Erythropotamos, Arda, their tributaries and Orestiadas Regional Moat.
2. This Service shall exercise all responsibilities of the management in accordance with the legal provisions for the design and construction of Public Constructions.
3. Registered office of EYDE EVROY is set Soufli and its running period is specified at eight years after the entry of this Decree into force.

Fig. 1. Article 1 of presidential decree 2011/54

Articles are the basic units in the main body of a legal document, numbered using Arabic numerals (1, 2, 3, ...). An article may consist of a list of paragraphs that are numbered using Arabic numerals as well. *Paragraphs* may have a list of indents (cases). *Indents* are numbered using lower-case Greek letters (α, β, γ, ...) and may have sub-indents which are numbered using double lower-case Greek letters ($\alpha\alpha$, $\beta\beta$, $\gamma\gamma$, ...). *Passages* are the elementary fragments of legal documents and are written contiguously, i.e., without any line breaks between them. Passages are the building blocks of indents and paragraphs. The main body of legal documents may be subdivided according to their size in larger units, such as *sections, chapters, parts* or *books*, which are numbered using upper-case Greek letters. Larger units and articles may have a title, which must be general and concise in order to reflect their content, and is used in the systematic classification of the substance of legal documents.

3.2   Metadata of Greek Legislation

In addition to the aforementioned structural elements, legal documents are accompanied by metadata. This primarily includes the *title* of the legal document, which must be general enough but concise so as to reflect its content, the *type* (e.g., Law, Presidential Decree), the *year* of publication, which can be inferred by the *publication date*, and the serial *number*. These latter elements (type, year, number) of metadata information serve also as a unique identifier of the legal document. Of equal importance are also the *issue* and the *sheet number* of the *government gazette* in which the legal document is published.

When reference to other legislation or public administration decisions is necessary, this is done with *citations*. For purposes of accuracy, citations must include the type of the legal document, its number, and its year of publication.

There is a large number of secondary metadata information, which could be related to a legal document. Currently in our model we consider three types of such metadata (named entities): the *signatories* (government members) that introduced the legislation, the *administrative areas* (locations), and the *organizations* mentioned in the text.

3.3   Legislative Modifications

Legislation is an event-driven process. Legal documents are passed based on an appropriate legislative procedure and then published in the government gazette. They may be modified by later legal documents with respect to their content (*amendments*) or they may be repealed (see Fig. 2). In the course of this process, we need to capture the structure of a legal document and the evolution of its content through time, given by the *legislative modifications* applied on the primary legal document or its versional successors. Although there is a standard encoding of Greek legislation (as presented in Sect. 3.1), there is no standard encoding for the codification of amendments or repeals. This is a challenge for any work [4, 10] like ours to which we offer the following solution.

PRESIDENTIAL DECREE No. 10
Article 1

1. After the end of paragraph 1 of Article 1 of Presidential Decree 54/2011, a paragraph is added as follows: "The maintenance of these projects remain within the remit of the region of E. Macedonia and Thrace in accordance with the provisions of Law 3852/2010 (G.G. A 87)."
2. Paragraph 3 of Article 1 of Presidential Decree 54/2011 is replaced as follows: "Registered office of EYDE EVROY is set Alexandroupoli and its running period is specified at eight years after the entry into force of this Decree."

Fig. 2. Article 1 of presidential decree 2012/10

By the analysis of a large corpus of Greek legislation and the consultation of Greek government officials, we have defined three main types of legislative modifications: *insertion*, *repeal*, and *substitution*. All three operations are applied with respect to an enacted and published legal document and an enacted but not yet published one. We explain these operations below and refer to the two documents by the identifiers **D1** and **D2**, respectively.

Insertion. Document **D2** specifies the exact structure and content of a new fragment that is to be inserted verbatim at a certain place of document **D1**.

Repeal. Document **D2** revokes a specific fragment of document **D1**.

Substitution. Document **D2** specifies the exact structure and content of a new fragment as well as a fragment of document **D1** that is to be replaced in document **D1** by the new fragment.

These three kinds of modifications (amendments) produce new *versions* of the original, as enacted, legal document (see Figs. 1, 2 and 3).

PRESIDENTIAL DECREE No. 54
Article 1
Establishment, Title, Responsibilities

1. A Civil Service entitled "Civil Service of Public Constructions for Flood Protection of river Evros' valley and its tributaries"(EYDE EVROY), with responsibilities relating to the special and important work of flood protection of the valley of the rivers: Evros, Erythropotamos, Arda, their tributaries and Orestiadas Regional Moat. The maintenance of these projects remain within the remit of the region of E. Macedonia and Thrace in accordance with the provisions of Law 3852/2010 (G.G. A 87).

2. This Service shall exercise all responsibilities of the management in accordance with the legal provisions for the design and construction of Public Constructions.

3. Registered office of EYDE EVROY is set Alexandroupoli and its running period is specified at eight years after the entry of this Decree into force.

Fig. 3. Article 1 of presidential decree 2011/54 (Current version)

4 Modeling Greek Legislation Using Semantic Web Technologies

In this section, we develop an OWL ontology for modeling the structural information of legal documents along with their accompanying metadata (i.e., title, gazette, publication date, etc.) and capturing how these documents may evolve through time in response to modifications. We call our ontology Nomothesia ontology and we discuss its current version (http://legislation.di.uoa.gr/nomothesia.owl) that adopts the ELI framework discussed in Sect. 2.

4.1 An OWL Ontology for Greek Legislation

Legal documents (LegalResource) are organized in fragments (LegalResourceSubdivision). Each legal document has (is_realized_by) subsequent versions (LegalExpression), which are differentiated by the previous ones

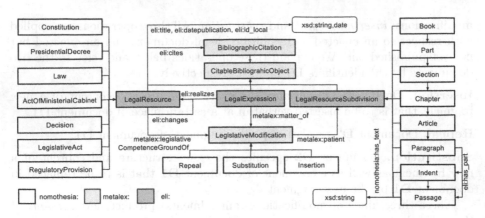

Fig. 4. The core of Nomothesia ontology

based on (matterf) some subsequent modifications (LegislativeModification) that are referred (legislativeCompetenceGroundOf) in subsequent legal documents and affect (patient) specific fragments. A legal document (LegalResource) changes an enacted legal document, which automatically inherits a new version (LegalExpression). Legal documents also include references (BibliographicCitation), which cite (cites) other legal documents or their fragments (CitableBibliographicObject).

Most of the above concepts are adopted from the ELI ontology discussed in Sect. 2. The rest of the concepts belong to the MetaLex ontology and there are no counterparts of them in ELI. Both ELI and MetaLex describe the event-driven process of legislation. We have extended this basis to describe also the structure, the content and the metadata of Greek legal documents (see Fig. 4).

As we already mentioned in Sect. 3, Greek legal documents are organized in specific types of fragments (e.g., Article, Paragraph, Indent). Indents and passages are the bottom elements of the main body's hierarchy, that define text (has_text). Legislative modifications are also classified in the three predefined types: Insertion, Repeal and Substitution. Given the above extensions, we can represent and store legislation natively only in the RDF format. Given the RDF encoding of a legal document, any additional format (e.g., PDF, XML) can be produced dynamically. Through this process, we eliminate the need to include special information about the different manifestations (Format) of any given legal document.

Primary metadata for identification of the legal documents are captured using both the ELI vocabulary and our own extensions: type of document (e.g., Law, Presidential Decree), title (title), serial number (id_local), date of publication (date_publication) and the Gazette issue (Gazette) in which a legal document has been published (published_in). We also capture a few secondary metadata (named entities) such as: signatories (Signatory), that is, the government members who signed (passed_by) a legal document, and possibly an administrative area (AdministrativeArea) or an organization (Organization) referred in (relevant_for) the document. There is also the appropriate means for the classification (is_about) of legal documents or divisions. As European legislation is adopted in Greek legislation, we should link the ELI URIs of directives, that are adopted (transposes) by national legal documents.

4.2   Persistent URIs in Nomothesia

In order to identify legal resources, we need appropriate URIs. Persistent URIs is a strong recommendation by EU [15] according to ELI. It is very important to have reliable means to identify legal documents, their fragments and generally any aspect related with a legal document. In Nomothesia, we structure the persistent URIs of legal documents according to template http://legislation. di.uoa.gr/eli/type/year/id. This pattern serves as a unique URI for each legal document based on its individual information: *type* of legislation (e.g., pd is the abbreviation for Presidential Decree), *year* of publication, and *identifier* (serial number). For example, the URI for Presidential Decree 54/2011 is http:// legislation.di.uoa.gr/eli/pd/2011/54.

Extending the basic URI of a legal document, we have URI extensions for its version. We define three different types: original (*as enacted*); *current* and *chronological* version given in the ISO 8601 format "YYYY-MM-DD". To retrieve Law 2014/4225 as of 21/10/2015, we use URI http://legislation.di.uoa. gr/eli/law/2014/4225/2015-10-21. Moving a step forward, we extend the versional URIs to capture the available manifestations of its version (e.g., PDF, XML, JSON). Despite the fact that our ontology does not capture such information, our RESTful API will serve such commodities, producing the appropriate representations on-the-fly. Through an HTTP GET request, an agent may request the JSON format of the enacted version of Ministerial Decision 2015/240 using URI http://legislation.di.uoa.gr/eli/md/2015/240/enacted/data.json. We can also refer to specific fragments of a legal document, following its nested structure from the top fragment level to our own point of interest. For instance, to refer to the first paragraph of the second article of Act of Ministerial Cabinet 2013/10 (as enacted version), we use URI http://legislation.di.uoa.gr/eli/ pd/2013/10/article/2/paragraph/1.

4.3   Population of Nomothesia Ontology

Population of Nomothesia ontology proved to be a demanding task that led us develop a tool tailored to the particularities of Greek legislation. We next list some of the challenges we faced, which highlight the need for an advanced computerized legislative system in Greece. First, the government gazette is only made available in PDF format and follows a two-column layout. It is well-known that transformation of PDF documents into plain text, written in a non Latin alphabet is error-prone. Second, although the encoding of Greek legislation is standardized, there are many instances of Greek legislation that do not conform with this encoding. This is mainly due to human errors during the typing process, a fact that contributes to the fragility of the task of ontology population.

Third, recognition of legislative modifications is even more intricate taking into account the lack of a standard encoding for the codification of amendments or repeals.

To deal with the above challenges, we built the Greek Government Gazette Parser (G3 Parser), which is a rule-based parsing tool designed to be flexible and robust towards frequent errors during the typing process of a legislative document, common deviations from the encoding of Greek legislation (see Sect. 3), as well as recognition of certain phrases denoting amendments and repeals (see Sect. 3.3). We successfully applied G3 Parser in almost all gazette issues during 2006–2015 (corresponding to 2,676 legal documents) producing approximately 1,85M RDF triples. The various stages of G3 Parser can be described as follows. (*a*) Transformation of double column PDF documents (gazette issues) in one-column plain text. (*b*) Split of text in individual legislative documents using appropriate regular expressions. (*c*) Parsing of the beginning of each legislative document for extracting primary metadata (i.e., title, issue number, etc.) and producing the corresponding RDF triples. (*d*) Parsing of the rest of the textual content of the document following a context free-grammar expressing the encoding of Greek legislation as described in Sect. 3. During this stage, G3 Parser is able to organize the textual content into fragments (e.g., articles and their cases, paragraphs, and passages) and generate the corresponding RDF triples following Nomothesia ontology. (*e*) Extraction of additional information from the bottom-level fragments (i.e., passages), like citations, legislative modifications, and named entities.

4.4   Linking Legislation with Other Open Data

The textual content of Greek legislation is very rich in references to named entities of three kinds: persons, places, and organizations. Nomothesia is able to capture these references and link them with the corresponding entities found in other public data sets, which provide additional information about them. To achieve this, for each of the above three kinds of entities, Nomothesia consults, respectively, the datasets of Greek DBpedia (http://el.dbpedia.org/), Greek Administrative Geography, and Greek Public Buildings. The last two datasets are part of a broader initiative started in the University of Athens for expressing various Greek geospatial data in RDF and publishing them as linked data on the Greek Linked Open Data portal (http://linkedopendata.gr/dataset).

As mentioned in the previous section, Stage (*e*) of G3 Parser is responsible for recognizing these entities. For each entity type and its corresponding dataset, G3 Parser has access to an inverted index, implemented using Apache Lucene (https://lucene.apache.org/), that associates the textual description of entities of this type with their corresponding URI. For example, there is an inverted index for entities of type dbpedia-owl:Person associating the lexical value of property prop-el:name to their URI. While in Stage (*e*), G3 Parser is able to recognize proper nouns and perform lookups on each one of the inverted indices. Whenever a lookup is successful over the inverted index for persons, G3 Parser creates a new entity linking it with the corresponding URI of DBpedia. On the other hand, if there is a successful lookup for a place or organization name, G3 Parser generates an RDF triple of the form ⟨*duri*, eli:relevant_for, *euri*⟩ where *duri* and *euri* correspond to the URIs of the current legislation document and named entity, respectively. As it is demonstrated in Sect. 5.3, interlinking Greek legislation with other relevant datasets allows for the formulation of sophisticated queries by indirectly accessing the additional information provided by these datasets.

5   Modeling Greek Legislation Using Nomothesia: A Short Example

In this section we give an example of how Nomothesia could be used to model Greek legislation. The example employs two legal documents, Presidential Decree 2011/54 and Presidential Decree 2012/10.

5.1   Representing P.D. 2011/54 in RDF

P.D. 2011/54 (see Fig. 1) is a Presidential Decree, which comes as a subclass of Legal Resource. It consists of 6 articles, each of which has its own nested subdivisions (paragraphs, passages, etc.). There are also primary metadata (title, publication date, etc.) related with the specific resource (see Fig. 5).

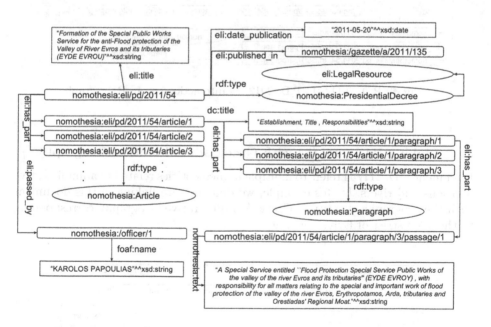

Fig. 5. P.D. 2011/54 structure and primary metadata as an RDF Graph

5.2 Capturing the Legislative Interaction between P.D. 2011/54 and P.D. 2012/10

P.D. 2012/10 (see Fig. 2) is a subsequent legal document that changes P.D. 2011/54 by applying legislative modifications to produce a new version that realizes the specific legal document. The legislative modifications are of a specific type (Insertion, Substitution). They have a specific structure and they refer to a specific part of the precedent legal document, which acts as the patient of the legislative modification (see Fig. 6).

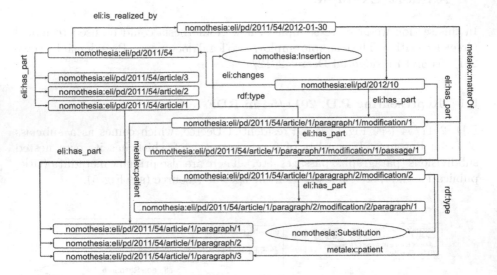

Fig. 6. P.D. 2011/54 - P.D. 2012/10 interactions as an RDF graph

5.3 Linking P.D. 2011/54 with Greek administration Geography

Document P.D. 211/54 refers to multiple named entities related to administrative divisions and rivers. In this example, we demonstrate linking of this document with the geographical entity of Greek Administrative Geography corresponding to the subdistrict of Evros (see Fig. 7).

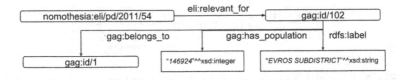

Fig. 7. Link P.D. 2011/54 and Evros subdistrict

5.4 Querying the Resulting RDF Data Using SPARQL

Based on the above, we have the ability to pose queries on the resulting RDF graph. Three queries are given in natural language together with their expression in SPARQL.

Q1. *Retrieve the structure of P.D. 2011/54, the type of each part (subdivision), and, optionally, its title and text.*

```
SELECT ?part ?text ?type ?title
WHERE {
    <http://legislation.di.uoa.gr/eli/pd/2011/54>
    eli:has_part+ ?part.
    ?part rdf:type ?type.
    OPTIONAL { ?part nomothesia:has_text ?text. }
    OPTIONAL { ?part eli:title ?title. } }
```

Q2. *Retrieve any modifications applied on P.D. 2011/54 accompanied with their type and the patient (affected) divisions.*

```
SELECT ?modification ?type ?patient
WHERE {
    <http://legislation.di.uoa.gr/pd/2011/54>
    eli:is_realized_by ?version.
    ?version metalex:matterOf ?modification.
    ?modification rdf:type ?type.
    ?modification metalex:patient ?part.
    <http://legislation.di.uoa.gr/eli/pd/2011/54>
    eli:has_part ?part }
```

Q3. *Retrieve any legal document that references geographical areas, which belong to the region of Thrace, and have population more than 100,000.*

```
SELECT ?doc ?municipality
WHERE {
    ?doc eli:relevant_for ?subdistrict.
    ?subdistrict gag:belongs_to ?region.
    ?subdistrict gag:has_population ?population.
    ?region gag:officialName "E.MACEDONIA-THRACE REGION"@en.
    FILTER(?population > 100000) }
```

The above queries are simple examples of the high level expressiveness (see Sect. 4) we can apply on Greek legislation, because we recorded a minimum number (only two) of legal documents in our example. Based on the above representation, we could build and present the current version of P.D. 2011/54 (see Fig. 3) through an automated process.

6   The Nomothesia Web Platform and RESTful API

We have implemented a web platform, using Java Spring framework, called Nomothesia (http://legislation.di.uoa.gr), which publishes legislation on the web. Legal documents are stored in RDF, using Sesame RDF store. Nomothesia comes with an intuitive interface for presenting legal documents. Among other capabilities, one may view how a certain legal document has evolved through time in response to modifications; view the actual modifications highlighted, or view the state of a legal document at a certain time point by applying all modifications to the initial document. With respect to search, Nomothesia offers an advanced search interface over the metadata of a legal document (e.g., title, enactment date) and the textual content of its structural elements. There are also pages presenting interesting statistics.

Last, Nomothesia offers a RESTful API serving legal documents in various formats (e.g., XML, JSON) and a SPARQL endpoint that can be used by more advanced users to form complex queries and easily retrieve and access legal documents as web resources. The latter functionality gives the opportunity to a programmer for direct open access to the knowledge base of Nomothesia, which can be employed to consume the legislative work of government into applications and combine it with other resources on the web so as to increase its value. Therefore, third party developers can build applications that utilize legislation for societal and business interests.

6.1 Querying Legislation Using SPARQL

Using Nomothesia SPARQL endpoint, we can answer complex queries as the following.

– *Retrieve the 5 most frequently modified legal documents during 2008–2013.*

```
SELECT ?doc (COUNT(DISTINCT ?expression) AS ?numver)
WHERE {
    ?doc eli:is_realized_by ?expression.
    ?expression metalex:matterOf ?modification.
    ?modification metalex:legislativeCompetenceGround
    ?doc2.
    ?doc2 eli:date_publication ?date.
    FILTER (?date >= "2008-01-01"^^xsd:date
    && ?date <= "2013-12-31"^^xsd:date)
} GROUP BY ?doc ?title ORDER BY DESC(?versions) LIMIT 5
```

?doc	?numver
nomothesia:eli/law/2010/3852	23
nomothesia:eli/law/2012/4093	19
nomothesia:eli/law/1994/2238	18
nomothesia:eli/law/1995/2362	13
nomothesia:eli/law/2007/3614	12

– *Retrieve the longest article that has been published in 2015.*

```
SELECT ?article (COUNT(?passage) as ?passages)
WHERE {
    ?doc eli:has_part+ ?article.
    ?doc eli:date_publication ?date.
    ?article rdf:type nomothesia:Article.
    ?article eli:has_part+ ?passage.
    ?passage rdf:type nomothesia:Passage.
    FILTER (?date >= "2015-01-01"^^xsd:date
    && ?date <= "2015-12-31"^^xsd:date)
} GROUP BY ?article ORDER BY DESC(?passages) LIMIT 1
```

?article	?passages
nomothesia:law/2015/4337/article/16	143

– *Find the 3 post government members that have signed the most legal documents during their service in 2008–2015.*

```
SELECT ?signatory_name (COUNT(?doc) AS ?docs)
WHERE {
    ?doc eli:passed_by ?signatory.
    ?doc eli:date_publication ?date.
    ?signatory foaf:name ?signatory_name.
    FILTER ( ?date >= "2008-01-01"^^xsd:date
    && ?date <= "2015-12-31"^^xsd:date)
} GROUP BY ?signatory_name ORDER BY DESC(?docs) LIMIT 4
```

?signatory_name	?docs
"C. ATHANASIOU"@en	227
"E. VENIZELOS"@en	211
"I. STOURNARAS"@en	175
"M. CHRISOHOIDIS"@en	167

7 Preliminary Feedback on Nomothesia

Nomothesia is a research prototype and it is still not well-known to the public. Thus, we have currently received limited feedback from our main target group, the citizens. In April of 2016, Nomothesia was presented and received an award in the 1st IT4GOV Contest, which was organized by the Greek Ministry of Administrative Reform & Electronic Governance. Among the contest jury members were 3 ministers and people from the industry, who judged Nomothesia positively. Following this contest, we have been in contact with officials from the House of Parliament and the Ministry of Administrative Reform & Electronic Governance, who have expressed interest in deploying Nomothesia in the governmental portal.

8 Conclusions and Future Work

In this paper we presented how Greek legislation can be published as open data using semantic web technologies. Our work is based on the latest European standards. We highlighted the significance of extending those standards in order to produce a semantic national legal publishing system and capture the particularities of Greek legislation. We highlighted the importance of interlinking legislation with other publicly available open data. Finally, we demonstrated the Nomothesia web platform, SPARQL endpoint and RESTful API, which offer advanced services to both citizens and programmers.

As an initial step of our future work we would like to proceed with an evaluation of our system based on user experience and study possible improvements. As a part of functional improvements, we would like to re-engineer the existing parsing system with natural language components for document zoning and entity recognition. Another direction would be the interlinking of Greek legislation with European legislation and case laws following the European Case Law Identifier scheme. Finally, we would like to extend our ontology to capture more complex legislative modifications.

References

1. Athan, T., Governatori, G., Palmirani, M., Paschke, A., Wyner, A.Z.: Legal-RuleML: design principles and foundations. In: Faber, W., Paschke, A. (eds.) Reasoning Web 2015. LNCS, vol. 9203, pp. 151–188. Springer, Cham (2015). doi:10.1007/978-3-319-21768-0_6
2. Barabucci, G., Cervone, L., Palmirani, M., Peroni, S., Vitali, F.: Multi-layer markup and ontological structures in Akoma Ntoso. In: Casanovas, P., Pagallo, U., Sartor, G., Ajani, G. (eds.) AICOL -2009. LNCS (LNAI), vol. 6237, pp. 133–149. Springer, Heidelberg (2010). doi:10.1007/978-3-642-16524-5_9
3. Boer, A., Hoekstra, R., Winkels, R., van Engers, T., Willaert, F.: ^{META}lex: legislation in XML. In: JURIX: The Fifteenth Annual Conference, London (2002)
4. Boer, A., Winkels, R., van Engers, T., de Maat, E.: Time and versions in ^{META}lex XML. In: Proceeding of the Workshop on Legislative XML, Kobaek Strand (2004)
5. Casanovas, P., Palmirani, M., Peroni, S., van Engers, T.M., Vitali, F.: Semantic web for the legal domain: the next step. Semant. Web 7(3), 213–227 (2016)
6. ELI Task Force. ELI - A technical implementation guide (2015)
7. E. C. for Standardization (CEN). CEN Workshop Agreement: Metalex (Open XML Interchange Format for Legal and Legislative Resources). Technical report (2006)
8. Francesconi, E., Küster, M.W., Gratz, P., Thelen, S.: The ontology-based approach of the publications office of the EU for document accessibility and open data services. In: Kő, A., Francesconi, E. (eds.) EGOVIS 2015. LNCS, vol. 9265, pp. 29–39. Springer, Cham (2015). doi:10.1007/978-3-319-22389-6_3

9. Frosterus, M., Tuominen, J., Wahlroos, M., Hyvönen, E.: The finnish law as a linked data service. In: Cimiano, P., Fernández, M., Lopez, V., Schlobach, S., Völker, J. (eds.) ESWC 2013. LNCS, vol. 7955, pp. 289–290. Springer, Heidelberg (2013). doi:10.1007/978-3-642-41242-4_46

10. Hallo Carrasco, M., Martínez-González, M.M., De La Fuente Redondo, P.: Data models for version management of legislative documents. J. Inf. Sci. **39**(4), 557–572 (2013)

11. Hoekstra, R.: The MetaLex document server – legal documents as versioned linked data. In: Aroyo, L., Welty, C., Alani, H., Taylor, J., Bernstein, A., Kagal, L., Noy, N., Blomqvist, E. (eds.) ISWC 2011. LNCS, vol. 7032, pp. 128–143. Springer, Heidelberg (2011). doi:10.1007/978-3-642-25093-4_9

12. Hoekstra, R., Breuker, J., Di Bello, M., Boer, A.: LKIF core: principled ontology development for the legal domain. In: Proceedings of the 2009 Conference on Law, Ontologies and the Semantic Web: Channelling the Legal Information Flood, Amsterdam, The Netherlands, pp. 21–52. IOS Press (2009)

13. Van Opijnen, M.: European case law identifier: indispensable asset for legal information retrieval. In: Biasiotti, M.A., Faro, S. (eds.) From Information to Knowledge. Frontiers in Artificial Intelligence and Applications, vol. 236, pp. 91–103. IOS Press (2011). doi:10.3233/978-1-60750-988-2-91. ISBN: 978-1-60750-987-5

14. Palmirani, M., Vitali, F.: Akoma-Ntoso for legal documents. In: Sartor, G., Palmirani, M., Francesconi, E., Biasiotti, M.A. (eds.) Legislative XML for the Semantic Web. Law, Governance and Technology Series, vol. 4, pp. 75–100. Springer, Heidelberg (2011)

15. Archer, P., Goedertier, S., Loutas, N.: Study on persistent URIs, with identification of best practices and recommendations on the topic for the MSs and the EC (2012)

16. PwC EU Services. Case study: How Linked Data is transforming eGoverment (2013)

Self-Enforcing Access Control
for Encrypted RDF

Javier D. Fernández[1,2], Sabrina Kirrane[1], Axel Polleres[1,2],
and Simon Steyskal[1,3(✉)]

[1] Vienna University of Economics and Business, Vienna, Austria
{javier.fernandez,sabrina.kirrane,axel.polleres,
simon.steyskal}@wu.ac.at
[2] Complexity Science Hub Vienna, Vienna, Austria
[3] Siemens AG Österreich, Vienna, Austria

Abstract. The amount of raw data exchanged via web protocols is
steadily increasing. Although the Linked Data infrastructure could
potentially be used to selectively share RDF data with different individuals or organisations, the primary focus remains on the unrestricted
sharing of public data. In order to extend the Linked Data paradigm to
cater for closed data, there is a need to augment the existing infrastructure with robust security mechanisms. At the most basic level both access
control and encryption mechanisms are required. In this paper, we propose a flexible and dynamic mechanism for securely storing and efficiently
querying RDF datasets. By employing an encryption strategy based on
Functional Encryption (FE) in which controlled data access does not
require a trusted mediator, but is instead enforced by the cryptographic
approach itself, we allow for fine-grained access control over encrypted
RDF data while at the same time reducing the administrative overhead
associated with access control management.

1 Introduction

The Linked Data infrastructure could potentially be used not only to distribut-
edly share public data, but also to selectively share data, perhaps of a sensitive
nature (e.g., personal data, health data, financial data, etc.), with specific individuals or organisations (i.e., closed data). In order to realise this vision, we
must first extend the existing Linked Data infrastructure with suitable security mechanisms. More specifically, encryption is needed to protect data in case
the server is compromised, while access control is needed to ensure that only
authorised individuals can access specific data. Apart from the need to protect
data, robustness in terms of usability, performance, and scalability is a major
consideration.

Supported by the Austrian Science Fund (FWF): M1720-G11, the Austrian Research
Promotion Agency (FFG) under grant 845638, and European Union's Horizon 2020
research and innovation programme under grant 731601.

E. Blomqvist et al. (Eds.): ESWC 2017, Part I, LNCS 10249, pp. 607–622, 2017.
DOI: 10.1007/978-3-319-58068-5_37

However, current encryption techniques for RDF are still very limited, especially with respect to the flexible maintenance and querying of encrypted data in light of user access control policies. Initial partial encryption techniques [16,17] focus on catering for both plain and encrypted data in the same representation and how to incorporate the metadata necessary for decryption. More recently, [21] proposed the generation of multiple ciphertexts per triple (i.e. each triple is encrypted multiple times depending on whether or not access to the subject, predicate and/or object is restricted) and the distribution of several keys to users. Although finer-grained access control is supported, the maintenance of multiple ciphertexts (i.e. encrypted triples) and keys presents scalability challenges. Additionally, such an approach, or likewise term-based encryption of RDF graphs, means that the *structure* of parts of the graph that should not be accessible could potentially be recovered, thus posing a security risk (cf. for instance [32]).

Beyond RDF, novel cryptography mechanisms have been developed that enable the flexible specification and enforcement of access policies over encrypted data. Predicate-based Encryption (PBE) [22] – which we refer to as Functional Encryption (FE) in order to avoid confusion with *RDF predicates* – enables searching over encrypted data, mainly for keywords or the conjunction of keyword queries, while alleviating the re-encryption burden associated with adding additional data.

Herein, we extend recent findings on FE to RDF, and demonstrate how FE can be used for fine-grained access control based on triples patterns over encrypted RDF datasets. Summarising our contributions, we: (i) adapt functional encryption to RDF such that it is possible to enforce access control over encrypted RDF data in a self-enforcing manner; (ii) demonstrate how encryption keys based on triple patterns can be used to specify flexible access control for Linked Data sources; and (iii) propose and evaluate indexing strategies that enhance query performance and scalability. Experiments show reasonable loading and query performance overheads with respect to traditional, non-encrypted data retrieval. The remainder of the paper is structured as follows: We discuss related work and potential alternatives to our proposal in *Sect.* 2. The details of our specific approach and optimisations are presented in *Sects.* 3 and 4 respectively, and evaluated in *Sect.* 5. Finally, we conclude and outline directions for future work in *Sect.* 6.

2   Related Work

When it comes to access control for RDF, broadly speaking researchers have focused on representing existing access control models and standards using semantic technology; proposing new access control models suitable for open, heterogeneous and distributed environments; and devising languages and frameworks that can be used to facilitate access control policy specification and maintenance. Kirrane et al. [23] provide a comprehensive survey of existing access control proposals for RDF. Unlike access control, encryption techniques for RDF has received very little attention to date. Giereth [17] demonstrate how public-key encryption techniques can be used to partially encryption RDF data represented using XML. While, Giereth [17] and Gerbracht [16] propose strategies for combining partially encrypted RDF data with the metadata that is necessary for decryption. Kasten et al. [21] propose a framework that can be used to query encrypted data. In order to support SPARQL queries based on triple patterns each triple is encrypted eight times according to the eight different binding possibilities. Limitations of the approach include the blowup associated with maintaining eight ciphers per triple and the fact that the structure of the graph is still accessible.

Searchable Symmetric Encryption (SSE) [9] has been extensively applied in database-as-a-service and cloud environments. SSE techniques focus on the encryption of outsourced data such that an external user can encrypt their query and subsequently evaluate it against the encrypted data. More specifically, SSE extracts the key features of a query (the data structures that allow for its resolution) and encrypts them such that it can be efficiently evaluated on the encrypted data. Extensive work has been done in basic SSE, which caters for a single keyword [6]. Recent improvements have been proposed to handle conjunctive search over multiple keywords [5], and to optimise the resolution to cater for large scale data in the presence of updates [4,20,30]. However, all of these works focus on keyword-based retrieval, whereas structured querying (such as SPARQL) over encrypted RDF datasets would require (at least) an unrestricted set of triple query patterns. In contrast, Fully Homomorphic Encryption (FHE) [15] allows any general circuit/computation over encrypted data, however it is prohibitively slow for most operations [7,28]. Thus, practical, encryption databases such as CryptDB [28] make use of lighter forms of encryption that still cater for computations (such as sums) over the encrypted data [27], at the cost of different vulnerability/feasibility trade-offs. Recently, predicate encryption [22], whereby predicates correspond to the evaluation of disjunctions, polynomial equations and inner products, enables security in light of unrestricted queries. Predicate encryption has a proven track record of efficiency in terms of conjunctive equality, range and subset queries.

The solution we propose builds on an existing work that defines access control policies based on RDF patterns that are in turn enforced over RDF datasets [23]. While existing proposals enforce access control over plain RDF data via data filtering (i.e., a query is executed against a dataset which is generated by removing the unauthorised data) or query rewriting (i.e., a query is updated so that unauthorised data will not be returned and subsequently executed over the unmodified dataset), we demonstrate how functional encryption can be used to enforce access control over encrypted RDF data in a self-enforcing manner (i.e., without the need for either data filtering or query rewriting nor a trusted mediator). Unlike previous approaches we store one cipher per triple and employ indexing strategies based on secure hashes (cf. PBKDF2 [19]) that can be used for efficient querying of encrypted RDF. In addition, we propose a mechanism to obfuscate the graph structure with real indexes and dummy ciphers that cannot be decrypted, making the dummy hashes and ciphers indistinguishable from real hashes and ciphers.

3   Secure and Fine-Grained Encryption of RDF

Common public-key encryption schemes usually follow an all-or-nothing approach (i.e., given a particular decryption key, a ciphertext can either be decrypted or not) which in turn requires users to manage a large amount of keys, especially if there is a need for more granular data encryption [2]. Recent advances in public-key cryptography, however, have led to a new family of encryption schemes called *Functional Encryption (FE)* which addresses aforementioned issue by making encrypted data self-enforce its access restrictions, hence, allowing for fine-grained access over encrypted information. In a functional encryption scheme, each decryption key is associated with a boolean function and each ciphertext is associated with an element of some attribute space Σ; a decryption key corresponding to a boolean function f is able to decrypt a particular ciphertext associated with $I \in \Sigma$ iff $f(I) = 1$. A functional encryption scheme is defined as a tuple of four distinct algorithms (**Setup, Enc, KeyGen, Dec**) such that:

Setup is used for generating a master public and master secret key pair.
Enc encrypts a plaintext message m given the master public key and an element $I \in \Sigma$. It returns a ciphertext c.
KeyGen takes as input the master secret key and generates a decryption key (i.e., secret key) SK_f for a given boolean function f.
Dec takes as input a secret key SK_f and a ciphertext c. It extracts I from c and computes $f(I)$.

3.1   A Functional Encryption Scheme for RDF

While there exist various different approaches for realising functional encryption schemes, we build upon the work of Katz et al. [22] in which functions correspond to the computation of inner-products over $\mathbb{Z}_N$ (for some large integer N). In their construction, they use $\Sigma = \mathbb{Z}_N^n$ as set of possible ciphertext attributes of length n and $\mathcal{F} = \{f_{\vec{x}} | \vec{x} \in \mathbb{Z}_N^n\}$ as the class of decryption key functions. Each ciphertext is associated with a (secret) attribute vector $\vec{y} \in \Sigma$ and each decryption key corresponds to a vector $\vec{x}$ that is incorporated into its respective boolean function $f_{\vec{x}} \in \mathcal{F}$ where $f_{\vec{x}}(\vec{y}) = 1$ iff $\sum_{i=1}^{n} y_i x_i = 0$.

In the following, we discuss how this encryption scheme can be utilised (i.e., its algorithms adopted[1]) to provide fine-grained access over encrypted RDF triples. Thus, allow for querying encrypted RDF using triple patterns such that a particular decryption key can decrypt all triples that satisfy a particular triple pattern (i.e., one key can open multiple locks). For example, a decryption key generated from a triple pattern (?,p,?) should be able to decrypt all triples with p in the predicate position.

---

[1] The **Setup** algorithm remains unchanged.

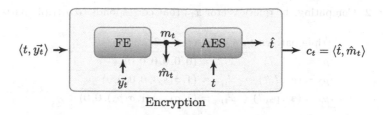

Fig. 1. Process of encrypting an RDF triple t.

Encrypting RDF Triples (Enc). To be able to efficiently encrypt large RDF datasets, we adopt a strategy commonly used in public-key infrastructures for securely and efficiently encrypting large amounts of data called *Key Encapsulation* [24]. Key encapsulation allows for secure but slow asymmetric encryption to be combined with simple but fast symmetric encryption by using asymmetric encryption algorithms for deriving a symmetric encryption key (usually in terms of a seed) which is subsequently used by encryption algorithms such as AES [11] for the actual encryption of the data. We illustrate this process in Fig. 1.

Thus, to encrypt an RDF triple $t = (s, p, o)$, we first compute its respective triple vector (i.e., attribute vector) $\vec{y}_t$ and functionally encrypt (i.e., compute **Enc** as defined in [22]) a randomly generated seed m_t using $\vec{y}_t$ as the associated attribute vector. Triple vector $\vec{y}_t$ where $\vec{y}_t = (y_s, y_s', y_p, y_p', y_o, y_o')$ for triple t is constructed as follows, where σ denotes a mapping function that maps a triple's subject, predicate, and object value to elements in $\mathbb{Z}_N$:

$$y_l := -r \cdot \sigma(l), y_l' := r, \text{with } l \in \{s, p, o\} \text{ and random } r \in \mathbb{Z}_N$$

Table 1 illustrates the construction of a triple vector $\vec{y}_t$ based on RDF triple t.

Table 1. Computing the triple vector $\vec{y}_t$ of an RDF triple t.

Triple t	Triple vector $\vec{y}_t$
$t_1 = (s_1, p_1, o_1)$	$\vec{y}_{t_1} = (-r_1 \cdot \sigma(s_1), r_1, -r_2 \cdot \sigma(p_1), r_2, -r_3 \cdot \sigma(o_1), r_3)$
$t_2 = (s_2, p_2, o_2)$	$\vec{y}_{t_2} = (-r_4 \cdot \sigma(s_2), r_4, -r_5 \cdot \sigma(p_2), r_5, -r_6 \cdot \sigma(o_2), r_6)$
...	...
$t_n = (s_n, p_n, o_n)$	$\vec{y}_{t_n} = (-r_{3n-2} \cdot \sigma(s_n), r_{3n-2}, -r_{3n-1} \cdot \sigma(p_n), r_{3n-1}, -r_{3n} \cdot \sigma(o_n), r_{3n})$

We use AES to encrypt the actual plaintext triple t with an encryption key derivable from our previously generated seed m_t and return both, the resulting AES ciphertext of t denoted by $\hat{t}$ and the ciphertext of the seed denoted by $\hat{m}_t$ as final ciphertext triple $c_t = \langle \hat{t}, \hat{m}_t \rangle$.

Generating Decryption Keys (KeyGen). As outlined above, decryption keys must be able to decrypt all triples that satisfy their inherent triple pattern

Table 2. Computing the query vector $\vec{x}_{tp}$ that corresponds to a triple pattern tp

Triple pattern tp	Query vector $\vec{x}_{tp}$
$tp_1 = (?, ?, ?)$	$\vec{x}_{tp_1} = (0, 0, 0, 0, 0, 0)$
$tp_2 = (\mathsf{s}_2, ?, ?)$	$\vec{x}_{tp_2} = (1, \sigma(s_2), 0, 0, 0, 0)$
$tp_3 = (\mathsf{s}_3, \mathsf{p}_3, ?)$	$\vec{x}_{tp_3} = (1, \sigma(s_3), 1, \sigma(p_3), 0, 0)$
$\ldots$	$\ldots$
$tp_n = (\mathsf{s}_n, \mathsf{p}_n, \mathsf{o}_n)$	$\vec{x}_{tp_n} = (1, \sigma(s_n), 1, \sigma(p_n), 1, \sigma(o_n))$

(i.e., one query key can open multiple locks). In order to compute a decryption key based on a triple pattern $tp = (s, p, o)$ with s, p, and o either bound or unbound, we define its corresponding vector $\vec{x}$ as $\vec{x}_{tp} = (x_s, x'_s, x_p, x'_p, x_o, x'_o)$ with:

$$\text{if } l \text{ is bound: } x_l := 1, x'_l := \sigma(l), \text{ with } l \in \{s, p, o\}$$
$$\text{if } l \text{ is not bound: } x_l := 0, x'_l := 0, \text{ with } l \in \{s, p, o\}$$

Again, σ denotes a mapping function that maps a triple pattern's subject, predicate, and object value to elements in $\mathbb{Z}_N$. Table 2 illustrates the construction of a query vector $\vec{x}_{tp}$ that corresponds to a triple pattern tp.

Decryption of RDF Triples (Dec). To verify whether an encrypted triple can be decrypted with a given decryption key, we compute the inner-product of their corresponding triple vector $\vec{y}_t$ and query vector $\vec{x}_{tp}$, with $t = (s_t, p_t, o_t)$ and $tp = (s_{tp}, p_{tp}, o_{tp})$:

$$\vec{y}_t \cdot \vec{x}_{tp} = y_{s_t} x_{s_{tp}} + y'_{s_t} x'_{s_{tp}} + y_{p_t} x_{p_{tp}} + y'_{p_t} x'_{p_{tp}} + y_{o_t} x_{o_{tp}} + y'_{o_t} x'_{o_{tp}}$$

Only when $\vec{y}_t \cdot \vec{x}_{tp} = 0$ is it possible to decrypt the encrypted seed $\hat{m}_t$, hence the corresponding symmetric AES key can be correctly derived and the plaintext triple t be returned. Otherwise (i.e., $\vec{y}_t \cdot \vec{x}_{tp} \neq 0$), an arbitrary seed $m' \neq m_t$ is generated hence encrypted triple c_t cannot be decrypted [26].

4 Optimising Query Execution over Encrypted RDF

The *secure data store* holds all the encrypted triples, i.e. $\{c_{t_1}, c_{t_2}, \cdots, c_{t_n}\}$, being n the total number of triples in the dataset. Besides assuring the confidentiality of the data, the data store is responsible for enabling the querying of encrypted data.

In the most basic scenario, since triples are stored in their encrypted form, a user's query would be resolved by iterating over all triples in the dataset, checking whether any of them can be decrypted with a given decryption key. Obviously, this results in an inefficient process at large scale. As a first improvement one can distribute the set of encrypted triples among different peers such that decryption could run in parallel. In spite of inherent performance improvements, such a solution is still dominated by the available number of peers and the – potentially large – number of encrypted triples each peer would have to process. Current efficient solutions for querying encrypted data are based on (a) using indexes to speed up the decryption process by reducing the set of potential solutions; or (b) making use of specific encryption schemes that support the execution of operations directly over encrypted data [13]. Our solution herein follows the first approach, whereas the use of alternative and directly encryption mechanisms (such as homomorphic encryption [28]) is complementary and left to future work.

In our implementation of such a secure data store, we first encrypt all triples and store them in a key-value structure, referred to as an `EncTriples Index`, where the keys are unique integer IDs and the values hold the encrypted triples (see Figs. 2 and 3 (right)). Note that this structure can be implemented with any traditional *Map* structure, as it only requires fast access to the encrypted value associated with a given ID. In the following, we describe two alternative approaches, i.e., one using *three individual indexes* and one based on *Vertical Partitioning (VP)* for finding the range of IDs in the `EncTriples Index` which can satisfy a triple pattern query. In order to maintain simplicity and general applicability of the proposed store, both alternatives consider key-value backends, which are increasingly used to manage RDF data [8], especially in distributed scenarios. It is also worth mentioning that we focus on basic triple pattern queries as (i) they are the cornerstone that can be used to build more complex SPARQL queries, and (ii) they constitute all the functionality to support the Triple Pattern Fragments [31] interface.

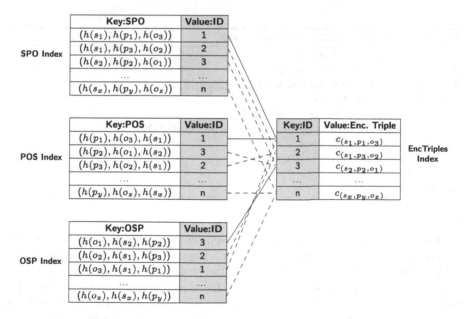

Fig. 2. 3-Index approach for indexing and retrieval of encrypted triples.

3-Index Approach. Following well-known indexing strategies, such as from CumulusRDF [25], we use three key-value B-Trees in order to cover all triple pattern combinations: SPO, POS and OSP Indexes. Figure 2 illustrates this organisation. As can be seen, each index consists of a *Map* whose keys are the securely hashed (cf. PBKDF2 [19]) subject, predicate, and object of each triple, and values point to IDs storing the respective ciphertext triples in the EncTriples Index.

Algorithm 1 shows the resolution of a (s,p,o) triple pattern query using the 3-Index approach. First, we compute the secure hashes h(s), h(p) and h(o) from the corresponding s, p and o provided by the user (Line 1). Our $hash(s, p, o)$ function does not hash unbounded terms in the triple pattern but treats them as a wildcard '?' term (hence all terms will be retrieved in the subsequent range queries). Then, we select the best index to evaluate the query (Line 2). In our case, the SPO Index serves (s,?,?) and (s,p,?) triple patterns, the POS Index satisfies (?,p,?) and (?,p,o), and the OSP Index index serves (s,?,o) and (?,?,o). Both (s,p,o) and (?,?,?) can be solved by any of them. Then, we make use of the selected index to get the range of values where the given h(s), h(p), h(o) (or 'anything' if the wildcard '?' is present in a term) is stored (Line 3). Note that this search can be implemented by utilising B-Trees [10,29] for indexing the keys. For each of the candidate ID values in the range (Line 4), we retrieve the encrypted triple for such ID by searching for this ID in the EncTriples Index (Line 5). Finally, we proceed with the decryption of the encrypted triple using the key provided by the user (Line 6). If the status of such decryption is *valid* (Line 7) then the decryption was successful and we output the decrypted triples (Line 8) that satisfy the query.

Algorithm 1. 3-Index_Search(s,p,o,key)

1: $(h(s), h(p), h(o)) \leftarrow hash(s, p, o);$
2: $index \leftarrow selectBestIndex(s, p, o);$ ▷ $index = \{SPO|POS|OSP\}$
3: $IDs[\ ] \leftarrow index.getRangeValues(h(s), h(p), h(o));$
4: **for each** $(id \in IDs)$ **do**
5:     $encryptedTriple \leftarrow EncTriples.get(id);$
6:     $< decryptedTriple, status > \leftarrow Decrypt(encryptedTriple, key);$
7:     **if** $(status = valid)$ **then**
8:         **output**$(decryptedTriple);$
9:     **end if**
10: **end for**

Thus, the combination of the three SPO, POS and OSP Indexes reduces the search space of the query requests by applying simple range scans over hashed triples. This efficient retrieval has been traditionally served through tree-based map structures guaranteeing $log(n)$ costs for searches and updates on the data, hence we rely on B-Tree stores for our practical materialisation of the indexes. In contrast, supporting all triple pattern combinations in 3-Index comes at the expense of additional space overheads, given that each $(h(s), h(p), h(o))$ of a triple is stored three times (in each SPO, POS and OSP Indexes). Note, however, that this is a typical scenario for RDF stores and in our case the triples are encrypted and stored just once (in EncTriples Index).

Vertical Partitioning Approach. Vertical partitioning [1] is a well-known RDF indexing technique motivated by the fact that usually only a few predicates are used to describe a dataset [14]. Thus, this technique stores one "table" per predicate, indexing (S,O) pairs that are related via the predicate. In our case, we propose to use one key-value B-Tree for each h(p), storing (h(s), h(o)) pairs as keys, and the corresponding ID as the value. Similar to the previous case, the only requirement is to allow for fast range queries on their map index keys. However, in the case of an SO index, traditional key-value schemes are not efficient for queries where the first component (the subject) is unbound. Thus, to improve efficiency for triple patterns with unbounded subject (i.e. $(?,p_y,o_z)$ and $(?,?,o_z)$), while remaining in a general key-value scheme, we duplicate the pairs and introduce the inverse $(h(o), h(s))$ pairs. The final organisation is shown in Fig. 3 (left), where the predicate maps are referred to as Pred_h(p_1), Pred_h(p_2),..., Pred_h(p_n) Indexes. As depicted, we add "so" and "os" keywords to the stored composite keys in order to distinguish the order of the key.

Algorithm 2 shows the resolution of a (s,p,o) triple pattern query with the VP organisation. In this case, after performing the variable initialisation (Line 1) and the aforementioned secure hash of the terms (Line 2), we inspect the predicate term h(p) and select the corresponding predicate index (Line 3), i.e., Pred_h(p). Nonetheless, if the predicate is unbounded, all predicate indexes are selected as we have to iterate through all tables, which penalises the performance of such queries. For each predicate index, we then inspect the subject term (Lines 5–9). If the subject is unbounded (Line 5), we will perform a ("os",h(o),?) range query over the corresponding predicate index (Line 6),

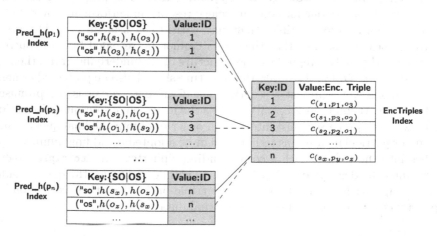

Fig. 3. Vertical Partitioning (VP) approach for indexing and retrieval of encrypted triples.

Algorithm 2. VerticalPartitioning_Search(s,p,o,key)

```
1:  IDs[ ] ← ();
2:  (h(s), h(p), h(o)) ← hash(s, p, o);
3:  Indexes[] ← selectPredIndex(h(p)); ▷ Indexes ⊆ {Pred_h(p₁), · · · , Pred_h(pₙ)Index}
4:  for each (index ∈ Indexes) do
5:     if (s =?) then
6:        IDs[ ] ← index.getRangeValues("os", h(o), ?);
7:     else
8:        IDs[ ] ← index.getRangeValues("so", h(s), h(o));
9:     end if
10:    for each (id ∈ IDs) do
11:       encryptedTriple ← EncTriples.get(id);
12:       < decryptedTriple, status >← Decrypt(encryptedTriple, key);
13:       if (status = valid) then
14:          output(decryptedTriple);
15:       end if
16:    end for
17: end for
```

otherwise we execute a ("so",h(s),h(o)) range query. Note that in both cases the object could also be unbounded. The algorithm iterates over the candidates IDs (Lines 10-end) in a similar way to the previous cases, i.e., retrieving the encrypted triple from EncTriples Index (Line 11) and performing the decryption (Lines 12–14).

Overall, VP needs less space than the previous 3-Index approach, since the predicates are represented implicitly and the subjects and objects are represented only twice. In contrast, it penalises the queries with unbound predicate as it has to iterate through all tables. Nevertheless, studies on SPARQL query logs show that these queries are infrequent in real applications [3].

Protecting the Structure of Encrypted Data. The proposed hash-based indexes are a cornerstone for boosting query resolution performance by reducing the encrypted candidate triples that may satisfy the user queries. The use of secure hashes [19] assures that the terms cannot be revealed but, in contrast, the indexes themselves reproduce the structure of the underlying graph (i.e., the in/out degree of nodes). However, the structure should also be protected as hash-based indexes can represent a security risk if the data server is compromised. State-of-the-art solutions (cf., [13]) propose the inclusion of spurious information, that the query processor must filter out in order to obtain the final query result.

In our particular case, this technique can be adopted by adding dummy triple hashes into the indexes with a corresponding ciphertext (in EncTriples Index) that cannot be decrypted by any key, hence will not influence the query results. Such an approach ensures that both the triple hashes and their corresponding ciphertexts are not distinguishable from real data.

5 Evaluation

We develop a prototypical implementation[2] of the proposed encryption and indexing strategies. Our tool is written in Java and it relies on the Java Pairing-Based Cryptography Library (JPBC [12]) to perform all the encryption/decryption operations. While, we use MapDB[3] as the supporting framework for the indexes. We provide an interface that takes as input a triple pattern query and a query key, and outputs the results of the query.

We evaluate our proposal in two related tasks: (i) performance of the data loading (encryption and indexing) and (ii) performance of different user queries (query execution on encrypted data). In both cases, we compare our proposed 3-Index strategy w.r.t the vertical partitioning (VP) approach. Finally, we measure the performance overhead associated with query resolution, introduced by the secure infrastructure, by comparing its results with a counterpart non-secure triplestore. For a fair comparison, we implement the non-secure triplestore with similar 3-Index and VP indexing strategies, storing the RDF data in plain. The approaches are referred to as 3-Index-plain and VP-plain respectively.

Table 3. Statistical dataset description.

Dataset	Triples	$|S|$	$|P|$	$|O|$	Size (MB)
Census	361,842	51,768	26	6,901	52
Jamendo	1,049,637	335,925	26	440,602	144
AEMET	3,547,154	394,289	23	793,664	726
	100,000	22,932	18	11,588	15
	200,000	39,244	18	23,749	29
	500,000	87,984	18	60,028	71
LUBM	1,000,000	169,783	18	120,464	139
	2,000,000	333,105	18	241,342	277
	5,000,000	820,185	18	604,308	694

Table 3 describes our experimental datasets, reporting the number of triples, different subjects ($|S|$), predicates ($|P|$) and objects ($|O|$), as well as the file size (in NT format). Note that there is no standard RDF corpus that can be used to evaluate RDF encryption approaches, hence we choose a diverse set of datasets that have been previously used to benchmark traditional RDF stores or there is a use case that indicates they could potentially benefit from a secure data store. On the one hand, we use the well-known Lehigh University Benchmark (LUBM [18]) data generator to obtain synthetic datasets of incremental sizes from 100K triples

---

[2] Source code and experimental datasets are available at: https://aic.ai.wu.ac.at/comcrypt/sld/.
[3] http://www.mapdb.org/.

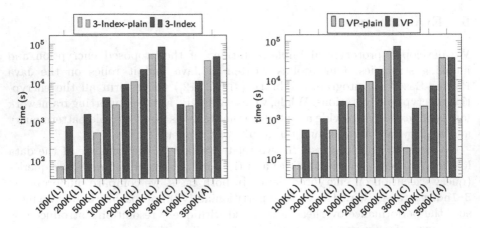

Fig. 4. Time for loading (encrypting+indexing) the entire dataset for `3-Index` and VP. We only report indexing time for the non-secure counterparts `3-Index-plain` and `VP-plain`.

to 5M triples. On the other hand, we choose real-world datasets from different domains: `Census` represents the 2010 Australian census, where sensitive data must be preserved and users could have different partial views on the dataset; `Jamendo` lists music records and artists, where some data can be restricted to certain subscribers; and `AEMET` includes sensor data from weather stations in Spain, which is a real use case where the old data is public but the most recent data is restricted to particular users. Tests were performed on a computer with 2 x Intel Xeon E5-2650v2 @ 2.6 GHz (16 cores), RAM 171 GB, 4 HDDs in RAID 5 config. (2.7 TB netto storage), Ubuntu 14.04.5 LTS running on a VM with QEMU/KVM hypervisor. All of the reported (elapsed) times are the average of three independent executions.

Data Loading. Figure 4 shows the dataset load times[4] for the `3-Index` and VP strategies. The reported time consists of the time to encrypt the triples using the aforementioned FE scheme, and the time to securely hash the terms and create the different indexes. In contrast, the non-secure triplestores, i.e. the `3-Index-plain` and `VP-plain` counterparts, only require the dataset to be indexed (we also make use of the hash of the terms in order to compare the encryption overhead).

The results show that the time of both the `3-Index` and the VP strategy scales linearly with the number of triples, which indicates that the representation can scale in the envisioned Linked Data scenario. It is worth noting that both strategies report similar performance results, where VP is slightly faster for loading given that only the subject and object is used to index each triple (the predicate is implicitly given by vertical partitioning). Finally, note that the

---

[4] We first list the LUBM datasets in increasingly order of triples, and use name abbreviations for LUBM (L), Census (C), Jamendo (J), and AEMET (A).

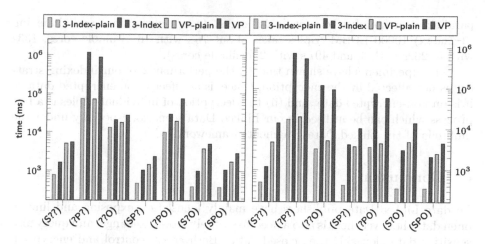

Fig. 5. Cold query times of LUBM with 5M triples (LHS) and Jamendo (RHS) for `3-Index`, `VP`, and their non-secure counterparts in ms (logarithmic y-axis).

comparison w.r.t the plain counterparts shows that the encryption overhead can be of one order of magnitude for the smaller datasets. In contrast, the encryption overhead is greatly reduced for larger datasets which is primarily due to the fact that the loading time for large datasets is the predominant factor, as the B-Tree indexes become slower the more triples are added (due to rebalancing).

Query Resolution. Figure 5 shows the query resolution time for two selected datasets[5], LUBM with 5M triples and Jamendo, considering all types of triple patterns. To do so, we sample 1,000 queries of each type and report the average resolution time. As expected, the `3-Index` reports a noticeable better performance than `VP` for queries with unbound predicates given that `VP` has to iterate though all predicate tables in this case. In turn, the `3-Index` and the `VP` approaches remain competitive with respect to their non-secure counterparts, if a look-up returns only a small amount of results as it is usually the case for (s,?,?), (s,?,o), (s,p,o) queries. However, the more query results that need to be returned the longer the decryption takes. At this point we also want to stress that due to the nature of our approach, each result triple can be returned as soon as its decryption has finished. This is in line with the incremental nature of the Triple Pattern Fragment [31] approach, which paginates the query results (typically including 100 results per page), allowing users to ask for further pages if required. For example, decrypting Jamendo entirely took about 2256 s for `VP` and 2808 s for `3-Index`, leading to respective triple decryption rates of 465 triples/s and 374 triples/s in a cold scenario, which already fulfils the performance requirements to feed several Triple Pattern Fragments per second.

Scalability. As mentioned in Sect. 4, our approach allows for parallel encryption/decryption of triples, thus scales with the system's supported level of

---

[5] Results are comparable for all datasets.

parallelisation/number of available cores (e.g., encrypting and indexing (3-Index) 10000 LUBM triples takes about 76s with 16 available cores, 133s with 8, 262s with 4, and 497s with 2 available cores).

Our experiments have shown that (i) the performance of our indexing strategy is not affected by the encryption, hence, is as effective on encrypted data as it is on non-encrypted data, and (ii) the decryption of individual triples is a fast process which can be utilised in our Linked Data scenario, especially under the umbrella of the Linked Data Fragments framework.

6   Conclusion

To date Linked Data publishers have mainly focused on exposing and linking open data, however there is also a need to securely store, exchange, and query also sensitive data alongside (i.e., closed data). Both access control and encryption mechanisms are needed to protect such data from unauthorised access, security breaches, and potentially untrusted service providers. Herein, we presented a mechanism to provide secure and fine-grained encryption of RDF datasets. First, we proposed a practical realisation of a functional encryption scheme, which allows data providers to generate query keys based on (triple-)patterns, whereby one decryption key can decrypt all triples that match its associated triple pattern. As such, our approach operates on a very fine level of granularity (i.e., triple level), which provides a high degree of flexibility and enables controlled access to encrypted RDF data. In existing literature, enforcing access control at the level of single statements or tuples is generally referred to as fine-grained access control (cf. [23]). Then, we presented two indexing strategies (implemented using MapDB) to enhance query performance, the main scalability bottleneck when it comes to serving user requests.

Our empirical evaluation shows that both indexing strategies on encrypted RDF data report reasonable loading and query performance overheads with respect to traditional, non-encrypted data retrieval. Our results also indicate that the approach is relatively slow for batch decryption, but this can be counteracted by the fact that it is suitable for serving incremental results, hence it is particularly suitable for Linked Data Fragments.

In future work, we plan to inspect different indexing strategies in order to optimise the loading time and query performance of large queries. We also consider extending our proposal to cater for named graphs, that is, encrypting quads instead of triples and generating keys based on quad patterns. Finally, we aim to integrate the proposed secure RDF store with a "policy" tier by employing Attribute-based Access Control (ABAC), which will manage the access/revocation to the query keys and serve as fully fledged security framework for Linked Data.

References

1. Abadi, D.J., Marcus, A., Madden, S.R., Hollenbach, K.: Scalable semantic web data management using vertical partitioning. In: Proceedings of Very Large Data Bases, pp. 411–422 (2007)
2. Abdalla, M., Bourse, F., Caro, A., Pointcheval, D.: Simple functional encryption schemes for inner products. In: Katz, J. (ed.) PKC 2015. LNCS, vol. 9020, pp. 733–751. Springer, Heidelberg (2015). doi:10.1007/978-3-662-46447-2_33
3. Arias, M., Fernández, J.D., Martínez-Prieto, M.A., de la Fuente, P.: An empirical study of real-world SPARQL queries. arXiv preprint arXiv:1103.5043 (2011)
4. Cash, D., Jaeger, J., Jarecki, S., Jutla, C.S., Krawczyk, H., Rosu, M.-C., Steiner, M.: Dynamic searchable encryption in very-large databases: data structures and implementation. IACR Cryptology ePrint Archive, 2014:853 (2014)
5. Cash, D., Jarecki, S., Jutla, C., Krawczyk, H., Roşu, M.-C., Steiner, M.: Highly-scalable searchable symmetric encryption with support for boolean queries. In: Canetti, R., Garay, J.A. (eds.) CRYPTO 2013. LNCS, vol. 8042, pp. 353–373. Springer, Heidelberg (2013). doi:10.1007/978-3-642-40041-4_20
6. Chang, Y.-C., Mitzenmacher, M.: Privacy preserving keyword searches on remote encrypted data. In: Ioannidis, J., Keromytis, A., Yung, M. (eds.) ACNS 2005. LNCS, vol. 3531, pp. 442–455. Springer, Heidelberg (2005). doi:10.1007/11496137_30
7. Chase, M., Shen, E.: Pattern matching encryption. IACR Cryptology ePrint Archive, 2014:638 (2014)
8. Cudré-Mauroux, P., et al.: NoSQL databases for RDF: an empirical evaluation. In: Alani, H., et al. (eds.) ISWC 2013. LNCS, vol. 8219, pp. 310–325. Springer, Heidelberg (2013). doi:10.1007/978-3-642-41338-4_20
9. Curtmola, R., Garay, J., Kamara, S., Ostrovsky, R.: Searchable symmetric encryption: improved definitions and efficient constructions. In: Proceedings of Computer and Communications Security, pp. 79–88 (2006)
10. da Rocha Pinto, P., Dinsdale-Young, T., Dodds, M., Gardner, P., Wheelhouse, M.J.: A simple abstraction for complex concurrent indexes. In: Proceedings of Object-Oriented Programming, Systems, Languages, and Applications, pp. 845–864 (2011)
11. Daemen, J., Rijmen, V.: The Design of Rijndael: AES - The Advanced Encryption Standard. Information Security and Cryptography. Springer, Heidelberg (2002)
12. De Caro, A., Iovino, V.: JPBC: Java pairing based cryptography. In: Proceedings of IEEE Symposium on Computers and Communications, pp. 850–855 (2011)
13. De Capitani di Vimercati, S., Foresti, S., Livraga, G., Samarati, P.: Practical techniques building on encryption for protecting and managing data in the cloud. In: Ryan, P.Y.A., Naccache, D., Quisquater, J.-J. (eds.) The New Codebreakers - Essays Dedicated to David Kahn on the Occasion of His 85th Birthday. LNCS, vol. 9100, pp. 205–239. Springer, Heidelberg (2016). doi:10.1007/978-3-662-49301-4_15
14. Fernández, J.D., Martínez-Prieto, M.A., Gutiérrez, C., Polleres, A., Arias, M.: Binary RDF representation for publication and exchange (HDT). J. Web Seman. 19, 22–41 (2013)
15. Gentry, C., et al.: Fully homomorphic encryption using ideal lattices. In: Proceedings of ACM Symposium on Theory of Computing, vol. 9, pp. 169–178 (2009)
16. Gerbracht, S.: Possibilities to encrypt an RDF-Graph. In: Proceedings of Information and Communication Technologies: From Theory to Applications, pp. 1–6 (2008)

17. Giereth, M.: On partial encryption of RDF-Graphs. In: Gil, Y., Motta, E., Benjamins, V.R., Musen, M.A. (eds.) ISWC 2005. LNCS, vol. 3729, pp. 308–322. Springer, Heidelberg (2005). doi:10.1007/11574620_24
18. Guo, Y., Pan, Z., Heflin, J.: LUBM: a benchmark for OWL knowledge base systems. J. Web Seman. **3**(2), 158–182 (2005)
19. Kaliski, B.: PKCS #5: Password-Based Cryptography Specification Version 2.0. RFC 2898 (Informational), September 2000
20. Kamara, S., Papamanthou, C.: Parallel and dynamic searchable symmetric encryption. In: Sadeghi, A.-R. (ed.) FC 2013. LNCS, vol. 7859, pp. 258–274. Springer, Heidelberg (2013). doi:10.1007/978-3-642-39884-1_22
21. Kasten, A., Scherp, A., Armknecht, F., Krause, M.: Towards search on encrypted graph data. In: Proceedings of the International Conference on Society, Privacy and the Semantic Web-Policy and Technology, pp. 46–57 (2013)
22. Katz, J., Sahai, A., Waters, B.: Predicate encryption supporting disjunctions, polynomial equations, and inner products. J. Cryptology **26**(2), 191–224 (2013)
23. Kirrane, S., Mileo, A., Decker, S.: Access control and the resource description framework: a survey. Seman. Web **8**(2), 311–352 (2017). doi:10.3233/SW-160236. http://dx.doi.org/10.3233/SW-160236
24. Kurosawa, K., Phong, L.T.: Kurosawa-desmedt key encapsulation mechanism, revisited. IACR Cryptology ePrint Archive, 2013:765 (2013)
25. Ladwig, G., Harth, A.: CumulusRDF: linked data management on nested key-value stores. In: Proceedings of Scalable Semantic Web Knowledge Base Systems, p. 30 (2011)
26. Lewko, A., Okamoto, T., Sahai, A., Takashima, K., Waters, B.: Fully secure functional encryption: attribute-based encryption and (hierarchical) inner product encryption. In: Gilbert, H. (ed.) EUROCRYPT 2010. LNCS, vol. 6110, pp. 62–91. Springer, Heidelberg (2010). doi:10.1007/978-3-642-13190-5_4
27. Paillier, P.: Public-key cryptosystems based on composite degree residuosity classes. In: Stern, J. (ed.) EUROCRYPT 1999. LNCS, vol. 1592, pp. 223–238. Springer, Heidelberg (1999). doi:10.1007/3-540-48910-X_16
28. Popa, R., Zeldovich, N., Balakrishnan, H.: Cryptdb: a practical encrypted relational dbms. Technical report, MIT-CSAIL-TR–005 (2011)
29. Sagiv, Y.: Concurrent operations on B*-trees with overtaking. J. Comput. Syst. Sci. **33**(2), 275–296 (1986)
30. Stefanov, E., Papamanthou, C., Shi, E.: Practical dynamic searchable encryption with small leakage. In: Proceedings of Network and Distributed System Security, vol. 14, pp. 23–26 (2014)
31. Verborgh, R., Vander Sande, M., Hartig, O., Van Herwegen, J., De Vocht, L., De Meester, B., Haesendonck, G., Colpaert, P.: Triple pattern fragments: a low-cost knowledge graph interface for the Web. J. Web Seman. **37–38**, 184–206 (2016)
32. Zheleva, E., Getoor, L.: To join or not to join: the illusion of privacy in social networks with mixed public and private user profiles. In: Proceedings of World Wide Web, pp. 531–540 (2009)

Removing Barriers to Transparency: A Case Study on the Use of Semantic Technologies to Tackle Procurement Data Inconsistency

Giuseppe Futia[1]([✉]), Alessio Melandri[2], Antonio Vetrò[1], Federico Morando[2], and Juan Carlos De Martin[1]

[1] DAUIN, Nexa Center for Internet and Society, Politecnico di Torino, Turin, Italy
{giuseppe.futia,antonio.vetro,demartin}@polito.it
[2] Synapta Srl, Turin, Italy
{alessio.melandri,federico.morando}@synapta.it
https://nexa.polito.it
https://synapta.it/

Abstract. Public Procurement (PP) information, made available as Open Government Data (OGD), leads to tangible benefits to identify government spending for goods and services. Nevertheless, making data freely available is a necessary, but not sufficient condition for improving transparency. Fragmentation of OGD due to diverse processes adopted by different administrations and inconsistency within data affect opportunities to obtain valuable information. In this article, we propose a solution based on linked data to integrate existing datasets and to enhance information coherence. We present an application of such principles through a semantic layer built on Italian PP information available as OGD. As result, we overcame the fragmentation of datasources and increased the consistency of information, enabling new opportunities for analyzing data to fight corruption and for raising competition between companies in the market.

Keywords: Public procurement · Linked data · Data integration · Data consistency

1 Introduction and Motivations

Transparency refers to the principle according to which data related to functioning of government can be accessed and interpreted, without being predefined, preprocessed, and manipulated. As stated by Janssen [6], "adherence to this principle requires that the mechanisms for creating transparency are integrated in the heart of the government functions". Open Government Data (OGD) - i.e. data which is made freely available by public institutions to everyone - is a concrete step in this direction, playing a fundamental role in promoting transparency and accountability [9]. In addition, OGD improves the relationship among the

© Springer International Publishing AG 2017
E. Blomqvist et al. (Eds.): ESWC 2017, Part I, LNCS 10249, pp. 623–637, 2017.
DOI: 10.1007/978-3-319-58068-5_38

government and citizens [11], who are enabled to be much more directly informed and involved in data driven decision-making[1].

Open Data on Public Procurement (PP), namely the procurement of goods or services on behalf of a public authority, is a specific area of the OGD characterized by big potential for increased openness of government information and incentives for supporting business activities. As reported by the Organisation for Economic Co-operation and Development (OECD), around US\$ 9.5 trillion of public money is spent each year by governments procuring goods and services for citizens[2]. Furthermore, PP transparency is a crucial toolset to identify problems that arise from corruption, promoting competition and growth: according to the Transparency International Slovakia initiative[3] "reforms in procurement that included contract publication led to an increase in bids from an average of 2.3 per public tender in 2009 to 3.6 in 2013". In other words, as argued by Svátek [10], PP information is able to unify *public* needs and *commercial* offers: it enables a lively context to increase the interoperability between data models[4], methodologies, and sources independently designed within the two sectors.

Despite the tangible benefits that come with the publication of PP information as OGD, making data available does not automatically produce transparency. As underlined by Janssen [6], providing data alone is not sufficient: deep insights into the working of mechanisms to ensure that information can be easily accessible, processed, and interpreted are necessary to create transparency. Such reflections emerge from a conceptual framework called Big and Open Linked Data (BOLD), proposed by Janssen himself, that identify categories, dimensions, and sub-dimensions that influence transparency [6].

An obstacle to a comprehensive implementation of transparency through OGD is related to the fragmentation of existent open government datasets, in particular in the domain of procurement data[5]. As proposed by Tim Berners-Lee [2], linked data principles can be a *modular* and *scalable* solution to overcome the fragmentation in government data, increasing citizen awareness of government functions and enabling administrations to work more efficiently.

---

[1] The so-called "Participatory Governance" is one of the key aspect of the OGD mentioned by the Open Knowledge Foundation (OKFN). More information available at: http://opengovernmentdata.org/.

[2] More details available in the OECD blog post "Transparency in public procurement, moving away from the abstract": http://oecdinsights.org/2015/03/27/transparency-in-public-procurement-moving-away-from-the-abstract/.

[3] For more information, see: http://www.transparency.sk/.

[4] The ISA initiative of the European Commission represents a landmark for understanding different levels of data interoperability. More information available at http://ec.europa.eu/isa/documents/isa_annex_ii_eif_en.pdf.

[5] Only 1/5 of total public expenditure on goods and services is published with rules complying with the EU Directives, for an estimated value of 420 billion. It means that "the bulk of total public expenditure on goods, services, and works is not organised in accordance with EU procurement legislation". See http://ec.europa.eu/internal_market/publicprocurement/docs/modernising_rules/executive-summary_en.pdf.

The data integration process, in fact, allows to identify cases of discrepancies respect to the knowledge base [13] of Italian procurement data. Some of these inconsistencies, in fact, can be only detected and fixed exploiting the crucial element of linked data principles, namely the interlinks between different data sources: accessing and retrieving data from an authoritative and reliable source, it is possible to reduce contradictions in the legacy dataset and enable opportunities to get less misleading results when the database is queried.

This article presents a solution based on linked data and semantic technologies to overcome the fragmentation of existing datasets and to increase the consistency of information: we process and transform PP data published on different websites of Italian administrations - in compliance with the Italian anti-corruption Act (law n. 190/2012) - relying upon linked data principles. The objective is to demonstrate how this approach has a fundamental impact to increase transparency of public bodies in the context of procurement data.

The structure of this contribution is the following. Section 2 describes the Italian context and gives an overall view of public procurement data made available by administrations. Section 3 explains current problems in terms of data quality of such data. Section 4 illustrates our approach for processing, transforming, and publishing procurement information as linked data. Section 5 reports results of the analysis on data quality issues. Section 6 presents a discussion on obtained results and limitations. Section 7 describes related work in the field of public procurement and spending information published according to linked data principles. The last section summarizes the work and present future advancements of the research project.

2   Study Context

In this Section we present the Italian legislative context according to which procurement data is published by public bodies and we describe the key characteristics of such data.

2.1   The Italian Legislative Context

The Italian Legislative Decree n. 33/2013 (DL33/2013) of March 14th, 2013[6] re-ordered obligations of disclosure, transparency, and dissemination of information by public administrations. According to specific requirements defined by the decree (Article 9 - DL33/2013), each body is required to create a specific section on its website called "Amministrazione Trasparente" (Transparent Administration). In this section, administrations provide details related to public procurement, with particular emphasis on procedures for the award and execution of

---

[6] See: http://www.normattiva.it/uri-res/N2Ls?urn:nir:stato:decreto.legislativo:2013-03-14;33!vig=. Notice that DL33/2013 has been recently amended by D.Lgs. 25 maggio 2016, n. 97 (DL97/2916), with a general tendency toward a more centralized publication of data - see, in particular, Article 9-bis - but no immediate impact on the publication requirements discussed in the paper at hand. See below footnote 7 for additional comments. Last visit on Nov. 2016.

public works, services, and supplies (Article 37 - DL33/2013[7]). Such data is published on the basis of a precise XML Schema Definition[8] (XSD) provided by *ANAC - Autorità Nazionale Anticorruzione* (Italian National Anti-Corruption Authority)[9], which has supervisory duties. After the publication on their websites, administrations transmit the link of the dataset to ANAC via certified mail. ANAC, at this point, performs a preliminary check and releases an index file (in JSON format), containing details related to the availability of data[10].

2.2 Source Data

Public bodies can publish and transmit to ANAC two types of XML files. The first type contains the actual data on contracts until the publication date (January 31st of each year). In order to facilitate the consistency of publications and the comparison of information, the structure of the document is defined by a precise XSD Schema[11]. The main structure of the XML file includes a section with the dataset metadata and a section containing multiple contracts, each of whom can be identified by the XML tag "lotto". The metadata section lists some information, including the first publication date and the last update of the dataset, the business name[12] of the contracting authority that spreads the dataset, the url of the dataset, and the license. The section containing data on contracts includes the following information: the identification code of the tender notice or CIG (that stands for Codice Identificativo Gara), the description of the tender, the procedure type for the selection of the beneficiary, the identification code and the business name of bidders (tender participants), the identification code and the business name of the beneficiary, the awarded amount, the paid amount, the dates of commencement and completion of works.

The second type of XML, instead, is an index that collects links to other XML files containing actual public procurement data[13].

3 Data Quality Problems

As described in Sect. 2.2, public contracts information is generated and spread on Italian public bodies websites. Due to diverse processes and tools adopted

---

[7] Following the aforementioned amendments by DL97/2016, a National Public Contracts Data Base -the BDNCP- is forthcoming, as described in the new Annex B of DL33/2013, but its creation will face all the problems described in the paper at hand - and possibly benefit from the approach that we suggest.

[8] XSD is a W3C recommendation that specifies how to describe an XML document.

[9] See: http://www.anticorruzione.it/.

[10] The JSON index is available at https://dati.anticorruzione.it/#/l190, by clicking on the "Esporta" (Export) button.

[11] A representation of the XSD schema is available at http://dati.anticorruzione.it/schema/datasetAppaltiL190.xsd.

[12] A pseudonym used by companies to perform their business under a name that may differs from their legal name.

[13] A representation of the XSD schema is available at http://dati.anticorruzione.it/schema/datasetIndiceAppaltiL190.xsd.

by administrations, the quality of PP data is extremely variable depending on the single case and it is impaired in terms of *accuracy, completeness,* and *consistency*[14].

Accuracy is defined as *the degree to which a data value conforms to its actual or specified value.* We distinguish between syntactical accuracy and semantic accuracy, which are defined in the following way:

– *Syntactical accuracy is defined as the closeness of the data values to a set of values defined in a domain considered syntactically correct.*
– *Semantic accuracy is defined as the closeness of the data values to a set of values defined in a domain considered semantically correct.*

The definition of completeness is dependent on the perspective used:

– Computer system's point of view: *completeness is the extent to which all necessary values have been assigned and stored in the computer system.*
– End-user point of view: *completeness is the extent by which the data consumer's need is met.*

Consistency *refers to the absence of apparent contradictions within data.* Inconsistency can be verified on the same or different entities. In the context of XML data that refers to a schema, integrity constraints are properties that must be satisfied by all instances of a database schema. Although integrity constraints are typically defined on schemas, they can at the same time be checked on a specific instance of the schema that presently represents the extension of the database.

3.1   Interdependence Between Quality Metrics

Although there are different metrics to assess the quality of data as shown in the previous Section, such metrics are closely interdependent. In the procurement domain, for instance, a contract could present issues like bad comma position in a payment value (accuracy) or the lack of the payment field (completeness). Both errors have a direct impact on the consistency of information: contradictions inevitably occur when we analyze the total amount of expenditure resulting by several XML files that report data of an ongoing contract.

For these reasons, although the focus of the article concerns the consistency of the information, Sect. 5.2 proposes also a comprehensive analysis of the data quality in terms of accuracy and completeness.

3.2   Focus on Data Consistency

Certain types of consistency problems directly emerge analyzing contracts data collected in a single XML file. We report here 3 examples of such inconsistencies:

---

[14] Such data quality metrics are defined by the International Organization for Standardization: ISO/IEC 25012.

- contracts in which the beneficiary is more than one;
- contracts in which the amount of money is paid, but no recipient is present in the data;
- contracts in which the sum reported as paid is greater than the sum initially awarded to the beneficiary.

Other types of inconsistencies manifest themselves only after merging data contained in different sources. The following cases are real examples of inconsistencies with Italian PP data:

- business entities with more than one business name;
- CIGs that identify more than one contract;
- incoherent payments among different versions of an ongoing contract.

The aforementioned issues represent a significant barrier to achieve transparency, because the results obtained by querying the dataset are likely to be inconsistent and misleading. For example, consider a citizen trying to access the effective business name of a contracting authority identified by the id "00518460019" (this is VAT number of the "Politecnico di Torino"). If the data quality in terms of semantic accuracy is poor, such id could be associated with wrong business names like "Politecnico di Milano" and/or "Politecnico di Bari". So, when the citizen performs a search using the business name as search key, he obtains an inconsistent result. The same problem happens when he wants to get details of a contract identified by a CIG: in this case, he is likely to get discordant values. Moreover, incoherent values of payments are not deductible from a single XML file, because errors emerge by analyzing the evolution of the contract data published in different years.

Section 5.2 show results of the analysis from the first kind of inconsistencies, that directly emerge analyzing contracts data collected in a single XML file.

In Sect. 6, instead, we demonstrate why reducing fragmentation of data with semantic technologies is an essential element to tackle the second type of inconsistencies, and we explain how some of these errors can be fixed exploiting linked data.

4   Applying the Linked Data Approach

In this Section we illustrate all stages of the approach to publish Italian PP according to linked data principles. Each stage is accomplished by means of different software components and resources that are shown in Fig. 1.

4.1   Harvesting of XML Files

The task of data harvesting is assigned to the Downloader component (3), that exploits the index file provided by ANAC[15] (1). On the basis of URLs contained in this file, the data fetching process from public bodies' websites (2) is able

---

[15] The index file provided by ANAC is available at http://dati.anticorruzione.it.

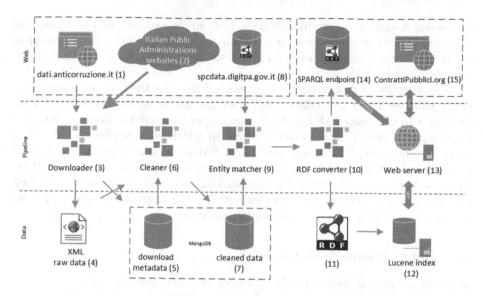

Fig. 1. Architecture for processing and publishing Italian PP as linked data

to manage two different cases. In the first case, the component gets XML files containing PP data and store them locally (4). Download metadata and the local path of each XML are stored in a MongoDB NoSQL database (5), in order to facilitate accessibility to such information and keep track of any duplicate file. In the second case, in which the XML contains links to other XML files, the component is able to cross the links chain[16] and performs the download process shown in the first case. When the component is not able to recognize the expected XML schema (as actual procurement data or as index) the file is stored in a dedicated directory of the file system for a later manual check. This problem occurs when the resource is not published according to an accepted format (e.g., it is a PDF file). In the worst cases, XML indexes are recursive, since they contain URLs that reference to the XML index itself. For these reasons, we implement some features in order to manage this critical issue that threatens to undermine the entire pipeline. Moreover, during the download operation, a lot of servers do not reply: we collected more than 10 different HTTP responses, which reveal how the quality of service over the 15,000 infrastructures of the Italian public administration might not be reliable. Results of the download process are reported in Sect. 5.1.

4.2   Cleaning of Procurement Data

The next step to the data harvesting is performed by the Cleaner (6). During this stage procurement information is extracted from XML files and each contract

---

[16] In some cases an index points to another index that finally might point to a file, or to another index.

is processed and stored as unique document in an instance of the MongoDB database (7). Analyzing such data we evaluate the magnitude of data quality in terms of accuracy, completeness, and consistency (results of such evaluation are available in Sect. 5.2).

When we detect new errors we progressively improve the Cleaner component. Every time we apply a specific fix on data regarding to a contract, we add some metadata to the related MongoDB document: we preserve the original data and we compile a specific field called "errors" in which we describe the identified issue. For instance, if we encounter a bad format for the date value (dd-mm-yyyy), we transform such value in the correct format according to ISO 8601 (in our case yyyy-mm-dd), preserving the original data and saving in the "errors" field the following string: "bad date format" (Sect. 5.2 reports adopted solutions for the most common procurement data quality issues).

4.3 Public Contracts Ontology

In order to publish Italian public procurement according to linked data principles, we use the Public Contracts Ontology (PCO) developed in the context of the Czech OpenData.cz initiative[17]. According to its authors, this ontology describes "information which is available in existing systems on the Web" and "which will be usable for matching public contracts with potential suppliers" [4]. Therefore, the goal of the PCO is to offer a generic model for describing public contracts, without providing details of the public procurement domain, that are specific to fields and countries.

In the PCO domain, a call for tenders is submitted for the award of a public procurement contract. Therefore, we map XML fields and data described in Sect. 2.2 into entities, classes, and relations provided by the PCO. Figure 2 shows the data model adopted for publishing Italian PP as linked data. Although there is a significant degree of overlap between the XSD that describes the schema of source data and the PCO, we have to introduce additional elements to better describe our domain. For instance, the concept of tender was not fully expressed in the data model adopted in XML files, since there are only information about participants, but not details related to offering services and prices. Nevertheless, the tender is one of the most important entity in the PCO to link participants to the public contract. For these reasons, during the conversion to linked data (Sect. 4.4), we create tender entities using as identifier the identification code of the participant and the CIG of the contract.

4.4 Triplification and Interlinking

After the cleaning stage, we convert contracts data stored in the MongoDB instance into RDF using the N-Triples serialization[18]. The component that

---

[17] The Public Contracts Ontology is available on GitHub platform at: https://github.com/opendatacz/public-contracts-ontology.

[18] More information available at: https://www.w3.org/TR/n-triples/.

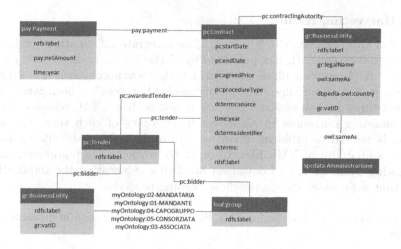

Fig. 2. Schema that describes Italian PP in the linked data domain

performs this task is called RDF converter (10), that maps fields and data values of the contracts into properties, entities, and literals defined by the PCO.

Before completing the triplification process, the Entity matcher component (9) performs the so-called *interlinking* stage. For our application, to improve the consistency of the information, we interlink public bodies listed in Italian PP to public bodies gathered from the SPCData database[19], provided by the *Agenzia per l'Italia Digitale* (8), that contains the index of Italian public administrations. The matching between entities are created using the identification code contained in both datasets. After the interlinking step, the final RDF file (11) is pushed into a Triple Store that exposes data via a SPARQL endpoint (14).

As shown in Fig. 1, the RDF file is also published in a Lucene[20] index (12) to enable full-text search features and data published within the endpoint can be queried by a Web server (13) to populate a Web interface[21] (15).

5   Results

In this Section we report results obtained with stages described previously in order to reduce data fragmentation and to improve data quality, in particular the consistency of information.

---

[19] More information available at: http://spcdata.digitpa.gov.it/index.html.
[20] Apache Lucene is a high-performance, full-featured text search engine library written entirely in Java. More information available at: http://lucene.apache.org/core/.
[21] The ContrattiPubblici.org project developed by Synapta Srl aims at demonstrating the opportunities for transparency and business of public procurement spread according to linked data.

5.1 Harvesting Results

With the approach described in Sect. 4, we integrate information coming from more than 300,000 XML files published by 15,000 public bodies. Table 1 reports details of the harvesting phase (see Sect. 4.1) that was accomplished in 4 different periods, starting from May 2015[22]. The download process has been carried out at different times for two reasons. The first reason is that ANAC releases the index file containing references to XML files in February of each year. The second reason is related to problems about servers uptime, which inevitably impacts on the availability of XML files. In order to tackle the last problem, on each harvesting cycle we try to download all XMLs, generating the duplication of files, that we manage during the data cleaning process.

Table 1. Number of downloaded XML files of procurement data in different periods of time

	May 2015	Nov 2015	Feb 2016	Nov 2016
URL requested	-	-	207.674	271.664
Downloaded files	205.415	184.738	201.451	252.246
Valid XMLs	199.341	180.609	197.338	247.881

5.2 Quality Problems Addressed

During the cleaning phase of PP (see Sect. 4.2), each contract is stored as single document within a MongoDB instance, enabling a first level of analysis on quality issues in terms of accuracy, completeness, and consistency of data. Table 2 shows, for each field of the contract, the type of data quality issue, the occurrence of such issue (in percent), the adopted solution (where available). The 41,65% of all contracts (almost 6 million in total) presents at least one of these issues. Analyzing in particular data inconsistency issues mentioned in Sect. 3.2, we discover that contracts in which the beneficiary is more than one correspond to 1.78% of all contracts; contracts in which the amount of money is paid, but no recipient is present in the data correspond to 4.30% of all contracts; contracts in which the sum reported as paid is greater than the sum initially awarded to the beneficiary correspond 5.96% of all contracts.

Exploiting the linked data principles, we build a semantic layer on procurement data to reduce fragmentation and to identify further inconsistencies. The dimension of the dataset built according to linked data principles is available in Table 3. Such dataset is published using the *Virtuoso Triple Store*[23] and can be queried via SPARQL endpoint[24].

---

[22] Some information is missing because in 2015 we did not store the number of requested URLs.

[23] More information available at: https://virtuoso.openlinksw.com/.

[24] The SPARQL endpoint on public procurement data is available at: https://contrattipubblici.org/sparql.

Table 2. Accuracy, completeness, and consistency degree in PP data

Field	Error	Occ. (%)	Solution
Completeness			
Start date	Missing	12.25	Nothing
End date	Missing	21.61	Nothing
Agreed price	Missing	0.06	Nothing
Payment	Missing	0.20	Nothing
Procedure type	Missing	0.11	Nothing
Business entity ID	Missing	1,05	Hash value
Accuracy			
Identifier	Syntactic errors	0.96	String cleaned
	Semantic errors	5.83	Hash value
Start date	Semantic errors	1.36	Nothing
End date	Semantic errors	2.00	Nothing
Agreed price	Syntactic errors	0.94	String cleaned
	Semantic errors	0.23	Nothing
Payment	Syntactic errors	0.76	String cleaned
	Semantic errors	0.65	Nothing
Procedure type	Syntactic errors	2.81	Optimal string match
Business entity ID	Semantic errors	1,08	Hash value
Consistency			
Start date	Non standard format	5.63	Uniformed to ISO 8601
End date	Non standard format	5.20	Uniformed to ISO 8601
Beneficiary	More than one beneficiary	1.78	Nothing
Payment	Payment without winner	4.30	Nothing
	Greater than awarded price	5.96	Nothing

Table 3. Characteristics of Italian procurement information published as linked data

Dimension	Value
RDF triples	168,961,163
Entities	22,436,784
Contracts	5,783,968
Public bodies	16,593
Companies	652,121
Links to external datasets	13,486

6   Discussion on Inconsistency Issues

As explained in Sect. 3.2, there are some inconsistencies that are visible only after completing a data integration process. We resume 3 different reported cases:

– business entities with more than one business name;
– CIGs that identify more than one contract;
– incoherent payments among different versions of an ongoing contract.

The first of this case can be detected with the following SPARQL query:

```
PREFIX rdfs: <http://www.w3.org/2000/01/rdf-schema#>
PREFIX gr: <http://purl.org/goodrelations/v1#>

SELECT (COUNT(DISTINCT ?be)) WHERE {
  {
    SELECT DISTINCT(?be) WHERE {
    ?be rdfs:label ?label .
    ?be a gr:BusinessEntity .
    }
    GROUP BY ?be HAVING (count(*)>1)
  }
}
```

This issue can be fixed exploiting the most important feature of linked data, namely the interlink among different datasets. In fact, we obtain a unique business name on a subset of business entities (in our case the contracting authorities), building links to the Italian public administration index of SPCData, shown in Sect. 4.4. From this dataset, exposed as linked data, we can get the official business name of contracting authorities, using as primary key their identification code (in our domain the VAT number of the contracting authority). In this way, we improve the consistency of information for a subset of business entities and we enable the opportunity to obtain valuable results.

The second case consists in the duplication of CIG for different contracts and can be detected with the following SPARQL query:

```
PREFIX dcterms: <http://purl.org/dc/terms/>
PREFIX pc: <http://purl.org/procurement/public-contracts#>

SELECT (COUNT(DISTINCT ?contract)) WHERE {
  {
    SELECT DISTINCT(?contract) WHERE {
    ?contract dcterms:identifier ?CIG .
    ?contract a pc:Contract .
    }
    GROUP BY ?contract HAVING (count(*)>1)
  }
}
```

The solution to this issue is generating a hash value, avoiding ambiguity due to duplicate CIGs, to build contracts URIs. In this way, we separate different contracts, misidentified by the same CIG, in different entities: we build the URI through a hash value generated combining the identity code of the contracting authority, the awarded amount, and the procedure type mentioned in the contract. The user is therefore able to detect this kind of error and he can semantically distinguish different contracts identified by the same CIG. Nevertheless, we need more context information to establish which is the correct CIG attribution for a specific contract.

The last problem is tracking incoherent payments published in different XML files of ongoing contracts. This problem is currently not solvable with our approach, because we lack information on provenance and reliability of different sources. In Sect. 6.1 we deepen this kind of limitation.

6.1   Limitations

The limitation of our approach is related to the absence of context information (in particular the reliability of the provenance) published in a structured way as linked data. In fact, we resolve inconsistency cases related to multiple business names for contracting authorities, because we can entrust an authoritative source (the SPCdata repository) that eliminates wrong business names in terms of semantic accuracy. Conversely, in the case of duplicate CIGs for different contracts and incoherent payments among different versions of an ongoing contract, we are not able to detect which is the invalid data because we can not support our analysis with an authoritative source. In other words, we are not in possession of the necessary data to adopt informed decision to govern the processing of information and consequently improve the quality of data. These limitations negatively impact on the level of achievement of transparency, because the citizen access to incomplete and controversial information.

7   Related Work

In this Section, we report contributions of procurement and spending data transformed and published according to linked data principles. This domain has already been addressed by several research projects, however a comprehensive work on Italian procurement data is not addressed yet. Furthermore, at the best of our knowledge, an analysis on procurement data consistency exploiting semantic technologies to improve transparency has not yet been accomplished.

One of the most important contributions in this domain is the LOD2 project, since it systematically addresses many phases of procurement linked data processing [10]. There are several other notable initiatives: the TWC Data-Gov Corpus [3], Publicspending.gr [8], The Financial Transparency System (FTS) project [7], Linked Spending [5], LOTED [12] and MOLDEAS [1].

In particular, the TWC Data-Gov Corpus gathers linked government data on US financial transactions from the Data.gov project[25]. This project exploits

---

[25] Data.gov project website: https://www.data.gov/.

a semantic-based approach in order to incrementally generate and enhance data via crowdsourcing. Publicspending.gr has the objective of interconnecting and visualizing Greek public expenditure with linked data to promote clarity and enhance citizen awareness through easily-consumed visualization diagrams. The FTS project of the European Commission contains information about grants for EU projects starting from 2007 to 2011, and publishes such data as linked data. Exploring this dataset, users are able to get an overview on EU funding, including data on beneficiaries as well as the amount and type of expenditure. Linked Spending is a project for the conversion to linked data of information published by the OpenSpending.org, an open platform that releases public finance information from governments around the world. The project uses the DataCube vocabulary[26] to model data in order to represent multidimensional statistical observations. LOTED[27] is focused on extracting data from single procurement acts and aggregating it over a SPARQL endpoint. Finally, MOLDEAS presents some methods to expand user queries to retrieve public procurement notices in the e-Procurement sector using linked data.

8   Conclusion and Future Works

In this article we present an approach to tackle fragmentation of Italian procurement data and to improve consistency of such data. Both these issues represent a significant barrier to achieve a full transparency, because user and robots that query the datasets risk to obtain partial, inconsistent, and misleading results. As shown by our use case, some inconsistency problems already emerge analyzing data in XML files released by administrations. Among these issues we mention cases in which a contract presents more than one beneficiary; contracts in which the amount of money is paid, but the beneficiary is not present; contracts in which the payment is greater than the initially-awarded amount.

Nevertheless, in order to reduce data fragmentation and detect inconsistency cases that only emerge by integrating information, linked data and semantic technologies play a fundamental role. As demonstrated in this article, our approach allow user to interact with a single access point to information (the SPARQL endpoint), to detect inconsistencies cases, giving the opportunity to resolve some of them. Among these issues we mention: business entities with more than one business name; CIGs that identify more than one contract; incoherent payments among different versions of an ongoing contract. Some of these issues can be fixed exploiting peculiar characteristics of linked data principles, according to which is possible to create interlinks between different datasets. Business names of contracting authorities can be fixed exploiting an authoritative data source for public administrations, such as the SPCData portal released by the Agency for Digital Italy. For what concerns duplicate CIG values, we build URI generating a hash value to distinguish contracts, avoiding problems of data loss or data overlapping. In the last case (incoherent payments in ongoing contract) we are

---

[26] DataCube vocabulary information: https://www.w3.org/TR/vocab-data-cube/.
[27] LOTED project website: http://www.loted.eu/.

not able to apply any fix because we do not have enough information to establish the correctness of the data according to the provenance.

Among the future works, we mention the development of automatic tools to detect and fix consistency problems among contracts published in different files. We want to explore possible ways to evaluate the provenance of data in order to improve the data processing stage and improve data consistency. Lastly, we want to develop advanced methods and dashboards in order to monitor the consistency degree of the data.

References

1. Álvarez, J.M., Labra, J.E., Calmeau, R., Marín, Á., Marín, J.L.: Query expansion methods and performance evaluation for reusing linking open data of the european public procurement notices. In: Lozano, J.A., Gámez, J.A., Moreno, J.A. (eds.) CAEPIA 2011. LNCS (LNAI), vol. 7023, pp. 494–503. Springer, Heidelberg (2011). doi:10.1007/978-3-642-25274-7_50
2. Berners-Lee, T.: Putting government data online (2009)
3. Ding, L., DiFranzo, D., Graves, A., Michaelis, J.R., Li, X., McGuinness, D.L., Hendler, J.A.: Twc data-gov corpus: incrementally generating linked government data from data. gov. In: Proceedings of the 19th International Conference on World Wide Web, pp. 1383–1386. ACM (2010)
4. Distinto, I., d'Aquin, M., Motta, E.: Loted2: an ontology of european public procurement notices. Semant. Web 7(3), 267–293 (2016)
5. Höffner, K., Martin, M., Lehmann, J.: Linkedspending: openspending becomes linked open data. Semant. Web 7(1), 95–104 (2016)
6. Janssen, M., van den Hoven, J.: Big and Open Linked Data (BOLD) in government: a challenge to transparency and privacy? Gov. Inf. Q. 32(4), 363–368 (2015)
7. Martin, M., Stadler, C., Frischmuth, P., Lehmann, J.: Increasing the financial transparency of european commission project funding. Semant. Web J. Spec. Call Linked Dataset Descr. 2(2), 157–164 (2013)
8. Vafolopoulos, M., Meimaris, M., Alexiou, G., et al.: Publicspending. gr: interconnecting and visualizing Greek public expenditure following Linked Open Data directives. In: Using Open Data: Policy Modeling, Citizen Empowerment, Data Journalism. W3C, The European Commission (2012)
9. Stiglitz, J.E., Orszag, P.R., Orszag, J.M.: Role of government in a digital age (2000)
10. Svátek, V., Mynarz, J., Węcel, K., Klímek, J., Knap, T., Nečaský, M.: Linked open data for public procurement. In: Auer, S., Bryl, V., Tramp, S. (eds.) Linked Open Data. LNCS, vol. 8661, pp. 196–213. Springer, Cham (2014). doi:10.1007/978-3-319-09846-3_10
11. Ubaldi, B.: Open government data. OECD Working Papers on Public Governance (2013)
12. Valle, F., dAquin, M., Di Noia, T., Motta, E.: Loted: exploiting linked data in analyzing european procurement notices. In: Proceedings of the 1st Workshop on Knowledge Injection into and Extraction from Linked Data - KIELD 2010 (2010)
13. Zaveri, A., Rula, A., Maurino, A., Pietrobon, R., Lehmann, J., Auer, S.: Quality assessment for linked data: a survey. Semant. Web 7(1), 63–93 (2016)

NdFluents: An Ontology for Annotated Statements with Inference Preservation

José M. Giménez-García[1]([✉]), Antoine Zimmermann[2], and Pierre Maret[1]

[1] Université de Lyon, CNRS, UMR 5516,
Laboratoire Hubert-Curien, Saint-Étienne, France
jose.gimenez.garcia@univ-st-etienne.fr
[2] Université de Lyon, MINES Saint-Étienne, CNRS, Laboratoire Hubert Curien
UMR 5516, 42023 Saint-Étienne, France

Abstract. RDF provides the means to publish, link, and consume heterogeneous information on the Web of Data, whereas OWL allows the construction of ontologies and inference of new information that is implicit in the data. Annotating RDF data with additional information, such as provenance, trustworthiness, or temporal validity is becoming more and more important in recent times; however, it is possible to natively represent only binary (or dyadic) relations between entities in RDF and OWL. While there are some approaches to represent metadata on RDF, they lose most of the reasoning power of OWL. In this paper we present an extension of Welty and Fikes' 4dFluents ontology—on associating temporal validity to statements—to any number of dimensions, provide guidelines and design patterns to implement it on actual data, and compare its reasoning power with alternative representations.

Keywords: Annotations · Contexts · Metadata · Ontologies · OWL · RDF · Reasoning · Reification

1 Introduction

The Resource Description Framework (RDF) represents statements as triples that typically match phrases with a subject, a verb and a complement. However, it is often the case that more complex information has to be encoded, such as qualifying a statement with its origin, its validity within a time frame, its degree of certainty, and so on. In this case, one may have to represent statements about a statement. We describe this as an annotated statement. However, with the RDF model it is only possible to represent binary (or dyadic) relations between subject and object [10]. In order to represent additional data about statements it is usually needed to use external annotations, extend either the data model [1] or the semantics of RDF [4,11], or use design patterns to represent that information [2,12].

On the other hand, RDF Schema (RDFS) and the Web Ontology Language (OWL) add formal semantics to RDF, making it possible to infer new statements

© Springer International Publishing AG 2017
E. Blomqvist et al. (Eds.): ESWC 2017, Part I, LNCS 10249, pp. 638–654, 2017.
DOI: 10.1007/978-3-319-58068-5_39

from pre-existing knowledge. However, when data is annotated using the previous approaches, the inferences in the original dataset are no longer possible, or the new inferred data is missing part of the annotations. For instance, OWL allows to define a relation between two resources as transitive. In that case, if a resource A is related to another resource Busing that property, and B is in turn related with another resource C with the same property, then it is inferred that A and C are also related. This inference is not preserved when using Reification, a classic approach to reference a triple and annotate it with metadata, that removes the original triple and replaces it with four new triples to identify the statement and describe the position of each element of the original triple.

Along these lines, Welty and Fikes [16] proposed an ontology for representing temporally changing information using a perdurantist view, where statements are asserted over temporal slices of entities, retaining most reasoning capabilities. This approach can be generalized to annotate data not only with temporal information, but with information from any dimension [15]. However, modeling several context dimensions for a statement is not straightforward and presents some challenges. In this work, we propose a generalization of Welty and Fikes model in the form of a generic ontology that can be extended to implement any number of concrete metadata dimensions, while preserving reasoning capacity relative to each dimension.

The rest of the paper is structured as follows: Sect. 2 presents the 4dFluents ontology for annotating statements with temporal data; Sect. 3 introduces NdFluents, the generalization of 4dFluents to annotate statements with any number of context dimensions; Sect. 4 describes three design patterns that can be used to model a combination of context dimensions; Sect. 5 discusses issues and possible solutions when representing metadata with NdFluents; Sect. 6 compares the reasoning capabilities of NdFluents with other current approaches to represent metadata about statements in RDF; Sect. 7 portrays related work; finally, we present some conclusions in Sect. 8.

2   Welty and Fikes' 4dFluents Ontology

Welty and Fikes [16] address the problem of representing *fluents*, *i.e.*, relations that hold within a certain time interval and not in others. They address the issue from the perspective of diachronic identity (that is, how an entity looks to be different at different times), showcasing the two ways of tackling it:

- The *endurantist* (*3D*) view maintains a differentiation between *endurants*, entities that are present at all times during its whole existence, and *perdurants*, events affecting an entity during a definite period of time during the entity's existence.
- The *perdurantist* (*4D*) view argues that entities themselves have to be handled as perdurants, *i.e.*, temporal parts of a four dimensional meta-entity. Instead of making an assertion about some entities, such as *"Paris is the capital of France"*, one should make the assertion about their temporal parts: *"A temporal part of Paris (since 508 up to now) is the capital of a temporal part of France (since 508 up to now)"*.

Welty and Fikes adopt the perdurantist approach to create the *4dFluents* ontology, representing *entities at a time* and using them as resources for their statements. The 4dFluents ontology expressed in OWL2 Functional Syntax is shown in Ontology 1.

```
Prefix( 4d:=<http://www.example.com/4dFluents#> )
Ontology( <http://www.example.com/4dFluents>
      Declaration( Class( 4d:Interval ) )
      Declaration( Class( 4d:TemporalPart ) )
      DisjointClasses( 4d:Interval 4d:TemporalPart )

      Declaration( ObjectProperty( 4d:fluentProperty ) )
      ObjectPropertyDomain( 4d:fluentProperty 4d:TemporalPart )
      ObjectPropertyRange( 4d:fluentProperty 4d:TemporalPart )

      Declaration( ObjectProperty( 4d:temporalExtent ) )
      FunctionalObjectProperty( 4d:temporalExtent )
      ObjectPropertyDomain( 4d:temporalExtent 4d:TemporalPart )
      ObjectPropertyRange( 4d:temporalExtent 4d:Interval )

      Declaration( ObjectProperty( 4d:temporalPartOf ) )
      FunctionalObjectProperty( 4d:temporalPartOf )
      ObjectPropertyDomain( 4d:temporalPartOf 4d:TemporalPart )
      ObjectPropertyRange( 4d:temporalPartOf ObjectComplementOf( 4d:Interval ))
)
```

Ontology 1. 4dFluents ontology (from [16])

In order to use the ontology for describing fluents, one has to introduce axioms at the terminological level (TBox) as well as assertions in the knowledge base (ABox). For instance, if one wants to say that *"Paris is the capital of France"* since 508, the relation "capital of" has to be a subproperty of `fluentProperty` and new individuals have to be introduced for the temporal part of Paris and of France, as shown in Ontology 2.

```
Declaration( ObjectProperty( ex:capitalOf ) )
SubObjectPropertyOf( ex:capitalOf 4d:fluentProperty )
ClassAssertion( 4d:TermporalPart ex:Paris@508 )
ClassAssertion( 4d:TermporalPart ex:France@508 )
ClassAssertion( 4d:Interval ex:year508) )
ObjectPropertyAssertion( ex:capitalOf ex:Paris@508 ex:France@508 )
ObjectPropertyAssertion( 4d:temporalExtent ex:Paris@508 ex:year508 )
ObjectPropertyAssertion( 4d:temporalExtent ex:France@508 ex:year508 )
ObjectPropertyAssertion( 4d:temporalPartOf ex:Paris@508 ex:Paris )
ObjectPropertyAssertion( 4d:temporalPartOf ex:France@508 ex:France )
```

Ontology 2. Expressing a fact about a fluent entity with the 4dFluents ontology

In this way, temporal information can be represented with standard OWL semantics, preserving reasoning capabilities.

3   The NdFluents Ontology

A temporal part of an entity can be viewed as an individual context dimension of the entity. A similar approach can then be used to represent different dimensions, such as provenance or confidence. Continuing with our running example, if Wikipedia states that *"Paris is the capital of France"*, we can articulate that fact as *"Paris as defined by Wikipedia is the capital of France as defined by Wikipedia"*. Different context dimensions of an entity could then be combined if applicable, allowing the representation of complex information, such as: *"A temporal part of Paris as defined by Wikipedia is the capital of a temporal part of France as defined by Wikipedia"*.

We use this idea to extend the 4dFluents ontology for an arbitrary number of context dimensions in the *NdFluents* ontology. The ontology, shown in Ontology 3, and published in http://www.emse.fr/~zimmermann/ndfluents.html, is a generalization from temporal parts to contextual parts.

```
Prefix( nd:=<http://purl.org/NET/NdFluents#> )
Ontology( <http://purl.org/NET/NdFluents>
     Declaration( Class( nd:Context ) )
     Declaration( Class( nd:ContextualPart ) )
     DisjointClasses( nd:Context nd:ContextualPart )

     Declaration( ObjectProperty( nd:contextualProperty ) )
     ObjectPropertyDomain( nd:contextualProperty nd:ContextualPart )
     ObjectPropertyRange( nd:contextualProperty nd:ContextualPart )

     Declaration( ObjectProperty( nd:contextualExtent ) )
     ObjectPropertyDomain( nd:contextualExtent nd:ContextualPart )
     ObjectPropertyRange( nd:contextualExtent nd:Context )

     Declaration( ObjectProperty( nd:contextualPartOf ) )
     FunctionalObjectProperty( nd:contextualPartOf )
     ObjectPropertyDomain( nd:contextualPartOf nd:ContextualPart )
     ObjectPropertyRange( nd:contextualPartOf ObjectComplementOf( nd:Context ))
)
```

Ontology 3. The NdFluents ontology

Note that `FunctionalObjectProperty( nd:contextualExtent )` axiom is not present in the ontology. This axiom should appear if the ontology was a direct translation from temporal dimension to a generic context dimension, but it is no longer applicable when we have more than one dimension simultaneously.

The NdFluents ontology is meant to be implemented for different context dimensions in a modular way. In this sense, the 4dFluents ontology can be seen as a concrete implementation of NdFluents, as we show in Ontology 4. In Fig. 1a we show the representation of a statement with temporal annotations using this ontology. The non-dashed parts are equivalent to the original 4dFluents ontology, while the dashed parts correspond to the NdFluents extension. Other dimensions, such as provenance, can be modeled similarly to the temporal dimension by replacing `TemporalPart` with `ProvenancePart`, `temporalExtent` with `provenanceExtent`, `Interval` with `Provenance`, and `temporalPartOf` with `provenancePartOf`. Additionally, an assertion like *"Paris is the capital of France, according to Wikipedia"* can be modeled following the same pattern as in Ontology 2, replacing the property and class names with their counterparts in the provenance dimension.

```
Prefix( nd:=<http://purl.org/NET/ndfluents#> )
Prefix( 4d:=<http://purl.org/NET/ndfluents/4dFluents#>)
Ontology( <http://www.example.com/4dFluentsV2>
       Import( <http://www.example.com/NdFluents> )

       Declaration( Class( 4d:Interval ) )
       SubClassOf( 4d:Interval nd:Context )
       Declaration( Class( 4d:TemporalPart ) )
       SubClassOf( 4d:TemporalPart nd:ContextualPart )

       Declaration( ObjectProperty( 4d:temporalExtent ) )
       SubObjectPropertyOf( 4d:temporalExtent nd:contextualExtent )
       ObjectPropertyDomain( 4d:temporalExtent 4d:TemporalPart )
       ObjectPropertyRange( 4d:temporalExtent 4d:Interval )

       Declaration( ObjectProperty( :temporalPartOf ) )
       SubObjectPropertyOf( 4d:temporalExtent nd:contextualPartOf )
       ObjectPropertyDomain( 4d:temporalPartOf 4d:TemporalPart )
)
```

Ontology 4. 4dFluents ontology as implementation of NdFluents

4   Design Patterns

An important scenario where NdFluents becomes relevant is when the necessity of combining two or more context dimensions arises, such as *"According to Wikipedia, Paris is the capital of France since 508"*. In this section we present three design patterns to combine different dimensions, along with added axioms that can be necessary depending on the modeling needs. Methodological support for choosing and implementing a design pattern can be found at Giménez-García et al. [3]

4.1   Contexts in Context

One possible model to represent information using different context dimensions is to relate a `ContextualPart` to another `ContextualPart`. This approach can be taken when the "first level" annotations are relevant facts of the knowledge base, and the intention is to state additional information about them. To be able to reason about different annotation levels of any entity, it is desirable for the contextualPartOf property to be transitive, which can be achieved by adding the axiom of Ontology 5.

While data about different dimensions can be more fine-grained using this model, it also grows in complexity. For example, in Fig. 1b the statement capitalOf is related to the ProvenancePart Paris@1.1. This information is in no way related to the TemporalPart Paris@1. While we could have this statement duplicated in the example, this can become unfeasible when we start adding more contextual parts to the data. We believe that this pattern can be useful in some specific cases, but it is usually too cumbersome.

```
Prefix( nd:=<http://purl.org/NET/ndfluents#> )
Ontology( <http://purl.org/NET/ndfluents/transitivecontextualpartof>
    TransitiveObjectProperty( nd:contextualPartOf )
)
```

Ontology 5. Transitive axiom for NdFluents ontology

4.2   Use Multiple Contextual Extents on Each Contextual Part

A generic approach for representing entities with more than one context dimension is to have ContextualParts with more than one contextual extent. Using this model, only one ContextualPart is created for a combination of context dimensions. This ContextualPart is then related to all related contextual extents, as shown in Fig. 1c. This pattern is easier to model: Relating the ContextualPart with the context dimensions is straightforward. It also avoids ambiguity when modeling annotations related to more than one dimension, and reduces the number of resources in the ontology (*i.e.*, while the previous model needed one ContextualPart for each dimension involved, this approach only requires one). Note that contextualPartOf is a functional property, which means that there cannot be a contextualPartOf of more than one entity.

4.3   Combine Different Contexts on One Contextual Extent

Finally, a third possibility is to create compound Contexts, and enforce a limit of only one Context per ContextualPart. This model adds a layer of complexity to the previous approach, but it can be useful to require a specific combination of dimensions on a set of ContextualParts. This can be achieved by adding the axiom in Ontology 6.

We show an example of this approach on Fig. 1d. Note that the combined classes and properties are subclasses and subproperties of the corresponding classes and properties of the two context dimensions they are combining (*e.g.*, Temporal+ProvenancePart is subclass of TemporalPart and ProvenancePart). As a result, querying and reasoning can be performed in an identical way as the previous approach.

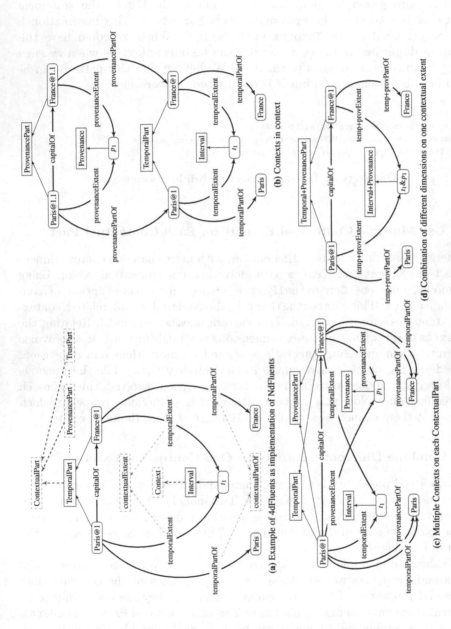

(a) Example of 4dFluents as implementation of NdFluents

(b) Contexts in context

(c) Multiple Contexts on each ContextualPart

(d) Combination of different dimensions on one contextual extent

Fig. 1. NdFluents ontology and design patterns

```
Prefix( nd:=<http://purl.org/NET/ndfluents#> )
Ontology( <http://purl.org/NET/ndfluents/functionalcontextualExtent>
        FunctionalObjectProperty( nd:contextualExtent )
)
```

Ontology 6. Functional contextual extents axiom for NdFluents ontology

5 Additional Considerations

In this section we discuss issues that may arise when modeling annotations using fluents, and possible approaches to deal with them if they exist. While the first one is common to the original 4dFluents ontology, the second is only relevant when dealing with more than one context dimension.

5.1 Dealing with Datatype Properties

The original 4dFluents ontology does not provide any information for modeling datatype properties. While there is nothing that prevents using regular datatype properties with ContextualParts of an entity, it may be desirable to declare explicit axioms for annotation properties to facilitate reasoning on that information. In that case, the statements of Ontology 7 need to be added to the NdFluents ontology. Figure 2 shows an example where a annotated property is used to state the population of Paris in a specific temporal interval. Note that it is also possible to create specific contextualProperty subproperties for different context dimensions (*i.e.*, temporalProperty for TemporalPart) for properties related to concrete context dimensions.

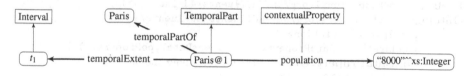

Fig. 2. Example of annotated datatype property

```
Prefix( nd:=<http://purl.org/NET/ndfluents#> )
Ontology( <http://purl.org/NET/ndfluents/annotatedDatatypeProperty>
        Declaration( DataProperty( nd:annotatedDatatypeProperty ) )
        DataPropertyDomain ( nd:annotatedDataProperty nd:
            ContextualPart )
)
```

Ontology 7. Datatype axioms for NdFluents ontology

5.2   Relations Between ContextualParts of Different Dimensions

The NdFluents ontology presented thus far allows the modeling of relations among different ContextualParts of different dimensions (*i.e.*, a TemporalPart of Paris could be the capital of a ProvenancePart of France). While this can be convenient for individual cases, it is often needed for an contextualProperty to be related to ContextualParts of the same dimension. In this case, it is necessary to add the appropriate axioms to the ontology. In Ontology 8. we show the needed axioms to include this restriction on the TemporalParts. Conversely, if there are datatype properties related to specific dimensions, axioms from Ontology 9 should be added.

```
Prefix( nd:=<http://purl.org/NET/ndfluents#> )
Prefix( 4d:=<http://purl.org/NET/ndfluents/4dFluents#>)
Ontology( <http://purl.org/NET/ndfluents/4dFluents/
    temporalpartrestriction>
        Declaration( ObjectProperty( 4d:fluentProperty ) )
        SubObjectPropertyOf( 4d:fluentProperty nd:contextualProperty )
        ObjectPropertyDomain( 4d:fluentProperty 4d:TemporalPart )
        ObjectPropertyRange( 4d:fluentProperty 4d:TemporalPart )
)
```

Ontology 8. Temporal restriction on object properties 4dFluents ontology

```
Prefix( nd:=<http://purl.org/NET/ndfluents#> )
Prefix( 4d:=<http://purl.org/NET/ndfluents/4dFluents#>)
Ontology( <http://purl.org/NET/ndfluents/4dFluents/
    temporalpartrestriction>
        Declaration( DataProperty( 4d:fluentDataTypeProperty ) )
        SubDataPropertyOf( 4d:fluentDataTypeProperty nd:
            contextualProperty )
        DataPropertyDomain( 4d:fluentProperty 4d:TemporalPart)
)
```

Ontology 9. Temporal restriction on datatype properties 4dFluents ontology

In a similar fashion, it is usually desirable that ContextualParts of the same dimension relate to the same Context. That is, if a Provenance Part of Paris relates to a ProvenancePart of France, their provenanceExtent properties should have the same ProvenancePart object. However, this restriction cannot be expressed in OWL. If needed, a rule language (such as SWRL [7] or RIF [8]) can be used for this purpose, but this case goes beyond the scope of this paper.

6 Reasoning with Annotated Data

In this section, we compare the reasoning capabilities of the NdFluents ontology with other approaches to annotate statements, namely RDF reification, N-ary relations, and singleton property. The interest is to know what RDFS and OWL entailments are preserved wrt the original unannotated data. For that, we need to formally define what annotations and entailment preservation mean. We assume that annotated statements can be described as a pair (G, A) where G is the graph corresponding to the statements that are annotated, and A denotes the annotations on G. The structure of A could be arbitrarily complex (e.g., containing dates, creator, provenance) but for the sake of this section and to simplify the presentation, we simply assume that the annotation structure is identified with an IRI. Thus, we approximate the notion of annotated statements with the concept of named graphs, i.e., pairs (n, G) where n is an IRI and G is an RDF graph. However, there is no standard way of reasoning with named graphs [19]. Our objective then is to compare approaches that convert annotated statements into RDF graphs. We name such approaches *RDF representation of annotated statements* and formalize it as follows.

Definition 1 (RDF representation of annotated statements). An *RDF representation of annotated statements* is a function f that maps annotated statements (in our simplified model, named graphs) (n, G) to an RDF graph $f(n, G)$.

For examples of this function, refer to Subsect. 6.1, where we describe four existing models to annotate statements and present their corresponding functions.

We want to assess to what extent each representation is preserving entailment with the notions of entailment preservation (when the entailment preserves also the annotations) and non-contextual entailment preservation (when only the original entailment is preserved) defined as follows.

Definition 2 (Entailment preservation). Let G_1 and G_2 be two RDF graphs such that $G_1 \models G_2$ and f be an RDF representation of annotated statements.[1] We say that f *preserves the entailment* between G_1 and G_2 iff for all annotation IRI n, $f(n, G_1) \models f(n, G_2)$.

Definition 3 (Non-Contextual Entailment preservation). Let G_1 and G_2 be two RDF graphs such that $G_1 \models G_2$ and f be an RDF representation of annotated statements.[1] We say that f *non-contextually preserves the entailment* between G_1 and G_2 iff for all annotation IRI n, $f(n, G_1) \models G_2$.

We generalize these notions to the case of entailment rules of the form $P(\mathbf{x}) \leftarrow Q(\mathbf{x}, \mathbf{y})$, where P and Q are graph patterns and $\mathbf{x}, \mathbf{y}$ are tuples of variables used in the patterns.

---

[1] This definition can apply to any entailment regime so that it is not necessary to specify what the relation $\models$ exactly is.

Definition 4 (Rule preservation). Let $R = P(\mathbf{x}) \leftarrow Q(\mathbf{x}, \mathbf{y})$ be a rule and f an RDF representation of annotated statements. We say that f *preserves the rule* R iff for all mappings μ from variables in $\mathbf{x}$ and $\mathbf{y}$ to RDF terms, $f(n, Q(\mu(x), \mu(y))) \models f(n, P(\mu(x)))$.

Definition 5 (Non-Contextual Rule preservation). Let $R = P(\mathbf{x}) \leftarrow Q(\mathbf{x}, \mathbf{y})$ be a rule and f an RDF representation of annotated statements. We say that f *non-contextually preserves the rule* R iff for all mappings μ from variables in $\mathbf{x}$ and $\mathbf{y}$ to RDF terms, $f(n, Q(\mu(x), \mu(y))) \models P(\mu(x))$.

For example, if we have an inference rule that allows us to infer that (France, hasCapital,Paris) from the triple (Paris,capitalOf,France), and we have an representation of annotated statements for (Paris,capitalOf,France), (508,now), rule preservation would allow us to infer (France, hasCapital, Paris),(508,now), while non-contextual rule preservation would allow to infer (France,hasCapital,Paris) from the annotated triple. Note this kind of inferences function annotates triples of the condition but the conclusion is not annotated are not always desirable. This will be further explained in Subsect. 6.2.

In the following subsections we first present the *RDF representation of annotated statements* (see Definition 1) for the representation approaches, and then proceed to compare the rule preservation for each one of them.

6.1   RDF Representation Approaches

- **Reification**[2] is the standard W3C model to represent information about an statement, proposed in 2004. A triple is represented as an instance of rdf: Statement, that relates to the original triple with the properties rdf: subject, rdf:predicate and rdf:object. Then, a triple (s, p, o) is replaced by the following set: $\{(i,\texttt{rdf:type},\texttt{rdf:Statement}), (i,\texttt{rdf:subject}, s), (i,\texttt{rdf:predicate}, p), (i,\texttt{rdf:object}, o)\}$, and annotations are related to i.
- **N-Ary relations** [12] were proposed in 2006 to represent relations between more than two individuals, or to describe the relation themselves. In this model, an individual is created to represent the relation, which can be used as the subject for new statements. Thus, a triple (s, p, o) is replaced by the following set: $\{(s, p'_1, r), (r, p'_2, o)\}$, and annotations are related to r.
- The **Singleton Property** [11] is a recent proposal to represent information about statements in RDF. A particular instance of the predicate is created for every triple. This instance is related to the original predicate by the singletonPropertyOf property. Then, each statement can be unequivocally referenced using its predicate for attaching additional information. Therefore, a triple (s, p, o) is replaced by the set: $\{(s, p', o), (p',\texttt{sp:singletonPropertyOf}, p)\}$, and annotations are related to p'.
- **NdFluents**, the approach presented in this paper, creates a contextualized individual for both subject and object (in case it is a URI or blank node) of the triple. The triple is the replaced by a new one that uses the contextualized individuals. These two new resources are related to the original individuals

---

[2] https://www.w3.org/TR/2004/REC-rdf-primer-20040210/#reification.

and with a Context, where the annotations are attached. Hence, the original triple (s, p, o) is replaced by the following set of triples $\{(s_c, p, o_c), (s_c, \text{nd: contextualPartOf}, s), (o_c, \text{nd:contextualPartOf}, o), (s_c, \text{nd:contextual Extent}, c), (o_c, \text{nd: contextual Extent}, c)\}$, where c is a function of the context. Annotations are related to c.

6.2   Comparison of Rule Preservation

For comparing how entailment is preserved in each of the 4 approaches presented in Sect. 6.1, we analyze which rules from the pD* fragment of OWL ter Horst [14] are preserved. This fragment is a modified subset of RDFS and OWL that can be expressed as a complete set of rules and is computationally feasible. For each rule, we check if is in accordance with *Rule Preservation* and *Non-Contextual Rule Preservation* (*i.e.*, for the former, if the inference rule holds when we apply the *RDF representation of annotated statements* function to both condition and conclusion; for the latter, if it holds when we apply the function only to the condition). It is important to note that the representation approaches are usually used to annotate data on relations between resources. For this reason, we decide to implement the representations on triples that do not include RDF, RDFS, or OWL vocabularies.

Table 1 shows the D* (modified RDFS) entailment rules and rule preservations for each one of the approaches, whereas Table 2 presents the same information for P entailments (modified subset of OWL). Note that we remove those rows where both condition and conclusion include only triples with RDF, RDFS, or OWL vocabularies. A P indicates that there is rule preservation for the corresponding approach, while a P_{NC} denotes non-contextual rule preservation. As mentioned in Sect. 6, it is worth noting that not all rule preservations are desirable. When the preserved rule entails new knowledge on the non-annotated graph, and the annotated triples are not universally true, then the inferences can lead to conclusions that do not conform with real-world knowledge. This

Table 1. Preserved D* entailments (P = Rule preservation, P_{NC} = Non-contextual rule preservation, ! = Risk of undesirable inference)

Rule	Condition	Constraint	Conclusion	Reif.	N-Ary	S.P.	NdF
lg	$v\,p\,l$	$l \in L$	$v\,p\,b_l$	P	P	P	P
gl	$v\,p\,b_l$	$l \in L$	$v\,p\,l$	P	P	P	P
rdf1	$v\,p\,w$		p type Property			P	P
rdf2-D	$v\,p\,l$	$l = (s,a) \in L_D^+$	b_l type a	P	P	P	P
rdfs1	$v\,p\,l$	$l \in L_p$	b_l type Literal	P	P	P	P
rdfs2	p domain u						
	$v\,p\,w$		v type u			P!	
rdfs3	p range u						
	$v\,p\,w$	$w \in U \cup B$	w type u			P!	
rdfs4a	$v\,p\,w$		v type Resource	P	P	P	P
rdfs4b	$v\,p\,w$	$w \in U \cup B$	w type Resource	P	P	P	P
rdfs7x	p subPropertyOf q						
	$v\,p\,w$	$q \in U \cup B$	$v\,q\,w$			P_{NC}!	P

Table 2. Preserved P-Entailments (P = Rule preservation, P_{NC} = Non-contextual rule preservation, ! = Risk of undesirable inference)

Rule	Condition	Constraint	Conclusion	Reif.	N-Ary	S.P.	NdF
rdfp1	p type FunctionalProperty u p v u p w	$v \in U \cup B$	v sameAs w			P!	
rdfp2	p type InverseFunctionalProperty u p w v p w		v sameAs w			P!	
rdfp3	p type SymmetricProperty v p w	$w \in U \cup B$	w p v			P_{NC}!	P
rdfp4	p type TransitiveProperty u p v v p w		u p w			P_{NC}!	P
rdfp5a	v p w		v sameAs v	P	P	P	P
rdfp5b	v p w	$w \in U \cup B$	w sameAs w	P	P	P	P
rdfp8ax	p inverseOf q v p w	$w, q \in U \cup B$	w q v			P_{NC}!	P
rdfp8bx	p inverseOf q v q w	$w \in U \cup B$	w p v			P_{NC}!	P
rdfp11	u p v u sameAs u' v sameAs v'	$u' \in U \cup B$	u' p v'	P	P	P_{NC}!	
rdfp14a	v hasValue w v onProperty p u p w		u type v			P!	
rdfp14bx	v hasValue w v onProperty p u type v	$p \in U \cup B$	u p w	P_{NC}	P_{NC}	P_{NC}	P_{NC}
rdfp15	v someValuesFrom w v onProperty p u p x x type w		u type v			P!	
rdfp16	v allValuesFrom w v onProperty p u type v u p x	$x \in U \cup B$	x type w			P!	

happens when the *RDF representation of annotated statements* function annotates at least one triple of the condition, and either we have non-contextual rule preservation, or we have rule preservation but the function does not annotate the triple in the conclusion. This is actually what happens with the Singleton Property for the rules `rdfs2`, `rdfs3`, and `rdfs7x` from the D*-entailments ruleset, and rules `rdfp1`, `rdfp2`, `rdfp3`, `rdfp4`, `rdfp8ax`, `rdfp8bx`, `rdfp11`, `rdfp14a`, `rdfp15`, and `rdfp16` (identified in the table with an exclamation mark), due to the RDFS interpretation that considers the singleton property as belonging to the extension of the original property [11, Sect. 3]. While there is no problem if the annotated fact is universally true (*i.e.*, we just want to provide additional information about a fact), it leads to undesirable conclusions when the context of the annotation is related with the identity of the resources (such as provenance or trust contexts), where we want to express that something is true only according to a source, or with a degree of confidence. For instance, let us suppose a functional property `birthplace` that we want to use in the context of provenance. It can be desirable to model that Barack Obama was born in the United

States according to a source, but in Kenya according to a different source. In this case the rule `rdfp1` would infer that the United States and Kenya are the same place in the non-annotated graph when using the Singleton Property.

It can be seen that Reification and N-Ary relations show poor preservation of rules, where most of those rules could be considered tautologies. The Singleton Property provides a mixture of rule preservation and non-contextual rule preservation for all the rules, that can be useful when we want to annotate universally true facts, but it is not usable when we want to have contextual information that is not universally true. NdFluents, by contrast, has neither non-contextual rule preservation nor rule preservation that can lead to undesirable inferences for any rule. There is only one rule where NdFluents is surpassed by the other approaches. Rule `rdfp11` presents *Rule Preservation* for Reification and N-Ary relations, but no rule preservation at all for NdFluents.

In addition, for the rules where NdFluents has no rule preservation, we observe that different conclusions hold, where we entail contextual knowledge. In Table 3 we see the conclusions for that set of rules with their conclusions. We can observe that the individual used in the annotation is entailed in the conclusion. For instance, let us suppose a property `capitalOf` with a domain of `PopulatedPlace`; if we state that Babylon was the capital of the Babylonian empire between 609 BC and 539 BC, instead of inferring that Babylon *is* a populated place (as a universal truth), we entail that Babylon between 609 BC and 539 BC was a populated place.

Table 3. Conclusions for rules with no rule preservation for NdFluents

Rule	Conclusion	Rule	Conclusion	Rule	Conclusion
rdfs2	v_c type u	rdfp1	v_c sameAs w_c	rdfp14a	u_c type v
rdfs3	w_c type u	rdfp2	v_c sameAs w_c	rdfp15	u_c type v

7 Related Work

In the original 4dFluents paper there were some issues not addressed by the authors. Later works have tried to identify and address those issues. Zamborlini and Guizzardi [17] present an alternative work to 4dFluents, where they present two different alternatives to represent temporally changing information in OWL. Both approaches have a similar model to Welty and Fikes's, where the entities are sliced for different times. The main difference is that in the first one, *Individual Concepts and Rigidity*, the original individuals are considered as classes. Thus, they are not described by any property, and a new slice has to be created every time that a property changes. On the other hand the second approach, "Objects and Moments", is based on *Relators* and *Qua-individuals* [9], where the individuals are represented by an entity, and their slices inherit its properties. Then, any time a property changes, it is reflected in the original entity. The first approach is more prone to the proliferation of timeslices, and can only guarantee the immutability of original properties only by repetition on every timeslice. The second approach solves those issues at the cost of blurring the details of the changes of individual properties, and it is not clear how inheritance works in OWL. In a later work [18], Zamborlini and Guizzardi focus on solving the issues of the prior approaches for representing events and properties of individuals. They maintain the fluent-like representation for events, but move to an N-ary representation for properties. However, they still not address the possibility to have more that one domain relation, nor address how inheritance is performed in OWL.

There are also other works that compare the different approaches to represent contextual information. Gangemi and Presutti [2] present and compare a number of design patterns to represent N-Ary relations, including Reification and Context Slices [15], to represent additional information on binary relations. The comparison is done in four qualitative dimensions (DL reasoning support, polymorphism support, relation footprint, and intuitiveness) and five quantitative dimensions (number of needed axioms, expressivity, consistency checking time, classification time, and amount on newly generated constants). However, they only provide a brief outline of the reasoning power of each approach, while we are interested in more fine-grained comparison of entailment preservations. Scheuermann et al. [13], on the other side, perform a qualitative research that compares user preferences and ability for using different design patterns. In their study the fluents pattern is regarded as the most complicated and less used to model, while making a temporal slice of the predicate (which could be represented using the Singleton Property in RDF) seems more intuitive. The N-ary pattern is the model most frequently used. The model regarded as the most user-friendly is not representable using OWL, because it requires having a predicate as an argument of another (an approximation in RDF could be using N-Quads, though). Hernández et al. [5] compare Reification, N-Ary relations, Singleton Properties and Named Graphs to encode Wikidata in practice. They provide space requirements and query performance for each approach in 4store[3], BlazeGraph[4], GraphDB[5], Jena TDB[6] and Virtuoso[7]. They report that Singleton Properties provide the most concise representation on a triple level, while N-Ary predicates is the only model with built-in support for SPARQL property paths. In addition, the Singleton Property usually lacks performance due to the number of predicates, whereas there is no clear winner among the other approaches. Virtuoso exhibits the best performance, while Jena and 4store show the worst results. Later, Hernández et al. [6] extend their previous work to compare Virtuoso, BlazeGraph, Neo4J[8], PostgreSQL[9] with a set of new experiments, based on the idea of performing sets of lookups for atomic patterns with exhaustive combinations of constants and

---

[3] https://github.com/garlik/4store.
[4] https://www.blazegraph.com.
[5] http://graphdb.ontotext.com.
[6] https://jena.apache.org/documentation/tdb.
[7] https://virtuoso.openlinksw.com.
[8] https://neo4j.com.
[9] https://www.postgresql.org.

variables, in order to give an idea of the low-level performance of each configuration. In this set of experiments standard reification and named graphs performed best, with N-Ary relations following in third, and singleton properties not being well-supported.

8 Conclusions

Representing annotations on multiple dimensions is a current challenge in RDF and OWL. We have proposed the NdFluents ontology, a multi-dimension annotation ontology, based on 4dFluents. To the best of our knowledge, this is the first generic extension of 4dFluents for an arbitrary combinations of context dimensions. This representation is intended to be extended in a modular way for each desired dimension. In addition, we have presented three design patterns and additional considerations to keep in mind when modeling data with NdFluents. We study how many of the original inference rules are preserved when annotating the data with NdFluents and compare with the main approaches to annotate data: Reification, N-Ary Relations, and Singleton Property. The results show that NdFluents preserves more desirable entailments, while omitting undesirable entailments, than any alternative. The Singleton property presents non-contextual rule preservation for many of the rules, and can lead to undesirable entailments when the annotated facts are not universally true. Reification and N-Ary relations preserve the fewest number of entailment rules.

Lines of future work are manifold: First, we want to apply this model to real world datasets. Our goal is to exploit the context of information to make the datasets fit for question answering, as well as determine the most relevant data sources. This includes providing additional information based on the context and helping to find the most trustworthy data for the answer. Second, we intend to look deeper into the entailment preservations for different approaches using bigger subsets of OWL 2, such as OWL LD and OWL 2 RL/RDF, and possible reformulations of the approaches that could improve the results. Third, we plan to perform an experimental evaluation of the different annotation models using different triple stores wrt different factors, such as size, loading time, query response time, and query formulation complexity.

Acknowledgements. This work is supported by funding from the EU H2020 research and innovation program under the Marie Skłodowska-Curie grant No 642795, and from ANR grant 14-CE24-0029 for project OpenSensingCity. Authors would like to thank Chris Welty for his supportive comments, and Amro Najjar for his suggestions.

References

1. Carothers, G.: RDF 1.1 N-Quads: a line-based syntax for RDF datasets. W3C Recommendation (2014). https://www.w3.org/TR/n-quads
2. Gangemi, A., Presutti, V.: A multi-dimensional comparison of ontology design patterns for representing n-ary relations. In: Emde Boas, P., Groen, F.C.A., Italiano, G.F., Nawrocki, J., Sack, H. (eds.) SOFSEM 2013. LNCS, vol. 7741, pp. 86–105. Springer, Heidelberg (2013). doi:10.1007/978-3-642-35843-2_8

3. Giménez-García, J.M., Zimmermann, A., Maret, P.: NdFluents: A multidimensional contexts ontology. Technical report, Université Jean Monnet (2016)
4. Hartig, O., Thompson, B.: Foundations of an alternative approach to reification in RDF. CoRR (2014). http://arxiv.org/abs/1406.3399x
5. Hernández, D., Hogan, A., Krötzsch, M., Reifying, R.D.F.: What works well with wikidata? In: 14th International Semantic Web Conference (ISWC) (2015)
6. Hernández, D., Hogan, A., Riveros, C., Rojas, C., Zerega, E.: Querying wikidata: comparing SPARQL, relational and graph databases. In: Groth, P., et al. (eds.) ISWC 2016. LNCS, vol. 9982, pp. 88–103. Springer, Cham (2016). doi:10.1007/978-3-319-46547-0_10
7. Horrocks, I., Patel-Schneider, P.F., Boley, H., Tabet, S., Grosof, B., Dean, M., et al.: SWRL: A semantic web rule language combining OWL and RuleML. W3C Member Submission (2004). http://www.w3.org/TR/2004/REC-owl-ref-20040210/
8. Kifer, M., Boley, H.: RIF overview. W3C Working Draft, W3C, October 2009 (2013). http://www.w3.org/TR/rif-overview
9. Masolo, C., Guizzardi, G., Vieu, L., Bottazzi, E., Ferrario, R.: Relational roles and qua-individuals. In: AAAI Fall Symposium on Roles, an Interdisciplinary Perspective (2005)
10. Nardi, D., Brachman, R.J., et al.: An introduction to description logics. In: The Description Logic Handbook: Theory, Implementation, and Applications (2003)
11. Nguyen, V., Bodenreider, O., Sheth, A.P.: Don't like RDF reification? Making statements about statements using singleton property. In: 23rd International Conference on World Wide Web (2014)
12. Noy, N., Rector, A., Hayes, P., Welty, C.: Defining n-ary relations on the semantic web. W3C Working Group Note (4) (2006). https://www.w3.org/TR/swbp-n-aryRelations/
13. Scheuermann, A., Motta, E., Mulholland, P., Gangemi, A., Presutti, V.: An empirical perspective on representing time. In: 7th International Conference on Knowledge Capture (2013)
14. ter Horst, H.J.: Completeness, decidability and complexity of entailment for RDF schema and a semantic extension involving the OWL vocabulary. J. Web Seman. 3(2–3), 79–115 (2005)
15. Welty, C., Slices, C.: Representing contexts in OWL. In: Workshop on Ontology Patterns (2010)
16. Welty, C., Fikes, R.: A reusable ontology for fluents in OWL. In: 1st International Conference of Formal Ontology in Information Systems (FOIS) (2006)
17. Zamborlini, V., Guizzardi, G.: On the representation of temporally changing information in OWL. In: Enterprise Distributed Object Computing Conference Workshops (EDOCW) (2010)
18. Zamborlini, V., Guizzardi, G.: An ontologically-founded reification approach for representing temporally changing information in OWL. In: 11th International Symposium on Logical Formalizations of Commonsense Reasoning (COMMONSENSE) (2013)
19. Zimmermann, A.: RDF 1.1: On semantics of RDF datasets (2014). https://www.w3.org/TR/rdf11-datasets/

Adopting Semantic Technologies for Effective Corporate Transparency

Maria Mora-Rodriguez[1](✉), Ghislain Auguste Atemezing[2], and Chris Preist[1]

[1] System Centre, University of Bristol, Bristol, UK
{maria.mora,chris.preist}@bristol.ac.uk
[2] Mondeca S.A, 35 boulevard de Strasbourg, Paris, France
ghislain.atemezing@mondeca.com

Abstract. A new transparency model with more and better corporate data is necessary to promote sustainable economic growth. In particular, there is a need to link factors regarding non-financial performance of corporations - such as social and environmental impacts, both positive and negative - into decision-making processes of investors and other stakeholders. To do this, we need to develop better ways to access and analyse corporate social, environmental and financial performance information, and to link together insights from these different sources. Such sources are already on the web in non-structured and structured data formats, a big part of them in XBRL (Extensible Business Reporting Language). This study is about promoting solutions to drive effective transparency for a sustainable economy, given the current adoption of XBRL, and the new opportunities that Linked Data can offer. We present (1) a methodology to formalise XBRL as RDF using Linked data principles and (2) demonstrate its usefulness through a use case connecting and making the data accessible.

Keywords: XBRL · Transparency · Linked data · Open government data · Interoperability

1 Introduction

Transparency is increasingly used as a means for holding to account organisations with power, both in the public and private sector. In this paper, we focus primarily on the latter - the role of transparency and open data to promote good governance in the private sector, and trust between the private sector and its diverse stakeholders. The main tool for this is corporate reporting - the self-disclosure of information by a company in a well-defined and regular way to a set of stakeholders - primarily, though not exclusively, existing and potential investors. Such stakeholders need to satisfy themselves as to the financial performance and good governance of the company they invest in. Many governments require such reports to be in specific formats, and increasingly are making them

© Springer International Publishing AG 2017
E. Blomqvist et al. (Eds.): ESWC 2017, Part I, LNCS 10249, pp. 655–670, 2017.
DOI: 10.1007/978-3-319-58068-5_40

publically available through open data initiatives – such as the EDGAR[1] program of the U.S. Security and Exchange Commission and the data repository of the Spanish Security Exchange Commission (CNMV)[2].

Such data is often submitted and made available in XBRL (Extensible Business Reporting Language), an XML format adopted to make corporate data more standardised and exchangeable. XBRL is currently in use in more than 60 countries, implemented by over 100 regulators covering some 10 million companies worldwide.

In addition to mandatory reporting, there are a number of voluntary initiatives encouraging corporations to disclose information regarding their performance in areas of economic, social and environmental impact. These include the CDP (Carbon Disclosure Project), GRI (Global Reporting Initiative), SASB (Sustainability Accounting Standards Board) and the IIRC (International Integrated Reporting Council). These are often international non-governmental organisations, with representatives of both corporations and other stakeholders including investors, academics, environmental NGOs and policymakers in their governance structures. Some of these, are promoting the disclosure and use of their data through XBRL, in a similar way to governmental open data initiatives.

Such initiatives encourage engagement with this data: Data journalists can investigate corporate behaviour; investors can integrate future risks associated with factors such as climate change in their assessment of companies; companies can benchmark themselves against others; academics can explore wider trends and correlations in financial performance and non-financial behaviours.

However, such transparency alone is not enough to support companies and their stakeholder's decisions and actions. There are several barriers such as:

- The lack of easily accessible corporate data to quickly and accurately inform the management about issues pertinent during decision-making-processes.
- The unfamiliarity of how sustainable aspects have an impact on financial outcomes and vice versa.
- Inadequate levels of integration of financial and non-financial information within the internal performance, strategy and operational frameworks of an organisation.
- A dearth of consistency, comparability, reliability and clarity of climate change information emerging from organisations globally. Standardisation and mainstreaming of disclosures needs to be facilitated.

The exposure of XBRL reports as open data is a first step to overcome some of these barriers. XBRL brings access to standardised data in an open format about corporate financial, sustainability and environmental performance, in independent silos. However, XBRL offers limited interconnection between them, In particular, XBRL exhibits the following weaknesses:

---

[1] https://www.sec.gov/xbrl/site/xbrl.shtml.
[2] http://www.cnmv.es/ipps/Default.aspx.

- It is primarily structured around documents and entities, rather than data making links between data elements difficult.
- The same data structures can be modelled in different ways, generating different technical implementations.

The Open Information Model (OIM) is being developed by the XBRL community in an effort to increase the adoption of existing XBRL data by reducing the heterogeneity of the XBRL reports generated by different modelling practices and enabling a better integration with Information Systems. OIM aims to facilitate the serialisation of XBRL reports in CSV, JSON and XML formats. In this paper, we use the XBRL Open Information Model as a template for extracting linked open data from XBRL documents, and so allowing them to be combined more easily and to be used alongside other open data sources to enrich analysis by stakeholders. We demonstrate this by reasoning with data from the Spanish Security and Exchange Commission (CNMV) alongside data from the CDP.

Our paper makes a contribution in the field of the Semantics and Corporate Transparency, by more effectively integrating data sources using the most common corporate standard - XBRL - into the linked data environment. We show that this in turn makes contributions in the following areas:

- The generation and integration of Linked Open Government data made available from government sources (CNMV) in a different format (XBRL).
- Provenance and Accountability of companies, showing how financial and non-financial data from different sources can be linked using our approach to hold companies more environmentally accountable.
- Trust, Data Traceability and Fact Checking of corporate data sources, enabled by using the proposed ontology together with SPARQL to facilitate cross-checking of corporate data with alternate sources.

This paper is organised as follows. In Sects. 2 and 3, we provide background on XBRL and prior work connecting XBRL with Linked Data formats. Section 4 presents a methodology to transform XBRL to RDF following Linked Data principles and the Open Information Model[3]. We describe a case study implementing our proposal, using financial and environmental data in XBRL format from the Spanish companies Repsol S.A and Amadeus IT Group, published by the CNMV and CDP respectively. Having XBRL data in RDF format, we proceed to link with others data sources from LOD using LIMES[4] as Link Discovery Framework. Then, we make the RDF data accessible and queryable via SPARQL endpoint, and we proceed to evaluate some queries to show the potential benefits for data users. Section 6 presents and discusses the results. The paper closes with conclusions and lessons learned which, if addressed by the XBRL and Linked data community, would further promote transparency.

---

[3] http://www.xbrl.org/Specification/oim/PWD-2016-01-13/oim-PWD-2016-01-13.html.

[4] http://aksw.org/Projects/LIMES.html.

2   XBRL Fundamentals

XBRL aims to overcome the limitations in traditional and mainly paper-based disclosures [7]. Through data standardisation in an open digital format, XBRL can help enhance data quality and data analysis through:

- An open mechanism to represent contextualised business facts under defined business requirements (presentation, period, legal references, calculation) and data quality;
- Enabling data-driven decision management;
- Improved accessibility and integration of the information to any application or management process;
- Standardised validation and comparability of information.

For financial disclosure regimes like International Financial Reporting Standards (IFRS) and US Generally Accepted Accounting Principles (US GAAP) or non-financial disclosure regimes like the CDP and GRI, and its corresponding statements and reports, a single XBRL taxonomy is created. The taxonomy is where the rules and data definitions are organised. It is comprised of a set of elements (i.e., Key Performance Indicators and narratives) and all the presentation, calculation, labels in different languages, and standard logic rules (linkbases) which provide semantic meaning. The taxonomy also includes mechanisms for defining reporting requirements in a multi-lingual context and allowing filers to use their language of choice while allowing consumers to review it in their own.

Once created, the XBRL taxonomy is published online. Then, for a given firm, software can be used to create an XBRL instance (the report itself), containing facts and figures for a certain period (Fig. 1). The XBRL instance can be checked against the taxonomy by all parties (reporting entity, a regulator, or even the public) in order to guarantee its data quality and reliability, as the taxonomy contains data quality checks that any XBRL engine can validate. The validation rules supported in XBRL allow a good level of data quality, from basic rules to validate data types (number, text, precision), to more complex rules relating to elements that have been disclosed. For example, rules can be implemented to check if a breakdown of carbon emissions is equal or not to the total emissions reported, or a CO_2 intensity figure (tCO_2/revenue) is actually in line with revenue and emissions figures reported.

The XBRL core specification (XBRL 2.1) has evolved since its creation in 2001 to enrich its dimensional data representation and validation rules. The XBRL Dimensions 1.0^5 in 2006 and Formulas 1.0^6 in 2009 provide optional incremental syntax to the core specification. Dimensions 1.0 enriches rules and procedures for constructing dimensional taxonomies and therefore instance documents. Taxonomies using Dimensions specification can define new dimensional

---

[5] http://www.xbrl.org/specification/dimensions/rec-2012-01-25/dimensions-rec-2006-09-18+corrected-errata-2012-01-25-clean.html.

[6] http://www.xbrl.org/wgn/xbrl-formula-overview/pwd-2011-12-21/xbrl-formula-overview-wgn-pwd-2011-12-21.html.

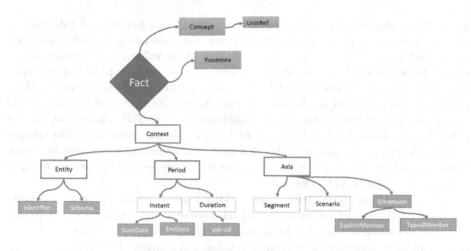

Fig. 1. Fact representation in XBRL

contexts, specifying valid values ("domains") for dimensions, using a mechanism called hypercube to define which dimensions apply to which business concept. There are two types of dimensions:

- Explicit dimensions: These have a fixed number of dimension members. For example, in a two-dimensional table, the number of row and columns are known.
- Typed dimensions: the number of dimension members is unknown. For example, in a two-dimensional table, it means that the number of columns is known, but the number of rows depends on user reporting needs.

However, the Dimensions specification in XBRL 2.1 has certain limitations. It does not fully support calculation rules (defined in calculation linkbase); calculations cannot be executed across different contexts. In other words, in a simple two-dimensional table, calculations can be executed by columns and not by rows. In part, these limitations, and the need to have strong validation capabilities in XBRL resulted in the Formulas 1.0 specification in 2009. This module enhances the XBRL validation capabilities, using XPath to validate instances and to calculate new XBRL facts.

XBRL has evolved to support both global regulatory environments and emerging domains such as sustainability reporting. The flexibility needed to do this has resulted in difficulties with regard to standardisation. Primarily, this is because of the diversity of technical implementations produced by different modelling practices during the taxonomy development phase. Though XBRL is a standard, it offers different ways to model data structures. For example, tables can be represented in XBRL using tuples or dimensions. This is true of the two taxonomies we use in our work later in the paper: The CNMV taxonomy represents financial facts as items and tuples and only makes use of the XBRL 2.1 core specification. Items are facts holding a simple value represented by a single XML element with the value as its content and period and information about reporting entity as a context attribute. An example item could be equity in the last quarter. Tuples are facts holding multiple values, and they are represented by a single XML element containing items or other tuples. For example, preferred stock is always defined by the combination of different stocks. Thus, preferred stock is a tuple defined by two items: the preferred stock-nominal value and the preferred stock-shares authorised. The CDP taxonomy, on the other hand, represents (environmental) facts as simple items and dimensional structures and makes use of XBRL 2.1 core and Dimensions 1.0 specifications. Dimensional structures are represented as items where context attributes also include dimensional XML elements.

The new Open Information Model (OIM) specification proposes an independent XBRL model to represent XBRL business facts; it focuses on XBRL instance documents instead of taxonomy definitions. This overcomes the taxonomy modelling difficulties and keeps the value and context of the data. However, the semantic richness of the taxonomy disappears, such as advanced validation rules, human labels in different languages and how the data should be presented.

In our study, we propose an ontology based on OIM specification to transform XBRL data to RDF, using as a case of study financial and environmental company data from the CNMV and CDP.

3   Related Work

Previous efforts to make XBRL data more interoperable with other data sources and formats have used RDF and OWL ontologies, as well as linking and publishing solutions. The majority of these base their examples on transforming financial XBRL taxonomies models into RDF from well-known open government data initiatives, such as XBRL filings available from the SECs EDGAR program. However, none of these studies covers the full XBRL specifications including XBRL 2.1 and Dimensions 1.0. In other words, these studies do not offer a general solution to convert any XBRL report to RDF. For example, Garcia and Gil [2], propose a solution to transform XBRL filings available from the EDGAR program to RDF. Their approach is generic to the XBRL 2.1 specification: simple items, scenarios, segments and tuples data structures. They use US-GAAP reports from 2006 as a case study, which do not use Dimensions 1.0 specification. On the other hand, Kampgen et al. [5], propose RDF Data Cube Vocabulary to model XBRL reports as a multidimensional dataset. They exemplified their methods by using 2009 and 2011 US-GAAP reports, whose taxonomy uses XBRL 2.1 and Dimensions 1.0 specification. However, it is unclear how that solution can be generic to other dimensional taxonomies and how the ontology proposed covers tuples, simple items and contextual information modelled with scenarios and segments. There is an experimental initiative, called the Edgar Linked Data wrapper[7], that provides access to XBRL filings from the SEC as Linked Data. The approach is to publish US-GAAP taxonomies

---

[7] http://edgarwrap.ontologycentral.com/.

into RDF as vocabularies. In fact, each new US-GAAP taxonomy version means a new semantic vocabulary. This represents a solution to convert US-GAAP reports into Linked data, but it is not a solution for any other type of XBRL reports, such as CNMV reports. Closer to the sustainability domain, Madlberger *et al.* [6] presented an ontology-based approach using GRI-XBRL taxonomy to build a Corporate Sustainability ontology. However, the result is content-based approach instead of metadata conversion, meaning that the solution proposed is not generic to transform any XBRL report to RDF, only GRI reports to RDF.

Authors agree that there are some limitations when representing XBRL data in RDF graphs and as Linked data, due to the lack of formal semantics and inference mechanisms, and difficulties to find correspondences with well-known vocabularies (SKOS, FOAF, etc.). Furthermore, complete architectures for evolving information systems enabling a better financial data integration using Linked data are proposed in [4,5]. Basically, these solutions integrate XBRL financial data with DBpedia and Yahoo!Finance Web API. For the purpose of this study, we also consider their requirements necessary to boost effective corporate transparency:

- to break the barriers which hold XBRL data in isolated data silos of information and vendor lock-in of proprietary XBRL tools,
- to reach a better level of data coverage and data quality and
- to facilitate a comprehensive picture of company performance.

We distinguish our study from previous work by proposing a solution to enhance an effective corporate transparency, increasing the adoption of financial and non-financial data and generating impact on decisions. Our central thesis is that in order to create that impact, two components are necessary: (1) Foster interoperability across economic, social and environmental data published in XBRL format and others; and (2) better integration of these combined data in information systems that are part of the decision-making processes of companies, their stakeholders, regulators and supervisory entities. In order to turn corporate data into valuable information for decision-makings, we focus on the following tasks:

- A generic ontology to transform any XBRL report into Linked data.
- Interlinking with existing data available in the LOD cloud.
- Data publication via SPARQL query endpoint.
- Enabling data contextualization, cross-data-source analyses and data accuracy.

4   Transforming XBRL into Linked Data

4.1   Lightweight Vocabulary for XBRL (XBRLL)

We develop a lightweight ontology using the Web Ontology Language (OWL) based on XBRL standard (XBRL 2.1 and Dimensions 1.0). The goal of implementing a lightweight vocabulary for XBRL is threefold: (1) Easy identification of the key concepts of the XBRL standard; (2) Reuse of existing vocabularies to describe XBRL datasets; (3) Enrichment and linking of data with relevant ones in the Linked Open Data cloud.

Unlike previous efforts, we base our ontology proposal on OIM, which is a syntax-independent model of the content of an XBRL report instead of taxonomy definition. OIM defines 4 components:

Namespace: Representation of XML namespace prefixes.
DTS Reference: Reference to XML documents and schema linked to an XBRL report.
Report: Top-level component that encapsulates the data of an XBRL report.
Fact: Representation of a business fact in an XBRL report. As explained in section one, a fact can be a simple item, a tuple fact or a dimension. All facts have the following common properties (id, aspects and footnotes). Id is a unique identifier; Aspects are properties which represent:
  - the entity and period which a business fact is referred to (oim:entity and oim:period),
  - the unit of measure, such as "USD" and "MWh" (oim:unit),
  - the reporting item (oim:concept),
  - tuples definition, which represents a grouping container for other facts.
  - dimensional structure, axis and members.
  - the footnotes of the fact (oim:footnotes)

Our XBRL-Lightweight (XBRLL) ontology is composed of 15 classes, 12 object properties and 12 data properties, with DL expressivity: ALC(D). The ontology follows best practice in the semantic web by reusing existing ontologies to improve data interoperability [1], through the Linked Open Vocabularies initiative (LOV[8]).

We provide a hierarchical structure following OIM, mapping XBRL components to Semantic web vocabularies. The class Report is a subclass of schema:Report[9]. The class Fact is used to represent XBRL business facts (items, tuples and dimensions) that in turn refer to entity (hasEntity), concept, period, scenario, value and footnote modelled as object properties. In the case of numeric and currency values, the number of decimals and unit type are also represented. These properties are enough to represent a simple XBRL item.

As required from XBRL, a fact can hold multiple values in the form of tuples and dimensional structures. For that purpose, we defined the has Tuple and hasDimension properties as part of the Fact class. They point out to Tuple and Dimension classes respectively. Tuple class is composed by concept, and hasTuple properties. The latter means a tuple can be embedded as part of another tuple, defining the context of the main item. Dimension class is composed of Axis and Member properties, representing the axis and member per axis which define the context of the main item. We decided not to differentiate whether the axis is explicit or typed, as our model is focused on instance documents instead of

---

[8] https://lov.okfn.org/.
[9] http://schema.org.

taxonomy definition. It means, that as we are working with XBRL reports, the dimensional members are defined.

Unlike the OIM model, we decided to define Entity as a class instead of an object property of the class Fact that links to the schema identifier of the entity that is part of the XBRL report. In that way, we extend the class rov:RegisteredOrganization[10], (1) keeping the correspondence with well-known vocabularies and (2) allowing the full information of the reporting firm facilitating the discovering link process. Figure 2 shows classes and relationships defined in the XBRLL ontology, which is available at https://w3id.org/vocab/xbrll.

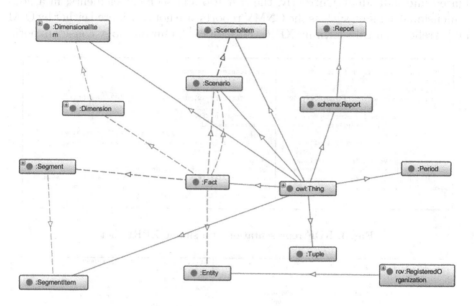

Fig. 2. Classes and relationships XBRLL ontology

4.2   From XBRL Data to Linked Data

As a next step, we demonstrate how XBRL data can be mapped to the ontology and how it can be published using Linked data principles. We demonstrate our method using financial and environmental XBRL data from the Spanish companies Repsol and Amadeus IT Group, chosen as ones which are published in both CNMV and CDP[11]. For the transformation process, we developed a script in Python (https://goo.gl/VqgJQZ) to transform the XBRL reports using JSON files generated by the Arelle[12] open source platform. Note that the JSON data used as our input is generated according to OIM.

---

[10] http://www.w3.org/ns/regorg.
[11] The CDP data used is publically available, but not yet in XBRL format. This is currently only available internally to CDP, and made available to this project.
[12] http://arelle.org/.

A simple fact in XBRL (Fig. 3), is the representation of a concept (cdp:IntroductionCompany) and its value, where the context consists of the period, unit and information about the reporting company (entity).

A tuple fact in XBRL (Fig. 4) represents facts with multiple values. In this case, the concepts ifrs-gp:IntagibleAssetsNet and ifrs-gp:GoodwillNet are the elements that compose the tuple ipp-gen:BalanceIndividual. The CNMV allows reporting companies to use xbrli:scenario element to determine if the value is part of an individual or consolidated financial statement. Hence in CNMV reports we find the same tuple hierarchy and concepts linked to two different contexts: consolidated and individual. Currently, the scenario and segment elements, in a non-dimensional domain, such as the CNMV reports, are not considered either by OIM and Arelle when transforming XBRL into JSON[13]. Our ontology considers both.

```
<http://data.mondeca.com/id/fact/f001> a
nsl:Fact ;
    nsl:hasEntity
<http://data.mondeca.com/id/entity/000084
7838> ;
    nsl:concept cdp:IntroductionCompany;
    nsl:period [ nsl:startPeriod "2015-
01-01"^^xsd:date ;
            nsl:endPeriod "2015-12-
31"^^xsd:date] ;
    nsl:value "Repsol is international
integrated Oil and Gas Company whose main
activity consists of the upstream and
downstream business." ;
```

```
<xbrli:context id="ctx_00">
    <xbrli:entity>
        <xbrli:identifier
scheme="http://www.cdp.net/CIK">0000847838</xbrli:identifie
r>
    </xbrli:entity>
    <xbrli:period>
        <xbrli:startDate>2015-01-01</xbrli:startDate>
        <xbrli:endDate>2015-12-31</xbrli:endDate>
    </xbrli:period>
</xbrli:context>
<cdp:IntroductionCompany contextRef="ctx_00">Repsol is
international integrated Oil and Gas Company whose main
activity consists of the upstream and downstream
business.<cdp:IntroductionCompany>
```

Fig. 3. RDF representation of a simple XBRL fact

```
<http://data.mondeca.com/id/fact/f0004> a nsl:Fact ;
    nsl:concept ifrs-gp:IntangibleAssetsNet;
    nsl:entity <http://data.mondeca.com/id/entity/A7837472>;
    nsl:period [ nsl:instant "2015-12-31"^^xsd:date] ;
    nsl:hasTuple <http://data.mondeca.com/id/tuple/t89493> ;
    nsl:unitRef < http://dbpedia.org/resource/EUR> ;
    nsl:value 94588000.0 ;
    nsl:decimals 0 .
    nsl:hasScenario
<http://data.mondeca.com/id/scenario/s0001> .
 <http://data.mondeca.com/id/fact/f0004> a nsl:Fact ;
    nsl:concept ifrs-gp:GoodwillNet ;
    nsl:entity <http://data.mondeca.com/id/entity/A7837472>;
    nsl:period [ nsl:instant "2015-12-31"^^xsd:date] ;
    nsl:hasTuple <http://data.mondeca.com/id/tuple/t89493> ;
    nsl:unitRef < http://dbpedia.org/resource/EUR>;
    nsl:value 350.0 ;
    nsl:decimals 0 .
    nsl:hasScenario
<http://data.mondeca.com/id/scenario/s0001> .
 <http://data.mondeca.com/id/scenario/s0001> a nsl:Scenario
    nsl:scenarioItem [nsl:concept ipp:Modelo; nsl:value "GE"]
    nsl:scenarioItem [nsl:concept ipp:Apartado; nsl:value
"Individual"]

 <http://data.mondeca.com/id/tuple/t89493> a nsl:Tuple ;
    nsl:concept ipp-gen:BalanceIndividual;
    nsl:hasTuple <http://data.mondeca.com/id/tuple/t89494>
```

```
<xbrli:context id="S22015_A-78374725_ici">
    <xbrli:entity>
        <xbrli:identifier
scheme="http://www.cnmv.es/xbrl/ipp/A-
78374725">REPSOL, S.A</xbrli:identifier>
    </xbrli:entity>
    <xbrli:period>
        <xbrli:instant>2015-12-31</xbrli:instant>
    </xbrli:period>
    <xbrli:scenario>
        <ipp:Modelo>GE</ipp:Modelo>
        <ipp:Apartado>Individual</ipp:Apartado>
    </xbrli:scenario>
</xbrli:context>
 <xbrli:unit id="euro">
    <xbrli:measure>iso4217:EUR</xbrli:measure>
</xbrli:unit>
 <ipp-gen:InformacionFinancieraSeleccionada>
    <ipp-gen:BalanceIndividual>
        <ifrs-gp:IntangibleAssetsNet
decimals="0" contextRef="S22015_A-78374725_ici"
unitRef="euro">94588000</ifrs-
gp:IntangibleAssetsNet>
        <ifrs-gp:GoodwillNet decimals="0"
contextRef="S22015_A-78374725_ici"
unitRef="euro">350</ifrs-gp:GoodwillNet>
    </ipp-gen:BalanceIndividual>
</ipp-gen:InformacionFinancieraSeleccionada>
```

Fig. 4. RDF representation of a tuple fact

[13] The lack of segment and scenario representation in a non-dimensional domain, was informed to the OIM working group and Arelle's authors.

Dimensional facts in XBRL (Fig. 5), can also represent multiple values. For example, the concepts cdp:EmissionValueGross and cdp:Scope are linked to the same Axis cdp:TotalEmissionDataAxis and related member(cdp:GreenhouseInventoryBoundariesID). This structure allows disclosing the total Emissions gross values per type of scope (Scope 1, Scope 2 location-based, Scope 2 market-based and Scope 3). Here the value of 21068516 CO2e corresponds to the Scope 1 emissions.

```
<http://data.mondeca.com/id/fact/f0005> a
ns1:Fact ;
   ns1:concept  cdp:EmissionValueGrossCO2e
   ns1:entity
<http://data.mondeca.com/id/entity/00008478
38> ;
   ns1:period [ ns1:startPeriod "2015-01-
01"^^xsd:date; ns1:endPeriod "2015-12-
31"^^xsd:date] ;
   ns1:hasDimension
<http://data.mondeca.com/id/dimension/d0001
> ;
   ns1:unitRef
<http://dbpedia.org/page/Carbon_dioxide_equ
ivalent>;
   ns1:value 21068516;
   ns1:decimals 0 .

<http://data.mondeca.com/id/dimension/d0001
> a ns1:Dimension ;
   ns1:dimensionItem [ns1:axis
TotalEmissionDataAxis; ns1:value "id01"]
```

```
<xbrli:context id="ctx_8_2_2015_id01">
   <xbrli:entity>
     <xbrli:identifier
scheme="http://www.cdp.net/CIK">0000847838</xbrli:identifier>
     <xbrli:segment>
       <xbrldi:typedMember
dimension="cdp:TotalEmissionDataAxis">
<cdp:GreenhouseInventoryBoundariesID>id01</cdp:GreenhouseInvento
ryBoundariesID>
       </xbrldi:typedMember>
     </xbrli:segment>
   </xbrli:entity>
   <xbrli:period>
     <xbrli:startDate>2015-01-01</xbrli:startDate>
     <xbrli:endDate>2015-12-31</xbrli:endDate>
   </xbrli:period>
</xbrli:context>
 <xbrli:unit id="CO2e">
   <xbrli:measure>cdp:CO2e</xbrli:measure>
 </xbrli:unit>
   <cdp:EmissionValueGrossCO2e decimals="0"
contextRef="ctx_8_2_2015_id01"
unitRef="CO2e">21068516</cdp:EmissionValueGrossCO2e>
   <cdp:Scope contextRef="ctx_8_2_2015_id01">cdp-
enum:Scope1</cdp:Scope>
```

Fig. 5. RDF representation of a dimensional fact

During the transformation of the CNMV report, we found certain XBRL elements with content about persons and activities, which belongs to the imported XBRL taxonomy called Data of General Identification (DGI)[14]. We map these using Friend Of a Friend (FOAF) vocabulary[15] (Fig. 6), in line with best practice of reusing existing vocabularies in specific contexts to increase the level of interoperability.

As in XML [8], XBRL Namespaces specifications do not need to reference a real location, just be unique. However, in RDF, the namespace URI must identify the location of the schemas. As certain XBRL namespaces from the CNMV reports were not valid we had to store the schemas in our server and point out the namespaces to real locations. In many cases, we decided to map the XBRL units and currencies to well-known DBpedia links, connecting related data that were not previously linked. For example:

<xbrli:measure>iso4217:EUR</xbrli:measure>   to   http://dbpedia.org/resource/EUR

---

[14] https://joinup.ec.europa.eu/asset/data_of_general_identification/home.
[15] http://xmlns.com/foaf/spec/.

```
                              <ipp-com:DetallePersonasResponsables>
                                 <dgi-est-gen:PersonName>
                                    <dgi-est-gen:FormattedName contextRef="S12015_A-
                              78374725_ici">D. Ángel Durández Adeva</dgi-est-
<http://data.mondeca.com/id/xbrl/person/016   gen:FormattedName>
06701J> a ns5:Person ;                </dgi-est-gen:PersonName>
   ns7:identifier "01606701J" ;       <dgi-est-gen:Identifier>
   ns6:role "Consejero" ;             <dgi-est-gen:IdentifierCode>
   ns5:name "D. Javier Echenique   <dgi-lc-es:Xcode_IDC.CIF contextRef="S12015_A-
Landiribar"                         78374725_ici">CIF</dgi-lc-es:Xcode_IDC.CIF>
   ns2:codeID "CIF" .               <dgi-est-gen:IdentifierValue contextRef="S12015_A-
                              78374725_ici">01606701J</dgi-est-gen:IdentifierValue>
                                 </dgi-est-gen:IdentifierCode>
                                 </dgi-est-gen:Identifier>
                              <dgi-est-gen:PositionType contextRef="S12015_A-
                              78374725_ici">Consejero</dgi-est-gen:PositionType>
                              </ipp-com:DetallePersonasResponsables>
```

Fig. 6. DGI representation in RDF using FOAF vocabulary

<xbrli:measure>cdp:CO2e</xbrli:measure>   to http://dbpedia.org/page/
Carbon_dioxide_equivalent

As XML, an XBRL document forms a tree structure ready to be consumed as a full report [3]. The move to data consumption requires the use of dereferenceable URIs to denote facts in a unique way, keeping its context. We use the following URI conventions to denote related facts and classes:

Fact: http://data.mondeca.com/id/fact/f[0-9]* ->http://data.mondeca.com/id/fact/f88557

Entity: http://data.mondeca.com/id/entity/(company identifier) ->http://data.mondeca.com/id/entity/A-78374725

Through the dereferenceable URIs, facts can be visualised using open source tools like LodLive[16]. We provide an example here: https://goo.gl/iFVE0B.

4.3 Linking XBRL Data to Other Data

If transparency is to be enabled, it is very important to convert the independent XBRL silos of information into pieces connected with existing Linked Data sources available on the web.

For that purpose, we use LIMES, which is a tool that allows detecting similar Linked datasets. LIMES works specifying the search criteria and the target endpoint to search in. Our search criteria is the company name contained in the Entity fact from the generated RDF. The DBPedia endpoint is the target source that we choose to gather the links, restricting the search by sch:Organization. LIMES requires a metric and acceptance condition setting a threshold value. We use the trigrams metric offered by LIMES to mapping correspondences between the ns6:legalName of our local RDFs and the sch:Organization from DBpedia. For the purpose of this paper, we only accept results with a minimum of 0.90 level acceptance. The final results were the following URLs http://dbpedia.org/resource/Repsol and http://dbpedia.org/page/Amadeus_IT_Group, included as a SameAs relationship in our local RDFs files.

---

[16] http://en.lodlive.it/.

5   Validation

For validation purposes, we run queries against the final ontology generated, evaluating its quality and accuracy by checking whether they contain enough information to cover three goals to promote effective transparency:

Data coverage: through better data contextualization.
Better data analysis: enabling cross-data-source analyses.
Data accuracy: facilitating data cross-checking contained in different sources.

For that, firstly we implemented an endpoint[17] using Apache Jena Fuseki[18], available here: http://data.mondeca.com/dataset.html?tab=query& ds=/xbrl-data. We used SPARQL (SImple Protocol and RDF Query language) because it allows us to express queries across diverse data. We conduct three queries with each of the two companies (Repsol and Amadeus IT Group) data. We illustrate each query below with one of the companies. Full results are available at the SPARQL queries provided.

Goal 1. Data Coverage Using DBpedia

Question: What is the context of the company Repsol?
Data: Abstract, subsidiary and industry.
SPARQL query: https://goo.gl/if8ydG
Output: presented in Table 1.

Table 1. Data coverage: information about the context of Repsol S.A

URL	Abstract	Subsidiary	Industry
http://dbpedia.org/resource/Repsol	Repsol S.A. is an integrated global energy company based in Madrid, Spain. It carries out upstream and downstream activities throughout the entire world. It has more than 24,000 employees worldwide	http://dbpedia.org/resource/Petronor	http://dbpedia.org/resource/Petroleum_industry

Goal 2. Cross Data Source Analysis Using CNMV and CDP Data

Question: What was the emission intensity of Repsol in 2015?
Data: Scope 1 emissions (CDP) divided by Consolidated sales (CNMV).
SPARQL query: https://goo.gl/7bIE9m
Output: presented in Table 2.

Table 2. Data analysis: emission intensity of Repsol S.A in 2015

CO2 Emissions	Consolidated sales	Emission intensity
21068516	39737000000	0.00053

---

[17] http://data.mondeca.com/xbrl-data/sparql.
[18] https://jena.apache.org/.

Goal 3. Data Accuracy Using DBPedia and CNMV Data

Question: How reliable is the equity figure presented in DBpedia?
Data: Equity (DBPedia) and equity (CNMV) in the year 2013.
SPARQL query: https://goo.gl/LGb53s
Output: presented in Table 3.

Table 3. Data consistency: reliability of equity figure presented in DBpedia

Entity name	Equity(DBPedia) dbo:equity	Equity(CNMV) ipp-gen:PatrimonioNetoNiif	Difference
Repsol S.A	2.792E10	27920000000	0.0001
Amadeus IT holding	€1,840.1 million@en	1840066000.0	-

6 Discussion

This study demonstrates that Linked data can be used to integrate financial and non-financial data and can facilitate transparency among diverse stakeholders. Our work does this by converting corporate XBRL reports into RDF and linking them to other relevant financial and non-financial data (e.g., environmental, DBpedia). A generic ontology to transform any XBRL report into Linked data has been proposed, along with ways to resolve the lack of formal correspondences with well-known vocabularies. This solution overcomes the XBRL challenges related to the diversity of technical implementations produced by different modelling practices, and so goes beyond prior related works.

We demonstrate that using linked data with well-adopted standards, such as XBRL, improves the interoperability and access to existing corporate datasets, as well as straightforward integration with related data in other formats. The validation exercise demonstrates that the solution proposed offers three benefits for data users: data coverage, better data analysis and data consistency. The results in Table 3 present an interesting point for discussion. It shows that the DBPedia data (dbo: equity) lacks context and numeric precision. For example, there is no year associated with the equity figure nor consistent use of datatype. Amadeus equity is a string €1,840.1 million while Repsol equity, which has the same tag (dbo:equity), is a number. The data from the CNMV in XBRL format does not have any of those problems.

Given these results, we believe that XBRL is a better format than RDF to standardise corporate information. However, it is less able to connect different data silos in various formats. For that, the publication of data using Linked data principles is the most appropriate solution.

This study proposes the combination of both solutions, XBRL and Linked data, to improve corporate transparency. Below, we enumerate a set of technical requirements to apply on XBRL schemas, definitions and reports to converge towards a Linked data approach. Adoption of these best practices in XBRL modelling would enhance interoperability and transparency of corporate data.

- Use of common data structures must be encouraged in XBRL taxonomies. For example, taxonomies such as DGI to represent common corporate information such as company name, unique identification number, activities and sectors. This would not only enhance the interoperability between XBRL data from different taxonomies but also ease the mapping process with well-known vocabularies in the Linked data world.
- The use of namespaces notation that point out to real locations should be promoted in XBRL. This would ease the transformation of XBRL data into Linked data, facilitating better inference mechanisms.
- Using dereferenceable URIs to denote and identify XBRL facts provides a way to access and link relevant information to those objects across the web. It enables better interoperability between data in XBRL format and other data sources.
- Reusing existing RDF data on units and currencies already published in LOD brings more contextual information than the current ISO and XBRL units reference.
- Using tools like LIMES can help to increase the coverage of information by continuously integrating data sources.

We made all scripts and tools available to let academia and industry evaluate and contribute to this work.

7   Conclusions and Further Research

In this study, we show the role of Linked Data and XBRL in bringing new opportunities for effective transparency in corporate reporting. Linked data principles can encourage better corporate data publication and therefore data analysis, defining the interconnection across financial and non-financial data (such as sustainability data) and documents publicly available in open government data initiatives and voluntary reporting initiatives. XBRL enables a standard and accurate representation of corporate data with advanced validation rules. We present a solution to convert independent silos of XBRL data into interconnected pieces. Lessons learned during the process and benefits are presented. While our work demonstrates the potential of this approach, it would benefit from extension in the following ways; (i) incorporate data sources beyond environmental, financial and DBpedia; (ii) incorporate non public-domain data, and address associated data protection issues necessary to do this; (iii) consider scalability and performance issues in the transformations necessary. In future work, we intend to evaluate and integrate sustainability reporting in XBRL, such as GRI data, and extend the ontology proposed using RDF Data Cube. We believe this study encourages scholars, regulators, data publishers and users to promote and use both XBRL and Linked data, as each solution has a different role to play. Combined use of them enables non-financial factors such as environmental and social performance of companies to be integrated into reasoning, allowing improved transparency and accountability by a diverse group of stakeholders.

Acknowledgments. This work was supported by the Systems Centre at the University of Bristol, the EPSRC funded Industrial Doctorate Centre in Systems (Grant EP/G037353/1) and the CDP Worldwide, London, UK.

References

1. Breslin, J., Passant, A., Decker, S.: The Social Semantic Web. Springer Science and Business Media, Heidelberg (2009)
2. Garcia, A., Gil, R.: Publishing XBRL as linked open data. In: World Wide Web Workshop: Linked Data on the Web (LDOW 2009), vol. 538 (2009)
3. Garcia, R., Gil, R.: Triplificating and linking XBRL financial data. In: Proceedings of the 6th International Conference on Semantic Systems. ACM (2010)
4. Goto, M., Hu, B., Naseer, A., Vandenbussche, P.: Linked data for financial reporting. In: Proceedings of the 4th International Conference on Consuming Linked Data, vol. 1034, pp. 123–135. CEUR-WS (2013)
5. Kämpgen, B., Weller, T., O'Riain, S., Weber, C., Harth, A.: Accepting the XBRL challenge with linked data for financial data integration. In: Presutti, V., d'Amato, C., Gandon, F., d'Aquin, M., Staab, S., Tordai, A. (eds.) ESWC 2014. LNCS, vol. 8465, pp. 595–610. Springer, Cham (2014). doi:10.1007/978-3-319-07443-6_40
6. Madlberger, L., Thöni, A., Wetz, P., Schatten, A., Tjoa, A.: Ontology-based data integration for corporate sustainability information systems. In: Proceedings of International Conference on Information Integration and Web-Based Applications. ACM (2013)
7. Mora-Gonzalbez, J., Mora-Rodriguez, M.: XBRL and integrated reporting: the Spanish accounting association taxonomy approach. Int. J. Dig. Acc. Res. **12**, 59–91 (2012)
8. Namespaces in XML 1.1, 2nd (edn.) (2016). goo.gl/UEMDG8. Accessed March

Author Index

Printed in the United States
By Bookmasters